BIOCHAR PRODUCTION FOR GREEN ECONOMY

BIOCHAR PRODUCTION FOR GREEN ECONOMY

Agricultural and Environmental Perspectives

Edited by

SHIV VENDRA SINGH
College of Agriculture, Rani Lakshmi Bai Central Agricultural, University, Jhansi, Uttar Pradesh, India

SANDIP MANDAL
Agricultural Energy and Power Division (AEPD), ICAR – Central Institute of Agricultural Engineering, Bhopal, Madhya Pradesh, India

RAM SWAROOP MEENA
Department of Agronomy, Institute of Agricultural Sciences, Banaras Hindu University, Varanasi, Uttar Pradesh, India

SUMIT CHATURVEDI
Department of Agronomy, G. B. Pant University of Agriculture & Technology, Pantnagar, Uttarakhand, India

K. GOVINDARAJU
Sathyabama Institute of Science and Technology, Chennai, India

Academic Press is an imprint of Elsevier
125 London Wall, London EC2Y 5AS, United Kingdom
525 B Street, Suite 1650, San Diego, CA 92101, United States
50 Hampshire Street, 5th Floor, Cambridge, MA 02139, United States

Copyright © 2024 Elsevier Inc. All rights are reserved, including those for text and data mining, AI training, and similar technologies.

Publisher's note: Elsevier takes a neutral position with respect to territorial disputes or jurisdictional claims in its published content, including in maps and institutional affiliations.

No part of this publication may be reproduced or transmitted in any form or by any means, electronic or mechanical, including photocopying, recording, or any information storage and retrieval system, without permission in writing from the publisher. Details on how to seek permission, further information about the Publisher's permissions policies and our arrangements with organizations such as the Copyright Clearance Center and the Copyright Licensing Agency, can be found at our website: www.elsevier.com/permissions.

This book and the individual contributions contained in it are protected under copyright by the Publisher (other than as may be noted herein).

Notices

Knowledge and best practice in this field are constantly changing. As new research and experience broaden our understanding, changes in research methods, professional practices, or medical treatment may become necessary.

Practitioners and researchers must always rely on their own experience and knowledge in evaluating and using any information, methods, compounds, or experiments described herein. In using such information or methods they should be mindful of their own safety and the safety of others, including parties for whom they have a professional responsibility.

To the fullest extent of the law, neither the Publisher nor the authors, contributors, or editors, assume any liability for any injury and/or damage to persons or property as a matter of products liability, negligence or otherwise, or from any use or operation of any methods, products, instructions, or ideas contained in the material herein.

ISBN: 978-0-443-15506-2

For Information on all Academic Press publications visit our website at https://www.elsevier.com/books-and-journals

Publisher: Nikki Levy
Acquisitions Editor: Nancy Maragioglio
Editorial Project Manager: Dan Egan
Production Project Manager: Bharatwaj Varatharajan
Cover Designer: Miles Hitchen

Typeset by MPS Limited, Chennai, India

Contents

List of contributors xi
About the editors xvii
Preface xix

1. Crop waste conversion into biochar: an overview

Rini Labanya, Parmanand Sahu, Sandip Mandal, Shiv Vendra Singh and Ram Swaroop Meena

1.1 Introduction 1
1.2 Properties of biochar 2
1.3 Biochar production technologies 3
1.4 Effect of biochar production process parameters on biochar yield and quality 6
1.5 Feedstock for biochar production 7
1.6 Applications of biochar 8
1.7 Policy and legislative framework 18
1.8 Overview of current biochar markets 18
1.9 Conclusion 19
References 19

A
Biochar production and modification

2. Advanced pyrolysis reactors for energy efficient production of biochar

Rajat Kumar Sharma, Sandip Mandal, Mohammad Ali Nazari, Juma Haydary and Akarsh Verma

2.1 Introduction 27
2.2 Types of pyrolysis reactors 28
2.3 Factors affecting the design of pyrolysis reactor 35
2.4 Conclusion 38
Acknowledgment 38
References 39

3. Microwave-assisted hydrothermal carbonization for biochar production: potential application and limitations 43

R. Divyabharathi, P. Komalabharathi and P. Subramanian

3.1 Introduction 43
3.2 Microwave-assisted hydrothermal carbonization 44
3.3 Applications of microwave-assisted hydrothermal carbonization 50
3.4 Limitations of microwave-assisted hydrothermal carbonization 52
3.5 Conclusion and future prospects 53
References 53

4. Sustainable management and diversification of problematic wastes: prospects and challenges 57

Anamika Barman, Sougata Roy, Priyanka Saha, Saptaparnee Dey, Shashank Patel, Deepak Kumar Meena, Anurag Bera, Shiv Vendra Singh, Sandip Mandal, Suprava Nath and Shreyas Bagrecha

4.1 Introduction 57
4.2 Types of problematic waste 58
4.3 Agro-ecological hazards from problematic wastes 61
4.4 Current technologies for waste management 63
4.5 Types of pyrolysis reactors used for treating problematic wastes 67
4.6 Diversification of problematic wastes through thermal degradation 71
4.7 Challenges of pyrolysis of problematic wastes 72
4.8 Conclusion and future thrust 74
References 74

5. Modern tools and techniques of biochar characterization for targeted applications 81

Rajat Kumar Sharma, T.P. Singh, Juma Haydary, Deepshikha Azad and Akarsh Verma

5.1 Introduction 81
5.2 Properties and characterization of biochar 82
5.3 Conclusion 92
Acknowledgment 92
References 92

6. Modeling the surface chemistry of biochar for efficient and wider applicability: opportunities and limitations 97

Adnan Shakeel, Riya Sawarkar, Suhel Aneesh Ansari, Shrirang Maddalwar and Lal Singh

6.1 Introduction 97
6.2 Modifications in surface functional groups of biochar 98
6.3 Modification techniques for improving the surface chemistry of biochar 99
6.4 Role of feedstock in enhancing the surface chemistry of biochar 100
6.5 Role of pyrolysis temperature in enhancing the surface chemistry of biochar 103
6.6 Application of nanoparticles to enhance the surface chemistry of biochar 105
6.7 Opportunities, limitations, and conclusion 107
Acknowledgment 108
References 108

B
Biochar for soil improvements

7. Biochar-based carbon farming: a holistic approach for crop productivity and soil health improvement 117

Debarati Bhaduri, Bibhash Chandra Verma, Soumya Saha, Trisha Roy and Rubina Khanam

7.1 Introduction 117
7.2 Determinants of biochar stability 122
7.3 Biochar for yield sustenance: facts and prospects from different soils 123
7.4 Effect of biochar application on soil health parameters: an insight of soil physical-fertility-biological factors 125
7.5 Biochar for mitigation of climate change in different soils 128
7.6 Contribution of biochar in economics of carbon farming 129
7.7 Conclusion 130
References 131

8. Biochar as a soil amendment: effects on microbial communities and soil health 137

Tanmaya K. Bhoi, Ipsita Samal, Anuj Saraswat, H.C. Hombegowda, Saubhagya K. Samal, Amit K. Dash, Sonal Sharma, Pramod Lawate, Vipula Vyas and Md Basit Raza

8.1 Introduction 137
8.2 Biochar, its preparation methods, and properties 138
8.3 Biochar qualities associated with microbial diversity 139
8.4 Biochar–microbe interactions 140
8.5 Significance of soil microbial diversity on soil–plant–atmospheric continuum 141
8.6 The potential advantages of biochar and its application in enhancing soil health and promoting plant growth 142
8.7 Mechanism of plant growth stimulation by plant growth promoting rhizobacteria 143
8.8 Interaction of biochar and plant growth promoting rhizobacterias 144
8.9 Factors influencing the effectiveness of biochar 151
8.10 Considerations and difficulties 152
8.11 Conclusion and future direction 153
References 154

9. Biochar mediated carbon and nutrient dynamics under arable land 161

Adeel Abbas, Rashida Hameed, Aitezaz A.A. Shahani, Wajid Ali Khattak, Ping Huang and Daolin Du

9.1 Introduction 161
9.2 Characteristics and production of biochar 163

9.3 Influence of biochar on agriculture productivity 164
9.4 Biochar effects on soil carbon dynamics and soil fertility 168
9.5 Mechanisms of biochar-mediated carbon sequestration 169
9.6 Biochar effects on different soil nutrient dynamics 170
9.7 Uses and limitations of biochar in agricultural arable land 175
9.8 Conclusion 177
References 177

10. Role of biochar in acidic soils amelioration 185

Nidhi Luthra, Shakti Om Pathak, Arham Tater, Samarth Tewari, Pooja Nain, Rashmi Sharma, Daniel Prakash Kushwaha, Manoj Kumar Bhatt, Susheel Kumar Singh and Ashish Kaushal

10.1 Introduction 185
10.2 Properties of the biochar 187
10.3 Extent of acid soils: world and India 188
10.4 Biochar interactions in acidic soils 188
10.5 Biochar and soil carbon sequestration 195
10.6 Future challenges and perspective 197
10.7 Conclusions 198
References 198

11. Biochar as a soil amendment for saline soils reclamation: mechanisms and efficacy 205

Rashida Hameed, Adeel Abbas, Guanlin Li, Aitezaz A.A. Shahani, Beenish Roha and Daolin Du

11.1 Introduction 205
11.2 Biochar can serve as a viable amendment for soil with high salinity 207
11.3 Role of biochar on soil properties 209
11.4 Conclusion 220
References 220

12. Biochar imparting abiotic stress resilience 227

Debarati Datta, Sourav Ghosh, Kajal Das, Shiv Vendra Singh, Sonali Paul Mazumdar, Sandip Mandal and Yogeshwar Singh

12.1 Introduction 227
12.2 Drought 229
12.3 Flood 233
12.4 Salinization 235
12.5 Heat stress 238
12.6 Heavy metals 240
12.7 Conclusion and future perspective 242
References 242

13. Biochar application in sustainable production of horticultural crops in the new era of soilless cultivation 249

Ashish Kaushal, Rajeev Kumar Yadav and Neeraj Singh

13.1 Introduction 249
13.2 Biochar for horticultural rooting media 250
13.3 Biochar-mediated suppression of abiotic stresses 255
13.4 Biochar-mediated suppression of biotic stresses 259
13.5 Enhanced seed germination using biochar 261
13.6 Biochar-mediated enhanced root growth of horticultural crops 262
13.7 Conclusion and future perspectives 262
References 263

14. Biochar-based slow-release fertilizers toward sustainable nutrition supply 269

Xiuxiu Zhang, Dan Luo and Chongqing Wang

14.1 Introduction 269
14.2 Biochar-based slow-release fertilizers 270
14.3 Preparation methods of biochar-based slow-release fertilizers 271
14.4 Applications of biochar-based slow-release fertilizers 277
14.5 Challenges and future perspectives 278
14.6 Conclusions 279
References 280

15. Microbial dynamics and carbon stability under biochar-amended soils 285

Shreyas Bagrecha, Kadagonda Nithinkumar, Nilutpal Saikia, Ram Swaroop Meena, Artika Singh and Shiv Vendra Singh

15.1 Introduction 285

- 15.2 Microbial dynamics in biochar-amended soils 287
- 15.3 Biochar and carbon stability 295
- 15.4 Interactions between microbial dynamics and carbon stability 301
- 15.5 Conclusion and future perspectives 303
- References 304

16. Nutrient enriched and co-composted biochar: system productivity and environmental sustainability 311

Cícero Célio de Figueiredo, Leônidas Carrijo Azevedo Melo, Carlos Alberto Silva, Joisman Fachini, Jefferson Santana da Silva Carneiro, Everton Geraldo de Morais, Ornelle Christiane Ngo Ndoung, Shiv Vendra Singh and Tony Manoj Kumar Nandipamu

- 16.1 Introduction 311
- 16.2 Biochar as a matrix to produce sustainable fertilizers 312
- 16.3 Slow-release mechanisms of biochar-based fertilizers 314
- 16.4 Cocomposted biochar as an enriched nutrient source 323
- 16.5 Conclusion 326
- References 326

C
Environmental sustainability and bioremediation

17. Biochar-led methanogenic and methanotrophic microbial community shift: mitigating methane emissions 335

Tony Manoj Kumar Nandipamu, Prayasi Nayak, Sumit Chaturvedi, Vipin Chandra Dhyani, Rashmi Sharma and Nishanth Tharayil

- 17.1 Introduction 335
- 17.2 Overview of the biochar's potential to mitigate methane emissions 338
- 17.3 Methanogenesis and methane emissions from cropping systems 340
- 17.4 Methanotrophs and methane oxidation from cropping systems 342
- 17.5 Biochar-led methanogenesis and methanotrophy 343
- 17.6 Biochar stimulates the methanogenic and methanotrophic microbial shift 345
- 17.7 Physicochemical properties of biochar relevant to methane mitigation 348
- 17.8 Conclusion and future perspectives 350
- References 351

18. Biochar enhances carbon stability and regulates greenhouse gas flux under crop production systems 359

Anamika Barman, Anurag Bera, Priyanka Saha, Saptaparnee Dey, Suman Sen, Ram Swaroop Meena, Shiv Vendra Singh and Amit Kumar Singh

- 18.1 Introduction 359
- 18.2 Surplus crop residue vis-a-vis biochar production 361
- 18.3 Biochar and its significant properties 362
- 18.4 Carbon storage potential of biochar 364
- 18.5 Biochar mitigating greenhouse gases emission 368
- 18.6 Challenges and future prospects 382
- References 382

19. The use of biochar to reduce carbon footprint: toward net zero emission from agriculture 389

Anurag Bera, Ram Swaroop Meena, Anamika Barman and Priyanka Saha

- 19.1 Introduction 389
- 19.2 Carbon footprint and its significance in the changing global environment 390
- 19.3 Carbon footprint 391
- 19.4 Greenhouse gas emissions and the role of agriculture's production system 395
- 19.5 Potential of biochar to cut down on greenhouse effects 400
- 19.6 Related favorable impacts of biochar on climate change mitigation 403
- 19.7 Conclusion 406
- References 406

20. Biochar as climate-smart strategy to address climate change mitigation and adoption in 21st century 413

Dipita Ghosh, Subodh Kumar Maiti, Sk Asraful Ali, Sayantika Bhattacharya, Tony Manoj Kumar Nandipamu, Biswajit Pramanick and Manpreet Singh Preet

20.1 Introduction 413
20.2 Factors affecting the ability of biochar to fix carbon in the soil 415
20.3 Biochar and climate change 415
20.4 Impact of biochar application on carbon footprint under different landforms 421
20.5 Biochar-based carbon crediting 424
20.6 Conclusions and future recommendations 427
References 427

21. Biochar for pollutants bioremediation from soil and water ecosystem 433

Amit K. Dash, Saloni Tripathy, A. Naveenkumar, Tanmaya K. Bhoi, Arpna Kumari, Divya, Ashish M. Latare, Tony Manoj Kumar Nandipamu, Virendra Singh, Md. Basit Raza, Anuj Saraswat and Jehangir Bhada

21.1 Introduction 433
21.2 Occurrence and fates of emerging contaminants in the ecosystem 436
21.3 Properties of biochar affecting removal of emerging contaminants 437
21.4 Biochar application for removal of emerging contaminants from aqueous system 438
21.5 Biochar application for removal of emerging contaminants from the soil system 442
21.6 Challenges and future perspective 445
21.7 Conclusions 445
References 446

22. Tailored biochar: a win–win strategy to remove inorganic contaminants from soil and water 453

Saptaparnee Dey, T.J. Purakayastha and Anurag Bera

22.1 Introduction 453
22.2 Occurrence and fates of inorganic pollutants in the soil and water ecosystem 454
22.3 Modification methods for the preparation of engineered biochar 455
22.4 Chemical modification of biochar and their effect on contaminant bioremediation from soil and water 459
22.5 Physical modification of biochar and their effect on contaminant bioremediation from soil and water 466
22.6 Biological modification of biochar and their effect on contaminant bioremediation from soil and water 468
22.7 Regeneration of used modified biochar 469
22.8 Future perspective 470
22.9 Conclusion 470
References 471

23. Bioremediation of organic pollutants soil and water through biochar for a healthy ecosystem 479

Diksha Pandey, Nikhil Savio, Nishtha Naudiyal, R.K. Srivastava, Prayasi Nayak, Beatriz Cabañas, Andrés Moreno and Shiv Vendra Singh

23.1 Introduction 479
23.2 Bioremediation of organic contaminants in water using biochar 481
23.3 Mechanisms for organic pollutant removal using biochar 490
23.4 Factors affecting biochar and its impact on soil and water management 494
23.5 Challenges, opportunities, and future research directions 495
23.6 Conclusion 496
References 496

24. Briquetting and pelleting of pine biochar: an income source for small-scale farmers of the hilly area 507

Hemant Kumar Sharma, T.P. Singh and T.K. Bhattacharya

24.1 Introduction 507
24.2 Pine needle biomass 508
24.3 Pine needle biochar 511
24.4 Characteristics of pine needle biochar 516
24.5 Briquetting and pelleting of pine biochar 516

24.6 Briquetting and pelleting as an income source 524
24.7 Rural development and employment for hilly region farmers 526
24.8 Conclusion 527
References 527

25. Engineered biochar: potential application toward agricultural and environmental sustainability 531

Asik Dutta, Abhik Patra, Pooja Nain, Surendra Singh Jatav, Ram Swaroop Meena, Sayon Mukharjee, Ankita Trivedi, Kiran Kumar Mohapatra and Chandini Pradhan

25.1 Introduction 531
25.2 Production technology of engineered biochar 533
25.3 Properties of engineered biochar 537
25.4 Critical factors to maximize the advantages of engineered biochar 539
25.5 Application of engineered biochar 540
25.6 Conclusion 549
References 550

Index 557

List of contributors

Adeel Abbas School of the Environment and Safety Engineering, Institute of Environment and Ecology, Jiangsu University, Zhenjiang, P.R. China

Sk Asraful Ali Division of Agronomy, ICAR—Indian Agricultural Research Institute, Pusa, New Delhi, India

Suhel Aneesh Ansari Environmental Biotechnology & Genomics Division, CSIR—National Environmental Engineering Research Institute, Nagpur, India

Deepshikha Azad Department of Farm Machinery and Power Engineering, G.B. Pant University of Agriculture & Technology, Pantnagar, Uttarakhand, India

Shreyas Bagrecha Agronomy Section, ICAR—National Dairy Research Institute, Karnal, Haryana, India

Anamika Barman Division of Agronomy, ICAR—Indian Agricultural Research Institute, Pusa, New Delhi, India

Anurag Bera Department of Agronomy, Institute of Agricultural Sciences, Banaras Hindu University, Varanasi, Uttar Pradesh, India; Department of Agronomy, Tea Research Association, North Bengal Regional R & D Center, Jalpaiguri, West Bengal, India

Jehangir Bhada Ecosystem Sciences Department, University of Florida, FL, United States

Debarati Bhaduri Crop Production Division, ICAR—National Rice Research Institute, Cuttack, Odisha, India

Manoj Kumar Bhatt Department of Soil Science, Govind Ballabh Pant University of Agriculture & Technology, Pantnagar, Uttarakhand, India

Sayantika Bhattacharya School of Agriculture, Graphic Era Hill University, Dehradun, Uttarakhand, India

T.K. Bhattacharya Department of Farm Machinery and Power Engineering, G. B. Pant University of Agriculture & Technology, Pantnagar, Uttarakhand, India

Tanmaya K. Bhoi Forest Protection Division, ICFRE—Arid Forest Research Institute (ICFRE—AFRI), Jodhpur, Rajasthan, India

Beatriz Cabañas Universidad de Castilla-La Mancha, Instituto de Combustión y Contaminación Atmosférica (ICCA). Ciudad Real, Spain

Jefferson Santana da Silva Carneiro School of Technological Education, Xinguara, Pará, Brazil

Sumit Chaturvedi Department of Agronomy, G. B. Pant University of Agriculture & Technology, Pantnagar, Uttarakhand, India

Kajal Das Division of Germplasm Evaluation, ICAR—National Bureau of Plant Genetic Resources, New Delhi, Delhi, India

Amit K. Dash ICAR—Indian Institute of Seed Science, Mau, Uttar Pradesh, India

Debarati Datta Division of Crop Production, ICAR—Central Research Institute for Jute and Allied Fibres, Barrackpore, West Bengal, India

Saptaparnee Dey Division of Soil Science and Agricultural Chemistry, ICAR—Indian Agricultural Research Institute, Pusa, New Delhi, India

Vipin Chandra Dhyani Department of Agronomy, G. B. Pant University of Agriculture & Technology, Pantnagar, Uttarakhand, India

Divya Department of Agronomy, College of Agriculture, Govind Ballabh Pant University of Agriculture & Technology, Pantnagar, Uttarakhand, India

R. Divyabharathi Department of Renewable Energy Engineering, AEC&RI, Tamil Nadu Agricultural University, Coimbatore, India

Daolin Du School of the Environment and Safety Engineering, Institute of Environment and Ecology, Jiangsu University, Zhenjiang, P.R. China

Asik Dutta Crop Production Division, ICAR—Indian Institute of Pulse Research, Kanpur, Uttar Pradesh, India

Joisman Fachini Department of Agronomy, School of Agronomy and Veterinary Medicine, University of Brasilia, Brasilia, Brazil

Cícero Célio de Figueiredo Department of Agronomy, School of Agronomy and Veterinary Medicine, University of Brasilia, Brasilia, Brazil

Dipita Ghosh Department of Environmental Science & Engineering, Indian Institute of Technology (Indian School of Mines), Dhanbad, Jharkhand, India; Forest Operations and Biomass Utilization, Ecological Restoration Institute, Northern Arizona University, Flagstaff, AZ, United States

Sourav Ghosh Division of Crop Production, ICAR-Central Research Institute for Jute and Allied Fibres, Barrackpore, West Bengal, India

Rashida Hameed School of the Environment and Safety Engineering, Institute of Environment and Ecology, Jiangsu University, Zhenjiang, P.R. China

Juma Haydary Department of Chemical and Biochemical Engineering, Faculty of Chemical and Food Technology, Slovak University of Technology, Bratislava, Slovakia

H.C. Hombegowda ICAR-Indian Institute of Soil and Water Conservation, Dehradun, Uttarakhand, India

Ping Huang School of the Environment and Safety Engineering, Institute of Environment and Ecology, Jiangsu University, Zhenjiang, P.R. China

Surendra Singh Jatav Department of Soil Science and Agricultural Chemistry, Institute of Agricultural Sciences, Banaras Hindu University, Varanasi, Uttar Pradesh, India

Ashish Kaushal Department of Horticulture, School of Agriculture, Graphic Era Hill University, Dehradun, Uttarakhand, India

Rubina Khanam Crop Production Division, ICAR-National Rice Research Institute, Cuttack, Odisha, India

Wajid Ali Khattak School of the Environment and Safety Engineering, Institute of Environment and Ecology, Jiangsu University, Zhenjiang, P.R. China

P. Komalabharathi Department of Agricultural Engineering, Nandha Engineering College, Erode, India

Arpna Kumari Department of Applied Biological Chemistry, The University of Tokyo, Tokyo, Japan

Daniel Prakash Kushwaha Krishi Vigyan Kendra, Rohtas, Bihar Agricultural University, Sabour, Bhagalpur, Bihar, India

Rini Labanya College of Agriculture, Sri Sri Sri University, Cuttack, Odisha, India

Ashish M. Latare Institute of Agricultural Sciences, Banaras Hindu University, Varanasi, Uttar Pradesh, India

Pramod Lawate ICAR—Indian Institute of Soil and Water Conservation, Dehradun, Uttarakhand, India

Guanlin Li School of the Environment and Safety Engineering, Institute of Environment and Ecology, Jiangsu University, Zhenjiang, P.R. China

Dan Luo School of Chemical Engineering, Zhengzhou University, Zhengzhou, P.R. China

Nidhi Luthra Department of Soil Science and Agricultural Chemistry, C.C.R. (PG) College,

Maa Shakumbhari University, Saharanpur, Uttar Pradesh, India

Shrirang Maddalwar Environmental Biotechnology & Genomics Division, CSIR–National Environmental Engineering Research Institute, Nagpur, India

Subodh Kumar Maiti Department of Environmental Science & Engineering, Indian Institute of Technology (Indian School of Mines), Dhanbad, Jharkhand, India

Sandip Mandal Agricultural Energy and Power Division (AEPD), ICAR—Central Institute of Agricultural Engineering, Bhopal, Madhya Pradesh, India

Sonali Paul Mazumdar Division of Crop Production, ICAR-Central Research Institute for Jute and Allied Fibres, Barrackpore, West Bengal, India

Deepak Kumar Meena Division of Agronomy, ICAR—Indian Agricultural Research Institute, Pusa, New Delhi, India

Ram Swaroop Meena Department of Agronomy, Institute of Agricultural Sciences, Banaras Hindu University, Varanasi, Uttar Pradesh, India

Leônidas Carrijo Azevedo Melo Department of Soil Science, School of Agricultural Sciences, Federal University of Lavras, Lavras, Minas Gerais, Brazil

Kiran Kumar Mohapatra Department of Soil Science and Agricultural Chemistry, College of Agriculture, Odisha University of Agriculture and Technology, Bhubaneswar, Odisha, India

Everton Geraldo de Morais Department of Soil Science, School of Agricultural Sciences, Federal University of Lavras, Lavras, Minas Gerais, Brazil

Andrés Moreno Universidad de Castilla-La Mancha, Instituto de Combustión y Contaminación Atmosférica (ICCA). Ciudad Real, Spain

Sayon Mukherjee Department of Soil Science and Agricultural Chemistry, Institute of Agricultural Sciences, Banaras Hindu University, Varanasi, Uttar Pradesh, India

Pooja Nain Department of Soil Science, Govind Ballabh Pant University of Agriculture & Technology, Pantnagar, Uttarakhand, India

Tony Manoj Kumar Nandipamu Department of Agronomy, G. B. Pant University of Agriculture & Technology, Pantnagar, Uttarakhand, India

Suprava Nath Department of Agronomy, M.S. Swaminathan School of Agriculture, Centurion University of Technology and Management, Paralakhemundi, Odisha, India

Nishtha Naudiyal College of Basic Sciences and Humanities, Govind Ballabh Pant University of Agriculture & Technology, Pantnagar, Uttarakhand, India

A. Naveenkumar Division of Soil Science and Agricultural Chemistry, ICAR—Indian Agricultural Research Institute, Pusa, New Delhi, India

Prayasi Nayak College of Basic Sciences and Humanities, Govind Ballabh Pant University of Agriculture & Technology, Pantnagar, Uttarakhand, India; Department of Agronomy, G. B. Pant University of Agriculture & Technology, Pantnagar, Uttarakhand, India

Mohammad Ali Nazari Department of Chemical and Biochemical Engineering, Faculty of Chemical and Food Technology, Slovak University of Technology, Bratislava, Slovakia

Ornelle Christiane Ngo Ndoung Department of Agronomy, School of Agronomy and Veterinary Medicine, University of Brasilia, Brasilia, Brazil

Kadagonda Nithinkumar Division of Agronomy, ICAR—Indian Agricultural Research Institute, Pusa, New Delhi, India

Diksha Pandey College of Basic Sciences and Humanities, Govind Ballabh Pant University of Agriculture & Technology, Pantnagar, Uttarakhand, India

Shashank Patel Division of Agronomy, ICAR—Indian Agricultural Research Institute, Pusa, New Delhi, India

Shakti Om Pathak Department of Soil Science and Agricultural Chemistry, S.G.T. University, Gurugram, Haryana, India

Abhik Patra Department of Soil Science and Agricultural Chemistry, Institute of Agricultural Sciences, Banaras Hindu University, Varanasi, Uttar Pradesh, India; Krishi Vigyan Kendra, Narkatiaganj, West Champaran, Dr. Rajendra Prasad Central Agricultural University, Samastipur, Pusa, Bihar, India

Chandini Pradhan Department of Soil Science and Agricultural Chemistry, Institute of Agricultural Sciences, Banaras Hindu University, Varanasi, Uttar Pradesh, India

Biswajit Pramanick Department of Agronomy, College of Agriculture, Dr. Rajendra Prasad Central Agricultural University, Samastipur, Bihar, India

Manpreet Singh Preet School of Agriculture, Graphic Era Hill University, Dehradun, Uttarakhand, India

T.J. Purakayastha Division of Soil Science and Agricultural Chemistry, ICAR—Indian Agricultural Research Institute, Pusa, New Delhi, India

Md. Basit Raza ICAR-Indian Institute of Soil and Water Conservation, Dehradun, Uttarakhand, India; ICAR-Directorate of Floricultural Research, Pune, Maharashtra, India

Beenish Roha School of Water Resources and Environment, China University of Geosciences (Beijing), Beijing, China

Sougata Roy Division of Agronomy, ICAR—Indian Agricultural Research Institute, Pusa, New Delhi, India

Trisha Roy Division of Soil Science and Agronomy, ICAR-Indian Institute of Soil & Water Conservation, Dehradun, Uttarakhand, India

Priyanka Saha Division of Agronomy, ICAR—Indian Agricultural Research Institute, Pusa, New Delhi, India

Soumya Saha Central Rainfed Upland Rice Research Station (CRURRS), ICAR-NRRI, Hazaribag, Jharkhand, India

Parmanand Sahu ICAR-Central Institute of Agricultural Engineering, Nabibagh, Bhopal, Madhya Pradesh, India

Nilutpal Saikia Division of Agronomy, ICAR–Indian Agricultural Research Institute, Pusa, New Delhi, India

Ipsita Samal ICAR-National Research Centre on Litchi, Muzaffarpur, Bihar, India

Saubhagya K. Samal ICAR-Indian Institute of Soil and Water Conservation, Dehradun, Uttarakhand, India

Anuj Saraswat Department of Soil Science, College of Agriculture, Govind Ballabh Pant University of Agriculture & Technology, Pantnagar, Uttarakhand, India

Nikhil Savio College of Basic Sciences and Humanities, Govind Ballabh Pant University of Agriculture & Technology, Pantnagar, Uttarakhand, India

Riya Sawarkar Environmental Biotechnology & Genomics Division, CSIR–National Environmental Engineering Research Institute, Nagpur, India

Suman Sen Division of Agronomy, ICAR—Indian Agricultural Research Institute, Pusa, New Delhi, India

Aitezaz A.A. Shahani Key Laboratory of Crop Sciences and Plant Breeding Genetics, College of Agriculture, Yanbian University, Yanji, Jilin, P.R. China

Adnan Shakeel Environmental Biotechnology & Genomics Division, CSIR–National Environmental Engineering Research Institute, Nagpur, India

Hemant Kumar Sharma Department of Farm Machinery and Power Engineering, G. B.

Pant University of Agriculture & Technology, Pantnagar, Uttarakhand, India

Rajat Kumar Sharma Department of Farm Machinery and Power Engineering, G.B. Pant University of Agriculture & Technology, Pantnagar, Uttarakhand, India; Department of Chemical and Biochemical Engineering, Faculty of Chemical and Food Technology, Slovak University of Technology, Bratislava, Slovakia

Rashmi Sharma Department of Agronomy, School of Agriculture, Graphic Era Hill University, Dehradun, Uttarakhand, India

Sonal Sharma Department of Soil Science, Maharana Pratap University of Agriculture and Technology, Udaipur, Rajasthan, India

Carlos Alberto Silva Department of Soil Science, School of Agricultural Sciences, Federal University of Lavras, Lavras, Minas Gerais, Brazil

Amit Kumar Singh Department of Agronomy, College of Agriculture, Rani Lakshmi Bai Central Agriculture University, Jhansi, Uttar Pradesh, India

Artika Singh Department of Agronomy, College of Agriculture, Rani Lakshmi Bai Central Agricultural University, Jhansi, Uttar Pradesh, India

Lal Singh Environmental Biotechnology & Genomics Division, CSIR–National Environmental Engineering Research Institute, Nagpur, India; Academy of Scientific and Innovative Research (AcSIR), Ghaziabad, India

Neeraj Singh Department of Horticulture, Ch. Shivnath Singh Shandilya PG College Meerut, Meerut, Uttar Pradesh, India

Shiv Vendra Singh Department of Agronomy, College of Agriculture, Rani Lakshmi Bai Central Agricultural University, Jhansi, Uttar Pradesh, India

Susheel Kumar Singh Department of Soil Science, College of Agriculture, Rani Lakshmi Bai Central Agricultural University, Jhansi, Uttar Pradesh, India

T.P. Singh Department of Farm Machinery and Power Engineering, G.B. Pant University of Agriculture & Technology, Pantnagar, Uttarakhand, India

Virendra Singh Department of Agronomy, College of Agriculture, Govind Ballabh Pant University of Agriculture & Technology, Pantnagar, Uttarakhand, India

Yogeshwar Singh Department of Agronomy, College of Agriculture, Rani Lakshmi Bai Central Agricultural University, Jhansi, Uttar Pradesh, India

R.K. Srivastava College of Basic Sciences and Humanities, Govind Ballabh Pant University of Agriculture & Technology, Pantnagar, Uttarakhand, India

P. Subramanian Department of Renewable Energy Engineering, AEC&RI, Tamil Nadu Agricultural University, Kumulur, India

Arham Tater Division of Soil Science and Agricultural Chemistry, ICAR—Indian Agricultural Research Institute, Pusa, New Delhi, India

Samarth Tewari College of Agriculture Sciences, Teerthanker Mahaveer University, Moradabad, Uttar Pradesh, India

Nishanth Tharayil Plant Ecophysiology, Plant and Environmental Sciences Department, Clemson University, SC, United States

Saloni Tripathy Division of Soil Science and Agricultural Chemistry, ICAR—Indian Agricultural Research Institute, Pusa, New Delhi, India

Ankita Trivedi Division of Soil Science and Agricultural Chemistry, ICAR—Indian Agricultural Research Institute, Pusa, New Delhi, India

Akarsh Verma Department of Mechanical Engineering, University of Petroleum and Energy Studies, Dehradun, Uttarakhand, India

Bibhash Chandra Verma Central Rainfed Upland Rice Research Station (CRURRS), ICAR-NRRI, Hazaribag, Jharkhand, India

Vipula Vyas Forest Protection Division, ICFRE-Arid Forest Research Institute (ICFRE-AFRI), Jodhpur, Rajasthan, India

Chongqing Wang School of Chemical Engineering, Zhengzhou University, Zhengzhou, P.R. China

Rajeev Kumar Yadav Department of Horticulture, School of Agriculture, Graphic Era Hill University, Dehradun, Uttarakhand, India

Xiuxiu Zhang School of Chemical Engineering, Zhengzhou University, Zhengzhou, P.R. China

About the editors

Dr. Shiv Vendra Singh his earned doctoral degree in agronomy from G.B. Pant University of Agriculture and Technology, Pantnagar, Uttarakhand, India. Currently, he is working as an Assistant Professor of agronomy at Rani Lakshmi Bai Central Agricultural University, India. His research works concentrate on the aspects of biochar production and diversification, biochar enrichment, soil quality enhancement, carbon sequestration, and carbon footprint mitigation under changing climatic scenarios. He has published more than 30 scientific publications, including full-length articles, books, book chapters, conference papers, and popular articles. Besides, he has received reputed awards and fellowships from state and national agencies, including Vice-Chancellor Gold Medal; INSPIRE Fellowship, DST, GOI; Young Scientist Award, UCOST-Uttarakhand; and Outstanding Doctoral Research Award by Fertilizer Association of India.

Dr. Sandip Mandal is working as a Senior Scientist at ICAR-Central Institute of Agricultural Engineering, Bhopal, India. He possesses nearly 17 years of research experience in the field of biochar and bioenergy, design of machines and equipment for pyrolysis and gasification, biochar production, enrichment, and carbon sequestration. He has published more than 70 scientific publications, including full-length articles, book chapters, and other technical and scientific articles in national and international journals of repute. He has been selected as NAAS Associate in the year 2021. He was also awarded with the prestigious NRDC Societal Innovation Award 2018 and Jawaharlal Nehru Award for Outstanding Doctoral Thesis Research in Agricultural and Allied Sciences in 2019 by ICAR. As a scientist, he has been associated with the development of machines and equipment for bioenergy generation, biomass management, and conversion of biomass to energy and value-added products through thermochemical and biochemical routes. Dr. Mandal is working on several projects to develop biochar equipment and high-end biochar-based products with NTPC, WWF, and other national agencies.

Dr. Ram Swaroop Meena works in the Department of Agronomy at the Institute of Agricultural Sciences, BHU, Varanasi, Uttar Pradesh. Dr. Meena has secured first division in all the classes with triple NET, JRF, and SRF from the ICAR, and he has awarded with RGNF from the UGC, GOI. Dr. Meena is a Fellow of the National Academy of Agricultural Sciences (FNAAS), National Academy of Biological Sciences (FNABS), and Society for Rapeseed-Mustard Research. He was a Visiting Scientist at IRRI-ISARC under

an INSA Fellowship. Dr. Meena has also been awarded the Raman Research Fellowship by the Ministry of Education, GOI. He has been listed in the World's Top 2% Scientists by the Stanford University, United States, and Elsevier report. He has working experience in 10 externally funded projects as PI/Co-PI, including running projects from IoE, BHU, SERB-DST, ICAR, MOE; GOI. He has the right to one patent on low-cost biochar preparation. Dr. Meena has published more than 120 international research and review papers in peer-reviewed reputed journals with an H-index of 63, I-10 index of 173, citations of more than 11,500, and a total impact factor of 398.37, with the highest 16 and average of 3.6 impact factor.
He has published 4 books at the national level and another 24 books (Springer, Elsevier, and so on) at the international level, and he contributed to the books with 15 chapters at the national level and 70 at the international level. His research aim is to reduce soil organic carbon oxidation and enhance stability in agroecosystems for farmers' benefit through carbon credits.

Dr. Sumit Chaturvedi has over 15-year-long experience in research, teaching, and extension education in the field of Agronomy and Agroforestry. He is working as a Professor in the Department of Agronomy, GBPUAT, Pantnagar. He has more than 100 publications, including research papers in national and international repute journals, book chapter, books/proceedings, manuals/technical bulletins, and poplar articles. He has received various awards to his credits; notable among those are Chancellor and Vice-Chancellor Gold Medal (1999), Best Oral Presentation Award, Best Editor Award (2017) in international event, Faculty Excellence Award, ISA-Indian Associateship Award, and Dr. P. S. Deshmukh Young Agronomist Award, ISA, New Delhi. Presently, he
is associated with ~20 research projects from different government and nongovernment agencies; notable among those are designing controlled release fertilizer utilizing agriwaste biochar, funded by NICRA, ICAR; AICRP-IFS; AICRP-Agroforestry; morpho-physiological and biochemical characterization and mitigation of terminal heat in wheat, UCOST; and FIST-DST, GOI.

Dr. K. Govindaraju is currently working as a Professor (Research) at the Centre for Ocean Research, Sathyabama Institute of Science and Technology, Chennai, India. He has 16 years of research experience in the field of nanotechnology, climate change, nanofertilizer, and biochar-based nanomicronutrient applications. He has published 142 SCI journal articles (Scopus H-index 36 with ~6000 citations), 14 book chapters, and 4 patents. He has received ~20 awards at the national (Young Scientist Award, 2020; National Excellence Award, 2015) and international levels (The IET Premium Award 2017; Outstanding contribution in Reviewing Awards). He has acted
as reviewers for ~50 journals. He has completed 14 research projects funded by DST, DBT, ICAR, NICRA, MoES, and DST-TIFAC.

Preface

Over the past decade, the potential of biochar derived through thermal decomposition has been intensively and excessively embraced by the scientific community for agricultural and environmental applications through improved productivity, soil fertility, water and soil bioremediation, GHG emission mitigation, and green and renewable energy production. However, more considerable continuous efforts are being made to upgrade the pyrolysis process and increase the targeted biochar applicability on a larger scale, still to be acknowledged and compiled as one. The lack of collective literature with in-depth scientific information is the need of the hour. This book covers the advancements and developments of biochar derived from the efficient diversification of crop waste, weeds, agroforestry, agroindustry, municipal waste, and so on toward achieving stunning agricultural and environmental benefits. The book is divided into three sections: biochar production and modification, biochar for soil improvements, and environmental sustainability and bioremediation. Advancements in pyrolysis reactors, hydrothermal carbonization, biochar surface tailoring, and biochar characterization are very well structured in the first section. The second section comprises aspects of soil applications such as carbon farming, biochar-led soil carbon–nutrient dynamics, problematic soil amendment, stress mitigation, and nutrient composite development for improving agricultural productivity. Biochar-mediated GHG mitigation with a detailed understanding of the mechanism, carbon footprint mitigation under changing climatic scenarios, and biofuel productions are covered in the third section. The surface-modified or enriched or tailored biochar is a novel concept utilized for soil amendment, GHG mitigation, bioremediation of metals and organic, inorganic, and other emerging pollutants, degraded and problematic soil restoration, and so on, which have been critically reviewed and compiled. The book also covers the trending multidisciplinary approaches of nanotechnology, surface chemistry, and thermal decomposition advancements being adopted for capacity enhancement and versatile applicability in agriculture, environment science, and green energy synthesis. Hence, the broad-spectrum subject knowledge covered in this book makes it a valuable resource for graduates, research scholars, academicians, the scientific community, industrialists, policymakers, university courses, subject specialization, and industrial perspectives of novel and sustainable agricultural and environmental sustainability.

Shiv Vendra Singh
Sandip Mandal
Ram Swaroop Meena
Sumit Chaturvedi
K. Govindaraju

CHAPTER 1

Crop waste conversion into biochar: an overview

Rini Labanya[1], Parmanand Sahu[2], Sandip Mandal[2], Shiv Vendra Singh[3] and Ram Swaroop Meena[4]

[1]College of Agriculture, Sri Sri Sri University, Cuttack, Odisha, India [2]ICAR-Central Institute of Agricultural Engineering, Nabibagh, Bhopal, Madhya Pradesh, India [3]Department of Agronomy, College of Agriculture, Rani Lakshmi Bai Central Agricultural University, Jhansi, Uttar Pradesh, India [4]Department of Agronomy, Institute of Agriculture Sciences, Banaras Hindu University, Varanasi, Uttar Pradesh, India

1.1 Introduction

The natural balance in the four carbon pools was disturbed when the indiscriminate use of coal and petroleum fuel was started due to Industrial Revolution. The way to reduce the level of carbon dioxide in the atmospheric carbon pool is to convert the carbon dioxide to solid carbon and store it for thousands of years. Biochar is one of these paths taken for balancing the carbon pools (Lehmann & Joseph, 2009; Singh, Chaturvedi, & Datta, 2019). When plant biomass is subjected to the thermo-chemical conversion process (pyrolysis) at a temperature range of 450°C–600°C in the presence of little to no oxygen, the result is a fine-grained, carbon-rich, porous substance known as biochar (Amonette & Joseph, 2009). In contrast to pure carbon, biochar is a mixture of various amounts of carbon (C), hydrogen (H), oxygen (O), nitrogen (N), sulfur (S), and ash (Masek, 2009; Chaturvedi, Singh, Dhyani, Govindaraju, & Mandal, 2021). Biochar is a type of charcoal or black carbon, but unlike black carbon made from fossil fuels or non-biomass waste, it is produced from biomass. Due to its porous structure, biochar improves water retention and increases soil surface area, which makes it a significant asset as a soil amendment.

It is believed that applying charcoal to increase soil fertility has been in practice for several thousand years. Research on the soils of the Amazon basin, known as Terra Preta de Indio, which contains varying amounts of organic black carbon thought to be of anthropogenic origin, has played a significant role in stoking this interest. The present research interest in biochar for carbon sequestration is growing as political and scientific awareness of climate change and issues with increases in residue burning. India produces 500 MT of crop waste per year, of which 141 MT are surplus. The three crops that produce the most agricultural leftovers in India are rice (154 MT), wheat (131 MT), and sugarcane (111 MT). Maize, cotton, and mustard are the other three crops producing significant amounts of crop residues. Due to various restrictions, these residues are either completely unutilized or only partially used. When left untreated, surplus and unneeded agricultural leftovers frequently disturb field preparation, crop establishment, and early crop growth. As a result, they are typically burned on farms, which release air pollutants, including greenhouse gases (GHGs), and results in significant nutrient losses (Singh, Chaturvedi, Dhyani, & Govindaraju, 2020; Singh et al., 2023). The conversion of crop wastes and other unusable biomass into biochar through a thermo-chemical process (slow pyrolysis) is becoming more and more important as a cutting-edge and financially viable alternative method of managing surplus and unusable crop residues and biomass. The surplus residues present in India can be converted into a beneficial material for boosting soil health and crop productivity by converting crop residues and on-site agroforestry residues to biochar and applying it to the soil as a soil amendment. The creation of biochar and its application to soil could be advantageous and provide benefits beyond carbon sequestration. This entails enhancing the physical qualities of soil that are advantageous to crops, enhancing the retention and availability of soil nutrients, enhancing biological activity and resulting in higher crop yields, as well as societal benefits through the reduction of non-CO_2-GHG emissions and the mitigation of global warming through carbon sequestration. Biochar has also become increasingly important for environmental applications in recent years. Due to its favorable structural and surface properties, simple production process, and widely available feedstocks, biochar is considered an efficient, cost-effective, and environmentally friendly way to remove a variety of pollutants. The heterogeneity of biochar's physicochemical properties allows it to be optimized for specific applications.

1.2 Properties of biochar

Grossly biochar is characterized by its carbon content, surface area, pore sizes, and pore volume as these properties influence the soil properties the most. Biochar is typically lightweight due to its porous nature. It can be produced in various sizes, depending on the feedstock and processing methods. Carbon content is considered important in terms of fuel value and soil carbon sequestration. Higher carbon content ensures better stability of biochar in soil. Biochar prepared at higher temperatures contains more carbon and is considered recalcitrant against decaying. High surface area and pore volume make it suitable for the adsorption of contaminants from soil, conserve moisture and also to hold soil nutrients. pH, CEC, and ash content influence the chemical environment of soil and also the crop productivity. Biochar pH can range from slightly acidic to alkaline, depending on the feedstock and pyrolysis conditions. CEC refers to the ability of biochar to attract and hold positively charged ions (cations) like calcium, magnesium, and potassium. This can improve soil fertility. There are many other chemical and physical properties of biochar which are relevant to diverse uses of biochar. These properties and their effects will be discussed in subsequent chapters.

1.3 Biochar production technologies

Numerous technologies have been developed for the conversion of biomass into biochar and similar products, such as combustion, gasification, torrefaction, and pyrolysis (Singh et al., 2016). These technologies are known as thermo-chemical conversion technologies. The process of each technology is distinguished in the aspect of ramp rate, residence time, bio compositions of feedstock, amount of air supply and many more parameters (Sahu, Gangil, & Kumar, 2023).

The majority of technologies have been concentrated on using biomass as a source of energy (Polygeneration et al., 2020). The use of woody biomass to make charcoal through combustion was one of the oldest approaches. However, due to low yields and excessive levels of air pollution, the production method was eventually abandoned. With the passage of time, better technologies were developed to extract energy from organic materials. Many technologies are inefficient in yielding quality products and less emissive gas reduction (Pei-dong, Guomei, & Gang, 2007). In contrast, many studies have shown that pyrolysis is a prominent and viable way to convert recalcitrant carbon (e.g., lignin) into carbon rich product with a satisfactory emission level (Arenas, Navarro, & Martínez, 2019; Durga et al., 2022).

Biochar is typically produced at relatively medium temperatures (400°C–700°C) with a lower ramping rate since the carbon in biomass rapidly degrades into gas/liquid at a high temperature. Biochar is generated during the pyrolysis of feedstock through a sequence of processes including isomerization, dehydration, decarboxylation, repolymerization, and charring (Rangabhashiyam & Balasubramanian, 2019; Zhou, Broadbelt, & Vinu, 2016). Moreover, due to the thermal breakdown of the carbon linked with the cellulose and hemicellulosic chain converted into liquid hydrocarbons via the random dissociation of the polymeric backbone (Zhou et al., 2016), whereas biochar formation is the result of the mid-chain dehydration and repolymerization of carbon linked with lignin polymeric chain. Higher amounts of lignin content in biomass feedstock impart the mass of biochar because, at the elevated temperature range, the presence of $-O-CH_3$ and aliphatic substituents functional group promotes the polymerization into ring chain (charring). The complex structure of the lignin improves thermal stability and the formation of a porous structure of biochar (Kim et al., 2019).

1.3.1 Pyrolysis

The pyrolysis process involves the thermal degradation of the organic matter in a closed chamber (absence of oxygen) at an elevated temperature range. The three main products of pyrolysis are char, oil, and gas, in varying proportions (Shahbaz et al., 2022). However, the reactor conditions could be changed to enhance the production of any one of these three products. Based on the process parameters, pyrolysis is categorized into three major categories: slow, fast, and flash pyrolysis. Modern days' technologies evolved microwave pyrolysis, catalytic pyrolysis, catalytic hydro-pyrolysis, and hydrous and hydro-pyrolysis (Zaman et al., 2017). An in-depth description of pyrolysis types and their important aspects are presented in the subsequent sections. These parameters significantly affected the properties, char yield and the level of carbon in produced biochar (Fig. 1.1).

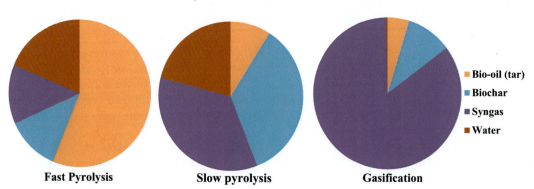

FIGURE 1.1 Product yield in various pyrolysis processes.

1.3.1.1 Types of pyrolysis

To create biochar, a variety of heat conversion techniques can be applied (Deal et al., 2012). The four types of pyrolysis systems used to convert unneeded and surplus crop and agroforestry residues for the creation of biochar are slow pyrolysis, fast pyrolysis, flash pyrolysis, and gasification. A significant output of biochar (35%), produced via slow pyrolysis carried out at lower temperatures (400°C–500°C) and with lengthy contact durations, is common (De Meyer et al., 2011). Faster gasification or pyrolysis operates at higher temperatures (800°C) and produces a high yield of combustible gases in comparison to solid charcoal (12%) (Laird, Fleming, Wang, Horton, & Karlen, 2010). The process of slow pyrolysis is the one used the most frequently to produce biochar. Under oxygen-deprived circumstances in a closed reactor, this technique uses direct thermo-chemical decomposition (exothermic reaction) to convert low-density residue matrix into biochar (Jeffery et al., 2011; Gul et al., 2015). Three steps make up the process in commercial biochar pyrolysis systems: first, moisture and some volatiles are removed; second, unreacted wastes are transformed into volatiles, gases, and biochar; and third, the biochar undergoes a gradual chemical rearrangement (Demirbas, 2004).

1.3.1.1.1 Slow pyrolysis

This technique of pyrolysis was utilized in the early 1900s in industries to produce ethanol, methanol, acetic acid, and coal. Generally, the heating of organic matter takes place comparatively for longer time period, it may be from minutes to over days, at the temperature range of 400°C–650°C with a slower rate of heating ranging from 5°C to 100°C/minute (Chen et al., 2021; Goyal, Seal, & Saxena, 2008). Meanwhile, the thermal cracking of complex structure of the organic matter possessed and restructured with different functional groups to produce char as a solid residue. The combined effect of lower heating rate, lower temperature range, and higher residence time can lead to higher quality char by retarding the formation of liquid and gaseous products (Bertero & Sedran, 2015; Bridgwater, Meier, & Radlein, 1999). Hence, researchers and policymakers have adopted slow pyrolysis as a well-stabilized technology at a pilot scale for biochar production.

1.3.1.1.2 Fast pyrolysis

Fast pyrolysis, often known as rapid pyrolysis, is a decomposition method that promotes the production of superior fluid oil. The aforementioned oil is a medium-density fuel source and can be processed into hydrocarbons in both diesel and gasoline (Strahan, Mullen, & Boateng, 2011). Fast pyrolysis includes the thermal treatment of the organic matter in the absence of oxygen at a temperature range of 600°C–650°C, at a higher rate of heating up to 1000°C/second. This leads to the rapid decomposition of the organic matter, resulting in the building prominently vapors and aerosols with trace amounts of gas and coal. This technique has aroused the interest of researchers and developers worldwide. During the fast pyrolysis process, it ensures the primary degradation of the molecular structure of organic matter. In addition, the technique is extremely versatile, therefore being financially viable. Most importantly, gas generation is higher in fast pyrolysis than in slow pyrolysis (Chen et al., 2021).

1.3.1.1.3 Flash pyrolysis

The method is referred to as flash pyrolysis because it is extremely rapid. However, speed is a key consideration in this process, despite that the methods by which heat and mass are transferred across the feedstock are also crucial aspects (Amenaghawon, Anyalewechi, Okieimen, & Kusuma, 2021). Flash pyrolysis also produces dark brown bio-oil similar to fast pyrolysis due to the decomposition of feedstock, condensation of vapor gasses, and cooling of pyrolysis oil. In this method of pyrolysis, the heating temperature ranges to 1000°C and the heating rate is maintained at more than 700°C/second for a fraction of a second (less than 0.5 seconds) (Bridgwater, 2007). As a result, the process is meticulously optimized to generate high quantities of high-quality oil (Bridgwater, 2007). Significant differentiating aspects include the necessity of heating the feedstock to the appropriate temperatures to allow reaction, the employing of small-sized particles in fluidized bed reactors in order to avoid coke formation and offering faster heat distribution in-depth of the particle surface, as accomplished during ablation (Bridgwater et al., 1999). Flash pyrolysis can produce about 75% bio-oil (Amenaghawon et al., 2021). However, flash pyrolysis has weaker thermal stability. Furthermore, the char has a catalytic effect, resulting in the generation of a more viscous oil with solid residue, rendering it economically unfeasible due to the additional expenditure for quality enhancement (Zaman et al., 2017).

1.3.1.1.4 Catalytic pyrolysis and hydro-pyrolysis

Catalytic pyrolysis and hydro-pyrolysis are two variations of the pyrolysis process. Certain catalysts can be used throughout the pyrolysis process to generate and enhance the outcomes in catalytic pyrolysis methods (Kan et al., 2020). Catalysts must be introduced to the feedstock prior to the pyrolysis process in order to get the requisite volatile and gaseous materials. According to research, ZSM-5 can be employed as a catalyst to produce hydrocarbons (Naqvi, Uemura, Yusup, Sugiura, & Nishiyama, 2015). However, the process enhances coke production, resulting in low efficiency and waste. Recent advances in technology have enabled the production of excellent oil from wood chips, with yields exceeding 60 gallons per ton of biomass (Jullian & Montagne, 2013). In contrast, the hydro-pyrolysis process employs a fluidized bed reactor to conduct pyrolysis under a hydrogen flow, but instead of a fluidized bed, a meta catalyst can be used. The presence of water helps in moderating the temperature and, in some cases, acts as a catalyst to enhance the reactions. The studies suggest that the pressure should be applied ranging from 7 to 34 bars during pyrolysis (Haider et al., 2022). Moreover, due to the intricacy of feeding organic materials into a pressurized pyrolyzer and the presence of hydrogen in the process, the technology is still in its nascent stages. The intricacy raises the expense of construction, lowering its economic viability (Zaman et al., 2017).

The summary of the section is that the pyrolysis process can be categorized as slow and fast pyrolysis based on the rate of heat supply to the biomass particles. In general, slow pyrolysis takes a longer time to attain the pyrolysis temperature and retains the temperature for a longer time. The physical properties and chemical properties of biochar are also influenced by its retention time (Janus et al., 2015; Yuan, Wang, Pan, Shen, & Wu, 2019). The fast pyrolysis of biochar requires rapid heating rates and short residence times, which, decreases secondary reactions and yields. In contrast, long residence time and slow heating rates result in higher biochar yield (Mohan, Rajput, Singh, Steele, & Pittman, 2011). According to Amarasinghe et al. (2016), high carbon content is obtained when raw material is pyrolyzed for 45–60 minutes at 450°C–500°C. Biochar pore size is significantly affected by variations in the retention time of pyrolysis (Indrawati, Ma'as, Utami, & Hanuddin, 2017). The process involved in slow pyrolysis promotes the vapor residence time in organic matter that reflects the production of a better-quality solid product which is known as biochar. Thus, slow pyrolysis is the most adaptable method of pyrolysis for biochar production from a wide range of feedstock. While both the method of pyrolysis slow and fast is performed under anaerobic conditions, whereas the hydro-pyrolysis executed in the hydrogen environment and hydrous pyrolysis is in the presence of water.

1.4 Effect of biochar production process parameters on biochar yield and quality

The effect of biochar production process parameters on biochar yield and quality is a complex issue that has been the subject of much research. The following are some of the most important process parameters that affect biochar production:

- **Pyrolysis temperature:** As mentioned above, the pyrolysis temperature has a significant effect on biochar yield and quality. In general, the higher the pyrolysis temperature, the lower the biochar yield and the higher the carbon content of the biochar. This is because the higher temperatures cause the volatile compounds in the biomass to be vaporized and released, leaving behind a carbon-rich residue. However, the pyrolysis temperature also affects the biochar's physical and chemical properties, such as its porosity, surface area, and reactivity. For example, biochar produced at higher temperatures is typically more porous and has a higher surface area, which can make it more effective for some applications, such as soil amendment.
- **Feedstock:** The type of biomass used as feedstock also affects biochar yield and quality. Biomass with a high lignin content, such as wood, will produce biochar with a higher yield and carbon content than biomass with a low lignin content, such as straw. This is

because lignin is a relatively stable compound that is not easily decomposed during pyrolysis. As a result, biochar produced from wood will typically have a higher carbon content than biochar produced from straw.
- **Particle size:** The particle size of the biomass feedstock also affects biochar yield and quality. Smaller particles will produce biochar with a higher yield and carbon content than larger particles. This is because smaller particles have a larger surface area, which allows them to react more easily with the heat and produce biochar.
- **Heating rate:** The heating rate is the rate at which the biomass is heated during pyrolysis. A faster heating rate will produce biochar with a lower yield and carbon content than a slower heating rate. This is because a faster heating rate does not allow the volatile compounds in the biomass to escape, resulting in a lower yield of biochar.
- **Residence time:** The residence time is the amount of time that the biomass is held in the pyrolysis reactor. A longer residence time will produce biochar with a higher yield and carbon content than a shorter residence time. This is because a longer residence time allows the heat to have more time to react with the biomass, resulting in a higher yield of biochar.
- **Pressure:** The pressure in the pyrolysis reactor can also affect biochar yield and quality. Higher pressures will produce biochar with a higher yield and carbon content than lower pressures. This is because higher pressures prevents the volatile compounds in the biomass from escaping, resulting in a higher yield of biochar.

1.5 Feedstock for biochar production

Biomass feedstock refers to organic materials used for biochar production. Organic waste materials, agricultural residues, forestry residues, and energy crops are acceptable feedstocks for biochar production. Listed below are some of the most commonly used biomass feedstocks for biochar production.

1. Agricultural residues:
   - Crop residues (e.g., wheat straw, paddy straw, corn cob, pigeon pea stalk, etc.);
   - Animal manure and bedding materials (e.g., dairy manure, poultry litter, etc.);
   - Food processing residues (e.g., sugarcane bagasse, fruit pomace, etc.).
2. Forestry residues:
   - Forest residue and logging slash;
   - Sawdust and wood chips;
   - Bark and tree trimmings.
3. Energy crops:
   - Fast-growing plant specially cultivated for bioenergy (e.g., switchgrass, miscanthus);
   - Dedicated energy crop residues.
4. Organic waste materials:
   - Municipal solid waste and sewage sludge;
   - Food waste and crop processing waste;
   - Animal by-products.

Several factors need to be considered when choosing biomass feedstock, such as availability, cost, local regulations, and desired biochar properties. Depending on the feedstock used, biochars may have different porosity, surface area, and nutrient content, making them more suitable for specific purposes. The characteristics and implications of different feedstocks impact biochar production and application. Study shows that Cd can be effectively removed from contaminated soil by applying biochar derived from municipal sewage sludge, with the removal efficiency increasing with a higher application rate (Chen, Peng, & Bi, 2015). Also, it is needed to consider that feedstock selection should not create adverse environmental impacts. To determine the best feedstock for specific biochar production, it is recommended to use scientific literature and relevant studies. Tong, Li, Yuan, and Xu (2011) found that biochar produced from leguminous biomass has a greater removal capacity of Cu than the biochar derived from other plants. A variety of high-molecular lignocellulosic materials can be used to make biochar (Thomsen, Nielsen, Bruun, & Ahrenfeldt, 2011).

1.6 Applications of biochar

Biochars are diverse materials with distinctive physical and chemical properties, including a stable carbon fraction and high surface area. They are used for a variety of applications, including soil amendment, plant and animal agriculture, filtration, and soil rehabilitation.

1.6.1 Effect of biochar on soil physical and chemical properties

The application of biochar may help degraded soils' physicochemical characteristics. Soil porosity is defined as the proportion of soil pores to total soil volume. Important soil characteristics have a significant impact on plant growth. The three types of pores in the soil differ in size. The soil pores function as a transport network in the soil, allowing for aeration, nutrient movement, and retention, as well as providing a home for a variety of soil microorganisms. Depending on the various characteristics of the biochar utilized and the soil being amended with the biochar, biochar can help to some extent in enhancing the porosity of the soil (Hseu, Jien, Chien, & Liou, 2014; Roy et al., 2022). This increase in soil porosity caused by the addition of biochar may be due to the material's high porosity. Biochar can hold onto soil water because of its surface functioning and porous interior structure (Suliman et al., 2017). Several researches revealed that applying biochar can enhance total pore volume, decrease bulk density of soil, and improve water retention capacity (Abel et al., 2013; Chan, van Zwieten, Meszaros, Downie, & Joseph, 2007).

Bulk density is a term used to indicate how tightly soil particles are linked together. It is the proportion of the mass of oven-dried soil to the soil's overall volume (volume of soil particles plus pore space). Bulk density has a significant impact on the soil's characteristics, which in turn influences plant growth and development. A soil with a higher bulk density is more resistant to root encroachment and less able to absorb water, which has an impact on both the soil and plant system. When biochar is applied, the porosity of the soil increases, which amply demonstrates that the bulk density of the soil reduces (Mukherjee & Lal, 2013).

As a result, biochar improves soil aeration, nutrient movement, and retention, which benefits plant development and soil health.

The ability of soil particles to stick together is referred to as soil aggregation. It is clear that well-aggregated soil has good soil structure, which makes it an ideal medium for the transport of water and nutrients in the soil and their subsequent uptake by plants. Beneficial bacteria are given a safe environment to live in by adding biochar to the soil, protecting them from predators and desiccation. Certain polysaccharides secreted by these microorganisms improve soil aggregation (Herth, Arbestain, & Hedley, 2013). However, very high rate of biochar addition may make survival difficult for soil bacteria , which would impede their activity and ultimately reduce soil aggregation (Paz-Ferreiro, Gasco, Gutierrez, & Mendez, 2012).

The water-holding capacity of soil refers to the greatest amount of water the soil can hold or maintain. This aspect of the soil is crucial for both plant growth and the farmer. In order to support plant growth with the amount of water stored in the soil, soil that can hold a lot of water does not require irrigation to be applied frequently. The addition of biochar to the soil can significantly increases the water holding capacity of soil (Rasa et al., 2018).

Addition of biochar to the soil improves its many qualities, paving the way for better plant growth and development. Due to various pyrolysis parameters, such as pyrolysis temperature, time, etc., that are involved in the creation of biochar from feedstock, the outcomes obtained by applying various biochars may vary. Soil pH and organic carbon content are the two main soil variables to consider when evaluating the impact of biochar on the soil. The departure or enhancement in soil characteristics following the addition of biochar was minimal and very varied, but statistically significant (Biederman & Harpole, 2013). Therefore, biochar ultimately contributes to some enhancement in soil fertility. The pH of the soil can be raised with biochar, although it takes longer time (Haefele et al., 2011).

By raising the pH of acidic soils with biochar, the issue of Al toxicity may be resolved, which would raise the soils' CEC and increase the availability of P and other basic cations that are essential for soil fertility (Peng, Ye, Wang, Zhou, & Sun, 2011). The structural porosity of the biochar would increase the availability of nutrients in the soil. Improvements in total N and soil organic carbon (SOC) have also been noted as a result of applying biochar to the soil (Beck, Johnson, & Spolek, 2011).

As it exhibits hydrophilic, hydrophobic, acidic, and basic properties as well, biochar has a very heterogeneous nature. Biochar may collect soil nutrients in solution form and hold them on its surface for later use thanks to its heterogeneous features. This would lessen the amount of nutrients that are lost from the soil through leaching or other gaseous losses (Yao et al., 2012). This nutrient adsorption or decrease by biochar is dependent on a number of biochar parameters, including feedstock, residence duration, and pyrolysis temperature (Yao et al., 2012; Zheng, Wang, Deng, Herbert, & Xing, 2013). In addition to this, the soil characteristics and the amount of biochar applied to the soil have an impact on how well biochar controls nutrient adsorption and leaching (Laird et al., 2010; Xu, Sun, Shao, & Chang, 2014). Biochar's adsorption ability would enable it to function as a gradual release nitrogen source in soil. This pattern of nutrient release would be reliant on the biochar's desorption capabilities (Ding et al., 2022).

1.6.2 Effect of biochar on soil biological properties

The cycling of nutrients is greatly aided by the vital ecosystem service that soil microorganisms perform. When compared to the addition of fresh organic matter, applying biochar to soil has various impacts on the soil biota, which may have an impact on the diversity, activity, and abundance of biotic communities in the soil (Table 1.1). Instead of giving microorganisms a primary source of nutrients, biochar is expected to alter the physical and chemical conditions of soils to create a more appropriate home for them (Lehmann et al., 2011; Singh et al., 2023). Still little is known about the precise impacts of biochar on the microbial population in soil.

According to research by Lehmann et al. (2011), biochar has the potential to be a source of organic matter and inorganic nutrients for microorganisms and can serve as a refuge to keep them safe from predators (Warnock, Lehman, Kuype, & Rillig, 2007). When inoculated with soil, biochar is a porous material that provides a home for a variety of microorganisms, ultimately increasing their population in the soil system (Hardy et al., 2019; Latini et al., 2019). However, because of the benefits of biochar inoculation, there are some threats. The inoculation of biochar may have negative impacts on several microorganisms resulting from a number of leftover bio-oil molecules and their products of recondensation. Due to its vast surface area, biochar can absorb nutrients when they are present. The different intra- and interspecific interactions of soil bacteria are likewise impacted by biochar. The signaling molecules that are employed by different microbes for cell-to-cell communication readily bind to biochar. The biochar that was pyrolyzed at a higher temperature and had a greater specific surface area adsorbs more signaling chemicals than the biochar that was pyrolyzed at a lower temperature, and it more thoroughly inhibited the transfer of signal between cells. The change of microbe-to-plant communication in the rhizospheric soil can also be aided by biochar. This alteration may have an impact on both harmful and helpful pathogens, enabling plants to defend themselves against soil-borne diseases. Because it lessens the toxicity of many soil pollutants to soil microbes, biochar also functions as a soil ameliorant. This decrease in hazardous chemicals for microorganisms results in a decrease in their mortality rate and healthy microorganism growth in the soil system. Depending on the type of biochar used and the type of soil, the effects of biochar on the soil microbial community have been observed to be negative, null, or favorable.

1.6.3 Biochar and heavy metal contamination

For the treatment of soils contaminated with heavy metals, biochar is a possible choice. The concentration of soil contaminants, physicochemical properties, and the stability of the metal-biochar complex, which may be correlated with the type of pyrolysis process, all influence how contaminants and metals are retained and released. The primary mechanisms that are typically taken into consideration for how biochar adheres to heavy metal pollution are as follows: (1) ion exchange, (2) coprecipitation, (3) complexation, and (4) electrostatic absorption (Wang, Xu, Norbu, & Wang, 2018). According to Cui et al. (2011), Cd mobility in soil was significantly reduced after biochar application in both years of the experiment. Under 10, 20, and 40 t ha^{-1} biochar treated plots, respectively, the content of CaCl$_2$-extractable Cd in soil was considerably reduced by 32.0%, 39.2%, and 52.5% in 2009 and by 5.3%, 43.4%, and 39.8% in 2010. The study also showed that rice's overall Cd

TABLE 1.1 Effects of application of biochar on physical, chemical, and biological properties of soils from different field experiments.

Properties	Impact		Biochar feedstock, application rate	Soil type	References
	Positive/ negative/ neutral	% Change			
Bulk density ($kg\,ha^{-1}\,m^{-3}$)	Reduced from 1520 to 1490	−21.68	Corn, 4 t ha^{-1}	Calcareous sandy loam	Saffari, Hajabbasi, Shirani, Mosaddeghi, & Owens (2021)
	Reduced from 1660 to 1300		Birch, 9 t ha^{-1}		Karhu, Mattila, Bergström, and Regina (2011)
Porosity (%)	Increased from 50.9 to 52.8	+1.9	Corn, 4 t ha^{-1}	Calcareous sandy loam	Saffari et al. (2021)
	Increased from 37 to 50	+13	Willow wood, 10 t ha^{-1}	Red soil	Bass, Bird, Kay, and Muirhead, (2016)
Water content (%)	Increased from 7.8 to 19	+11.2	Grain husk, 15 t ha^{-1}	Acidic sandy soil	Farkas et al. (2020)
	Increased from 26 to 31	+5			
Chemical properties					
pH	Increased from 5.9 to 6.2	+5.08	Corn cob, 10 t ha^{-1}	Sandy loam	Amoakwah, Frimpong, Okae-Anti, and Arthur (2017)
	Increased from 8.2 to 8.3	+1.21	Orange peel, 10 t ha^{-1}	Silty clay loam	Sial et al. (2019)
CEC ($cmol\,ha^{-1}\,kg^{-1}$)	Increased from 11.19 to 45.66	+308.04	Woody shrub, 4 t ha^{-1}	Silty loam	Pandit, Mulder, Hale, Schmidt, and Cornelissen, (2017)
	Increased from 6.27 to 9.69	+54.54	Peanut shells & wheat straw	Subtropical landfill cover soil	Lu et al. (2020)
SOC ($g\,ha^{-1}\,kg^{-1}$)	Increased from 3 to 7.5	+150	Corn	Calcareous sandy loam	Saffari et al. (2021)
	Increased from 8.89 to 16.7	+87.85	Orange peel	Silty clay loam	Sial et al. (2019)

(Continued)

TABLE 1.1 (Continued)

Properties	Impact		Biochar feedstock, application rate	Soil type	References
	Positive/ negative/ neutral	% Change			
Total nitrogen (N) (cmol ha^{-1}kg^{-1})	Increased from 0.18 to 0.21	+16.67	Hardwood, 30 t ha^{-1}	Sandy Loam	Adekiya et al. (2020)
	Increased from 1.04 to 1.14	+9.62	Wheat straw, 40 t ha^{-1}	Aqualluvic Primisol	Liu et al. (2014)
Available P (mg ha^{-1}kg^{-1})	Increased from 11.1 to 17.7	+59.46	Hardwood, 30 t ha^{-1}	Sandy Loam	Adekiya et al. (2020)
Available K (mg ha^{-1}kg^{-1})	Increased from 0.17 to 0.21	+23.53	Wheat straw, 40 t ha^{-1}	Aqualluvic Primisol	Liu et al. (2014)
Biological properties					
Acidobacteria	Increased from 4.86 to 5.91%	+1.05	Peanut shells and wheat straw, 10% (v/v)	Subtropical landfill cover soil	Lu et al. (2020)
Proteobacteria	Increased from 21.96 to 24.1%	+2.14	Peanut shells and wheat straw, 10% (v/v)	Subtropical landfill cover soil	Lu et al. (2020)

uptake was dramatically reduced when the biochar amendment was present. If the temperature, contact time, pH, and initial concentration of heavy metals were held constant, according to Komkiene and Baltrenaite (2016), then the amount of adsorbent or biochar used enhanced the number of heavy metals that were able to be adsorbed. Karami et al. (2011) introduced biochar to soil that had been contaminated by Pb and Cu as a result of mining. These workers observed that adding biochar decreased the Pb concentrations in pore water to a level that was half that of the mining soil. The levels of Pb concentrations in the pore water were 20 times lower when biochar and composted green waste were added.

1.6.4 Biochar and carbon sequestration

By using photosynthesis to create organic matter, atmospheric CO_2 is removed from the atmosphere and eventually stored in the soil as long-lasting, stable forms of carbon. Terrestrial, atmospheric, oceanic, and geological systems are significant sources of C. Rapid movement of carbon between pools occurs in the active C pool (Lehmann, 2006; Singh et al., 2023). Moving C into a passive pool with stable or inert C is important to minimize the amount of C in the atmosphere. C may easily flow from the active pool to the passive pool with the help of biochar. Controlled carbonization, as opposed to burning, produces stable C pools from even greater amounts of organic biomass, which are thought to last for generations (Schmidt & Noack, 2000; Glaser et al., 2001). In contrast to the little amounts retained after burning (3%) and biological degradation (less than 10%–20% after 5–10 years), the conversion of biomass carbon to biochar results in the sequestration of roughly 50% of the initial carbon (Lehmann et al., 2006). Different organic materials that cover the soil, such as mulches, manures, and cover crops, can be employed to increase crop production and the soil's carbon pool. However, these additions are only transient in nature since the components quickly disintegrate and mineralize into carbon dioxide (CO_2). In order to maintain a consistent supply of C in the soil, these organic amendments must be applied on a regular basis. Another way to keep carbon in the soil is to stop the ground from being tilled, although this strategy only keeps carbon in the soil for as long as the land is not tilled. Therefore, these crop and soil management techniques only temporarily store soil carbon. So, adding biochar to the soil system may be a useful way to increase the soil's carbon content. When added to soil, biochar would maintain its stability for a very long time. So, biochar has the potential to be an extremely effective source of soil carbon sequestration. A relatively stable soil carbon sink is biochar.

In an experiment using Brandt silty clay loam and maize stover biochar, CO_2-C emissions were reduced by 12% and 14%, respectively, in the low and high WFP treatments (Cheng et al., 2016). The kind of feedstock has a considerable impact on how effectively biomass is converted to biochar, whereas the pyrolysis temperature (which is typically between 350°C and 500°C) has little impact. According to Mohan et al. (2011), when stover or husk biomass is used as soil amendments, it decomposes and releases the majority of its carbon (C) back into the atmosphere primarily as CO_2. However, some of the C may end up in incremental plant growth biomass, which may once again get mineralized in the soil and contribute to atmospheric carbon dioxide within a few years (Lehmann et al., 2006). Contrarily, biochar is a much more durable type of carbon that can last in the soil for hundreds or even thousands of years (Lehmann et al., 2006). Therefore, if used frequently and widely in agriculture, biochar has a significant potential to sequester carbon (Fig. 1.2).

FIGURE 1.2 Biochar—a potential option for carbon buildup under agricultural soils.

1.6.5 Effect of biochar application on crop yield and quality

It has been demonstrated that adding biochar to the soil increases agricultural output and quality, either directly or indirectly. By serving as a source of nutrients, it immediately aids in the growth of the plant. The favorable impact of biochar on the physical, chemical, and biological characteristics of soil may help to explain the indirect effect. According to Tian et al. (2018), the interaction between the biochar application and year had a substantial (p 0.01) impact on cotton yields. With increasing rates of biochar amendments and increasing numbers of applications made at the same rate, the cotton yield rose. When compared to the control in 2014, the seed yield for the plots treated with 5, 10, and 20 t ha^{-1} biochar rose by 9.6%, 12.2%, and 13.5%, respectively. With three dosages of biochar application in 2015, the equivalent increases were 8.1%, 15.4%, and 18.6%, respectively. Cotton lint yield patterns were likewise consistent across all treatments. Applications of biochar and seasons had a considerable impact on plant height, but there was no discernible relationship between them. The biochar amendment and years had significant (p 0.01) effects on cotton fiber length and fiber strength, according to the analysis of variance in this study. According to Mutezo (2013), crop emergence was at its maximum (99.6%) when biochar was applied with half the recommended amount of fertilizer, followed by biochar applied alone. Because charcoal can increase nutrient availability, treatments with biochar and full fertilizer application rates had the longest coleoptiles. According to the research done by Gebremedhin et al. (2015), the treatment that included biochar as a soil amendment was superior to other treatments in terms of increasing grain, straw, and root yields over NP application alone by a factor of 15.7%, 16.5%, and 20%, respectively. According to (Hien, Shinogi, & Taniguchi, 2017), the fruit quality characteristics of sweet potatoes, including dry matter and total sugar content, were significantly improved at the level of p 0.05 with biochar application at both rates of 2% and 4%. Control was determined to have the lowest dry matter content and total sugar content, measuring 32.7%–1.7% and 13.6–0.2 (g/L), respectively. The dry matter content was 39.6% ± 0.5% and the total sugar content was 26.7 ± 0.4 (g/L) in the samples with the highest biochar content (4%).

1.6.6 Control of greenhouse gas emissions

Biochar is a carbon-rich material that can be produced from biomass through pyrolysis. It has been hailed as a potential solution to mitigate GHG emissions by sequestering carbon in soils. The Kyoto Protocol and the Paris Agreement are two international agreements that aim to reduce GHG emissions. These agreements list nitrous oxide (N_2O), carbon dioxide (CO_2), methane (CH_4), hydrofluorocarbons (HFCs), perfluorocarbons (PFCs), and sulfur hexafluoride (SF_6) as the main GHGs to be mitigated. The global warming potentials of different GHGs vary greatly. HFCs, PFCs, and SF6 have global warming potentials that are many thousands of times greater than CO_2, while CH_4 and N_2O have global warming potentials that are 25 and 298 times greater than CO_2, respectively. However, CO_2 accounts for around 55% of total emissions due to its high abundance. Studies have shown biochar addition to soil can lead to significant reductions in N_2O emissions, ranging from 10% to up to 90% in laboratory settings. The exact reasons behind this reduction are still being explored, but several mechanisms are likely involved. Biochar may act as a physical barrier, hindering N_2O diffusion from the soil. It can also increase the soil's capacity to store nitrogen, limiting the availability of precursors for N_2O formation. Additionally, biochar might enhance the activity of microbes that consume N_2O before it escapes into the atmosphere. Biochar addition to soil can lead to a decrease in CH_4 emissions, with an average reduction of around 7%. Biochar's porous structure can enhance soil aeration, reducing the creation of anaerobic environments where methane-producing microbes thrive.

1.6.7 Wastewater treatment

Wastewater is the polluted form of water generated as a byproduct of industrial, commercial, domestic, and agricultural activities and it has become a global concern. Wastewater contains heavy metals, organic pollutants, and nitrogen and phosphorous as a major source of contaminants. After proper treatment, wastewater can be utilized in the industrial sector as a source of cooling water, agricultural sector as irrigation water, and in domestic and aquaculture thus reducing the burden on overexploitation of fresh water from rivers, lakes, streams, etc. Wastewater treatment utilizing biochar has a lot of potential due to its high specific surface area and abundance in surface functional groups which help in the adsorption of various contaminants present in wastewater. Heavy metals are mostly positive in nature. Due to biochar's enormous surface functional groups, it has been widely used to remove heavy metals like arsenic, cadmium and lead from wastewater (Table 1.2). Paper mill biochar when applied to wastewater contaminated with arsenic, it was observed that biochar adsorbed arsenic and the maximum adsorptive capacity was 34.1 mg g^{-1} (Cho et al., 2017). Heavy metal adsorption capacity increased by increasing the pyrolysis temperature as it increased the specific surface area as well as the functional groups. Increasing pyrolysis temperature from 350°C to 650°C increased cadmium

TABLE 1.2 Adsorption of heavy metals and organic contaminants by biochar in wastewater.

	Biochar feedstock	Pyrolysis temperature (°C)	Biochar dose (g/l)	Heavy metals/ organic contaminants	Initial concentration (mg/L)	Adsorption capacity (mg/g)	Removal mechanism	Reference
1	Pinewood	600	2.5	As^{3+}	20	4.38	Electrostatic attraction and surface complexation with hydroxyl groups	Wang, Gao, Li, Zimmerman, and Cao (2016)
2	Rice husk	450–500	1	Cr^{6+}	100	435.7	Introduction of amino group facilitate chemical reduction of Cr^{6+}	Rajapaksha et al. (2016)
3	Banana peels	600	2.5	Cu^{+2}	200	75.99	Electrostatic attraction	Ahmad et al. (2018)
4	Maple wood	500	50	Pb^{+2}	50	43.3	Complexation by oxygen functional groups	Wang, Wang, Lee, Lehmann, and Gao (2018)
5	Switchgrass	–	1	Metribuzin herbicide	100	39.6	Electrostatic attraction	Essandoh, Wolgemuth, Pittman, Mohan, and Mlsna (2017)
6	Sawdust	–	–	Tetracycline	150	86	Site recognition	Zhou et al. (2017b)

removal percentage (Higashikawa, Conz, Colzato, Cerri, & Alleoni, 2016). Organic contaminants, such as pesticides, herbicides and antibiotics constitute another prominent group of pollutants in the aquatic environment. The aromaticity index, polarity index, specific surface area and number of oxygen functional groups of biochar have a direct impact on the effectiveness with which biochar will adsorb organic pollutants (Braghiroli, Bouafif, Neculita, & Koubaa, 2018). Biochars are employed to remove antibiotics (sulfonamides and tetracyclines) from wastewater (Sun et al., 2018) in which attracting groups present on biochar surface plays a significant role (Peiris, Gunatilake, Mlsna, Mohan, & Vithanage, 2017). The primary forms of reactive nitrogen, that is, ammonium, nitrate, and phosphorus in wastewater lead to eutrophication in water bodies. Biochar is employed to remove ammonium and phosphorus from wastewater (Xue et al., 2016).

1.6.8 Other emerging applications

The potential applications of biochar are still being explored, and new applications are being discovered all the time. Biochar is a versatile material with a wide range of potential benefits, and it is likely to play an increasingly important role in a sustainable future.

Here are some specific examples of emerging applications of biochar:

- **Biochar as a catalyst:** Biochar can be used as a catalyst for a variety of reactions, including the production of fuels and chemicals. Biochar can be used to catalyze the water-splitting reaction, which produces hydrogen gas. This reaction is typically catalyzed by metal catalysts, but biochar can also be used as carrier material of base material. Biochar is a more sustainable option than metal catalysts, as it is made from biomass and does not require the mining of metals. Biochar can also be used to catalyze the fermentation of biomass to produce ethanol, biodiesel, methanol and acetic acid.
- **Biochar as a biostimulant:** Biochar has been shown to have a biostimulant effect on plants, meaning that it can improve plant growth and development. The biostimulant effect of biochar is thought to be due to a number of factors, including its high surface area, its ability to retain water, and its ability to chelate nutrients. There is a growing body of research on the use of biochar as a biostimulant. Some studies have shown that biochar can increase crop yields, while others have shown that it can improve the resistance of plants to pests and diseases. More research is needed to fully understand the biostimulant effect of biochar and to optimize its use for plant growth and development. A study in Brazil found that biochar increased the yields of corn, soybean, and sugarcane by 10%−20%. Biochar increased the resistance of wheat to drought stress by 20% whereas, it reduced the incidence of root rot in tomatoes by 50%.
- **Biochar as a biofilter:** Biochar is effective at removing a variety of pollutants from water and air, including heavy metals, organic compounds, bacteria, volatile organic compounds (VOCs), nitrogen oxides, and particulate matter. The use of biochar as a biofilter is a promising technology for improving environmental quality. It is a sustainable, cost-effective, and environmentally friendly alternative to traditional filter materials. It has been found to be effective at removing heavy metals from water, such as lead, copper, and zinc, VOCs from the air, such as benzene and toluene, nutrients from wastewater, such as nitrogen and phosphorus, and storing CO_2 from atmosphere.

- **Biochar as filler material for cement:** Several researchers have explored its potential application as green filler in order to reduce the carbon footprint both of cement production and cement-based construction materials. The 2 wt.% of biochar particles were sufficient to increase the strength and toughness of the cement and mortar composites and in place of the cement in the mixture can maintain the mechanical properties equal to those of the reference samples (Suarez-Riera et al., 2020).

1.7 Policy and legislative framework

The Government of India is currently in the process of developing a National Biochar Policy. The policy is expected to provide a framework for the production, use, and trade of biochar in India. The policy is also expected to address the environmental and social impacts of biochar production and use. The National Biomass Action Plan (NBAP), which was launched in 2018, aims to promote the use of biomass for energy generation, heat and power, and other applications. The NBAP includes a section on biochar, which states that biochar can be used to improve soil quality, reduce GHG emissions, and improve water quality. The Solid Waste Management Rules, 2016, regulate the management of solid waste in India. The rules include provisions for the composting of organic waste, which could also be used to produce biochar. The Environmental Impact Assessment (EIA) Notification, 2006, requires that certain projects undergo an environmental impact assessment (EIA) before they can be approved. The EIA notification includes a list of projects that are considered to be potentially polluting, and biochar production could be considered a potentially polluting activity.

1.8 Overview of current biochar markets

The global biochar market is projected to grow at a CAGR of 11.14% from 2023 to 2032, reaching a value of US$633.31 million by 2032. The growth of the market is driven by the increasing demand for biochar from the agricultural, environmental, and energy sectors. The agricultural sector is the largest end-user of biochar, accounting for over 60% of the market in 2023. Biochar is used in agriculture as a soil amendment to improve soil quality, water retention, and crop yields. It can also be used to reduce the need for chemical fertilizers and pesticides. The environmental sector is the second largest end-user of biochar, accounting for over 25% of the market in 2023. Biochar can be used to mitigate climate change by capturing and storing carbon dioxide from the atmosphere. It can also be used to treat wastewater and contaminated soil. The energy sector is the third largest end-user of biochar, accounting for about 15% of the market in 2023. Biochar can be used as a fuel in biomass power plants and other renewable energy applications. The major regions in the biochar market are Asia Pacific, North America, Europe, and South America. Asia Pacific is the largest market for biochar, accounting for over 65% of the market due to the growing demand for biochar from the agricultural sector in the region. The biochar market is facing some challenges, such as the high cost of production and the lack of awareness about the benefits of biochar. However, the market is expected to grow in the coming years due to the increasing demand from the agricultural, environmental, and energy sectors.

1.9 Conclusion

In addition to being an excellent source of carbon, biochar also possesses a number of qualities that can enhance soil health and boost agricultural productivity. Significant environmental and agricultural benefits may be attained if agricultural biomass wastes, which are currently burned in India and other countries, were pyrolyzed into biochars and used to enrich the soil. First, producing biochar would produce less CO_2 than burning trash and waste outdoors. As a result, rather than just being ash from open burning, a greater portion of the carbon in these wastes would be returned to the soil as biochar. All of the essential micronutrients that are lacking in many agricultural soils are found in biochar, which could provide them in a form that is less likely to be fixed in the soil. Since a sizeable portion of the carbon in biochar does not decompose, it remains trapped in the soil for extended periods of time, preventing global warming and fostering microbial activity. Biochar technology can ensure carbon negativity in food grain production in countries like India and China by adopting an integrated pyrolysis electricity generation model owing to the versatility of the application of biochar.

References

Abel, S., Peters, A., Trinks, S., Schonsky, H., Facklam, M., & Wessolek, G. (2013). Impact of biochar and hydrochar addition on water retention and water repellency of sandy soil. *Geoderma*, 202–203, 183–191.

Adekiya, A. O., Agbede, T. M., Olayanju, A., Ejue, W. S., Adekanye, T. A., Adenusi, T. T., et al. (2020). Effect of biochar on soil properties, soil loss, and cocoyam yield on a tropical sandy loam alfisol. *Scientific World Journal*, 2020, 1–9.

Ahmad, Z., Gao, B., Mosa, A., Yu, H., Yin, X., Bashir, A., ... Wang, S. (2018). Removal of Cu(II), Cd(II) and Pb(II) ions from aqueous solutions by biochars derived from potassium-rich biomass. *Journal of Cleaner Production*, 180, 437–449.

Amenaghawon, A. N., Anyalewechi, C. L., Okieimen, C. O., & Kusuma, H. S. (2021). Biomass pyrolysis technologies for value-added products: A state-of-the-art review. Environment, development and sustainability. Netherlands: Springer. Available from https://doi.org/10.1007/s10668-021-01276-5.

Amoakwah, E., Frimpong, K. A., Okae-Anti, D., & Arthur, E. (2017). Soil water retention, air flow and pore structure characteristics after corn cob biochar application to a tropical sandy loam. *Geofisica International*, 307, 189–197.

Amonette, J., & Joseph, S. (2009). Characteristics of biochar Micro-chemical properties. In *Biochar for environmental management* (pp. 33–52). Science and Technology Earth Scan: London.

Arenas, C. N., Navarro, M. V., & Martínez, J. D. (2019). Pyrolysis kinetics of biomass wastes using isoconversional methods and the distributed activation energy model. *Bioresource Technology*, 288, 121485. Available from https://doi.org/10.1016/j.biortech.2019.121485.

Bass, A. M., Bird, M. I., Kay, G., & Muirhead, B. (2016). Soil properties, greenhouse gas emissions and crop yield under compost, biochar and co-composted biochar in two tropical agronomic systems. *The Science of the Total Environment*, 550, 459–470.

Beck, D. A., Johnson, G. R., & Spolek, G. A. (2011). Amending greenroof soil with biochar to affect runoff water quantity and quality. *Environmental Pollution (Barking, Essex: 1987)*, 159(8–9), 2111–2118.

Bertero, M., & Sedran, U. (2015). Coprocessing of bio-oil in fluid catalytic cracking. *Recent advances in thermochemical conversion of biomass*. Elsevier B.V. Available from https://doi.org/10.1016/B978-0-444-63289-0.00013-2.

Biederman, L. A., & Harpole, W. S. (2013). Biochar and its effects on plant productivity and nutrient cycling: A meta-analysis. *GCB Bioenergy*, 5(2), 202–214.

Braghiroli, F. L., Bouafif, H., Neculita, C. M., & Koubaa, A. (2018). Activated biochar as an effective sorbent for organic and inorganic contaminants in water. *Water, Air, and Soil Pollution*, 229230.

Bridgwater, A. V. (2007). The production of biofuels and renewable chemicals by fast pyrolysis of biomass. *International Journal of Global Energy Issues*, 27, 160–203. Available from https://doi.org/10.1504/IJGEI.2007.013654.

Bridgwater, A. V., Meier, D., & Radlein, D. (1999). An overview of fast pyrolysis of biomass. *Organic Geochemistry, 30*, 1479–1493. Available from https://doi.org/10.1016/S0146-6380(99)00120-5.

Chan, K., van Zwieten, L., Meszaros, I., Downie, A., & Joseph, S. (2007). Agronomic values of greenwaste biochar as a soil amendment. *Australian Journal of Soil Research, 45*, 629–634.

Chaturvedi, S., Singh, S. V., Dhyani, V. C., Govindaraju, K., & Mandal, S. (2021). Characterization, bioenergy value and thermal stability of biochar derived from diverse agriculture and forestry lignocellulosic waste. *Biomass Conversion and Biorefinery*. Available from https://doi.org/10.1007/s13399-020-01239-2.

Chen, W. H., Farooq, W., Shahbaz, M., Naqvi, S. R., Ali, I., Al-Ansari, T., & Saidina Amin, N. A. (2021). Current status of biohydrogen production from lignocellulosic biomass, technical challenges and commercial potential through pyrolysis process. *Energy, 226*120433. Available from https://doi.org/10.1016/j.energy.2021.120433.

Chen, W. H., Peng, J., & Bi, X. T. (2015). A state-of-the-art review of biomass torrefaction, densification and applications. *Renewable Sustainable Energy Review, 44*, 847–866. Available from https://doi.org/10.1016/j.rser.2014.12.039.

Cheng, Y., Cai, Z., Chang, S.X., Wang, J., Zhang, J. (2016). Wheat straw and its biochar have contrasting effects on inorganic N retention and N_2O production in a cultivated Black Chernozem. *Biology and Fertility of Soils, 48*, 941–946. https://doi.org/10.1007/s00374-012-0687-0.

Cho, D.-W., Kwon, G., Yoon, K., Tsang, Y. F., Ok, Y. S., Kwon, E. E., & Song, H. (2017). Simultaneous production of syngas and magnetic biochar via pyrolysis of paper mill sludge using CO_2 as reaction medium. *Energy Conversion. Management, 145*, 1–9.

De Meyer, A., Poesen, J., Isabirye, M., Deckers, J., & Rates, D. (2011). Soil erosion rate in tropical villages: A case study from Lake Victoria Basin, Uganda. *Catena, 84*, 89–98. Available from https://doi.org/10.1016/j.catena.2010.10.001.

Demirbas, A. (2004). Combustion characteristics of different biomass fuels. *Progress in Energy and Combustion Science, 30*, 219–230. Available from https://doi.org/10.1016/j.pecs.2003.10.004.

Deal, C., Brewer, C., Brown, R., Okure, M., & Amoding, A. (2012). Comparison of kiln-derived and gasifier-derived biochars as soil amendments in the humid tropics. *Biomass & Bioenergy, 37*, 161–168. Available from https://10.1016/j.biombioe.2011.12.017.

Demirbaş, A. (2004). Effects of temperature and particle size on bio-char yield from pyrolysis of agricultural residues. *Journal of Analytical and Applied Pyrolysis, 72*, 243–248. Available from https://doi.org/10.1016/j.jaap.2004.07.003.

Ding, Y., Liu, Y. G., Liu, S. B., Li, Z. W., Tan, X. F., Huang, X. X., ... Bhargav, V. K. (2022). Thermal influx induced biopolymeric transitions in paddy straw. *Renewable Energy, 199*, 1024–1032. Available from https://doi.org/10.1016/j.renene.2022.09.054.

Essandoh, M., Wolgemuth, D., Pittman, C. U., Mohan, D., & Mlsna, T. (2017). Adsorption of metribuzin from aqueous solution using magnetic and nonmagnetic sustainable low-cost biochar adsorbents. *Environmental Science and Pollution Control Series., 24*, 4577–4590.

Farkas, É., Feigl, V., Gruiz, K., Vaszita, E., Fekete-Kertész, I., Tolner, M., et al. (2020). Long-term effects of grain husk and paper fibre sludge biochar on acidic and calcareous sandy soils - A scale-up field experiment applying a complex monitoring toolkit. *The Science of the Total Environment, 731*138988.

Glaser, B., Haumaier, L., Guggenberger, G., & Zech, W. (2001). The 'Terra Preta' phenomenon: a model for sustainable agriculture in the humid tropics. *Naturwissenschaften, 88*, 37–41.

Goyal, H. B., Seal, D., & Saxena, R. C. (2008). Bio-fuels from thermochemical conversion of renewable resources: A review. *Renewable Sustainable Energy Review, 12*, 504–517. Available from https://doi.org/10.1016/j.rser.2006.07.014.

Gul, S., Whalen, J. K., Thomas, B. W., Sachdeva, V., & Deng, H. (2015). Physico-Chemical Properties and Microbial Responses in Biochar-Amended Soils: Mechanisms and Future Directions. *Agriculture, Ecosystems & Environment, 206*, 46–59. Available from https://doi.org/10.1016/j.agee.2015.03.015.

Haefele, S., Konboon, Y., Wongboon, W., Amarante, S., Maarifat, A. A., Pfeiffer, E. M., & Knoblauch, C. (2011). Effects and fate of biochar from rice residues in rice-based systems. *Field Crops Research, 121*, 430–440.

Haider, F. U., Coulter, J. A., Cai, L., Hussain, S., Cheema, S. A., Wu, J., & Zhang, R. (2022). An overview on biochar production, its implications, and mechanisms of biochar-induced amelioration of soil and plant characteristics. *Pedosphere, 32*, 107–130. Available from https://doi.org/10.1016/S1002-0160(20)60094-7.

Hardy, B., Sleutel, S., Dufey, J., & Cornelis, J.-T. (2019). The long-term effect of biochar on soil microbial abundance, activity and community structure Is overwritten by land management. *Frontiers in Environmental Science, 7*. Available from https://doi.org/10.3389/fenvs.2019.00110.

Herth, H. M. S. K., Arbestain, M. C., & Hedley, M. (2013). Effect of biochar on soil physical propertiesin two contrasting soils: An Alfisol and an Andisol. *Journal of Geoderma, 209–210*, 188–197.

References

Hien, T., Shinogi, Y., & Taniguchi, T. (2017). The different expressions of draft cherry tomato growth, yield, quality under bamboo and rice husk biochars application to clay loamy soil. *Agricultural Sciences, 8*(9), 934–948. Available from https://doi.org/10.4236/as.2017.89068.

Higashikawa, F. S., Conz, R. F., Colzato, M., Cerri, C. E. P., & Alleoni, L. R. F. (2016). Effects of feedstock type and slow pyrolysis temperature in the production of biochars on the removal of cadmium and nickel from water. *Journal of Cleaner Production, 137*, 965–972.

Hseu, Z. Y., Jien, S. H., Chien, W. H., & Liou, R. C. (2014). Impacts of biochar on physical properties and erosion potential of a mudstone slopeland soil. *The Scientific World Journal, 2014*, 1–10.

Indrawati, U. S. Y. V., Ma'as, A., Utami, S. N. H., & Hanuddin, E. (2017). Characteristics of three biochar types with different pyrolysis time as ameliorant of peat soil. *Indian Journal of Agricultural Research, 51*, 458–462. Available from https://doi.org/10.18805/IJARe.A-274.

Janus, A., Pelfrêne, A., Heymans, S., Deboffe, C., Douay, F., & Waterlot, C. (2015). Elaboration, characteristics and advantages of biochars for the management of contaminated soils with a specific overview on Miscanthus biochars. *Journal of Environmental Management, 162*, 275–289. Available from https://doi.org/10.1016/j.jenvman.2015.07.056.

Jeffery, S., Verheijen, F. G. A, Van Der Velde, M., & Bastos, A. C. (2011). A quantitative review of the effects of biochar application to soils on crop productivity using meta-analysis. *Agriculture, Ecosystems & Environment, 144*, 175–187. Available from https://doi.org/10.1016/j.agee.2011.08.015.

Jullian, S., & Montagne, X. (2013). A short historical review of fast pyrolysis of biomass. *Oil Gas Science Technology, 68*, 621–631. Available from https://doi.org/10.2516/ogst/2013177.

Kan, T., Strezov, V., Evans, T., He, J., Kumar, R., & Lu, Q. (2020). Catalytic pyrolysis of lignocellulosic biomass: A review of variations in process factors and system structure. *Renewable Sustainable Energy Review., 134*110305. Available from https://doi.org/10.1016/j.rser.2020.110305.

Karami, N., Clemente, R., Morenojimenez, E., Lepp, N.W., & Beesley, L. (2011). Efficiency of green waste compost and biochar soil amendments for reducing lead and copper mobility and uptake to ryegrass. *Journal of Hazardous Materials, 191*(1), 41–48.

Karhu, K., Mattila, T., Bergström, I., & Regina, K. (2011). Biochar addition to agricultural soil increased CH_4 uptake and water holding capacity – Results from a short-term pilot field study. *Agriculture, Ecosystems & Environment, 140*, 309–313.

Kim, Y., Ok, J. I., Vithanage, M., Park, Y. K., Lee, J., & Kwon, E. E. (2019). Modification of biochar properties using CO_2. *Chemical Engineering Journal, 372*, 383–389. Available from https://doi.org/10.1016/j.cej.2019.04.170.

Komkiene, J., & Baltrenaite, E. (2016). Biochar as adsorbent for removal of heavy metal ions Cadmium(II), Copper(II), Lead(II), Zinc(II) from aqueous phase. *International Journal of Environmental Science and Technology, 13*, 471–482.

Laird, D., Fleming, P., Wang, B. Q., Horton, R., & Karlen, D. (2010). Biochar impact on nutrient leaching from a Midwestern agricultural soil. *Geoderma, 158*, 436–442.

Lehmann, J., Gaunt, J., & Rondon, M. (2006). Bio-char sequestration in terrestrial ecosystems – A review. *Mitigation and Adaptation Strategies for Global Change, 11*, 395–419.

Lehmann, J., & Joseph, S. (2009). Biochar for environmental management: An introduction. *Biochar for environmental management: Science and technology*. Earthscan Publishers Ltd.

Lehmann, J., Rillig, M. C., Thies, J., Masiello, C. A., Hockaday, W. C., & Crowley, D. (2011). Biochar effects on soil biota – A review. *Soil Biology & Biochemistry, 43*, 1812–1836.

Liu, X., Ye, Y., Liu, Y., Zhang, A., Zhang, X., Li, L., et al. (2014). Sustainable biochar effects for low carbon crop production: A 5-crop season field experiment ona low fertility soil from Central China. *Agricultural Systems, 129*, 22–29.

Lu, H., Yan, M., Wong, M. H., Mo, W. Y., Wang, Y., Chen, X. W., et al. (2020). Effects of biochar on soil microbial community and functional genes of a landfill cover three years after ecological restoration. *The Science of the Total Environment, 717*137133.

Masek, O. 2009. Biochar production technologies, http://www.geos.ed.ac.uk/sccs/biochar/documents/BiocharLaunch-OMasek.pdf. (2009). *BiocharLaunch.* (2009).

Mohan, D., Rajput, S., Singh, V. K., Steele, P. H., & Pittman, C. U. (2011). Modeling and evaluation of chromium remediation from water using low cost bio-char, a green adsorbent. *Journal of Hazardous Materials, 188*, 319–333. Available from https://doi.org/10.1016/j.jhazmat.2011.01.127.

Mukherjee, A., & Lal, R. (2013). Biochar impacts on soil physical properties and greenhouse gas emissions. *Journal of Agronomy, 3*(2), 313–339.

Mutezo, W. T. (2013). Early crop growth and yield responses of maize (Zea mays) to biochar applied on soil NAF, IRN International Working Paper Series. Paper no. 13/03.

Naqvi, S. R., Uemura, Y., Yusup, S., Sugiura, Y., & Nishiyama, N. (2015). In situ catalytic fast pyrolysis of paddy husk pyrolysis vapors over MCM-22 and ITQ-2 zeolites. *Journal of Analytical Applications of Pyrolysis, 114*, 32–39. Available from https://doi.org/10.1016/j.jaap.2015.04.003.

Pandit, N. R., Mulder, J., Hale, S. E., Schmidt, H. P., & Cornelissen, G. (2017). Biochar from "Kon Tiki" flame curtain and other kilns: Effects of nutrient enrichment and kiln type on crop yield and soil chemistry. *PLoS One, 12*.

Paz-Ferreiro, J., Gasco, G., Gutierrez, B., & Mendez, A. (2012). Soil biochemical activities and the geometric mean of enzyme activities after application of sewage sludge and sewage sludge biochar to soil. *Biology and Fertility of Soils, 48*(5), 511–517.

Pei-dong, Z., Guomei, J., & Gang, W. (2007). Contribution to emission reduction of CO_2 and SO_2 by household biogas construction in rural China. *Renewable Sustainable Energy Review., 11*, 1903–1912. Available from https://doi.org/10.1016/j.rser.2005.11.009.

Peiris, C., Gunatilake, S. R., Mlsna, T. E., Mohan, D., & Vithanage, M. (2017). Biochar based removal of antibiotic sulfonamides and tetracyclines in aquatic environments: A critical review. *Bioresource Technology, 246*, 150–159.

Peng, X., Ye, L., Wang, C., Zhou, H., & Sun, B. (2011). Temperature and duration-dependent rice straw-derived biochar: Characteristics and its effects on soil properties of an Ultisol in southern China. *Soil Tillage Research, 112*(2), 159–166.

Polygeneration, I., Plant, P., Firing, B. S. S., Alghassab, M., Samuel, O. D., Khan, Z. A., ... Farooq, M., 2020. Exergoeconomic and environmental modeling of energies 13, 6018.

Rajapaksha, A. U., Chen, S. S., Tsang, D. C. W., Zhang, M., Vithanage, M., Mandal, S., ... Ok, Y. S. (2016). Engineered/designer biochar for contaminant removal/immobilization from soil and water: Potential and implication of biochar modification. *Chemosphere, 148*, 276–291.

Rangabhashiyam, S., & Balasubramanian, P. (2019). The potential of lignocellulosic biomass precursors for biochar production: Performance, mechanism and wastewater application—A review. *Industrial Crops Production, 128*, 405–423. Available from https://doi.org/10.1016/j.indcrop.2018.11.041.

Rasa, K., Heikkinen, J., Hannula, M., Arstila, K., Sampo Kulju, A., & Hyväluoma, J. (2018). How and why does willow biochar increase a clay soil water retention capacity? *Biomass and Bioenergy, 119*, 346–353, Available fromhttps://doi.org/10.1016/j.biombioe.2018.10.004.

Roy, A., Chaturvedi, S., Singh, S., Govindaraju, K., Dhyani, V. C., & Pyne, S. (2022). Preparation and evaluation of two enriched biochar-based fertilizers for nutrient release kinetics and agronomic effectiveness in direct-seeded rice. *Biomass Conversion and Biorefinery*. Available from https://doi.org/10.1007/s13399-022-02488-z.

Saffari, N., Hajabbasi, M. A., Shirani, H., Mosaddeghi, M. R., & Owens, G. (2021). Influence of corn residue biochar on water retention and penetration resistance in a calcareous sandy loam soil - ScienceDirect. *Geoderma, 383*.

Sahu, P., Gangil, S., & Kumar, V. (2023). Biopolymeric transitions under pyrolytic thermal degradation of Pigeon pea stalk. *Renewable Energy, 206*, 157–167. Available from https://doi.org/10.1016/j.renene.2023.02.012.

Schmidt, M. W. I., & Noack, A. G. (2000). Black carbon in soils and sediments: Analysis, distribution, implications, and current challenges. *Global Biogeochemical Cycles, 14*, 777–793.

Shahbaz, M., AlNouss, A., Parthasarathy, P., Abdelaal, A. H., Mackey, H., McKay, G., & Al-Ansari, T. (2022). Investigation of biomass components on the slow pyrolysis products yield using Aspen Plus for techno-economic analysis. *Biomass Conversion of Biorefinery, 12*, 669–681. Available from https://doi.org/10.1007/s13399-020-01040-1.

Sial, T. A., et al. (2019). Evaluation of orange peel waste and its biochar on greenhouse gas emissions and soil biochemical properties within a loess soil. *Waste Management, 87*, 125–134.

Singh, J., Kumar, M., Sharma, A., Pandey, G., Chae, K., & Lee, S. (2016). Influence of process parameters on synthesis of biochar by pyrolysis. *Intech, 11*, 13.

Singh, S., Chaturvedi, S., & Datta, D. (2019). Biochar: An eco-friendly residue management approach. *Indian Farming, 69*(08), 27–29.

Singh, S., Chaturvedi, S., Nayak, P., Dhyani, V. C., Nandipamu, T. M., Singh, D. K., ... Govindaraju, K. (2023). Carbon offset potential of biochar based straw management under rice–wheat system along Indo-Gangetic Plains of India. *Science of the Total Environment, 897*165176. Available from https://doi.org/10.1016/j.scitotenv.2023.165176.

Singh, S., Luthra, N., Mandal, S., Kushwaha, D. P., Pathak, S. O., Datta, D., ... Pramanik, B. (2023). Distinct behavior of biochar modulating biogeochemistry of salt-affected and acidic soil: A review. *Journal of Soil Science and Plant Nutrition*. Available from https://doi.org/10.1007/s42729-023-01370-9.

References

Singh, S. V., Chaturvedi, S., Dhyani, V. C., & Govindaraju, K. (2020). Pyrolysis temperature influences the characteristics of rice straw and husk biochar and sorption/desorption behavior of their biourea composite. *Bioresource Technology*, 314123674. Available from https://doi.org/10.1016/j.biortech.2020.123674.

Strahan, G. D., Mullen, C. A., & Boateng, A. A. (2011). Characterizing biomass fast pyrolysis oils by 13C NMR and chemometric analysis. *Energy and Fuels*, 25, 5452–5461. Available from https://doi.org/10.1021/ef2013166.

Suliman, W., Harsh, J. B., Abu-Lail, N. I., Fortuna, A.-M., Dallmeyer, I., & Garcia-Pérez, M. (2017). The role of biochar porosity and surface functionality in augmenting hydrologic properties of a sandy soil. *Science of the Total Environment*, 574, 139–147.

Sun, P., Li, Y., Meng, T., Zhang, R., Song, M., & Ren, J. (2018). Removal of sulphonamide antibiotics and human metabolite by biochar and biochar/H_2O_2 in synthetic urine. *Water Research*, 147, 91–100.

Thomsen, T., Nielsen, H. H., Bruun, E. W., & Ahrenfeldt, J. (2011). The potential of pyrolysis technology in climate change mitigation: Influence of process design and–parameters, simulated in SuperPro designer software. Riso-R-1764 (EN) National Laboratory for Sustainable Energy Technical University of Denmark. ISBN 978-87-550–3877-6, p. 112.

Tian, X., Li, C., Zhang, M., Wan, W., Xies, Z., Chen, B., & Li, W. (2018). Biochar derived from corn straw affected availability and distribution of soil nutrients and cotton yield. *PLoS One*, 13(1), 2–19.

Tong, X. J., Li, J. Y., Yuan, J. H., & Xu, R. K. (2011). Adsorption of Cu(II) by biochars generated from three crop straws. *Chemical Engineering Journal*, 172, 828–834. Available from https://doi.org/10.1016/j.cej.2011.06.069.

Wang, Q., Wang, B., Lee, X., Lehmann, J., & Gao, B. (2018). Sorption and desorption of Pb(II) to biochar as affected by oxidation and pH. *The Science of the Total Environment*, 634, 188–194.

Wang, S., Gao, B., Li, Y., Zimmerman, A. R., & Cao, X. (2016). Sorption of arsenic onto Ni/ Fe layered double hydroxide (LDH)-biochar composites. *RSC Advances*, 6, 17792–17799.

Wang, S., Xu, Y., Norbu, N., & Wang, Z. (2018). Remediation of biochar on heavy metal polluted soils. *IOP Conference. Series: Earth and Environmental Science*, 108, 042113.

Warnock, D. D., Lehman, J., Kuype, T. W., & Rillig, M. C. (2007). Mycorrhizal responses to biochar in soil e concepts and mechanisms. *Plant and Soil*, 300, 9–20.

Xu, G., Sun, J. N., Shao, H. B., & Chang, S. X. (2014). Biochar had effects on phosphorus sorption and desorption in three soils with differing acidity. *Ecological Engineering*, 62, 54–60.

Xue, L. H., Gao, B., Wan, Y. S., Fang, J. N., Wang, S. S., Li, Y. C., ... Yang, L. Z. (2016). High efficiency and selectivity of MgFe-LDH modified wheatstraw biochar in the removal of nitrate from aqueous solutions. *Journal of the Taiwan Institute of Chemical Engineers*, 63, 312–317.

Yao, Y., Gao, B., Zhang, M., Inyang, M., & Zimmerman, A. R. (2012). Effect of biochar amendment on sorption and leaching of nitrate, ammonium, and phosphate in a sandy soil. *Chemosphere*, 89, 1467–1471.

Yuan, P., Wang, J., Pan, Y., Shen, B., & Wu, C. (2019). Review of biochar for the management of contaminated soil: Preparation, application and prospect. *The Science of the Total Environment*, 659, 473–490. Available from https://doi.org/10.1016/j.scitotenv.2018.12.400.

Zaman, C. Z., Pal, K., Yehye, W. A., Sagadevan, S., Shah, S. T., Adebisi, G. A., ... Johan, R. Bin (2017). Pyrolysis: A sustainable way to generate energy from waste. *Pyrolysis*. Available from https://doi.org/10.5772/intechopen.69036.

Zheng, H., Wang, Z. Y., Deng, X., Herbert, S., & Xing, B. S. (2013). Impacts of adding biochar on nitrogen retention and bioavailability in agricultural soil. *Geoderma*, 206, 32–39.

Zhou, X., Broadbelt, L. J., & Vinu, R. (2016). Mechanistic understanding of thermochemical conversion of polymers and lignocellulosic biomass. *Advances in Chemical Engineering* (1st ed). Elsevier Inc.. Available from https://doi.org/10.1016/bs.ache.2016.09.002.

Zhou, Y., Liu, X., Xiang, Y., Wang, P., Zhang, J., Zhang, F., ... Tang, L. (2017b). Modification of biochar derived from sawdust and its application in removal of tetracycline and copper from aqueous solution: Adsorption mechanism and modelling. *Bioresource Technology*, 245, 266–273.

Cui, H.J., Wang, M.K., Fu, M.L. and Ci, E. 2011. Enhancing phosphorus availability in phosphorus-fertilized zones by reducing phosphate adsorbed on ferrihydrite using rice straw-derived biochar. (2011). *Journal of Soils Sediments*, 11, 1135–1141.

Latini, A., Bacci, G., Teodoro M, Gattia DM, Bevivino A, Trakal L (2019) The impact of soil-applied biochars from different vegetal feedstocks on durum wheat plant performance and rhizospheric bacterial microbiota in low metal-contaminated soil. *Frontiers in Microbiology* 10, 2694. https://doi.org/10.3389/fmicb.2019.02694.

SECTION A

Biochar production and modification

CHAPTER 2

Advanced pyrolysis reactors for energy efficient production of biochar

Rajat Kumar Sharma[1], Sandip Mandal[2], Mohammad Ali Nazari[3], Juma Haydary[4] and Akarsh Verma[5]

[1]Department of Farm Machinery and Power Engineering, G.B. Pant University of Agriculture & Technology, Pantnagar, Uttarakhand, India [2]Division of Agricultural Energy and Power, ICAR—Central Institute of Agricultural Engineering, Bhopal, Madhya Pradesh, India [3]Institute of Chemical and Environmental Engineering, Slovak University of Technology, Bratislava, Slovak Republic [4]Department of Chemical and Biochemical Engineering, Faculty of Chemical and Food Technology, Slovak University of Technology, Bratislava, Slovaki [5]Department of Mechanical Engineering, University of Petroleum and Energy Studies, Dehradun, Uttarakhand, India

2.1 Introduction

Considering the challenges posed by climate change, population expansion, environmental pollution and energy deficit, there is a continuous quest to develop more efficient value chains and sustainable substitutes for traditional energy resources such as coal and fossil fuel (Mandal & Sharma, 2023; Mandal, Kumar Sharma, & Kumar Bhattacharya, 2023). Technologies like solar power and wind energy hold promise for a more environmentally friendly approach to meeting the energy needs of both industries and households. In addition to exploring innovative methods of energy generation, another viable option is the expanded utilization of biomass on an industrial scale (Sharma, Bhattacharya, & Kumain, 2020). Specifically, harnessing agricultural and silvicultural leftovers, whether for energy or recycling purposes, represents a sensible and pragmatic approach (Mandal & Sharma, 2023; Mandal et al., 2023; Sharma, Nazari, & Haydary, 2023).

According to Lal (2005), global crop residue production amounted to approximately 3.8 pentagrams per year, excluding forestry residues. This suggests significant untapped potential if these residual materials can be effectively repurposed (Lal, 2005). One such approach is the application of pyrolysis, which involves the thermochemical conversion of biomass in an oxygen-free environment. This ancient technology has been known to humanity for millennia. What sets pyrolysis apart is its ability to efficiently convert a wide range of organic residues into valuable resources, requiring minimal additional energy input, which presents a practical option compared to conventional thermochemical conversion techniques like combustion or gasification (Kloss, Zehetner, & Dellantonio, 2012; Tsai, Chang, & Lee, 1997).

Biochar, generated from pyrolysis, serves diverse sectors, enhancing soil fertility, mitigating climate change through carbon sequestration, aiding wastewater treatment, acting as a livestock feed additive, supporting renewable energy production, and facilitating bioremediation and odor control (Azad, Pateriya, & Sharma, 2023; Mandal, Sharma, & Bhattacharya, 2022). Additionally, biochar finds use in forestry, landscaping, livelihood enhancement, carbon credit accrual, eco-friendly construction materials, animal husbandry enhancement, fire risk reduction, and an expanding array of industrial and environmental roles (Chaturvedi, Singh, & Dhyani, 2023; Roy, Chaturvedi, & Singh, 2022; Singh, Chaturvedi, et al., 2023; Singh, Luthra, et al., 2023; Vendra Singh, Chaturvedi, Dhyani, & Kasivelu, 2020). Its versatility and sustainability drive its incorporation into numerous fields. Meeting biochar's varying application needs necessitates the development of advanced carbonization reactors (Bhatt, Rene, & Kumar, 2021; Sharma, Singh, & Mandal, 2022).

Importantly, while combustion and gasification are primarily oriented towards using biomass for energy generation, pyrolysis opens additional possibilities by converting biomass residues into more valuable products, which can include chemicals, liquid fuels, and valuable biochar (Mandal, Haydary, & Bhattacharya, 2019). Various types of pyrolysis reactors have been utilized to produce different grades of biochar. In pyrolysis, there are two frequently used reactor designs: rotary kilns and screw reactors. Both of these setups consist of extended tubular chambers and utilize some form of rotational motion to move the feedstock through the reactor (Mandal et al., 2022). Despite considerable research having been conducted on both technologies, there has been limited effort to assess and compare their advantages and disadvantages specifically within the context of biochar production.

Considering these considerations, this chapter explains various pyrolysis reactors used for carbonizing biomass and also sheds light on the pros and cons associated with different reactor types.

2.2 Types of pyrolysis reactors

The central component of any pyrolysis procedure is its reactor, and in recent decades, various reactor designs have been examined to enhance heat transfer efficiency. These reactor arrangements have demonstrated their capability to achieve significant yields of liquid and solid products, constituting around 70%—80% of the initial dry biomass weight (Garcia-Nunez, Pelaez-Samaniego, & Garcia-Perez, 2017). There are five key categories of pyrolysis reactors, encompassing bubbling fluidized-bed reactors, circulating fluidized-bed reactors (comprising both dilute and dense versions), ablative reactors (consisting of cyclonic and plate-type designs), vacuum pyrolysis reactors, and screw reactors, primarily employed for slow pyrolysis. A brief overview of each of these reactor types is illustrated below.

2.2.1 Bubbling fluidized bed

A "bubbling fluidized bed" represents a category of fluidized bed reactors widely utilized in diverse industrial applications. In this configuration, solid particles are suspended and fluidized by a flowing gas or liquid within the reactor. The term "bubbling" describes the movement of these solid particles, causing them to generate small bubbles within the fluidized bed (Garcia-Nunez et al., 2017; Liu, Song, & Ran, 2020). In a bubbling fluidized bed reactor, the fluid's velocity is typically controlled at a level where solid particles move individually or in small groups, creating a bubbling or churning effect. This design finds common use in various processes, including combustion, gasification, and specific chemical reactions. It offers notable advantages by promoting efficient mixing, heat transfer, and mass transfer between the solid particles and the surrounding gas or liquid (Patel, Kundu, & Halder, 2019). The specific structure and application of bubbling fluidized bed reactors can vary depending on the industrial process and the desired outcomes.

Mohan, Pittman, and Steele (2006) detailed the configuration of bubbling fluidized bed reactors, often referred to as fluidized bed reactors, as illustrated in Fig. 2.1. These reactors are characterized by their simplicity in operation and construction (Mohan et al., 2006). They provide effective temperature control and highly efficient heat transfer to biomass particles. Sand is commonly used as a heat transfer medium. The residence time of solids and vapors is regulated by the flow rate of the fluidizing gas. Notably, char particles have a longer residence time compared to vapors and serve as effective catalysts for vapor cracking (Richardson, Drobek, & Julbe, 2015). Thus, rapid and efficient char separation is crucial, typically achieved through cyclone separators. For the separation of very fine char particles, a hot gas filtration system is employed at the outlet.

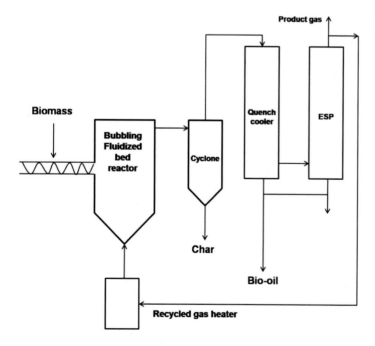

FIGURE 2.1 Bubbling fluidized bed reactor.

2.2.2 Circulating fluidized bed

A "circulating fluidized bed (CFB) reactor" is a type of reactor used in various chemical and industrial processes. In this reactor design, solid particles, often catalysts or reactants, are suspended and constantly kept in motion by a flowing gas or liquid within the reactor. The term "circulated" is associated with the continuous movement of particles within the fluidized bed. In a CFB reactor, solid particles remain in constant circulation within the bed. This dynamic behavior offers numerous benefits in processes like chemical reactions, combustion, and gasification (Garcia-Nunez et al., 2017; Kartal, Sezer, & Özveren, 2022). The ongoing circulation of particles enables efficient mixing, heat transfer, and mass transfer, thereby enhancing the reactor's overall performance. These reactor systems are widely applied in industries such as petroleum and chemicals, serving roles in catalytic cracking, fluidized catalytic cracking units, and other processes that demand effective interaction between solid particles and gases or liquids (Garcia-Nunez et al., 2017).

Mohan et al. (2006) also noted that CFB reactors closely resemble bubbling fluidized beds, except that the residence time for char is nearly equivalent to that for the vapors (as shown in Fig. 2.2) (Mohan et al., 2006). The gas velocity in CFB reactors is relatively high, resulting in more char condensing in the bio-oil. However, the heat transfer rate is not particularly high, primarily relying on gas-solid convective transfer. In certain CFB designs, a common twin-bed reactor is used, with the second reactor serving as a char combustor to reheat the circulating solids (Meyer, Glaser, & Quicker, 2011). This leads to the accumulation of ash in the circulating solids. The ash acts as a cracking catalyst for the organic molecules in the volatile pyrolysis products, resulting in a lower yield of bio-oil and a higher yield of product gas. The primary advantage of CFBs is their suitability for very high throughputs (Mohan et al., 2006).

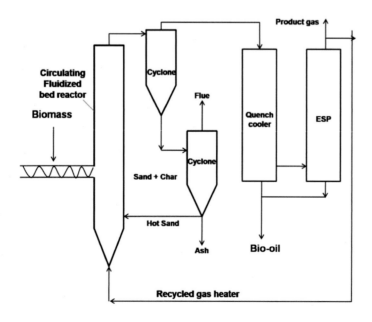

FIGURE 2.2 Circulating fluidized bed reactor.

2.2.3 Ablative reactors

A specialized reactor known as the "vortex ablative bed reactor" is employed in specific industrial processes. This reactor type uses the term "vortex" to indicate the creation of a swirling or rotating motion within it. The "ablative bed" is composed of solid material, often a refractory substance or a bed made up of small solid particles and plays a crucial role in the reaction process (Gupta, Savla, & Pandit, 2022). Within the vortex ablative bed reactor, a high-velocity flow of gas or liquid is directed into the reactor, resulting in a swirling vortex within the ablative bed. This design is typically chosen for processes that require high temperatures and significant mixing or interaction among reactants (Vuppaladadiyam, Vuppaladadiyam, & Sahoo, 2023). It finds applications in various scenarios, such as combustion, gasification, and specific chemical reactions, where it enhances effective heat transfer, mixing, and chemical transformations. The specific design and applications of vortex ablative bed reactors may vary based on the industrial process they are used for.

Ringer, Putsche, and Scahill (2006) introduced ablative reactors, where biomass particles experienced either melting or vaporization from one side of the reactor plate, as illustrated in Fig. 2.3. This innovative method had the potential to accommodate larger particle sizes, reaching up to 20 mm, a significant departure from the 2 mm particle size requirement of fluidized bed designs (Ringer et al., 2006). The process involved propelling biomass particles at high speeds using an inert carrier gas like steam or nitrogen and then introducing them tangentially into the vortex (tubular) reactor. Under these conditions, the particles were compelled to glide along the inner surface of the reactor at elevated speeds, driven by the combination of high velocities and centrifugal force, which pressed the particles

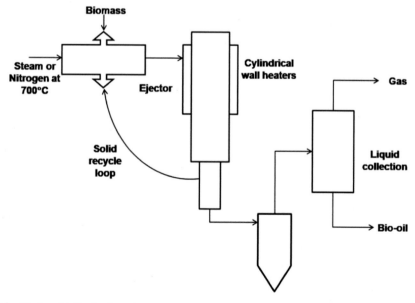

FIGURE 2.3 Vortex ablative bed reactor.

against the reactor wall. The reactor wall's temperature was meticulously maintained at 625°C, effectively causing the particles to liquefy, resembling the process of butter melting in a hot skillet (Ringer et al., 2006). Vapors produced at the surface were swiftly expelled from the reactor by the carrier gases, resulting in vapor residence times of 50–100 ms. As a result, this design was also well-suited for fast pyrolysis and yielded a substantial 65% bio-oil yield.

2.2.4 Vacuum pyrolysis reactors

Vacuum pyrolysis reactors are specialized equipment used in a variety of industrial applications. These reactors operate under reduced pressure conditions, creating a vacuum or low-pressure environment inside the reactor. Vacuum pyrolysis is a thermochemical process where organic materials breakdown without the presence of oxygen, often yielding valuable byproducts like biochar, bio-oil, or syngas (Yadav, Ansari, & Simha, 2016). In a vacuum pyrolysis reactor, organic materials are exposed to high temperatures in a low-pressure environment. This unique condition can influence the pathways of reactions and the quantities of materials produced. These reactors are employed in diverse applications, including waste management, biomass conversion, and the production of biofuels and specialized chemicals (Carrier, Hardie, & Uras, 2012). The specific design and uses of vacuum pyrolysis reactors can vary depending on the industrial process and the desired end products or goals.

Yang et al. (2000) introduced an industrial vacuum pyrolysis reactor measuring 14.6 meters in length and 2.2 meters in diameter, which operated successfully, achieving a feed rate of 3000 kg h^{-1} on a dry biomass basis. Vacuum pyrolysis is typically conducted at low pressures, ranging from 2 to 20 kPa, and at temperatures within the 400°C–500°C range (Yang, Malendoma, & Roy, 2000). The vacuum conditions enable the rapid extraction of pyrolysis products from the hot reaction chamber, preserving the primary fragments resulting from the thermal decomposition reactions. However, one drawback of vacuum pyrolysis technology concerns heat transfer (Yadav et al., 2016). A distinctive feature of the vacuum pyrolysis process is its ability to achieve very short residence times for volatile components. The rapid vaporization of biomass fragments under vacuum conditions minimizes the extent of secondary decomposition reactions (Bardestani & Kaliaguine, 2018). Consequently, the chemical composition of the pyrolysis products more faithfully reflects the original structures of the complex biomolecules present in the initial organic material. Fig. 2.4 depicts the schematic of a vacuum pyrolysis reactor.

2.2.5 Screw reactors

Auger or screw pyrolysis reactors, also referred to as auger reactors, are specialized devices designed for the thermal transformation of organic materials, often biomass, into valuable end products like biochar, bio-oil, and syngas. These reactors are characterized by their elongated, tube-like structure that houses a rotating screw (auger) and a heat source. The principal function of the auger is to convey the biomass feedstock through the reactor, facilitating the pyrolysis process (Brassard, Godbout, & Raghavan, 2017). Auger

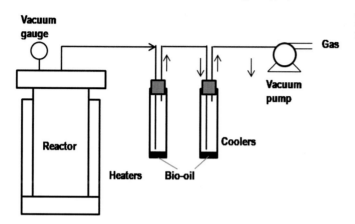

FIGURE 2.4 Vacuum pyrolysis reactors (Yang et al., 2000).

reactors are distinguished by their continuous operation, making them well-suited for maintaining a steady feedstock flow and ensuring efficient and consistent pyrolysis. Operating at relatively lower temperatures, typically around 400°C, auger reactors are especially suitable for gradual pyrolysis processes. Their compact design, efficient heat transfer system, reduced reliance on carrier gases, and the ability to control residence time makes them particularly valuable in research and development contexts and for applications that require precision and operation on a smaller scale (Mandal et al., 2019).

Mohan et al. (2006) presented screw reactors, as depicted in Fig. 2.5, characterizing them as compact systems with minimal carrier gas requirements. They function at lower process temperatures, typically hovering around 400°C, operating continuously (Mohan et al., 2006). The screws efficiently convey biomass feedstock through an oxygen-free cylindrical tube, subjecting it to the requisite pyrolysis temperature, resulting in devolatilization and gasification. This process generates char, and it condenses gases into bio-oil, with noncondensable gases being collected as biogas. As a result, this design leads to reduced energy costs. Furthermore, it offers the flexibility to tailor the residence time of vapor components by extending the length of the heated zone through which the vapors traverse before entering the condensation stage.

2.2.6 Rotary klin

A rotary kiln, sometimes known as a rotary drum, is essentially a long, hollow tube that rotates along its longitudinal axis. Typically, this tube is slightly inclined to facilitate the movement of feedstock from the entrance to the exit. The degree of inclination usually ranges just a few degrees and can be adjusted as required. Rotary kilns are typically made from steel, although on a smaller scale, laboratory kilns can be constructed from quartz glass. In certain instances, rotary kilns may incorporate an inner protective layer of fire-resistant material, referred to as refractory. Fig. 2.6 presents a schematic representation of a basic rotary kiln. Rotary kilns can be operated either continuously or in batch mode. While batch rotary kilns offer advantages in terms of easier operation and simpler

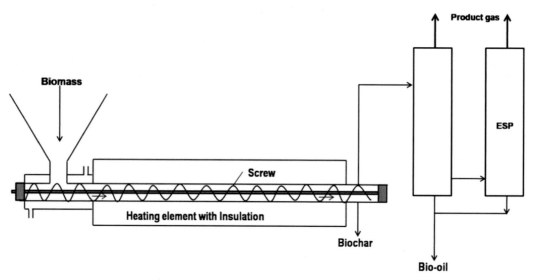

FIGURE 2.5 Screw or screw pyrolysis reactor.

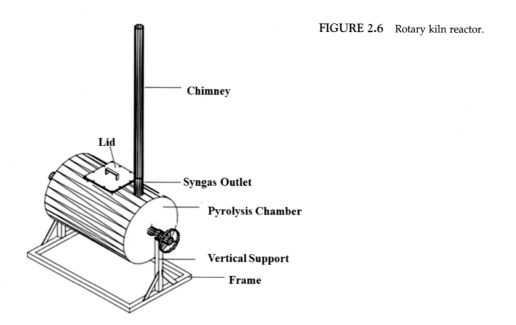

FIGURE 2.6 Rotary kiln reactor.

construction, this paper primarily focuses on continuous rotary kilns because they have a more significant role in research.

Mandal et al. (2022) developed a rotary kiln for the production of biochar from pine needles. During the experiment, the horizontal charring drum exhibited the ability to recover 29% of premium-quality char from pine needles (Mandal et al., 2022). To achieve uniform heat distribution, the drum requires periodic rotation. It's worth emphasizing that the properties of the resulting char vary significantly depending on the specific conditions. Ensuring that there is adequate headspace above the biomass surface is vital for consistent heat distribution and the creation of high-quality char. The charring drum used in the study can process up to 48 kg of shredded pine needles and generate char with characteristics resembling those produced at 400°C.

2.3 Factors affecting the design of pyrolysis reactor

The consideration of a suitable reactor for the carbonization of biochar is a function of different variables such as temperature, residence time, heating rate, type of biomass, particle size of biomass etc. These factors are discussed in a detailed manner in the following section.

2.3.1 Temperature

Elevating the temperature during pyrolysis has an adverse impact on the production of biochar. This temperature increase encourages the thermal decomposition of long-chain hydrocarbons such as lignin and hemicellulose materials, leading to higher yields of bio-oil and syngas byproducts which results in a reduction in biochar production. For instance, raising the temperature from 400°C to 700°C resulted in a 10% reduction in the biochar yield obtained from hazelnut shells (Pütün, Özcan & Pütün, 1999) and a 17% decrease for sesame stalks (Ateş, Pütün & Pütün, 2004). Recent research by Choi, Choi, and Park (2012) corroborates this trend, showing a decline in biochar yield as the pyrolysis temperature increases (Choi et al., 2012). Various published studies on the temperature's impact on the biochar yield during biomass pyrolysis, summarized in Table 2.1, consistently demonstrate a decrease in biochar yield as temperature rises. At higher temperatures, the biochar formed in the initial pyrolysis reactions undergoes secondary reactions, leading to increased formation of liquid and gaseous products at the expense of solid char. In contrast, lower temperatures are more conducive to achieving higher biochar yields because excessive energy supplied to the biomass may be more than the energy needed to shatter the chemical bonds supporting the release of volatile biomass components (Mandal et al., 2022). These volatile components within the biomass are transformed into gases during pyrolysis, resulting in reduced biochar yield. While there is a substantial number of literature addressing the influence of temperature on the yield of biochar, evaluating the optimal temperature for each biomass, still it remains a complex problem due to the significant influence of the specific nature, composition, and type of biomass involved in the process.

2.3.2 Residence time

A long residence time during which vapor resides at a lower temperature is a critical factor when aiming to maximize the production of biochar (Encinar, Beltrán, & Bernalte, 1996). This prolonged residence time provides an ample window for the thorough repolymerization of biomass components, enabling them to engage in comprehensive reactions.

TABLE 2.1 Effect of temperature on biochar yield.

Feedstock	Temperature (°C)	Biochar yield	Reference
Pine needles	300	39.5	Mandal et al. (2022)
Pine needles	400	35.4	Mandal et al. (2022)
Pine needles	500	32.8	Mandal et al. (2022)
Pine needles	600	30.1	Mandal et al. (2022)
Rice husk	400	33	Williams and Nugranad (2000)
Rice husk	600	25.5	Williams and Nugranad (2000)
Corncob	450	30.6	Demirbas (2004a)
Corncob	1250	5.7	Demirbas (2004a)

On the other hand, when the residence period is decreased, the repolymerization process of the biomass's contents is left unfinished, which results in a lower production of biochar (Park, Park, & Kim, 2008). It's worth noting that when the residence period is extended during the quick pyrolysis of materials like poplar wood (Kim, Eom, & Lee, 2011) and yellow-brown coal (Yeasmin, Mathews, & Ouyang, 1999) it has been shown to slightly enhance the char production. Nevertheless, research conducted by Mohamed, Hamzah, Daud, and Zakaria (2013) and Tsai, Lee, and Chang (2007) suggests that while residence time does have an impact on the composition of liquid and gaseous products, its influence on char yield is not particularly substantial. Moreover, by promoting the development of both micro and macro pores, the time that the material stays in the system not only impacts the production of biochar but also has a substantial effect on the characteristics and attributes of the material. Studies have shown that prolonged residence times lead to an enlargement of the pores within the char structure (Tsai et al., 1997). Understanding the relationship between residence time and biochar production is intricate and is greatly shaped by numerous factors, such as temperature, heating rate, and various other parameters. Providing a straightforward description of how residence time functions in biochar synthesis is challenging due to these intricate interconnections.

2.3.3 Particle size

Considering its immediate influence on the speed of heat transfer to the initial biomass in the pyrolysis procedure, particle size is a crucial aspect that warrants thorough investigation. As the particle size increases, so does the gap between the surface and the core of the input biomass, resulting in a slower heat transfer from the high-temperature region to the low-temperature region. This temperature contrast fosters the development of char (Encinar, González, & González, 2000). Furthermore, larger particle sizes compel the vapor generated during the thermal breakdown of biomass to traverse a greater distance within the char layer. This extended travel leads to additional secondary reactions and the generation of extra char. When materials including olive husk, corncobs, and tea trash were pyrolyzed at a temperature of 677°C, Demirbas (2004a, 2004b) examined the effect of particle size on biochar output (Demirbas, 2004b). (Demirbas, 2004a). The findings indicated that as the particle size increased from 0.5 to 2.2 mm, the biochar yield also increased, ranging from 19.4% to 35.6% for olive husk and 5.7% to 16.6% for corncob. A similar pattern of increased biochar production with larger particle sizes was observed in the case of tea waste. Additional studies conducted by Ateş et al. (2004) and Choi et al. (2012) also reported heightened biochar yield when increasing the particle size during the pyrolysis of various biomass materials. In an experiment with wheat straw pyrolysis (Ateş et al., 2004; Choi et al., 2012). Mani, Murugan, Abedi, & Mahinpey (2010) noted that raising the particle size from 0.25 to 0.475 mm led to a considerable biochar yield increase, specifically from 11.85% to 23.28% (Mani et al., 2010). Interestingly, they observed that within the particle size range of 0.475 to 1.35 mm, there was no significant increase in biochar yield. Contrasting to other studies, Onay and Kockar (2003) reported varying findings in their study on rapeseed pyrolysis, observing a decrease in biochar yield as particle size increased from 0.425 to 0.85 mm. However, when the particle size exceeded 0.85 mm, the biochar yield began to increase (Onay & Kockar, 2003). These diverse findings, which present different trends, underscore the fact that the impact of particle size on biochar yield in pyrolysis remains not entirely understood, emphasizing the necessity for further investigations to include these criteria in the design of a reactor.

2.3.4 Heating rate

In the biomass pyrolysis process, the heating rate is crucial because, to a certain extent, it affects the type and content of the final product. A low heating rate avoids thermal cracking of biomass and reduces the possibility of additional pyrolysis processes, increasing the output of biochar. However, a high heating rate promotes the fragmentation of the biomass, which reduces the creation of biochar and increases the generation of gaseous and liquid byproducts. Several studies, such as those by Angın (2013) and Şensöz and Angın (2008), reported a decrease in biochar yield when the heating rate was increased from 30°C to 50°C/minute during the pyrolysis of various biomass materials at different temperatures ranging from 400°C to 600°C. Demirbas (2004a, 2004b) examined the biochar output from the pyrolysis of beech wood bark while adjusting the heating rates at different levels of temperature (Demirbas, 2004b). The formation of char is eventually impeded by high heating rates, which tend to encourage the depolymerization of biomass into its basic volatile components. Secondary pyrolysis processes establish dominance at rapid heating rates and help to produce gaseous byproducts. At lower temperatures, the effect of the heating rate on the production of biochar is more significant.

2.3.5 Feedstocks bed height

Pyrolysis reactors can be broadly categorized into two primary types: fixed-bed reactors and moving-bed reactors. Moving bed reactors are characterized by the motion of the biomass during the pyrolysis process, in contrast to fixed bed reactors where the biomass remains stationary. This mobility in moving bed reactors can be achieved through mechanical means, as seen in rotary bed reactors, or through fluid flow, as observed in fluidized bed reactors, entrained bed reactors, spouted bed reactors, and various other reactor designs (Mandal & Sharma, 2023). In pyrolysis, heat is transferred either by solid-solid or gas-solid heat transfer. Moving bed reactors enable heat to travel from the heat source to the biomass during pyrolysis by combining conduction and convection heat transfer mechanisms. Fixed bed reactors, on the other hand, primarily rely on solid-solid heat transfer processes. Both reactor types of bed heights are crucial variables that have a big impact on product yields. Despite differences between fluidized and fixed bed reactors' liquid and gas output, both reactor types' char yields are mostly constant. Increasing the bed height from 5 to 10 cm has been shown to decrease the biochar yield, lowering it from 28.48% to 25.04%. Zhang, Xiao, Huang, and Xiao (2009) also found that raising the bed height from 5 to 10 cm resulted in a reduction in biochar yield (Zhang et al., 2009). However, successive elevations of the bed caused a rise in char output. Longer vapor residence time, which directly affects biochar production, is a result of the bed height expansion. Lower char yield is the result of insufficient vapor residence time for the repolymerization of volatile product particles when the bed height is reduced. A longer vapor residence time, on the other hand, allows for the repolymerization of biomass particles and, as a result, increases char production.

2.4 Conclusion

The proper technology selection is a complex choice influenced by many different aspects. This chapter examines key components to clarify the key distinctions and parallels between the two technologies to aid in making this decision. All processes offer viable choices for the manufacture of biochar, although the suitability of each depends on the individual feedstock or objectives. The primary conclusions are as follows, to put it clearly. When there is a requirement for very large capacity in industrial applications, choosing rotary kilns is advised. This is because screw reactors may have upper-size capacity restrictions, but rotary kilns may be well-established at an industrial scale for a variety of applications. The large freeboard area of rotary kilns is another benefit as it enables better feedstock and gas interaction in studies utilizing reactive gases, such as steam activation of biochar. Screw reactors, in contrast, have a long history of utilizing solid heat carriers, making them a more practical choice for quick pyrolysis operations. Screw reactors' immovable reactor walls also make it simpler to access products at any time during the process. Screw reactors also take up less room since they often run at greater load factors, which means that they require less capacity to process the same quantity of material.

Acknowledgment

The authors gratefully recognize the assistance offered during the conduct of this work, the National Scholarship Program of Slovakia (NSP-SAIA) and ICAR-AICRP on EAAI. The Slovak Research and Development Agency, grant number APVV-19−0179, is also thanked by the authors for financing the study.

References

Angın, D. (2013). Effect of pyrolysis temperature and heating rate on biochar obtained from pyrolysis of safflower seed press cake. *Bioresource Technology, 128*, 593–597. Available from https://doi.org/10.1016/j.biortech.2012.10.150.

Ateş, F., Pütün, E., & Pütün, A. E. (2004). Fast pyrolysis of sesame stalk: Yields and structural analysis of bio-oil. *Journal of Analytical Applications of Pyrolysis, 71*, 779–790. Available from https://doi.org/10.1016/j.jaap.2003.11.001.

Azad, D., Pateriya, R. N., & Sharma, R. K. (2023). Chemical activation of pine needle and coconut shell biochar: Production, characterization and process optimization. *International Journal of Environmental Science and Technology*. Available from https://doi.org/10.1007/s13762-023-04913-w.

Bardestani, R., & Kaliaguine, S. (2018). Steam activation and mild air oxidation of vacuum pyrolysis biochar. *Biomass Bioenergy, 108*, 101–112. Available from https://doi.org/10.1016/j.biombioe.2017.10.011.

Bhatt, P., Rene, E. R., Kumar, A. J., et al. (2021). Fipronil degradation kinetics and resource recovery potential of Bacillus sp. strain FA4 isolated from a contaminated agricultural field in Uttarakhand, India. *Chemosphere, 276*130156. Available from https://doi.org/10.1016/j.chemosphere.2021.130156.

Brassard, P., Godbout, S., & Raghavan, V. (2017). Pyrolysis in auger reactors for biochar and bio-oil production: A review. *Biosystems Engineering, 161*, 80–92. Available from https://doi.org/10.1016/j.biosystemseng.2017.06.020.

Carrier, M., Hardie, A. G., Uras, Ü., et al. (2012). Production of char from vacuum pyrolysis of South-African sugar cane bagasse and its characterization as activated carbon and biochar. *Journal of Analytical Applications of Pyrolysis, 96*, 24–32. Available from https://doi.org/10.1016/j.jaap.2012.02.016.

Chaturvedi, S., Singh, S. V., Dhyani, V. C., et al. (2023). Characterization, bioenergy value, and thermal stability of biochars derived from diverse agriculture and forestry lignocellulosic wastes. *Biomass Conversion Biorefinement, 13*, 879–892. Available from https://doi.org/10.1007/s13399-020-01239-2.

Choi, H. S., Choi, Y. S., & Park, H. C. (2012). Fast pyrolysis characteristics of lignocellulosic biomass with varying reaction conditions. *Renewable Energy, 42*, 131–135. Available from https://doi.org/10.1016/j.renene.2011.08.049.

Demirbas, A. (2004a). Effects of temperature and particle size on bio-char yield from pyrolysis of agricultural residues. *Journal of Analytical Applications of Pyrolysis, 72*, 243–248. Available from https://doi.org/10.1016/j.jaap.2004.07.003.

Demirbas, A. (2004b). Determination of calorific values of bio-chars and pyro-oils from pyrolysis of beech trunkbarks. *Journal of Analytical Applications of Pyrolysis, 72*, 215–219. Available from https://doi.org/10.1016/j.jaap.2004.06.005.

Encinar, J. M., Beltrán, F. J., Bernalte, A., et al. (1996). Pyrolysis of two agricultural residues: Olive and grape bagasse. Influence of particle size and temperature. *Biomass Bioenergy, 11*, 397–409. Available from https://doi.org/10.1016/S0961-9534(96)00029-3.

Encinar, J. M., González, J. F., & González, J. (2000). Fixed-bed pyrolysis of *Cynara cardunculus* L. Product yields and compositions. *Fuel Processing Technology, 68*, 209–222. Available from https://doi.org/10.1016/S0378-3820(00)00125-9.

Garcia-Nunez, J. A., Pelaez-Samaniego, M. R., Garcia-Perez, M. E., et al. (2017). Historical developments of pyrolysis reactors: A review. *Energy & Fuels, 31*, 5751–5775. Available from https://doi.org/10.1021/acs.energyfuels.7b00641.

Gupta, M., Savla, N., Pandit, C., et al. (2022). Use of biomass-derived biochar in wastewater treatment and power production: A promising solution for a sustainable environment. *Science of The Total Environment, 825*153892. Available from https://doi.org/10.1016/j.scitotenv.2022.153892.

Kartal, F., Sezer, S., & Özveren, U. (2022). Investigation of steam and CO_2 gasification for biochar using a circulating fluidized bed gasifier model in Aspen HYSYS. *Journal of CO_2 Utilization, 62*102078. Available from https://doi.org/10.1016/j.jcou.2022.102078.

Kim, K. H., Eom, I. Y., Lee, S. M., et al. (2011). Investigation of physicochemical properties of biooils produced from yellow poplar wood (Liriodendron tulipifera) at various temperatures and residence times. *Journal of Analytical Applications of Pyrolysis, 92*, 2–9. Available from https://doi.org/10.1016/j.jaap.2011.04.002.

Kloss, S., Zehetner, F., Dellantonio, A., et al. (2012). Characterization of slow pyrolysis biochars: Effects of feedstocks and pyrolysis temperature on biochar properties. *Journal of Environmental Quality, 41*, 990–1000. Available from https://doi.org/10.2134/jeq2011.0070.

Lal, R. (2005). Forest soils and carbon sequestration. *For Ecological Management, 220,* 242−258. Available from https://doi.org/10.1016/j.foreco.2005.08.015.

Liu, Y., Song, Y., Ran, C., et al. (2020). Pyrolysis of furfural residue in a bubbling fluidized bed reactor: Biochar characterization and analysis. *Energy, 211*118966. Available from https://doi.org/10.1016/j.energy.2020.118966.

Mandal, S., Haydary, J., Bhattacharya, T. K., et al. (2019). Valorization of pine needles by thermal conversion to solid, liquid and gaseous fuels in a screw reactor. *Waste Biomass Valorization, 10,* 3587−3599. Available from https://doi.org/10.1007/s12649-018-0386-7.

Mandal, S., Kumar Sharma, R., & Kumar Bhattacharya, T. (2023). Deriving fuel from pine needles through pyrolysis, charring and briquetting and their GHG emission potential. *Current Science, 124,* 1210−1215.

Mandal, S., & Sharma, R. K. (2023). *Application of gasification technology in agriculture for power generation. In:* Handbook of energy management in agriculture (pp. 1−22). Singapore: Springer Nature Singapore.

Mandal, S., Sharma, R. K., Bhattacharya, T. K., et al. (2022). Charring of pine needles using a portable drum reactor. *Chemical Papers, 76,* 1239−1252. Available from https://doi.org/10.1007/s11696-021-01893-4.

Mani, T., Murugan, P., Abedi, J., & Mahinpey, N. (2010). Pyrolysis of wheat straw in a thermogravimetric analyzer: Effect of particle size and heating rate on devolatilization and estimation of global kinetics. *Chemical Engineering Research and Design, 88,* 952−958. Available from https://doi.org/10.1016/j.cherd.2010.02.008.

Meyer, S., Glaser, B., & Quicker, P. (2011). Technical, economical, and climate-related aspects of biochar production technologies: A literature review. *Environmental Science & Technology, 45,* 9473−9483. Available from https://doi.org/10.1021/es201792c.

Mohamed, A. R., Hamzah, Z., Daud, M. Z. M., & Zakaria, Z. (2013). The effects of holding time and the sweeping nitrogen gas flowrates on the pyrolysis of EFB using a fixed−bed reactor. *Procedia Engineering, 53,* 185−191. Available from https://doi.org/10.1016/j.proeng.2013.02.024.

Mohan, D., Pittman, C. U., & Steele, P. H. (2006). Pyrolysis of wood/biomass for bio-oil: A critical review. *Energy & Fuels, 20,* 848−889. Available from https://doi.org/10.1021/ef0502397.

Onay, O., & Kockar, O. M. (2003). Slow, fast and flash pyrolysis of rapeseed. *Renewable Energy, 28,* 2417−2433. Available from https://doi.org/10.1016/S0960-1481(03)00137-X.

Park, H. J., Park, Y.-K., & Kim, J. S. (2008). Influence of reaction conditions and the char separation system on the production of bio-oil from radiata pine sawdust by fast pyrolysis. *Fuel Processing Technology, 89,* 797−802. Available from https://doi.org/10.1016/j.fuproc.2008.01.003.

Patel, S., Kundu, S., Halder, P., et al. (2019). Slow pyrolysis of biosolids in a bubbling fluidised bed reactor using biochar, activated char and lime. *Journal of Analytical Applications of Pyrolysis, 144*104697. Available from https://doi.org/10.1016/j.jaap.2019.104697.

Pütün, A. E., Özcan, A., & Pütün, E. (1999). Pyrolysis of hazelnut shells in a fixed-bed tubular reactor: Yields and structural analysis of bio-oil. *Journal of Analytical Applications of Pyrolysis, 52,* 33−49. Available from https://doi.org/10.1016/S0165-2370(99)00044-3.

Richardson, Y., Drobek, M., Julbe, A., et al. (2015). *Biomass gasification to produce syngas. In:* Recent advances in thermo-chemical conversion of biomass (pp. 213−250). Elsevier.

Ringer M., Putsche V., Scahill J. (2006) Large-scale pyrolysis oil production: A technology assessment and economic analysis. Golden, CO.

Roy, A., Chaturvedi, S., Singh, S. V., et al. (2022). Preparation and evaluation of two enriched biochar-based fertilizers for nutrient release kinetics and agronomic effectiveness in direct-seeded rice. *Biomass Conversion Biorefinement*. Available from https://doi.org/10.1007/s13399-022-02488-z.

Şensöz, S., & Angın, D. (2008). Pyrolysis of safflower (*Charthamus tinctorius* L.) seed press cake: Part 1. The effects of pyrolysis parameters on the product yields. *Bioresource Technology, 99,* 5492−5497. Available from https://doi.org/10.1016/j.biortech.2007.10.046.

Sharma, R. K., Bhattacharya, T. K., Kumain, A., et al. (2020). Energy use pattern in wheat crop production system among different farmer groups of the Himalayan Tarai region. *Current Science, 118,* 448. Available from https://doi.org/10.18520/cs/v118/i3/448-454.

Sharma, R. K., Nazari, M. A., Haydary, J., et al. (2023). A review on advanced processes of biohydrogen generation from lignocellulosic biomass with special emphasis on thermochemical conversion. *Energies (Basel), 16,* 6349. Available from https://doi.org/10.3390/en16176349.

Sharma, R. K., Singh, T. P., Mandal, S., et al. (2022). *Chemical treatments for biochar modification: Opportunities, limitations and advantages. Engineered biochar* (pp. 65−84). Singapore: Springer Nature Singapore.

Singh, S., Chaturvedi, S., Nayak, P., et al. (2023a). Carbon offset potential of biochar based straw management under rice- wheat system along Indo-Gangetic Plains of India. *Science of The Total Environment, 897*165176. Available from https://doi.org/10.1016/j.scitotenv.2023.165176.

Singh, S., Luthra, N., Mandal, S., et al. (2023b). Distinct behavior of biochar modulating biogeochemistry of salt-affected and acidic soil: A review. *Journal of Soil Science and Plant Nutrition, 23*, 2981–2997. Available from https://doi.org/10.1007/s42729-023-01370-9.

Tsai, W. T., Chang, C. Y., & Lee, S. L. (1997). Preparation and characterization of activated carbons from corn cob. *Carbon, 35*, 1198–1200. Available from https://doi.org/10.1016/S0008-6223(97)84654-4.

Tsai, W., Lee, M., & Chang, Y. (2007). Fast pyrolysis of rice husk: Product yields and compositions. *Bioresource Technology, 98*, 22–28. Available from https://doi.org/10.1016/j.biortech.2005.12.005.

Vendra Singh, S., Chaturvedi, S., Dhyani, V. C., & Kasivelu, G. (2020). Pyrolysis temperature influences the characteristics of rice straw and husk biochar and sorption/desorption behaviour of their biourea composite. *Bioresource Technology, 314*123674. Available from https://doi.org/10.1016/j.biortech.2020.123674.

Vuppaladadiyam, A. K., Vuppaladadiyam, S. S. V., Sahoo, A., et al. (2023). Bio-oil and biochar from the pyrolytic conversion of biomass: A current and future perspective on the trade-off between economic, environmental, and technical indicators. *Science of The Total Environment, 857*159155. Available from https://doi.org/10.1016/j.scitotenv.2022.159155.

Williams, P. T., & Nugranad, N. (2000). Comparison of products from the pyrolysis and catalytic pyrolysis of rice husks. *Energy, 25*, 493–513. Available from https://doi.org/10.1016/S0360-5442(00)00009-8.

Yadav, A., Ansari, K. B., Simha, P., et al. (2016). Vacuum pyrolysed biochar for soil amendment. *Resource-Efficient Technologies, 2*, S177–S185. Available from https://doi.org/10.1016/j.reffit.2016.11.004.

Yang, J., Malendoma, C., & Roy, C. (2000). Determination of the overall heat transfer coefficient in a vacuum pyrolysis moving and stirred bed reactor. *Chemical Engineering Research and Design, 78*, 633–642. Available from https://doi.org/10.1205/026387600527581.

Yeasmin, H., Mathews, J. F., & Ouyang, S. (1999). Rapid devolatilisation of Yallourn brown coal at high pressures and temperatures. *Fuel, 78*, 11–24. Available from https://doi.org/10.1016/S0016-2361(98)00119-7.

Zhang, H., Xiao, R., Huang, H., & Xiao, G. (2009). Comparison of non-catalytic and catalytic fast pyrolysis of corn-cob in a fluidized bed reactor. *Bioresource Technology, 100*, 1428–1434. Available from https://doi.org/10.1016/j.biortech.2008.08.031.

CHAPTER 3

Microwave-assisted hydrothermal carbonization for biochar production: potential application and limitations

R. Divyabharathi[1], P. Komalabharathi[2] and P. Subramanian[3]

[1]Department of Renewable Energy Engineering, AEC&RI, Tamil Nadu Agricultural University, Coimbatore, India [2]Department of Agricultural Engineering, Nandha Engineering College, Erode, India [3]Department of Renewable Energy Engineering, AEC&RI, Tamil Nadu Agricultural University, Kumulur, India

3.1 Introduction

Industrialization of countries by enormous usage of electricity generated from coal-based power plants has significantly contributed to the emission of CO_2 and other greenhouse gases (GHGs) to the atmosphere ultimately evolving as the prime reason for global climate change. An alternative solution to mitigate the emissions from fossil-based power plants and provide a pollution-free power sector is to opt for renewable energy. Over the past few years, biomass energy-based industrial research has gained more interest (Hesam et al., 2020). Biomass energy especially in the form of solid biofuel serves the need by not only producing energy in power plants but also helping in carbon sequestration indirectly benefitting soil health. Overall, utilizing all forms of renewable energy will provide a promising way to reduce overall energy consumption and environmental pollution.

Char, a solid biofuel produced from the thermochemical conversion of biomass attracts more attention from researchers and industrialists as a viable alternative to coal. Solid biochar produced from biomass is available abundantly worldwide and is renewable, and carbon-neutral with less sulfur, nitrogen and ash (Singh et al., 2023). This helps in reducing the emissions of SO_x, NO_x and other particles. It also helps in cutting down the costs required for handling and disposal of ash and maintenance of ash collection equipment. Solid char is produced conventionally by two processes: pyrolysis and hydrothermal carbonization (HTC), both of which involve anaerobic thermochemical conversion at longer residence time (Yan, Perez, & Sheng, 2017). The difference is that pyrolysis is a dry (moisture in biomass up to 10%), high temperature (200°C—600°C) and low-pressure (1—2 MPa) process (Surenderan, Saad, Zhou, Neshaeimoghaddam, & Rahman, 2018) whereas HTC is a wet (moisture in biomass up to 80%), low temperature (150°C—300°C) and high pressure (up to 20 MPa) process (Román, Nabais, Laginhas, Ledesma, & González, 2012). Comparatively, pyrolysis has a drawback of higher energy consumption and costs required for drying of feedstock, whereas HTC eliminates the drying step, and it can directly process wet biomass under thermochemical conversion to yield solid biofuel (char) thereby cutting the energy costs (Divyabharathi & Subramanian, 2022; Zhao, Shen, Ge, Chen, & Yoshikawa, 2014).

Most of the HTC research followed the conventional method of heating using an electric furnace, however, microwave heating using heat radiation is a recent advancement in the field. Microwave-assisted heating helps in direct heat transfer inside the feedstock by convection rather than conduction, however, it has a problem of forming hotspots (Huang, Chiueh, & Lo, 2016). There are limited studies on the application of microwave-assisted hydrothermal carbonization (MHTC) in the production of solid char and its applications over a larger scale. This chapter gives a technical overview of the MHTC process, suitable feedstock for MHTC, effect of process parameters such as temperature, solids loading, residence time, and catalyst on the yield and quality of biochar (also termed hydrochar when produced through HTC), biochar applications and limitations of MHTC.

3.2 Microwave-assisted hydrothermal carbonization

Over recent years, microwave-assisted heating technique has been employed over conventional electric heating, since it is comparatively advantageous. The advantages include quick processing time due to its molecular level heat dissipation and uniform controllable heat advancement, which makes it economic in terms of both energy and cost (Xia et al., 2020). On the other hand, conventional heating has longer residence time and heat loss due to unnecessary side reactions (Tripathi, Sahu, Ganesan, & Dey, 2015). The char produced from MHTC is comparatively enriched in energy content, due to its higher processing rate and avoidance of side processes (Silitonga et al., 2020). The major difference between the two methods is the mode of heat generation and passage. Conventional heating works on the principle of conduction/convection and the heat passes from outside to inside of the reactor thereby increasing the temperature required for reaction after attaining stabilization. Microwave heating is quite opposite, involving microwave radiation penetrating the molecules to convert them into internal energy and heat passes inside-out.

Microwave energy contains nonionizing electromagnetic radiation of shorter wavelength, higher frequency and higher energy (Gong & Bassi, 2016). It activates movement between molecules through ions and dipoles, targeting only the object to be heated without affecting its molecular structure. The presence of polar units helps in the microwave heating process, especially under a water medium, the dispersiveness and permittivity are enhanced, increasing heating rates (Santos & Rodriguez-Esquerre, 2020). The negatively charged oxygen atom in the water molecule gets attracted to the positively charged end of the microwave and similarly, the positively charged hydrogen atoms in water move toward a negatively charged end, thereby rotating the water molecules upon microwave oscillation. This develops heat due to the friction force between molecules and is easily transferred to the substance by conduction, convection and radiation (Kalla & Devaraju, 2017). Thus, microwave-assisted heating is a promising and preferable thermochemical conversion technology in a water medium. Furthermore, many studies have compared char produced from conventional heating with microwave heating and stated that biochar is produced at higher processing rates, that is, around 10 times reduction in reaction time from MHTC and has enhanced properties of that produced from conventional method. Thus, it proves that MHTC is a potentially effective way to produce biochar than conventional HTC. There are also studies on the utilization of bio-digestate produced from anaerobic biochemical conversion in HTC to produce char (Sharma, Panigrahi, Sarmah, & Dubey, 2020). This depicts the biorefinery concept in which a circular bioeconomy will be achieved by the reuse and recycling of materials ensuring waste reduction.

3.2.1 Feedstock for microwave-assisted hydrothermal carbonization

The organic biowastes rich in lignocellulose such as agricultural residues, manure, municipal solid wastes, and energy crops are potential feedstock to derive bioenergy (Chaturvedi, Singh, Dhyani, Govindaraju, & Mandal, 2023), yet direct utilization of this feedstock without proper pretreatment is not recommended due to its low energy content and high moisture. Lignocellulosic biomass contains about 40%–60% cellulose, 20%–40% hemicellulose, and 10%–20% lignin (Acharya, Dutta, & Minaret, 2015). The structural intricacy of lignocellulosic biomass together with inhibitors and intermediates formed during the process throws a major challenge for its utilization. To minimize the complexity, hydrothermal processing may be opted to degrade the lignocellulosic portions. At subcritical conditions of hydrothermal processing, hemicellulose, cellulose, and lignin components disintegrates at 160°C, 200°C, and 220°C, respectively (Grønli, Várhegyi, & Di Blasi, 2002). MHTC process undergoes a chemistry of complex chemical reactions such as dehydration, hydrolysis, devolatilization, polymerization, and decarboxylation (Zhang et al., 2018). Hydrolysis leads to the formation of monomers and oligomers in aqueous media. The cellulose disintegrates into hydroxyl organic derivatives. The lignin and hemicellulose undergo major dehydration and decarboxylation producing primary char, the liquid substances undergo polymerization producing secondary char and the overall decarboxylation reactions lead to the production of syngas (Joshi, Ramola, Singh, Anerao, & Singh, 2022; Titirici, 2013). MHTC process in subcritical water-soluble media aids in ion transfer by increasing the ionization constant and accelerating hydrolysis reactions to break down and rearrange the biomass structure. This eventually transforms the lignocellulosic components into low molecular polymeric organic compounds.

Currently, there are studies related to MHTC of coconut shells, paddy straw, husk, corn stalk, sugarcane bagasse, bamboo-sawdust, etc. (Dai et al., 2018) which reported that a high residence time and temperature leads to high-energy biochar production, which can be used as a solid biofuel due to its high calorific value (19 MJ kg^{-1}). High residence time and temperature favor aromatics disintegration and increase the thermal stability of the product. Catalyst addition during the MHTC is proven to be effective in increasing the carbonizing effect of biomass leading to solid yield, thereby increasing the heating value and energy density. This shows that the biochar produced from MHTC is preferable for utilization than that produced from conventional pyrolysis.

3.2.2 Chemistry of microwave-assisted hydrothermal carbonization

During MHHTC, the feedstock submerged in a hot compressed water environment is transformed into solid, liquid, and gaseous products through a number of chemical reactions that occur simultaneously. Among the reactions involved, hydrolysis had the lowest activation energy. So, hydrolysis predominates the HTC process and the ester and ether bonds present in the feedstock matrix are broken down into smaller molecules. The reduction of the H/C ratio is governed by dehydration reactions by the release of water molecules in the process water. The O/C reduction is driven by the removal of carbon dioxide via decarboxylation reactions. The fragmented smaller molecules are combined into larger molecules and release the water molecule into process water via condensation and polymerization reactions. The aromatic polymer structure is preserved via aromatization reaction. The order and severity of each reaction have not yet discovered (Heidari, Dutta, Acharya, & Mahmud, 2019). One step kinetic equation is applicable for biochar combustion since it involves solid-state decomposition (Islam, Kabir, Asif, & Hameed, 2015; Zulkornain et al., 2021). The kinetic equation for the combustion of biochar is given in Eq. (3.1).

$$\frac{d\alpha}{dt} = k(T)f(\alpha) \tag{3.1}$$

where $k(T)$ is the rate constant (temperature dependent) and α is the thermal conversion of combustible matter. The process chemistry behind the MHTC of biomass is depicted in Fig. 3.1.

3.2.3 Reaction profile

The severity of hydrothermal process conditions determines the product formation (solid/liquid/gas). Hydrothermal process temperature, reaction time, solids loading, and

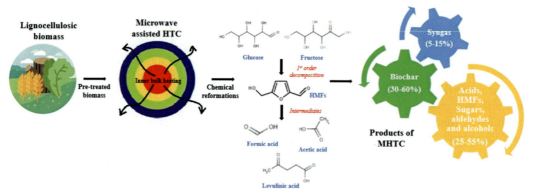

FIGURE 3.1 Chemistry of microwave-assisted hydrothermal carbonization process.

selection of solvent and addition of catalysts mainly influence the biochar yield and quality such as fuel value, surface functional groups, surface charge and physicochemical characteristics (Komalabharathi et al., 2022). Apart from the above-said factors, microwave power and feedstock properties also influence the process to a lesser extent. Fig. 3.2 shows the effect of an increase in temperature, residence time and solid content on char yield. The effect of each parameter on char yield and its characteristics are explained below.

3.2.3.1 Temperature

Hydrothermal process temperature has a significant effect on product distribution and quality. At lower temperatures ($\approx 160°C-180°C$), higher solid yields can be attained with lower liquid and gaseous products. The lesser energy availability at lower temperature is insufficient for destroying the feedstock matrix and results in higher char yield (Singh, Chaturvedi, Dhyani, & Govindaraju, 2020). With the increase in temperature up to a certain level ($\approx 180°C-250°C$), solid yields tend to decrease and increase the liquid and gaseous yield. When the temperature rises, the hydrogen and oxygen from the feedstock matrix are stripped off due to dehydration and deoxygenation which results in improved physicochemical properties and fuel value of biochar. At higher temperatures ($>250°C$), the yield of gaseous products will be higher than that of solid and liquid products because of the intensified decarboxylation reactions.

The Arrhenius equation depicts the relationship between the reaction rate and the process temperature (Kumar, Kuang, Liang, & Sun, 2020).

$$K = Ae^{-E_a/RT}$$

where K is a reaction rate constant, E_a is the activation energy, R is the universal gas constant, T is the absolute temperature, and A is the preexponential factor.

In microwave heating, the rate of reaction depends on the change in process temperature. An increase in process temperature under microwave heating tends to increase the number of collisions between molecules and the rate of different chemical reactions which results in the improved physicochemical properties of biochar. With the increase in process temperature

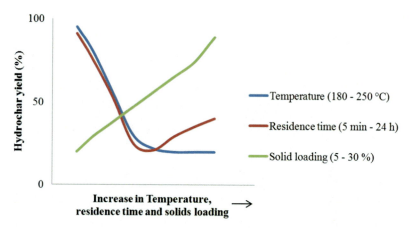

FIGURE 3.2 Effect of process parameters on hydrochar yield.

from 180°C to 220°C, the gross calorific value (GCV) and content of carbon were improved to 17.6 MJ kg^{-1} and 48.8%, respectively in the rice husk biochar when compared to raw rice husk which had a GCV of 12.3 MJ kg^{-1} with 37.19% carbon content (Nizamuddin et al., 2019). Similarly in another study, the GCV of coconut shell improved from 15.06 to 19.76 MJ kg^{-1} (in coconut shell biochar) at 200°C (Elaigwu & Greenway, 2019).

3.2.3.2 Reaction time

At low reaction time, biochar yield will be higher due to the lesser interaction between the feedstock matrix and microwave energy. Depending on the feedstock's nature and characteristics, increasing the reaction time can cause two types of actions. For feedstocks like palm shell, increasing the reaction time tends to increase the liquid product due to enhanced dehydration and decarboxylation reactions which promotes intermediate product formation such as acids, ketones and furfurals (Nizamuddin et al., 2016). Whereas for dairy manure and other animal wastes, an increase in reaction time increased the char yield due to the condensation and polymerization reactions. Different feedstocks react differently with an increase in reaction time. So, it is necessary to study the material-specific reaction profile to get a deep understanding of MHTC.

Compared to conventional HTC, the same quality of char can be attained at the same temperature with lesser residence time in MHTC due to the fast process kinetics (i.e., rapid increment in the rate of chemical reactions). In conventional heating, it may take several hours to days to produce biochar. However, in microwave-assisted method, it will take several minutes to a few hours (Kumar et al., 2020). Since MHTC involves direct heating of the material itself, the entire volume can be heated uniformly at a lesser residence time without much energy losses. The difference between conventional pyrolysis/HTC and MHHTC is shown in Fig. 3.3. The optimum reaction time for microwave-assisted biochar synthesis may be in the range of 20–60 minutes. Initially, an increase in reaction time tends to decrease the yield of char to a certain limit. Later, there is no significant change in char yield with an increase in residence time due to the complete degradation of hemicellulose and cellulose.

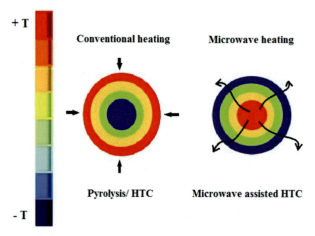

FIGURE 3.3 Conventional pyrolysis/hydrothermal carbonization (HTC) versus microwave-assisted HTC.

3.2.3.3 Solids loading

A lower solids loading is preferred in MHTC, since it allows sufficient water to absorb microwave energy to interact with the solid matter. Due to the enhanced interaction, hydrothermal reactions are accelerated and improved quality of char (increased heating value) can be produced with lower yield. Higher solid loading resists degradation due to the lesser penetration of microwave energy into the available water. Thus, higher biochar yield can be achieved at higher solid content. The solid content in the water can be varied from 5% to 50%, depending on the type, structure, density and hydrophobic nature of biomass. Biomass slurry with lower solid content resulted in higher biochar yield for corn stalks and red seaweed (Cao et al., 2019; Kang et al., 2019).

3.2.3.4 Selection of solvent

Selecting the appropriate solvent determines the amount of microwave energy absorbed by the reaction mixture and the process efficiency. Solvents are classified into three types based on loss tangent (tan δ). Loss tangent determines the ability of a solvent to convert electromagnetic energy into heat energy at a given frequency of microwave and temperature. The first type is solvents having tan δ higher than 0.5 which shows improved absorption for microwaves. Solvents having middle tan δ in the range of 0.1−0.5 come under the second category that are having medium absorbance for microwaves. Lastly, poor microwave absorbing solvents come under the third category which are having tan δ less than 0.5. The dielectric constant (ε') and dielectric losses (ε'') of the solvent are the parameters that influence the loss tangent of the solvent (Kumar et al., 2020). Using low-cost and clean solvents eliminates the hazardous compound generation during the process. In MHTC, water is used as a solvent since it possesses a higher dielectric constant (80.4) with medium losses (9.889) at a microwave frequency of 2.45 GHz and room temperature which makes it a suitable candidate for microwave absorption (Schmidt, Prado-Gonjal, & Morán, 2015). Other solvents such as NaOH and ethylene glycol are used and the process is known as solvothermal synthesis (Xia et al., 2022; Zambzickaite et al., 2022).

3.2.3.5 Catalyst

In microwave-assisted hydrothermal synthesis, hydrolysis reaction aids in the formation of intermediate products like acids which will auto-catalyze the hydrothermal process by enhancing the hydrolysis of the feedstock matrix. The addition of external catalysts like acid, alkali and metal salts will further help to enhance the degradation of feedstock and the formation of functional groups on the surface of biochar. Depending on the applications, in situ surface modification of biochar can be achieved using suitable catalysts (Zhang et al., 2020). Selection of catalyst and concentration influences the product formation during MHTC. For example, the addition of an acidic catalyst increases the char formation due to the enhanced hydrolysis reactions. At lower concentrations, acidic catalysts perform well without affecting the char yield. On the other hand, the addition of basic catalysts improved the liquid formation due to the enhanced volatilization reactions. Catalyst addition can also control the emission of NO_x during the combustion of biochar by enhancing the removal of cell-bound nitrogen during MHHTC (Nizamuddin et al., 2018). The addition of sulfuric acid (H_2SO_4) increased the biochar yield for red seaweed (Cao et al., 2019). Liu et al. (2021) reported that the addition of H_3PO_4 increased the liquid yield and improved the higher heating value of biochar.

3.2.3.6 Process water recycling

Recirculation of process water is more economical and environmentally friendly since it reduces the water requirement for the process. Process water is enriched with numerous chemicals such as acids, aldehydes, ketones and furfurals. Utilizing the process water as a reaction medium in consecutive cycles enhances the polymerization reaction and results in improved yield of biochar because of the increased acidity of the process water and feedstock mixture (Heidari et al., 2019). Biochar yield was increased by the presence of reactive compounds in the recirculated process water such as furfurals. Overall, process water recirculation enhanced the formation of protein, lipid, carbohydrate and lignin-originated compounds. Dissolved nitrogen content was high in the process water. This recirculation of process water in the temperature range of 200°C to 240°C, improved the formation of heterocycles of nitrogen or quaternary nitrogen in the biochar produced from brewer's spent grains (Dominik et al., 2022). Mass and energy yields were increased in the biochar produced from grape pomace with process water recirculation (Catalkopru, Kantarli, & Yanik, 2017).

3.2.3.7 Other factors

The structure of feedstock can have a significant influence on the product formation during MHHTC. Higher cellulose and hemicellulose-containing feedstock release more intermediate products and result in decreased yield of biochar (Singh et al., 2023). Whereas feedstock with a higher amount of lignin results in higher biochar yield due to the fact that lignin is unaffected during hydrothermal process conditions (Nizamuddin et al., 2017). Regarding the particle size of feedstock, lower particle size offers higher penetration of water into the feedstock matrix owing to the larger surface area and destruction of cellulose crystallinity. Higher penetration of water enhances the hydrolysis, decarboxylation and dehydration reactions which results in decreased yield and improved quality of char (Meehnian, Jana, & Jana, 2016).

Microwave power is also another influential factor in HTC. Increase in microwave power enables the easier degradation of the feedstock matrix. The improved degradation results in the enormous release of volatile and gaseous compounds thus decreasing the yield of biochar at increased microwave power. A higher yield of char can be achieved at lower microwave power levels (Satpathy, Tabil, Meda, Naik, & Prasad, 2014). The effect of various parameters on biochar yield and properties are given in Table 3.1.

3.3 Applications of microwave-assisted hydrothermal carbonization

Biomass being a renewable energy source has limited applications for use as direct fuel due to its low density and calorific value and high moisture and volatile content. Hence biomass needs to be bioprocessed via thermochemical conversion to produce solid fuel for potential use. Biochar produced from MHTC is high in density, carbon and energy content with similar properties comparable to conventional coal. In addition, the microstructure

TABLE 3.1 Effect of process parameters on properties of biochar.

S. no.	Feedstock	Temperature (°C)	Residence time (min)	Solid to water ratio	Maximum yield (%)	Remarks	References
1.	Sewage sludge (SS) and Human biowaste (HFS)	160–200	30–120	1:19	60 (SS) and 72 (HFS)	Increase in temperature increased the darkness of biochar due to melanoidin generation. Higher biochar production rate is observed at lower residence time. Due to selective heating of microwave, biochar has porous structure	Afolabi, Sohail, and Thomas (2015)
2.	Prosopis Africana shell	200	5–20	1:15	36	Increase in residence time decreased the biochar yield	Elaigwu and Greenway (2016)
3.	Dairy manure	180–260	30–840	1:20	68	Increase in residence time improved the surface morphology of biochar. Increase in temperature improved biochar energy content	Gao et al. (2018)
4.	Rice straw	180–220	20–60	1:5–1:15	NA	Higher yield of biochar is observed at lower temperature, lower residence time and lower solid to liquid ratio	Nizamuddin et al. (2019)
5.	Groundnut shell	180–220	20	1:20	61	Increase in temperature decreased the biochar yield	Komalabharathi and Subramanian (2022)
6.	Palm kernel shell	150–300	10	NA	94	Microwave reactor combined with steam purging increased the biochar surface area	Yek et al. (2022)
7.	Apple, wheat, barley, oats and pea straw	180–220	10–60	1:10 and 1:50	35–45	Decrease in the solid to water ratio improved the yield, carbon recovery and energy yield	Holliday, Parsons, and Zein (2022)
8.	Seaweed	200	60	1:5	52	Compared to conventional method, improved oxygen containing functional groups were present in biochar produced through microwave-assisted carbonization	Soroush et al. (2023)
9.	Pomegranate peel	200	60	1:10	NA	In situ chemical modification to enhance the oxygen functionality of the biochar	Hessien (2023)

NA: Not available.

and porosity are transformed through various chemical reactions viz. depolymerization, devolatilization, dehydration, and decomposition of lignocellulosic structure (Funke & Ziegler, 2010). This mechanism increases the biochar grindability and decreases the adhering nature of biomass components. Due to the dehydration reaction, the hydroxyl groups which contribute to the hygroscopic nature of biomass are removed thereby making the biofuel hydrophobic in nature and convenient to store, transport and use.

One of the recent interests is the co-combustion of biochar produced from MHTC with coal, however, there are only limited studies on biochar co-combustion. Conventional co-combustion faces problems of early burnout, low efficiency, slagging and fouling. These problems will be eliminated in the utilization of biochar since it is easily grindable and has high fixed carbon, which is compatible with coal (Kambo & Dutta, 2015). The microcarbon structure and porous nature make the biochar blend easily with coal and enhance the reaction profile (Zhang et al., 2020). This in turn increases the activation energy of the reaction and combustion efficiency. Few studies have shown that the use of biochar in co-combustion with coal would decrease the burnout temperature and higher blending ratios will decrease the ignition temperature comparatively (Parshetti, Kent Hoekman, & Balasubramanian, 2013). There are also studies on Co-MHTC with other biomasses where two different biomasses are blended and MHTC is applied. Co-MHTC is applied on lignocellulosic biomass, woody matter, food waste and animal manure and the biochar obtained has higher carbon content, low ash and better mechanical properties (Wang et al., 2018).

3.4 Limitations of microwave-assisted hydrothermal carbonization

Though MHHTC is advantageous in terms of shortened reaction time and instant heating, there are a few limitations on economic, technical and environmental characteristics that need to be addressed to enhance the process efficiency and product quality. In terms of technical aspects, the results from various experimental studies on MHTC may not be similar, due to the feedstock species variation, analysis techniques, standardization and variability of instruments. The chemistry of microwave heating may also be disturbed by random external factors like plasma formation, discharge of metals, formation of thermal hotspots, etc. due to instant nonuniform heating and varied electric fields (Kostas, Beneroso, & Robinson, 2017). Biomass on the other hand is a poor absorber of microwaves due to its dielectric nature affecting the permittivity of microwaves. This leads to the setting up of microwave absorbers adding more energy costs (Li et al., 2016). The dielectric properties must be clearly defined by more studies so that there is no failure in the design and development of MHTC reactors with particular temperature and frequency. Large-scale applications of MHTC are challenging due to the limited knowledge of the distribution of electromagnetic fields under microwave environment and the biomass-microwave interaction. Hence, the ideal solution to uplift the knowledge and understanding of MHTC would be the promotion of multidisciplinary research.

Microwave-assisted studies on energy efficiency, balance and economics are countable. The techno-economic analysis to find the feasibility involves a diversity of information viz. type of feedstock, process parameters, product output, capacity, capital investment, operational cost and returns generated. Hence, only less information on these aspects is found in existing studies on MHTC (Treichel et al., 2020; Wang, Lei, & Ruan, 2015) With respect to the environment, MHTC has a comparatively good environmental impact in terms of waste utilization, however, the liquid and gaseous discharge may cause water pollution and air pollution, respectively, which definitely needs to be addressed by proper waste management strategies.

Though MHTC has these challenges, a high potential for energy recovery from biomass, enhanced reaction chemistry and higher process efficiency is attained due to its specific microwave properties. Hence, with proper technical knowledge and more research, a better understanding of the process, high quality and quantity of products and industry-level mass-scale production can be achieved in the near future.

3.5 Conclusion and future prospects

Globally, GHG emissions have become a major challenge due to the continued usage of fossil fuel-based resources. One of the promising alternatives is the use of renewables especially in the form of solid biofuel. Hydrothermal conversion is one of the recent and fast-growing biomass energy conversions as it has higher process efficiency and the ability to use a wide range of feedstock. Biomass underwater media offers great advantages in terms of reaction chemistry, process time, product output and energy efficiency specifically when microwave heating over conventional heating is employed. The biochar produced via MHTC is rich in carbon content and calorific value which can be substituted for coal. The grindable nature of biochar also makes it a potential fuel for combustion in existing boilers and power plants. One of the future prospects is the utilization of biochar produced via MHTC in co-combustion with coal as the biochar from MHTC is less volatile in nature with more carbon making it compatible with coal for co-combustion. Another way is direct co-MHTC which has the potential to enhance the biochar quality by increasing the carbon content, reducing ash content, increasing durability and improving the mechanical properties of biochar by strengthening the inter-particle bonding. However, the dielectric nature of biomass, formation of hotspots during MHTC, environmental discharges and lack of information on economics are the major limitations that are to be remedied through more research and knowledge sharing.

References

Acharya, B., Dutta, A., & Minaret, J. (2015). Review on comparative study of dry and wet torrefaction. *Sustainable Energy Technologies and Assessments, 12*, 26−37.

Afolabi, O. O., Sohail, M., & Thomas, C. P. L. (2015). Microwave hydrothermal carbonization of human biowastes. *Waste and Biomass Valorization, 6*, 147−157.

Cao, L., Iris, K. M., Cho, D. W., Wang, D., Tsang, D. C., Zhang, S., ... Ok, Y. S. (2019). Microwave-assisted low-temperature hydrothermal treatment of red seaweed (*Gracilaria lemaneiformis*) for production of levulinic acid and algae hydrochar. *Bioresource Technology, 273*, 251−258.

Catalkopru, A. K., Kantarli, I. C., & Yanik, J. (2017). Effects of spent liquor recirculation in hydrothermal carbonization. *Bioresource Technology, 226*, 89−93.

Chaturvedi, S., Singh, S. V., Dhyani, V. C., Govindaraju, K., & Mandal, S. (2023). Characterization, bioenergy value and thermal stability of biochar derived from diverse agriculture and forestry lignocellulosic waste. *Biomass Conversion and Biorefinery, 13*, 879–892.

Dai, L., He, C., Wang, Y., Liu, Y., Ruan, R., Yu, Z., … Zhao, Y. (2018). Hydrothermal pretreatment of bamboo sawdust using microwave irradiation. *Bioresource Technology, 247*, 234–241.

Divyabharathi, R., & Subramanian, P. (2022). Biocrude production from orange (*Citrus reticulata*) peel by hydrothermal liquefaction and process optimization. *Biomass Conversion and Biorefinery, 12*(1), 183–194.

Dominik, W., Pablo, A., Sonja, H., Fernando, C., Luca, F., & Andrea, K. (2022). Process water recirculation during hydrothermal carbonization as a promising process step towards the production of nitrogen-doped carbonaceous materials. *Waste and Biomass Valorization*, 1–25.

Elaigwu, S. E., & Greenway, G. M. (2016). Microwave-assisted and conventional hydrothermal carbonization of lignocellulosic waste material: Comparison of the chemical and structural properties of the hydrochars. *Journal of Analytical and Applied Pyrolysis, 118*, 1–8.

Elaigwu, S. E., & Greenway, G. M. (2019). Characterization of energy-rich hydrochars from microwave-assisted hydrothermal carbonization of coconut shell. *Waste and Biomass Valorization, 10*(7), 1979–1987.

Funke, A., & Ziegler, F. (2010). Hydrothermal carbonization of biomass: A summary and discussion of chemical mechanisms for process engineering. *Biofuels, Bioproducts and Biorefining, 4*(2), 160–177.

Gao, Y., Liu, Y., Zhu, G., Xu, J., Xu, H., Yuan, Q., … Ji, L. (2018). Microwave-assisted hydrothermal carbonization of dairy manure: Chemical and structural properties of the products. *Energy, 165*, 662–672.

Gong, M., & Bassi, A. (2016). Carotenoids from microalgae: A review of recent developments. *Biotechnology Advances, 34*(8), 1396–1412.

Grønli, M. G., Várhegyi, G., & Di Blasi, C. (2002). Thermogravimetric analysis and devolatilization kinetics of wood. *Industrial & Engineering Chemistry Research, 41*(17), 4201–4208.

Heidari, M., Dutta, A., Acharya, B., & Mahmud, S. (2019). A review of the current knowledge and challenges of hydrothermal carbonization for biomass conversion. *Journal of the Energy Institute, 92*(6), 1779–1799.

Hesam, N. M., Saad, J. M., Shamsuddin, A. H., Zamri, M. F. M. A., Rahman, A. A., & Ghazali, A. F. (2020). Pyrolysis kinetic study of homogenized waste plastic and date blend. *IOP Conference Series: Earth and Environmental Science, 476*(1).

Hessien, M. (2023). Methylene blue dye adsorption on iron oxide-hydrochar composite synthesized via a facile microwave-assisted hydrothermal carbonization of pomegranate peels' waste. *Molecules (Basel, Switzerland), 28*(11), 4526.

Holliday, M. C., Parsons, D. R., & Zein, S. H. (2022). Microwave-assisted hydrothermal carbonisation of waste biomass: The effect of process conditions on hydrochar properties. *Processes, 10*(9), 1756.

Huang, Y.-F., Chiueh, P.-T., & Lo, S.-L. (2016). A review on microwave pyrolysis of lignocellulosic biomass. *Sustainable Environment Research, 26*(3), 103–109.

Islam, M. A., Kabir, G., Asif, M., & Hameed, B. H. (2015). Combustion kinetics of hydrochar produced from hydrothermal carbonisation of Karanj (*Pongamia pinnata*) fruit hulls via thermogravimetric analysis. *Bioresource Technology, 194*, 14–20.

Joshi, S., Ramola, S., Singh, B., Anerao, P., & Singh, L. (2022). *Waste to wealth: Types of raw materials for preparation of biochar and their characteristics*. Engineered Biochar (pp. 21–33). Singapore: Springer.

Kalla, A. M., & Devaraju, R. (2017). Microwave energy and its application in food industry: A reveiw. *Asian Journal of Dairy and Food Research, 36*, 37–44.

Kambo, H. S., & Dutta, A. (2015). Comparative evaluation of torrefaction and hydrothermal carbonization of lignocellulosic biomass for the production of solid biofuel. *Energy Conversion and Management, 105*, 746–755.

Kang, K., Nanda, S., Sun, G., Qiu, L., Gu, Y., Zhang, T., … Sun, R. (2019). Microwave-assisted hydrothermal carbonization of corn stalk for solid biofuel production: Optimization of process parameters and characterization of hydrochar. *Energy, 186*, 115795.

Komalabharathi, P., Karuppasamy Vikraman, V., Praveen Kumar, D., Boopathi, G., & Subramanian, P. (2022). *Hydrochar: Sustainable and low-cost biosorbent for contaminant removal*. Encyclopedia of green materials (pp. 1–8). Singapore: Springer Nature Singapore.

Komalabharathi, P., & Subramanian, P. (2022). Energy densification of groundnut shell through microwave-assisted hydrothermal carbonization. *Madras Agricultural Journal, 109*(special), 1.

Kostas, E. T., Beneroso, D., & Robinson, J. P. (2017). The application of microwave heating in bioenergy: A review on the microwave pre-treatment and upgrading technologies for biomass. *Renewable and Sustainable Energy Reviews, 77*, 12–27.

Kumar, A., Kuang, Y., Liang, Z., & Sun, X. (2020). Microwave chemistry, recent advancements, and eco-friendly microwave-assisted synthesis of nanoarchitectures and their applications: A review. *Materials Today Nano, 11*, 100076.

Li, J., Dai, J., Liu, G., Zhang, H., Gao, Z., Fu, J., ... Huang, Y. (2016). Biochar from microwave pyrolysis of biomass: A review. *Biomass and Bioenergy, 94*, 228–244.

Liu, J., Zhong, F., Niu, W., Zhao, Y., Su, J., Feng, Y., & Meng, H. (2021). Effects of temperature and catalytic methods on the physicochemical properties of microwave-assisted hydrothermal products of crop residues. *Journal of Cleaner Production, 279*, 123512.

Meehnian, H., Jana, A. K., & Jana, M. M. (2016). Effect of particle size, moisture content, and supplements on selective pretreatment of cotton stalks by *Daedalea flavida* and enzymatic saccharification. *3 Biotech, 6*(2), 235.

Nizamuddin, S., Baloch, H. A., Griffin, G. J., Mubarak, N. M., Bhutto, A. W., Abro, R., ... Ali, B. S. (2017). An overview of effect of process parameters on hydrothermal carbonization of biomass. *Renewable and Sustainable Energy Reviews, 73*, 1289–1299.

Nizamuddin, S., Baloch, H. A., Siddiqui, M. T. H., Mubarak, N. M., Tunio, M. M., Bhutto, A. W., ... Srinivasan, M. P. (2018). An overview of microwave hydrothermal carbonization and microwave pyrolysis of biomass. *Reviews in environmental science and bio/technology, 17*, 813–837.

Nizamuddin, S., Mubarak, N. M., Tiripathi, M., Jayakumar, N. S., Sahu, J. N., & Ganesan, P. (2016). Chemical, dielectric and structural characterization of optimized hydrochar produced from hydrothermal carbonization of palm shell. *Fuel, 163*, 88–97.

Nizamuddin, S., Qureshi, S. S., Baloch, H. A., Siddiqui, M. T. H., Takkalkar, P., Mubarak, N. M., & Tanksale, A. (2019). Microwave hydrothermal carbonization of rice straw: Optimization of process parameters and upgrading of chemical, fuel, structural and thermal properties. *Materials, 12*(3), 403.

Parshetti, G. K., Kent Hoekman, S., & Balasubramanian, R. (2013). Chemical, structural and combustion characteristics of carbonaceous products obtained by hydrothermal carbonization of palm empty fruit bunches. *Bioresource Technology, 135*, 683–689.

Román, S., Nabais, J. M. V., Laginhas, C., Ledesma, B., & González, J. F. (2012). Hydrothermal carbonization as an effective way of densifying the energy content of biomass. *Fuel Processing Technology, 103*, 78–83.

Santos, F.S., & Rodriguez-Esquerre, V.F. (2020). Water-based broadband metamaterial absorbers operating at microwave frequencies. In *Proceedings: Metamaterials, metadevices, and metasystems*, 11460.

Satpathy, S. K., Tabil, L. G., Meda, V., Naik, S. N., & Prasad, R. (2014). Torrefaction of wheat and barley straw after microwave heating. *Fuel, 124*, 269–278.

Schmidt, R., Prado-Gonjal, J., & Morán, E. (2015). *Microwaves: Microwave assisted hydrothermal synthesis of nanoparticles. CRC Concise encyclopedia of nanotechnology* (pp. 561–572). CRC Press Taylor & Francis Group.

Sharma, H. B., Panigrahi, S., Sarmah, A. K., & Dubey, B. K. (2020). Downstream augmentation of hydrothermal carbonization with anaerobic digestion for integrated biogas and hydrochar production from the organic fraction of municipal solid waste: A circular economy concept. *The Science of the Total Environment, 706*, 135907.

Silitonga, A. S., Shamsuddin, A. H., Mahlia, T. M. I., Milano, J., Kusumo, F., Siswantoro, J., ... Ong, H. C. (2020). Biodiesel synthesis from *Ceiba pentandra* oil by microwave irradiation-assisted transesterification: ELM modeling and optimization. *Renewable Energy, 146*, 1278–1291.

Singh, S., Chaturvedi, S., Nayak, P., Dhyani, V. C., Nandipamu, T. M., Singh, D. K., ... Govindaraju, K. (2023). Carbon offset potential of biochar based straw management under rice – wheat system along Indo-Gangetic Plains of India. *Science of the Total Environment, 897*, 165176.

Singh, S., Luthra, N., Mandal, S., Kushwaha, D. P., Pathak, S. O., Datta, D., ... Pramanik, B. (2023). Distinct behavior of biochar modulating biogeochemistry of salt-affected and acidic soil: A review. *Journal of Soil Science and plant Nutrition*.

Singh, S. V., Chaturvedi, S., Dhyani, V. C., & Govindaraju, K. (2020). Pyrolysis temperature influences the characteristics of rice straw and husk biochar and sorption/desorption behavior of their biourea composite. *Bioresource Technology, 314*, 123674.

Soroush, S., Ronsse, F., Park, J., Ghysels, S., Wu, D., Kim, K. W., & Heynderickx, P. M. (2023). Microwave assisted and conventional hydrothermal treatment of waste seaweed: Comparison of hydrochar properties and energy efficiency. *Science of the Total Environment, 878*, 163193.

Surenderan, L., Saad, J. M., Zhou, H., Neshaeimoghaddam, H., & Rahman, A. A. (2018). Characterization studies on waste plastics as a feedstock for energy recovery in Malaysia. *International Journal of Engineering and Technology (UAE), 7*(4), 534–537.

Titirici, M. M. (2013). *Sustainable carbon materials from hydrothermal processeso title*. John Wiley & Sons, Ltd.

Treichel, Helen, Fongaro, Gislaine, Scapini, T., Camargo, A. F., Stefanski, F. S., & V, B. (2020). *Utilising biomass in biotechnology* (1st ed.). cham: Springer.

Tripathi, M., Sahu, J. N., Ganesan, P., & Dey, T. K. (2015). Effect of temperature on dielectric properties and penetration depth of oil palm shell (OPS) and OPS char synthesized by microwave pyrolysis of OPS. *Fuel, 153,* 257–266.

Wang, L., Lei, H., & Ruan, R. R. (2015). *Techno-economic analysis of microwave-assisted pyrolysis for production of biofuels. Production of biofuels and chemicals with microwave. biofuels and biorefineries* (pp. 251–263). Springer.

Wang, T., Zhai, Y., Li, H., Zhu, Y., Li, S., Peng, C., ... Li, C. (2018). Co-hydrothermal carbonization of food waste-woody biomass blend towards biofuel pellets production. *Bioresource Technology, 267,* 371–377.

Xia, K., Liu, X., Liu, H., Li, W., Li, Y., Han, F., ... Hou, Z. (2022). Microwave-assisted solvothermal synthesis of hollow mesoporous SiOC ceramics in NaOH solution. *Ceramics International, 48*(13), 19232–19239.

Xia, Y., Luo, H., Li, D., Chen, Z., Yang, S., Liu, Z., ... Gai, C. (2020). Efficient immobilization of toxic heavy metals in multi-contaminated agricultural soils by amino-functionalized hydrochar: Performance, plant responses and immobilization mechanisms. *Environmental Pollution, 261,* 114217.

Yan, W., Perez, S., & Sheng, K. (2017). Upgrading fuel quality of moso bamboo via low temperature thermochemical treatments: Dry torrefaction and hydrothermal carbonization. *Fuel, 196,* 473–480.

Yek, P. N. Y., Liew, R. K., Mahari, W. A. W., Peng, W., Sonne, C., Kong, S. H., ... Lam, S. S. (2022). Production of value-added hydrochar from single-mode microwave hydrothermal carbonization of oil palm waste for dechlorination of domestic water. *Science of The Total Environment, 833,* 154968.

Zambzickaite, G., Talaikis, M., Dobilas, J., Stankevic, V., Drabavicius, A., Niaura, G., & Mikoliunaite, L. (2022). Microwave-assisted solvothermal synthesis of nanocrystallite-derived magnetite spheres. *Materials, 15*(11), 4008.

Zhang, J., An, Y., Borrion, A., He, W., Wang, N., Chen, Y., & Li, G. (2018). Process characteristics for microwave-assisted hydrothermal carbonization of cellulose. *Bioresource Technology, 259,* 91–98.

Zhang, N., Wang, G., Zhang, J., Ning, X., Li, Y., Liang, W., & Wang, C. (2020). Study on co-combustion characteristics of hydrochar and anthracite coal. *Journal of the Energy Institute, 93*(3), 1125–1137.

Zhao, P., Shen, Y., Ge, S., Chen, Z., & Yoshikawa, K. (2014). Clean solid biofuel production from high moisture content waste biomass employing hydrothermal treatment. *Applied Energy, 131,* 345–367.

Zulkornain, M. F., Normanbhay, S., Saad, J. M., Zhang, Y. S., Samsuri, S., & Ghani, W. A. W. A. K. (2021). Microwave-assisted hydrothermal carbonization for solid biofuel application: A brief review. *Carbon Capture Science & Technology, 1,* 100014.

CHAPTER

4

Sustainable management and diversification of problematic wastes: prospects and challenges

Anamika Barman[1], Sougata Roy[1], Priyanka Saha[1], Saptaparnee Dey[2], Shashank Patel[1], Deepak Kumar Meena[1], Anurag Bera[3], Shiv Vendra Singh[4], Sandip Mandal[5], Suprava Nath[6] and Shreyas Bagrecha[7]

[1]Division of Agronomy, ICAR—Indian Agricultural Research Institute, Pusa, New Delhi, India [2]Division of Soil Science and Agricultural Chemistry, ICAR—Indian Agricultural Research Institute, Pusa, New Delhi, India [3]Department of Agronomy, Institute of Agricultural Sciences, Banaras Hindu University, Varanasi, Uttar Pradesh, India [4]Department of Agronomy, College of Agriculture, Rani Lakshmi Bai Central Agriculture University, Jhansi, Uttar Pradesh, India [5]Agricultural Energy and Power Division, ICAR—Central Institute of Agricultural Engineering, Nabibagh, Bhopal, Madhya Pradesh, India [6]Department of Agronomy, M.S. Swaminathan School of Agriculture, Centurion University of Technology and Management, Paralakhemundi, Odisha, India [7]Agronomy Section, ICAR—National Dairy Research Institute, Karnal, Haryana, India

4.1 Introduction

Owing to the rapid growth of the population in modern society, the generation of waste is also increasing significantly. This rise in waste generation has severe ecological consequences (Singh, Duan, & Tang, 2020). Traditional methods of waste disposal, such as incineration, dumping, and landfilling, when not managed properly, lead to pollution of air, water, and soil (Ahamed et al., 2020; Azar & Azar, 2016). Problematic waste types encompass municipal solid waste (MSW), agricultural and industrial waste, plastics, e-waste, and biomedical waste. Mismanaged handling of these wastes results in the emission of greenhouse gases (GHGs), contributing to environmental pollution. Despite recognizing the nonrenewable nature of fossil fuels, society continues to rely on them for energy, causing global warming and shifts in climate (Raihan & Tuspekova, 2022; Wang & Yan, 2022). Various sectors, including industry, agriculture, mining, and municipalities, contribute to the annual generation of solid waste in the country. It is projected that global waste production will reach 27 billion tonnes annually by 2050. At present, Asia is responsible for one-third of total waste, with China (0–0.49 kg capita^{-1} day^{-1}) and India (0.50–0.9 kg capita^{-1} day^{-1}) making significant contributions (Kumar & Agrawal, 2020). Moreover, the world produces approximately 300 million tonnes of plastic waste each year, of which only 9% is recycled, about 14% is collected for recycling, and the remainder ends up in the oceans annually (Rezania et al., 2019). The persistent nature of plastics poses a global threat, as microplastics (MPs) infiltrate water bodies, polluting rivers and oceans (Hira et al., 2022; Wojnowska-Baryła, Bernat, & Zaborowska, 2022). These MPs originate from larger plastic items that are not properly disposed of, including agricultural plastic films, municipal plastic debris such as bags and bottles, as well as electronic waste (Gangwar & Pathak, 2021). The management of waste from various sources has become paramount. Pyrolysis has emerged as a novel technique to convert waste into solid, liquid, and gaseous products by adjusting the temperature (Bhatnagar, Khatri, Krzywonos, Tolvanen, & Konttinen, 2022). This process enables the transformation of low-energy-density materials into high-energy-density biofuels and valuable chemicals (Zhai et al., 2022). An advantage is the versatility of raw material sources, encompassing industrial and household residues (Al-Mrayat et al., 2022). Agricultural and industrial wastes like crop residues, press mud, synthetic oil, and MSW can be effectively converted to valuable energy sources through pyrolysis, contributing to global sustainability (Cheng et al., 2022). The utilization of agro-industrial waste for biofuel production offers an eco-friendly and renewable alternative to fossil fuels. This shift towards pyrolysis-based energy generation reduces reliance on fossil fuels and mitigates environmental contamination and climate change (Nair, Agrawal, & Verma, 2022). As a result, the aim is to achieve a zero-waste society while promoting sustainable bioenergy production (Sarkar, Butti, & Mohan, 2018). This chapter explores problematic waste types, their ecological risks, current waste management practices, diversification of wastes through pyrolysis, properties of the pyrolyzed products, advancements in pyrolysis technology for waste management, and challenges in handling problematic wastes.

4.2 Types of problematic waste

The types of problematic wastes which can be pyrolyzed are seaweed, agro-industrial waste, MSW, plastic waste, e-waste, etc. Each type of problematic waste can be used as an energy source by replacing conventional fossil fuels (Czajczyńska et al., 2017). The reduction of detrimental emissions from fossil fuels necessitates the application of renewable resources and the effective utilization of waste materials for generating power (Sultan et al., 2021). Recent research into the biochar generated through pyrolysis has shown that it serves as a valuable addition with diverse uses. These applications span from enhancing soil and water quality, boosting agricultural productivity, and enabling technologies like supercapacitors and fuel cells, to aiding catalysts and sustainable chemistry, as well as contributing to carbon sequestration efforts (Barrett, Alexander, Robinson, & Bragg, 2016). In the following paragraphs, we will discuss the different types of problematic wastes which can be pyrolyzed to reduce environmental hazards.

4.2.1 Agro-industrial waste

Agro-industrial waste refers to the residue or by-products generated from agricultural and food processing activities. These wastes often consist of materials that are leftover after the primary products have been extracted or processed, and they can include plant parts, peels, shells, stems, and other materials that are not utilized for their original intended purpose. Agro-industrial waste can originate from various stages of the food production chain, such as farming, harvesting, processing, and packaging (Ogbu & Okechukwu, 2023). These waste materials can have negative environmental impacts if not managed properly, leading to issues like pollution, GHG emissions, and habitat degradation. However, they also present opportunities for sustainable resource utilization through various approaches, such as composting, recycling, and conversion into valuable products like biofuels, animal feed, and biochar (Sadh, Duhan, & Duhan, 2018). Agro-industrial waste materials can be broadly categorized into three distinct groups (Ogbu & Okechukwu, 2023). The first group consists of naturally occurring agricultural and agro-industrial wastes that are either recyclable or compostable. These materials can be reused on farms or processed at recycling facilities. Examples of primary residues within this category include pruning residues, straw, leaves, stover, stalks, bagasse, cobs, and animal dung or manure. These residues originate directly from crop cultivation and animal husbandry activities. The second category comprises nonrecyclable and noncompostable agricultural and agro-industrial wastes. These materials do not possess the potential for recycling or composting. The third group encompasses hazardous agricultural and agro-industrial wastes, which may pose risks to the environment and human health due to their potentially harmful properties.

4.2.2 Municipal solid waste

MSW commonly referred to as "garbage" or "trash," consists of various discarded materials generated from households, businesses, institutions, and other nonindustrial sources within a community or municipality. MSW includes a diverse range of items such as paper, cardboard, plastic, glass, textiles, food scraps, yard waste, electronics, and more. It's essentially the waste generated by individuals and organizations in daily life (Huang et al., 2018). MSW does not include hazardous waste (such as chemicals, medical waste, or radioactive materials) or industrial waste (generated by manufacturing processes), as these types of waste have specific disposal regulations. A study conducted by Guermoud et al. (2009), revealed the typical composition of MSW to be approximately 64.6% organic matter, 15.9% paper-cardboard, 10.5% plastic, 2.8% glass, 1.9% textiles, and 2.3% diverse waste (Table 4.1).

TABLE 4.1 Sources and types of municipal solid waste (Hester & Harrison, 2002).

Sources	Typical waste generators	Types of solid waste
Residential	Single and multifamily dwellings	Food wastes, paper, cardboard, plastics, textiles, glass, metals, ashes, special wastes (bulky items, consumer electronics, batteries, oil, tyres) and household hazardous wastes
Commercial	Stores, hotels, restaurants, markets, office buildings	Paper, cardboard, plastics, wood, food wastes, glass, metals, special wastes, hazardous wastes
Institutional	Schools, government centers, hospitals, Prisons	Paper, cardboard, plastics, wood, food wastes, glass, metals, special wastes, hazardous wastes
Municipal services	Street cleaning, landscaping, parks, beaches, recreational areas	Street sweepings, landscape and tree trimmings, general wastes from parks, beaches, and other recreational areas

4.2.3 Plastic waste

Plastic waste refers to discarded or unwanted plastic materials that are no longer being used and have reached the end of their useful life. Plastic waste can encompass a wide range of plastic items, including packaging materials, containers, bottles, bags, utensils, toys, electronics, and more (Cleetus, Thomas, & Varghese, 2013). It is a major environmental concern due to its nonbiodegradable nature, which means that plastic waste can persist in the environment for a long time, causing pollution and harm to ecosystems (Ratnasari, Nahil, & Williams, 2017). Disposable plastics like masks, gloves, containers, medical packaging, and utensils associated with the ongoing COVID-19 pandemic are undeniably influencing the management of plastic waste (Korley, Epps, Helms, & Ryan, 2021). They are indispensable materials used across various applications that simplify our daily tasks in activities like household chores, packaging in stores, marketing, construction, and healthcare. Their lightweight, chemical stability (resistance to rust and decay), ready availability, and potential for reuse make them highly versatile (Ratnasari et al., 2017). The United States holds the global lead in per capita plastic consumption, reaching 142 kg year^{-1} (Cleetus et al., 2013). In 2015, a staggering 6.3 billion metric tons of plastics were produced, with 79% of this total being disposed of in landfills, 12% undergoing incineration, and 9% being recycled (Jia et al., 2021). In the European Union, half of the plastics produced annually are eventually discarded, transforming into the third-largest source of MSW following food and paper waste (Cleetus et al., 2013). The substantial quantity of plastic waste resulting from the exponential growth in production and consumption has sparked serious concerns. These plastics do not naturally decompose, remaining within municipal waste for extended periods. Compared to other organic waste, plastic waste occupies more space, adding strain to increasingly scarce and costly landfill areas. While incineration might seem a solution, it presents its own issues due to the emission of substantial volumes of harmful gases such as HCl, dioxins, SO_x, NO_x, and CO_2. Prolonged exposure to these toxic gases can lead to respiratory issues and exacerbate global warming and acid rain concerns (Akpanudoh, Gobin, & Manos, 2005).

4.2.4 E-waste

E-waste, short for electronic waste, refers to discarded electrical and electronic devices that have reached the end of their useful life or are no longer in working condition. This category includes a wide range of electronic products, such as computers, laptops, mobile phones, televisions, refrigerators, washing machines, digital cameras, and other electronic equipment. E-waste is characterized by its complex composition, as it contains a mixture of valuable and hazardous materials, including metals, plastics, glass, and chemicals (Popescu, 2015). The improper disposal or mishandling of e-waste can lead to environmental pollution and health hazards due to the release of toxic substances into the air, soil, and water. Proper management of e-waste involves recycling and safe disposal methods to recover valuable materials and reduce environmental impact Environmental pollutants such as heavy metals (HMs) (namely Pb, Sb, Ba, Cd, and As) present in e-waste possess hazardous properties attributed to their resistance to degradation, toxic nature, and tendency to accumulate within ecosystems (Ali, Khan, & Ilahi, 2019). Printed circuit boards (PCBs) constitute a fundamental component in nearly all electronic devices. Waste printed circuit boards (WPCBs) encompass a wealth of valuable resources intertwined with potentially harmful elements. This category of waste encompasses HMs, such as lead and cadmium, as well as hazardous substances that pose an environmental risk if not adequately managed. Nonetheless, the considerable presence of valuable materials within WPCBs renders them worthy of recycling efforts (Cui & Zhang, 2008). The waste stream of electrical and electronic equipment (WEEE) is experiencing swift expansion. It attained a global volume of around 44.7 million metric tons (Mt) in 2016, and projections anticipate an increase to approximately 53.9 Mt by 2025 (Baldé, Forti, Gray, Kuehr, & Stegmann, 2017; Kitila, 2018). According to recent estimates by the International Telecommunication Union, approximately 53.6 million metric tons of e-waste were generated in 2019. Shockingly, only 17.4% of this global e-waste was properly collected and recycled. The recycling and collection rates of e-waste across continents varied in 2019: 0.9% in Africa, 8.8% in Oceania, 42.5% in Europe, 9.4% in the Americas, and 11.7% in Asia (Forti, Balde, Kuehr, & Bel, 2020). Nevertheless, it remains a concerning fact that a significant portion of e-waste worldwide still lacks adequate management. Effective e-waste recycling necessitates a precise understanding of the composition and structure of each individual e-waste item. Consequently, we meticulously analyze the structural characteristics of five pivotal e-waste categories: PCBs, liquid crystal displays, light emitting diodes, lithium-ion batteries, and thermocouples. This scrutiny aims to uncover the rationale behind the pyrolysis of each specific type (Cui & Zhang, 2008).

4.3 Agro-ecological hazards from problematic wastes

Agro-ecological systems are highly sensitive to various forms of waste generated by human activities. Problematic wastes, such as seaweed plants, agro-industrial products, MSW, plastic waste, and electronic waste (e-waste), pose significant threats to agricultural ecosystems and the environment. These wastes can lead to soil and water pollution, reduced agricultural productivity, biodiversity loss, and negative impacts on human health. Here's a breakdown of the potential hazards associated with each type of waste:

4.3.1 Hazards from agro-industrial products

Agricultural waste from industries can pose hazards when not managed properly. Agricultural-based industries produce a vast amount of residue every year. If these residues are released into the environment without proper disposal procedure they may cause environmental pollution and harmful effects on human and animal health. Most of the agro-industrial wastes are untreated and underutilized, therefore in maximum reports it disposed of either by burning, dumping or unplanned landfilling. These untreated wastes create different problems with climate change by increasing the number of GHGs (Sadh et al., 2018). These wastes cause a serious disposal problem (Rodríguez Couto, 2008). For example, the juice industries produced a huge amount of waste as peels, the coffee industry produced coffee pulp as waste, and the cereal industry produced husks. All over the world approximately 147.2 million metric tons of fiber sources are found, whereas 709.2 and 673.3 million metric tons of wheat straw residues and rice straws were estimated, respectively, in the 1990s (Belewu & Babalola, 2009).

4.3.2 Hazards from municipal solid waste

Unmanaged MSW can lead to several agro-ecological hazards. Improper disposal of MSW, including organic and Inorganic materials, can result in soil and water pollution. Organic waste decomposition produces methane gas, a potent GHG that contributes to climate change. The leachate from landfills can infiltrate soil and contaminate groundwater, affecting both agricultural productivity and human health. Open burning of solid waste releases harmful pollutants into the air, further impacting agro-ecological systems. One of the major environmental problems is the collection, management and disposal of the MSW in the urban areas. Lack of MSW management and disposal is leading to significant environmental problems. This includes soil, air water, and esthetic pollution. Such environmental problems are associated with human health disorders, due to the increase in GHG emissions (Weigand, Fripan, Przybilla, & Marb, 2003). In addition, most countries worldwide are facing a serious challenge in managing domestic food waste. It is wet, put in random ways, and sometimes mixed with impurities of inorganic waste and metals. Primarily, the composition of such domestic food waste is very complex because it includes papers, water, and oil, as well as spoiled and leftover foods from kitchen wastes and markets. All these waste substances are chemically comprised of fats, cellulose, starch, lipids, protein, and other organic matter. The moisture and salt contents lead to rapid decomposition of the organic contents in the wastes thus producing unpleasant odors. This condition can attract bugs and flies which are vectors for several diseases. Apart from being perishable, these MSWs including household kitchen waste as well as domestic food waste from restaurants and markets consist of high lignocellulosic materials that could be decomposed and exploited to produce valuable bio-products (Abdel-Shafy & Mansour, 2018).

4.3.3 Hazards from plastic waste

Plastic waste has detrimental effects on agro-ecosystems. Plastic waste is a global concern due to its persistence in the environment and its detrimental effects on ecosystems.

MPs, small plastic particles, can accumulate in soil and water, altering soil structure and reducing water retention capacity. Plastics can also block waterways, leading to flooding and disrupting irrigation systems. Chemicals leached from plastics can contaminate agricultural products and pose health risks to consumers. Once plastics are integrated with the soil matrix, they can change the soil porosity and soil (Guo et al., 2020) binding properties by affecting soil aggregation and water dynamics of soil. Plastic wastes, especially MPs, can interact with multiple soil properties. MPs contain toxic additives and hazardous contaminants such as polybrominated diphenyl ether, perfluorochemicals, HMs like copper, zinc, and lead and the high dispersion capabilities of MPs allow them to mix up with soil and degrade soil fertility (Brennecke, Duarte, Paiva, Caçador, & Canning-Clode, 2016; Gaylor, Harvey, & Hale, 2013; Hodson, Duffus-Hodson, Clark, Prendergast-Miller, & Thorpe, 2017). Moreover, plastic also hinders the growth of earthworms and other useful microorganisms, which consequently lead to the infertility of soil. The appearance of plastics as ecosystem stressors affects soil health and changes soil biophysical properties, which creates a complex variation in the environmental behavior of other pollutants in soil (Alimi, Farner Budarz, Hernandez, & Tufenkji, 2018; Wang et al., 2018).

4.3.4 Hazards from e-waste

Electronic waste contains hazardous substances that can be harmful when not managed properly e-waste, composed of discarded electronic devices, contains hazardous materials such as lead, mercury, cadmium, and flame retardants. Improper recycling or disposal of e-waste can release these toxic substances into the environment, contaminating soil and water. These pollutants can be taken up by crops, entering the food chain and posing health risks to humans and animals. E-waste contains numerous hazardous components in the form of halogenated compounds like polychlorinated biphenyls, tetrabromobisphenol A, polybrominated biphenyl, etc. along with other toxic materials which cause an adverse impact on the plants, microbes and human beings. One of the major toxic components of e-waste are HMs like As, Cr, Cd, Cu, and Hg, which need to be handled carefully at the time of dismantling the e-waste, being managed by the informal sector in developing countries compounds the problem, also, the available disposal/treatment technologies of e-waste are inadequate, and they have a direct as well as indirect impact on human health and the environment. As a result of long-term exposure, these contaminants persist in soil for a longer period causing numerous impacts on the soil microbe and health of higher organisms. They also lead to changes in soil microbiology, phylogenetically and functionally (Tang, Huang, Hao, & Zhao, 2013; Zhang et al., 2010).

4.4 Current technologies for waste management

Efficient handling and reutilization of substantial waste quantities are necessary to alleviate their environmental impact and mitigate health risks to humans. These waste materials represent an overlooked resource with significant potential for energy generation. The existing methods for waste management encompass thermochemical techniques

A. Biochar production and modification

(such as incineration, thermal gasification, and pyrolysis), biochemical processes [including fermentation, anaerobic digestion (AD), and landfilling with gas capture], and chemical approaches (like esterification). The thermal treatment of waste leads to the liberation of a noteworthy quantity of heat, contingent on the calorific value of the treated waste. This discourse delves into the competitive advantages and concurrent challenges inherent in these technologies.

4.4.1 Thermochemical technology

4.4.1.1 Incineration

Incineration stands as a crucial technique within the realm of thermochemical technology, involving the combustion of waste materials at temperatures exceeding 1000°C. This method encompasses the combustion of waste masses as well as simultaneous co-burning with coal and biomass. The outcome of the incineration process encompasses the generation of heat, power, or a combination of both. Incineration serves the purpose of eliminating solid wastes arising from industrial, urban, and agricultural sources, including hazardous medical waste. This approach exhibits the potential to reduce waste volume by 80%–95% and minimize the land area required for efficient waste disposal (Paulraj, Bernard, Raju, & Abdulmajid, 2019; Pham, Kaushik, Parshetti, Mahmood, & Balasubramanian, 2015). However, the hesitance of certain nations to fully adopt waste incineration stems from concerns about the release of harmful airborne pollutants such as dioxins, nitrogen oxides, particulates, and HMs. These pollutants are by-products of the combustion process using earlier equipment and technologies (Katami, Yasuhara, & Shibamoto, 2004). Consequently, the implementation of measures for controlling air pollution becomes imperative. Furthermore, the residue remaining after combustion, commonly in the form of incombustible ash, represents a concentrated inorganic waste that necessitates proper disposal. This aspect has previously cast waste incineration in a negative light, leading to bans on the practice in some countries. However, advancements in air emission control systems have paved the way for the development of a new generation of facilities that adhere to more stringent environmental regulations. This evolution significantly mitigates potential adverse effects on human health.

4.4.1.2 Pyrolysis and thermal gasification

Pyrolysis and gasification represent thermal methods for converting waste into energy. Unlike combustion, which primarily yields heat, both pyrolysis and gasification processes generate valuable compounds. Pyrolysis involves converting waste into biochar, gases, aerosols, and syngas under higher pressure and low oxygen conditions. Conversely, gasification transforms waste materials into a combustible gas mixture through partial oxidation at elevated temperatures, typically around 800°C–900°C. Notably, the gasification process produces syngas rich in hydrogen, serving as a foundational component for manufacturing valuable items such as chemicals and fuels (Young, 2010). Compared to incineration, both pyrolysis and gasification require lower temperatures. By controlling the extent of pyrolytic reactions, the production of syngas (comprising carbon monoxide, hydrogen, and methane) or other recovery products leads to an immediate reduction in volume and

4.4 Current technologies for waste management

weight. Since petroleum was the primary raw material for producing plastic, the conversion of plastic back into liquid oil through a process called pyrolysis held significant potential. This was because the resulting oil had a high energy content similar to commercial fuels. Pyrolyzing plastic waste resulted in the production of oil, gas, and solid residue. Polyethylene terephthalate (PET) became a popular choice for packaging food items, especially beverages like bottled water, soft drinks, and fruit juices. Cepeliogullar and Putun investigated the use of PET in pyrolysis. They conducted experiments using a fixed-bed reactor at a temperature of 500°C and found that the pyrolysis of PET yielded liquid oil. Another study by Belewu and Babalola (2009) discovered that polystyrene pyrolysis also produced a high percentage of liquid oil yield, reaching 97.0 wt.% at an optimal temperature of 425°C. Bernando (2011) focused on producing char by co-pyrolyzing various plastic wastes, pine biomass, and used tyres (Fig. 4.1).

E-waste was also subjected to pyrolysis to recover polymers and concentrate metals in a solid residue. Different methods of e-waste pyrolysis exist, such as vacuum pyrolysis,

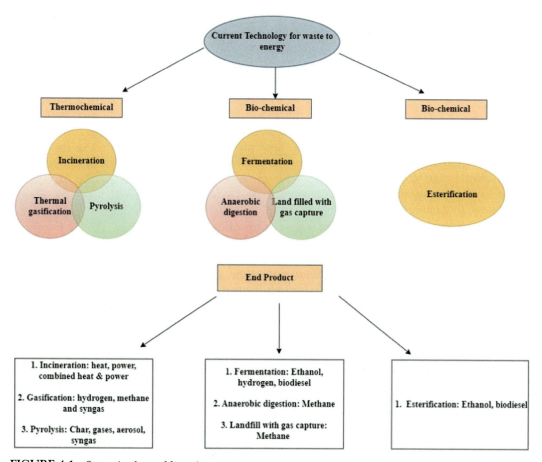

FIGURE 4.1 Strategies for problematic waste management.

A. Biochar production and modification

catalytic pyrolysis, co-pyrolysis, microwave pyrolysis, and plasma pyrolysis. The thermal cracking (pyrolysis) of e-waste proved beneficial for energy and material recovery, following a waste-to-energy approach. However, most studies on e-waste pyrolysis were limited to small-scale tests due to the lack of comprehensive information about the kinetics, activation energy, and component yield. Microwave-assisted pyrolysis showed promise for e-waste recycling, offering better potential for metal recovery compared to conventional pyrolysis, as reported by Su et al. (2022). Solid residues from MSW pyrolysis mainly comprised carbon, followed by inorganic materials, metals, or transition elements. Extremely high pyrolysis temperatures hindered solid residue formation. For instance, pyrolyzing MSW at 800°C resulted in a 39.2% char yield, according to Tokmurzin et al. (2020). Additionally, when food waste was co-pyrolyzed with algae biomass, it increased gas yield but reduced pyrolysis oil yield, as demonstrated by Chen, Yu, Fang, Dai, and Ma (2018).

4.4.2 Biochemical technology

4.4.2.1 Fermentation

The process of fermentation encompasses dark fermentation and photo-fermentation, resulting in the production of ethanol, hydrogen, and biodiesel. Dark fermentation involves the treatment of organic waste with bacteria in the absence of light. On the other hand, photo-fermentation achieves the same by employing a light source during the treatment (Kumar & Agrawal, 2020). The fermentation procedure employs various methods, including batch, fed-batch, continuous processes, submerged fermentation, and continuous immobilized fermentation. The selection of these methods hinges on factors like the composition of lignocellulosic hydroxylates, microbial behavior, and economic considerations (John, Muthukumar, & Arunagiri, 2017). This biological transformation can be executed with or without the involvement of enzymatic hydrolysis. Recognized by its cost-effectiveness, low energy requirements, and minimal wastewater generation, this approach holds promise for converting organic waste into valuable commodities (Tlais, Fiorino, Polo, Filannino, & Di Cagno, 2020). Multiple studies suggest that agri-food waste substrates can yield various value-added products subsequent to fermentation. These encompass antibiotics, pigments, biosurfactants, hydrolytic enzymes, plastics, pesticides, and bioactive compounds (Cerda et al., 2019; Lizardi-Jiménez & Hernández-Martínez, 2017).

4.4.2.2 Anaerobic digestion

In recent times, there has been a growing focus on addressing the issue of detrimental disposal of organic waste and simultaneously recovering clean bioenergy (Lu, Lin, Wu, & Zhang, 2021). AD has emerged as a promising strategy for managing organic waste and harnessing bioenergy (He, Guo, Lu, & Zhang, 2021). AD offers numerous environmental advantages, encompassing the creation of a renewable energy platform, potential nutrient recycling, and reduction in waste volumes (Kosseva, 2009). According to a report, the energy content of 1 cubic meter of biogas obtained from AD is estimated to be equivalent to 21 MJ. Considering a generation efficiency of 35%, this amount of biogas could potentially produce approximately 2.04 kWh of electricity (Murphy, McKeogh, & Kiely, 2004).

When organic waste undergoes AD in landfills, it generates biogas primarily composed of CH_4 and CO_2, alongside small quantities of other gases such as nitrogen (N_2), oxygen (O_2), and hydrogen sulfide (H_2S), which can escape into the atmosphere and contribute to environmental pollution. Notably, AD holds an edge over conventional aerobic processes due to its minimal energy requirements for operation, low initial investment costs, and limited sludge production (Zhao et al., 2021). Furthermore, the AD process results in the production of biogas, offering a clean and renewable energy source (Vyas, Prajapati, Shah, Srivastava, & Varjani, 2022). From both an economic and environmental standpoint, AD stands as an attractive option, generating cost-effective clean energy without emitting GHGs. Despite the existence of multiple technologies, contemporary preferences worldwide tend to lean towards incineration and landfilling methods due to their high energy production capabilities (Hussain, Mishra, & Vanacore, 2020; Tozlu, Özahi, & Abuşoğlu, 2016). In the realm of MSW management, AD is recognized as a valuable technology (Jain, Jain, Wolf, Lee, & Tong, 2015). Nevertheless, there is a current imperative to critically examine operational parameters and available pretreatment technologies (including mechanical, thermal, chemical, and biological methods) for treating the substrate. This exploration is essential for maximizing output and enhancing the efficiency of waste AD.

4.5 Types of pyrolysis reactors used for treating problematic wastes

Pyrolysis, a thermochemical conversion process, has emerged as a promising avenue for transforming such waste materials into valuable resources. Through controlled heating in specialized reactors, pyrolysis breaks down complex compounds into biochar, bio-oil, and gases. Different types of pyrolytic reactors play important roles in treating hazardous wastes. It really indicates the technological advancements in shaping a sustainable future for waste management.

4.5.1 Fixed-bed reactor and batch reactor

Fixed-bed reactors are one of the simplest and oldest designs for pyrolysis. The waste feedstock is loaded into a stationary vessel, and heat is applied to initiate thermal decomposition. Prior to conducting the experiment, the reactor is purged using an inert gas like nitrogen (N_2) or argon (Ar), and this gas flow is sustained throughout the entire process to establish an oxygen-free environment. While gases and vapors produced are released from the reactor during pyrolysis, the solid char residue is typically extracted after the process concludes (Jouhara, Nannou, Anguilano, Ghazal, & Spencer, 2017). The fixed-bed reactor is notable for its relatively slow heating rate. While easy to construct and operate, fixed-bed reactors can suffer from temperature gradients due to uneven heat distribution, potentially leading to incomplete pyrolysis and varied product quality (Jouhara et al., 2017; Lewandowski, Januszewicz, & Kosakowski, 2019). Research has shown that proper insulation and improved design can mitigate these issues, resulting in more uniform heating and better yields of valuable products like bio-oil (Li et al., 2016).

Ordinarily, batch reactors operate in closed setups, maintaining consistent reactants and products throughout, resulting in notable conversion rates. Conversely, semibatch reactors enable the addition of reactants and removal of products during the reaction process. Nevertheless, ensuring uniform product yields across various batches is a limitation, and scaling up presents difficulties. Furthermore, extended solid retention times and the intricate removal of char compounds are some limitations of this reactor category (Jouhara et al., 2017).

4.5.2 Fluidized-bed reactor

Fluidized-bed reactors involve suspending the waste feedstock in an upward-flowing stream of gas or air. The fluidized particles help distribute heat more evenly and promote mixing, enhancing the overall pyrolysis process. Research has focused on optimizing fluidized-bed conditions, such as gas flow rates and particle size, to achieve higher conversion rates and improved product quality (Eke, Onwudili, & Bridgwater, 2020). This reactor design appears to offer a viable answer for the pyrolysis of discarded polymers. The utilization of a fluidized-bed reactor for polymer pyrolysis can offer significant benefits compared to other reactor methods, where heat transfer is less effective for breaking down polymers because of their low thermal conductivity and high viscosity. Moreover, Pandey et al. (2020) reported it is one of the most promising reactors for pyrolysis of plastic wastes.

However, fluidized-bed reactors, while potentially advantageous for waste utilization, come with significant challenges. The need for finely processed raw materials to allow them to float in the fluid and the formidable task of efficiently separating char residues from bed materials are pivotal hurdles (Iannello, Morrin, & Materazzi, 2020). Consequently, the adoption of this reactor type remains limited within larger-scale projects. However, the intricate design of their system poses challenges for potential scaling up. Moreover, the laborious process of sample preparation further complicates matters, casting doubts on the practicality and feasibility of industrial-scale implementation (Jouhara et al., 2017).

4.5.3 Spouted bed reactor

The spouted bed reactor (SBR) is an innovative and efficient technology that holds significant promise, particularly for the treatment of plastic waste. In an SBR, solid particles are continuously circulated within a central column by gas injection at the bottom, creating a spout-like movement. This dynamic bed behavior results in enhanced mixing and heat transfer, making SBRs highly effective for various processes, including pyrolysis (Jouhara et al., 2017). The spout-like motion of particles ensures thorough mixing of the plastic feedstock, promoting uniform heating and enhancing the contact between the plastics and the heat carrier. This effect is crucial for achieving high pyrolysis conversion rates and consistent product quality. Nevertheless, the intense gas-solid contact in an SBR promotes efficient heat transfer, allowing for rapid and controlled heating of the plastic waste. This feature is particularly beneficial for plastic pyrolysis, where achieving the optimal temperature range is vital for breaking down complex polymer structures into valuable products (Lopez et al., 2010). Moreover, the continuous circulation of particles in the SBR helps prevent particle agglomeration and reactor fouling. This characteristic is especially valuable when dealing with plastic waste, which can be prone to forming clumps that hinder efficient pyrolysis reactions in traditional fixed-bed reactors (Lopez et al., 2010). Prior studies reported the use of this reactor in treating plastic wastes (Orozco et al., 2021), polystyrene (Artetxe et al., 2015), polypropylene and PET (Niksiar, Faramarzi, & Sohrabi, 2015), etc.

4.5.4 Rotary kiln reactor

Rotary kiln reactors consist of a rotating cylinder where waste feedstock is continuously fed in at one end and moves through various temperature zones as it progresses along the cylinder's length. Research has shown that rotational movement aids in achieving uniform heating, resulting in consistent product quality. Continuous operation in rotary kilns allows for higher throughput, making them suitable for large-scale waste treatment. Residence time holds significant importance in the pyrolysis process, as it governs the amount of energy absorbed by the material at a particular heating rate. Prior studies evaluated the relationship between residence time and factors like the mean volumetric flow and rotational speed of rotary kilns (Fantozzi, Colantoni, Bartocci, & Desideri, 2007). However, controlling residence times and optimizing heat distribution are areas of ongoing research to maximize pyrolysis efficiency (Li, Yan, Li, Chi, & Cen, 2002).

4.5.5 Microwave pyrolysis reactor

Microwave reactors apply microwave radiation to heat the waste feedstock directly, rapidly inducing pyrolysis. Research in this area has focused on optimizing microwave energy absorption by waste materials, enhancing reaction kinetics, and improving product yields. The most important advantages provided by microwaves are uniform and rapid internal heating of large biomass particles, rapid initiation and cessation of processes, elimination of agitation requirements, and the ability to maintain precise control (Jouhara et al., 2017). Moreover, microwave pyrolysis can achieve faster reaction rates and higher energy efficiency compared to conventional heating methods (Molino, Nanna, Ding, Bikson, & Braccio, 2013). Through microwave pyrolysis, agricultural waste can be converted to biochar and as by-products syngas and bio-oil will be produced (Ge et al., 2021). Moreover, it has been used for pyrolysis of hazardous plastic wastes in hospitals (Mahari et al., 2022) paper (Khongkrapan, Thanompongchart, Tippayawong, & Kiatsiriroat, 2013), biomass (Li et al., 2016; Wu et al., 2014) and plastics (Putra, Rozali, Patah, & Idris, 2022). However, scaling up microwave reactors while maintaining uniform heating remains a challenge. Moreover, the introduction of catalysts in microwave reactors can specifically modify the output and characteristics of the pyrolysis products. Although catalytic microwave pyrolysis comes with significant initial and operational expenses, it proves economically viable by generating high-quality jet fuels, biochar, and hydrogen (Ge et al., 2021).

4.5.6 Screw auger pyrolysis reactor

Screw auger reactors utilize a rotating screw mechanism to transport the waste feedstock through a heated chamber. This controlled movement ensures prolonged exposure to heat and allows for better mixing, resulting in improved product yields (Funke, Henrich, Dahmen, & Sauer, 2017). Research has demonstrated that adjusting the screw's speed and pitch can influence the residence time and temperature profile, leading to the production of specific pyrolysis products. This reactor design has gained attention for its ability to achieve consistent and controllable outcomes (Campuzano, Brown, & Martínez, 2019).

4.5.7 Plasma reactor

Plasma is often described as the fourth state of matter, distinct from solids, liquids, and gases. It consists of a high-energy, ionized gas in which electrons are separated from their parent atoms, resulting in a mixture of free electrons and positively charged ions. This ionized gas can conduct electricity and respond to electromagnetic fields (Jouhara et al., 2017). A plasma reactor is an advanced technology used for the pyrolysis of hazardous waste materials. In this method, a high-energy plasma torch is employed to generate an extremely hot and ionized gas (plasma), which reaches temperatures exceeding those of traditional pyrolysis methods. This intense heat breaks down the complex molecular structures of hazardous waste into simpler molecules, reducing them to their basic constituents. Plasma pyrolysis offers several benefits for hazardous waste treatment. Firstly, the extremely high temperatures achieved by the plasma torch facilitate the decomposition of even the most stubborn and hazardous compounds, biomedical wastes ensuring a thorough breakdown of the waste material. This leads to the destruction of toxic and hazardous substances, reducing their harmful impact on the environment (Harussani, Sapuan, Rashid, Khalina, & Ilyas, 2022). Secondly, the rapid and precise heating provided by the plasma torch minimizes the formation of undesirable by-products that could be generated through slower heating processes. This can lead to higher efficiency and a cleaner end product (Harussani et al., 2022). Furthermore, the versatility of plasma reactors allows for the treatment of a wide range of hazardous waste materials, including those that are highly persistent and resistant to other treatment methods (Tang et al., 2013). However, plasma pyrolysis comes with certain challenges, such as high energy consumption and equipment costs. Additionally, the design and operation of plasma reactors require specialized expertise (Sanlisoy & Carpinlioglu, 2017).

4.5.8 Solar reactor

A solar reactor is a technology that uses concentrated sunlight to initiate pyrolysis, breaking down materials like biomass and waste into valuable products. It offers environmental benefits through renewable energy use and reduced emissions. Solar reactors are especially suitable for sunny regions (Joardder, Halder, Rahim, & Masud, 2017). However, challenges like intermittent sunlight and complex engineering need to be addressed. Despite this, solar reactors hold promise for sustainable waste treatment by harnessing clean energy (Sobek & Werle, 2019).

4.6 Diversification of problematic wastes through thermal degradation

4.6.1 Biofuel production

Both developing and developed nations heavily rely on oil and gas derived from finite fossil fuel deposits. The growth in global population and industrialization has triggered an energy crisis, leading to increased fuel costs and demand. Consequently, there's a pressing need to explore alternative renewable fuels like biofuels. Agricultural waste resulting from human activities is less harmful than chemical waste, but improper disposal can cause environmental problems, accumulating and posing pollution risks. The rich organic content in agricultural waste fosters microbial growth, presenting health and environmental hazards. Paradoxically, these very characteristics make such waste suitable for biofuel production. Biogas, a gaseous biofuel like bio-methane, can be generated through AD of agro-industrial waste. This method is cost-effective, produces minimal residual waste, and is recognized as a renewable resource (Molino et al., 2013). Various biofuels, such as bioethanol, biogas, biodiesel, and bio-hydrogen, can be derived from agro-industrial waste. Biogas, composed of CH_4 and CO_2, is produced from the organic portion of crop residue or biomass. AD converts biomass, like sugar cane residue, rice residue, and cow dung, into biogas (Devi et al., 2022). Biodiesel serves as an alternative to diesel, produced from substrates like rice bran, waste vegetable oil, and fruit waste (Leiva-Candia et al., 2014; Ngoie, Oyekola, Ikhu-Omoregbe, & Welz, 2020). Various oils like corn, soybean, palm, and sunflower oil can be used, with waste cooking oil being a significant source (Gouran, Aghel, & Nasirmanesh, 2021; Hosseinzadeh-Bandbafha et al., 2022). Ethanol, sourced from plants, is another biofuel. Starch can be enzymatically converted to glucose and then fermented into ethanol, often using the yeast species S. cerevisiae. The lignocellulosic structure of spent grain aids ethanol production, and this content can be converted to ethanol post-enzymatic hydrolysis and fermentation. MSW is a promising cellulose-rich source for sustainable bioethanol production (Rezania et al., 2019), demanding effective utilization of this resource for fuel production due to its abundance.

4.6.2 Biochar production

In recent times, there has been a growing interest in biochar due to its manifold advantages. Biochar can be easily produced from MSW and agricultural residues. Various sources, such as lignocellulose-based biomass, microalgae, and food waste, serve as raw materials for generating biochar. Additionally, the potential of sewage sludge and organic components from landfills as biochar precursors has been investigated. Biochar possesses significant physicochemical attributes like high porosity, surface area, cation exchange capacity, and a substantial presence of organic and mineral substances. The practical applications of biochar in global agriculture have ancient origins. Its use varies based on its specific physicochemical characteristics. Biochar can function as a substrate for activated carbon production, a solid biofuel in the power industry, or an adsorbent for pollutants. The utility of biochar in environmental preservation yields multiple benefits. It enhances soil properties by augmenting soil carbon content and water retention capacity. Furthermore, it offers the potential to substitute fossil fuels with renewable alternatives, reduce reliance on synthetic fertilizers and pesticides, and thereby decrease the risk of surface and groundwater contamination. Biochar's physicochemical attributes also make it valuable in composting. Incorporating biochar as a structural element and an additive in composting mitigates ammonia emissions. Composting materials with a narrow carbon-to-nitrogen ratio tend to release more ammonia during composting, leading to lower nitrogen content in the final compost. By introducing biochar, the overall porosity, air permeability, and water-holding capacity of the composting mixture can be enhanced. Moreover, biochar contributes to addressing climate change by curbing GHG emissions due to its substantial carbon content (Sri Shalini, Palanivelu, Ramachandran, & Raghavan, 2021). It emerges as a promising alternative for fossil fuels, waste management, bioenergy, and combating climate change.

4.6.3 Composting of problematic wastes

The decay of organic refuse gives rise to unpleasant odors and olfactory concerns. Therefore, effective management and reduction of its pollution levels are essential to ensure safe disposal. Multiple techniques are accessible for addressing such waste-related issues. Among these approaches, composting of organic waste stands out as a well-established method. It proves particularly suitable for various organic wastes like biodegradable MSW, agricultural by-products, garden waste, slaughterhouse waste, and livestock waste. Composting involves the controlled microbial breakdown of organic matter into a relatively stable, amorphous, brown to dark-brown humus-like material. Agricultural wastes and MSW undergo thorough decomposition under regulated conditions of either thermophilic or mesophilic temperatures, usually in heaps, windows, or peat, with appropriate moisture levels. While the decomposition and stabilization of organic waste occur naturally, the systematic production of composted manure through the orchestrated utilization of this natural process is categorized as composting. This method can significantly reduce waste volume by 50%–85% (Malav et al., 2020). The transformation of food waste and agricultural residues into compost, followed by its application in agriculture, presents a successful waste management strategy. Compost application leads to augmented soil carbon content, improved infiltration rates, greater water retention capacity, enhanced sorptivity, and decreased evaporation (Al-Omran, Ibrahim, & Alharbi, 2019). It can substantially replace synthetic fertilizers by supplying essential macro and micronutrients. Additionally, composting can contribute to carbon credits by preventing landfill deposition and incineration, resulting in reduced GHG emissions.

4.7 Challenges of pyrolysis of problematic wastes

Plastics' high volatile content makes them suitable for pyrolysis, yielding substantial liquid oil (Dogu et al., 2021). However, challenges arise when processing commercial polymers that include additives like fillers, plasticizers, colorants, and flame retardants through pyrolysis. In these cases, there is an increase in the production of char due to the inherent rise in fixed carbon content (Zhou, Broadbelt, & Vinu, 2016). Furthermore, the presence of contaminants in solid plastic waste (such as food residues) can significantly alter the outcomes of pyrolysis and gasification processes. Plastic waste comes from diverse sources like WEEE, end of life vehicle, construction and demolition waste, packaging, and agriculture, resulting in a varied mix. These waste streams can contain harmful substances, including halogens and biohazardous materials. Thorough feedstock characterization before pyrolysis is essential for product quality and operational safety, as toxic materials could escape during recycling, posing environmental and health risks. Additionally, pyrolysis transforms plastics back into hydrocarbons, including potentially dangerous aromatic compounds classified as possible carcinogens. Therefore, conducting a detailed risk assessment before pyrolysis is crucial (Qureshi et al., 2020). During the pyrolysis of plastics, notably polyolefins, a major product obtained is wax. To recover these waxes effectively, recovery systems need to be tailored for wax handling. Inadequate condensing systems can lead to wax build-up on inner surfaces, complicating the recovery process. An innovative approach, as demonstrated by Scheirs and Kaminsky (2006), involves incorporating heated impact precipitators within the condensing setup. This method keeps waxes in a liquid state, streamlining the efficiency of product recovery. An integral challenge in the development of a continuous polymer pyrolysis process revolves around the consistent introduction of solid feedstock into the reactor (Wong, Ngadi, Abdullah, & Inuwa, 2015). Fluid catalytic cracking provides a resolution to this issue by dissolving plastic feedstock in a suitable solvent, followed by subjecting the solution to pyrolysis (Lopez, Artetxe, Amutio, Bilbao, & Olazar, 2017; Ragaert, Delva, & Van Geem, 2017; Wong et al., 2015). The quality of pyrolysis liquids depends on their stability and how they age over time. Pyrolysis liquids have a tendency to undergo repolymerization due to their thermodynamic instability. As a result, it's necessary to implement posttreatment methods to maintain their quality over an extended period. These posttreatment approaches might involve techniques like blending and dewaxing, among others (Butler, Devlin, & McDonnell, 2011). Pyrolysis offers flexibility through adjustments in operational factors such as temperature, residence time, and catalyst usage. Optimizing these parameters for different types of e-waste materials is imperative. However, many studies primarily focus on specific components or fractions of e-waste, partly due to the hazardous nature of e-waste and the challenges of analyzing its diverse products. To make pyrolysis processes more practical, researchers are exploring methods for pyrolyzing entire e-waste instead of specific components. It has been extensively studied at laboratory scales. However, the absence of operational industrial-scale e-waste pyrolysis plants is evident. Future research efforts should focus on upscaling the latest pyrolysis technologies to pilot and industrial levels. It's important to implement laboratory-developed future-generation technologies as practical industrial solutions. This approach could significantly contribute to addressing e-waste disposal challenges on both local and global scales. Besides technological progress, societal demand and government regulations play a vital role in promoting the use of pyrolysis processes in e-waste management (Joo, Kwon, & Lee, 2021). MSW processing requires significant capital investment. Pyrolysis transforms it into various byproducts: syngas mixture, liquid bio-oil or tar, and solid residue known as char. However, it has drawbacks, such as the complexity of byproduct streams that can't be directly released into the environment. Due to high CO gas concentrations, harmful gases, and elevated temperatures, further treatment is necessary (Kumar & Agrawal, 2020).

4.8 Conclusion and future thrust

The persistent rise in the volume of waste generated by society underscores the pressing requirement for the creation of novel and improved approaches to waste disposal. Conventional waste management techniques like composting, landfilling, and incineration have become outdated and should be substituted with contemporary, efficient, and user-friendly alternatives, such as pyrolysis. The escalating cost of energy derived from conventional sources has rendered the extraction of power from waste progressively more crucial in today's context. These waste management methodologies not only offer cost-effective solutions but also adhere to environmental standards and gain societal approval. The advantages of waste management are increasingly apparent, encompassing the reduction of GHG emissions, waste volume reduction, revenue generation through energy sales, and the repurposing of waste materials. In contrast to commonly employed biological waste disposal methods, pyrolysis permits the utilization of all carbon-based materials, both organic and inorganic. This technique yields readily usable fuels in a straightforward and secure manner. Typically, gas and char are employed as energy sources due to the convenience of utilizing and marketing energy products. The liquid by-products resulting from the pyrolysis of problematic waste materials tend to be intricate and frequently contain water. It is crucial to permit researchers and scientists to explore waste-to-energy strategies comprehensively. This exploration should encompass efficient waste management approaches and technologies, all underpinned by thorough environmental impact assessments. Furthermore, there is merit in adopting emerging laboratory-scale technologies as industrial solutions, alongside seeking novel alternatives that could altogether eradicate domestic waste disposal challenges.

References

Abdel-Shafy, H. I., & Mansour, M. S. (2018). Solid waste issue: Sources, composition, disposal, recycling, and valorization. *Egyptian Journal of Petroleum*, 27(4), 1275–1290.

Ahamed, A., Veksha, A., Yin, K., Weerachanchai, P., Giannis, A., & Lisak, G. (2020). Environmental impact assessment of converting flexible packaging plastic waste to pyrolysis oil and multi-walled carbon nanotubes. *Journal of Hazardous Materials*, 390, 121449.

Akpanudoh, N. S., Gobin, K., & Manos, G. (2005). Catalytic degradation of plastic waste to liquid fuel over commercial cracking catalysts: Effect of polymer to catalyst ratio/acidity content. *Journal of Molecular Catalysis A: Chemical*, 235(1–2), 67–73.

Ali, H., Khan, E., & Ilahi, I. (2019). Environmental chemistry and ecotoxicology of hazardous heavy metals: Environmental persistence, toxicity, and bioaccumulation. *Journal of Chemistry*, 2019.

Alimi, O. S., Farner Budarz, J., Hernandez, L. M., & Tufenkji, N. (2018). Microplastics and nanoplastics in aquatic environments: Aggregation, deposition, and enhanced contaminant transport. *Environmental Science & Technology*, 52(4), 1704–1724.

Al-Mrayat, T., Al-Hamaiedeh, H., El-Hasan, T., Aljbour, S. H., Al-Ghazawi, Z., & Mohawesh, O. (2022). Pyrolysis of domestic sewage sludge: Influence of operational conditions on the product yields using factorial design. *Heliyon*, 8(5).

Al-Omran, A., Ibrahim, A., & Alharbi, A. (2019). Evaluating the impact of combined application of biochar and compost on hydro-physical properties of loamy sand soil. *Communications in Soil Science and Plant Analysis*, 50(19), 2442–2456.

Artetxe, M., Lopez, G., Amutio, M., Barbarias, I., Arregi, A., Aguado, R., ... Olazar, M. (2015). Styrene recovery from polystyrene by flash pyrolysis in a conical spouted bed reactor. *Waste Management*, 45, 126–133.

Azar, S. K., & Azar, S. S. (2016). Waste related pollutions and their potential effect on cancer incidences in Lebanon. *Journal of Environmental Protection, 7*(6), 778–783.

Baldé, C.P., Forti, V., Gray, V., Kuehr, R., & Stegmann, P. (2017). *The global e-waste monitor 2017: Quantities, flows and resources*. United Nations University, International Telecommunication Union, and International Solid Waste Association.

Barrett, G. E., Alexander, P. D., Robinson, J. S., & Bragg, N. C. (2016). Achieving environmentally sustainable growing media for soilless plant cultivation systems—A review. *Scientia Horticulturae, 212*, 220–234.

Belewu, M. A., & Babalola, F. T. (2009). Nutrient enrichment of waste agricultural residues after solid state fermentation using Rhizopus oligosporus . *Journal of Applied Bioscience* (13, pp. 695–699).

Bernando, M. (2011). *Physico-chemical characterization of chars produced in the copyrolysis of wastes and possible routes of valorisation* (pp. 27–36). Portugal: Chemical Engineering. Universidade Nova de Lisboa.

Bhatnagar, A., Khatri, P., Krzywonos, M., Tolvanen, H., & Konttinen, J. (2022). Techno-economic and environmental assessment of decentralized pyrolysis for crop residue management: Rice and wheat cultivation system in India. *Journal of Cleaner Production, 367*, 132998.

Brennecke, D., Duarte, B., Paiva, F., Caçador, I., & Canning-Clode, J. (2016). Microplastics as vector for heavy metal contamination from the marine environment. *Estuarine, Coastal and Shelf Science, 178*, 189–195.

Butler, E., Devlin, G., & McDonnell, K. (2011). Waste polyolefins to liquid fuels via pyrolysis: Review of commercial state-of-the-art and recent laboratory research. *Waste and Biomass Valorization, 2*, 227–255.

Campuzano, F., Brown, R. C., & Martínez, J. D. (2019). Auger reactors for pyrolysis of biomass and wastes. *Renewable and Sustainable Energy Reviews, 102*, 372–409.

Cerda, A., Artola, A., Barrena, R., Font, X., Gea, T., & Sánchez, A. (2019). Innovative production of bioproducts from organic waste through solid-state fermentation. *Frontiers in Sustainable Food Systems, 3*, 63.

Chen, L., Yu, Z., Fang, S., Dai, M., & Ma, X. (2018). Co-pyrolysis kinetics and behaviors of kitchen waste and chlorella vulgaris using thermogravimetric analyzer and fixed bed reactor. *Energy Conversion and Management, 165*, 45–52.

Cheng, S., Meng, M., Xing, B., Shi, C., Nie, Y., Xia, D., ... Xia, H. (2022). Preparation of valuable pyrolysis products from poplar waste under different temperatures by pyrolysis: Evaluation of pyrolysis products. *Bioresource Technology, 364*, 128011.

Cleetus, C., Thomas, S., & Varghese, S. (2013). Synthesis of petroleum-based fuel from waste plastics and performance analysis in a CI engine. *Journal of Energy, 2013*.

Cui, J., & Zhang, L. (2008). Metallurgical recovery of metals from electronic waste: A review. *Journal of Hazardous Materials, 158*(2–3), 228–256.

Czajczyńska, D., Anguilano, L., Ghazal, H., Krzyżyńska, R., Reynolds, A. J., Spencer, N., & Jouhara, H. (2017). Potential of pyrolysis processes in the waste management sector. *Thermal Science and Engineering Progress, 3*, 171–197.

Devi, M. K., Manikandan, S., Oviyapriya, M., Selvaraj, M., Assiri, M. A., Vickram, S., ... Awasthi, M. K. (2022). Recent advances in biogas production using agro-industrial Waste: A comprehensive review outlook of techno-economic analysis. *Bioresource Technology*, 127871.

Dogu, O., Pelucchi, M., Van de Vijver, R., Van Steenberge, P. H., D'hooge, D. R., Cuoci, A., ... K., M. (2021). The chemistry of chemical recycling of solid plastic waste via pyrolysis and gasification: State-of-the-art, challenges, and future directions. *Progress in Energy and Combustion Science, 84*, 100901.

Eke, J., Onwudili, J. A., & Bridgwater, A. V. (2020). Influence of moisture contents on the fast pyrolysis of trommel fines in a bubbling fluidized bed reactor. *Waste and Biomass Valorization, 11*, 3711–3722.

Fantozzi, F., Colantoni, S., Bartocci, P., & Desideri, U. (2007). Rotary kiln slow pyrolysis for syngas and char production from biomass and waste—Part I: Working envelope of the reactor.

Forti, V., Balde, C.P., Kuehr, R., & Bel, G. (2020). The global e-waste monitor 2020: Quantities, flows and the circular economy potential.

Funke, A., Henrich, E., Dahmen, N., & Sauer, J. (2017). Dimensional analysis of auger-type fast pyrolysis reactors. *Energy Technology, 5*(1), 119–129.

Gangwar, S., & Pathak, V. K. (2021). A critical review on tribological properties, thermal behavior, and different applications of industrial waste reinforcement for composites. *Proceedings of the Institution of Mechanical Engineers, Part L: Journal of Materials: Design and Applications, 235*(3), 684–706.

Gaylor, M. O., Harvey, E., & Hale, R. C. (2013). Polybrominated diphenyl ether (PBDE) accumulation by earthworms (*Eisenia fetida*) exposed to biosolids-, polyurethane foam microparticle-, and penta-BDE-amended soils. *Environmental Science & Technology, 47*(23), 13831–13839.

Ge, S., Yek, P. N. Y., Cheng, Y. W., Xia, C., Mahari, W. A. W., Liew, R. K., ... Lam, S. S. (2021). Progress in microwave pyrolysis conversion of agricultural waste to value-added biofuels: A batch to continuous approach. *Renewable and Sustainable Energy Reviews, 135*, 110148.

Gouran, A., Aghel, B., & Nasirmanesh, F. (2021). Biodiesel production from waste cooking oil using wheat bran ash as a sustainable biomass. *Fuel, 295*, 120542.

Guo, J. J., Huang, X. P., Xiang, L., Wang, Y. Z., Li, Y. W., Li, H., ... Wong, M. H. (2020). Source, migration and toxicology of microplastics in soil. *Environment International, 137*, 105263.

Harussani, M. M., Sapuan, S. M., Rashid, U., Khalina, A., & Ilyas, R. A. (2022). Pyrolysis of polypropylene plastic waste into carbonaceous char: Priority of plastic waste management amidst COVID-19 pandemic. *Science of The Total Environment, 803*, 149911.

He, X., Guo, Z., Lu, J., & Zhang, P. (2021). Carbon-based conductive materials accelerated methane production in anaerobic digestion of waste fat, oil and grease. *Bioresource Technology, 329*, 124871.

Hester, R. E., & Harrison, R. M. (Eds.), (2002). *Environmental and health impact of solid waste management activities* (18). Royal Society of Chemistry.

Hira, A., Pacini, H., Attafuah-Wadee, K., Vivas-Eugui, D., Saltzberg, M., & Yeoh, T. N. (2022). Plastic waste mitigation strategies: A review of lessons from developing countries. *Journal of Developing Societies, 38*(3), 336–359.

Hodson, M. E., Duffus-Hodson, C. A., Clark, A., Prendergast-Miller, M. T., & Thorpe, K. L. (2017). Plastic bag derived-microplastics as a vector for metal exposure in terrestrial invertebrates. *Environmental Science & Technology, 51*(8), 4714–4721.

Hosseinzadeh-Bandbafha, H., Nizami, A. S., Kalogirou, S. A., Gupta, V. K., Park, Y. K., Fallahi, A., ... Tabatabaei, M. (2022). Environmental life cycle assessment of biodiesel production from waste cooking oil: A systematic review. *Renewable and Sustainable Energy Reviews, 161*, 112411.

Huang, B., Wang, X., Kua, H., Geng, Y., Bleischwitz, R., & Ren, J. (2018). Construction and demolition waste management in China through the 3R principle. *Resources, Conservation and Recycling, 129*, 36–44.

Hussain, Z., Mishra, J., & Vanacore, E. (2020). Waste to energy and circular economy: the case of anaerobic digestion. *Journal of Enterprise Information Management, 33*(4), 817–838.

Iannello, S., Morrin, S., & Materazzi, M. (2020). Fluidised bed reactors for the thermochemical conversion of biomass and waste. *KONA Powder and Particle Journal, 37*, 114–131.

Jain, S., Jain, S., Wolf, I. T., Lee, J., & Tong, Y. W. (2015). A comprehensive review on operating parameters and different pretreatment methodologies for anaerobic digestion of municipal solid waste. *Renewable and Sustainable Energy Reviews, 52*, 142–154.

Jia, C., Xie, S., Zhang, W., Intan, N. N., Sampath, J., Pfaendtner, J., & Lin, H. (2021). Deconstruction of high-density polyethylene into liquid hydrocarbon fuels and lubricants by hydrogenolysis over Ru catalyst. *Chem Catalysis, 1*(2), 437–455.

Joarddar, M. U. H., Halder, P. K., Rahim, M. A., & Masud, M. H. (2017). Solar pyrolysis: Converting waste into asset using solar energy. *Clean energy for sustainable development* (pp. 213–235). Academic Press.

John, I., Muthukumar, K., & Arunagiri, A. (2017). A review on the potential of citrus waste for D-Limonene, pectin, and bioethanol production. *International Journal of Green Energy, 14*(7), 599–612.

Joo, J., Kwon, E. E., & Lee, J. (2021). Achievements in pyrolysis process in e-waste management sector. *Environmental Pollution, 287*, 117621.

Jouhara, H., Nannou, T. K., Anguilano, L., Ghazal, H., & Spencer, N. (2017). Heat pipe based municipal waste treatment unit for home energy recovery. *Energy, 139*, 1210–1230.

Katami, T., Yasuhara, A., & Shibamoto, T. (2004). Formation of dioxins from incineration of foods found in domestic garbage. *Environmental Science & Technology, 38*(4), 1062–1065.

Khongkrapan, P., Thanompongchart, P., Tippayawong, N., & Kiatsiriroat, T. (2013). Fuel gas and char from pyrolysis of waste paper in a microwave plasma reactor. *International Journal of Energy and Environment (Print), 4*.

Kitila, A.W. (2018). *Waste electrical and electronic equipment (e-waste) management and disposal methods in the city of Addis Ababa, Ethiopia* (Doctoral dissertation).

Korley, L. T., Epps, T. H., Helms, B. A., & Ryan, A. J. (2021). Toward polymer upcycling—adding value and tackling circularity. *Science (New York, N.Y.), 373*(6550), 66–69.

Kosseva, M. R. (2009). Processing of food wastes. *Advances in Food and Nutrition Research, 58*, 57–136.

Kumar, A., & Agrawal, A. (2020). Recent trends in solid waste management status, challenges, and potential for the future Indian cities—A review. *Current Research in Environmental Sustainability, 2*, 100011.

Leiva-Candia, D. E., Pinzi, S., Redel-Macías, M. D., Koutinas, A., Webb, C., & Dorado, M. P. (2014). The potential for agro-industrial waste utilization using oleaginous yeast for the production of biodiesel. *Fuel, 123,* 33–42.

Lewandowski, W. M., Januszewicz, K., & Kosakowski, W. (2019). Efficiency and proportions of waste tyre pyrolysis products depending on the reactor type—A review. *Journal of Analytical and Applied Pyrolysis, 140,* 25–53.

Li, J., Dai, J., Liu, G., Zhang, H., Gao, Z., Fu, J., ... Huang, Y. (2016). Biochar from microwave pyrolysis of biomass: A review. *Biomass and Bioenergy, 94,* 228–244.

Li, S. Q., Yan, J. H., Li, R. D., Chi, Y., & Cen, K. F. (2002). Axial transport and residence time of MSW in rotary kilns: Part I. Experimental. *Powder technology, 126*(3), 217–227.

Lizardi-Jiménez, M. A., & Hernández-Martínez, R. (2017). Solid state fermentation (SSF): diversity of applications to valorize waste and biomass. *3 Biotech, 7*(1), 44.

Lopez, G., Artetxe, M., Amutio, M., Bilbao, J., & Olazar, M. (2017). Thermochemical routes for the valorization of waste polyolefinic plastics to produce fuels and chemicals. A review. *Renewable and Sustainable Energy Reviews, 73,* 346–368.

Lopez, G., Olazar, M., Aguado, R., Elordi, G., Amutio, M., Artetxe, M., & Bilbao, J. (2010). Vacuum pyrolysis of waste tires by continuously feeding into a conical spouted bed reactor. *Industrial & Engineering Chemistry Research, 49*(19), 8990–8997.

Lu, J., Lin, Y., Wu, J., & Zhang, C. (2021). Continental-scale spatial distribution, sources, and health risks of heavy metals in seafood: Challenge for the water-food-energy nexus sustainability in coastal regions? *Environmental Science and Pollution Research,* 1–14.

Mahari, W. A. W., Awang, S., Zahariman, N. A. Z., Peng, W., Man, M., Park, Y. K., ... Lam, S. S. (2022). Microwave co-pyrolysis for simultaneous disposal of environmentally hazardous hospital plastic waste, lignocellulosic, and triglyceride biowaste. *Journal of Hazardous Materials, 423,* 127096.

Malav, L. C., Yadav, K. K., Gupta, N., Kumar, S., Sharma, G. K., Krishnan, S., ... Bach, Q. V. (2020). A review on municipal solid waste as a renewable source for waste-to-energy project in India: Current practices, challenges, and future opportunities. *Journal of Cleaner Production, 277,* 123227.

Molino, A., Nanna, F., Ding, Y., Bikson, B., & Braccio, G. (2013). Biomethane production by anaerobic digestion of organic waste. *Fuel, 103,* 1003–1009.

Murphy, J. D., McKeogh, E., & Kiely, G. (2004). Technical/economic/environmental analysis of biogas utilisation. *Applied Energy, 77*(4), 407–427.

Nair, L. G., Agrawal, K., & Verma, P. (2022). An overview of sustainable approaches for bioenergy production from agro-industrial wastes. *Energy Nexus, 6,* 100086.

Ngoie, W. I., Oyekola, O. O., Ikhu-Omoregbe, D., & Welz, P. J. (2020). Valorisation of edible oil wastewater sludge: Bioethanol and biodiesel production. *Waste and Biomass Valorization, 11,* 2431–2440.

Niksiar, A., Faramarzi, A. H., & Sohrabi, M. (2015). Mathematical modeling of polyethylene terephthalate pyrolysis in a spouted bed. *AIChE Journal, 61*(6), 1900–1911.

Ogbu, C. C., & Okechukwu, S. N. (2023). *Agro-industrial waste management: The circular and bioeconomic perspective.* Onwudili.

Orozco, S., Alvarez, J., Lopez, G., Artetxe, M., Bilbao, J., & Olazar, M. (2021). Pyrolysis of plastic wastes in a fountain confined conical spouted bed reactor: Determination of stable operating conditions. *Energy Conversion and Management, 229,* 113768.

Pandey, U., Stormyr, J. A., Hassani, A., Jaiswal, R., Haugen, H. H., & Moldestad, B. M. E. (2020). Pyrolysis of plastic waste to environmentally friendly products. *Energy Produc Manage in the 21st century IV: The Quest for Sustain Energy, 246,* 61–74.

Paulraj, C. R. K. J., Bernard, M. A., Raju, J., & Abdulmajid, M. (2019). Sustainable waste management through waste to energy technologies in India-opportunities and environmental impacts. *International Journal of Renewable Energy Research (IJRER), 9*(1), 309–342.

Pham, T. P. T., Kaushik, R., Parshetti, G. K., Mahmood, R., & Balasubramanian, R. (2015). Food waste-to-energy conversion technologies: Current status and future directions. *Waste Management, 38,* 399–408.

Popescu, M. L. (2015). Waste electrical and electronic equipment management in Romania. Harmonizing national environmental law with the UE legislation. *Procedia-Social and Behavioral Sciences, 188,* 264–269.

Putra, P. H. M., Rozali, S., Patah, M. F. A., & Idris, A. (2022). A review of microwave pyrolysis as a sustainable plastic waste management technique. *Journal of Environmental Management, 303,* 114240.

Qureshi, M. S., Oasmaa, A., Pihkola, H., Deviatkin, I., Tenhunen, A., Mannila, J., ... Laine-Ylijoki, J. (2020). Pyrolysis of plastic waste: Opportunities and challenges. *Journal of Analytical and Applied Pyrolysis, 152,* 104804.

Ragaert, K., Delva, L., & Van Geem, K. (2017). Mechanical and chemical recycling of solid plastic waste. *Waste Management, 69*, 24–58.

Raihan, A., & Tuspekova, A. (2022). Nexus between economic growth, energy use, agricultural productivity, and carbon dioxide emissions: New evidence from Nepal. *Energy Nexus, 7*, 100113.

Ratnasari, D. K., Nahil, M. A., & Williams, P. T. (2017). Catalytic pyrolysis of waste plastics using staged catalysis for production of gasoline range hydrocarbon oils. *Journal of Analytical and Applied Pyrolysis, 124*, 631–637.

Rezania, S., Oryani, B., Park, J., Hashemi, B., Yadav, K. K., Kwon, E. E., ... Cho, J. (2019). Review on transesterification of non-edible sources for biodiesel production with a focus on economic aspects, fuel properties and by-product applications. *Energy Conversion and Management, 201*, 112155.

Rodríguez Couto, S. (2008). Exploitation of biological wastes for the production of value-added products under solid-state fermentation conditions. *Biotechnology Journal: Healthcare Nutrition Technology, 3*(7), 859–870.

Sadh, P. K., Duhan, S., & Duhan, J. S. (2018). Agro-industrial wastes and their utilization using solid state fermentation: A review. *Bioresources and Bioprocessing, 5*(1), 1–15.

Sanlisoy, A., & Carpinlioglu, M. O. (2017). A review on plasma gasification for solid waste disposal. *International Journal of Hydrogen Energy, 42*(2), 1361–1365.

Sarkar, O., Butti, S. K., & Mohan, S. V. (2018). Acidogenic biorefinery: Food waste valorization to biogas and platform chemicals. *Waste Biorefinery*, 203–218.

Scheirs, J., & Kaminsky, W. (2006). *Feedstock recycling and pyrolysis of waste plastics*. Chichester, UK; Hoboken, NJ: J. Wiley & Sons.

Singh, N., Duan, H., & Tang, Y. (2020). Toxicity evaluation of E-waste plastics and potential repercussions for human health. *Environment International, 137*, 105559.

Sobek, S., & Werle, S. (2019). Solar pyrolysis of waste biomass: Part 1 reactor design. *Renewable Energy, 143*, 1939–1948.

Sri Shalini, S., Palanivelu, K., Ramachandran, A., & Raghavan, V. (2021). Biochar from biomass waste as a renewable carbon material for climate change mitigation in reducing greenhouse gas emissions—A review. *Biomass Conversion and Biorefinery, 11*, 2247–2267.

Su, G., Ong, H. C., Cheah, M. Y., Chen, W.-H., Lam, S. S., & Huang, Y. (2022). Microwave-assisted pyrolysis technology for bioenergy recovery: Mechanism, performance, and prospect. *Fuel, 326*, 124983.

Sultan, U., Zhang, Y., Farooq, M., Imran, M., Khan, A. A., Zhuge, W., ... Ali, Q. (2021). Qualitative assessment and global mapping of supercritical CO_2 power cycle technology. *Sustainable Energy Technologies and Assessments, 43*, 100978.

Tang, L., Huang, H., Hao, H., & Zhao, K. (2013). Development of plasma pyrolysis/gasification systems for energy efficient and environmentally sound waste disposal. *Journal of Electrostatics, 71*(5), 839–847.

Tlais, A. Z. A., Fiorino, G. M., Polo, A., Filannino, P., & Di Cagno, R. (2020). High-value compounds in fruit, vegetable and cereal byproducts: An overview of potential sustainable reuse and exploitation. *Molecules (Basel, Switzerland), 25*, 2987.

Tokmurzin, D., Kuspangaliyeva, B., Aimbetov, B., Abylkhani, B., Inglezakis, V., Anthony, E. J., & Sarbassov, Y. (2020). Characterization of solid char produced from pyrolysis of the organic fraction of municipal solid waste, high volatile coal and their blends. *Energy, 191*, 116562.

Tozlu, A., Özahi, E., & Abuşoğlu, A. (2016). Waste to energy technologies for municipal solid waste management in Gaziantep. *Renewable and Sustainable Energy Reviews, 54*, 809–815.

Vyas, S., Prajapati, P., Shah, A. V., Srivastava, V. K., & Varjani, S. (2022). Opportunities and knowledge gaps in biochemical interventions for mining of resources from solid waste: a special focus on anaerobic digestion. *Fuel, 311*, 122625.

Wang, F., Wong, C. S., Chen, D., Lu, X., Wang, F., & Zeng, E. Y. (2018). Interaction of toxic chemicals with microplastics: A critical review. *Water Research, 139*, 208–219.

Wang, X., & Yan, L. (2022). Driving factors and decoupling analysis of fossil fuel related-carbon dioxide emissions in China. *Fuel, 314*, 122869.

Weigand, H., Fripan, J., Przybilla, I., & Marb, C. (2003) Composition and contaminant load of household waste in Bavaria, Germany: Investigating effects of settlement structure and waste management practice. In *Proceedings of the 9th international waste management and landfill symposium*, Cagliari, CD-ROM.

Wojnowska-Baryła, I., Bernat, K., & Zaborowska, M. (2022). Plastic waste degradation in landfill conditions: The problem with microplastics, and their direct and indirect environmental effects. *International Journal of Environmental Research and Public Health, 19*(20), 13223.

Wong, S. L., Ngadi, N., Abdullah, T. A. T., & Inuwa, I. M. (2015). Current state and future prospects of plastic waste as source of fuel: A review. *Renewable and Sustainable Energy Reviews, 50*, 1167–1180.

Wu, C., Budarin, V. L., Gronnow, M. J., De Bruyn, M., Onwudili, J. A., Clark, J. H., & Williams, P. T. (2014). Conventional and microwave-assisted pyrolysis of biomass under different heating rates. *Journal of Analytical and Applied Pyrolysis, 107*, 276–283.

Young, G. (2010). Municipal solid waste to energy conversion processes: Economic. *Technical and renewable comparisons*. J. Wiley & Sons Inc.

Zhai, P., Zhao, Y., Yang, S., Jin, X., Liang, Z., Yuan, H., ... Li, C. (2022). Comparison of yield and physicochemical characteristics of tropical crop residue biochar under different pyrolysis temperatures. *Biomass Cosnversion and Biorefinery*, 1–13.

Zhang, W., Wang, H., Zhang, R., Yu, X. Z., Qian, P. Y., & Wong, M. H. (2010). Bacterial communities in PAH contaminated soils at an electronic-waste processing center in China. *Ecotoxicology (London, England), 19*, 96–104.

Zhao, D., Yan, B., Liu, C., Yao, B., Luo, L., Yang, Y., ... Zhou, Y. (2021). Mitigation of acidogenic product inhibition and elevated mass transfer by biochar during anaerobic digestion of food waste. *Bioresource Technology, 338*, 125531.

Zhou, X., Broadbelt, L. J., & Vinu, R. (2016). Mechanistic understanding of thermochemical conversion of polymers and lignocellulosic biomass. *Advances in Chemical Engineering, 49*, 95–198.

CHAPTER 5

Modern tools and techniques of biochar characterization for targeted applications

Rajat Kumar Sharma[1,2], T.P. Singh[1], Juma Haydary[2], Deepshikha Azad[1] and Akarsh Verma[3]

[1]Department of Farm Machinery and Power Engineering, G.B. Pant University of Agriculture & Technology, Pantnagar, Uttarakhand, India [2]Department of Chemical and Biochemical Engineering, Faculty of Chemical and Food Technology, Slovak University of Technology, Bratislava, Slovakia [3]Department of Mechanical Engineering, University of Petroleum and Energy Studies, Dehradun, Uttarakhand, India

5.1 Introduction

The field of biochar is gaining attention due to its versatility, cost-effectiveness, environmentally friendly impact, and potential to address global issues such as climate change, pollution, and soil degradation (Mandal, Sharma, & Bhattacharya, 2022). The International Biochar Initiative (IBI) defines biochar as a solid material obtained from the carbonization of organic feedstocks, such as wood, plant leaves, and crop residue, through thermal decomposition under low oxygen conditions at temperatures below 700°C (IBI Biochar Standards Version 2.0. 2014). Applications for biochar include producing solid fuel, removing pollutants, sequestering carbon, improving soil quality, reducing odors from livestock operations, and treating organic waste through methane fermentation and composting. It also has potential uses in producing supercapacitors, high-yield sorbent materials, and geoengineering. As a result, biochar with varying properties is needed to fulfill these different applications (Lehmann et al., 2011; Sharma, Singh, Mandal, Azad, & Kumar, 2022; Singh, Chaturvedi, et al., 2023).

It is important to characterize biochar prior to using it in a specific application in order to optimize its effectiveness. There are various methods used around the world to identify

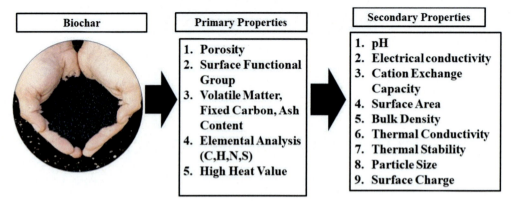

FIGURE 5.1 Different properties and analysis for biochar.

and quantify biochar properties. The IBI has published "Standard Product Definition and Product Testing Guidelines for Biochar that is used in soil" (International Biochar Initiative, 2014) which gives some guidance on biochar characterization (IBI Biochar Standards Version 2.0. 2014). The IBI strongly recommends characterizing biochar before using it as a soil amendment. Additionally, the IBI's mission is to test and use biochar as a tool for soil fertility enhancement and climate change mitigation and therefore these guidelines do not cover characterization required in other fields. The European Biochar Foundation has also published the European Biochar Certificate (EBC) which provides guidelines for the selection of proper methods for biochar characterization which are crucial to assessing biochar properties accurately and consistently (Hans-Peter, 2013). However, these guidelines may not be applicable to other fields of biochar use. Therefore, it is important to use appropriate methods for biochar characterization based on the intended application and to consider factors such as accessibility to technology and knowledge of the user. Improper characterization methods can compromise the validity and reliability of the data, leading to suboptimal outcomes for the intended application.

In this chapter, conventional methods and expected results for the evaluation and comparison of biochar are explored. The techniques discussed range from straightforward molecular investigations to more intricate standard procedures like proximal analysis. Fig. 5.1 illustrates the diverse types of analysis and the qualities of biochar, with the recognition that certain methods or attributes may overlap across multiple categories. The chapter concludes with a review of key findings and potential avenues for future research in the realm of biochar study.

5.2 Properties and characterization of biochar

5.2.1 Proximate analysis

Proximate analysis is a commonly used characterization method that can provide information on the basic properties of biochar such as moisture content, volatile matter, fixed carbon (or resident matter), and ash content (Joseph, Peacocke, Lehmann, & Munroe, 2009). The standard test method ASTM D1762-84 is widely used for this purpose and is recognized by organizations such as the IBI as a reliable method for determining these properties (ASTM International, 2007). The results of the proximate analysis can be used to determine the quality and suitability of biochar for different applications, such as soil amendment or as a solid fuel (Chaturvedi, Singh, Dhyani, Govindaraju, & Mandal, 2021; Singh, Chaturvedi, Dhyani, & Govindaraju, 2020). However, it is important to note that proximate analysis only provides information on a few basic properties of biochar and does not provide a complete characterization of the material.

Proximate analysis involves carefully heating a sample in controlled conditions to determine its composition. The process begins with obtaining a representative biochar sample, ensuring it's well-mixed and free from any extraneous materials. This sample is then ground into a uniform powder with particle sizes ranging from 250 μm to 500 μm. The moisture content is assessed by weighing a clean and dry crucible, typically made of porcelain or platinum. A known amount of the prepared biochar sample, usually between 1 and 3 g, is placed into the crucible, which is then heated in an oven at 104°C ± 2°C (219°F ± 4°F) for one hour. After cooling in a desiccator, the crucible is weighed again, and the moisture content is calculated as a percentage of the original sample weight. The analysis continues with the determination of volatile matter. The previously dried crucible containing the biochar sample is transferred to a furnace preheated to 950°C ± 25°C (1742°F ± 45°F) and heated for around 7 minutes. After cooling in a desiccator, the crucible is weighed to calculate the volatile matter content as a percentage of the original sample weight. Further, the ash content is determined by heating the crucible with the sample from the previous steps in a muffle furnace at 750°C ± 25°C (1382°F ± 45°F) for at least 1 hour or until complete oxidation occurs. After cooling in a desiccator, the crucible is weighed to calculate the ash content as a percentage of the original sample weight. In the end, to assess the fixed carbon content, the difference between 100% and the sum of moisture, volatile matter, and ash content is calculated. This value represents the proportion of biochar that remains after the volatile matter has been driven off during the heating process.

According to Torquato, Crnkovic, Ribeiro, and Crespi (2017), variations in the conditions of pyrolysis and the category of raw materials employed can have a considerable impact on the features of proximate analysis for biochar. Pyrolysis is the process of heating organic matter without oxygen to create biochar. Pyrolysis normally occurs at temperatures between 200°C and 800°C, and it can last anywhere from a few minutes to many hours. Biochar can be derived from a variety of organic sources, including wood, agricultural waste, sewage sludge, and algae. The properties of the produced biochar might vary depending on the selection of source material. According to Jafri, Wong, and Doshi (2018), greater temperatures during pyrolysis and longer residence durations are likely to elevate the amount of fixed carbon and ash in biochar. Ash is a representation of the biochar's inorganic mineral composition, and fixed carbon is the percentage of the material still remaining after the volatile matter has been removed. This is because numerous volatile chemicals are produced, and moisture is removed during pyrolysis. More volatile compounds are produced when the process is more vigorous, and some of them may become fixed carbon if they remain in the reactor for an extended period of time due to the additional reactions. The ash content in biochar can be different, ranging from 1% to 20%.

Nonwood biochar usually has higher ash levels compared to biochar made from wood. The amount of volatile content in biochar can be between 6.0% and 88.0%, and the fixed carbon content can be between 11% and 86% (Safdari, Rahmati, & Amini, 2018). The fuel ratio (FR), which is the ratio of fixed carbon to volatile matter, can be calculated using these results. The FR for different types of coal ranges from ≥10 for anthracite to <3 for bituminous. The FR of biomass biochar usually falls within the range of approximately 3:1 to 12:1, showing that there is a higher amount of fixed carbon compared to volatile matter. This higher FR is a consequence of the pyrolysis process, where moisture and volatile elements are eliminated, resulting in a relatively higher concentration of fixed carbon in the biochar (Azad, Pateriya, & Sharma, 2024; Mandal et al., 2022; Thrush, 1968). In order to anticipate NO_x emission during the burning of biochar and coal, a relationship has been established between FR and NO_x emissions from coal. The total solids (dry mass) and volatile solids (VSs) content (a lack of combustion or organic matter content) of biochar are additional crucial characteristics (Wang & Howard, 2018). These variables are critical for biochar, which is utilized as an additive in biological processes such as methane generation, anaerobic digestion, and composting (Azad, Pateriya, Arya, & Sharma, 2022).

5.2.2 Ultimate analysis and elemental composition

The elemental composition of a sample of biochar can be determined through a process called ultimate analysis or elemental analysis. The standard CHNS analysis procedure for biochar, following ASTM D5373-16, involves obtaining a representative sample of biochar and placing it into a tin capsule or combustion boat. The sample is then subjected to high-temperature combustion (around 1000°C) using pure oxygen. During this combustion process, the carbon, hydrogen, and nitrogen present in the biochar is transformed into carbon dioxide, water vapor, and nitrogen oxides, respectively. The resulting combustion gases are purified and passed through detectors to measure the concentrations of carbon dioxide, water vapor, and nitrogen oxides. These measurements are then used to calculate the percentages of carbon (C), hydrogen (H), and nitrogen (N) in the biochar sample, which provides valuable insights into its elemental composition (Caillat & Vakkilainen, 2013). The resulting combustion products are subsequently directed through a heated copper tube filled with helium gas, operating at approximately 600°C. During this phase, residual oxygen is eliminated, and nitrogen oxides are transformed into nitrogen gas. By employing absorbent traps, undesirable gases like hydrogen chloride are additionally eliminated. The subsequent step involves accurately measuring the purified gases, comprising CO_2, H_2O, nitrogen, and SO_x, through techniques such as gas chromatography, Flame ionization detection, thermal conductivity analysis, and infrared analysis (Thompson, 2008). Biochar with a high carbon content is preferred from an energy efficiency perspective since it improves the heating value. However, a sizable percentage of the sample's hydrogen is often released as volatile matter during the pyrolysis process used to produce charcoal, and sulfur can contribute to unfavorable emissions such sulfur oxides (SO_x). Additionally, nitrogen in the fuel might result in the production of nitrogen oxide (NO_x) emissions. Therefore, it is important to consider the balance between carbon content and emissions when producing biochar. The ultimate analysis is a useful tool for evaluating the performance of biochar in various applications. It also gives insight into the process of carbonization and the potential stability of the biochar. Additionally, as the pyrolysis temperature increases, the content of hydrogen and oxygen in biochar decreases, this results in a decrease in the ratios of H/C and O/C, indicating an improvement in the aromaticity and degree of carbonization of the biochar, as well as a reduction in surface polarity (Caillat & Vakkilainen, 2013; Mandal et al., 2020; Mandal et al., 2022).

5.2.3 Physiochemical characterization

5.2.3.1 Porosity

The increase in temperature during pyrolysis can also affect the porosity of the biochar. Studies have shown that biochar produced at a temperature of 350°C has an average porosity of $\leq 10\ \mu m$ (Weber & Quicker, 2018). Porosity refers to the percentage of the biochar particle that is not filled by solid material and includes all pore volumes that open directly to the exterior of the biochar particle but are smaller than the DryFlo particle standard size of 75–8000 μm (Brewer, Chuang, & Masiello, 2014). This information is important as it can provide insight into how the biochar will perform in different applications and can help predict its stability and reactivity. Eq. (5.1) can be used to measure to determine the porosity of biochar

$$\text{Porosity} = 1 - \frac{\rho}{\rho_s} \times 100 \tag{5.1}$$

There are some factors that have been linked to porosity, including pore accessibility and the use of mercury porosimetry. Pore accessibility refers to the ability for all pores within a given substance to be accessible from the outside (Wade, Martin, & Long, 2015). Mercury porosimetry is a method that can be used to measure the size and distribution of pores in biochar. It involves applying pressure to a sample submerged in mercury, and the pressure required to penetrate the pores is inversely proportional to the size of the pores (Anderson, 2001). This can be determined using a device such as the Auto Pore IV mercury porosimeter.

5.2.3.2 Density

Density is a fundamental characteristic of a substance, representing the relationship between its mass and volume. For wood biochar, density becomes a significant aspect to take into account, particularly following the pyrolysis procedure. The density of wood biochar as well as other biomass biochar can vary depending on the type of biomass used as raw material and the temperature employed during pyrolysis (Brewer et al., 2014, Plötze & Niemz, 2011). Density is influenced by a number of variables, including envelope density, skeletal density, and real density/true particle density. A set of procedures is used to calculate the envelope density of biochar, which describes its bulk or apparent density. Initially, a representative sample of biochar is acquired, ensuring it is thoroughly mixed and devoid of impurities. The sample's volume is then measured using a graduated cylinder or pycnometer, considering both the solid material and pore spaces. Following this, the biochar sample is accurately weighed using a laboratory balance. Subsequently, the envelope density is calculated by dividing the sample's mass by its volume, typically expressed in units like g/cm^3 or kg/m^3. This measurement provides valuable insights into the overall density and packing characteristics of the biochar. The envelope density considers the entire structure, including solid components, pores, and surface irregularities. The GeoPyc 1360 Envelope Density Analyzer is also used for the determination of envelope density. Higher porosity results in a lighter biochar per unit volume, and this porosity can be measured using ASTMD5004. The skeleton density, on the other hand, considers both the volume and mass of solid material contained inside each distinct component, including closed pores. The skeletal density may be calculated using ASTMD3766 as the ratio of these two variables. The mass of the wood material after pyrolysis divided by its volume, omitting open and closed pores, yields the actual density, also known as true particle density, of a biochar particle (Plötze & Niemz, 2011; Webb, 2001).

5.2.3.3 pH

The pH of biochar is an important consideration in determining how it impacts the environment, agriculture, and biology. The pH of biochar can range from acidic to alkaline, however, it is most generally recognized to be alkaline. Furthermore, the pH of biochar tends to rise as the temperature of carbonization rises (Al-Wabel, Al-Omran, & El-Naggar, 2013). Several ways have been tried to figure out the pH of biochar, however, the most common method is to mix it with deionized water in a specified ratio, which is often employed in biochar assessment. In 2015, the IBI proposed a technique for testing the pH of biochar that includes combining it with deionized water at a 1:20 ratio and stirring it for 1.5 hours (IBI, 2014). Another investigation, carried out in 2012 by Rajkovich et al., utilized a similar procedure of combining biochar and deionized water in a 1:20 proportion and stirring it to ensure that the pH is appropriately assessed by providing a sufficient amount of time for the biochar and deionized water to achieve equilibrium (Rajkovich et al., 2012).

5.2.3.4 Electrical conductivity

Biochar is known to contain water-soluble ions that can contribute to its electrical conductivity (EC), much like soil (Rajkovich et al., 2012). The IBI (2014) states that the EC of biochar is affected by feedstock properties and production conditions (IBI, 2014). The capacity of biochar to transmit electrons has been evaluated for supercapacitor development, although this is not the same as the tests used to calculate biochar EC. These metrics are affected by the porosity structure, surface area, and the amount of crystalline carbon structures present in the biochar (Jiang et al., 2013). The EC of biochar is measured using a glass electrode coupled to an electronic pH meter. Biochar is often blended with either deionized water or KCl, then well agitated and left to settle for a certain amount of time. Various biochar-to-water/KCl ratios have been proposed and examined, including 1:10, 1:20, and 1 g of sample with 60 mL of distilled water (Al-Wabel et al., 2013, Jiang et al., 2013; Rodriguez, Lustosa Filho, & Melo, 2021). The latter is created by stirring the mixture for 1 hour and then permitting it to cool to the ambient temperature.

Understanding biochar EC is important for applications such as agriculture and soil/water remediation. Fang, Singh, Singh, and Krull (2014) used a 1:5 ratio of biochar to deionized water to measure the EC of their sample (Fang et al., 2014). Kizito et al. (2015) took it a step further and boiled the biochar with deionized water for two hours at 90°C prior to determining the EC. Their method was intended to increase the dissolution of soluble biochar components (Kizito et al., 2015). Pituello et al. (2015) opted for a 1:20 ratio of biochar to deionized water without the heating step, which is also recommended by the IBI (2014). This suspension can be used to measure the EC as well as the pH of the sample (Pituello et al., 2015).

5.2.3.5 Cation and anion exchange capacity

Biochar's potential to exchange cations is evaluated through its cation exchange capacity (CEC), indicating the number of cations that can be swapped in a specific soil quantity, including $K+$, $Na+$, $NH4+$, $Mg2+$, $Ca2+$, $Fe3+$, $Al3+$, $Ni2+$, and $Zn2+$ (Das, Singh, & Adolphson, 2010). CEC is usually expressed in centimoles or millimoles of total or specific cations per kilogram of soil. Soils with CEC below 50 mmol kg^{-1} are considered deficient, while the typical CEC range falls between 50 and 250 mmol kg^{-1}, with values exceeding 250 mmol kg^{-1} being unusually high. The modified AOAC method 973.09 is utilized to measure CEC. As stated by Rizwan, Ali, and Qayyum (2016), a higher CEC in biochar indicates enhanced metal adsorption. However, when the pyrolysis temperature surpasses 350°C, the CEC decreases (Rizwan et al., 2016).

Biochar's CEC is essential information for assessing its potential as a pollutant adsorbent, as it involves the interaction of protons and ionized cations with salts dissolved on the biochar's surface (Singh, Luthra, et al., 2023). The biochar's ability to adsorb heavy metals depends on factors such as pollutant size and the biochar's surface functional groups (Roy et al., 2022; Singh et al., 2020). For instance, Trakal, Veselská, and Šafařík (2016) investigated the removal of Cd and Pb using biochar derived from various raw materials like grape husk, grape stalk, nutshell, plum stone, and wheat straw. Their findings indicated that biochar produced from raw materials containing iron oxides exhibited higher effectiveness in removing Pb and Cd, as the inclusion of iron in the raw material increased the biochar's CEC (Trakal et al., 2016).

Biochar exhibiting a higher CEC has demonstrated a greater capacity for retaining nutrients. These biochars often contain a higher concentration of hydrophilic oxygen-containing groups like phenolic and carboxylic groups, which enhance cation exchangeability. While we have a good understanding of the processes affecting changes in CEC in biochar, there is still much to be discovered regarding biochar AEC pathways (Cheng, Lehmann, & Engelhard, 2008). Biomass that has been carbonized with a high AEC has the potential to effectively remove anionic contaminants in water treatment, making it a valuable characteristic. Nevertheless, the anion exchange capacity (AEC) of biochar is limited in neutral and alkaline conditions.

AEC is determined by conducting a $Cl-$ for $Br-$ spontaneous exchange on 1-gram biochar samples in water. The quantity of available Br- is then measured using an ion chromatography system consisting of a Dionex 1100 ion chromatograph and an ASRS 300×4 mm conductance detector. Lawrinenko, Jing, and Banik (2017) have demonstrated that the AEC of biochar is significantly influenced by the pyrolysis temperature and the distribution of metal oxyhydroxides within the biochar, particularly when the biomass is pretreated with Al or Fe trichlorides (Lawrinenko et al., 2017). Biochar produced at 700°C has been found to possess a higher AEC compared to biochar produced at 500°C.

5.2.3.6 FTIR spectroscopy

FTIR spectroscopy offers a swift and cost-effective means of analyzing material surfaces. Employing FTIR and Raman spectroscopy in the laboratory assists in gaining a deeper comprehension of the chemical processes occurring during biochar synthesis and the distribution of chemical phases on both small and large scales. The behavior of infrared absorption is typically represented by linear regression functions, which commonly yield determination coefficients within the range of 60%–90% (Chia, Gong, & Joseph, 2012). Instruments like the Thermo Scientific Nicolet iZ10 FTIR or DRIFTS (using the Bruker Tensor 37) can be utilized for FTIR analysis. The creation of a spectrum typically involves combining 32–64 images within the 400–4000 cm^{-1} range at a resolution of 4 cm^{-1}.

The temperature of pyrolysis significantly impacts the functional groups in biochar. Numerous studies indicate that as the pyrolysis temperature increases, the quantity of polar functional groups decreases. This leads to the formation of hydrophobic biochar with fewer functional groups at higher pyrolysis temperatures, especially around 600°C. In such cases, the carbon structure in the biochar becomes well-organized. This approach provides a rapid assessment of various biochar's suitability for immobilizing specific environmental pollutants or for long-term carbon storage (Janu, Mrlik, & Ribitsch, 2021).

5.2.3.7 *Water-holding capacity*

The functional groups of biochar influence its water attraction, and its porous structure influences how much water it can hold. When compared to biochar generated at elevated temperatures (600°C–800°C), biochar produced at low temperatures (400°C) has more hydrophobic characteristics and reduced water retention capacity (Brewer et al., 2014; Singh, Chaturvedi, & Datta, 2019). The hydrophobicity and water-holding capacity of biochar may be measured using techniques such as mercury intrusion porosimetry, plant-available water capacity, and water-retention curve. Furthermore, the surface hydrophobicity of biochar samples may be determined using the molarity of an ethanol drop test (Kinney, Masiello, & Dugan, 2012).

5.2.3.8 *Thermal conductivity*

The variety of biomass utilized, the temperature at which it is pyrolyzed, and the direction of heat flow all impact the properties and usefulness of biochar as a thermal medium. The higher the density of the biomass, the higher the thermal conductivity, but a porous structure lowers the conductivity relative to the original material (Weber & Quicker, 2018). Thermal conductivity varies with heat flow direction, with the maximum values obtained when heat travels parallel to the biomass sample, which is 1.5–2.7 times more than when heat flows perpendicular to the specimen (Weber & Quicker, 2018). To measure thermal conductivity and heat flow in biochar, one can use the NETZSCH HFM 446 Heat Flow Meter, following the ASTM C518 standard (Jeon, Kim, & Park, 2021).

5.2.4 High heating value

Biochar's higher heating value (HHV) is a significant component in energy assessment since it measures the energy emitted per unit of mass. It differs depending on the type of biomass and the pyrolysis temperature employed. Yang, Wang, and Strong (2017) investigated biochar produced from eight different biomass sources pyrolyzed at 350°C and 500°C. The study discovered that HHV values for Miscanthus and bamboo sawdust were greater when pyrolyzed at 500°C compared to 350°C, but rice straw and bamboo leaves were contrary (Yang et al., 2017). Excluding pecan shells and paddy straw, which showed lower values (15.85 and 16.95 MJ/kg), the HHV of most materials was near (18.44–20.10 MJ/kg). The HHV may be calculated using a bomb calorimeter and ASTM D 240-19. A 1 g sample is placed in a bomb chamber and connected to an ignition wire during the procedure. After that, oxygen is injected to achieve a pressure of 30 bar. After burning the sample, the increase in core temperature during the combustion of organic material may be monitored, allowing the heat value to be determined (Basu, 2018).

5.2.4.1 Iodine number

The iodine number plays a pivotal role in the thorough characterization of biochar, offering valuable insights into its adsorption capacity and surface area. This metric quantifies the amount of iodine, typically denoted in milligrams per gram (mg/g^{-1}), adsorbed by 1 g of biochar under defined conditions. Typically, the procedure involves immersing the biochar specimen in an iodine solution, commonly potassium iodide, to enable the adsorption of iodine molecules onto its surface. This adsorption process is facilitated by the presence of pores and functional groups on the biochar's surface, acting as binding sites for iodine molecules. Subsequently, the residual iodine concentration in the solution is determined after a specified adsorption period using titration or spectrophotometric techniques. The disparity between the initial and final iodine concentrations permits the calculation of iodine adsorption per unit mass of biochar (Azad et al., 2024). A higher iodine number signifies a greater iodine adsorption, indicative of increased surface area and porosity within the biochar. Consequently, this suggests enhanced adsorption capabilities for a range of contaminants, including organic pollutants, heavy metals, and other substances found in water, soil, or gas streams (Mandal et al., 2023). The iodine number proves particularly beneficial in assessing the efficacy of biochar as an adsorbent in environmental remediation applications, such as wastewater treatment, soil enhancement, and air purification. Furthermore, it furnishes essential data for refining biochar production methods to tailor its characteristics for specific uses, thus ensuring the effective utilization of this sustainable and adaptable material across diverse environmental and industrial settings.

5.2.5 Morphology and surface functionality

5.2.5.1 Scanning electron microscopy

The morphology of biochar is studied using a scanning electron microscope, which gives detailed information about the microporous and mesoporous distributions and pore arrangements. SEM allows us to predict the surface changes that occur during the adsorption process (Yaashikaa, Kumar, Varjani, & Saravanan, 2020). Before analysis, samples are put on copper supports and plated with a thin coating of gold (Munar-Florez, Varón-Cardenas, Ramírez-Contreras, & García-Núñez, 2021). The microscope utilized has a high resolution of 3 nm with a zoom range of 10 60,000. SEM pictures show that pyrolysis alters the surface shape of biochar particles. Higher temperatures can also increase biochar's pore characteristics, although an extreme temperature of 700°C destroys its porous structure (Yaashikaa et al., 2020).

5.2.5.2 Brunauer–Emmett–Teller

In the pyrolysis process, the development of a pore structure takes place as volatile substances are separated from the biomass. This carbonization process results in the formation of a distinct surface area and pore volume within the resulting biochar, as indicated by Kong et al. (2021). The surface area of the biochar can be quantified using BET analysis. The BET technique utilizes the physisorption of a gas to calculate the surface area of a given sample. By allowing gas molecules to permeate through particles and penetrate all cracks, pores and irregularities on the surface, this measurement accurately assesses the entire microscopic surface area of the sample. Typically, the sample is in the form of powder or granules, and the outcome is expressed as a Specific Surface Area, denoting the area per unit mass. Alternatively, it can be presented as the area per unit volume or as the absolute area for a given object. This feature is critical for eliminating contaminants from soil and water, and greater temperatures result in a larger surface area (Howell, Pimentel, & Bhattacharia, 2021). High carbonization temperatures of 450°C, 550°C, or 650°C often result in a greater specific surface area (Kong, Zhang, & Liu, 2021). However, other investigations have indicated that the surface area of the biochar-mineral composite peaks around 500°C and subsequently drops over 700°C, suggesting that an optimal temperature is required for maximal surface area. This is considered to be due to the recondensation of volatilized organic molecules on the biochar surface, which plugs pores and lowers surface area (Howell et al., 2021).

5.2.5.3 Thermogravimetric analysis

Thermogravimetric analysis (TGA) is a method used to examine the physical and chemical properties of materials as they are subjected to a temperature increase (Mandal, Haydary, & Gangil, 2020). This type of analysis is commonly used to study the thermal behavior of different samples. The purpose of this study is to analyze the combustion properties of biochar and biochar/biomass blends using TGA. Additionally, the average weight of each component was evaluated to determine if a synergistic effect existed between the blend components. The results can help to better understand the thermal properties and behavior of the samples from a general perspective and for test analysis. During the TGA process, biochar is heated from room temperature to 1000°C, with heating rates reported to be 10°C and 20°C/minute, 10 K/minute, or less than 1000°C by various researchers (Ahmad et al., 2014; Mandal et al., 2020).

5.2.5.4 NMR spectroscopy

NMR spectroscopy may be used to investigate the structure of biochar. NMR studies particle structure using a magnetic field and radio waves based on atom resonance frequencies. The presence of carbon functional groups, the level of aromatic ring formation, and the overall molecular structure of biochar may all be determined using solid-state NMR approaches (Choung, Um, Kim, & Kim, 2013). NMR may also be used to determine the hydrocarbon concentration, including aliphatic and aromatic hydrocarbons. NMR may also be used to compare the stability and carbonization of different biochars. However, the use of NMR is limited because of ferromagnetic mineral interference in biochar and the poor signal-to-noise ratio of biochar formed at high temperatures by pyrolysis (Melligan et al., 2012).

5.2.5.4.1 X-ray diffraction technique

X-ray diffraction (XRD) is an indispensable tool for comprehensively characterizing biochar, providing detailed insights into its structural attributes with exceptional precision. When biochar specimens are subjected to X-ray irradiation, the resulting diffraction patterns offer valuable information regarding the arrangement and crystallinity of constituent materials embedded within the biochar matrix. Through meticulous analysis of diffraction peaks and their associated intensities, the presence of diverse crystalline phases including graphite, minerals, and various inorganic constituents can be determined. Moreover, XRD facilitates the determination of pivotal parameters such as crystallite dimensions, interlayer spacing, and the extent of graphitization (Yaashikaa et al., 2020). These parameters are crucial for elucidating the structural modifications inherent in the pyrolysis or activation processes undergone by biochar. Consequently, XRD emerges as an indispensable analytical instrument for advancing our understanding of biochar's multifaceted physicochemical attributes, thereby fostering its versatile applications across environmental and industrial domains (Usman et al., 2015). In addition, versatility of XRD extends to the assessment of biochar's crystalline characteristics and configuration. Within XRD analysis, identifiable features of amorphous material, developed at temperatures exceeding 350°C, are consistently discernible via diffractograms. Modern computerized XRD setups integrate critical components such as a monochromator, radiation source, and stepping motor, thereby enhancing analytical precision and efficacy. The distinctive and well-defined XRD peaks observed serve as indicators of the crystalline nature of the resultant nanocrystals, with particle dimensions demonstrating a gradual augmentation over time. Consequently, XRD patterns play a pivotal role in facilitating the production of biochar of superior quality, marked by expedited processing and nondestructive evaluation, thereby amplifying its sorption capabilities and broadening its utility across diverse applications.

5.2.5.4.2 Raman spectroscopy

Raman scattering emerges as a fundamental and extensively utilized molecular spectroscopy technique, leveraging the vibrational dynamics of atoms under electromagnetic radiation. This phenomenon entails the scattering of light with altered frequency from the incident radiation due to the absorption or loss of vibrational energy within the molecule. The aim is to establish a robust methodology for precisely evaluating the degree of chemical or nanostructural changes occurring during biomass carbonization. This methodology seeks to enable rapid determination of the heat treatment temperatures (HTTs) employed in producing a specific biochar sample. Raman spectroscopy offers notable advantages, including heightened sensitivity, minimal sample preparation requirements, and reduced interference, thus rendering it conducive to biochar characterization (Yaashikaa et al., 2020). Nevertheless, its cost may impede its widespread adoption. Analysis of resultant Raman spectra reveals discernible peaks indicative of vibrational modes specific to carbonaceous materials within the biochar, furnishing insights into its chemical composition, structural attributes, and level of graphitization. Quantitative assessment entails evaluating parameters such as peak intensity, width, and position to discern chemical or nanostructural modifications in the biochar sample. Validation involves comparing acquired spectra with reference standards to ensure result accuracy. Finally, comprehensive documentation of experimental procedures, findings, and interpretations is essential for dissemination and future reference, highlighting the efficacy of Raman spectroscopy in elucidating biochar properties (Vithanage et al., 2015).

5.3 Conclusion

There are several methods for measuring biochar qualities, ranging from simple physical to complex molecular analysis, each with its own purpose and significance. However, because of the great variety of biochar and its many uses, it is critical to consider the specific tests required for the ultimate application. Some recommendations for biochar analysis have been made, but due to the wide range of biochar types, uses, and laboratory equipment available, it is doubtful that a worldwide standard will be produced in the near future. Instead, present analytical methods are expected to be changed further, resulting in the production of enhanced or innovative descriptions of biochar characteristics. With prospective breakthroughs in analytical tools for a thorough understanding of its chemical composition and behavior, the future of biochar research holds considerable promise. Real-time monitoring is being investigated for in-situ biochar analysis, while nanoscale characterization leads to tailored biochar with specific capabilities. Biochar composites and hybrid materials are gaining popularity as possible advances and established processes assure uniformity, dependability, and safety in their manufacturing and use.

Acknowledgment

The authors are grateful to the National Scholarship Program of Slovakia (NSP-SAIA) and the ICAR-AICRP on EAAI for their assistance during this project. The authors also thank the Slovak Research and Development Agency, grant number APVV-19-0179, for financing the study.

References

Ahmad, M., Rajapaksha, A. U., Lim, J. E., Zhang, M., Bolan, N., Mohan, D., ... Ok, Y. S. (2014). Biochar as a sorbent for contaminant management in soil and water: A review. *Chemosphere, 99,* 19–33. Available from https://doi.org/10.1016/j.chemosphere.2013.10.071.

Al-Wabel, M. I., Al-Omran, A., & El-Naggar, A. H. (2013). Pyrolysis temperature induced changes in characteristics and chemical composition of biochar produced from Conocarpus wastes. *Bioresource Technology, 131,* 374–379. Available from https://doi.org/10.1016/j.biortech.2012.12.165.

Anderson, J. (2001). *Mercury porosimetry.* <https://www.mri.psu.edu/materials-characterization-lab/characterization-techniques/mercury-porosimetry> Accessed 22.06.21.

ASTM International. (2007). *ASTM D1762-84. Standard test method for chemical analysis of wood charcoal.* West Conshohocken: ASTM International 1–2. Available from http://doi.org/10.1520/D1762-84R07.

Azad, D., Pateriya, R. N., Arya, R., & Sharma, R. K. (2022). Biological treatment for biochar modification: Opportunities, limitations, and advantages. In S. Ramola, D. Mohan, O. Masek, A. Méndez, & T. Tsubota (Eds.), *Engineered biochar.* Singapore: Springer. Available from https://doi.org/10.1007/978-981-19-2488-0_6.

Azad, D., Pateriya, R. N., & Sharma, R. K. (2024). Chemical activation of pine needle and coconut shell biochar: Production, characterization and process optimization. *International Journal of Environmental Science and Technology, 21*(1), 757–772.

Basu, P. (2018). Biomass combustion and cofiring. *Biomass gasification, pyrolysis and torrefaction: Practical design and theory* (pp. 393−413). Elsevier.

Brewer, C. E., Chuang, V. J., & Masiello, C. A. (2014). New approaches to measuring biochar density and porosity. *Biomass and Bioenergy, 66*, 176−185. Available from https://doi.org/10.1016/j.biombioe.2014.03.059.

Caillat, S., & Vakkilainen, E. (2013). Large-scale biomass combustion plants: An overview. *Biomass combustion science, technology and engineering* (pp. 189−224). Elsevier.

Chaturvedi, S., Singh, S., Dhyani, V. C., Govindaraju, K., & Mandal, S. (2021). Characterization, bioenergy value and thermal stability of biochar derived from diverse agriculture and forestry lignocellulosic waste. *Biomass Conversion and Biorefinery*. Available from https://doi.org/10.1007/s13399-020-01239-2.

Cheng, C. H., Lehmann, J., & Engelhard, M. H. (2008). Natural oxidation of black carbon in soils: changes in molecular form and surface charge along a climosequence. *Geochimica et Cosmochimica Acta, 72*, 1598−1610. Available from https://doi.org/10.1016/j.gca.2008.01.010.

Chia, C. H., Gong, B., & Joseph, S. (2012). Imaging of mineral-enriched biochar by FTIR, Raman and SEM-EDX. *Vibrational Spectroscopy, 62*, 248−257. Available from https://doi.org/10.1016/j.vibspec.2012.06.006.

Choung, S. W., Um, W. Y., Kim, M. Y., & Kim, M. G. (2013). Uptake mechanism for iodine species to black carbon. *Environmental Science and Technology*. Available from https://doi.org/10.1021/es401570a, 130827075129003.

Das, K. C., Singh, K., & Adolphson, R. (2010). Steam pyrolysis and catalytic steam reforming of biomass for hydrogen and biochar production. *Applied Engineering in Agriculture, 26*, 137−146.

Fang, Y., Singh, B., Singh, B. P., & Krull, E. (2014). Biochar carbon stability in four contrasting soils. *European Journal of Soil Science, 65*, 60−71. Available from https://doi.org/10.1111/ejss.12094.

Hans-Peter, S. S. A. (2013). European biochar certificate − Guidelines for a sustainable production of biochar. European Biochar Foundation. <http://www.bio-inspecta.ch/htm/dl_detail.htm?sprache = e&id = 105&p = 3> Accessed 17.03.22.

Howell, N., Pimentel, A., & Bhattacharia, S. (2021). Material properties and environmental potential of developing world-derived biochar made from common crop residues. *Environmental Challenges, 4*, 100137. Available from https://doi.org/10.1016/j.envc.2021.100137.

IBI Biochar Standards Version 2.0. (2014). Standardized product definition and product testing guidelines for biochar that is used in soil. The International Biochar Initiative. http://www.biochar-international.org/characterizationstandard.

International Biochar Initiative. 2014. State of the biochar industry: a survey of commercial activity in the biochar field: report overview [online]. Available from: http://www.biochar-international.org/sites/de-fault/files/StateoftheBiocharIndustry_2013_4pager_final.pdf (Accessed 11.08.23).

Jafri, N., Wong, W. Y., & Doshi, V. (2018). A review on production and characterization of biochars for application in direct carbon fuel cells. *Process Safety and Environmental Protection, 118*, 152−166. Available from https://doi.org/10.1016/j.psep.2018.06.036.

Janu, R., Mrlik, V., & Ribitsch, D. (2021). Biochar surface functional groups as affected by biomass feedstock, biochar composition and pyrolysis temperature. *Carbon Resources Conversion, 4*, 36−46. Available from https://doi.org/10.1016/j.crcon.2021.01.003.

Jeon, J., Kim, H. I., & Park, J. H. (2021). Evaluation of thermal properties and acetaldehyde adsorption performance of sustainable composites using waste wood and biochar. *Environmental Research, 196*. Available from https://doi.org/10.1016/j.envres.2021.110910.

Jiang, J., Zhang, L., Wang, X., Holm, N., Rajagopalan, K., Chen, F., & Ma, S. (2013). Highly ordered macroporous woody biochar with ultra-high carbon content as supercapacitor electrodes. *Electrochimica Acta, 113*, 481−489. Available from https://doi.org/10.1016/j.electacta.2013.09.121.

Joseph, S., Peacocke, C., Lehmann, J., & Munroe, P. (2009). Developing a biochar classification and test methods. In: *Biochar for environmental management: Science and technology* (vol. 1, pp. 107−126).

Kinney, T. J., Masiello, C. A., & Dugan, B. (2012). Hydrologic properties of biochars produced at different temperatures. *Biomass and Bioenergy, 41*, 34−43. Available from https://doi.org/10.1016/j.biombioe.2012.01.033.

Kizito, S., Wu, S., Kipkemoi Kirui, W., Lei, M., Lu, Q., Bah, H., & Dong, R. (2015). Evaluation of slow pyrolyzed wood and rice husks biochar for adsorption of ammonium nitrogen from piggery manure anaerobic digestate slurry. *Science of the Total Environment, 505*, 102−112. Available from https://doi.org/10.1016/j.scitotenv.2014.09.096.

Kong, W., Zhang, M., & Liu, Y. (2021). Physico-chemical characteristics and the adsorption of ammonium of biochar pyrolyzed from distilled spirit lees, tobacco fine and Chinese medicine residues. *Journal of Analytical and Applied Pyrolysis, 156*. Available from https://doi.org/10.1016/j.jaap.2021.105148.

Lawrinenko, M., Jing, D., & Banik, C. (2017). Aluminium and iron biomass pretreatment impacts on biochar anion exchange capacity. *Carbon, 118*, 422–430. Available from https://doi.org/10.1016/j.carbon.2017.03.056.

Lehmann, J., Rillig, M. C., Thies, J., Masiello, C. A., Hockaday, W. C., & Crowley, D. (2011). Biochar effects on soil biota—A review. *Soil Biology & Biochemistry, 43*, 1812–1836. Available from https://doi.org/10.1016/j.soilbio.2011.04.022.

Mandal, S., Haydary, J., & Gangil, S. (2020). Inferences from thermogravimetric analysis of pine needles and its chars from a pilot-scale screw reactor. *Chemical Papers, 74*, 689–698. Available from https://doi.org/10.1007/s11696-019-00998-1.

Mandal, S., Jena, P. C., Gangil, S., Pal, S., Haydary, J., Sharma, R. K., & Verma, A. (2023). Ni-supported pigeon pea stalk biochar as a catalyst for ex situ tar cracking in biomass gasification. *Biomass Conversion and Biorefinery*, 1–11.

Mandal, S., Sharma, R. K., & Bhattacharya, T. K. (2022). Charring of pine needles using a portable drum reactor. *Chemical Papers, 76*, 1239–1252. Available from https://doi.org/10.1007/s11696-021-01893-4.

Melligan, F., Dussan, K., Auccaise, R., Novotny, E. H., Leahy, J. J., Hayes, M. H. B., & Kwapinski, W. (2012). Characterisation of the products from pyrolysis of residues after acid hydrolysis of Miscanthus. *Bioresource Technology, 108*, 258–263. Available from https://doi.org/10.1016/j.biortech.2011.12.110.

Munar-Florez, D. A., Varón-Cardenas, D. A., Ramírez-Contreras, N. E., & García-Núñez, J. A. (2021). Adsorption of ammonium and phosphates by biochar produced from oil palm shells: effects of production conditions. *Results in Chemistry, 3*. Available from https://doi.org/10.1016/j.rechem.2021.100119.

Pituello, C., Francioso, O., Simonetti, G., Pisi, A., Torreggiani, A., Berti, A., & Morari, F. (2015). Characterization of chemical–physical, structural and morphological properties of biochars from biowastes produced at different temperatures. *Journal of Soils and Sediments, 15*, 792–804. Available from https://doi.org/10.1007/s11368-014-0964-7.

Plötze, M., & Niemz, P. (2011). Porosity and pore size distribution of different wood types as determined by mercury intrusion porosimetry. *European Journal of Wood and Wood Products, 69*, 649–657. Available from https://doi.org/10.1007/s00107-010-0504-0.

Rajkovich, S., Enders, A., Hanley, K., Hyland, C., Zimmerman, A. R., & Lehmann, J. (2012). Corn growth and nitrogen nutrition after additions of biochars with varying properties to a temperate soil. *Biology and Fertility of Soils, 48*, 271–284. Available from https://doi.org/10.1007/s00374-011-0624-7.

Rizwan, M., Ali, S., & Qayyum, M. F. (2016). Mechanisms of biochar-mediated alleviation of toxicity of trace elements in plants: A critical review. *Environmental Science and Pollution Research, 23*, 2230–2248. Available from https://doi.org/10.1007/s11356-015-5697-7.

Rodriguez, J. A., Lustosa Filho, J. F., & Melo, L. C. A. (2021). Co-pyrolysis of agricultural and industrial wastes changes the composition and stability of biochars and can improve their agricultural and environmental benefits. *Journal of Analytical and Applied Pyrolysis, 155*. Available from https://doi.org/10.1016/j.jaap.2021.105036.

Roy, A., Chaturvedi, S., Singh, S., Govindaraju, K., Dhyani, V. C., & Pyne, S. (2022). Preparation and evaluation of two enriched biochar-based fertilizers for nutrient release kinetics and agronomic effectiveness in direct-seeded rice. *Biomass Conversion and Biorefinery*. Available from https://doi.org/10.1007/s13399-022-02488-z.

Safdari, M. S., Rahmati, M., & Amini, E. (2018). Characterization of pyrolysis products from fast pyrolysis of live and dead vegetation native to the Southern United States. *Fuel, 229*, 151–166. Available from https://doi.org/10.1016/j.fuel.2018.04.166.

Sharma, R. K., Singh, T. P., Mandal, S., Azad, D., & Kumar, S. (2022). Chemical treatments for biochar modification: Opportunities, limitations and advantages. In S. Ramola, D. Mohan, O. Masek, A. Méndez, & T. Tsubota (Eds.), *Engineered Biochar*. Singapore: Springer. Available from https://doi.org/10.1007/978-981-19-2488-0_5.

Singh, S., Chaturvedi, S., & Datta, D. (2019). Biochar: An eco-friendly residue management approach. *Indian Farming, 69*(08), 27–29.

Singh, S., Chaturvedi, S., Dhyani, V. C., & Govindaraju, K. (2020). Pyrolysis temperature influences the characteristics of rice straw and husk biochar and sorption/desorption behavior of their biourea composite. *Bioresource Technology, 314*, 123674. Available from https://doi.org/10.1016/j.biortech.2020.123674.

Singh, S., Chaturvedi, S., Nayak, P., Dhyani, V. C., Nandipamu, T. M., Singh, D. K., ... Govindaraju, K. (2023a). Carbon offset potential of biochar-based straw management under rice–wheat system along Indo-Gangetic Plains of India. *Science of the Total Environment, 897*, 165176. Available from https://doi.org/10.1016/j.scitotenv.2023.165176.

Singh, S., Luthra, N., Mandal, S., Kushwaha, D. P., Pathak, S. O., Datta, D., ... Pramanik, B. (2023b). Distinct behavior of biochar modulating biogeochemistry of salt-affected and acidic soil: A review. *Journal of Soil Science and Plant Nutrition*. Available from https://doi.org/10.1007/s42729-023-01370-9.

Thompson, M. (2008). *CHNS elemental analysers.* <https://www.rsc.org/images/CHNS-elemental-analysers-technical-brief-29_tcm18-214833.pdf> Accessed 17.03.23.

Thrush, P. W. (1968). Definition of coal fuel ratio. In: Bureau of mines. Department of the Interior. <https://www.mindat.org/glossary/coal_fuel_ratio> Accessed 29.05.21.

Torquato, L. D. M., Crnkovic, P. M., Ribeiro, C. A., & Crespi, M. S. (2017). New approach for proximate analysis by thermogravimetry using CO_2 atmosphere: Validation and application to different biomasses. *Journal of Thermal Analysis and Calorimetry, 128*, 1–14. Available from https://doi.org/10.1007/s10973-016-5882-z.

Trakal, L., Veselská, V., & Šafařík, I. (2016). Lead and cadmium sorption mechanisms on magnetically modified biochars. *Bioresource Technology, 203*, 318–324. Available from https://doi.org/10.1016/j.biortech.2015.12.056.

Usman, A. R., Abduljabbar, A., Vithanage, M., Ok, Y. S., Ahmad, M., Ahmad, M., ... Al-Wabel, M. I. (2015). Biochar production from date palm waste: Charring temperature induced changes in composition and surface chemistry. *Journal of Analytical and Applied Pyrolysis, 115*, 392–400.

Vithanage, M., Rajapaksha, A. U., Ahmad, M., Uchimiya, M., Dou, X., Alessi, D. S., & Ok, Y. S. (2015). Mechanisms of antimony adsorption onto soybean stover-derived biochar in aqueous solutions. *Journal of Environmental Management, 151*, 443–449.

Wade, J. B., Martin, G. P., & Long, D. F. (2015). An assessment of powder pycnometry as a means of determining granule porosity. *Pharmaceutical Development and Technology, 20*, 257–265. Available from https://doi.org/10.3109/10837450.2013.860550.

Wang, P., & Howard, B. H. (2018). Impact of thermal pretreatment temperatures on woody biomass chemical composition, physical properties and microstructure. *Energies, 11*, 25. Available from https://doi.org/10.3390/en11010025.

Webb, P. A. (2001). Volume and density determinations for particle technologists. *Micromeritics Instrument Corp, 2*(16), 01.

Weber, K., & Quicker, P. (2018). Properties of biochar. *Fuel, 217*, 240–261. Available from https://doi.org/10.1016/j.fuel.2017.12.054.

Yaashikaa, P. R., Kumar, P. S., Varjani, S., & Saravanan, A. (2020). A critical review on the biochar production techniques, characterization, stability and applications for circular bioeconomy. *Biotechnology Reports, 28*, e00570.

Yang, X., Wang, H., & Strong, P. J. (2017). Thermal properties of biochars derived from waste biomass generated by agricultural and forestry sectors. *Energies, 10*. Available from https://doi.org/10.3390/en10040469.

CHAPTER

6

Modeling the surface chemistry of biochar for efficient and wider applicability: opportunities and limitations

Adnan Shakeel[1], Riya Sawarkar[1], Suhel Aneesh Ansari[1], Shrirang Maddalwar[1] and Lal Singh[1,2]

[1]Environmental Biotechnology & Genomics Division, CSIR—National Environmental Engineering Research Institute, Nagpur, India [2]Academy of Scientific and Innovative Research (AcSIR), Ghaziabad, India

6.1 Introduction

The adverse effects of fossil fuel combustion for energy production on the environment signifies the need for immediate alternative energy sources which can produce minimum greenhouse gases and other pollutants (Barreto, 2018; Das, Majhi, Mohanty, & Pant, 2014; Shakeel, Khan, & Ahmad, 2019). The application of waste biomass products is gaining interest in this field. One such waste biomass product is biochar, which is produced when an organic feedstock is subjected to pyrolysis under limited or no oxygen conditions at very high temperatures (Kumar et al., 2021). The feedstock for biochar varies from agriculture, food, animal manure, municipal solid waste, and forestry (Chaturvedi, Singh, Dhyani, Govindaraju, & Mandal, 2021; Kumar et al., 2011; Kumar, Senth Lauamarai, Sai Deepthi, & Bharani, 2013). Biochar's unique qualities, such as high cation exchange capacity (CEC), surface area, functional groups, high porosity, and stability, make it appropriate for a wide range of applications (Abhishek et al., 2022; Singh et al., 2021). The advantages of biochar include its quick and simple preparation, eco-friendliness, reusability, and affordability (Gayathri, Gopinath, & Kumar, 2021; Hemavathy, Kumar, Kanmani, & Jahnavi, 2020). Current studies are mainly focusing on biochar to determine how effective it is at removing various impurities from aqueous and other media. Also, its promising results in soil amelioration, and carbon sequestration have opened new doors toward its application in agriculture on a broader scale (Oni, Oziegbe, & Olawole, 2019; Singh et al., 2023). The key factors affecting biochar qualities are the process parameters like temperature, biomass types, residence time, heating rate, and pressure (Babu & Chaurasia, 2003). Among all the physicochemical properties of biochar, its surface chemistry is the most important characteristic which influences its application and effectiveness in various pollution mitigation processes (Abbas et al., 2018).

Biochars are not all made equal. Its surface chemistry and the functional groups that enable surface complexation of cationic contaminants or aromatic matter contaminants via pi-pi-electron interactions are determined by different input materials and pyrolysis conditions (Atinafu, Yun, Kim, Wi, & Kim, 2021). The most crucial groups for polar pollutants of concern's surface binding are hydroxyl, aldehyde, and ketone groups (Liu et al., 2019). Recent studies demonstrate that though biochar is rich in various polar functional groups, to enhance its wider applicability, various modifications in feedstock, pyrolysis temperature, and pre—posttreatments are being explored (Ambika et al., 2022; Arif et al., 2021). Surface area, an important biochar attribute, is often observed to increase with biochar production temperature while ion exchange, CEC, and surface charge characteristics also vary among different biochar (de Mendonça, da Cunha, Soares, Tristão, & Lago, 2017). All these characteristics are very important for the surface adsorption capacity of biochar. This chapter, therefore, discusses the new insights into the modifications of surface chemistry in biochar which are being explored and exploited to widen the applicability of biochar.

6.2 Modifications in surface functional groups of biochar

The presence of functional groups like CO—, C=O, —OH on biochar is principally responsible for its specificity toward the removal of heavy metals and organic pollutants from various environmental substrates (Gupta et al., 2022). Therefore, there is a lot of research going on to manipulate and increase the number of functional groups on biochar surfaces. Also, different treatment processes during pyrolysis are being practiced introducing additional functional groups in biochar. These processes include surface amination, surface oxidation, surface sulfonation, and P—O—P insertion.

Typically, the amination of the biochar surface is started by oxidation of HNO_3 with functional nitrate and oxygen groups introduced on the biochar surface (Li et al., 2014). This process aims to further reduce the nitrate group which has been reported with enhanced selective adsorption of Cu (II). The $-NH_2$ group may also be created by heating $FeCl_3$-loaded agar in a single step under the NH_3 effect (Mian et al., 2018). This occurs due to the reaction of —OH present in agar with NH_3 to form amino groups. Under an acidic environment, $-NH_2$ undergoes protonation to NH_3^+ which shows improved electrostatic interactions with $HCrO_4^-$ (Yu, Lian, Cui, & Liu, 2018). For instance, at pyrolysis temperatures between 600°C and 700°C, the predominant N species were oxidized-N, pyrrolic-N, pyridine-N, and whereas graphitic-N was present at temperatures above 800°C (Ma et al., 2018). Inside the biochar matrix, graphitic-N might significantly improve cation—π interactions encourage electron transport and enhance heavy metal or organic pollutant removal (Wan et al., 2020). It is crucial to mention that highly N-concentrated biochar can also be used to make N-doped biochar, however this process would generate hazardous hydrogen cyanide (HCN) gas (Ren et al., 2011). There are unquestionably significant environmental concerns from this poisonous gas. Treatment of biochar with H_3PO_4 could introduce P–O–P bonds in C which can result in amplification of the amorphous forms and carbon dilation. These changes improve the micropore structure of the biochar. Research has shown that H_3PO_4-treated biochar has ten times more specific surface area and pore volume (Zhao et al., 2017). Also, the P–O–P bonding with C improves the carbon retention in the organic structure and thus can enhance the yield and efficiency of biochar. Treatment of biochar with H_2SO_4 and gaseous SO_3 introduces sulfonic acid groups (Xiong et al., 2017). This sulfonated biochar has been effectively used in the transesterifications and esterification of vegetables and fatty acids which leads to the fast production of biodiesel (Kastner et al., 2012).

6.3 Modification techniques for improving the surface chemistry of biochar

There have been continuous efforts to enhance the surface chemistry of biochar as improved surface properties will expose more functional groups present on biochar (Fig. 6.1).

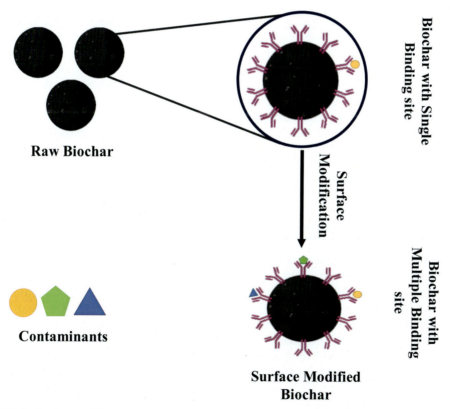

FIGURE 6.1 Surface modification of biochar to improve surface functional groups.

In general, several treatment processes have been reported to enhance the surface properties of biochar before, during, and after pyrolysis (Anerao, Salwatkar, Kumar, Pandey, & Singh, 2022; Wang et al., 2020). Different techniques have been explored to alter biochar properties such that it would respond to specific application needs and broad applications (Table 6.1). Many modification methods have been devised and employed by researchers to maximize the surface functional properties of biochar to suit various contaminant removal purposes. However, these alterations may change several features of biochar, which may result in some faults and limit its long-term application (Zhang, Duan, Peng, Pan, & Xing, 2021) (Fig. 6.2).

Iriarte-Velasco, Sierra, Zudaire, and Ayastuy (2016) reported that by treating biochar with H_2SO_4 and H_3PO_4, it was possible to see an increase in surface area and micropore volume. For instance, compared to the control, bone biochar treated with H_2SO_4 increased its surface area by 83% possibly due to the removal of inorganic components like hydroxyapatite and $CaCO_3$ in the bone char. Also, the bond formation (P–O–P) in biochar after treating with H_3PO_4 can reduce the energy needed for the decomposition of biomass in addition to removing inorganic components. It can also enhance the amplification of the amorphous structure of biochar by increasing its surface area and micropore volume (Mohammad, Shaibu-Imodagbe, Igboro, Giwa, & Okuofu, 2015; Zhao et al., 2017). In another study, Jin et al. (2014) reported that before pyrolysis, the feedstock could be alkali-treated to increase the surface area of the biochar. During pyrolysis, the preadded NaOH and KOH may react with the surface carbon of biochar which results in the opening of blocked pores and thus, increases the diameter of pores. This activation with KOH prior to pyrolysis increases the cadmium adsorption capacity of biochar via improved functional groups and surface area. Peng, Lang, and Wang (2016) observed that biochar surface area may be increased with HCl treatment. This happens since the inorganic content and soluble organic C linked to these inorganic constituents are removed during pyrolysis. Another study reported that CO_2 purging during pyrolysis may also enhance the surface area of the biochar because of the possibility of CO_2 reacting with the pore and surface carbon structure, together with CO release micropore formation (Xiong et al., 2013). Biochar can be treated at prepyrolysis, postpyrolysis and during pyrolysis stages to improve surface area as summarized in Table 6.1 (Fig. 6.3).

6.4 Role of feedstock in enhancing the surface chemistry of biochar

Feedstock plays a major role in the surface chemistry of biochar (Rowan et al., 2022). In fact, the selection of feedstock while preparing biochar depends solely on its applications and requirements of functional groups as well as surface physiochemical attributes (da Silva et al., 2022; Manasa, Sambasivam, & Ran, 2022; Tan et al., 2022; Wu et al., 2022). During pyrolysis, the phenomenon of carbonization takes place which converts wooden biomass into biochar (Kang, Nanda, & Hu, 2022). Hence, the chemical composition of wooden biomass is useful for retaining and understanding the surface properties, chemical properties as well as physical properties of biochar (Madadian & Simakov, 2022; Yang, Kang, Xu, & Yu, 2023). Biochar has a wide range of applications in the fields of agriculture for soil amendment and applications, in energy storage for making supercapacitors as well as hydrogen storage applications and in environment conservation practices to increase

TABLE 6.1 Modification techniques for improving the surface area of biochar.

	Modification technique	Surface area before treatment	Surface area after treatment	Enhancement times in surface area	References
Before pyrolysis	Clay impregnation	2.3	26.2	11.4	Chen et al. (2017)
	H_3PO_4 acid treatment	51	802	15.7	Mohammad et al. (2015); Zhao et al. (2017)
	H_2SO_4 acid treatment	76	140	1.8	Iriarte-Velasco et al. (2016); Liu et al. (2012)
	KOH alkali treatment	589	2183	3.7	Fu, Shen, Zhang, Ge, and Chen (2019)
After pyrolysis	HCl treatment	58.7	88.4	1.5	Peng et al. (2016)
During pyrolysis	CO_2 purging	486	1365	2.8	Alvarez, Lopez, Amutio, Bilbao, and Olazar (2015)

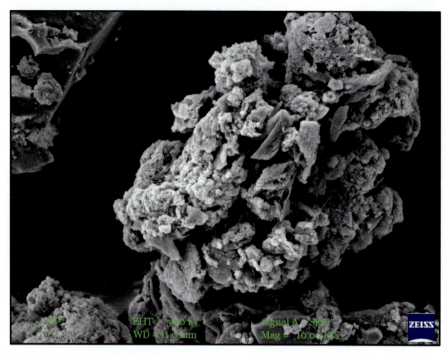

FIGURE 6.2 Scanning electron micrograph image of the surface of biochar.

the rate of degradation or any other aspects (Uday, Harikrishnan, Deoli, Mahlknecht, & Kumar, 2022; Zishan & Manzoor, 2022). So, each of these applications requires biochar with different properties and is produced from different substrates (Joshi, Ramola, Singh, Anerao, & Singh, 2022; Saif et al., 2022).

Generally, biochar produced from rice husk is widely used for a variety of applications due to its wide availability, specificity and standardized mode of production (Zhu et al., 2022). But other substrates like crop residues from food crops as well as commercial crops are also useful for preparing biochar along with biogas (Babu et al., 2022; Mishra et al., 2021). Various value-added products like cosmetics, surface cleaning agents etc. are made using biochar due to their high porosity, higher surface area as well as excellent bioadaptability (Chopra, Rangarajan, & Sen, 2022). In the case of chemical requirements, the biochar from certain feedstocks retains elements and functional groups on its surface from its natural materials (Leng, Liu, et al., 2022; Leng, Zhang, et al., 2022). Biomass is mainly composed of three constituents, namely cellulose, hemicellulose, and lignin. Out of these three components, hemicellulose weighs for 15%–30%, cellulose weighs for 40%–50% and lignin weighs for 15%–30% (Zheng et al., 2022). So, cellulose has a major stake in the overall composition of biochar. These components also vary in their structure and characteristics. Due to these differences in the chemical composition of these components of biomass, they all have different decomposition rates and behave differently in the process of pyrolysis of biomass (Chaudhary, Jain, & Jaiswar, 2022; Dias et al., 2022; Farias et al., 2022; Lopez et al., 2022). Hemicellulose shows decomposition at 200°C–260°C, while cellulose

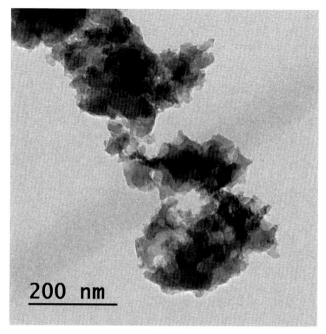

FIGURE 6.3 Transmitting electron micrograph of the surface of biochar.

decomposes at 240°C–400°C and lignin decomposes at 280°C–500°C during the pyrolysis of biomass for preparation of biochar (Zou et al., 2022). So, the temperature of decomposition is one of the useful parameters for the standardization of biochar for specific applications.

6.5 Role of pyrolysis temperature in enhancing the surface chemistry of biochar

Biochar is a multipurpose adsorbent due to the presence of a variety of adsorption groups on its surface, which include CHO, C=C, COOH, C–C, OH, aromatic carbon structure, mineral crystal stages, and other O_2-containing functional groups (Singh, Chaturvedi, Dhyani, & Govindaraju, 2020; Zhang et al., 2020). Surface area, hydrophobicity, functional groups, zeta potential, pH, stability, and substrate sources are the primary determinants of the physicochemical characters of biochar (Pariyar, Kumari, Jain, & Jadhao, 2020; Xiao, Chen, Chen, Zhu, & Schnoor, 2018). According to Maschio, Koufopanos, and Lucchesi (1992), in accordance with the temperature range of the pyrolysis process, there are three classes of pyrolysis techniques for the synthesis of biochar: carbonization/conventional pyrolysis/ (550°C–950°C), fast pyrolysis (850°C–1250°C), and flash pyrolysis (1050°C–1300°C). Depending on the feedstock materials and required biochar qualities, the control of biochar characteristics has been achieved by optimizing the temperature of pyrolysis ranging from 200°C–800°C. Aromaticity, Carbon content, surface area, pH, ash content, stability, and pore size all increase with pyrolysis temperature while biochar yield, hydrogen concentration, oxygen concentration, O/C, and H/C ratios decrease (Neris, Luzardo, da Silva, & Velasco, 2019; Ottman et al., 2019). Due to variations in the concentrations of cellulose, hemicelluloses, lignin, and inorganic minerals in different feedstock types, they respond to particular pyrolysis conditions accordingly (Clemente, Beauchemin, Thibault, MacKinnon, & Smith, 2018).

The content and intensity of the functional groups are influenced by the sources of the feedstock, high-temperature treatment (HTT) and the pyrolysis source (Maddalwar, Nayak, Kumar, & Singh, 2021). The pyrolysis temperature in particular is crucial for achieving the required functional groups. The common functional groups of biochar include C–C, C–O, CHO–, OH–, C=C–, CHO–, CH–-, and carbon structures. Various functional groups undergo a variety of state transitions in response to HTT and pyrolysis duration (Hassan et al., 2020). According to Lian and Xing (2017) in the first stage (<200°C–300°C), the feedstocks start to disintegrate and in second stage (300°C–600°C), hemicellulose and cellulose and many aliphatic functional groups start to develop. Lian and Xing (2017) estimated that, at temperature 300°C–350°C the cellulose and hemicellulose starts disintegrating and forms symmetric aromatic stretching of C–C (1475–1600 cm^{-1}), bending of CH_2 and CH_3 alkanes (1375 and 1465 cm^{-1}), stretching of C–H (855 cm^{-1}), stretching of C–O (1110 cm^{-1}), C–O stretching of conjugated ketonic groups (1600 cm^{-1}), O–H bonding of phenol (1375 cm^{-1}) and OH bending of alcohol (3300–3400 cm^{-1}). In the last stage (600°C –900°C), polycyclic structure and benzene derivatives (700 cm^{-1}) start to be created by transformation. Oxygen-bearing functional groups gradually decreased with increasing pyrolysis temperatures (Chen et al., 2011). High temperatures fully break down aliphatic structures like carboxyl, C=O, ketone, and esters (Lian & Xing, 2017). Consequently, biochar pyrolyzed at low temperatures tends to be hydrophilic because it contains various types of functional groups that contain oxygen (Lian & Xing, 2017). On the other hand, biochars produced at comparatively higher temperatures mostly include aromatic structures (Kloss et al., 2012). These aromatic ring structures of biochar result in hydrophobic properties (Lian & Xing, 2017). As a result, aromatic structures are primarily generated at higher temperatures, while functional groups comprising oxygen are more common at lower temperatures (Hou et al., 2022). Hassan et al. (2020) studied scanning electron micrographs of sawdust biochar at varying temperatures (200°C–600°C). This study did not find any visible pores at 200°C in SEM micrographs, but at 600°C honeycomb structured pores derived from lignocellulosic biomass were observed. This occurs because the surface of the biomass began to degrade and volatilize as the temperature increased. Biochar's phenotype at 300°C and 400°C showed multiple visible pore structures. Hassan et al. (2020) also studied the TEM micrograph of biochar produced from sugarcane bagasse and observed an amorphous structure. However, the breakdown of cellulose and hemicellulose structures caused the amorphous carbon to begin to reorganize at low pyrolysis temperatures (250°C–350°C). The biochar surface formed a consistent crystalline structure at a high pyrolysis temperature (600°C), which was most likely constituted of graphite. In general, hydrophobic interactions and intraparticle diffusion in graphitic carbon and the porous biochar increase the adsorption of organic contaminants but decrease the adsorption of inorganic pollutants because of reduced oxygen-containing functional groups.

The influence of pyrolysis temperature on the main elemental components of feedstock sources like C, O, and H has been demonstrated by Hassan et al. (2020). A summary of how temperature changes influence the functional groups of biochar is given in Table 6.2. Different biomass types have different element quantities in feedstocks (Xiao et al., 2018). Compared to equivalent raw feedstock materials, the concentration of total nitrogen in the biochar has nearly been unchanged (Hassan, Naidu, Du, Liu, & Qi, 2019). High pH and low CEC of biochar were observed at high pyrolysis temperatures, whereas low pH and high CEC were observed at low pyrolysis temperatures. This difference possibly occurs due to the presence of acidic functional groups like −OH and O_2 and COOH− (Banik, Lawrinenko, Bakshi, & Laird, 2018). Raheem et al. (2022) demonstrated that the number of heavy metals (Zn, Cr, Cu, As, and Pb) were reduced in the highly pyrolyzed (800°C) waste wood which caused metal emissions. Amorphous silicon can be converted into crystalline SiO_2 and silicic acid at low (< 250°C) and high (>500°C) HTT, through the pyrolysis of salicaceous feedstocks such as wheat straw, rice straw, and sugarcane bagasse (Wang, Xiao, Xu, & Chen, 2019). Silicon plays a vital role during the inorganic phase of metal co-precipitation (Hassan et al., 2020). Additionally, silicon can shield the organic phases from heat breakdown during pyrolysis (Wang et al., 2019). As a result, silicon found in biochar serves as a co-precipitator of adsorbed heavy metals (loids) (Fang et al., 2020).

6.6 Application of nanoparticles to enhance the surface chemistry of biochar

Nanoparticles (NPs) can be applied as a coating or formed in situ to modify biochar (Tan et al., 2022; Zhang et al., 2021; Zhao et al., 2017). The coated NPs consist of elemental metals (such as zerovalent iron), metal oxides (such as MgO, Fe_2O_3, $CaCO_3$, and MnO_2), carbonaceous NPs (such as GO, CNTS, and C60), and ore sulfides (e.g., FeS) (Wang et al., 2020). The effective surface area is significantly increased by the well-distributed NPs on the biochar surface because of its porous nature. Yan, Han, Gao, Xue, and Chen (2015) reported that persulfate and trichloroethylene can be activated up to 62.4% by using zero-valent iron NPs. Also, persulfate can be activated by biochar and biochar brings about the degradation of trichloroethylene (TCE) by 25.3%. Utilization of nZVI/BC composite increases the TCE degradation to 99.3%. When compared to the total of the two separate particles, the degradation ratio for nZVI/BC composites was clearly larger (Yan et al., 2015). The evenly disseminated particles of nZVI throughout the biochar surface were credited with the comparatively high reactivity of nZVI/BC. Biochar complexes with NP demonstrated effective adsorption of heavy metals, nitrate, organic contaminants, phosphate, and inorganic nutrients for the same reason (Ahmaruzzaman, 2021).

In addition to helping NPs disperse, biochar's redox characteristics can enhance the NP's catalytic reactivity. For instance, biochar containing nZVI showed strong removal of Cr (VI) (96.2 mg g^{-1}) throughout a variety of pH (Zhu et al., 2018). Cr (VI) removal would be greatly facilitated by the ability of nZVI-coated biochar to reduce toxic Cr (VI) to form Cr (III) which is comparatively less harmful. A mixture of TiO_2 and biochar was shown by Lu, Shan, Shi, Wang, and Yuan (2019), to have more catalytic activity than pure TiO_2. The synergistic interaction between biochar and TiO_2 was responsible for this increased catalytic performance. To promote the separation of charges and hinder the recombination of

TABLE 6.2 Effect of pyrolysis temperature on surface functional groups of biochar.

Pyrolysis temperature (°C)	Low pyrolysis (100°C–300°C)	Mild pyrolysis (300°C–500°C)	High pyrolysis (500°C–700/800°C)	References
Structural degradation	Cellulose and hemicellulose starts to degrade	Lignin begins to degrade	Formation of benzene polymer, carbon ring structure	Fang et al. (2020)
Active sites for sorption	$C=C-$, $-C-C-$, $-C-O-$, $-O-H$	$OH-$, $COH-$ $COOH-$	Benzene ring, aromatic structure, $COOH-$, and nonpolar groups	Zhao et al. (2020)
pH of biochar	Acidic to slightly acidic	Alkaline to slightly alkaline	Alkaline to slightly alkaline	Hassan et al. (2020)
Adsorption performance	Higher for ionic pollutant but lower for organic pollutants	Relatively lower for ionic pollutants but higher for organic pollutants	Maximum for organic contaminants and minimum for ionic pollutants	Hu et al. (2020)
Inorganic contaminants adsorption mechanism	Ion exchange, electrostatic interaction, complexation	Ion exchange, hydrogen bonding, electrostatic interaction, and complexation	Ion exchanges, co-precipitation, co-ordination	Xiao et al. (2018)
Organic contaminants adsorption mechanism	H-bond hydration and hydrophobic interaction, partition	Hydrogen bonding, hydrophobic interaction, and π-bond	$\pi-\pi$ bond interaction, Hydrophobic interaction, and pore filling	Ashoori et al. (2019)

hole-electron pairs, biochar served as a way to convert photoelectrons. Another study found that through the trapping of electrons in the conduction band, the immobilization of CuO and Cu_2O in biochar might reduce the photocatalysts' band gap (Khataee et al., 2019; Mostofa et al., 2022). The likelihood of interaction between impurities and NPs on the surface of the biochar is unquestionably increased since most chemicals incline and concentrate through adsorption at the interface. This promotes the subsequent reactions between the contaminant and the decorated NPs (Wang et al., 2019). By changing the point of zero charges, the reactions at the contact may be further improved. The positive charge on NPs could promote the adsorption of organic compounds or ions having negative charges like nitrate, phosphate and drugs onto the negatively charged biochar. These pollutants that have been adsorbed may be effectively removed by the following separation or redox reactions (Zhang et al., 2020; Zhang, et al., 2022).

6.7 Opportunities, limitations, and conclusion

The problem of growing environmental pollution has widened the scope of modified biochar globally. However, for the development of efficient modification procedures, thorough investigations into the links between the structure of biochar and reaction mechanisms are crucial, particularly the reactivity of carbon structures as impacted by the implanted modification substances. Numerous studies have demonstrated the effectiveness of using biochar to clean polluted media such as wastewater and contaminated soils. However, the majority of them concentrated on short-term, in-lab or small-scale in situ trials. Research on the widespread use of functionalized biochar is still in its early stages. Although surface-modified biochar has demonstrated considerable promise in the removal and management of various pollutants, some issues need to be carefully considered. The surface-decorated NPs may partially dissolve or disassociate. It is, therefore, very useful to have a parameter to evaluate the stability of biochar decorated with NPs. To achieve this metric, researchers may need to create detailed wear of the aging process. Biochar application also affects the pH of soil which influences the soil microbes. Hence, biochar should be functionalized and neutralized to optimum pH before soil applications. In the case of energy applications, biochar is preferred for hydrogen gas adsorption on its surface due to its higher surface area and lower cost of production as compared to activated carbon. But, when it comes to output, the hydrogen storage applications from activated carbon are more efficient than biochar-based applications. In the case of surface cleaning agents, biochar is an efficient and affordable alternative to traditional ones. However, the process of making these surface cleaning agents using biochar and other components like surfactants, stabilizers, etc should be properly standardized.

Although many methods have been looked into for biochar modification, the expense was rarely considered. It is not advised to modify biochar using pricey ingredients because the majority of biochar is made from biological waste. It is recommended to judiciously evaluate the production cost of biochar and the goods that result from their modification. A great approach toward recycling solid wastes, according to some researchers, is to modify biochar using industrial solid wastes including red mud, fly ash, and coal gangue. For a better understanding of biochar yield and function. The mechanism of thermal reaction kinetics of biomass in the presence of these solid wastes needs to be carefully assessed for a better understanding of biochar yield and functions. Additionally, the cost will be greatly reduced if biochar can be collected and renewed, as will be addressed later. The other problem that needs additional research is the recycling or regeneration of the functionalized biochar. The most popular recycling method is the creation of magnetic biochar, which is made by chemically combining Fe_3^+ and Fe_2^+ or by adding tiny magnetic particles. In the aqueous phase, magnetic biochar can typically be separated from one another and recovered. However, in wastewater, the recovery by magnetic force should be especially assessed when magnetic biochar is coated by a significant amount of organic matter or even microbes. The stability of these loaded magnetic substances on biochar is another issue. The loaded magnetic components may be released as a result of biochar ageing, turbulence, pH change, and microbial activity. Prior to application, it is important to assess how long biochar functionalities will last. When used in water and soil media, which are also known for the ageing process, the functionalized biochar's surface structure/properties and adsorption properties would eventually alter. The long-term effects of the magnetically functionalized biochar on the eradication of pollutants in water and soil must be investigated. Biochar may age as a result of chemical and biological activities, some of which may be advantageous to biochar function and others of which may not.

Acknowledgment

The authors are very thankful to the Director, CSIR-NEERI for providing facilities to accomplish this work. The manuscript has been checked by CSIR-NEERI KRC with the number: CSIR-NEERI/KRC/2022/SEP/EBGD/1.

References

Abbas, Q., Liu, G., Yousaf, B., Ali, M. U., Ullah, H., Munir, M. A. M., ... Liu, R. (2018). Contrasting effects of operating conditions and biomass particle size on bulk characteristics and surface chemistry of rice husk derived-biochars. *Journal of Analytical and Applied Pyrolysis, 134*, 281−292.

Abhishek, K., Srivastava, A., Vimal, V., Gupta, A. K., Bhujbal, S. K., Biswas, J. K., ... Kumar, M. (2022). Biochar application for greenhouse gas mitigation, contaminants immobilization and soil fertility enhancement: A state-of-the-art review. *Science of The Total Environment*, 158562.

Ahmaruzzaman, M. (2021). Biochar based nanocomposites for photocatalytic degradation of emerging organic pollutants from water and wastewater. *Materials Research Bulletin, 140*111262.

Alvarez, J., Lopez, G., Amutio, M., Bilbao, J., & Olazar, M. (2015). Physical activation of rice husk pyrolysis char for the production of high surface area activated carbons. *Industrial & Engineering Chemistry Research, 54*(29), 7241−7250.

Ambika, S., Kumar, M., Pisharody, L., Malhotra, M., Kumar, G., Sreedharan, V., ... Bhatnagar, A. (2022). Modified biochar as a green adsorbent for removal of hexavalent chromium from various environmental matrices: Mechanisms, methods, and prospects. *Chemical Engineering Journal*, 135716.

Anerao, P., Salwatkar, G., Kumar, M., Pandey, A., & Singh, L. (2022). *Physical treatment for biochar modification: Opportunities, limitations and advantages. Engineered Biochar* (pp. 49−64). Singapore: Springer.

Arif, M., Liu, G., Yousaf, B., Ahmed, R., Irshad, S., Ashraf, A., ... Rashid, M. S. (2021). Synthesis, characteristics and mechanistic insight into the clays and clay minerals-biochar surface interactions for contaminants removal—A review. *Journal of Cleaner Production, 310*, 127548.

Ashoori, N., Teixido, M., Spahr, S., LeFevre, G. H., Sedlak, D. L., & Luthy, R. G. (2019). Evaluation of pilot-scale biochar-amended woodchip bioreactors to remove nitrate, metals, and trace organic contaminants from urban stormwater runoff. *Water Research, 154*, 1−11.

References

Atinafu, D. G., Yun, B. Y., Kim, Y. U., Wi, S., & Kim, S. (2021). Introduction of eicosane into biochar derived from softwood and wheat straw: Influence of porous structure and surface chemistry. *Chemical Engineering Journal, 415*, 128887.

Babu, B. V., & Chaurasia, A. S. (2003). Modeling, simulation and estimation of optimum parameters in pyrolysis of biomass. *Energy Conversion and Management, 44*(13), 2135–2158.

Babu, S., Rathore, S. S., Singh, R., Kumar, S., Singh, V. K., Yadav, S. K., ... Wani, O. A. (2022). Exploring agricultural waste biomass for energy, food and feed production and pollution mitigation: A review. *Bioresource Technology, 127566*.

Banik, C., Lawrinenko, M., Bakshi, S., & Laird, D. A. (2018). Impact of pyrolysis temperature and feedstock on surface charge and functional group chemistry of biochars. *Journal of Environmental Quality, 47*(3), 452–461.

Barreto, R. A. (2018). Fossil fuels, alternative energy and economic growth. *Economic Modelling, 75*, 196–220.

Chaturvedi, S., Singh, S., Dhyani, V. C., Govindaraju, K., & Mandal, S. (2021). Characterization, bioenergy value and thermal stability of biochar derived from diverse agriculture and forestry lignocellulosic waste. *Biomass Conversion and Biorefinery*. Available from https://doi.org/10.1007/s13399-020-01239-2.

Chaudhary, S., Jain, V. P., & Jaiswar, G. (2022). The composition of polysaccharides: Monosaccharides and binding, group decorating, polysaccharides chains. *Innovation in Nano-Polysaccharides for Eco-sustainability* (pp. 83–118). Elsevier.

Chen, L., Chen, X. L., Zhou, C. H., Yang, H. M., Ji, S. F., Tong, D. S., ... Chu, M. Q. (2017). Environmental-friendly montmorillonite-biochar composites: Facile production and tunable adsorption-release of ammonium and phosphate. *Journal of Cleaner Production, 156*, 648–659.

Chen, X., Chen, G., Chen, L., Chen, Y., Lehmann, J., McBride, M. B., ... Hay, A. G. (2011). Adsorption of copper and zinc by biochars produced from pyrolysis of hardwood and corn straw in aqueous solution. *Bioresource Technology, 102*(19), 8877–8884.

Chopra, J., Rangarajan, V., & Sen, R. (2022). Recent developments in oleaginous yeast feedstock based biorefinery for production and life cycle assessment of biofuels and value-added products. *Sustainable Energy Technologies and Assessments, 53*, 102621.

Clemente, J. S., Beauchemin, S., Thibault, Y., MacKinnon, T., & Smith, D. (2018). Differentiating inorganics in biochars produced at commercial scale using principal component analysis. *ACS Omega, 3*(6), 6931–6944.

da Silva, M. D., da Boit Martinello, K., Knani, S., Lütke, S. F., Machado, L. M., Manera, C., ... Dotto, G. L. (2022). Pyrolysis of citrus wastes for the simultaneous production of adsorbents for Cu (II), H2, and d-limonene. *Waste Management, 152*, 17–29.

Das, S. K., Majhi, S., Mohanty, P., & Pant, K. K. (2014). CO-hydrogenation of syngas to fuel using silica supported Fe–Cu–K catalysts: Effects of active components. *Fuel Processing Technology, 118*, 82–89.

de Mendonça, F. G., da Cunha, I. T., Soares, R. R., Tristão, J. C., & Lago, R. M. (2017). Tuning the surface properties of biochar by thermal treatment. *Bioresource Technology, 246*, 28–33.

Dias, A. P. S., Rijo, B., Ramos, M., Casquilho, M., Rodrigues, A., Viana, H., ... Rosa, F. (2022). Pyrolysis of burnt maritime pine biomass from forest fires. *Biomass and Bioenergy, 163*, 106535.

Fang, Z., Gao, Y., Bolan, N., Shaheen, S. M., Xu, S., Wu, X., ... Wang, H. (2020). Conversion of biological solid waste to graphene-containing biochar for water remediation: A critical review. *Chemical Engineering Journal, 390*, 124611.

Farias, D., de Mélo, A. H., da Silva, M. F., Bevilaqua, G. C., Ribeiro, D. G., Goldbeck, R., ... Maugeri-Filho, F. (2022). New biotechnological opportunities for C5 sugars from lignocellulosic materials. *Bioresource Technology Reports, 100956*.

Fu, Y., Shen, Y., Zhang, Z., Ge, X., & Chen, M. (2019). Activated bio-chars derived from rice husk via one-and two-step KOH-catalyzed pyrolysis for phenol adsorption. *Science of the Total Environment, 646*, 1567–1577.

Gayathri, R., Gopinath, K. P., & Kumar, P. S. (2021). Adsorptive separation of toxic metals from aquatic environment using agro waste biochar: Application in electroplating industrial wastewater. *Chemosphere, 262*, 128031.

Gupta, M., Savla, N., Pandit, C., Pandit, S., Gupta, P. K., Pant, M., ... Thakur, V. K. (2022). Use of biomass-derived biochar in wastewater treatment and power production: A promising solution for a sustainable environment. *Science of the Total Environment, 153892*.

Hassan, M., Liu, Y., Naidu, R., Parikh, S. J., Du, J., Qi, F., ... Willett, I. R. (2020). Influences of feedstock sources and pyrolysis temperature on the properties of biochar and functionality as adsorbents: A meta-analysis. *Science of the Total Environment, 744*, 140714.

Hassan, M., Naidu, R., Du, J., Liu, Y., & Qi, F. (2019). Critical review of magnetic biosorbents: Their preparation, application, and regeneration for wastewater treatment. *Science of the Total Environment, 702*, 134893.

Hemavathy, R. V., Kumar, P. S., Kanmani, K., & Jahnavi, N. (2020). Adsorptive separation of Cu (II) ions from aqueous medium using thermally/chemically treated *Cassia fistula*-based biochar. *Journal of Cleaner Production, 249*, 119390.

Hou, J., Pugazhendhi, A., Sindhu, R., Vinayak, V., Thanh, N. C., Brindhadevi, K., ... Yuan, D. (2022). An assessment of biochar as a potential amendment to enhance plant nutrient uptake. *Environmental Research, 214*, 113909.

Hu, X., Zhang, X., Ngo, H. H., Guo, W., Wen, H., Li, C., ... Ma, C. (2020). Comparison study on the ammonium adsorption of the biochars derived from different kinds of fruit peel. *Science of the Total Environment, 707*, 135544.

Iriarte-Velasco, U., Sierra, I., Zudaire, L., & Ayastuy, J. L. (2016). Preparation of a porous biochar from the acid activation of pork bones. *Food and Bioproducts Processing, 98*, 341−353.

Jin, H., Capareda, S., Chang, Z., Gao, J., Xu, Y., & Zhang, J. (2014). Biochar pyrolytically produced from municipal solid wastes for aqueous As (V) removal: Adsorption property and its improvement with KOH activation. *Bioresource Technology, 169*, 622−629.

Joshi, S., Ramola, S., Singh, B., Anerao, P., & Singh, L. (2022). Waste to wealth: Types of raw materials for preparation of biochar and their characteristics. *Engineered biochar* (pp. 21−33). Singapore: Springer.

Kang, K., Nanda, S., & Hu, Y. (2022). Current trends in biochar application for catalytic conversion of biomass to biofuels. *Catalysis Today*.

Kastner, J. R., Miller, J., Geller, D. P., Locklin, J., Keith, L. H., & Johnson, T. (2012). Catalytic esterification of fatty acids using solid acid catalysts generated from biochar and activated carbon. *Catalysis Today, 190*(1), 122−132.

Khataee, A., Rad, T. S., Nikzat, S., Hassani, A., Aslan, M. H., Kobya, M., ... Demirbaş, E. (2019). Fabrication of NiFe layered double hydroxide/reduced graphene oxide (NiFe-LDH/rGO) nanocomposite with enhanced sonophotocatalytic activity for the degradation of moxifloxacin. *Chemical Engineering Journal, 375*, 122102.

Kloss, S., Zehetner, F., Dellantonio, A., Hamid, R., Ottner, F., Liedtke, V., ... Soja, G. (2012). Characterization of slow pyrolysis biochars: Effects of feedstocks and pyrolysis temperature on biochar properties. *Journal of Environmental Quality, 41*(4), 990−1000.

Kumar, M., Dutta, S., You, S., Luo, G., Zhang, S., Show, P., ... Tsang, D. C. (2021). A critical review on biochar for enhancing biogas production from anaerobic digestion of food waste and sludge. *Journal of Cleaner Production, 305*, 127143.

Kumar, S. P., Abhinaya, R. V., Gayathri Lashmi, K., Arthi, V., Pavithra, R., Sathyaselvabala, V., ... Sivanesan, S. (2011). Adsorption of methylene blue dye from aqueous solution by agricultural waste: Equilibrium, thermodynamics, kinetics, mechanism and process design. *Colloid Journal, 73*(5), 651−661.

Kumar, S. P., Senth Lauamarai, C., Sai Deepthi, A. S. L., & Bharani, R. (2013). Adsorption isotherms, kinetics and mechanism of Pb (II) ions removal from aqueous solution using chemically modified agricultural waste. *The Canadian Journal of Chemical Engineering, 91*(12), 1950−1956.

Leng, L., Liu, R., Xu, S., Mohamed, B. A., Yang, Z., Hu, Y., ... Li, H. (2022a). An overview of sulfur-functional groups in biochar from pyrolysis of biomass. *Journal of Environmental Chemical Engineering*, 107185.

Leng, L., Zhang, W., Chen, Q., Zhou, J., Peng, H., Zhan, H., ... Li, H. (2022b). Machine learning prediction of nitrogen heterocycles in bio-oil produced from hydrothermal liquefaction of biomass. *Bioresource Technology*, 127791.

Li, Y., Shao, J., Wang, X., Deng, Y., Yang, H., & Chen, H. (2014). Characterization of modified biochars derived from bamboo pyrolysis and their utilization for target component (furfural) adsorption. *Energy & Fuels, 28*(8), 5119−5127.

Lian, F., & Xing, B. (2017). Black carbon (biochar) in water/soil environments: Molecular structure, sorption, stability, and potential risk. *Environmental Science & Technology, 51*(23), 13517−13532.

Liu, P., Liu, W. J., Jiang, H., Chen, J. J., Li, W. W., & Yu, H. Q. (2012). Modification of bio-char derived from fast pyrolysis of biomass and its application in removal of tetracycline from aqueous solution. *Bioresource Technology, 121*, 235−240.

Liu, Y., Paskevicius, M., Wang, H., Parkinson, G., Veder, J. P., Hu, X., ... Li, C. Z. (2019). Role of O-containing functional groups in biochar during the catalytic steam reforming of tar using the biochar as a catalyst. *Fuel, 253*, 441−448.

Lopez, L., Alagna, F., Bianco, L., De Bari, I., Fasano, C., Panara, F., ... Perrella, G. (2022). Plants: A sustainable platform for second-generation biofuels and biobased chemicals. *Handbook of Biofuels* (pp. 47−72). Academic Press.

Lu, L., Shan, R., Shi, Y., Wang, S., & Yuan, H. (2019). A novel TiO_2/biochar composite catalysts for photocatalytic degradation of methyl orange. *Chemosphere, 222*, 391−398.

Ma, W., Wang, N., Du, Y., Xu, P., Sun, B., Zhang, L., ... Lin, K. Y. A. (2018). Human-hair-derived N, S-doped porous carbon: An enrichment and degradation system for wastewater remediation in the presence of peroxymonosulfate. *ACS Sustainable Chemistry & Engineering, 7*(2), 2718–2727.

Madadian, E., & Simakov, D. S. (2022). Thermal degradation of emerging contaminants in municipal biosolids: The case of pharmaceuticals and personal care products. *Chemosphere*, 135008.

Maddalwar, S., Nayak, K. K., Kumar, M., & Singh, L. (2021). Plant microbial fuel cell: opportunities, challenges, and prospects. *Bioresource Technology, 341*, 125772.

Manasa, P., Sambasivam, S., & Ran, F. (2022). Recent progress on biomass waste derived activated carbon electrode materials for supercapacitors applications—A review. *Journal of Energy Storage, 54*, 105290.

Maschio, G., Koufopanos, C., & Lucchesi, A. (1992). Pyrolysis, a promising route for biomass utilization. *Bioresource Technology;(United Kingdom), 42*(3).

Mian, M. M., Liu, G., Yousaf, B., Fu, B., Ullah, H., Ali, M. U., ... Ruijia, L. (2018). Simultaneous functionalization and magnetization of biochar via NH_3 ambiance pyrolysis for efficient removal of Cr (VI). *Chemosphere, 208*, 712–721.

Mishra, A., Kumar, M., Bolan, N. S., Kapley, A., Kumar, R., & Singh, L. (2021). Multidimensional approaches of biogas production and up-gradation: Opportunities and challenges. *Bioresource Technology, 338*, 125514.

Mohammad, Y. S., Shaibu-Imodagbe, E. M., Igboro, S. B., Giwa, A., & Okuofu, C. A. (2015). Effect of phosphoric acid modification on characteristics of rice husk activated carbon. *Iranica Journal of Energy and Environment, 6*(1), 20–25.

Mostofa, M. G., Rahman, Md. M., Ghosh, T. K., Kabir, A. H., Abdelrahman, M., Rahman Khan, Md. A., ... Tran, L. S. P. (2022). Potassium in plant physiological adaptation to abiotic stresses. *Plant Physiology and Biochemistry, 186*, 279–289. Available from https://doi.org/10.1016/J.PLAPHY.2022.07.011.

Neris, J. B., Luzardo, F. H. M., da Silva, E. G. P., & Velasco, F. G. (2019). Evaluation of adsorption processes of metal ions in multi-element aqueous systems by lignocellulosic adsorbents applying different isotherms: A critical review. *Chemical Engineering Journal, 357*, 404–420.

Oni, B. A., Oziegbe, O., & Olawole, O. O. (2019). Significance of biochar application to the environment and economy. *Annals of Agricultural Sciences, 64*(2), 222–236.

Ottman, N., Ruokolainen, L., Suomalainen, A., Sinkko, H., Karisola, P., Lehtimäki, J., ... Fyhrquist, N. (2019). Soil exposure modifies the gut microbiota and supports immune tolerance in a mouse model. *Journal of allergy and clinical immunology, 143*(3), 1198–1206.

Pariyar, P., Kumari, K., Jain, M. K., & Jadhao, P. S. (2020). Evaluation of change in biochar properties derived from different feedstock and pyrolysis temperature for environmental and agricultural application. *Science of the Total Environment, 713*, 136433.

Peng, P., Lang, Y. H., & Wang, X. M. (2016). Adsorption behavior and mechanism of pentachlorophenol on reed biochars: pH effect, pyrolysis temperature, hydrochloric acid treatment and isotherms. *Ecological Engineering, 90*, 225–233.

Raheem, A., He, Q., Mangi, F. H., Areeprasert, C., Ding, L., & Yu, G. (2022). Roles of heavy metals during pyrolysis and gasification of metal-contaminated waste biomass: a review. *Energy & Fuels, 36*(5), 2351–2368.

Ren, Q., Zhao, C., Chen, X., Duan, L., Li, Y., & Ma, C. (2011). NO_x and N_2O precursors (NH_3 and HCN) from biomass pyrolysis: Co-pyrolysis of amino acids and cellulose, hemicellulose and lignin. *Proceedings of the Combustion Institute, 33*(2), 1715–1722.

Rowan, M., Umenweke, G. C., Epelle, E. I., Afolabi, I. C., Okoye, P. U., Gunes, B., ... Okolie, J. A. (2022). Anaerobic co-digestion of food waste and agricultural residues: An overview of feedstock properties and the impact of biochar addition. *Digital Chemical Engineering*, 4.

Saif, I., Thakur, N., Zhang, P., Zhang, L., Xing, X., Yue, J., ... Li, X. (2022). Biochar assisted anaerobic digestion for biomethane production: Microbial symbiosis and electron transfer. *Journal of Environmental Chemical Engineering*, 107960.

Shakeel, A., Khan, A. A., & Ahmad, G. (2019). The potential of thermal power plant fly ash to promote the growth of Indian mustard (*Brassica juncea*) in agricultural soils. *SN Applied Sciences, 1*(4), 1–5.

Singh, S., Chaturvedi, S., Dhyani, V. C., & Govindaraju, K. (2020). Pyrolysis temperature influences the characteristics of rice straw and husk biochar and sorption/desorption behavior of their biourea composite. *Bioresource Technology, 314*, 123674.

Singh, S., Chaturvedi, S., Nayak, P., Dhyani, V. C., Nandipamu, T. M., Singh, D. K., ... Govindaraju, K. (2023). Carbon offset potential of biochar based straw management under rice−wheat system along Indo-Gangetic Plains of India. *Science of the Total Environment*, 897, 165176.

Singh, E., Kumar, A., Mishra, R., You, S., Singh, L., Kumar, S., ... Kumar, R. (2021). Pyrolysis of waste biomass and plastics for production of biochar and its use for removal of heavy metals from aqueous solution. *Bioresource Technology*, 320, 124278.

Tan, S., Narayanan, M., Huong, D. T. T., Ito, N., Unpaprom, Y., Pugazhendhi, A., ... Liu, J. (2022). A perspective on the interaction between biochar and soil microbes: A way to regain soil eminence. *Environmental Research*, 113832.

Uday, V., Harikrishnan, P. S., Deoli, K., Mahlknecht, J., & Kumar, M. (2022). Current trends in production, morphology, and real-world environmental applications of biochar for the promotion of sustainability. *Bioresource Technology*, 127467.

Wan, Z., Sun, Y., Tsang, D. C., Khan, E., Yip, A. C., Ng, Y. H., ... Ok, Y. S. (2020). Customised fabrication of nitrogen-doped biochar for environmental and energy applications. *Chemical Engineering Journal*, 401, 126136.

Wang, L., Ok, Y. S., Tsang, D. C., Alessi, D. S., Rinklebe, J., Wang, H., ... Hou, D. (2020). New trends in biochar pyrolysis and modification strategies: Feedstock, pyrolysis conditions, sustainability concerns and implications for soil amendment. *Soil Use and Management*, 36(3), 358−386.

Wang, Y., Xiao, X., Xu, Y., & Chen, B. (2019). Environmental effects of silicon within biochar (Sichar) and carbon−silicon coupling mechanisms: A critical review. *Environmental Science & Technology*, 53(23), 13570−13582.

Wu, Y., Wang, H., Li, H., Han, X., Zhang, M., Sun, Y., ... Xu, X. (2022). Applications of catalysts in thermochemical conversion of biomass (pyrolysis, hydrothermal liquefaction and gasification): A critical review. *Renewable Energy*, 196, 462−481.

Xiao, X., Chen, B., Chen, Z., Zhu, L., & Schnoor, J. L. (2018). Insight into multiple and multilevel structures of biochars and their potential environmental applications: a critical review. *Environmental Science & Technology*, 52(9), 5027−5047.

Xiong, X., Iris, K. M., Cao, L., Tsang, D. C., Zhang, S., & Ok, Y. S. (2017). A review of biochar-based catalysts for chemical synthesis, biofuel production, and pollution control. *Bioresource Technology*, 246, 254−270.

Xiong, Z., Shihong, Z., Haiping, Y., Tao, S., Yingquan, C., & Hanping, C. (2013). Influence of NH_3/CO_2 modification on the characteristic of biochar and the CO_2 capture. *BioEnergy Research; a Journal of Science and its Applications*, 6(4), 1147−1153.

Yan, J., Han, L., Gao, W., Xue, S., & Chen, M. (2015). Biochar supported nanoscale zerovalent iron composite used as persulfate activator for removing trichloroethylene. *Bioresource Technology*, 175, 269−274.

Yang, Y., Kang, Z., Xu, G., & Yu, Y. (2023). Enhanced adsorption performance of bensulfuron methyl with B doping biochar: Mechanism and density functional theory calculations. *Bioresource Technology*, 128657.

Yu, W., Lian, F., Cui, G., & Liu, Z. (2018). N-doping effectively enhances the adsorption capacity of biochar for heavy metal ions from aqueous solution. *Chemosphere*, 193, 8−16.

Zhang, P., Duan, W., Peng, H., Pan, B., & Xing, B. (2021). Functional biochar and its balanced design. *ACS Environmental Au*, 2(2), 115−127.

Zhang, X., Zhang, Y., Ngo, H. H., Guo, W., Wen, H., Zhang, D., ... Qi, L. (2020). Characterization and sulfonamide antibiotics adsorption capacity of spent coffee grounds-based biochar and hydrochar. *Science of the Total Environment*, 716, 137015.

Zhang, Y., Cao, L., Fu, H., Zhang, M., Meng, J., Althakafy, J. T., ... Guo, Z. (2022). Effect of sulfamethazine on anaerobic digestion of manure mediated by biochar. *Chemosphere*, 306, 135567.

Zhao, L., Zheng, W., Mašek, O., Chen, X., Gu, B., Sharma, B. K., ... Cao, X. (2017). Roles of phosphoric acid in biochar formation: Synchronously improving carbon retention and sorption capacity. *Journal of Environmental Quality*, 46(2), 393−401.

Zhao, M., Dai, Y., Zhang, M., Feng, C., Qin, B., Zhang, W., ... Qiu, R. (2020). Mechanisms of Pb and/or Zn adsorption by different biochars: Biochar characteristics, stability, and binding energies. *Science of the Total Environment*, 717, 136894.

Zheng, T., Yang, L., Li, J., Cao, M., Shu, L., Yang, L., ... Yao, J. (2022). Lignocellulose hydrogels fabricated from corncob residues through a green solvent system. *International Journal of Biological Macromolecules*, 217, 428−434.

Zhu, S., Huang, X., Ma, F., Wang, L., Duan, X., & Wang, S. (2018). Catalytic removal of aqueous contaminants on N-doped graphitic biochars: Inherent roles of adsorption and nonradical mechanisms. *Environmental Science & Technology, 52*(15), 8649–8658.

Zhu, X., Labianca, C., He, M., Luo, Z., Wu, C., You, S., ... Tsang, D. C. (2022). Life-cycle assessment of pyrolysis processes for sustainable production of biochar from agro-residues. *Bioresource Technology,* 127601.

Zishan, M., & Manzoor, U. (2022). Promoting crop growth with symbiotic microbes in agro-ecosystems—II. *Microbes and microbial biotechnology for green remediation* (pp. 135–148). Elsevier.

Zou, R., Qian, M., Wang, C., Mateo, W., Wang, Y., Dai, L., ... Lei, H. (2022). Biochar: from by-products of agro-industrial lignocellulosic waste to tailored carbon-based catalysts for biomass thermochemical conversions. *Chemical Engineering Journal,* 135972.

SECTION B

Biochar for soil improvements

CHAPTER 7

Biochar-based carbon farming: a holistic approach for crop productivity and soil health improvement

Debarati Bhaduri[1], Bibhash Chandra Verma[2], Soumya Saha[2], Trisha Roy[3] and Rubina Khanam[1]

[1]Crop Production Division, ICAR—National Rice Research Institute, Cuttack, Odisha, India
[2]Central Rainfed Upland Rice Research Station (CRURRS), ICAR-NRRI, Hazaribag, Jharkhand, India [3]Division of Soil Science and Agronomy, ICAR-Indian Institute of Soil & Water Conservation, Dehradun, Uttarakhand, India

7.1 Introduction

Like other farming practices, carbon farming is the management option where carbon fluxes are monitored to improve the carbon content in the soil. This process is well known as soil carbon sequestration. The atmospheric carbon dioxide level is regulated by continuous flow within the carbon pools of air (atmosphere), ocean (hydrosphere), and the earth's geological and biological systems (lithosphere). As long as the amount of carbon in-flow and out-flow into the atmosphere (as CO_2) are in balance, the carbon level will remain constant in the atmosphere. Carbon is a central element which governs almost all the soil properties, hence long-term storage of carbon is very important for sustainable development as well as also an important option for climate change mitigation by reducing the greenhouse gases (GHGs). Different agricultural management options are tried and recommended for soil carbon sequestration; among them, biochar has also emerged as one of the promising options for improving carbon content in soil. Environmental management through biochar application can be employed for different purposes like improvement of soil, waste management, energy production and climate change mitigation (Purakayastha, Bhaduri, & Singh, 2021). Biochar is a rich carbonaceous material composed of recalcitrant carbon compounds which are predominantly stable as well as and the stored carbon remains in the soil for several years (Karhu, Mattila, Bergström, & Regina, 2011; Lehmann, 2007). The nature and stability of biochar depend upon several factors like chemical characteristics of biochar itself, the feedstock used for the preparation of biochar, duration and temperature of the pyrolysis process, the soil management options, soil environment, interaction of soil matrix with biochar, soil microbial diversity, temperature and moisture of soil, etc. (Chaturvedi, Singh, Dhyani, Govindaraju, & Mandal, 2021; Singh, Singh, & Cowie, 2010). The different researchers had different opinions related to biochar use and its benefits, however, it was well established that biochar affects the physical, chemical and biological properties of soil by influencing soil pH, CEC, water holding capacity, carbon content, microbial activity, etc. Application of biochar in soil proved to have positive effects on net primary crop production, grain yield and dry matter in several instances.

7.1.1 What is biochar?

Biochar is a carbonaceous material obtained after the burning of biological materials (agricultural residues, wood residues, paper waste, etc.) mainly composed of cellulose, hemicellulose, and lignin under limited supply or in the absence of oxygen at a high temperature (300°C–1000°C). This process is well-known as pyrolysis (Lehmann, 2007). During the pyrolysis process around 50%–80% of biomass is converted into combustible liquids, bio-oils, vapors, gases, etc. which can be used to produce renewable fuel or bioenergy (Gray, Johnson, Dragila, & Kleber, 2014; Laird, 2008) and the remaining by product is called biochar. Calculations revealed that, depending upon charring temperature and duration, the mean residence time of biochar may vary from hundreds to thousands of years. "Terra-Preta" which means the application of biochar (charred biomass) to soils for enhancing soil productivity was practiced by the primitive Amazonians. Emerging from these old concepts, modern agriculturists are using biochar as a potent soil amendment. Biochar application can be viewed from different dimensions as for environmental management (to improve water quality by using it as a low-cost adsorbent to remove contaminants), soil improvement (enhancing soil quality, increasing pH and CEC, moisture retention, reducing leaching of nutrients) waste management (recycling of organic waste in different product), energy production (biogas, bio-oil etc.), and climate change mitigation (reduce emission of GHGs) (Purakayastha et al., 2021; Whitfield, 2009). The production and application goals to upscale the biochar technology were presented in a schematic diagram (Fig. 7.1)

The following criteria were fixed for the suitability of biochar as per the European Biochar Certificate (2012):

- Carbon content in biochar must be >50% of the dry mass. Any organic matter after pyrolysis with <50% C may be classified as pyrogenic carbonaceous material.
- The molar H/Corg ratio and O/Corg ratio must be <0.7 and <0.4, respectively.
- The quantity of volatile organic compounds must be available and listed.
- The nutrient contents in biochar in terms of N, P, K, Mg, and Ca must be available and listed.
- The thresholds for heavy metals (Pb, Cd, Hg, Zn, Cu, Ni, Cr, As) must be kept.

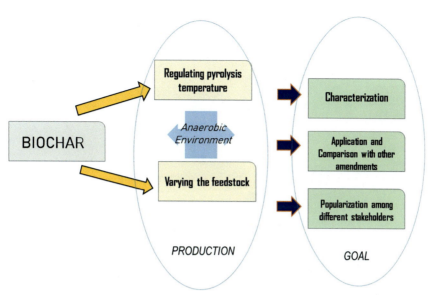

FIGURE 7.1 The production and application goals to upscale the biochar technology.

- The bulk density, pH, water and ash content of biochar as well as its specific surface area must be specified.

7.1.2 Why biochar is an option for carbon farming?

"Carbon farming" refers to the management of C fluxes through agricultural practices which essentially contributes to the increase of C and increasing C content (C sequestration) in living biomass, humus, and soil (Fig. 7.1). This is an essential "green business model" and has the capacity to combat the perils of climate change and achieve carbon neutrality (Appunn, 2022). The carbon farming initiatives have started across the globe and the European Union strongly recommends C farming practices as a part of a sustainable C cycle to reduce GHG emissions by 2050.

For agriculture practitioners and land managers C farming practices will lead to one or more of the following outcomes (McDonald et al., 2021) as illustrated in Fig. 7.2. There are multiple approaches to practice C farming. However, it does not involve a single solution for all land managers as the practices yielding positive results are location-specific.

The Natural Resource Conservation Services has identified almost 35C farming practices that help sequester C and/or reduce the emission of GHGs. The options for C farming are lucrative to the farming community as it is incentivised and focuses on rewarding an individual based on the carbon credit earned. One carbon credit is equivalent to one metric tonne of carbon dioxide. The estimation and pricing of carbon credits involve very careful measurement, reporting and verification procedures. The C credits earned by the farmers can be traded in the C market with firms or companies looking to offset their emissions. However,

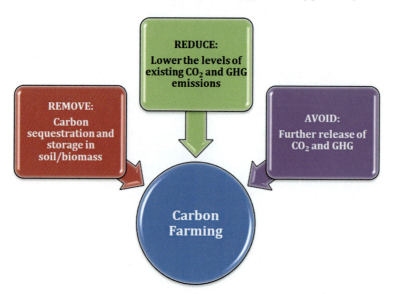

FIGURE 7.2 The positive outcomes of carbon farming for sustainable C cycle. Source: Modified from McDonald, H., Frelih-Larsen, A., Keenleyside, C., Lóránt, A., Duin, L., Andersen, S. P., ... Hiller, N. (2021). Carbon farming, Making agriculture fit for 2030.

the present value of carbon credit is variable depending on the market and the present rate of carbon credit is anywhere between $10 and $90 (source: eagaronom.com).

Biochar is known for the recalcitrant carbon in it and store carbon for a longer period of time, hence, it is one of the solutions to maintain the sustainable C cycle and is considered an integral part of C farming. The total C content in biochar is generally more than 40% (Ding et al., 2016) irrespective of the origin, source material and preparation methodology. The biochar C is characterized by extraordinarily high stability against chemical and microbial degradation (Cheng, Lehmann, Thies, & Burton, 2008), making it a potential source of C farming. Besides being the concentrated C source, it also has various synergistic impacts on the soil environment and plant growth (Ding et al., 2016; Grossman et al., 2010; Laird et al. 2010).

The process of pyrolysis is "carbon negative" as the absence of oxygen reduces the mineralization process of C to CO_2 and the placement of biochar in soil can help to bury the C for a long period of time (Das, Avasthe, Singh, & Babu, 2014) and help in achieving C credits. An estimated amount of 1.8 Pg CO_2-C equivalents per year (almost 12% of the annual net emission) can be curbed by adopting the potential of biochar without negatively impacting food security and sustainability (Woolf, Amonette, Street-Perrott, Lehmann, & Joseph, 2010). Biochar is a more potent climate change mitigation option compared to an equivalent amount of biomass combusted for bio-energy. The same quantity of biomass when used for bio-energy production can offset only 10% of the anthropogenic CO_2-C emission (Woolf et al., 2010). Thus, by all means, the use of biochar offers a promising solution to reduce CO_2 emissions.

The C present in biochar is mainly in the recalcitrant category which renders it more stable. Almost 97% of the total C in biochar is present in this form with a mean residence time of 556 years whereas the rest 3% belongs to the labile fraction which is prone to mineralization and losses and has a residence time of only around 108 days (Wang, Xiong, &

Kuzyakov, 2016). The median decomposition rate of biochar in soil is 0.0046% day^{-1} (Fang, Singh, Singh, & Krull, 2014; Gupta et al., 2020) and this stability is mainly attributed to the higher concentration of aromatic substances which makes the biochar chemically nonresponsive toward microbial decomposition. However, not all biochar incorporation in soil will allow the sequestration of C at the same rate as the quantity of C in different biochar varies depending on the residue composition, pyrolysis temperature etc. Comparing biochar produced from rice straw and apple and oak tree branches at pyrolysis temperatures 400°C–800°C, the former had 29%–50% C while the latter had 70%–85% (Jindo, Mizumoto, Sawada, Sanchez-Monedero, & Sonoki, 2014). To understand the C credit gain from biochar let us try to understand this very simple illustration in Table 7.1.

This simplified example indicates the C creditworthiness of different biochars applied to soil. Though the entire amount of applied biochar would not lead to C sequestration, but to make things more lucid the calculation process has been kept comprehensive. The application rate may vary depending on the availability of biochar and definitely, the amount of C stored will vary depending on the soil environment, the climatic conditions and management practices adopted by the farmers. The price of the C credit is also variable depending on the market. However, compared to other C sources added to the soil for C sequestration, the stability of biochar is noteworthy, and farmers can gain maximum C credit from its use.

India has the potential to produce 162.2 million tons of biochar from the crop residue with an average C content of 70% which amounts to 113.5 million tonnes C/Mg C (Gupta et al., 2020). At the policy level, the country has recently banned the export of C credits and passed the Energy Conservation (Amendment) Bill in August 2022 with an increased focus on the share of green energy. This is a welcome step for engaging the farmers in the C credit market, where they can do the trading within the country to reap the benefits. Still, a few bottlenecks remain, as with the adoption of any technology at a large scale. Besides the use of biochar for carbon sequestration, other related aspects of biochar use

TABLE 7.1 Carbon credit gain from biochar with the example of rice biochar and apple/oak tree branch biochar.

	Rice biochar	Apple/oak Biochar
C% (Jindo et al., 2014)	29–50	70–85
Avg C%	39.5	77.5
Application rate (t ha^{-1})	10	10
Total C (t ha^{-1})	3.95	7.75
Assuming 97% recalcitrant C	3.83	7.51
C credit earned	3.83	7.51
Monetary value (US$)	15.82	31.04
Monetary value (INR ha^{-1})	1303.7	2557.9
C credit value (Nature-based solutions) (https://carboncredits.com/carbon-prices-today/accessed on January 3, 2023)	$ 4.13	

must be considered. The production of biochar is a technically intense process, trained manpower and a production facility are required. To make it lucrative and economically beneficial, biochar production facilities at the village or block level may help the farmers utilize their potential in the best manner.

7.2 Determinants of biochar stability

7.2.1 Biochar's inherent chemical characteristics

The stability of biochar as an individual product and after it has been applied to the soil is an interplay of several factors. By virtue of this stability, it has found a place as a supply-side mitigation option for greenhouse gas emissions in the Intergovernmental Panel on Climate Change Report (Smith, 2016). The composition of biochar is principally guided by the feedstock from which it is derived. The biochar derived from crop residues is less stable with a decomposition rate of 0.025% day^{-1} (Wang et al., 2016) compared to other feedstocks. The chemical changes occurring during the process of pyrolysis are mainly responsible for the stability of biochar. As the temperature is increased in the absence of oxygen, the hydrogen and oxygen are preferentially removed from the biomass allowing the C to concentrate. This causes an increase in the aromaticity of C, and less aliphatic compounds are produced thus making the biochar highly resistant to microbial degradation (Liang et al., 2008). The removal of H and O leads to a decrease in the H/C and H/O ratios, which serves as an important indicator for the stability of biochar and conveys a fair idea about the biochar maturity, degree of aromaticity and carbonization (Lehmann & Joseph, 2009a, 2009b). The threshold value of the H/C ratio is 0.7 while for O/C ratio is 0.4 (EBC, 2012). Biochar with higher ratios than the threshold is of inferior quality and poorer stability. The H/C value of 0.7 is recommended by the International Biochar Initiative to draw a line between biochar and biomass. The nature of feedstock impacts the stability of biochar and studies have shown that wood-derived biochar is more stable compared to crop residue-derived biochar (Hilscher, Heister, Siewert, & Knicker, 2009; Singh et al., 2010) by virtue of the presence of more aryl C in the former (Wang et al., 2016). The C in the biochar is characterized by some amount of lability and not all the C is completely resistant to decomposition. Almost 3% of the biochar C is classified as labile (Gupta et al., 2020). Incubation studies with less than 0.5 years have shown higher decomposition rates (@0.023% day^{-1}) of biochar while studies conducted over more than 1 year have a decomposition rate of 0.005% day^{-1} (Wang et al., 2016). Thus, despite the stability in the structure and its chemical composition the biochar contains C which has skipped the condensation process and is susceptible to decomposition.

7.2.2 Interaction with minerals and soil organic matte

The soil properties play an important role in deciding the stability of the added biochar. The biochar when added to different soil will behave differently related to soil properties. Generally, the soil Fe content and clay content are negatively correlated with the decomposition rate of biochar while soil pH and Ca^{2+} exhibit a positive correlation. Therefore, acid soils with high clay and Fe content can better stabilize the applied biochar compared to soils with higher pH (Yang, Zuo, Zhou, Wu, & Wang, 2022). High soil temperature and moisture content, intense tillage practices, higher oxygen availability etc. stimulate the decomposition of biochar in the soil which is similar to the soil organic matter content (Sohi, Lopez-Capel, Krull, & Bol, 2009). Contrary to the observations by Yang et al. (2022) indicating slower biochar decomposition in acid soils, Rechberger et al. (2017) indicated an increase in the H and O-containing functional group and the formation of more hydrophilic functional groups in the soil surface under acidic conditions which enhances the decomposition of biochar. However, the conditions are reversible with the change in soil pH.

7.2.3 Feedstock type, temperature and duration of pyrolysis

The type of feedstock, pyrolysis temperature and duration all have a very significant impact on the characteristics of the biochar produced. Generally, it is observed that an increase in the pyrolysis temperature leads to the production of more stable biochar, irrespective of the feedstock used (Conz, Abbruzzini, Andrade, Milori, & Cerri, 2017). Lu et al. (2021) inferred that biochars produced at a temperature of 550°C or above produce a negative priming effect on soil and can sequester more C. The feedstock composition particularly the origin of plant materials and its chemical composition influence the C content in biochar. The feedstock rich in lignin results in biochar production with more aromatic compounds while those predominant in cellulose and hemicelluloses result in more aliphatic compounds (Collard & Blin, 2014).

7.3 Biochar for yield sustenance: facts and prospects from different soils

Worldwide, several field experiments have been carried out to study the effect of biochar on crop production. Biochar application was found more promising in soils prone to varying levels of degradation and nutrient deficiency (Laghari et al., 2015; Zhang et al., 2012) than in fertile soils (Hussain et al., 2017; Van Zwieten et al., 2010) in improving crop productivity. Many times crop yield improvement was not witnessed immediately after field application of biochar but rather from the subsequent years. It has been found that a one-time application of biochar may provide benefits over many cropping seasons, although long-term studies are still lacking to get a concrete result (Frimpong, Phares, Boateng, Abban-Baidoo, & Apuri, 2021; Xu et al., 2022).

The most common understanding of biochar application is that crop yield improves due to improvement of C content and soil structure which leads to improvement in nutrient availability and influencing soil biological properties (Table 7.2). Several metaanalysis studies on the application of biochar on crop performance have been undertaken to provide a better understanding of the effect of biochar in diverse soil types throughout the world. A metaanalysis by Jeffery et al. (2017) revealed that the application of biochar resulted in considerable improvement in crop productivity in tropical regions where a 25% crop yield increase was found, whereas the same effect was not observed in the temperate areas. The same study also found that biochar experiments conducted in pot culture were found with a higher degree of

TABLE 7.2 Effect of biochar on crop yield across different parts of the world.

Soil type	Country	Biochar rate (t ha^{-1})	Crop	Effect on crop yield	Reference
Sandy soil	Japan	15	Maize	150% yield increased over control	Uzoma et al. (2011)
Inceptisol	Colombia	30	Rice	294% yield increase over control	Noguera et al. (2010)
Oxisol	Colombia	88	Rice	21% yield reduction over control	Noguera et al. (2010)
Paddy soil	China	40	Rice	19% yield increase over control	Zhang et al. (2010)
Oxisol	Colombia	20	Maize	28% yield increase over control	Major, Rondon, Molina, Riha, and Lehmann (2010)
Silt loam Inceptisol	Chile	20	Barley	31% yield improvement over sole NPK application	Curaqueo, Meier, Khan, Cea, and Navia (2014)
Acidic sandy Oxisol	Zambia	4	Maize	Mean grain yield increases 45% ± 14% over fertilized control	Abiven, Hund, Martinsen, and Cornelissen (2015)
Sandy loam	Belgium	20	Barley	Nonsignificant yield improvement	Nelissen et al. (2015)
Acidic Red Ferrosol,	Australia	10	Maize	29% yield improvement over sole NPK application	Agegnehu, Bass, Nelson, and Bird (2016)
Acidic clay soil	Nepal	5	Radish, soybean, chilly, garlic	Nonsignificant yield improvement	Gautam, Bajracharya, and Sitaula (2017)
Highly acidic sandy loam Ultisol	Indonesia	5, 15	Maize	Significantly higher grain yield	Cornelissen et al. (2018)
Cambisol, Sandy loam	Germany	7.7	Wheat, Rye, Maize	No significant effect over 3-year cultivation period	Sänger et al. (2017)
Sandy loam	United States	29	Onion	No significant improvement	Gao et al. (2020)
Sandy loam	Ghana	10	Maize	56.9% yield improvement over sole NPK application	Frimpong et al. (2021)

positive outcome in terms of crop yield than in the field trials; the same is also true in case of acidic soils (pH < 5) than in soils of neutral reaction, as well as in soils with sandy nature than loamy and silty soils (Liu et al., 2013). These metaanalyses tried to figure out the average overall crop yield improvement after the application of biochar which varied from 11% to 13% (Jeffery, Verheijen, van der Velde, & Bastos, 2011; Liu et al., 2013). Lower response of biochar application in the temperate may be due to existing higher nutrient availability/fertility of the soils along with better water holding capacity which leaves very little scope to improve the crop yield level. On the contrary, in tropical regions, improved crop productivity following biochar application was mostly due to enhanced nutrient availability in combination with pH balance in acidic soil and increased water retention capacity.

Besides this, the application of biochar was found to have detrimental effects on crop growth and nutrient availability which hampered crop productivity (Jeffery et al., 2017; Sänger et al., 2017; Zhang, Ding, Wang, Su, & Zhao, 2020). The unfavorable effects of biochar addition on crop yields have widely been seen in temperate regions (Sänger et al., 2017; Schmidt et al., 2015; Wei et al., 2020) which may be due to the rising of the soil pH due to over-liming effect resulting in immobilization of nutrient elements such as P, Mg, B, and Fe. In the subtropical climate of China, uptake of three macronutrients (N, P, and K) by rice and wheat was reduced after 6 years of field experimentation (Wang et al., 2018). Similarly, in the upland red soil of China, Jin et al. (2019) observed an increase in rapeseed yield during the first year of wheat straw biochar application (at the rate >10 t ha^{-1}) subsequently the effect of biochar started to diminish due to its effect on soil pH, bulk density, available Phosphorus, soil organic C, and available soil water decreased. Crop yields may also be reduced upon biochar application without additional N fertilizer in soils having low inherent N (Asai et al., 2009). Generally, the application of freshly prepared biochar reduces crop growth due to temporarily disturbing the soil pH levels and/or nutrient immobilization in the soil. Mobile substances such as tars, and resins present on the surface of biochar particulates may also inhibit plant growth (Zhang et al., 2010).

From field studies conducted in different soils and climates throughout the world, it is clearly understood that the promising effect of biochar on crop yield is mostly witnessed in acidic, nutrient-deficient, degraded/weathered tropical soils which is due to its influence on soil pH and improvement in soil physical, biological and chemical parameters. It is very important to take note of the fact that, more than 30% of the world's soils are acidic which includes about half of potential cultivable soil (Galinato, Yoder, & Granatstein, 2011). However, soil pH and texture, production methodologies and feedstocks used for biochar preparation and application rates, agronomy of crops, and agro-climatic conditions are responsible for the variations found in the growth and yield of crops in different studies. The variable response of crops in different soils upon biochar application demands an advanced understanding of the underlying mechanisms of biochar in soil systems for improving crop performance.

7.4 Effect of biochar application on soil health parameters: an insight of soil physical-fertility-biological factors

When we talk about soil health (or more precisely soil quality), the role of biochar is yet to get any prime importance. The reason may be biochar is not popular so far, neither easily available nor its function in soils is well-explained to the crop growers. However, when biochar (from different feedstock/agricultural residues) was introduced to the scientific community, they could realize the immense potentiality of biochar. So far, biochar has been more known as a stable source of carbon, hence upon addition to soil it helps soil C build up for a longer time. Therefore, long-term sequestration of C into soil leads to climate change mitigation by reducing CO_2 emission from soil. However, the role of biochar in improving soil health has been less discussed to date. Soil health comprises of integration of soil's physical, chemical and biological attributes, and eventually, biochar being a raw source of carbon, impacts positively on all three aspects, thus exhibiting its direct role in soil health. In this section, an elaborate discussion will be made focussing the role of biochar, in managing and restoring soil health.

7.4.1 Biochar for protecting soil physical health

Being a porous structure, biochar helps in forming stable aggregates and improves water retention capacity. Few studies were directives in this particular aspect. Githinji (2014) reported that there was a positive impact on the soil's physical (decreased bulk density, increased total soil porosity) and hydraulic properties (increased volumetric water content and water retention) caused by biochar amendment in tomato cultivated soil which might have helped better plant growth since plant roots got an access of sufficient moisture and oxygen. Busscher et al. (2010) mentioned a decrease in soil penetration resistance due to biochar addition while Brockhoff, Christians, Killorn, Horton, and Davis (2010) highlighted an improved water retention but declined saturated hydraulic conductivity added with biochar for sand-based root zones; and Novak et al. (2012) reported increased moisture storage capacity beneficiated by biochar in both Ultisols and Aridisols. This shows biochar has a specific role in protecting soil physical parameters, which are turned as very crucial for optimum plant growth.

7.4.2 Biochar for improving nutrient cycling in soil

Being a stable C-source, biochar is often considered as a rich source of C; which is rather interlinked with other available nutrients and nutrient dynamics operating in soil. Yang, Cao, Gao, Zhao, and Li (2015) studied the fate of NH_4^+-N relevant to the N cycle, by an understanding of biochar-affected sorption-emission-transformation processes of NH_4^+-N in Anthrosol and Ferrosol. It was reported that biochar-applied soil possesses chemical sorption as a dominant process where NH_4^+-N sorption was significantly increased by the application of rice straw biochar, but it reduced NH_4^+-N biotransformation into NO_3^--N. However, an increment in soil pH leads to higher ammonia emissions from biochar-applied soils compared to nonamended ones.

7.4.3 Biochar for enriching soil biological health

The application of biochar for enriching soil biological health is closely linked with their improvement in soil physical and fertility parameters, which causes healthy plant root growth, promotes "rhizosphere effect" and abundance of microbial population and activity per se. All over the world, many studies are being carried out in a similar direction, which unanimously concluded the proliferation of soil microbes under biochar application. Liu et al. (2011) experimented with the effects of rice straw biochar and bamboo char on methanogenic activity in the paddy soil in relation to CH_4 emissions which was found to be reduced (even by 91%) from the soils amended with biochar. Comparatively rice straw biochar responded better than bamboo char. Among the four different biochars tested (maize stover, pearl millet stalk, rice straw and wheat straw), Purakayastha, Kumari, and Pathak (2015) found that dehydrogenase activity and microbial biomass carbon were maximized due to rice straw biochar application and showed greater microbial activities because of its higher lability over other biochar.

7.4.4 Biochar reclaiming health of disturbed soil

This facet of biochar science gained attraction since there are lots of discussions going on these days regarding managing the disturbed/deteriorated soil and how to harness the potentiality. Bhaduri, Saha, Desai, and Meena (2016) studied if the less-explored groundnut shell biochar could potentially restore the saline soil. They observed a positive effect on soil C achieved up to 10% w/w biochar addition in saline soil, while the lowest dose (2.5% w/w BC) influenced dehydrogenase, acid and alkaline phosphatase activities in saline soil and the highest dose (10% w/w BC) stimulated the urease and fluorescein di-acetate hydrolyzing activities. An increase in SOC, TOC and MBC indicated less C mineralization and the possibility of long-term C-sequestration in soil (Singh et al., 2023).

In other studies, different rates of biochar (B0: 0% biochar, B1: 1%, B2: 2%, B4: 4% w/w) prepared from rice residue were applied into different soils having EC2 (control), EC8 (8 dS m^{-1}), and EC16 (16 dS m^{-1}) were incubated for 8 weeks. Salinity depressed cumulative respiration and MBC. 2%–4% biochar application restored the cumulative respiration and MBC by 21% while 1% biochar in EC8 or EC16 soil did not show any effect on microbial properties. An increment of NO_3^--N was observed with incubation time and biochar rates, whereas 2%–4% biochar application was most effective in ameliorating soil salinity (Singh, Mavi, & Choudhary, 2019). Another study compared the prominence of biochar under two different stresses: while Biochar addition showed a stronger effect under nutrient-stressed conditions over nutrient-rich conditions; however, biochar application was revealed to be more effective for crop growth at abundant soil moisture than at moisture-stressed conditions under maize-grown soil (Pandit et al., 2018).

7.4.5 Biochar for overall soil health improvement

Lately, few studies have been undertaken on the holistic improvement of soil quality by applying biochar and quantifying the same with the soil quality index (SQI) approach (Munda et al., 2023). SQI was estimated for assessing/quantifying the long-term effects of biochar amendment on productivity in Alfisol when experimented with rice husk biochar (pyrolysed at 350°C–400°C) sequentially applied as treatments 0, 3, 6, and 12 t ha^{-1}. The positive effects of biochar amendments on soil pH, CEC, bulk density (decreased), water stable aggregates and water holding capacity remained consistently enhanced. In the long term, the soil physical quality parameters were more positively influenced than chemical indicators upon biochar application. Biochar amendment at 6–12 t ha^{-1} was found with best SQI after three consecutive years of application. Three years of biochar application showed a diminishing trend in terms of different soil quality indicators viz. electrical conductivity, organic C, total N and C:N (Oladele, 2019). In another study, effect of rice straw biochar (0, 10, 20, and 40 t ha^{-1}) in contrasting textured soils (loamy sand and sandy clay loam) was studied in a wheat-maize cropping sequence, and there was an enhancement of total and soil organic C stocks, expressed as ΔTC and ΔSOC (up to 4.8 t ha^{-1}) which indicates greater input and stabilization of belowground root-derived C in the biochar amended soils. In an excerpt from Bai et al. (2019) biochar amendment could potentially maximize the C sequestration in croplands (39%), followed by cover cropping, and conservation tillage by 6% and 5%, respectively, thus, playing an important role to enrich soil health. Soil properties such as EC, pH, DOC, MBC, N, P, and K also showed positive responses toward increasing rates of biochar addition (Mavi Singh et al., 2018). In a different field study in a semiarid climate, biochars sourced from feedstocks like pistachio shell, corn cobs and cotton straws were applied at 0, 4, and 8 Mg ha^{-1} rates under normal and limited irrigation conditions (65% FC) revealed that pistachio shells biochar led to the highest soil quality variables and indices, and showed more increment of SQI at limited irrigation conditions. Biochar had increased soil quality indicators like soil aggregate stability, porosity, plant available water content and total nitrogen. Further, SQI had a visible impact on crop yield which was proportional to biochar application (Bilgili et al., 2019).

7.5 Biochar for mitigation of climate change in different soils

Applying biochar as a soil amendment is among the several sustainable agricultural practices that can help farmers abate climate change by sequestering carbon in the soil. Further, the addition of biochar also reduces the need for synthetic fertilizers, which can reduce greenhouse gas emissions and simultaneously increase crop and forage yields. Since 2000, anthropogenic carbon dioxide (CO_2) emissions have increased by more than 3% yearly, placing Earth's ecosystems on a trajectory toward fast, hazardous, and irreversible climate change. One potential method of lowering the level of CO_2 in the atmosphere has been suggested as the application and storing of biochar in soils. Biochar application in soil may affect GHG emissions by modifying soil physico-bio-chemical properties viz., gas diffusivity, aggregation, water retention, pH, EC, C, N cycling, microbial enzymes, etc. Biochar can act as a long-term storage of C by capturing the carbon from the atmosphere into the soil. It builds organic carbon in soil by 20% and also controls nitrogen (N) cycling (Clough, Condron, Kammann, & Müller, 2013) and can reduce nitrous oxide (N_2O) emissions from the soil by 28%—54% (Joseph et al., 2021). Researchers showed that the application of biochar significantly improves soil health, and plant growth and reduces GHG emission from croplands in the low-nutrient acidic soils of the north coast of New South Wales and Queensland (Joseph et al., 2021). Sandy soils in Western Australia, Victoria, and South Australia, especially those in dryland areas where drought is becoming a bigger problem due to climate change, would also substantially benefit from the use of biochar. Biochar not only sequestrate C in the soil but also minimizes the release of methane (CH_4) gas from paddy fields and marshy lands (Singh et al., 2023). Karhu et al. (2011) showed that biochar application in fields (9 t ha^{-1}) significantly decreased CH_4 emission. Biochar application has reduced nitrogen loss as denitrifying bacterial communities increase their N_2O-reducing activity with increasing pH and may reduce N_2O emissions from soils due to the alkaline nature of biochar which increases the soil pH after application in acid soil (Lehmann & Joseph, 2009a, 2009b). Negative charge on biochar surfaces (Cheng, Lehmann, Thies, Burton, & Engelhard, 2006) as well as subsequent interactions with organic matter and clay minerals may further contribute to the mitigation of soil N_2O emissions. Biochar having a higher number of carboxyl groups may adsorb NH_3, lowering the NH_4 concentration in soils (Karhu et al., 2011). The escape of three potent GHGs (CO_2, CH_4, and N_2O) into the atmosphere can be reduced by incorporating biochar into the soil. Biochar is being considered as a new option for reducing the impact of climate change as it converts carbon from an active to a passive pool. They are present in soil in highly stabilized form and store carbon for several years and act as major carbon sinks (Schmidt & Noack, 2000). Carbon sequestration through biochar application enhances storage time as compared to other sequestration approaches like afforestation or reforestation (Wang et al., 2016). Thus, the application of biochar in soil may have a significant role in reducing carbon from the atmosphere and simultaneously reducing the GHGs emissions and climate change impact.

7.6 Contribution of biochar in economics of carbon farming

The aim of carbon farming is to optimize carbon capture by implementing practices which take-up CO_2 from the atmosphere and facilitate long-term storage in plant material and/or soil organic matter which results in a net loss of C in the atmosphere. Similarly, carbon farming practices are management practices which sequester C and/or reduce GHG emissions. Agronomic and economic benefits of biochar amendment must be clearly visible for effective large-scale production of biochar, which indicates that the crop selection, soil type, and biochar application rate and quality are very crucial as they influence the economics of the whole system (Baidoo, Sarpong, Bolwig, & Ninson, 2016; Shareef, Zhao, 2016). Moreover, globally the demand for energy is increasing due to the growth of every sector and population. Biomass is a promising source of renewable energy which can be converted through mechanical, biochemical, physical, and thermos-chemical processes (Oni, Oziegbe, & Olawole, 2019). Three main products viz. biochar, bio-oil, and syngas are obtained from biomass pyrolysis, that is, thermo-chemical conversion. Oil and syngas can be used as a renewable energy source. Ultimately, biochar is an economic bio-commodity, which can be used in agriculture as well as the energy sector.

Profitable biochar production can be achieved by introducing C credits or payments on CO_2 sequestration. Sequestering carbon in biochar is an added advantage to soil application. Sequestered carbon, that is, carbonized materials in biochar derived from biomass after pyrolysis is long-lasting with a half-life of 1400 years (Kuzyakov, Subbotina, Chen, Bogomolova, & Xu, 2009). The degree of emissions avoidance from slow pyrolysis of green waste is calculated to the tune of 3.8 tons/CO_2e ton dry feedstock (Gaunt & Cowie, 2009). This value includes "avoided emissions" from landfills, fossil fuel substitution and other GHG benefits. Emissions of methane and nitrous oxide may be avoided during the pyrolysis of biomass, additionally, the energy produced during pyrolysis displaces fossil fuels and may also account for "avoided emission." Biochar need not be applied every time to soil as an amendment which helps in improving the water retention capacity of soil and nutrient availability thus decreasing the cost of irrigation and fertilizer, respectively. While yield improvements lead to C sequestration, additionally, reduced fertilizer, irrigation and energy use in agriculture may lead to further reduction in C emissions. The cost of biochar is directly related to the cost of the feedstock, the processing method used and the economic utilization of the co-products. Biochar prepared from yard waste including livestock manures, is worth a net profit of $69 and $16 for the high and low revenue scenario of CO_2 (Roberts, Gloy, Joseph, Scott, & Lehmann, 2010). Transportation feedstock as well as biochar is also an important factor in the economics of biochar production.

Recent assessments on C sequestration by biochar mostly considered the availability of biomass for biochar production and the sequestration potential varies between 0.6 and 11.9 $GtCO_2$ per year (Fuss et al., 2018). Lenton (2010) assessed the sequestration rate of CO_2 to the tune of 2.8–3.3 $GtCO_2$ per year if all felling losses from the forest area, 50% of currently unused crop residues, and burned biomass from shifting cultivation fires were used to prepare biochar. Considering the annual unutilized waste biomass content of about 12.1 $GtCO_2$ per year, globally, the same estimate was found to be 6.1 $GtCO_2$ per year (Lee & Day, 2013). Roberts, Gloy, Joseph, Scott, and Lehmann (2010) estimated achievable annual carbon sequestration of around 0.7 $GtCO_2$ per year accounting for only the use of late stover as a feedstock for biochar. Fuss et al. (2018) opined a realistic lower range of 0.3–2 $GtCO_2$ per year by 2050 accounting for the limited availability of biomass as feedstock for biochar.

As long-term large-scale studies of biochar application in field crops are still lacking, therefore, practical feasibility, long-term mitigation capacity and disadvantages mostly remain unknown. Moreover, the estimation of CO_2 sequestration potentials by biochar at the world level does not consider the complex, locale-specific effects of biochar applications that vary with the quality and quantity of biochar, climate, nature of the soil, and management practices followed. However, offsetting biochar C in C trading for further offsetting the costs may be a complicated process as measurement of soil C is cumbersome, especially over large areas.

To improvise the choice of biochar over other soil amendments the following developmental needs can be tapped in future:

1. Building awareness among different stakeholders through capacity building in biochar production and application technologies.
2. Familiarizing biochar production and application technologies for awareness generation among the farmers by demonstrating its benefits for soil and crop yields.
3. Promoting biochar production as a start-up opportunity by encouraging unemployed youths and through forming self-help groups.
4. To impart training for selected progressive farmers, and their experience should be further used for improving technology and popularization.

7.7 Conclusion

Soil is a great source of carbon and holds approximately three times of atmosphere-equivalent amount of CO_2 or nearly four times the amount stored in living matter, as per estimates. But with the pace of time, agriculture and land conversion have decreased soil carbon in all over the world by 840 GtCO$_2$ in a span of the last 10,000 years globally. This problem was manifested when many arable soils deteriorated 50%−70% C from the original organic carbon. Biochar (recalcitrant carbon-containing compound) is one of the most potential and suitable technologies for removing carbon from atmosphere and capturing it into soil, that is, soil carbon sequestration (or alternatively called as carbon farming). It may also be categorized as a sustainable technology toward carbon-negative energy production because a large amount of organic carbon (agricultural residues and waste materials) can be converted into C-rich biochar using pyrolysis and stored in the soil for a long time. By virtue of good quality publications and research materials from all over the world, the role and importance of biochar are already well-known among the scientific community. It has a prominent role as an important soil amendment and improves soil properties by affecting mainly soil pH, CEC, water holding capacity, nutrient retention capacity, etc., which directly or indirectly regulates C-storage in soil. Carbon plays a pivotal role and probably the single most important element which is held responsible for maintaining all three facets, that is, soil physical−chemical−biological parameters. Moreover, the application of biochar can also be beneficial by reducing toxic components in soil.

Biochar can offer very definite benefits for soil C farming and improved soil quality, there are still some concerns among the scientific fraternity regarding biochar's role like saturation (up to what extent its role will be beneficial and certainly not to the infinite level), reversibility (if there's any possible consequences for release of C from biochar upon man-made or natural scenarios), and difficulty for measurement (in what quantity biochar can actually contribute for C-farming is still a tedious and costly method). Owing to these reasons, perhaps the potential and cost of using biochar on a large scale are less clear. A recent estimate calculated that biochar could sequester 0.5−2.0 Gt CO$_2$ per year by 2050 at a cost of $30−120 per ton of CO$_2$. More refined research is required to judge the estimates of biochar's potential. Despite this, Biochar is undoubtedly a suitable strategy for carbon sequestration, sustainable soil development, mitigation of climate change and remediation of environmental pollutants. More so when biochar is based on recyclable agro-waste and culminates as an example of a win−win strategy.

References

Abiven, S., Hund, A., Martinsen, V., & Cornelissen, G. (2015). Biochar amendment increases maize root surface areas and branching: A shovelomics study in Zambia. *Plant and Soil*, *395*, 45−55.

Agegnehu, G., Bass, A. M., Nelson, P. N., & Bird, M. I. (2016). Benefits of biochar, compost and biochar−compost for soil quality, maize yield and greenhouse gas emissions in a tropical agricultural soil. *Science of the Total Environment*, *543*, 295−306.

Appunn, K. (2022). https://www.cleanenergywire.org/factsheets/carbon-farming-explained-pros-cons-and-eus-plans.

Asai, H., Samson, B. K., Stephan, H. M., Songyikhangsuthor, K., Homma, K., Kiyono, Y., ... Horie, T. (2009). Biochar amendment techniques for upland rice production in Northern Laos: 1. Soil physical properties, leaf SPAD and grain yield. *Field Crops Research*, *111*(1−2), 81−84.

Bai, X., Huang, Y., Ren, W., Coyne, M., Jacinthe, P. A., Tao, B., ... Matocha, C. (2019). Responses of soil carbon sequestration to climate-smart agriculture practices: A meta-analysis. *Global Change Biology*, *25*(8), 2591−2606.

Baidoo, I., Sarpong, D. B., Bolwig, S., & Ninson, D. (2016). Biochar amended soils and crop productivity: A critical and meta-analysis of literature. *International Journal of Development and Sustainability*, *5*(9), 414−432.

Bhaduri, D., Saha, A., Desai, D., & Meena, H. N. (2016). Restoration of carbon and microbial activity in salt-induced soil by application of peanut shell biochar during short-term incubation study. *Chemosphere*, *148*, 86–98.

Bilgili, A. V., Aydemir, S., Altun, O., Sayğan, E. P., Yalçın, H., & Schindelbeck, R. (2019). The effects of biochars produced from the residues of locally grown crops on soil quality variables and indexes. *Geoderma*, *345*, 123–133.

Brockhoff, S. R., Christians, N. E., Killorn, R. J., Horton, R., & Davis, D. D. (2010). Physical and mineralnutrition properties of sand-based turfgrass root zones amended with biochar. *Agronomy Journal*, *102*, 1627–1631.

Busscher, W. J., Novak, J. M., Evans, D. E., Watts, D. W., Niandou, M. A. S., & Ahmedna, M. (2010). Influence of pecan biochar on physical properties of a Norfolk loamy sand. *Soil Science*, *175*(1), 10–14.

Chaturvedi, S., Singh, S., Dhyani, V. C., Govindaraju, K., & Mandal, S. (2021). Characterization, bioenergy value and thermal stability of biochar derived from diverse agriculture and forestry lignocellulosic waste. *Biomass Conversion and Biorefinery*. Available from https://doi.org/10.1007/s13399-020-01239-2.

Cheng, C. H., Lehmann, J., Thies, J. E., Burton, S. D., & Engelhard, M. H. (2006). Oxidation of black carbon by biotic and abiotic processes. *Organic Geochemistry*, *37*(11), 1477–1488.

Cheng, C. H., Lehmann, J., Thies, J. E., & Burton, S. D. (2008). Stability of black carbon in soils across a climatic gradient. *Journal of Geophysical Research: Biogeosciences*, *113*, 2027.

Clough, T. J., Condron, L. M., Kammann, C., & Müller, C. (2013). A review of biochar and soil nitrogen dynamics. *Agronomy*, *3*(2), 275–293.

Collard, F. X., & Blin, J. (2014). A review on pyrolysis of biomass constituents: Mechanisms and composition of the products obtained from the conversion of cellulose, hemicelluloses and lignin. *Renewable and Sustainable Energy Review*, *38*, 594–608.

Conz, R. F., Abbruzzini, T. F., Andrade, C. D., Milori, D. M., & Cerri, C. E. (2017). Effect of pyrolysis temperature and feedstock type on agricultural properties and stability and biochars. *Embrapa Instrumentação-Artigo em periódico indexado (ALICE)*.

Cornelissen, G., Nurida, N. L., Hale, S. E., Martinsen, V., Silvani, L., & Mulder, J. (2018). Fading positive effect of biochar on crop yield and soil acidity during five growth seasons in an Indonesian Ultisol. *Science of the Total Environment*, *634*, 561–568.

Curaqueo, G., Meier, S., Khan, N., Cea, M., & Navia, R. (2014). Use of biochar on two volcanic soils: Effects on soil properties and barley yield. *Journal of Soil Science and Plant Nutrition*, *14*(4), 911–924.

Das, S. K., Avasthe, R. K., Singh, R., & Babu, S. (2014). Biochar as carbon negative in carbon credit under changing climate. *Current Science*, *107*(7), 1090.

Ding, Y., Yunguo, L., Shaobo, L., Zhongwu, L., Xiaofei, T., Xixian, H., ... Zheng, B. (2016). Biochar to improve soil fertility. A review. *Agronomy for Sustainable Development*, *36*, 1–18.

EBC (2012) European biochar certificate – guidelines for a sustainable production of biochar. European Biochar Foundation (EBC). Available at: http://www.european-biochar.org/en/

Fang, Y., Singh, B., Singh, B. P., & Krull, E. (2014). Biochar carbon stability in four contrasting soils. *European Journal of Soil Science*, *65*, 60–71.

Frimpong, K. A., Phares, C. A., Boateng, I., Abban-Baidoo, E., & Apuri, L. (2021). One-time application of biochar influenced crop yield across three cropping cycles on tropical sandy loam soil in Ghana. *Heliyon*, *7*(2)e06267.

Fuss, S., Lamb, W. F., Callaghan, M. W., Hilaire, J., Creutzig, F., Amann, T., ... Minx, J. C. (2018). Negative emissions—Part 2: Costs, potentials and side effects. *Environmental Research Letters*, *13*(6), 063002.

Galinato, S. P., Yoder, J. K., & Granatstein, D. (2011). The economic value of biochar in crop production and carbon sequestration. *Energy Policy*, *39*(10), 6344–6350.

Gao, S., Wang, D., Dangi, S. R., Duan, Y., Pflaum, T., Gartung, J., ... Turini, T. (2020). Nitrogen dynamics affected by biochar and irrigation level in an onion field. *Science of the Total Environment*, *714*, 136432.

Gaunt, J., & Cowie, A. (2009). Biochar, greenhouse gas accounting and emission trading. In J. Lehmann, & S. Joseph (Eds.), *Biochar for Environmental Management: Science and Technology* (pp. 317–340). London: Earthscan.

Gautam, D. K., Bajracharya, R. M., & Sitaula, B. K. (2017). Effects of biochar and farm yard manure on soil properties and crop growth in an agroforestry system in the Himalaya. *Sustainable Agriculture Research*, *6*(4), 74–82.

Githinji, L. (2014). Effect of biochar application rate on soil physical and hydraulic properties of a sandy loam. *Archives of Agronomy and Soil Science*, *60*(4), 457–470.

Gray, M., Johnson, M. G., Dragila, M. I., & Kleber, M. (2014). Water uptake in biochars: The roles of porosity and hydrophobicity. *Biomass and Bioenergy*, *61*, 196–205.

Grossman, J. M., O'Neill, B. E., Tsai, S. M., Liang, B., Neves, E., Lehmann, J., ... Thies, J. E. (2010). Amazonian anthrosols support similar microbial communities that differ distinctly from those extant in adjacent, unmodified soils of the same mineralogy. *Microbial Ecology, 60,* 192–205.

Gupta, D. K., Gupta, C. K., Dubey, R., Fagodiya, R. K., Sharma, G., Mohamed, N., ... Shukla, A. K. (2020). *Role of biochar in carbon sequestration and greenhouse gas mitigation. Biochar applications in agriculture and environment management* (pp. 141–165). Cham: Springer.

Hilscher, A., Heister, K., Siewert, C., & Knicker, H. (2009). Mineralisation and structural changes during the initial phase of microbial degradation of pyrogenic plant residues in soil. *Organic Geochemistry, 40,* 332–342.

Hussain, M., Farooq, M., Nawaz, A., Al-Sadi, A. M., Solaiman, Z. M., Alghamdi, S. S., ... Siddique, K. H. (2017). Biochar for crop production: Potential benefits and risks. *Journal of Soils and Sediments, 17,* 685–716.

Jeffery, S., Abalos, D., Prodana, M., Bastos, A. C., Van Groenigen, J. W., Hungate, B. A., ... Verheijen, F. (2017). Biochar boosts tropical but not temperate crop yields. *Environmental Research Letters, 12*(5), 053001.

Jeffery, S., Verheijen, F. G. A., van der Velde, M., & Bastos, A. C. (2011). A quantitative review of the effects of biochar application to soils on crop productivity using meta-analysis. *Agriculture, Ecosystems & Environment, 144*(1), 175–187.

Jin, Z., Chen, C., Chen, X., Hopkins, I., Zhang, X., Han, Z., ... Billy, G. (2019). The crucial factors of soil fertility and rapeseed yield—A five year field trial with biochar addition in upland red soil, China. *Science of the Total Environment, 649,* 1467–1480.

Jindo, K., Mizumoto, H., Sawada, Y., Sanchez-Monedero, M., & Sonoki, T. (2014). Physical and chemical characterization of biochars derived from different agricultural residues. *Biogeosciences, 23,* 6613–6621.

Joseph, S., Cowie, A. L., Van Zwieten, L., Bolan, N., Budai, A., Buss, W., ... Luo, Y. (2021). How biochar works, and when it doesn't: A review of mechanisms controlling soil and plant responses to biochar. *GCB Bioenergy, 13*(11), 1731–1764.

Karhu, K., Mattila, T., Bergström, I., & Regina, K. (2011). Biochar addition to agricultural soil increased CH_4 uptake and water holding capacity—Results from a short-term pilot field study. *Agriculture, Ecosystems & Environment, 140*(1–2), 309–313.

Kuzyakov, Y., Subbotina, I., Chen, H., Bogomolova, I., & Xu, X. (2009). Black carbon decomposition and incorporation into soil microbial biomass estimated by ^{14}C labeling. *Soil Biology and Biochemistry, 41*(2), 210–219.

Laghari, M., Mirjat, M. S., Hu, Z., Fazal, S., Xiao, B., Hu, M., ... Guo, D. (2015). Effects of biochar application rate on sandy desert soil properties and sorghum growth. *Catena, 135,* 313–320.

Laird, D. A. (2008). The charcoal vision: A win–win–win scenario for simultaneously producing bioenergy, permanently sequestering carbon, while improving soil and water quality. *Agronomy Journal, 100,* 178–181.

Laird, D. A., Fleming, P., Davis, D. D., Horton, R., Wang, B., & Karlen, D. L. (2010). Impact of biochar amendments on the quality of a typical Midwestern agricultural soil. *Geoderma, 158,* 443–449.

Lehmann, J., & Joseph, S. (2009a). Biochar systems. In J. Lehmann, & S. Joseph (Eds.), *Biochar for environmental management* (pp. 147–168). London: Science and Technology, Earthscan.

Lee, J. W., & Day, D. M. (2013). Smokeless biomass pyrolysis for producing biofuels and biochar as a possible arsenal to control climate change. In *Advanced biofuels and bioproducts,* (pp. 23–34). New York: Springer.

Lehmann, J. (2007). Bio-energy in the black. *Frontier Ecology and Environment, 5*(7), 381–387.

Lehmann, J., & Joseph, S. (2009b). *Biochar for environmental management: Science and technology.* London: Earthscan.

Lenton, T. M. (2010). The potential for land-based biological $CO2$ removal to lower future atmospheric CO_2 concentration. *Carbon Management, 1*(1), 145–160.

Liang, B., Lehmann, J., Solomon, D., Sohi, S., Thies, J. E., Skjemstad, J. O., ... Wirick, S. (2008). Stability of biomass-derived black carbon in soils. *Geochimica et Cosmochimica Acta, 72,* 6096, 6078.

Liu, X., Zhang, A., Ji, C., Joseph, S., Bian, R., Li, L., ... Paz-Ferreiro, J. (2013). Biochar's effect on crop productivity and the dependence on experimental conditions—a meta-analysis of literature data. *Plant and Soil, 373,* 583–594.

Liu, Y., Yang, M., Wu, Y., Wang, H., Chen, Y., & Wu, W. (2011). Reducing CH_4 and CO_2 emissions from waterlogged paddy soil with biochar. *Journal of Soils and Sediments, 11*(6), 930–939.

Lu, X., Yin, Y., Li, S., Ma, H., Gao, R., & Yin, Y. (2021). Effects of biochar feedstock and pyrolysis temperature on soil organic matter mineralization and microbial community structures of forest soils. *Frontiers in Environmental Science, 313.*

Major, J., Rondon, M., Molina, D., Riha, S. J., & Lehmann, J. (2010). Maize yield and nutrition during 4 years after biochar application to a Colombian savanna oxisol. *Plant and Soil, 333*(1–2), 117–128.

Mavi Singh, M., Singh, G., Singh, B. P., Singh Sekhon, B., Choudhary, O. P., Sagi, S., ... Berry, R. (2018). Interactive effects of rice-residue biochar and N-fertilizer on soil functions and crop biomass in contrasting soils. *Journal of Soil Science and Plant Nutrition, 18*(1), 41–59.

McDonald, H., Frelih-Larsen, A., Keenleyside, C., Lóránt, A., Duin, L., Andersen, S.P., ... Hiller, N. (2021). Carbon farming, Making agriculture fit for 2030.

Munda, S., Nayak, A. K., Shahid, M., Bhaduri, D., Chatterjee, D., Mohanty, S., ... Jambhulkar, N. (2023). Soil quality assessment of lowland rice soil of eastern India: Implications of rice husk biochar application. *Heliyon, 9*(7).

Nelissen, V., Ruysschaert, G., Manka'Abusi, D., D'Hose, T., De Beuf, K., Al-Barri, B., ... Boeckx, P. (2015). Impact of a woody biochar on properties of a sandy loam soil and spring barley during a two-year field experiment. *European Journal of Agronomy, 62*, 65–78.

Noguera, D., Rondón, M., Laossi, K. R., Hoyos, V., Lavelle, P., de Carvalho, M. H. C., ... Barot, S. (2010). Contrasted effect of biochar and earthworms on rice growth and resource allocation in different soils. *Soil Biology and Biochemistry, 42*(7), 1017–1027.

Novak, J. M., Busscher, W. J., Watts, D. W., Amonette, J., Ippolito, J. A., Lima, I. M., Gaskin, J., Das, K. C., Steiner, C., Ahmedna, M., et al. (2012). Biochars impact on soil moisture storage in an Ultisol and two Aridisols. *Soil Science, 177*, 310–320.

Oladele, S. O. (2019). Changes in physicochemical properties and quality index of an Alfisol after three years of rice husk biochar amendment in rainfed rice–Maize cropping sequence. *Geoderma, 353*, 359–371.

Oni, B. A., Oziegbe, O., & Olawole, O. O. (2019). Significance of biochar application to the environment and economy. *Annals of Agricultural Sciences, 64*(2), 222–236.

Pandit, N. R., Mulder, J., Hale, S. E., Martinsen, V., Schmidt, H. P., & Cornelissen, G. (2018). Biochar improves maize growth by alleviation of nutrient stress in a moderately acidic low-input Nepalese soil. *Science of the Total Environment, 625*, 1380–1389.

Purakayastha, T. J., Bhaduri, D., & Singh, P. (2021). Role of biochar on greenhouse gas emissions and carbon sequestration in soil: Opportunities for mitigating climate change. *Soil Science: Fundamentals to Recent Advances*, 237–260.

Purakayastha, T. J., Kumari, S., & Pathak, H. (2015). Characterisation, stability, and microbial effects of four biochars produced from crop residues. *Geoderma, 239*, 293–303.

Rechberger, M. V., Kloss, S., Rennhofer, H., Tintner, J., Watzinger, A., Soja, G., ... Zehetner, F. (2017). Changes in biochar physical and chemical properties: Accelerated biochar aging in an acidic soil. *Carbon, 115*, 209–219.

Roberts, K. G., Gloy, B. A., Joseph, S., Scott, N. R., & Lehmann, J. (2010). Life cycle assessment of biochar systems: Estimating the energetic, economic, and climate change potential. *Environmental Science & Technology, 44*(2), 827–833.

Roberts, K. G., Gloy, B. A., Joseph, S., Scott, N. R., & Lehmann, J. (2010). Life cycle assessment of biochar systems: estimating the energetic, economic, and climate change potential. *Environmental Science & Technology, 44*(2), 827–833.

Sänger, A., Reibe, K., Mumme, J., Kaupenjohann, M., Ellmer, F., Roß, C. L., ... Meyer-Aurich, A. (2017). Biochar application to sandy soil: Effects of different biochars and N fertilization on crop yields in a 3-year field experiment. *Archives of Agronomy and Soil Science, 63*(2), 213–229.

Schmidt, H. P., Pandit, B. H., Martinsen, V., Cornelissen, G., Conte, P., & Kammann, C. I. (2015). Fourfold increase in pumpkin yield in response to low-dosage root zone application of urine-enhanced biochar to a fertile tropical soil. *Agriculture, 5*(3), 723–741.

Schmidt, M. W. I., & Noack, A. G. (2000). Black carbon in soils and sediments: Analysis, distribution, implications, and current challenges. *Global Biogeochemical Cycles, 14*(3), 777–793.

Shareef, T. M. E., & Zhao, B. (2016). The fundamentals of biochar as a soil amendment tool and management in agriculture scope: An overview for farmers and gardeners. *Journal of Agricultural Chemistry and Environment, 6*(1), 38–61.

Singh, S., Luthra, N., Mandal, S., Kushwaha, D. P., Pathak, S. O., Datta, D., ... Pramanik, B. (2023). Distinct behavior of biochar modulating biogeochemistry of salt-affected and acidic soil: A review. *Journal of Soil Science and Plant Nutrition*. Available from https://doi.org/10.1007/s42729-023-01370-9.

Singh, B., Singh, B. P., & Cowie, A. L. (2010). Characterisation and evaluation of biochars for their application as a soil amendment. *Soil Research, 48*(7), 516–525. Available from https://doi.org/10.1071/SR10058.

Singh, R., Mavi, M. S., & Choudhary, O. P. (2019). Saline soils can be ameliorated by adding biochar generated from rice-residue waste. *CLEAN–Soil, Air, Water, 47*(2)1700656.

Smith, P. (2016). Soil carbon sequestration and biochar as negative emission technologies. *Global Change Biology*, 22(3), 1315–1324.

Sohi, S., Lopez-Capel, E., Krull, E., & Bol, R. (2009). Biochar, climate change and soil: A review to 524 guide future research. *CSIRO Land Water Science Report*, 5(09), 17–31.

Uzoma, K. C., Inoue, M., Andry, H., Fujimaki, H., Zahoor, A., & Nishihara, E. (2011). Effect of cow manure biochar on maize productivity under sandy soil condition. *Soil Use and Management*, 27(2), 205–212.

Van Zwieten, L., Kimber, S., Morris, S., Chan, K. Y., Downie, A., Rust, J., … Cowie, A. (2010). Effects of biochar from slow pyrolysis of papermill waste on agronomic performance and soil fertility. *Plant and Soil*, 327, 235–246.

Wang, J., Xiong, Z., & Kuzyakov, Y. (2016). Biochar stability in soil: Meta-analysis of decomposition and priming effects. *Global Change Biology Bioenergy*, 8(3), 512–523.

Wang, L., Li, L., Cheng, K., Ji, C., Yue, Q., Bian, R., … Pan, G. (2018). An assessment of emergy, energy, and cost-benefits of grain production over 6 years following a biochar amendment in a rice paddy from China. *Environmental Science and Pollution Research*, 25, 9683–9696.

Wei, W., Yang, H., Fan, M., Chen, H., Guo, D., Cao, J., … Kuzyakov, Y. (2020). Biochar effects on crop yields and nitrogen loss depending on fertilization. *Science of the Total Environment*, 702, 134423.

Whitfield, R. (2009). Biochar: A way forward for India and the world, Paper II.3, http://www.climatecommunity.org/documents/PaperII3BiocharPaper.pdf

Woolf, D., Amonette, J. E., Street-Perrott, F. A., Lehmann, J., & Joseph, S. (2010). Sustainable biochar to mitigate global climate change. *Nature Communications*, 156.

Xu, Q., Wang, J., Liu, Q., Chen, Z., Jin, P., Du, J., … Wang, X. (2022). Long-term field biochar application for rice production: Effects on soil nutrient supply, carbon sequestration, crop yield and grain minerals. *Agronomy*, 12(8), 1924.

Yang, F., Cao, X., Gao, B., Zhao, L., & Li, F. (2015). Short-term effects of rice straw biochar on sorption, emission, and transformation of soil NH_4^+-N. *Environmental Science and Pollution Research*, 22(12), 9184–9192.

Yang, F., Zuo, X., Zhou, Y., Wu, S., & Wang, M. (2022). Stability of biochar in five soils: Effects from soil property. *Environmental Progress & Sustainable Energy*, 41(3), 13775.

Zhang, A., Bian, R., Pan, G., Cui, L., Hussain, Q., Li, L., … Yu, X. (2012). Effects of biochar amendment on soil quality, crop yield and greenhouse gas emission in a Chinese rice paddy: A field study of 2 consecutive rice growing cycles. *Field Crops Research*, 127, 153–160.

Zhang, A., Cui, L., Pan, G., Li, L., Hussain, Q., Zhang, X., … Crowley, D. (2010). Effect of biochar amendment on yield and methane and nitrous oxide emissions from a rice paddy from Tai Lake plain, China. *Agriculture, Ecosystems & Environment*, 139(4), 469–475.

Zhang, Y., Ding, J., Wang, H., Su, L., & Zhao, C. (2020). Biochar addition alleviate the negative effects of drought and salinity stress on soybean productivity and water use efficiency. *BMC Plant Biology*, 20, 1–11.

CHAPTER 8

Biochar as a soil amendment: effects on microbial communities and soil health

Tanmaya K. Bhoi[1], Ipsita Samal[2], Anuj Saraswat[3], H.C. Hombegowda[4], Saubhagya K. Samal[4], Amit K. Dash[5], Sonal Sharma[6], Pramod Lawate[4], Vipula Vyas[1] and Md. Basit Raza[4,7]

[1]Forest Protection Division, ICFRE—Arid Forest Research Institute (ICFRE—AFRI), Jodhpur, Rajasthan, India [2]ICAR-National Research Centre on Litchi, Muzaffarpur, Bihar, India [3]Department of Soil Science, College of Agriculture, Govind Ballabh Pant University of Agriculture & Technology, Pantnagar, Uttarakhand, India [4]ICAR—Indian Institute of Soil and Water Conservation, Dehradun, Uttarakhand, India [5]ICAR—Indian Institute of Seed Science, Mau, Uttar Pradesh, India [6]Department of Soil Science, Maharana Pratap University of Agriculture and Technology, Udaipur, Rajasthan, India [7]ICAR-Directorate of Floricultural Research, Pune, Maharashtra, India

8.1 Introduction

The production of enough food to feed a growing world population while maintaining the long-term health of soils and ecosystems is becoming an increasingly difficult task for agricultural systems around the world (United Nations, 2019). As a result, there is now more interest in sustainable farming methods like using biochar. Biochar is a refractory carbon molecule that is produced by the thermal degradation of biomass at 300°C–800°C in the absence or with limited oxygen (Chaturvedi, Singh, Dhyani, Govindaraju, & Mandal, 2021). It is a stable, high-carbon substance with the capacity to improve the condition of the soil and agricultural output (Malik et al., 2022). Biochar can improve soil's ability to hold water, retain nutrients, and support microbial activity which can improve plant development and lower greenhouse gas emissions (Singh et al., 2023). By acting as a slow-release source of nutrients and encouraging the development of PGPRs (plant growth-promoting rhizobacteria), biochar can improve soil fertility. PGPRs are beneficial microorganisms that colonize plant roots and drive plant development in a number of ways, including by creating plant growth hormones, improving nutrient uptake, and guarding plants against pests and diseases (Behera et al., 2021). Despite the potential advantages of using biochar to enhance PGPR activities and improve soil health, further research is still required to fully comprehend the underlying mechanism and tailor its use for various agricultural systems (de Medeiros et al., 2021; Majhi et al., 2022). The chapter attempts to give a summary of the most recent research on how applying biochar affects PGPR activity and soil health. The most recent research will be reviewed, and it will be discussed how biochar could be able to boost PGPR activities and enhance soil health. It will also draw attention to the challenges and opportunities presented by employing biochar in agricultural systems.

8.2 Biochar, its preparation methods, and properties

Biochar, a carbon-rich material created by regulated procedures such as pyrolysis and gasification, has received a lot of interest for its potential to solve environmental and agricultural problems (Širić et al., 2022). These production processes are crucial in molding biochar's properties, and other parameters such as feedstock, temperature, and residence duration contribute to its various features (Şirin, Ertürk, & Kazankaya, 2022). The physical, chemical, and biological properties of biochar all influence its applications and impacts in distinct ecosystems (Singh, Singh, Singh, Rai, & Kumar, 2016). The primary methods employed to produce biochar include pyrolysis and gasification. Pyrolysis refers to the thermal degradation of biomass in a controlled environment with limited oxygen, leading to the formation of char, volatile gases, and bio-oil. The procedure occurs at three distinct temperature levels, namely low, moderate, and high pyrolysis. The production of biochar with elevated levels of volatile matter concentration and surface area is achieved by the process of low-temperature pyrolysis, rendering it suitable for soil amendment purposes (Sani, Hasan, Uddain, & Subramaniam, 2020). The process of high-temperature pyrolysis yields biochar that exhibits enhanced carbon content and improved thermal stability, hence offering advantages for several uses, including carbon sequestration. In contrast, gasification entails the partial oxidation of biomass under elevated temperatures. The procedure yields syngas, a composite of hydrogen, carbon monoxide, methane, and char. The choice of gasifier type and operating conditions significantly influence both the content of the syngas and the properties of the resulting biochar. According to Sadegh Kasmaei et al. (2019), gasification offers a certain degree of flexibility as it allows for the regulation of the syngas composition and the potential for simultaneous generation of heat and power.

A variety of reasons contribute to biochar's diverse qualities. The chemical composition and physical structure of biochar are greatly influenced by the feedstock used (Riaz et al., 2017). Biochars with variable carbon-to-nitrogen ratios, ash content, and elemental composition are produced from various feedstocks such as wood chips, crop residues, and organic waste (Murtaza, Ditta, Ullah, Usman, & Ahmed, 2021). Temperature during manufacture is crucial;

higher temperatures promote carbonization and produce biochar with higher carbon content, more porosity, and greater stability. Another important issue is residence time, or how long biomass sits in the reactor. Longer residence times enable deeper pyrolysis, resulting in higher carbon content and improved biochar stability. Shorter residence durations, which are common in gasification, may result in biochars with diverse characteristics and possible applications (Mona et al., 2021).

Biochar's physical qualities include porosity, surface area, and particle size dispersion. The porous nature of biochar provides a home for microorganisms and improves soil water-holding capacity (WHC) (Negi, Bharat, & Kumar, 2021). Larger surface areas promote soil fertility and plant growth by facilitating nutrient adsorption and cation exchange. Furthermore, the size distribution of biochar particles affects its assimilation into soils and their efficiency as a soil conditioner (Nafees, Ullah, & Ahmed, 2022). Biochar's reactivity and interactions within ecosystems are chemically determined by its composition and surface functional groups. The variable carbon, hydrogen, oxygen, and nitrogen content affects its stability and potential to store carbon in soil over long periods of time (Ijaz et al., 2019). Nutrient retention, metal immobilization, and contaminant adsorption are all influenced by surface functional groups such as hydroxyl and carboxyl groups. The chemical characteristics of biochar have a strong influence on soil pH, cation exchange capacity (CEC), and nutrient availability (Hendrix, Crossley, Blair, & Coleman, 2020). The biological features of biochar are inextricably linked to soil ecology and plant—microbe interactions. It supports beneficial bacteria that regulate nutrient cycling and soil health (Kasak et al., 2018). Soils modified with biochar have enhanced microbial diversity, enzymatic activity, and plant growth. Biochar's involvement in encouraging microbial colonization, improving nutrient retention, and creating symbiotic connections between plants and soil microbes can be linked to these improvements (Kari et al., 2021).

8.3 Biochar qualities associated with microbial diversity

In addition to having a high level of hydrophobicity, alkalinity, and considerable concentrations of nutrients like nitrogen, phosphorus, and potassium, biochar is a substance that is rich in carbon. Additionally, it has a high level of surface porosity that enables it to interact with external organic and inorganic compounds, good water and nutrient retention capabilities, low thermal conductivity, high energy content, and low thermal conductivity (Hossain et al., 2020; Xiang et al., 2022). It has a vast surface area and is made of a very porous substance. It typically has a low ash level and a high carbon content (50%—90%). There are two categories of biochar porosity: macro-porosity and micro-porosity. The porosity of biochar is significant because it affects how well it can hold onto nutrients and water (Singh, Northup, Rice, & Prasad, 2022). The feedstock used to manufacture biochar affects its chemical properties. Biochar typically has a high CEC, ranging from 50 to 400 meq/100 g, and a pH often between 4.0 and 9.0. The ability of biochar to store cations like potassium, calcium, and magnesium, which can be released gradually over time to assist plant growth, is made possible by its CEC (Das, Sahoo, Raza, Barman, & Das, 2022). Biochar also contains functional groups, such as carboxyl, hydroxyl, and phenolic groups, which can influence its interactions with soil minerals and microbes (Panahi et al., 2020).

8.3.1 Woody biochar

Woody biomass is used to make woody biochar, including wood chips, sawdust, and bark. According to Rodriguez-Franco and Page-Dumroese (2021), this generally has a low ash level and a high carbon content (70%–90%). Woody biochar has a huge surface area and a high microporosity, making it useful for adsorbing contaminants and preserving nutrients. Before it may be utilized as a soil amendment, though, its low/high pH may require pH correction.

8.3.2 Agricultural biochar

Agricultural leftovers including rice husks, wheat straw, and maize stover are used to make agricultural biochar. It often contains a high ash percentage and a lower carbon content than woody biochar, ranging from 40% to 60%. Compared to woody biochar, agricultural biochar has a smaller surface area and lower microporosity, but it may also have a higher mesoporosity, which can improve its ability to retain water. According to Allohverdi, Mohanty, Roy, and Misra (2021), agricultural biochar can be useful for enhancing soil fertility and minimizing soil erosion.

8.3.3 Manure biochar

Animal dung, including those from cows, chickens, and pigs, is used to make manure biochar. It typically contains more nutrients and less carbon than woody biochar, ranging between 30% and 50%. Manure biochar is efficient at retaining nutrients and enhancing soil fertility due to its high porosity and strong CEC (Wani et al., 2021). However, before it can be utilized as soil amendments, its high pH may need to be adjusted and its high nutrient concentration may render it susceptible to nutrient leakage.

8.3.4 Municipal solid waste biochar

Municipal solid waste (MSW), including food waste and paper ash, is used to make MSW biochar. It typically contains a higher ash percentage and a lower carbon content than woody biochar, ranging from 20% to 40%. MSW biochar's function as soil amendments may be hampered by the presence of impurities such as heavy metals and organic pollutants. By storing carbon in soil, it can also be successful at lowering greenhouse gas emissions (Li et al., 2022).

8.4 Biochar—microbe interactions

Biochar—microbe interactions are a fascinating and intricate phenomenon that have a greater impact on ecosystem health (Gorovtsov et al., 2020). When biochar is added to the soil, a special microenvironment is created that encourages a variety of interactions with soil microbes (Asiloglu, 2022). Increased microbial diversity and abundance result from the substrate and refuge provided by the porous, stable structure of biochar. Microorganisms that live inside the pores of biochar participate in crucial activities like nutrient cycling, organic matter decomposition, and the release of compounds that aid in plant growth (Pathy, Ray, & Paramasivan, 2020). According to certain research, biochar can promote the development of advantageous microbial communities like mycorrhizal fungi, which work in symbiosis with plant roots to improve nutrient uptake (Siedt et al., 2021). Additionally, biochar can alter soil pH, making a particular microbial community's environment more conducive and affecting the activity and function of that community. For projects including carbon sequestration, soil remediation, and sustainable agriculture, it can be helpful to understand the complex interactions between biochar and microbes (Kimura, Uchida, & Madegwa, 2022).

However, not all the effects of biochar on soil bacteria are beneficial. The kind of soil, the feedstock used, the pyrolysis conditions, and the impact of biochar on microbial populations are just a few of the variables that can affect the outcome (Zhong et al., 2022). The availability of nutrients and total soil fertility may be impacted by some biochars' potential to hinder the growth and activity of specific microbial species. This suppression may be caused by the release of specific chemical compounds from biochar, modifications to the soil's structure, or adjustments to the dynamics of the nutrients, which have varied effects on different microbial communities (Wang, Zhang, Gan, Liu, & Mei, 2021). Understanding the complex mechanisms underlying these interactions is key to maximizing the benefits of biochar as a soil amendment. The management of agricultural and natural ecosystems can be done sustainably by modifying the properties of biochar to promote positive microbial activity while limiting any negative impacts. This can be done by conducting further study in this area.

8.5 Significance of soil microbial diversity on soil–plant–atmospheric continuum

According to Lohan et al. (2018), the soil–plant–atmospheric continuum functions in large part due to the diversity of its microbial inhabitants. The interrelated system that consists of the soil, the plants growing in it, and the atmosphere that surrounds it is referred to as the soil–plant–atmospheric continuum (Amitrano, Arena, Rouphael, De Pascale, & De Micco, 2019). These elements work together in an intricate network of chemical physical and biological processes that affect the amount of food water and energy available. The diversity of soil-dwelling bacteria, fungi, archaea, and viruses are referred to as soil microbiological diversity. These bacteria are essential for the cycling of nutrients in the soil. For decomposition, and for promoting plant growth. According to Hendrix et al. (2020), they are involved in a variety of ecosystem functions, including soil formation, nutrient cycling, carbon sequestration, and climate regulation. Numerous elements, including soil type, climate, land use, and management techniques, have an impact on soil microbial diversity. For instance, microbial diversity can be impacted by changes in soil pH, organic matter content, and nutrient availability caused by agricultural activities such as tillage, fertilizer application, and pesticide usage (Barriuso et al., 2008; Singh, Luthra et al., 2023). Through several processes, soil microbial diversity affects plant development and health. In order to release nutrients necessary for plant growth, organic matter must be broken down by microorganisms. By converting complicated chemicals into more digestible forms that plant roots can easily take up, they also aid plants in absorbing nutrients (Timmis & Ramos, 2021). In addition, some microbes create hormones and enzymes that aid in the growth and development of plants. In the process of storing carbon dioxide from the atmosphere in soil organic matter (soil carbon sequestration), microorganisms are also a key component. Due to its role in lowering atmospheric carbon dioxide concentrations, soil carbon sequestration is a crucial part of the mitigation of climate change (Sarfraz et al., 2019). The stability of soil carbon and the rate at which soil organic matter decomposes are both influenced by soil microbial diversity (Dash, Dwivedi, Dey, Meena, & Chakraborty, 2023). Soil microbial diversity controls greenhouse gas emissions, which have an impact on atmospheric processes. The diversity of soil microbes is essential to the operation of the continuity between soil, plants, and the atmosphere. It affects the cycling of nutrients, plant development, sequestration of carbon, and emissions of greenhouse gases, all of which are crucial elements of ecosystem services and climate change mitigation (Raza, Hombegowda, Kumar, & Kumar, 2023). To create sustainable land management methods that maintain soil health and foster ecosystem resilience, it is essential to comprehend the factors that determine soil microbial diversity and the mechanisms through which it affects ecosystem processes.

8.6 The potential advantages of biochar and its application in enhancing soil health and promoting plant growth

8.6.1 Soil fertility and plant growth

The biochar's notable CEC facilitates its ability to effectively absorb and retain essential plant nutrients, including nitrogen, phosphorus, and potassium, for prolonged durations. The progressive release of nutrients by biochar has the potential to enhance soil fertility and hence decrease reliance on synthetic fertilizers. According to Ding et al. (2016), the use of biochar has the potential to enhance soil structure, increase water retention capabilities, improve aeration, and thus facilitate plant development while mitigating erosion. The application of biochar can enhance plant growth by creating a stable environment for beneficial microorganisms, improving soil fertility, and reducing the incidence of soil-borne illnesses. The utilization of biochar has been found to have a positive impact on nutrient absorption through the promotion of mycorrhizal fungi growth. These fungi establish a mutually beneficial interaction with plant roots, leading to an increase in nutrient uptake (Kasak et al., 2018). Furthermore, the utilization of biochar has the potential to mitigate plant stress through its ability to improve soil retention, a crucial factor in regions characterized by aridity.

8.6.2 Reduces greenhouse gas emissions

By storing carbon in the soil, reducing the discharge of nitrous oxide, and improving soil structure, biochar application can reduce greenhouse emissions. By preserving carbon in the soil for an extended period of time, biochar can help mitigate climate change by reducing the amount of carbon dioxide released into the atmosphere (Sri Shalini, Palanivelu, Ramachandran, & Raghavan, 2021).

8.6.3 Reduces contaminant

Biochar reduces the bioavailability and mobility of soil pollutants such as heavy metals and pesticides. It has the ability to absorb these contaminants and reduce their bioavailability, thereby protecting plants and soil microorganisms from their deleterious effects (Behera et al., 2022; Samal et al., 2023).

8.7 Mechanism of plant growth stimulation by plant growth promoting rhizobacteria

Soil-dwelling bacteria known as "plant growth promoting rhizobacteria" (PGPR) are rhizobacteria that promote plant growth in a variety of situations. PGPR has gained popularity in recent years as a natural and sustainable strategy for increasing plant growth and yield (Basu et al., 2021; Singh, Bhoi, & Vyas, 2023). It is critical to first understand the principles of PGPR and their mode of action before digging into the mechanisms of PGPR-mediated plant growth promotion (Etesami & Adl, 2020). Associative and symbiotic PGPRs are frequently distinguished. Associative PGPR interacts loosely with plant roots, whereas symbiotic PGPR interacts mutualistically with plants, resulting in the production of nodules on legume roots. Plant development is facilitated by PGPRs through both direct and indirect mechanisms. Indirect strategies were reported to include the synthesis of siderophores and the solubilization of nutrients (Negi et al., 2021). PGPRs are known to synthesize essential plant growth regulators, including cytokinins, auxins, and gibberellins, which play crucial roles in the growth and development of plants. Auxin reported to promote cellular division and elongation, while cytokinins have been found to induce cellular division and differentiation. In contrast, gibberellins are known to promote elongation of stems, enlargement of leaves, and initiation of floral development. Plant growth regulators are synthesized by PGPRs, leading to enhancements in plant growth and development, including ABA and ethylene, which play a significant role in enhancing stress tolerance (Ha-Tran, Nguyen, Hung, Huang, & Huang, 2021; Renoud et al., 2022). The closing of stomata in response to water stress is regulated by ABA, while the development of plants in response to biotic and abiotic stress is controlled by ethylene. The production of siderophores by PGPR enhances the accessibility of iron to plants, leading to a subsequent augmentation in plant growth and development. Siderophores are a class of molecules that possess the ability to chelate iron, hence facilitating the process of iron absorption in plants. The presence of iron is crucial for the development and growth of plants, yet its accessibility in the soil is often restricted (Khatoon et al., 2020). Antibiotics, lytic enzymes, and antibacterial chemicals can all be synthesized by PGPR, which successfully halts the onset and spread of plant diseases. Plant glycine receptor (PGPR) has also been shown to trigger systemic resistance in plants, leading to increased plant disease resistance (Nozari, Ortolan, Astarita, & Santarém, 2021). Because it can reduce the use of chemical pesticides, the PGPR-mediated plant growth promotion mechanism is especially useful for sustainable agriculture (Bhat et al., 2020; Przybyłko, Kowalczyk, & Wrona, 2021). Although phosphorus is an essential macronutrient for plant growth and development, it is insoluble in soil. PGPR produce phosphatases, which are enzyme proteins.

Hydrolysis of insoluble phosphorus forms is aided by these enzymes, making them usable for plants. Potassium solubilizing enzymes are produced by PGPR, which aids in the solubilization of otherwise insoluble potassium compounds and similarly promotes plant growth and development. Mutualistic associations between fungi and plant roots are known as mycorrhizae, and they can be bolstered by the presence of PGPR (Şirin et al., 2022). Mycorrhizae are beneficial organisms because they increase a plant's ability to absorb nutrients, especially phosphorus (Fig. 8.1).

8.8 Interaction of biochar and plant growth promoting rhizobacterias

The potential of PGPR and biochar to promote soil health and stimulate plant development has led to a rise in their use in sustainable agriculture. There may be synergistic effects that increase plant growth and nutrient uptake as a result of the current interaction (Ajeng et al., 2020). The method through which PGPR interact with biochar can be studied from a variety of angles. Singh and Jha (2016) found that biochar may house PGPR and supply it with nutrients, hence fostering PGPR's growth and activity. Tomczyk, Sokołowska, and Boguta (2020) state that biochar's higher surface area makes it simpler for PGPRs and other beneficial soil

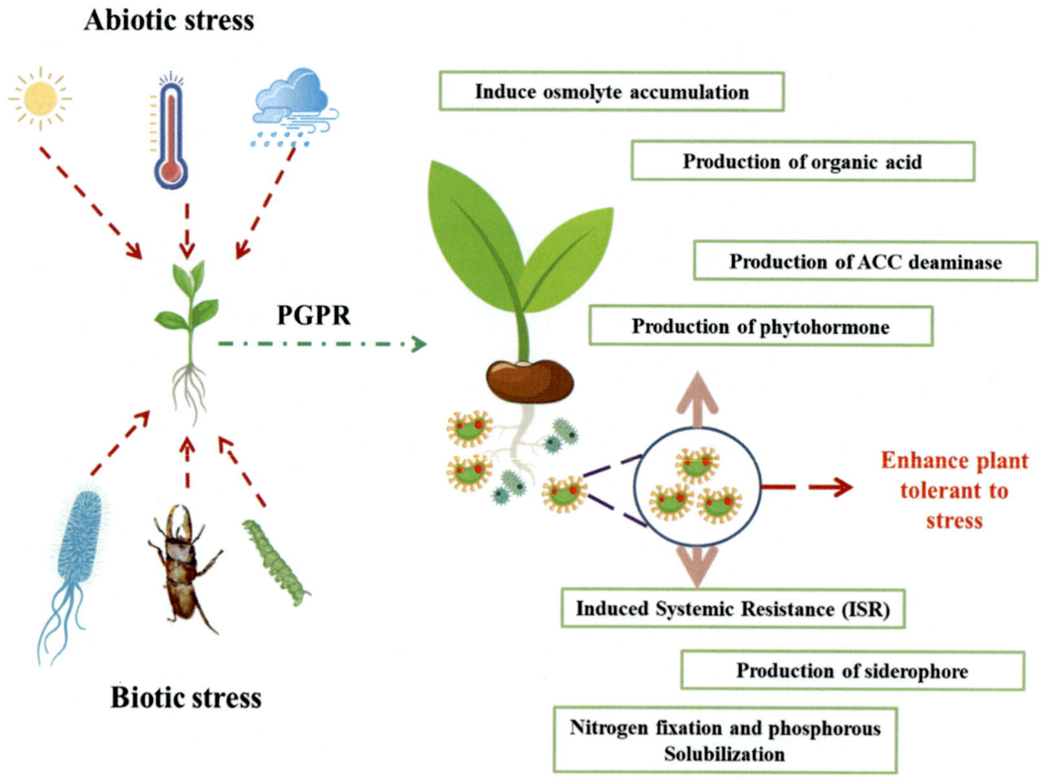

FIGURE 8.1 Mechanisms of plant growth stimulation by plant growth promoting rhizobacteria.

microbes to colonize and multiply in the soil. Biochar's ability to absorb nutrients and organic compounds and gradually release them aids PGPR colonization. According to the study by Riaz et al. (2017), biochar's activation process and nutrient retention ability can be enhanced by utilizing PGPR. According to Shemawar et al. (2021), PGPR can generate organic acids and enzymes that quickly break down biochar's complicated chemical components. Nutrients that plants can use are therefore liberated and made more accessible to the plant. Increased soil CEC thanks to PGPR can enhance biochar's nutrient retention and decrease leaching. When PGPR and biochar are used together, they boost plant growth and health. The production of plant growth hormones by PGPRs has been shown to improve nutrient uptake (El-Naggar, Shaheen, Ok, & Rinklebe, 2018). To further aid plant growth and water retention, biochar can enhance soil structure and fertility. Recent studies have shown that the positive effects of PGPR and biochar on plant growth and yield are significantly amplified when the two treatments are used together. For example, in one study on tomato plants, PGPR and biochar combined increased plant height, biomass, and fruit yield relative to the control group (Yuan, Wenqing, Binghai, & Hanhao, 2021). In comparison to the control, PGPR and biochar application increased plant height, leaf area, and grain yield in maize (Ullah et al., 2020). Kari et al. (2021) looked at how PGPR and biochar affected maize growth and nutrient uptake. The study discovered that combining PGPR and biochar boosted maize growth, yield, and nutrient uptake significantly more than each component alone. Adding biochar to PGPR also boosted nutrient availability and microbial activity in the soil, which assisted plant growth and health.

8.8.1 Effects of plant growth promoting rhizobacteria and biochar co-application on soil health

Biochar and PGPR have been proven to improve soil health via a variety of methods, including nutrient cycling, increased soil microbial activity, and improved soil structure. Co-application of biochar and PGPR boosted soil microbial activity, soil organic carbon, and available phosphorus in contrast to the control. Another study on damaged soil discovered that co-applying biochar and PGPR boosted soil fertility, soil structure, and microbial diversity, leading to an improvement in yield and plant development (Ren et al., 2021). Furthermore, applying biochar and PGPR together has been shown to increase the diversity and richness of beneficial soil microbes such as mycorrhizal fungi and nitrogen-fixing bacteria (Bamdad, Papari, Lazarovits, & Berruti, 2022).

8.8.2 Impact on soil pH and electrical conductivity

Soil electrical conductivity (EC) and pH are important soil parameters that influence nutrient availability, plant growth, and soil microbial activity (Khadem, Raiesi, Besharati, & Khalaj, 2021). In diverse cropping systems, it has been demonstrated that the addition of biochar along with PGPR can raise the pH of the soil and enhance its EC (Lalay, Ullah, & Ahmed, 2022). Recent studies have investigated how the addition of PGPR and biochar to soil affects both the pH and EC of the soil. Wang, Wang, and Song (2021) conducted an experiment in a greenhouse with cucumbers and found that the combined application of biochar and PGPR resulted in improved soil pH as well as reduced soil EC. The combination of PGPR and biochar can lead to increased plant growth as well as an increase in the accumulation of heavy metals in soil. It was determined that the decrease in soil EC and the rise in soil pH were responsible for this effect. This, in turn, leads to an improvement in the soil's health as well as an increase in plant growth. Fertilizer application and irrigation methods can change the soil's pH and EC; however, this can be efficiently handled by the utilization of biochar and PGPR. It's possible that the buffering effect reported in the study carried out by Niewiadomska et al. (2023) helps to keep soil pH and EC levels at values that are beneficial for microbial activity and plant growth. It is essential to recognize that the effect of biochar and PGPR on the pH and EC of soil is dependent on a number of parameters. These elements include the specific qualities of the soil, biochar, and PGPR that are used, as well as the dose and length of time that the application is carried out. It is necessary to undertake additional research in order to acquire a comprehensive understanding of the effects of this simultaneous application on the EC of the soil and the various agricultural systems. In a variety of agricultural settings, the incorporation of biochar with PGPR has been shown in a number of studies to have a beneficial effect on the soil's pH and EC. Maintaining ideal pH and EC levels in the soil, as well as the ability to absorb changes in these soil properties, can be beneficial for both the health of the soil and the growth of plants. This can be accomplished through a number of different methods. It is required to conduct additional research in order to determine the treatment rates and timing that are most effective for the various types of soil and crops. In general, the combination of biochar and PGPR were reported to show encouraging potential for the improvement of crop productivity over the long run as well as the quality of the soil (Fig. 8.2).

8.8.3 Impact on the availability of nutrients in soil

The amount of nutrients that are present in the soil has a crucial impact on the production and growth of plants. Danish, Zafar-Ul-Hye, Mohsin, and Hussain (2020) conducted a study on sandy loam soil and found that the combined application of PGPR and biochar had a significant impact on the distribution of phosphorous, nitrogen, and potassium. Further, changes in the distribution of nutrients lead to improved growth and yield of maize significantly. In addition, Murtaza et al. (2021) demonstrated that the co-integration of PGPR and biochar increased the availability of nitrogen and phosphorous in soil contaminated with heavy metals, which resulted in improved sorghum growth and higher nutrient uptake. The application of biochar and PGPR together has the potential to increase the availability of nutrients in the soil, which in turn will boost plant growth and productivity. The co-application of PGPR and biochar can also contribute to an increase in the organic matter content of the soil, which further results in higher nutrient availability in the soil. Biochar, on the other hand, has the potential to serve as a potential source of stable organic carbon (Vahedi, Rasouli-Sadaghiani, Barin, & Vetukuri, 2022). This is in contrast to the potential of PGPR to stimulate soil microbial activity and the decomposition of organic matter. Additionally, the combination of biochar and PGPR can improve nutrient availability for plant uptake by decreasing nutrient leaching and raising nutrient retention in the soil (Zafar-Ul-Hye et al., 2020) (Fig. 8.2).

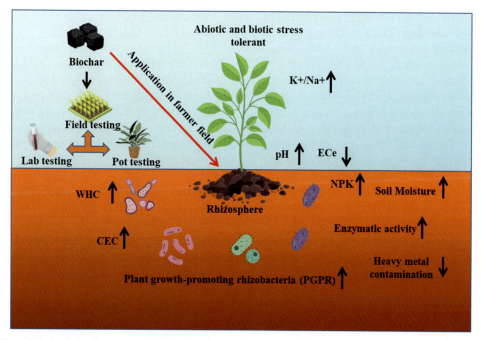

FIGURE 8.2 Impact of biochar and plant growth promoting rhizobacteria on soil quality and plant development.

Zinc (Zn), Fe, and Cu are three soil micronutrients that are crucial to the establishment and growth of plants (Rahman, Sofi, Javeed, Malik, & Nisar, 2020). Biochar and PGPR have been proposed as soil amendments that can boost the availability of micronutrients to plants. The availability of these micronutrients is often limited in soils, resulting in nutrient shortages and reduced agricultural yields. Adding PGPR and biochar to residual mushroom substrate dramatically enhanced the availability of Zn and Fe in cauliflower (Şirin et al., 2022). Additionally, the application of biochar and PGPR can boost soil microbial activity, which can aid in the mobilization of firmly bound micronutrients in the soil. Insoluble, forms of micronutrients like Fe and Zn, for instance, can be solubilized by PGPR, making them more readily available for plant uptake (Kaur et al., 2020). Additionally, according to Xiang et al. (2022). Biochar can operate as a habitat for soil bacteria, encouraging their development and activity (Fig. 8.2).

8.8.4 Effect on water holding capacity of soil

For crop growth and production, water is an essential resource, and its availability in soil is influenced by a variety of parameters, including soil texture, organic matter concentration and soil structure (Siedt et al., 2021). Two new agricultural techniques, PGPR and biochar, have been shown to increase soil WHC and water usage effectiveness (Ullah et al., 2021). Biochar, a porous carbon material produced by the pyrolysis of organic materials, can improve the physical properties of soil, such as its ability to store water and its porosity and rate of water infiltration (Ahmad Bhat et al., 2022). By boosting nutrient uptake and increasing stress tolerance, PGPRs are beneficial microorganisms that reside in plant roots and support plant development and production (Arora et al., 2020). According to Ahmad et al. (2020), adding wheat straw charcoal and *Bacillus* sp. considerably improved the soil's ability to hold water in a pot experiment. Co-applying biochar with PGPR can increase the soil's WHC, bulk density, and porosity, which can increase plant growth and soil water availability (Nafees et al., 2022). By forming a porous structure that improves soil water retention and decreases evaporation, biochar is known to increase soil porosity and water retaining capacity. Contrarily, PGPR can improve plant root development and nutrient uptake, which may increase water usage effectiveness and drought resistance. By encouraging the growth and activity of advantageous soil microorganisms such mycorrhizal fungi and nitrogen-fixing bacteria, the combined application of biochar and PGPR might boost these favorable effects. These microbes can aid in enhancing the nutrient cycle and soil structure, which can further increase the soil's capacity to hold water (Fig. 8.2) (Table 8.1).

8.8.5 Effect on soil microbial communities

By controlling nutrient cycling, the breakdown of organic matter, and interactions between plants and microbes, soil microbial communities are essential to the health of the soil and the operation of ecosystems (Table 8.1). Variations in microbial populations brought on by biochar addition may also impact their rate of C metabolism and functional diversity as measures of soil microbial function response. Biochar research increasingly uses the biological approach to assess microbial diversity and abundance as well as C utilization rates (Xu et al., 2018). Utilizing biochar has been recommended as a technique to increase the utilization of various polymeric group C substrates as a result of changes in the microbial population (Tian et al., 2016). The application of biochar has also been demonstrated to support functional diversity, evenness, and variation (Xu et al., 2018). Biochar may act as a structural refugium for microbial growth and a store of malleable or easily recoverable C, leading to an increase in microbial biomass C (Pawar et al., 2017). The impact of biochar on microbial biomass N (MBN) varies. Because the microbial community composition responds to changes in soil physicochemical parameters, the MBN can decrease with the addition of charcoal in some samples (Alburquerque et al., 2014). According to other studies, biochar has no effect on MBN; this could be due to a number of factors, such as the amount of nitrate in the soil or the existence of plants that compete with one another for nitrate. By offering a consistent source of carbon and nutrients for soil microorganisms and encouraging the growth and activity of beneficial microbes, the co-integration of PGPR and biochar can improve soil microbial diversity, abundance, and activity (Arif, Batool, & Schenk, 2020). By offering a consistent source of carbon that can promote microbial activity and growth, biochar can serve as a carbon sink (Osman et al., 2022). By secreting numerous plant growth-promoting substances such phytohormones, enzymes, and antibiotics that can boost the growth and activity of advantageous soil microbes, PGPR can increase microbial diversity and activity. By encouraging the growth

TABLE 8.1 Effect of co-application of plant growth promoting rhizobacteria and biochar on plant growth and soil quality.

PGPR	Biochar	Crop	Interactive Effect	Country	References
Bacillus subtilis and *Paenibacillus azotofixans*	Agricultural biochar	Einkorn wheat variety (*Triticum monococcum* L.)	Combination of biochar and PGPR applications has shown a positive effect in terms of soil properties, plant growth	Turkey	Çığ, Sönmez, Nadeem and Sabagh (2021)
Pseudomonas fluorescens	Azolla biochar	Rosemary (*Rosmarinus Officinalis*)	Significant enhancement in rosemary growth occurred due to the improved soil quality as a result of organic fertilizers application	Iran	Sadegh Kasmaei et al. (2019)
Bacillus deuterium and *Bacillus megatherian*	Wheat straw biochar	Eucalyptus DH32−29 (*Eucalyptus urophylla* × *E. grandis*)	co-application of PGPR and biochar increase soil sucrase activity, electrical conductivity (EC), and total potassium (K) concentration,	China	Ren et al. (2020)
Pseudomonas spp and *Bacillus* spp	Cotton stem biochar	Bread wheat (*Triticum aestivum* L.)	PGPR + biochar with half a dose of the fertilizer proved to be the best and improve the yield and profit of wheat crop with reduced synthetic fertilization.	India	Ijaz et al. (2019)
Pseudomonas spp., *Azotobacter chroococcum* and *Azospirillum brasilense*	Rice husk biochar	Rice (*Oryza sativa*)	Significantly increased rice yield and nutrients uptake	India	Singh et al. (2016)
Paenibacillus polymyxa and *Bacillus amyloliquefaciens*	Millet straw biochar	Tomato (*Lycopersicon esculentum* Mill.).	Biochar combined with PGPR increased the relative abundance of *Nitrospira* and *Bradyrhizobium* in the soil	China	Wang, Li, Du, and Li, (2021)
Arthrobacter crystallopoietes and *Azospirillum brasilense*	Organic grain husk and paper fiber sludge biochar	Maize (*Zea mays*)	High doses of biochar treatment influenced the abundance of bacterial group	Hungary	Kari et al. (2021)
Pseudomonas fluorescent	Pruning waste biochar	Bread wheat (*Triticum aestivum* L.)	Fe and Zn uptake rates increases	Iran	Vahedi et al. (2022)

(*Continued*)

TABLE 8.1 (Continued)

PGPR	Biochar	Crop	Interactive Effect	Country	References
Pseudomonas sp.	Poplar sawdust biochar	Maize (*Zea mays*)	Higher proline and peroxidase activity was observed in combined treatment with biochar and PGPR	Pakistan	Fazal and Bano (2016)
Pseudomonas koreensis MG209738, *Azotobacter chroococcum* SARS 10, *Azospirillum lipoferum* SP2, *Bacillus coagulans* NCAIM B 1086 and *Enterobacter cloacae* KX034162	Rice husk and corn stalk biochar	Rice (*Oryza sativa*)	Synergistic use of biochar and PGPR could be an effective strategy for improving plant growth and productivity	Egypt	Hafez et al. (2019)
Bacillus subtilis (MTCC 441) and *Pseudomonas fluorescence* (MTCC 103 T)	Spent mushroom substrate (SMS) biochar	Cauliflower (*Brassica oleracea* var. *botrytis*)	Positive correlation with growth, yield, and biochemical response of cauliflower,	India	Širić et al. (2022)
PG1 (*Pseudomonas* sp.) and PG2 (*Staphylococcus haemolyticus*)	*Morus alba* L. wood biochar	*Brassica napus* L.	Increase water use effeciency	Pakistan	Ullah, Nafees, and Ahmed (2023)

and activity of beneficial soil microbes like mycorrhizal fungi and bacteria that may fix nitrogen, which in turn promotes nutrient cycling and plant–microbe interactions, the application of PGPR and biochar combined can enhance these beneficial effects. By controlling nutrient availability, organic matter breakdown, plant development, and production (Lakshmi, Okafor, & Visconti, 2020) (Fig. 8.2), these microbes can support ecosystem health and function.

8.8.6 Effect on crops growth and productivity

It has been demonstrated that PGPR and biochar applied together boost crop yield and quality. In comparison to the control treatment, the co-application of PGPR and biochar enhanced the grain yield by 36%, the straw yield by 50%, and the biological yield by 40% in research on maize (Zafar-Ul-Hye et al., 2020). Similarly, when Trichoderma and charcoal were applied together, tomato fruit yield and quality rose (11.33%) in comparison to the control treatment. Higher tomato yields, antioxidant levels, and mineral content were produced as a result of the interaction between Trichoderma and biochar, which also increased soil fertility, nutrient uptake, and the expansion of rhizosphere fungal and bacterial populations (Sani et al., 2020). The results are mostly likely attributable to the enhanced soil fertility, nutrient availability, and water retention brought about by the co-application of PGPR and biochar.

8.9 Factors influencing the effectiveness of biochar

The efficiency of biochar as a soil amendment is influenced by a complex interplay of elements such as soil health, plant development, and microbial communities (Ahmad Bhat et al., 2022). These variables include a variety of soil, biochar, and environmental aspects that impact the consequences of biochar application (Ahmad et al., 2020). Understanding these aspects is critical for maximizing biochar's advantages and customizing its application to specific agricultural and environmental environments.

8.9.1 Soil type, texture, and mineralogy

The intrinsic qualities of the soil in which biochar is placed are critical to its success (Ajeng et al., 2020). The ability of biochar to adsorb and retain water, nutrients, and organic matter is influenced by soil type, which is defined by its composition of sand, silt, and clay. Sandy soils, which have a low WHC, can benefit from biochar's water retention qualities, which help plants cope with drought stress (Bamdad et al., 2022). Biochar's interactions with soil colloids, on the other hand, can improve nutrient retention and availability in clay-rich soils with increased CEC. Furthermore, soil mineralogy influences the possibility of chemical interactions between biochar and soil constituents, which influences nutrient release rates and pH adjustments (Asiloglu, 2022).

8.9.2 Biochar dosage and application methods

The amount of biochar used and the technique of application have a significant impact on its effectiveness (Basu et al., 2021). The appropriate dosage varies depending on soil type, crop variety, and intended results. Overuse of biochar can result in nutrient immobilization or changes in soil structure, whilst underuse may not produce substantial advantages (Bhat et al., 2020). Surface broadcasting, assimilation, or localized placement all have an impact on biochar-soil contact and subsequent interactions. Banding biochar near plant roots, for example, can maximize its influence on root zone conditions and nutrient uptake.

8.9.3 Climate and environmental conditions

Environmental elements such as climate, temperature, and precipitation all have an impact on biochar-soil interactions. The porosity and surface area of biochar can affect its WHC and resistance to breakdown (Çığ et al., 2021). Biochar can improve soil drainage and reduce waterlogging in wetter climates, reducing root asphyxiation. Biochar's impact on water retention, on the other hand, can be especially advantageous in arid environments, helping to preserve water and promote plant growth during drought periods (Das et al., 2022). Biochar's durability and nutrient release rates are affected by temperature, with higher temperatures generally resulting in more biochar mineralization and nutrient availability.

8.9.4 Long-term effects and aging

The persistence of biochar in soil, also known as its aging process, can have a substantial impact on its effectiveness over time. Freshly applied biochar may have a bigger impact on soil characteristics and microbial populations at first (de Medeiros et al., 2021). However, the surface chemistry of biochar can change over time due to weathering and microbial interactions, resulting in changes in its nutrient retention capability and interactions with soil organisms. Long-term research is essential for understanding how biochar's efficacy evolves and how its advantages remain or fade over successive growing seasons (Etesami & Adl, 2020).

8.9.5 Interactions with microbial communities

The interactions between biochar, soil, and microbes are critical determinants of its efficiency. Biochar can provide a home for a wide range of microbial populations, altering their makeup and activity (Gorovtsov et al., 2020). Specific biochar features, like as surface area and porosity, can provide microbial colonization niches, promoting the growth of beneficial bacteria that contribute to nutrient cycling and disease suppression. However, the impact on microbial communities is complicated because the impact of biochar on microbial diversity and functional groups varies depending on parameters such as biochar feedstock and soil properties (Hafez et al., 2019).

8.9.6 Synergies with other soil amendments

Using biochar in conjunction with other soil amendments such as compost, manure, or mineral fertilizers can result in synergistic effects that improve soil fertility and plant growth (El-Naggar et al., 2018). Biochar can act as a transporter for nutrients provided by other amendments, increasing retention and decreasing leaching (Fazal and Bano, 2016). Furthermore, biochar can alleviate the negative effects of certain amendments, such as lowering ammonia volatilization when used in conjunction with nitrogen-rich fertilizers. Understanding how biochar interacts with other soil amendments is critical for developing integrated and effective soil management techniques.

8.10 Considerations and difficulties

8.10.1 Biochar's potential drawbacks and limitations

While biochar has great potential as a soil supplement, it is not without restrictions. One significant problem is that biochar has the ability to change soil nitrogen dynamics (Adeniyi et al., 2023). Biochar may initially immobilize nutrients, causing temporary nutritional deficits in plants, depending on its source and production conditions (Koné and Galiegue, 2023). Furthermore, due to changes in soil physical qualities, poor application or excessive biochar dosage might result in reduced plant growth and impaired root development. Biochar can also have an effect on soil pH, which can result in unanticipated changes in soil acidity or alkalinity that alter nutrient availability and microbial activity.

8.10.2 Biochar quality control and standardization

Consistent biochar quality is critical for getting the desired results in various soil systems (Zafar-Ul-Hye et al., 2020). Biochar characteristics and performance can differ due to variances in biochar production processes, feedstock, and processing conditions. Standardizing biochar manufacturing techniques and quality control measures is critical for providing practitioners with accurate guidelines. The lack of standardized testing procedures for assessing biochar features including surface area, porosity, and nutrient content might stymie effective decision-making during application (Yuan et al., 2021).

8.10.3 Ecological implications and unintended consequences

The introduction of biochar into ecosystems can have far-reaching ecological consequences that should be carefully considered (Ullah et al., 2023). While biochar has the potential to improve soil carbon sequestration, its long-term influence on carbon cycling and soil microbial populations is unknown. Biochar's modification of microbial community composition could have a domino impact on nutrient cycling, plant–microbe interactions, and overall ecosystem dynamics (Vahedi et al., 2022). Comprehensive ecological studies are required to analyze the potential trade-offs and unforeseen consequences of large-scale biochar application.

8.10.4 Regulatory and policy issues regarding biochar application

The use of biochar as a soil amendment presents regulatory and policy issues that must be addressed in order to ensure responsible and sustainable deployment (Wang, Zhang et al., 2021). Regulations governing emissions, waste management, and land use may apply to biochar production. Concerns about potential impurities in biochar, such as heavy metals or polycyclic aromatic hydrocarbons, highlight the significance of quality control and monitoring (Wani et al., 2021). It is critical to develop clear rules for biochar application, storage, and transport in order to minimize environmental concerns and ensure safe and successful implementation.

8.11 Conclusion and future direction

The utilization of biochar as a soil supplement has demonstrated favorable outcomes in improving soil health and promoting the activities of PGPR. Research has demonstrated that the application of biochar has the potential to enhance soil fertility, water retention, and nutrient retention, while concurrently promoting a greater diversity and activity of microorganisms. Furthermore, empirical evidence has indicated that it can elevate the activity of PGPR, hence stimulating plant growth and improving soil quality. Despite these promising findings, there remains a dearth of understanding regarding the mechanisms that govern the interplay between biochar and PGPR, as well as the enduring effects of biochar on soil health. To use biochar most effectively in various agricultural systems and soil types, more research is needed. The use of biochar as a sustainable soil amendment has the potential to be extremely important in addressing issues like food security and climate change on a global scale. Because of its ability to improve soil health, promote plant growth, and enhance the activities of beneficial bacteria, it is a crucial tool for sustainable agriculture.

There are still many problems that need to be resolved despite the fact that our knowledge of how biochar affects PGPR activity and soil health has advanced significantly. The following are a few prospective future directions that could be specifically looked into:

- *The molecular mechanism underlying the interactions between biochar and PGPR is* needs to be studied: Even though some of the mechanisms through which biochar's advantageous impacts on PGPR activity have been established, more research is required to completely understand the molecular process behind these interactions. To further understand the changes in microbial communities that occur when biochar is applied, it would be necessary to use cutting-edge molecular techniques like transcriptomics and proteomics.
- *Evaluating the long-term effects of biochar on soil health*: Most of the research on biochar and soil health has been done during brief times, usually between a few months and a year. However, research that span years or even decades are necessary to completely comprehend the long-term consequences of biochar.
- *Evaluating the performance of biochar in various climatic and soil environments*: The effectiveness of biochar as a soil addition might vary depending on the kind of soil, climate, and crop. Therefore, further investigation should focus on the effectiveness of biochar in diverse soil types and climatic conditions.
- *Investigating the interactions between biochar and other soil amendments*: Compost and fertilizer are two common soil additives used in conjunction with biochar. Future research should look into how biochar interacts with other soil amendments and how this interaction influence PGPR activity and soil health.
- *Studying the impact of biochar on plant−microbe interactions*: Very little is known about how biochar impacts other microbial diversity in the soil environment. The majority of research has focused on how biochar affects PGPR activity. Future research should look into how biochar affects interactions between plants and microbes and how this can impact plant development and health.

References

Adeniyi, A., Iwuozor, K., Emenike, E., Amoloye, M., Aransiola, E., Motolani, F., & Kayode, S. (2023). Prospects and problems in the development of biochar-filled plastic composites: A review. *Functional Composites and Structures, 5*, 012002.

Ahmad Bhat, S., Kuriqi, A., Dar, M. U. D., Bhat, O., Sammen, S. S., Towfiqul Islam, A. R. M., & Heddam, S. (2022). Application of biochar for improving physical, chemical, and hydrological soil properties: A systematic review. *Sustainability, 14*(17), 11104.

Ahmad, M., Wang, X., Hilger, T. H., Luqman, M., Nazli, F., Hussain, A., & Mustafa, A. (2020). Evaluating biochar-microbe synergies for improved growth, yield of maize, and post-harvest soil characteristics in a semi-arid climate. *Agronomy, 10*(7), 1055.

Ajeng, A. A., Abdullah, R., Ling, T. C., Ismail, S., Lau, B. F., Ong, H. C., & Chang, J. (2020). Bioformulation of biochar as a potential inoculant carrier for sustainable agriculture. *Environmental Technology and Innovation, 20*, 101168.

Alburquerque, J. A., Calero, J. M., Barrón, V., Torrent, J., Del Campillo, M. C., Gallardo, A., & Villar, R. (2014). Effects of biochars produced from different feedstocks on soil properties and sunflower growth. *Journal of Plant Nutrition and Soil Science, 177*(1), 16−25.

Allohverdi, T., Mohanty, A. K., Roy, P., & Misra, M. (2021). A review on current status of biochar uses in agriculture. *Molecules (Basel, Switzerland), 26*(18), 5584.

Amitrano, C., Arena, C., Rouphael, Y., De Pascale, S., & De Micco, V. (2019). Vapour pressure deficit: The hidden driver behind plant morphofunctional traits in controlled environments. *Annals of Applied Biology, 175*(3), 313−325.

Arif, I., Batool, M., & Schenk, P. M. (2020). Plant microbiome engineering: Expected benefits for improved crop growth and resilience. *Trends in Biotechnology, 38*(12), 1385−1396.

Arora, N. K., Fatima, T., Mishra, J., Mishra, I., Verma, S., Verma, R., . . . Bharti, C. (2020). Halo-tolerant plant growth promoting rhizobacteria for improving productivity and remediation of saline soils. *Journal of Advanced Research, 26*, 69−82.

Asiloglu, R. (2022). Biochar−microbe interaction: More protist research is needed. *Biochar, 4*(1)72.

Bamdad, H., Papari, S., Lazarovits, G., & Berruti, F. (2022). Soil amendments for sustainable agriculture: Microbial organic fertilizers. *Soil Use and Management, 38*(1), 94−120.

Barriuso, J., Ramos Solano, B., Lucas, J. A., Lobo, A. P., García-Villaraco, A., & Gutiérrez Mañero, F. J. (2008). Ecology, genetic diversity and screening strategies of plant growth promoting rhizobacteria (PGPR). *Plant-Bacteria Interactions: Strategies and Techniques to Promote Plant Growth*, 1−7.

Basu, A., Prasad, P., Das, S. N., Kalam, S., Sayyed, R. Z., Reddy, M. S., & El Enshasy, H. (2021). Plant growth promoting rhizobacteria (PGPR) as green bioinoculants: Recent developments, constraints, and prospects. *Sustainability, 13*(3), 1140.

Behera, B., Das, T. K., Raj, R., Ghosh, S., Raza, M. B., & Sen, S. (2021). Microbial consortia for sustaining productivity of non-legume crops: Prospects and challenges. *Agricultural Research, 10*(1), 1−14.

Behera, B., Kancheti, M., Raza, M. B., Shiv, A., Mangal, V., Rathod, G., & Singh, B. (2022). Mechanistic insight on boron-mediated toxicity in plant vis-a-vis its mitigation strategies: A review. *International Journal of Phytoremediation*, 1−18.

Bhat, M. A., Kumar, V., Bhat, M. A., Wani, I. A., Dar, F. L., Farooq, I., & Jan, A. T. (2020). Mechanistic insights of the interaction of plant growth-promoting rhizobacteria (PGPR) with plant roots toward enhancing plant productivity by alleviating salinity stress. *Frontiers in Microbiology, 11*, 1952.

Chaturvedi, S., Singh, S., Dhyani, V. C., Govindaraju, K., & Mandal, S. (2021). Characterization, bioenergy value and thermal stability of biochar derived from diverse agriculture and forestry lignocellulosic waste. *Biomass Conversion and Biorefinery*. Available from https://doi.org/10.1007/s13399-020-01239-2.

Çığ, F., Sönmez, F., Nadeem, M. A., & Sabagh, A. E. (2021). Effect of biochar and PGPR on the growth and nutrients content of einkorn wheat (*Triticum monococcum* L.) and post-harvest soil properties. *Agronomy, 11*(12), 2418.

Danish, S., Zafar-Ul-Hye, M., Mohsin, F., & Hussain, M. (2020). ACC-deaminase producing plant growth promoting rhizobacteria and biochar mitigate adverse effects of drought stress on maize growth. *PLoS One, 15*(4), e0230615.

Das, D., Sahoo, J., Raza, M. B., Barman, M., & Das, R. (2022). Ongoing soil potassium depletion under intensive cropping in India and probable mitigation strategies. A review. *Agronomy for Sustainable Development, 42*(1), 4.

Dash, A. K., Dwivedi, B. S., Dey, A., Meena, M. C., & Chakraborty, D. (2023). Temperature sensitivity of soil organic carbon as affected by crop residue and nutrient management options under conservation agriculture. *Journal of Soil Science and Plant Nutrition*, 1−15.

de Medeiros, E. V., Lima, N. T., de Sousa Lima, J. R., Pinto, K. M. S., da Costa, D. P., Franco Junior, C. L., & Hammecker, C. (2021). Biochar as a strategy to manage plant diseases caused by pathogens inhabiting the soil: A critical review. *Phytoparasitica, 49*(4), 713−726.

Ding, Y., Liu, Y., Liu, S., Li, Z., Tan, X., Huang, X., Zeng, G., ... Zheng, B. (2016). Biochar to improve soil fertility. A review. *Agronomy for Sustainable Development, 36*, 1–8.

El-Naggar, A., Shaheen, S. M., Ok, Y. S., & Rinklebe, J. (2018). Biochar afects the dissolved and colloidal concentrations of Cd, Cu, Ni, and Zn and their phytoavailability and potential mobility in a mining soil under dynamic redox-conditions. *Science of the Total Environment, 624*, 1059–1071.

Etesami, H., & Adl, S. M. (2020). Plant growth-promoting rhizobacteria (PGPR) and their action mechanisms in availability of nutrients to plants. *Phytol.-Microbiome in Stress Regulation, 147*–203.

Fazal, A., & Bano, A. (2016). Role of plant growth-promoting rhizobacteria (PGPR), biochar, and chemical fertilizer under salinity stress. *Communications in Soil Science and Plant Analysis, 47*(17), 1985–1993.

Gorovtsov, A. V., Minkina, T. M., Mandzhieva, S. S., Perelomov, L. V., Soja, G., Zamulina, I. V., ... Yao, J. (2020). The mechanisms of biochar interactions with microorganisms in soil. *Environmental Geochemistry and Health, 42*, 2495–2518.

Hafez, E. M., Alsohim, A. S., Farig, M., Omara, A. E., Rashwan, E., & Kamara, M. M. (2019). Synergistic effect of biochar and plant growth promoting rhizobacteria on alleviation of water deficit in rice plants under salt-affected soil. *Agronomy, 9*(12), 847.

Ha-Tran, D. M., Nguyen, T. T. M., Hung, S. H., Huang, E., & Huang, C. C. (2021). Roles of plant growth-promoting rhizobacteria (PGPR) in stimulating salinity stress defense in plants: A review. *International Journal of Molecular Sciences, 22*(6), 3154.

Hendrix, P. F., Crossley, D. A., Blair, J. M., & Coleman, D. C. (2020). *Soil biota as components of sustainable agroecosystems*. In *Sustainable agricultural systems* (pp. 637–654). Boca Raton, FL: CRC Press.

Hossain, M. Z., Bahar, M. M., Sarkar, B., Donne, S. W., Ok, Y. S., Palansooriya, K. N., & Bolan, N. (2020). Biochar and its importance on nutrient dynamics in soil and plant. *Biochar, 2*(4), 379–420.

Ijaz, M., Tahir, M., Shahid, M., Ul-Allah, S., Sattar, A., Sher, A., & Hussain, M. (2019). Combined application of biochar and PGPR consortia for sustainable production of wheat under semiarid conditions with a reduced dose of synthetic fertilizer. *Brazilian Journal of Microbiology, 50*(2), 449–458.

Kari, A., Nagymáté, Z., Romsics, C., Vajna, B., Tóth, E., Lazanyi-Kovács, R., & Márialigeti, K. (2021). Evaluating the combined effect of biochar and PGPR inoculants on the bacterial community in acidic sandy soil. *Applied Soil Ecology, 160*, 103856.

Kasak, K., Truu, J., Ostonen, I., Sarjas, J., Oopkaup, K., Paiste, P., & Truu, M. (2018). Biochar enhances plant growth and nutrient removal in horizontal subsurface flow constructed wetlands. *Science of the Total Environment, 639*, 67–74.

Kaur, T., Rana, K. L., Kour, D., Sheikh, I., Yadav, N., Kumar, V., Yadav, A. N., ... Saxena, A. K. (2020). Microbe-mediated biofortification for micronutrients: Present status and future challenges. In *New and future developments in microbial biotechnology and bioengineering* (pp. 1–17). Amsterdam: Elsevier.

Khadem, A., Raiesi, F., Besharati, H., & Khalaj, M. A. (2021). The effects of biochar on soil nutrients status, microbial activity and carbon sequestration potential in two calcareous soils. *Biochar, 3*(1), 105–116.

Khatoon, Z., Huang, S., Rafique, M., Fakhar, A., Kamran, M. A., & Santoyo, G. (2020). Unlocking the potential of plant growth-promoting rhizobacteria on soil health and the sustainability of agricultural systems. *Journal of Environmental Management, 273*, 111118.

Kimura, A., Uchida, Y., & Madegwa, Y. M. (2022). Legume species alter the effect of biochar application on microbial diversity and functions in the mixed cropping system—Based on a pot experiment. *Agriculture, 12*(10), 1548.

Koné, S., & Galiegue, X. (2023). Potential development of biochar in Africa as an adaptation strategy to climate change impact on agriculture. *Environmental Management*, 1–15.

Lakshmi, G., Okafor, B. N., & Visconti, D. (2020). Soil microarthropods and nutrient cycling. *Environmental Climate Plant Vegetation Growth*, 453–472.

Lalay, G., Ullah, S., & Ahmed, I. (2022). Physiological and biochemical responses of *Brassica napus* L. to drought-induced stress by the application of biochar and plant growth promoting rhizobacteria. *Microscopy Research and Technique, 85*(4), 1267–1281.

Li, N., He, M., Lu, X., Yan, B., Duan, X., Chen, G., & Hou, L. (2022). Municipal solid waste derived biochars for wastewater treatment: Production, properties and applications. *Resources, Conservation and Recycling, 177*, 106003.

Lohan, S. K., Jat, H. S., Yadav, A. K., Sidhu, H. S., Jat, M. L., Choudhary, M., & Sharma, P. C. (2018). Burning issues of paddy residue management in north-west states of India. *Renewable and Sustainable Energy Reviews, 81*, 693–706.

Majhi, P. K., Raza, B., Behera, P. P., Singh, S. K., Shiv, A., Mogali, S. C., ... Behera, B. (2022). *Future-proofing plants against climate change: A path to ensure sustainable food systems. In* Biodiversity, functional ecosystems and sustainable food production (pp. 73–116). Springer International Publishing.

Malik, L., Sanaullah, M., Mahmood, F., Hussain, S., Siddique, M. H., Anwar, F., & Shahzad, T. (2022). Unlocking the potential of co-applied biochar and plant growth-promoting rhizobacteria (PGPR) for sustainable agriculture under stress conditions. *Chemical and Biological Technologies in Agriculture, 9*(1), 1–29.

Mona, S., Malyan, S. K., Saini, N., Deepak, B., Pugazhendhi, A., & Kumar, S. S. (2021). Towards sustainable agriculture with carbon sequestration, and greenhouse gas mitigation using algal biochar. *Chemosphere, 275*, 129856.

Murtaza, G., Ditta, A., Ullah, N., Usman, M., & Ahmed, Z. (2021). Biochar for the management of nutrient impoverished and metal contaminated soils: Preparation, applications, and prospects. *Journal of Soil Science and Plant Nutrition, 21*, 2191–2213.

Nafees, M., Ullah, S., & Ahmed, I. (2022). Modulation of drought adversities in Vicia faba by the application of plant growth promoting rhizobacteria and biochar. *Microscopy Research and Technique, 85*(5), 1856–1869.

Negi, S., Bharat, N. K., & Kumar, M. (2021). Effect of seed biopriming with indigenous PGPR, Rhizobia and Trichoderma sp. on growth, seed yield and incidence of diseases in French bean (*Phaseolus vulgaris* L.). *Legume Research An International Journal, 44*, 593–601.

Niewiadomska, A., Płaza, A., Wolna-Maruwka, A., Budka, A., Głuchowska, K., Rudziński, R., & Kaczmarek, T. (2023). Consortia of plant growth-promoting rhizobacteria and selected catch crops for increasing microbial activity in soil under spring barley grown as an organic farming system. *Applied Sciences, 13*(8), 5120.

Nozari, R. M., Ortolan, F., Astarita, L. V., & Santarém, E. R. (2021). *Streptomyces* spp. enhance vegetative growth of maize plants under saline stress. *Brazilian Journal of Microbiology, 52*(3), 1371–1383.

Osman, A. I., Fawzy, S., Farghali, M., El-Azazy, M., Elgarahy, A. M., Fahim, R. A., & Rooney, D. W. (2022). Biochar for agronomy, animal farming, anaerobic digestion, composting, water treatment, soil remediation, construction, energy storage, and carbon sequestration: A review. *Environmental Chemistry Letters, 20*(4), 2385–2485.

Panahi, H. K. S., Dehhaghi, M., Ok, Y. S., Nizami, A. S., Khoshnevisan, B., Mussatto, S. I., ... Lam, S. S. (2020). A comprehensive review of engineered biochar: production, characteristics, and environmental applications. *Journal of cleaner production, 270*, 122462.

Pathy, A., Ray, J., & Paramasivan, B. (2020). Biochar amendments and its impact on soil biota for sustainable agriculture. *Biochar, 2*, 287–305.

Pawar, A. B., Kumawat, C., Verma, A. K., Meena, R. K., Raza, M. B., Anil, A. S., & Trivedi, V. K. (2017). Threshold limits of soil in relation to various soil functions and crop productivity. *International Journal of Current Microbiology and Applied Sciences, 6*(5), 2293–2302.

Przybyłko, S., Kowalczyk, W., & Wrona, D. (2021). The effect of mycorrhizal fungi and PGPR on tree nutritional status and growth in organic apple production. *Agronomy, 11*(7), 1402.

Rahman, R., Sofi, J. A., Javeed, I., Malik, T. H., & Nisar, S. (2020). Role of micronutrients in crop production. *International Journal of Current Microbiology and Applied Sciences, 8*, 2265–2287.

Raza, M. B., Hombegowda, H. C., Kumar, P., & Kumar, S. (2023). Effect of conservation agriculture on soil health. *A Training Manual on Conservation Agriculture*, 41.

Ren, H., Huang, B., Fernández-García, V., Miesel, J., Yan, L., & Lv, C. (2020). Biochar and rhizobacteria amendments improve several soil properties and bacterial diversity. *Microorganisms, 8*(4), 502.

Ren, H., Lv, C., Fernández-García, V., Huang, B., Yao, J., & Ding, W. (2021). Biochar and PGPR amendments influence soil enzyme activities and nutrient concentrations in a eucalyptus seedling plantation. *Biomass Conversion and Biorefinery, 11*(5), 1865–1874.

Renoud, S., Abrouk, D., Prigent-Combaret, C., Wisniewski-Dyé, F., Legendre, L., Moënne-Loccoz, Y., & Muller, D. (2022). Effect of inoculation level on the impact of the PGPR *Azospirillum lipoferum* CRT1 on selected microbial functional groups in the rhizosphere of field maize. *Microorganisms, 10*(2), 325.

Riaz, M., Roohi, M., Arif, M. S., Hussain, Q., Yasmeen, T., Shahzad, T., & Khalid, M. (2017). Corncob-derived biochar decelerates mineralization of native and added organic matter (AOM) in organic matter depleted alkaline soil. *Geoderma, 294*, 19–28.

Rodriguez-Franco, C., & Page-Dumroese, D. S. (2021). Woody biochar potential for abandoned mine land restoration in the US: A review. *Biochar, 3*(1), 7–22.

Sadegh Kasmaei, L., Yasrebi, J., Zarei, M., Ronaghi, A., Ghasemi, R., Saharkhiz, M. J., & Schnug, E. (2019). Influence of plant growth promoting rhizobacteria, compost, and biochar of Azolla on rosemary (*Rosmarinus officinalis* L.)

growth and some soil quality indicators in a calcareous soil. *Communications in Soil Science and Plant Analysis, 50* (2), 119−131.

Samal, S. K., Datta, S. P., Dwivedi, B. S., Meena, M. C., Nogiya, M., Choudhary, M., . . . Raza, M. B. (2023). Phytoextraction of nickel, lead, and chromium from contaminated soil using sunflower, marigold, and spinach: Comparison of efficiency and fractionation study. *Environmental Science and Pollution Research*, 1−17.

Sani, M. N. H., Hasan, M., Uddain, J., & Subramaniam, S. (2020). Impact of application of Trichoderma and biochar on growth, productivity and nutritional quality of tomato under reduced NPK fertilization. *Annals of Agricultural Sciences, 65*(1), 107−115.

Sarfraz, R., Hussain, A., Sabir, A., Ben Fekih, I., Ditta, A., & Xing, S. (2019). Role of biochar and plant growth promoting rhizobacteria to enhance soil carbon sequestration—A review. *Environmental Monitoring and Assessment, 191*(4), 251.

Shemawar, M. A., Hussain, S., Mahmood, F., Iqbal, M., Shahid, M., Ibrahim, M., . . . Shahzad, T. (2021). Toxicity of biogenic zinc oxide nanoparticles to soil organic matter cycling and their interaction with rice-straw derived biochar. *Scientific Reports, 11*(1), 1−12.

Siedt, M., Schäffer, A., Smith, K. E. C., Nabel, M., Roß-Nickoll, M., & van Dongen, J. T. (2021). Comparing straw, compost, and biochar regarding their suitability as agricultural soil amendments to affect soil structure, nutrient leaching, microbial communities, and the fate of pesticides. *Science of the Total Environment, 751*, 141607.

Singh, S., Chaturvedi, S., Nayak, P., Dhyani, V. C., Nandipamu, T. M., Singh, D. K., . . . Govindaraju, K. (2023). Carbon offset potential of biochar based straw management under rice−wheat system along Indo-Gangetic Plains of India. *Science of the Total Environment, 897*, 165176.

Singh, S., Luthra, N., Mandal, S., Kushwaha, D. P., Pathak, S. O., Datta, D., . . . Pramanik, B. (2023). Distinct behavior of biochar modulating biogeochemistry of salt-affected and acidic soil: A review. *Journal of Soil Science and Plant Nutrition.* Available from https://doi.org/10.1007/s42729-023-01370-9.

Singh, A., Singh, A. P., Singh, S. K., Rai, S., & Kumar, D. (2016). Impact of addition of biochar along with PGPR on rice yield, availability of nutrients and their uptake in alluvial soil. *Journal of Pure and Applied Microbiology, 10*(3), 2181−2188.

Singh, H., Northup, B. K., Rice, C. W., & Prasad, P. V. V. (2022). Biochar applications influence soil physical and chemical properties, microbial diversity, and crop productivity: A meta-analysis. *Biochar, 4*(1), 8.

Singh, R. P., & Jha, P. N. (2016).). The multifarious PGPR Serratia marcescens CDP-13 augments induced systemic resistance and enhanced salinity tolerance of wheat (*Triticum aestivum* L.). *PLoS One, 11*(6), e0155026.

Singh, S., Bhoi, T. K., & Vyas, V. (2023). *Interceding microbial biofertilizers in agroforestry system for enhancing productivity. Inplant growth promoting microorganisms of arid region* (pp. 161−183). Singapore: Springer Nature Singapore.

Širić, I., Eid, E. M., Taher, M. A., El-Morsy, M. H. E., Osman, H. E. M., Kumar, P., & Kumar, V. (2022). Combined use of spent mushroom substrate biochar and PGPR improves growth, yield, and biochemical response of cauliflower (*Brassica oleracea* var. botrytis): A preliminary study on greenhouse cultivation. *Horticulturae, 8*(9), 830.

Şirin, E., Ertürk, Y., & Kazankaya, A. (2022). Effects of PGPR, AMF and Trichoderma applications on adaptation abilities to different biotic and abiotic conditions in medicinal and aromatic plants. *Turkish Journal of Agriculture − Food Science and Technology, 10*(2), 166−173.

Sri Shalini, S., Palanivelu, K., Ramachandran, A., & Raghavan, V. (2021). Biochar from biomass waste as a renewable carbon material for climate change mitigation in reducing greenhouse gas emissions—A review. *Biomass Conversion and Biorefinery, 11*(5), 2247−2267.

Tian, J., Wang, J., Dippold, M., Gao, Y., Blagodatskaya, E., & Kuzyakov, Y. (2016). Biochar affects soil organic matter cycling and microbial functions but does not alter microbial community structure in a paddy soil. *Science of the Total Environment, 556*, 89−97.

Timmis, K., & Ramos, J. L. (2021). The soil crisis: The need to treat as a global health problem and the pivotal role of microbes in prophylaxis and therapy. *Microbial Biotechnology, 14*(3), 769−797.

Tomczyk, A., Sokołowska, Z., & Boguta, P. (2020). Biochar physicochemical properties: Pyrolysis temperature and feedstock kind efects. *Reviews in Environmental Science and Bio/Technology, 19*(1), 191−215.

Ullah, N., Ditta, A., Imtiaz, M., Li, X., Jan, A. U., Mehmood, S., & Rizwan, M. (2021). Appraisal for organic amendments and plant growth-promoting rhizobacteria to enhance crop productivity under drought stress: A review. *Journal of Agronomy and Crop Science, 207*(5), 783−802.

Ullah, N., Ditta, A., Khalid, A., Mehmood, S., Rizwan, M. S., Ashraf, M., & Iqbal, M. M. (2020). Integrated effect of algal biochar and plant growth promoting rhizobacteria on physiology and growth of maize under deficit irrigations. *Journal of Soil Science and Plant Nutrition, 20*(2), 346−356.

Ullah, S., Nafees, M., & Ahmed, I. (2023). Resistance induction in *Brassica napus* L. against water deficit stress through application of biochar and plant growth promoting rhizobacteria. *Journal of the Saudi Society of Agricultural Sciences*.

United Nations. (2019). World population prospects: Highlights. Available at: http://un.org/development/desa/publications/world-population-prospects-2019.

Vahedi, R., Rasouli-Sadaghiani, M. H., Barin, M., & Vetukuri, R. R. (2022). Effect of biochar and microbial inoculation on P, Fe, and Zn bioavailability in a calcareous soil. *Processes*, 10(2), 343.

Wang, F., Wang, X., & Song, N. (2021). Biochar and vermicompost improve the soil properties and the yield and quality of cucumber (*Cucumis sativus* L.) grown in plastic shed soil continuously cropped for different years. *Agriculture, Ecosystems and Environment*, 315, 107425.

Wang, H., Zhang, K., Gan, L., Liu, J., & Mei, G. (2021). Expansive soil-biochar-root-water-bacteria interaction: Investigation on crack development, water management and plant growth in green infrastructure. *International Journal of Damage Mechanics*, 30(4), 595–617.

Wang, Y., Li, W., Du, B., & Li, H. (2021). Effect of biochar applied with plant growth-promoting rhizobacteria (PGPR) on soil microbial community composition and nitrogen utilization in tomato. *Pedosphere*, 31(6), 872–881.

Wani, I., Kumar, H., Rangappa, S. M., Peng, L., Siengchin, S., & Kushvaha, V. (2021). Multiple regression model for predicting cracks in soil amended with pig manure biochar and wood biochar. *Journal of Hazardous, Toxic, and Radioactive Waste*, 25(1), 04020061.

Xiang, L., Harindintwali, J. D., Wang, F., Redmile-Gordon, M., Chang, S. X., Fu, Y., & Xing, B. (2022). Integrating biochar, bacteria, and plants for sustainable remediation of soils contaminated with organic pollutants. *Environmental Science and Technology*, 56(23), 16546–16566.

Xu, W., Wang, G., Deng, F., Zou, X., Ruan, H., & Chen, H. Y. H. (2018). Responses of soil microbial biomass, diversity and metabolic activity to biochar applications in managed poplar plantations on reclaimed coastal saline soil. *Soil Use and Management*, 34(4), 597–605.

Yuan, W. A. N. G., Wenqing, L. I., Binghai, D. U., & Hanhao, L. I. (2021). Effect of biochar applied with plant growth-promoting rhizobacteria (PGPR) on soil microbial community composition and nitrogen utilization in tomato. *Pedosphere*, 31(6), 872–881.

Zafar-Ul-Hye, M., Tahzeeb-Ul-Hassan, M., Abid, M., Fahad, S., Brtnicky, M., & Danish, S. (2020). Potential role of compost mixed biochar with rhizobacteria in mitigating lead toxicity in spinach. *Scientific Reports*, 10(1), 12159.

Zhong, L., Li, G., Qing, J., Li, J., Xue, J., Yan, B., … Rui, Y. (2022). Biochar can reduce N_2O production potential from rhizosphere of fertilized agricultural soils by suppressing bacterial denitrification. *European Journal of Soil Biology*, 109, 103391.

CHAPTER

9

Biochar mediated carbon and nutrient dynamics under arable land

Adeel Abbas[1,*], Rashida Hameed[1,*], Aitezaz A.A. Shahani[2,*], Wajid Ali Khattak[1], Ping Huang[1] and Daolin Du[1]

[1]School of the Environment and Safety Engineering, Institute of Environment and Ecology, Jiangsu University, Zhenjiang, P.R. China [2]Key Laboratory of Crop Sciences and Plant Breeding Genetics, College of Agriculture, Yanbian University, Yanji, Jilin, P.R. China

9.1 Introduction

Biochar, commonly recognized as black carbon, is a by-product obtained through the pyrolysis of carbon-rich organic matter and is prevalent in soils as remarkably stable solid structures, frequently manifesting as sedimentary accumulations (Novotný et al., 2023). Biochar exhibits long-lasting soil persistence, enduring for extended durations, often spanning thousands of years, and distributed across varying depths. Recent studies have extensively investigated the characteristics and properties of biochar to ascertain its potential advantages and adverse effects, particularly concerning its application as agricultural amendments (Koné & Galiegue, 2023). The beneficial effects of biochar have been acknowledged for 1000 years, dating back to the era when the "slash-and-burn" agricultural method was prevalent. Furthermore, the formation of biochar was facilitated by natural forest fires, as well as historical and cultural practices, leading to the establishment of biochar-enriched soil deposits that have endured for millennia (Nieder & Benbi, 2023). Many biochar variants exist, contingent upon the source material from which they originate. Diverse carbon-rich materials yield distinct biochar types characterized by unique physical and chemical attributes inherited from the parent substance. For instance, biochar derived from various tree species or plant varieties yields disparate biochar varieties (Yu et al., 2023).

---

* These authors are equally contributed.

Biochar addition brings forth a diverse range of beneficial effects, including heightened soil microbial activity, improved plant nutrient assimilation, and enhanced agricultural productivity (Qian et al., 2023). Furthermore, biochar enhances bulk density, porosity, soil aeration, aggregate stability, infiltration rate, hydraulic conductivity and water-holding capacity. It also plays a vital role in the immobilization of heavy metals, thereby reducing their bioavailability in agriculturally challenging or low-quality soils. Additionally, biochar stimulates microbial abundance, mitigates the detrimental impacts of heat, drought, and salinity stress, and fosters crop growth, development, and agricultural productivity. It promotes the biological nitrogen fixation in leguminous plants and contributes to the sequestration of carbon (Musto, Swanepoel, & Strauss, 2023). Nevertheless, there is an immediate need for extensive research and comprehension of the intricate interactions among biochar, soil, and plants. This is crucial to unravel the diverse responses displayed by soils and plants to various biochar types. Factors encompassing feedstock composition, pyrolysis temperature, and techniques for biochar application and management should be thoroughly examined within diverse environmental contexts (Murtaza et al., 2023).

Biochar has attracted considerable attention due to its potential to modulate the natural carbon cycle within biogeochemical systems, primarily in soil ecosystems. It acts as an environmentally friendly adsorbent for carbon dioxide (CO_2). Furthermore, as a soil alteration, biochar improves soil abundance, enhances agricultural plant productivity, and provides additional benefits in the agricultural domain that can be optimized through efficient nutrient management strategies (Shareef & Zhao, 2016). As a soil amendment, biochar has the potential to amend the physio-chemical estates of soil, stimulate microbial processes, improve nutrient availability, and impact plant productivity. Biochar has been widely acknowledged for its efficacy in environmental remediation, as it exhibits adsorption capabilities toward diverse pollutants including pharmaceuticals, dyes, heavy metals, and eternal organic pollutants (Vithanage, Ashiq, Ramanayaka, & Bhatnagar, 2020) Furthermore, the utilization of biochar influences the soil's nutrient transport processes and normalizes fertilizer liberation owing to its notable sorption function. Multiple surface possessions, including specific functional groups, surface area and surface charge, significantly contribute to the adsorption capacity of biochar (Chausali, Saxena, & Prasad, 2021).

The utilization of biochar induces substantial modifications in diverse soil attributes, encompassing volume density, cation replacement activities, pH, water retention, and different biological activities. These transformations are expected to influence various nutrient interactions with microbial nutrient transformations and put impact on soil particles. The tender of biochar enhances soil fertility and promotes crop productivity through increased nutrient availability, enhanced nutrient mobility, stimulation of microbial activity, improved soil aeration and water retention, regulation of soil reactions, reduction of bulk density, and preservation of soil cumulative manufacture characteristics (Palansooriya et al., 2023). Additionally, biochar mitigates different nutrient losses through filtering and volatilization by modifying soil pH and improving ion movement activities. Additionally, it improves the composition of soil microbial communities, consequently impacting different nutrient uptake agriuculture plants (Banu, Rani, Kavya, & Nihala Jabin, 2023). Here we discuss the characteristic and methods of biochar production and its impact on agriculture arable land. How biochar influences agriculture productivity, the mechanisms by which it affects soil carbon and nutrient cycling, including its impact on nutrient availability and examine biochar's role in enhancing soil fertility and crop productivity through improved nutrient preservation and accessibility.

9.2 Characteristics and production of biochar

The term "char" encompasses the product formed through the decomposition of different matter. Although biochar has been used interchangeably, its specific applications can be differentiated. Charcoal is mainly utilized for energy production, while biochar is primarily employed for carbon sequestration and environmental purposes (Shao et al., 2023). Biochar is alternatively known as "pyrochar" due to its production through the pyrolysis process of biomass. The International Biochar Initiative provides a universally acknowledged definition of biochar as "a solid substance produced through chemical conversion of different biomass under less availability of oxygen availability" (Vedovato et al., 2023). Upon studying and analyzing the characteristics of terra-preta soils, the scientific community identified similarities between biochar and these soils. This discovery prompted extensive research on biochar and its soil application. Recent scientific research has provided evidence that biochar can be made through various microwave-assisted pyrolysis, carbonization, and alternative thermochemical methods, including hydrothermal, gasification and torrefaction. The unique attributes of biochar are persuaded by the selection of biomass feedstock and the specific specifications employed during the heating procedure (Fig. 9.1) and different biochar sources affect soil properties (Table 9.1). Additional considerations such as kiln pressure, boiling rate and the components of the kiln surroundings (whether nitrogen or carbon dioxide) also play a considerable role in determining the physio-chemical characteristics of the ensuing biochar. (Osman et al., 2023). Categorization of biochar into three primary classifications is possible based on its ash creation and characteristics (Joseph & Taylor, 2014).

1. Biochar obtained from low-ash biomass, nut shells, such as wood, specific seeds and bamboo, exhibits elevated porosity and surface area (SA), facilitating improved water retention in comparison to biochar derived from alternative sources. Moreover, biochar derived from these feedstocks typically exhibits enhanced hardness.
2. Agricultural residues, bark, and best-grade green waste commonly exhibit a moderate ash content, ranging from 5% to 13%. Biochar derived from these biomass sources falls into this particular classification.
3. Biochar is generated utilizing high-ash biomass feedstocks, such as animal manures, slurries, wastepaper material, rice husks and municipal waste.

Interest in biochar has surged, leading to increased production. Thermochemical conversion techniques like pyrolysis, hydrothermal carbonization, gasification, torrefaction, and hydrothermal liquefaction are commonly used (Saravanan, Karishma, Kumar, & Rangasamy, 2023). Pyrolysis involves heating biomass under limited oxygen conditions, generating biochar, bio-oil, and syngas (Amalina, Krishnan, Zularisam, & Nasrullah, 2023). Slow pyrolysis is a gentle process with a longer residence time, while fast pyrolysis is rapid with a shorter residence time, producing bio-oil and syngas. Gasification decomposes biomass at high

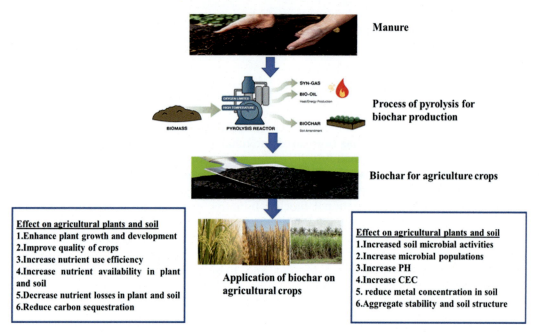

FIGURE 9.1 Basic and conceptual framework of biochar application, characteristics and impact for agriculture plant and soil.

temperatures, creating biochar and syngas (Singh, 2023). Torrefaction is oxygen-free pyrolysis at moderate temperatures, resulting in bio-oil or biochar (Khairy et al., 2023). Flash carbonization achieves rapid combustion of biomass under pressure, known as flash pyrolysis (Safarian, 2023).

9.3 Influence of biochar on agriculture productivity

Applying biochar as a soil amendment has been shown to enhance soil's physical and chemical properties, resulting in increased fertility and productivity. Ongoing research efforts aim to develop a comprehensive understanding of the mechanisms behind the beneficial effects of soil-applied biochar (Koné & Galiegue, 2023). Researchers are investigating various application methods and rates and documenting the advantages of biochar use in agriculture. Furthermore, recent studies have examined the wide-ranging impacts of biochar, including its potential role in mitigating climate change on a worldwide scale. Recent studies evaluate regional increases in soil carbon storage and suggest that by 2050, biochar conversion technologies could store an estimated 2.2 gigatons of carbon (Pant et al., 2023). In addition to its climate mitigation advantages, the incorporation of biochar into the soil can also mitigate emissions of nitrous oxide and methane, mitigate nutrient leaching into groundwater, and reduce soil contaminant levels, among various other benefits. The composition of biochar, as well as the method of production, can

TABLE 9.1 Effects of various biochar's on soil properties and their diverse applications in agriculture.

	Biochar Resource	Experiment Type	Soil Type	Application Rate	Crops Grown	Soil characteristic	Application Effects	Reference
1	Corn feedstock	Incubation study	Hapludand soil	0-, 11.3- and 10- ton biochar ha^{-1}	Soil Incubation	Aggregate stability, volumetric, water content (θ_V), bulk density, saturated hydraulic, conductivity (Ks) and soil water repellence.	1. Biochar application increased Ag. St. and θ_V, decreased bulk density (increased porosity). No effect on soil water repellence. 2. Biochar helped drainage capacity in poorly drained soils.	Herath (2012), Yu (2022)
2	Soft and hard wood	column, lab incubation, and field studies	Sandy, Silty clay	0, 0.5, 1 wt.%	6-year, corn-tomato rotation	Soil water retention (SWR), wilting point (WP), field capacity (FC) and Plant available water (PAW)	1. High surface area biochar increased FC and SWR (sandy soil), no significant effect on wilting pointer silt-clay soil. 2. Biochar increased the FC and PAW in coarse-textured soil.	Danso et al. (2019)
3	Eucalyptus wood stock	Pot type study	Sandy loam	0%, 1%, 2%, and 4%	3 consecutive corn crops	Soil organic carbon (SOC) stability (soil C remaining relative to initial soil C).	1. Ultisol: SOC stability increased up to 2% application, Oxisol: SOC stability decreased at 2% & 4% application of high temperature Biochar. 2. SOC stability affected by different buffering capacity and mineralogy of soils.	El-Naggar et al. (2019), Majumder, Neogi, Dutta, Powel, and Banik (2019)
4	Hard and soft types of wood	column, lab incubation, and field studies	sandy, Silty clay	0, 0.5, 1 wt.%	5-to-6-year, corn- tomato rotation	Soil water retention (SWR), wilting point (WP), field capacity (FC) and Plant available water.	1. High surface area biochar increased FC and SWR (sandy soil), no significant effect on wilting pointer silt-clay soil. 2. Biochar increased the FC and PAW in coarse-textured soil.	Lizcano Toledo et al. (2019)

(Continued)

TABLE 9.1 (Continued)

	Biochar Resource	Experiment Type	Soil Type	Application Rate	Crops Grown	Soil characteristic	Application Effects	Reference
5	Mixed woody feedstock	Column leaching	Sandy and Sandy loam	0%, 1%, 5%, 10%, and 20% (by volume)	Soil incubation	bulk density and water holding capacity	1. Biochar decreased soil BD and increase maximum WHC, more pronounced effect on sandy (coarse text. soil) and finer biochar. 2. 20% biochar to 15 cm depth increase soil water storage of sandy soil from 0.56 mm (control) to 0.83–0.91 (mm).	Emde (2021), Ndede, Kurebito, Idowu, Tokunari, and Jindo (2022)
6	Wheat straw biochar	Field experiment.	Upland red soil	0, 5, 10, 15, 25, 40 t ha^{-1}	Maize (Zea mays) -Mustard (Brassica sps.)	Plant available phosphorus (PAP) and water (PAW); Soil pH, CEC, base saturation, soil organic carbon.	1. PAP increased at 10 and 40 t ha^{-1}, PAW at 40 t ha^{-1}; increase in soil EC and BS; additive effect on SOC. 2. Increased crop yield of maize and mustard in year 2&3	Smit (2023)

influence its properties and potential advantages when utilized in land applications. Biochar exhibits the capacity to retain applied fertilizers and nutrients, gradually releasing them to crops over an extended period (Lei et al., 2023). This characteristic, in conjunction with its capacity to prolong water and nutrient retention in the upper soil layers, renders biochar a valuable asset in agricultural practices. By minimizing nutrient leaching from the plant root zone, biochar has the potential to enhance crop yields and diminish the need for excessive fertilizer application, consequently mitigating adverse environmental effects. It is crucial to differentiate between biochar and compost, as they serve distinct functions. While compost serves as a direct nutrient source through the decomposition of organic matter, biochar remains stable over time, eliminating the need for repeated applications (Qian et al., 2023).

Biochar's impact on nutrient management is paramount for sustainable agriculture (Pillala, Sowmya, Gajbhiye, & Upadhyay, 2023). Its sorption properties enable it to adsorb nutrients such as nitrogen, phosphorus, and potassium, preventing their leaching into groundwater and runoff into water bodies (Al-Hazmi et al., 2023). Moreover, biochar's slow-release characteristics gradually supply nutrients to plants over time, reducing the risk of nutrient imbalances and optimizing plant growth (Ofori, Titriku, Xiang, Ahmed, & Zheng, 2023; Yadav, Yadav, & Abd-Elsalam, 2023). By mitigating nutrient losses, biochar contributes to higher nutrient-use efficiency and overall agricultural productivity (Tyagi, Ashoka, Saikanth, & Ojha, 2023). Water scarcity is a significant challenge in agriculture, making water retention strategies crucial (Alotaibi et al., 2023). Biochar's ability to improve water-holding capacity can prove transformative in regions prone to drought (Ghorbani et al., 2023). Its porous structure absorbs and retains water, reducing surface runoff and enhancing soil moisture content (Rasool, Rasool, & Gani, 2023). This water-retaining property increases the resilience of crops during dry periods, ensuring sustained growth, yield, and agricultural productivity (Baloch et al., 2023; Tamboli, Chaurasiya, Upadhyay, & Kumar, 2023). Biochar's positive impact on soil structure is a cornerstone of its influence on agricultural productivity (Gao et al., 2023). Incorporating biochar enhances soil aggregation, which in turn improves soil porosity, aeration, and drainage (Khan et al., 2023; Ramezanzadeh et al., 2023). Improved soil structure fosters a healthier root environment, leading to increased nutrient uptake and ultimately contributing to higher crop productivity (Hajam, Kumar, & Kumar, 2023). Soil erosion poses a serious threat to agricultural productivity and environmental sustainability (Ali Warsame & Hassan Abdi, 2023). Biochar's application aids in soil conservation by enhancing soil stability and reducing erosion risk (Rakkar, Kaplan, & Blanco-Canqui, 2023). It acts as a binding agent, promoting aggregation and reducing surface runoff (Fernandes, da Silva, Ferraudo, & Fernandes, 2023). This effect is particularly beneficial in sloped landscapes or areas susceptible to erosion, safeguarding the topsoil layer essential for plant growth and maintaining agricultural productivity (Roy et al., 2023; Gupta, 2023).

Integrating biochar into agricultural practices holds the potential to enhance crop yields while mitigating environmental consequences (Xia et al., 2023). Despite the promising effects of biochar on agricultural production, certain studies have reported no noticeable benefits in crop yields or even negative outcomes in certain scenarios (Majumder et al., 2023). The observed lower crop yields in certain instances of biochar application can be attributed to factors such as limited nutrient availability for plant uptake, application of biochar on already nutrient-rich soils, or inadequate biochar application rates (Bilias,

Kalderis, Richardson, Barbayiannis, & Gasparatos, 2023). Conversely, the higher yields observed in some cases of biochar application lack a straightforward explanation, although they may be influenced by factors such as biochar characteristics, soil fertility levels, and the specific crop being cultivated (Roberts, Greene, & Nemet, 2023). Recent scientific investigations on biochar have predominantly concentrated on extensively weathered and nutrient-poor soils, where the positive effects of biochar application have frequently been observed (Anand, Gautam, & Ram, 2023).

Water scarcity poses a significant challenge to agricultural productivity, particularly in regions prone to drought (Sherzad et al., 2023). Biochar's water retention capacity mitigates this challenge by increasing the soil's water-holding capacity (Jie et al., 2023). Its porous structure absorbs and stores water, reducing surface runoff and enhancing soil moisture content (Rojano-Cruz et al., 2023). This property bolsters crops against water stress, maintaining growth and yield during dry periods and enhancing overall agricultural productivity (Vassileva, Georgieva, Zehirov, & Dimitrova, 2023). Agriculture is both a contributor to and a victim of climate change. Biochar's carbon sequestration potential offers a dual benefit—reducing atmospheric carbon dioxide levels while enhancing soil health (Brouziyne et al., 2023). Sequestered carbon enriches soil organic matter, supporting microbial activity and nutrient availability (Mockevičienė, Karčauskienė, & Repšienė, 2023). Healthy soils, in turn, foster resilient plants that can withstand environmental stressors, contributing to climate-resilient agriculture and sustained productivity (Rayanoothala, Hasibul Alam, Mahapatra, Gafur, & Antonius, 2023). Currently, researchers from the University of Florida are engaged in assessing the advantages of biochar application on low-fertility sandy soils in Florida. Their aim is to document any enhancements in crop growth and yield resulting from the incorporation of biochar into these specific soil types (McAmis et al.).

9.4 Biochar effects on soil carbon dynamics and soil fertility

Biochar, a carbon-rich material formed by the pyrolysis process of organic matter in the lack of oxygen, has attracted significant attention due to its potential for mitigating carbon emissions and promoting carbon sequestration. It has been anticipated as a soil amendment to amend soil assets and utilities, including enhancing fertility and stability, as well as facilitating carbon sequestration (Zakaria, Farid, Andou, Ramli, & Hassan, 2023). Agricultural plants sustain their growth and metabolic processes through the process of photosynthesis and the utilization of below-ground nutrient resources. Biochar has the potential to modulate the underground nutrient environment, thereby enhancing the plants' ability to assimilate atmospheric CO_2. A comprehensive analysis of multiple studies indicated that the remedy of biochar leads to an average increase in crop productivity of 10%. This increase can be accredited to the limiting effects of biochar on the soil, improved water retention capacity, and enhanced availability of soil nutrients (Mumivand, Izadi, Amirizadeh, Maggi, & Morshedloo, 2023). Biochar exerts a beneficial influence on plant growth by modulating soil water and nutrient availability, leading to notable upgrades in plants' different physiology and biochemical process. The presence of biochar has a significant impact on multiple parameters related to the photosynthetic performance of maize, such as leaf area index, chlorophyll content, leaf net photosynthetic rate, and dry matter growth (Das, Ghosh, & Avasthe, 2020).

Biochar's most apparent contribution to soil dynamics lies in its capacity to sequester carbon and its recalcitrant nature resists decomposition, leading to long-term carbon storage in soils (Ajmal, Ganapathi, & Chaithanya, 2023; Das, Choudhury, Hazarika, Mishra, & Laha, 2023). By introducing biochar to agricultural soils, carbon sequestration potential can be harnessed, mitigating atmospheric carbon dioxide levels and contributing to climate change mitigation efforts (Fagodiya, Verma, & Verma, 2023). Sequestered carbon enhances soil organic carbon stocks, supporting soil health, structure, and fertility (Chander et al., 2023). Biochar's porous structure and high surface area facilitate nutrient retention and release (Singh et al., 2023). Cations such as nitrogen, phosphorus, and potassium are adsorbed onto biochar surfaces, reducing leaching and enhancing nutrient availability (Utkarsh, Thakur, & Shukla, 2023). The cation exchange capacity (CEC) of biochar aids in nutrient retention, making it a valuable tool for nutrient management (Velásquez Anrango, 2023).

Biochar exhibits the capacity to augment soil aggregation through direct and indirect mechanisms. In a direct manner, biochar actively utilizes its physical and chemical properties to enhance the process of soil aggregation. Indirectly, biochar exerts its influence on soil accumulation by modulating bacterial interest, impacting the planning and secretion of plant roots, and fostering cooperative interactions. These combined effects contribute to the overall improvement of soil aggregation facilitated by biochar (Liu, Li, Chen, & Zhou, 2023). In a recent investigation, it was discovered that the inoculation of arbuscular mycorrhizal fungi (AMF) in different agricultural areas led to significant enhancements in soil particulate organic carbon and light fraction organic carbon. These improvements were primarily ascribed to the increased mycelial length and elevated glomalin content facilitated by AMF. Furthermore, the plant growth-indorsing abilities of AMF contributed to an augmented carbon response in the soil, further improving the overall carbon dynamics (Ren et al., 2023). Numerous investigations have provided additional evidence supporting the beneficial impacts of biochar and mycorrhizal fungi on soil aggregation and carbon sequestration. Soil aggregates, which act as biochemical reactors within the soil matrix, play a pivotal role in regulating greenhouse gas productions at the soil-air boundary. The functionality of these aggregates is intricately linked to their size, distribution, and the biotic and abiotic properties of the soil environment. However, there is currently a dearth of research focused on elucidating the intricate interplay between biochar, plant root systems, mycorrhizal fungi, and soil aggregation. This knowledge gap limits our comprehensive understanding of the intricate mechanisms through which biochar influences the soil carbon cycle (Atkinson, Fitzgerald, & Hipps, 2010; Li, Xiong, & Cai, 2021).

9.5 Mechanisms of biochar-mediated carbon sequestration

The long-term stability of soil carbon following biochar application can be ascribed to multiple factors, including the inherent recalcitrant characteristics of biochar. These properties stem from its aromatic carbon framework and the presence of crystalline silicon structures within silica—carbon complexes that are constructed during pyrolysis

(Gupta et al., 2022). The collaboration between surface oxygen-functional groups, such as carboxyl, present in biochar, and soil minerals plays a considerable function in maintaining the carbon stability of biochar during the ageing process. The adsorption of soil organic substance onto biochar also facilitates the formation of soil aggregates. The presence of AMF in the soil, which releases glomalin, a protein that promotes microaggregation by influencing hydration, contributes to the abundance of these fungi. Furthermore, these factors exert an influence on soil microbial cooperation and modify soil enzyme performances, which regulate the decomposition of soil organic carbon. Notably, the shelter of soil aggregates and the modulation of enzyme accomplishments are closely linked to microbial movement (Libra et al., 2011; Murtaza et al., 2023). Biochar serves as a cohesive driver for soil organic matter during the process of aggregate formation, promoting the formation of larger soil aggregates and contributing to their stability. When applied to sandy soil, biochar has been observed to increase the proportion of stable macroaggregates (>250 mm) by approximately 25% after an 80-day incubation period. This enhancement in macro aggregation has several advantages, including improved water retention within the soil pores, increased sequestration of carbon in the soil, and protection of microbial communities (Han et al., 2021). The formation of large soil aggregates is influenced by the adsorption of soil organic mixtures, such as humic acids, onto the surfaces of biochar. Changes in soil aggregation patterns can have implications for the structure of microbial communities, as gram-positive and gram-negative bacteria employ different approaches for dominating specific fractions of soil particles (Keswani, 2020). Gram-positive bacteria exhibit higher abundance in aged silt and clay fractions that contain extensively decomposed soil organic matter. In contrast, gram-negative bacteria are more dominant in fractions characterized by larger aggregates (>200 mm) as they preferentially utilize recently derived carbon sources from plants. Likewise, fungal populations decrease in smaller soil aggregates (<200 mm) due to the limited availability of carbon substrates in those fractions (Ortiz et al., 2022).

9.6 Biochar effects on different soil nutrient dynamics

Biochar holds promise as a valuable source of nutrients for diverse agricultural crops, and its nutrient composition is prejudiced by the possessions of the feedstock materials and the specific pyrolysis environments, such as temperature, residence time, and gaseous environment (Zakaria et al., 2023). The nutrient-rich feedstock materials yield biochar that is enriched with essential elements including nitrogen, phosphorus, potassium, other important secondary nutrients and soil carbon-nutrient cycling (Fig. 9.2). Ensuring a continuous supply of available nitrogen is crucial for maintaining agricultural soil productivity throughout the cropping seasons, and biochar holds promise as a nitrogen source for plants. Apart from organic nitrogen forms such as water-soluble-N, hydrolysable-N and nonhydrolysable-N, biochar also contains inorganic nitrogen arrangements (Javeed et al., 2023). As a result, not all nitrogen forms that appear in the feedstock are sequestered in biochar.

The impact of biochar on nitrogen dynamics in different soils is multifaceted (Lytand, Singh, & Katiyar, 2023). In sandy soils, characterized by poor water and nutrient retention, biochar can act as a sponge, absorbing and retaining nitrates that would otherwise leach out (Abuseif, 2023; Ogunwole & Ogunwole, 2023). This contributes to reduced nitrogen loss and

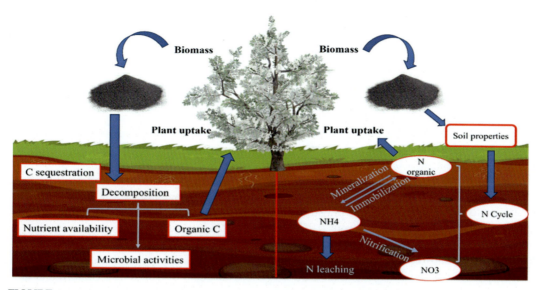

FIGURE 9.2 Mechanism and regulations of carbon and nitrogen between soil and plant.

enhanced plant access to this vital nutrient (He et al., 2023; Yang, Imran, & Ortas, 2023). However, in clay-rich soils, where nitrates may already be less prone to leaching, the influence of biochar on N dynamics might be less pronounced (Gunathunga, Gagen, Evans, Erskine, & Southam, 2023). The high surface area and CEC of biochar allows it to adsorb ammonium ions, making them available to plants over an extended period, effectively acting as a slow-release fertilizer (Rafique et al., 2023; Singh et al., 2023). For example, arginine, which contains amide clusters, is frequently transformed into ammonia or other gaseous nitrogen established during biomass pyrolysis, leading to their loss. Additionally, metallic elements present in the feedstock can affect the alteration of nitrogen-encompassing compounds, thus influencing the abundance and composition of nitrogen species in the resulting biochar properties (Yuan, Li, Shan, Zhang, & Chen, 2023). Even though most biochars have nitrogen substances below 1.5%, there are exceptions with significantly higher levels. For example, biochar derived from sewage sludge shows a nitrogen content of 6.8%, while poultry litter and grass waste contain 5.85% and 4.9% nitrogen content, respectively. Specifically, biochar obtained from sewage sludge at a pyrolysis temperature of 350°C exhibits a higher nitrogen content of 3.17% compared to biochar derived from sugarcane and eucalyptus wastes, which have nitrogen contents of 1.4% and 0.4%, respectively. Furthermore, increasing the pyrolytic temperature leads to a reduction in biochar's nitrogen content due to the renovation of specific amino acids into pyridine-N and pyrrolic-N forms (Abdulraheem, Hu, Ahmed, Li, & Naqvi, 2023). Volatilization-induced nitrogen loss occurs during the process of pyrolysis. For example, chicken manure biochar exhibited nitrogen contents of 2.79%, 2.45%, and 1.81% when produced at temperatures of 250°C, 350°C, and 550°C, respectively. Similarly, maize straw biochar showed a decrease in nitrogen content from 1.25% (300°C) to 1.20% (500°C), while biochar from elephant grass diminished from 3.87% (400°C) to 2.15% (600°C) due to higher pyrolytic temperatures. Prepyrolysis acidification of biochar resulted in a reduction in total nitrogen content, attributed to nitrogen loss through volatilization during pyrolysis. However, salt-impregnated biochar (achieved by prepyrolysis impregnation of the biomass with $CaCl_2$ mineral salts) retained nitrogen content in chicken manure biochar (Hossain et al., 2020). Biochar harbors nitrogen constituents on its surfaces and within its pores, encompassing nitrates, ammonium salts, and heterocyclic compounds. Notably, algal biochar exhibited significantly higher levels of these nitrogen components compared to conventional biochar derived from manure and biosolid/sewage sludge. Amongst the inorganic nitrogen methods, nitrate (NO_3), and organic nitrogen (O-N) exhibited substantial increases at a high pyrolysis temperature of 800°C, whereas ammonium (NH_4-N) experienced a pronounced decrease at 300°C. However, different nitrogen forms persisted stable at 600°C. Hence, careful consideration of the appropriate pyrolytic temperature and feedstock type is crucial when aiming to produce nitrogen-enriched biochar (Chang, Tsai, & Li, 2015).

Similar to nitrogen content, the phosphorus (P) content in different biochars demonstrates significant variability, spanning from 0.005% to 5.9%. While the nitrogen substance reduces with increasing pyrolytic different temperatures, the phosphorus exhibits a positive correlation with temperature. Biochar's effects on phosphorus availability depend on its properties and the inherent characteristics of the soil (Mbabazize, Mungai, & Ouma, 2023; Sharma et al., 2023). In soils with high phosphorus-fixing capacities, biochar's ability to adsorb phosphates can reduce fixation, making P more accessible to plants (Ghosh et al., 2023; Khoshru et al., 2023). This led to increased plant growth and productivity (Kumawat et al., 2023). However, in soils with low phosphorus content, the presence of biochar may have a more subtle effect on P dynamics (Hamidzadeh, Ghorbannezhad, Ketabchi, & Yeganeh, 2023; Yixuan et al., 2023). Biochar's alkaline nature can also influence P availability by altering soil pH, affecting the solubility of phosphates (Peng et al., 2023).

The augmented phosphorus content in biochar at higher pyrolytic temperatures can be recognized to the "absorption influence" which arises due to reduced biochar yield at elevated temperatures. For instance, chicken manure biochar generated at temperatures of 250°C, 350°C, and 550°C manifested subsequent phosphorus contents of 1.91%, 2.15%, and 2.96%, respectively (Yoon et al., 2023). The phosphorus content is also influenced by the specific biomass utilized. For instance, biochar derived from swine solid displayed a higher P content of 5.9%, followed by chicken manure with 2.96%, and poultry litter with 2.57%. In contrast, biochar obtained from rice husks and apple branches exhibited significantly lower P contents of 0.15% and 0.18%, respectively. Therefore, the selection of suitable feedstock plays a pivotal role in producing phosphorus-enriched biochar (Azeem et al., 2023). Furthermore, the investigation revealed a negative relationship between water-extractable phosphorus and pyrolytic temperature for both modified and unmodified biochars. Conversely, Olsen-P demonstrated a positive correlation with expanding temperature. The researchers also performed a decline in Olsen-P levels while chicken manure underwent pretreatment with different salts. Additionally, the analysis unveiled that acidified biochar obtained from maize straw at a temperature of 700°C possessed Olsen-P and water-soluble P contents of 775.45 and 495.21 mg/kg, respectively (Mousavi, Rasouli-Sadaghiani, Sepehr, Barin, & Vetukuri, 2023).

Biochar's influence on secondary nutrients (calcium, magnesium, sulfur) and micronutrients is noteworthy (Ghassemi-Golezani & Rahimzadeh, 2023). Its ability to enhance cation exchange and modify soil pH contributes to improved availability of secondary nutrients (Amin, 2023; Kang et al., 2023). This is particularly relevant in soils that exhibit deficiencies in these nutrients (Jiang et al., 2023). Moreover, biochar's unique properties, such as surface functional groups, can chelate micronutrients, enhancing their availability to plants (El-Gamal, Salem, Mahmoud, & Saleh, 2023; Shafiq, Anwar, Zhang, & Ashraf, 2023). Additionally, biochar can mitigate the adverse effects of certain trace elements (e.g., heavy metals) by reducing their mobility and bioavailability (Gondek, Mierzwa-Hersztek, & Jarosz, 2023). The effects of biochar on nutrient dynamics are profoundly influenced by soil characteristics (Du et al., 2023; Li & Tasnady, 2023). In sandy soils, biochar's capacity to improve water and nutrient retention prevents nutrient leaching, leading to improved plant access to N, P, and K (Brtnicky et al., 2023; Neththasinghe, Dissanayaka, & Karunarathna, 2023). In clay soils, biochar's enhancement of soil structure promotes root penetration and nutrient diffusion, thereby facilitating nutrient uptake by plants (Hartmann & Six, 2023; Wan et al., 2023). Furthermore, the pH-modifying properties of biochar can be tailored to address pH-related nutrient deficiencies (Guo et al., 2022; Msimbira & Smith, 2020). In alkaline soils, biochar can increase nutrient availability, while in acidic soils, it can act as a pH buffer, aiding nutrient accessibility (Neina, 2019).

The potassium content in biochar exhibits variation depending on the feedstock type and pyrolysis temperature. Biochar can be derived from types of manure including chicken and poultry manure, bamboo, and rice demonstrate higher K gist compared to biochar from rice husks, corn stalks, and apple branches. Similar to phosphorus, the potassium content in biochar increases with higher pyrolysis temperatures due to the "concentration effect." For instance, chicken manure biochar exhibited an increase in potassium content ranging from 4.1% to 5.9% as the pyrolysis temperature improved from 250°C to 550°C. Poultry litter-derived biochar displayed K contents of 3.88% and 5.88% at pyrolysis temperatures of 400°C and 600°C, respectively (Bilias et al., 2023). Biosolid-derived biochar was subjected to pyrolysis at temperatures of 300°C, 400°C, 500°C, 700°C, and 900°C, resulting in K contents of 3.9%, 3.98%, 4.1%, 4.02%, 8.12%, and 9.83%, respectively. In another investigation, the enrichment of K in banana peduncle biochar was examined using various gases and plasma, with processing durations of 3, 5, 7, and 9 minutes. The researchers observed that plasma processing for up to 7 minutes resulted in K enrichment in the biochar under different gas environments (Karim, Kumar, Mohapatra, Panda, & Singh, 2014).

Different type of manure-derived biochar exhibits elevated levels of secondary elements such as sulfur (S), calcium (Ca), and magnesium (Mg). The calcium content in biochar derived from animal manure varies between 0.40% and 6.15%, while that from industrial and municipal discarded ranges from 0.37% to 6.57%. Biochar gained from crop remainders contains calcium concentrations ranging from 0.20% to 1.57%, while woody biochar falls within the range of 0.05% to 2.42%. Notably, biochar produced from apple areas demonstrates a higher calcium content of 2.42% compared to other feedstocks as well including wheat and barley straw (0.20%), sugar maple sawdust (0.50%), and acacia (0.27%) (Issahaku, Tetteh, & Tetteh, 2023). Biochar harbors substantial amounts of different nutrients (Table 9.2), known as micro and macronutrients, including iron, copper, boron, zinc, manganese, and molybdenum. Among these micronutrients, the existing scientific literature primarily emphasizes Fe, Zn, and Cu, while Mn, Mo, and B receive comparatively less attention. Notably, biochar derived from animal manure exhibits higher Fe content,

TABLE 9.2 Carbon, nitrogen, potassium contents percentage in different biochar product.

Source	PT[1] (°C)	PH	C	N (TN)	C/N	P (TP Available P) (g kg^{-1})		K
Chicken manure	250	7.6	34.55	2.79	12	1.91	—	4.16
Chicken manure	350	8.9	29.21	2.45	12	2.12	—	4.93
Chicken manure	550	10.2	23.65	1.81	13	2.96	—	5.93
Chicken manure — Ca	250	7.8	30.00	2.85	11	2.21	—	4.87
Chicken manure — Ca	550	10.6	24.73	1.96	13	3.06	—	6.03
Chicken manure — Mg	250	7.3	26.40	2.43	11	2.05	—	3.92
Chicken manure — Mg	350	9.1	26.22	2.42	11	2.67	—	5.03
Chicken manure- Mg	550	10.3	27.04	2.06	13	3.03	—	5.88
Chicken manure — Fe	250	5.75	28.26	2.91	10	2.01	—	3.92
Chicken manure — Fe	550	6.68	27.13	2.17	13	3.10	—	5.95
Poultry litter	500–520	9.30	33.72	3.39	10	2.57	—	5.24
Poultry litter	600	10.4	52.1	4.0	13	1.54	—	5.88
Poultry litter	350	8.70	51.1	4.45	11	2.08	—	4.85
Poultry litter	700	10.3	45.9	2.07	22	3.12	—	1.0
Turkey litter	350	8.00	49.3	4.07	12	2.62	—	4.01
Cow manure	300	8.59	41.02	0.71	58	0.19	—	0.26
Bull manure	300	8.20	60.6	1.3	47	0.3	—	0.20
Bull manure	600	9.50	76.0	0.8	95	0.3	—	0.36
Digested dairy manure	400	9.22	57.7	0.24	240	0.65	—	1.66
Digested dairy manure	600	9.94	59.4	0.23	258	0,83	—	1.49
Dairy manure	350	9.92	55.8	1.51	37	1.0	—	1.43
Barely straw	400	8.02	71.50	1.3	55	—	—	—
Corn stalks	500–600	8.87	71.5	0.69	104	—	—	1.61
Maize straw	450	10.4	—	1.22	—	—	—	—
Maize straw	300	9.84	53.1	2.22	—	—	—	—
Wheat straw	500	10.2	83.40	1.5	56	—	—	—
Elephant grass	400	—	63.86	3.87	17	—	—	—
Elephant grass	500	—	74.85	2.08	36	—	—	—
Kunai grass	600	—	82.23	2.15	38	—	—	—
Switch grass	400	—	55.00	0.7	79	0.10	—	0.46

(*Continued*)

TABLE 9.2 (Continued)

Source	PT[1] (°C)	PH	C	N (TN)	C/N	P (TP Available P) (g kg^{-1})	K
Corn stover	300	7.33	59.5	1.16	54	–	–	0.32
Pearl millet	400	10.6	64	1.1	58	0.16	–	2.52
Sugar maple sawdust	450	7.22	80.00	0.32	250	0.02	–	0.32
Castor stalk	550	–	43.18	1.57	27	0.22	–	0.62
Bambo	600	9.84	70.90	0.41	173	0.11	–	2.78
Hardwood	550	7.80	76	0.22	345	0.02	–	–
Cashew wood residue	–	–	–	0.94	–	0.01	–	0.13
Eucalypt green Waste	650–750	7.30	79.00	0.26	303	0.04	–	0.03
Willow wood waste	550	8.30	47.50	0.38	125	–	–	–
Acacia	400–500	7.01	57.80	1.02	57	1.14	–	–
Sugarcane bagasse	450–500	8.79	63.27	0.67	94	0.07	–	–
Sewage sludge	350	8.15	34.56	2.7	13	1.70	–	0.26
Swine manure	400	7.6	54.9	2.23	24	–	–	–
Pig manure	550	9.90	42.7	–	–	1.55	–	1.62
Swine manure	300	9.11	32.58	2.80	12	–	–	–
Swine solids	350	8.40	51.5	3.54	15	3.89	–	1.78
Rice husk	450	8.53	39.90	0.54	74	0.16	–	0.58
Dairy manure	700	9.9	56.7	0.24	236	1.69	–	2.31
Rice husk	–	9.50	–	0.1	–	0.15	–	0.20
Rice straw	550–600	9.71	44.27	0.64	69	0.09	–	2.82
Maize straw	450	10.4	–	1.22	–	–	–	–

ranging from 311 to 7480 mg/kg, compared to biochar obtained from crop remainders and woody resources. Waste-derived biochar displays Fe content within the scale of 0.009 to 380 mg/kg (Cakmak et al., 2023).

9.7 Uses and limitations of biochar in agricultural arable land

Gaining insights into the consequences and implications of biochar in different agricultural soils holds significant scientific importance. Prior research endeavors have provided evidence of the remarkable potential of biochar as a soil amendment in agricultural landscapes. Regarding the physical properties of soil, biochar exerts beneficial effects on soil stability, promoting particle aggregation and enhancing water-holding capacity. These improvements are attributed to the amelioration of soil pore structure and the enhancement of water retention characteristics (Siedt et al., 2021).

The soil's texture and composition play a crucial role in determining the effects of biochar application. Notably, the advantages of biochar are more pronounced in soils with a coarse texture compared to those with a fine texture. Sandy soils exhibit higher responsiveness to biochar compared to clay-rich soils. The particle size of biochar and its function depth also influence the soil's water retention capability. For sandy soil, optimal water retention is achieved with biochar particles ranging from 0.5 to 1.0 mm in size, applied at a depth of 4 to 6 cm. Moreover, the concentration of biochar used has an impact on water retention. Concentrations exceeding 3% can enhance water-holding capacity in clay soils by up to 60%. However, the ideal biochar concentration may vary depending on specific conditions. In sandy loam soil, higher biochar concentrations ($>5\%$) lead to reduced pore sizes, affecting the soil's hydraulic conductivity (Zhang, Amonette, & Flury, 2021). The choice of biomass feedstock is a critical factor in biochar production. This difference can be ascribed to the wood chip biochar's greater porosity and surface area, enabling superior water adsorption capabilities. The accumulation of biochar to the soil also leads to increased soil fertility by promoting the biological cycling of different nutrients (nitrogen and phosphorus). Moreover, biochar provides organic matter and essential nutrients such as potassium, nitrogen and phosphorus to support plant growth. The mean dwelling time of biochar in the soil is predictable to exceed 3000 years, indicating its long-lasting stability and slow decomposition rate. However, it should be observed that biochar's phospholipid and glycolipid components are more susceptible to decomposition. In addition to the settlement, biochar also influences several other properties of soil, including changes in swelling and shrinkage behavior, tensile strength, surface area, and cracking density (Shakoor et al., 2020).

While most literature suggests the positive effects of biochar modifications, there are restrictions and limitations. Biochar inhibits soil ageing, requiring alternating accumulation of fresh biomass for nutrient cycling and soil—water circumstances. Aged biochar reduces root biomass in rice and *Solanum Lycopersicon* and lowers soil thermal diffusivity. The constructive influences of biochar are soil-specific, not universally positive. Weed problems arise with biochar application at high rates, hindering weed control. Biochar's impact on agricultural productivity varies by plant species and part (Alkharabsheh et al., 2021). It delays flowering and selectively adsorbs pollutants. High-temperature biochar may harm plants, adsorb nitrogen, and counteract essential nutrients. It competes with soil nutrients, reducing phosphorus availability (Schmidt, Hagemann, Draper, & Kammann, 2019). Moreover, biochar's effects can be variable, depending on factors such as feedstock type, production methods, and soil characteristics (Sharma et al., 2023). This necessitates tailoring biochar choices to specific soil conditions and desired outcomes (Kerner et al., 2023). Despite enhancing nutrient availability, biochar itself possesses low nutrient content, requiring it to be seen as a complement rather than a substitute for traditional fertilizers (Ersahin, Cicekalan, Cengiz, Zhang, & Ozgun, 2023). The long-term stability and persistence of biochar in soils remain under investigation, as its sustained benefits and potential impacts on soil health need thorough assessment (Irshad et al., 2023). Furthermore, biochar-soil interactions can impact microbial communities, necessitating careful consideration of site-specific adjustments (Li & Tasnady, 2023; Wang & Delavar, 2023). Lastly, there's a potential risk of contaminants, highlighting the importance of ensuring the biochar's source material is uncontaminated and compatible with the soil's ecosystem

(Phang, Mohammadi, & Mingyuan, 2023). Biochar disturbs organic matter decomposition, decreasing fungal abundance. Source matters: coconut husk biochar improves Zea mays biomass, while orange bagasse biochar has no important effect (Nascimento et al., 2023). Impurity harms plant growth. Biochar production costs vary with feedstock availability, ranging from around 148–389 British pounds per ton. Regulatory issues and biomass testing further increase costs (Espinosa, Moore, Rasmussen, Fehmi, & Gallery, 2020).

9.8 Conclusion

Biochar is a substantial repository of essential plant nutrients, providing micro, macro, and secondary nutrients. Its distinct physical and chemical properties profoundly influence the intricate dynamics of nutrient interactions within the soil by altering key soil attributes like pH and CEC. Biochar is a substantial reservoir of essential agricultural plant nutrients, encompassing micro, macro, and secondary nutrients. The availability of these vital elements within biochar-amended soil hinges upon various influential factors, including the specific feedstock utilized for biochar production, pyrolytic conditions, the rate at which biochar is integrated into the soil matrix, and the unique characteristics of the soil itself. Diverse types of manures and biochar derived from waste materials demonstrate elevated levels of N, P, and K in comparison availability of biochar.

Furthermore, manure and waste-derived biochar, both municipal and industrial, contain a greater abundance of micronutrients in comparison to crop residues and woody biochar. Due to volatilization losses, mainly nutrients correlate completely with pyrolytic temperature, excluding sulfur and nitrogen. By mitigating leaching losses, biochar effectively retains nutrients such as P, K, and others in the soil, enhancing nutrient use efficiency for agriculture and arable land. In the future, biochar is expected to play a crucial role in sustainable agriculture and soil fertility management. Its ability to act as a substantial reservoir of essential plant nutrients, including micro and macronutrients, makes it a valuable amendment for enhancing nutrient availability in the soil. Different types of biochar, derived from various feedstocks and produced under specific conditions, offer diverse nutrient profiles that can be tailored to meet specific agricultural needs. Additionally, biochar's capacity to retain nutrients and mitigate leaching losses can contribute to improved nutrient use efficiency and reduced environmental impacts in agricultural practices. Overall, biochar's potential in agricultural arable land is both promising and challenging. It offers a range of benefits, from improving soil fertility to aiding in carbon sequestration and water retention. Nevertheless, the limitations, including cost, variability, and potential long-term effects, must not be overlooked. Effective utilization of biochar requires informed decisions that consider local conditions, ensuring that its implementation aligns with sustainable agricultural practices.

References

Abdulraheem, M. I., Hu, J., Ahmed, S., Li, L., & Naqvi, S. M. Z. A. (2023). Advances in the use of organic and organomineral fertilizers in sustainable agricultural production.

Abuseif, M. (2023). Exploring influencing factors and innovative solutions for sustainable water management on green roofs: A systematic quantitative review. *Architecture (Washington, D.C.)*, 3(2), 294–327.

Ajmal, K., Ganapathi, S., & Chaithanya, J. (2023). Chapter-3 Soil management strategies to enhance carbon sequestration potential of degraded lands. *Agronomy*, 45.

Al-Hazmi, H. E., Mohammadi, A., Hejna, A., Majtacz, J., Esmaeili, A., Habibzadeh, S., ... Mąkinia, J. (2023). Wastewater treatment for reuse in agriculture: Prospects and challenges. *Environmental Research*, 116711.

Ali Warsame, A., & Hassan Abdi, A. (2023). Towards sustainable crop production in Somalia: Examining the role of environmental pollution and degradation. *Cogent Food & Agriculture, 9*(1), 2161776.

Alkharabsheh, H. M., Seleiman, M. F., Battaglia, M. L., Shami, A., Jalal, R. S., Alhammad, B. A., ... Al-Saif, A. M. (2021). Biochar and its broad impacts in soil quality and fertility, nutrient leaching and crop productivity: A review. *Agronomy, 11*(5), 993.

Alotaibi, K. D., Alharbi, H. A., Yaish, M. W., Ahmed, I., Alharbi, S. A., Alotaibi, F., & Kuzyakov, Y. (2023). Date palm cultivation: A review of soil and environmental conditions and future challenges. *Land Degradation & Development, 34*(9), 2431–2444.

Amalina, F., Krishnan, S., Zularisam, A., & Nasrullah, M. (2023). Effect of process parameters on bio-oil yield from lignocellulosic biomass through microwave-assisted pyrolysis technology for sustainable energy resources: Current status. *Journal of Analytical and Applied Pyrolysis*, 105958.

Amin, A. E.-E. A. Z. (2023). Effects of saline water on soil properties and red radish growth in saline soil as a function of co-applying wood chips biochar with chemical fertilizers. *BMC Plant Biology, 23*(1), 1–14.

Anand, A., Gautam, S., & Ram, L. C. (2023). Feedstock and pyrolysis conditions affect suitability of biochar for various sustainable energy and environmental applications. *Journal of Analytical and Applied Pyrolysis, 170*, 105881.

Atkinson, C. J., Fitzgerald, J. D., & Hipps, N. A. (2010). Potential mechanisms for achieving agricultural benefits from biochar application to temperate soils: A review. *Plant and Soil, 337*, 1–18.

Azeem, M., Jeyasundar, P. G. S. A., Ali, A., Riaz, L., Khan, K. S., Hussain, Q., ... Majrashi, A. (2023). Cow bone-derived biochar enhances microbial biomass and alters bacterial community composition and diversity in a smelter contaminated soil. *Environmental Research, 216*, 114278.

Baloch, M. Y. J., Zhang, W., Sultana, T., Akram, M., Al Shoumik, B. A., Khan, M. Z., & Farooq, M. A. (2023). Utilization of sewage sludge to manage saline-alkali soil and increase crop production: Is it safe or not? *Environmental Technology & Innovation*, 103266.

Banu, M., Rani, B., Kavya, S., & Nihala Jabin, P. (2023). Biochar: A black carbon for sustainable agriculture. *International Journal of Environment and Climate Change, 13*(6), 418–432.

Bilias, F., Kalderis, D., Richardson, C., Barbayiannis, N., & Gasparatos, D. (2023). Biochar application as a soil potassium management strategy: A review. *Science of The Total Environment, 858*, 159782.

Brouziyne, Y., El Bilali, A., Epule Epule, T., Ongoma, V., Elbeltagi, A., Hallam, J., ... McDonnell, R. (2023). Towards lower greenhouse gas emissions agriculture in North Africa through climate-smart agriculture: A systematic review. *Climate, 11*(7), 139.

Brtnicky, M., Mustafa, A., Hammerschmiedt, T., Kintl, A., Trakal, L., Beesley, L., ... Holatko, J. (2023). Pre-activated biochar by fertilizers mitigates nutrient leaching and stimulates soil microbial activity. *Chemical and Biological Technologies in Agriculture, 10*(1), 57.

Cakmak, I., Brown, P., Colmenero-Flores, J. M., Husted, S., Kutman, B. Y., Nikolic, M., ... Zhao, F.-J. (2023). Micronutrients. *Marschner's Mineral Nutrition of Plants* (pp. 283–385). Elsevier.

Chander, G., Singh, A., Abbhishek, K., Whitbread, A. M., Jat, M., Mequanint, M. B., ... Cuba, P. (2023). Consortium of management practices in long-run improves soil fertility and carbon sequestration in drylands of semi-arid tropics. *International Journal of Plant Production*, 1–14.

Chang, Y.-M., Tsai, W.-T., & Li, M.-H. (2015). Chemical characterization of char derived from slow pyrolysis of microalgal residue. *Journal of Analytical and Applied Pyrolysis, 111*, 88–93.

Chausali, N., Saxena, J., & Prasad, R. (2021). Nanobiochar and biochar based nanocomposites: Advances and applications. *Journal of Agriculture and Food Research, 5*, 100191.

Danso, E. O., Yakubu, A., Darrah, Y. O. K., Arthur, E., Manevski, K., Sabi, E. B., ... Andersen, M. N. (2019). Impact of rice straw biochar and irrigation on maize yield, intercepted radiation and water productivity in a tropical sandy clay loam. *Field Crops Research, 243*, 107628.

Das, S. K., Choudhury, B. U., Hazarika, S., Mishra, V. K., & Laha, R. (2023). Long-term effect of organic fertilizer and biochar on soil carbon fractions and sequestration in maize-black gram system. *Biomass Conversion and Biorefinery*, 1–14.

Das, S. K., Ghosh, G. K., & Avasthe, R. (2020). Valorizing biomass to engineered biochar and its impact on soil, plant, water, and microbial dynamics: A review. *Biomass Conversion and Biorefinery*, 1–17.

Du, Z., Zhang, Q., Sun, J., Hong, J., Xu, X., Pang, B., ... Wang, X. (2023). Combined effects of freezing-thawing cycles and livestock excreta deposition complicate soil nitrogen and phosphorus nutrient dynamics in an alpine steppe of northern Tibetan Plateau, China.

El-Gamal, E. H., Salem, L. R., Mahmoud, A. H., & Saleh, M. E. (2023). Evaluation of rice husk biochar as a micronutrients carrier on micronutrients availability in a calcareous sandy soil. *Journal of Soil Science and Plant Nutrition*, 23(2), 1633–1647.

El-Naggar, A., El-Naggar, A. H., Shaheen, S. M., Sarkar, B., Chang, S. X., Tsang, D. C., ... Ok, Y. S. (2019). Biochar composition-dependent impacts on soil nutrient release, carbon mineralization, and potential environmental risk: A review. *Journal of Environmental Management*, 241, 458–467.

Emde, D. (2021). *Understanding the impact of irrigated agricultural systems on soil organic carbon storage*. University of British Columbia.

Ersahin, M. E., Cicekalan, B., Cengiz, A. I., Zhang, X., & Ozgun, H. (2023). Nutrient recovery from municipal solid waste leachate in the scope of circular economy: Recent developments and future perspectives. *Journal of Environmental Management*, 335, 117518.

Espinosa, N. J., Moore, D. J., Rasmussen, C., Fehmi, J. S., & Gallery, R. E. (2020). Woodchip and biochar amendments differentially influence microbial responses, but do not enhance plant recovery in disturbed semiarid soils. *Restoration Ecology*, 28, S381–S392.

Fagodiya, R. K., Verma, K., & Verma, V. K. (2023). Climate resilient agricultural practices for mitigation and adaptation of climate change. *Social Science Dimensions of Climate Resilient Agriculture*.

Fernandes, M. M. H., da Silva, M. F., Ferraudo, A. S., & Fernandes, C. (2023). Soil structure under tillage systems with and without cultivation in the off-season. *Agriculture, Ecosystems & Environment*, 342, 108237.

Gao, F., Khan, R., Yang, L., Chi, Y. X., Wang, Y., & Zhou, X. B. (2023). Uncovering the potentials of long-term straw return and nitrogen supply on subtropical maize (Zea mays L.) photosynthesis and grain yield. *Field Crops Research*, 302, 109062.

Ghassemi-Golezani, K., & Rahimzadeh, S. (2023). Biochar-based nutritional nanocomposites: A superior treatment for alleviating salt toxicity and improving physiological performance of dill (*Anethum graveolens*). *Environmental Geochemistry and Health*, 45(6), 3089–3111.

Ghorbani, M., Neugschwandtner, R. W., Konvalina, P., Asadi, H., Kopecký, M., & Amirahmadi, E. (2023). Comparative effects of biochar and compost applications on water holding capacity and crop yield of rice under evaporation stress: A two-years field study. *Paddy and Water Environment*, 21(1), 47–58.

Ghosh, A., Biswas, D. R., Bhattacharyya, R., Das, S., Das, T. K., Lal, K., ... Biswas, S. S. (2023). Recycling rice straw enhances the solubilisation and plant acquisition of soil phosphorus by altering rhizosphere environment of wheat. *Soil and Tillage Research*, 228, 105647.

Gondek, K., Mierzwa-Hersztek, M., & Jarosz, R. (2023). Effect of willow biochar and fly ash-derived zeolite in immobilizing heavy metals and promoting enzymatic activity in a contaminated sandy soil. *CATENA*, 232, 107429.

Gunathunga, S. U., Gagen, E. J., Evans, P., Erskine, P. D., & Southam, G. (2023). Anthropedogenesis in coal mine overburden; the need for a comprehensive, fundamental biogeochemical approach. *Science of The Total Environment*, 164515.

Guo, Y., Song, B., Li, A., Wu, Q., Huang, H., Li, N., ... Yang, L. (2022). Higher pH is associated with enhanced co-occurrence network complexity, stability and nutrient cycling functions in the rice rhizosphere microbiome. *Environmental Microbiology*, 24(12), 6200–6219.

Gupta, M., Savla, N., Pandit, C., Pandit, S., Gupta, P. K., Pant, M., ... Nair, R. R. (2022). Use of biomass-derived biochar in wastewater treatment and power production: A promising solution for a sustainable environment. *Science of The Total Environment*, 153892.

Hajam, Y. A., Kumar, R., & Kumar, A. (2023). Environmental waste management strategies and vermi transformation for sustainable development. *Environmental Challenges*, 100747.

Hamidzadeh, Z., Ghorbannezhad, P., Ketabchi, M. R., & Yeganeh, B. (2023). Biomass-derived biochar and its application in agriculture. *Fuel*, 341, 127701.

Han, L., Zhang, B., Chen, L., Feng, Y., Yang, Y., & Sun, K. (2021). Impact of biochar amendment on soil aggregation varied with incubation duration and biochar pyrolysis temperature. *Biochar*, 3(3), 339–347.

Hartmann, M., & Six, J. (2023). Soil structure and microbiome functions in agroecosystems. *Nature Reviews Earth & Environment*, 4(1), 4–18.

He, X., Tian, J., Zhang, Y., Zhao, Z., Cai, Z., & Wang, Y. (2023). Attribution and driving force of nitrogen losses from the Taihu Lake Basin by the InVEST and GeoDetector models. *Scientific Reports, 13*(1), 7440.

Gupta, A.K. 2023 The role of soil conservation in plain and hilly areas.

Herath, H. M. S. K. (2012). *Stability of biochar and its influence on the dynamics of soil properties*. A thesis presented in partial fulfilment of the requirements for the degree of Doctor of Philosophy (PhD) in Soil Science, Institute of Natural Resources, College of Sciences, Massey University, Palmerston North, New Zealand Massey University.

Hossain, M. Z., Bahar, M. M., Sarkar, B., Donne, S. W., Ok, Y. S., Palansooriya, K. N., ... Bolan, N. (2020). Biochar and its importance on nutrient dynamics in soil and plant. *Biochar, 2*, 379–420.

Irshad, M. K., Zhu, S., Javed, W., Lee, J. C., Mahmood, A., Lee, S. S., ... Ali, A. (2023). Risk assessment of toxic and hazardous metals in paddy agroecosystem by biochar-for bio-membrane applications. *Chemosphere*, 139719.

Issahaku, I., Tetteh, I. K., & Tetteh, A. Y. (2023). Chitosan and chitosan derivatives: Recent advancements in production and applications in environmental remediation. *Environmental Advances*, 100351.

Javeed, H. M. R., Ali, M., Zamir, M. S. I., Qamar, R., Andleeb, H., Qammar, N., ... Tahir, M. (2023). Biochar application to soil for mitigation of nutrients stress in plants. *In* Sustainable Agriculture reviews 61: Biochar to improve crop production and decrease plant stress under a changing climate (pp. 189–216). Springer.

Jiang, Y., Huang, S., Zhu, F., Guo, X., Zhang, X., Zhu, M., ... Xue, S. (2023). Long-term weathering difference in soil-like indicators of bauxite residue mediates the multifunctionality driven by microbial communities. *Science of The Total Environment, 890*, 164377.

Joseph, S., & Taylor, P. (2014). The production and application of biochar in soils. In Advances in biorefineries, (pp. 525–555). Elsevier.

Kang, F., Lv, Q., Fan, J., Zhang, Y., Song, Y., Ren, X., & Hu, S. (2023). Ameliorative effect of calcium poly (aspartic acid)(PASP-Ca) and calcium poly-γ-glutamic acid (γ-PGA-Ca) on soil acidity in different horizons. *Environmental Science and Pollution Research*, 1–13.

Karim, A. A., Kumar, M., Mohapatra, S., Panda, B., & Singh, A. (2014). Banana peduncle biochar: characteristics and adsorption of hexavalent chromium from aqueous solution. *Methodology, 7*, 1–10.

Kerner, P., Struhs, E., Mirkouei, A., Aho, K., Lohse, K. A., Dungan, R. S., & You, Y. (2023). Microbial responses to biochar soil amendment and influential factors: A three-level meta-analysis. *bioRxiv*, 2023.2006. 2002.543269.

Keswani, C. (2020). *Intellectual property issues in nanotechnology*. CRC Press.

Khairy, M., Amer, M., Ibrahim, M., Ookawara, S., Sekiguchi, H., & Elwardany, A. (2023). The influence of torrefaction on the biochar characteristics produced from sesame stalks and bean husk. *Biomass Conversion and Biorefinery*, 1–22.

Khan, N., Bolan, N., Jospeh, S., Anh, M. T. L., Meier, S., Kookana, R., ... Solaiman, Z. M. (2023). Complementing compost with biochar for agriculture, soil remediation and climate mitigation. *Advances in Agronomy, 179*, 1–90.

Khoshru, B., Nosratabad, A. F., Mitra, D., Chaithra, M., Danesh, Y. R., Boyno, G., ... Pellegrini, M. (2023). Rock phosphate solubilizing potential of soil microorganisms: advances in sustainable crop production. *Bacteria, 2*(2), 98–115.

Koné, S., & Galiegue, X. (2023). Potential development of biochar in Africa as an adaptation strategy to climate change impact on agriculture. *Environmental Management*, 1–15.

Kumawat, K. C., Sharma, B., Nagpal, S., Kumar, A., Tiwari, S., & Nair, R. M. (2023). Plant growth-promoting rhizobacteria: Salt stress alleviators to improve crop productivity for sustainable agriculture development. *Frontiers in Plant Science, 13*, 1101862.

Lei, W., Ruizhi, L., Yan, W., Zongjun, G., Jiaheng, L., Hang, Y., & Xiaoyi, M. (2023). Laboratory and numerical modelling of irrigation infiltration and nitrogen leaching in homogeneous soils. *Pedosphere*..

Jie, H., Zhang, W., Yuan, Q., Ma, Z., Wu, H., Rao, W., ... Wang, D. (2023). Combined biochar and water-retaining agent application increased soil water-holding capacity and maize seedling drought resistance in fluvo-aquic soils.

Li, M., Xiong, Y., & Cai, L. (2021). Effects of biochar on the soil carbon cycle in agroecosystems: an promising way to increase the carbon pool in dryland. In *IOP conference series: Earth and environmental science*.

Li, S., & Tasnady, D. (2023). Biochar for soil carbon sequestration: current knowledge, mechanisms, and future perspectives. *C, 9*(3), 67.

Libra, J. A., Ro, K. S., Kammann, C., Funke, A., Berge, N. D., Neubauer, Y., ... Kern, J. (2011). Hydrothermal carbonization of biomass residuals: A comparative review of the chemistry, processes and applications of wet and dry pyrolysis. *Biofuels, 2*(1), 71–106.

Liu, Z., Li, Z., Chen, S., & Zhou, W. (2023). Enhanced phytoremediation of petroleum-contaminated soil by biochar and urea. *Journal of Hazardous Materials, 453*, 131404.

Lizcano Toledo, R., Lerda, C., Martin, M., Gorra, R., Mania, I., Moretti, B., ... Celi, L. (2019). Effects of inorganic and organic p availability on N fixing capacity of *Vicia villosa*. In *First joint meeting on soil and plant system sciences (SPSS 2019). Natural and human-induced impacts on the critical zone and food production*.

Lytand, W., Singh, B. V., & Katiyar, D. (2023). Biochar's influence on soil microorganisms: Understanding the impacts and mechanisms. *International Journal of Plant Soil Science, 35*(18), 455–464.

Majumder, D., Saha, S., Mukherjee, B., Das, S., Rahman, F., & Hossain, A. (2023). Biochar application for improving the yield and quality of crops under climate change. *Sustainable agriculture reviews 61: Biochar to improve crop production and decrease plant stress under a changing climate* (pp. 3–55). Springer.

Majumder, S., Neogi, S., Dutta, T., Powel, M. A., & Banik, P. (2019). The impact of biochar on soil carbon sequestration: meta-analytical approach to evaluating environmental and economic advantages. *Journal of Environmental Management, 250*, 109466.

Mbabazize, D., Mungai, N. W., & Ouma, J. P. (2023). Effect of biochar and inorganic fertilizer on soil biochemical properties in Njoro Sub-County, Nakuru County, Kenya. *Open Journal of Soil Science, 13*(7), 275–294.

McAmis, S., Bae, H., Ogram, A., Rathinasabapathi, B., & Spakes Richter, B. Living mulches present tradeoffs between soil nutrient cycling and competition during establishment of tea in an organic production system. Available at SSRN 4425602.

Mockevičienė, I., Karčauskienė, D., & Repšienė, R. (2023). The response of retisol's carbon storage potential to various organic matter inputs. *Sustainability, 15*(15), 11495.

Mousavi, R., Rasouli-Sadaghiani, M., Sepehr, E., Barin, M., & Vetukuri, R. R. (2023). Improving phosphorus availability and wheat yield in saline soil of the Lake Urmia Basin through enriched biochar and microbial inoculation. *Agriculture, 13*(4), 805.

Msimbira, L. A., & Smith, D. L. (2020). The roles of plant growth promoting microbes in enhancing plant tolerance to acidity and alkalinity stresses. *Frontiers in Sustainable Food Systems, 4*, 106.

Mumivand, H., Izadi, Z., Amirizadeh, F., Maggi, F., & Morshedloo, M. R. (2023). Biochar amendment improves growth and the essential oil quality and quantity of peppermint (*Mentha* × *piperita* L.) grown under waste water and reduces environmental contamination from waste water disposal. *Journal of Hazardous Materials, 446*, 130674.

Murtaza, G., Ahmed, Z., Eldin, S. M., Ali, B., Bawazeer, S., Usman, M., ... Tariq, A. (2023). Biochar-soil-plant interactions: A cross talk for sustainable agriculture under changing climate. *Frontiers in Environmental Science, 11*, 1059449.

Musto, G., Swanepoel, P., & Strauss, J. (2023). Regenerative agriculture v. conservation agriculture: Potential effects on soil quality, crop productivity and whole-farm economics in Mediterranean-climate regions. *The Journal of Agricultural Science*, 1–11.

Nascimento, I., Fregolente, L. G., Pereira, A. P. d A., Nascimento, C. D. V. d, Mota, J. C. A., Ferreira, O. P., ... Souza Filho, A. (2023). Biochar as a carbonaceous material to enhance soil quality in drylands ecosystems: A review. *Environmental Research*, 116489.

Ndede, E. O., Kurebito, S., Idowu, O., Tokunari, T., & Jindo, K. (2022). The potential of biochar to enhance the water retention properties of sandy agricultural soils. *Agronomy, 12*(2), 311.

Neina, D. (2019). The role of soil pH in plant nutrition and soil remediation. *Applied and environmental soil science, 2019*, 1–9.

Neththasinghe, N., Dissanayaka, D., & Karunarathna, A. (2023). Rhizosphere nutrient availability and nutrient uptake of soybean in response to biochar application. *Journal of Plant Nutrition*, 1–11.

Nieder, R., & Benbi, D. K. (2023). Potentially toxic elements in the environment—A review of sources, sinks, pathways and mitigation measures. *Reviews on Environmental Health* (0)).

Novotný, M., Marković, M., Raček, J., Šipka, M., Chorazy, T., Tošić, I., & Hlavínek, P. (2023). The use of biochar made from biomass and biosolids as a substrate for green infrastructure: A review. *Sustainable Chemistry and Pharmacy, 32*, 100999.

Ofori, A. D., Titriku, J. K., Xiang, X., Ahmed, M. I., & Zheng, A. (2023). Rice production in ghana: a multidimensional sustainable approach.

Ogunwole, A. A., & Ogunwole, O. D. (2023). Effects of water application rates and sawdust biochar on the physicochemical properties of soil and performance of five tree species used in urban landscaping in Ondo. *Nigeria. Cities and the Environment (CATE), 16*(2), 7.

Ortiz, C., Fernández-Alonso, M. J., Kitzler, B., Díaz-Pinés, E., Saiz, G., Rubio, A., & Benito, M. (2022). Variations in soil aggregation, microbial community structure and soil organic matter cycling associated to long-term afforestation and woody encroachment in a Mediterranean alpine ecotone. *Geoderma, 405*, 115450.

Osman, A. I., Farghali, M., Ihara, I., Elgarahy, A. M., Ayyad, A., Mehta, N., ... Hosny, M. (2023). Materials, fuels, upgrading, economy, and life cycle assessment of the pyrolysis of algal and lignocellulosic biomass: A review. *Environmental Chemistry Letters*, 1–58.

Palansooriya, K. N., Dissanayake, P. D., Igalavithana, A. D., Tang, R., Cai, Y., & Chang, S. X. (2023). Converting food waste into soil amendments for improving soil sustainability and crop productivity: A review. *Science of The Total Environment, 881*, 163311.

Pant, D., Shah, K. K., Sharma, S., Bhatta, M., Tripathi, S., Pandey, H. P., ... Bhat, A. K. (2023). Soil and ocean carbon sequestration, carbon capture, utilization, and storage as negative emission strategies for global climate change. *Journal of Soil Science and Plant Nutrition*, 1–17.

Peng, Y., Chen, Q., Guan, C.-Y., Yang, X., Jiang, X., Wei, M., ... Li, X. (2023). Metal oxide modified biochars for fertile soil management: Effects on soil phosphorus transformation, enzyme activity, microbe community, and plant growth. *Environmental Research*, 116258.

Phang, L.-Y., Mohammadi, M., & Mingyuan, L. (2023). Underutilised plants as potential phytoremediators for inorganic pollutants decontamination. *Water, Air, & Soil Pollution, 234*(5), 306.

Pillala, R., Sowmya, N., Gajbhiye, P., & Upadhyay, H. (2023). Impact of biochar in sustainable crop production and soil health management: A review.

Qian, S., Zhou, X., Fu, Y., Song, B., Yan, H., Chen, Z., ... Lai, C. (2023). Biochar-compost as a new option for soil improvement: Application in various problem soils. *Science of The Total Environment, 870*, 162024.

Rafique, M. I., Ahmad, J., Usama, M., Ahmad, M., Al-Swadi, H. A., Al-Farraj, A. S., & Al-Wabel, M. I. (2023). Clay-biochar composites: Emerging applications in soil. *Clay Composites: Environmental Applications*, 143–159.

Rakkar, M., Kaplan, S., & Blanco-Canqui, H. (2023). Sustainable soil management: Soil physical health. *Sustainable Soil Management: Beyond Food Production*, 136.

Ramezanzadeh, H., Zarehaghi, D., Baybordi, A., Bouket, A. C., Oszako, T., Alenezi, F. N., & Belbahri, L. (2023). The impacts of biochar-assisted factors on the hydrophysical characteristics of amended soils: A review. *Sustainability, 15*(11), 8700.

Rasool, S., Rasool, T., & Gani, K. M. (2023). Unlocking the potential of wetland biomass: Treatment approaches and sustainable resource management for enhanced utilization. *Bioresource Technology Reports*, 101553.

Rayanoothala, P., Hasibul Alam, S., Mahapatra, S., Gafur, A., & Antonius, S. (2023). Rhizosphere microorganisms for climate resilient and sustainable crop production. *Gesunde Pflanzen*, 1–19.

Ren, A.-T., Li, J.-Y., Zhao, L., Zhou, R., Ye, J.-S., Wang, Y.-B., ... Xiong, Y.-C. (2023). Reduced plastic film mulching under zero tillage boosts water use efficiency and soil health in semiarid rainfed maize field. *Resources, Conservation and Recycling, 190*, 106851.

Roberts, C., Greene, J., & Nemet, G. F. (2023). Key enablers for carbon dioxide removal through the application of biochar to agricultural soils: Evidence from three historical analogues. *Technological Forecasting and Social Change, 195*, 122704.

Rojano-Cruz, R., Martínez-Moreno, F. J., Galindo-Zaldívar, J., Lamas, F., González-Castillo, L., Delgado, G., ... Cárceles-Rodríguez, B. (2023). Impacts of a hydroinfiltrator rainwater harvesting system on soil moisture regime and groundwater distribution for olive groves in semi-arid Mediterranean regions. *Geoderma, 438*, 116623.

Roy, P., Pal, S. C., Chakrabortty, R., Islam, A. R. M. T., Chowdhuri, I., Saha, A., & Mosavi, A. (2023). The role of indigenous plant species in controlling the erosion of top soil in sub-tropical environment: In-situ field observation and validation. *Journal of Hydrology*, 129993.

Safarian, S. (2023). Performance analysis of sustainable technologies for biochar production: A comprehensive review. *Energy Reports, 9*, 4574–4593.

Saravanan, A., Karishma, S., Kumar, P. S., & Rangasamy, G. (2023). A review on regeneration of biowaste into bio-products and bioenergy: Life cycle assessment and circular economy. *Fuel, 338*, 127221.

Schmidt, H.-P., Hagemann, N., Draper, K., & Kammann, C. (2019). The use of biochar in animal feeding. *PeerJ*, 7, e7373.

Shafiq, F., Anwar, S., Zhang, L., & Ashraf, M. (2023). Nano-biochar: Properties and prospects for sustainable agriculture. *Land Degradation & Development*.

Shakoor, M. B., Ali, S., Rizwan, M., Abbas, F., Bibi, I., Riaz, M., ... Rinklebe, J. (2020). A review of biochar-based sorbents for separation of heavy metals from water. *International Journal of Phytoremediation*, 22(2), 111–126.

Shao, F., Xu, J., Chen, F., Liu, D., Zhao, C., Cheng, X., & Zhang, J. (2023). Insights into olation reaction-driven coagulation and adsorption: A pathway for exploiting the surface properties of biochar. *Science of The Total Environment*, 854, 158595.

Shareef, T. M. E., & Zhao, B. (2016). The fundamentals of biochar as a soil amendment tool and management in agriculture scope: An overview for farmers and gardeners. *Journal of Agricultural Chemistry and Environment*, 6(1), 38–61.

Sharma, S., Negi, M., Sharma, U., Kumar, P., Chauhan, A., Katoch, V., & Sharma, R. (2023). A critique of the effectiveness of biochar for managing soil health and soil biota. *Applied Soil Ecology*, 191, 105065.

Sherzad, R. A., Shinwari, H., Noor, N. A., Baber, B. M., Durani, A., Aryan, S., ... Sarobol, E. (2023). Improving water efficiency, nutrients utilization, and maize yield using super absorbent polymers combined with NPK during water deficit conditions. *NUIJB*, 2(02), 15–31.

Siedt, M., Schäffer, A., Smith, K. E., Nabel, M., Roß-Nickoll, M., & van Dongen, J. T. (2021). Comparing straw, compost, and biochar regarding their suitability as agricultural soil amendments to affect soil structure, nutrient leaching, microbial communities, and the fate of pesticides. *Science of The Total Environment*, 751, 141607.

Singh, A. (2023). Thermal and thermochemical conversion of solid biomass. In Bioenergy for power generation, transportation and climate change mitigation. IOP Publishing.

Singh, P., Rawat, S., Jain, N., Bhatnagar, A., Bhattacharya, P., & Maiti, A. (2023). A review on biochar composites for soil remediation applications: Comprehensive solution to contemporary challenges. *Journal of Environmental Chemical Engineering*, 110635.

Smit, L. (2023). The use of a starch-based superabsorbent polymer to support and optimise potato production in the sandy soils of the Sandveld production region in South Africa.

Tamboli, P., Chaurasiya, A. K., Upadhyay, D., & Kumar, A. (2023). Climate change impact on forage characteristics: An appraisal for livestock production. In. *Molecular interventions for developing climate-smart crops: A forage perspective* (pp. 183–196). Springer.

Tyagi, A. K., Ashoka, P., Saikanth, D., & Ojha, R. (2023). A smart way to increase crop productivity and soil health through biochar. *Advanced Farming Technology*, 15.

Utkarsh, K., Thakur, N., & Shukla, S. K. (2023). Biochar: a feasible and visible solution for agricultural sustainability. *Arabian Journal of Geosciences*, 16(4), 239.

Vassileva, V., Georgieva, M., Zehirov, G., & Dimitrova, A. (2023). Exploring the genotype-dependent toolbox of wheat under drought stress.

Vedovato, L. B., Carvalho, L., Aragão, L. E., Bird, M., Phillips, O. L., Alvarez, P., ... Castro, W. (2023). Ancient fires enhance Amazon forest drought resistance. *Frontiers in Forests and Global Change*, 6.

Velásquez Anrango, R. N. (2023). *Raw materials and usages of Biochar: A state of art* Universidad de Investigación de Tecnología Experimental Yachay.

Vithanage, M., Ashiq, A., Ramanayaka, S., & Bhatnagar, A. (2020). Implications of layered double hydroxides assembled biochar composite in adsorptive removal of contaminants: Current status and future perspectives. *Science of The Total Environment*, 737, 139718.

Wan, H., Liu, X., Shi, Q., Chen, Y., Jiang, M., Zhang, J., ... Hossain, M. A. (2023). Biochar amendment alters root morphology of maize plant: Its implications in enhancing nutrient uptake and shoot growth under reduced irrigation regimes. *Frontiers in Plant Science*, 14, 1122742.

Wang, J., & Delavar, M. A. (2023). Techno-economic analysis of phytoremediation: A strategic rethinking. *Science of The Total Environment*, 165949.

Xia, L., Cao, L., Yang, Y., Ti, C., Liu, Y., Smith, P., ... Butterbach-Bahl, K. (2023). Integrated biochar solutions can achieve carbon-neutral staple crop production. *Nature Food*, 4(3), 236–246.

Yadav, A., Yadav, K., & Abd-Elsalam, K. A. (2023). Nanofertilizers: Types, delivery and advantages in agricultural sustainability. *Agrochemicals*, 2(2), 296–336.

Yang, S., Imran., & Ortas, I. (2023). Impact of mycorrhiza on plant nutrition and food security. *Journal of Plant Nutrition*, 1–26.

Yixuan, C., Zhonghua, W., Jun, M., Zunqi, L., Jialong, W., Xiyu, L., ... Wenfu, C. (2023). The positive effects of biochar application on *Rhizophagus irregularis*, rice seedlings, and phosphorus cycling in paddy soil. *Pedosphere.*.

Yoon, K., Cho, D.-W., Kwon, G., Rinklebe, J., Wang, H., & Song, H. (2023). Practical approach of As (V) adsorption by fabricating biochar with low basicity from $FeCl_3$ and lignin. *Chemosphere*, 138665.

Yu, H., Li, C., Yan, J., Ma, Y., Zhou, X., Yu, W., ... Dong, P. (2023). A review on adsorption characteristics and influencing mechanism of heavy metals in farmland soil. *RSC Advances*, 13(6), 3505–3519.

Yu, Y. (2022). *Fate and transport of micro-and nanoparticles in the subsurface*. Washington State University.

Yuan, H., Li, C., Shan, R., Zhang, J., & Chen, Y. (2023). Nitrogen-containing species evolution during co-pyrolysis of gentamicin residue and biomass. *Journal of Analytical and Applied Pyrolysis*, 169, 105812.

Zakaria, M. R., Farid, M. A. A., Andou, Y., Ramli, I., & Hassan, M. A. (2023). Production of biochar and activated carbon from oil palm biomass: Current status, prospects, and challenges. *Industrial Crops and Products*, 199, 116767.

Zhang, J., Amonette, J. E., & Flury, M. (2021). Effect of biochar and biochar particle size on plant-available water of sand, silt loam, and clay soil. *Soil and Tillage Research*, 212, 104992.

CHAPTER 10

Role of biochar in acidic soils amelioration

Nidhi Luthra[1], Shakti Om Pathak[2], Arham Tater[3], Samarth Tewari[4], Pooja Nain[5], Rashmi Sharma[6], Daniel Prakash Kushwaha[7], Manoj Kumar Bhatt[5], Susheel Kumar Singh[8] and Ashish Kaushal[9]

[1]Department of Soil Science and Agricultural Chemistry, C.C.R. (PG) College, Maa Shakumbhari University, Saharanpur, Uttar Pradesh, India [2]Department of Soil Science and Agricultural Chemistry, S.G.T. University, Gurugram, Haryana, India [3]Division of Soil Science and Agricultural Chemistry, ICAR—Indian Agricultural Research Institute, Pusa, New Delhi, India [4]College of Agriculture Sciences, Teerthanker Mahaveer University, Moradabad, Uttar Pradesh, India [5]Department of Soil Science, Govind Ballabh Pant University of Agriculture & Technology, Pantnagar, Uttarakhand, India [6]Department of Agronomy, School of Agriculture, Graphic Era Hill University, Dehradun, Uttarakhand, India [7]Krishi Vigyan Kendra, Rohtas, Bihar Agricultural University, Sabour, Bhagalpur, Bihar, India [8]Department of Soil Science, College of Agriculture, Rani Lakshmi Bai Central Agricultural University, Jhansi, Uttar Pradesh, India [9]Department of Horticulture, School of Agriculture, Graphic Era Hill University, Dehradun, Uttarakhand, India

10.1 Introduction

The contemporary phenomenon of climate change has fascinated the scientific community towards carbon sequestration. Soil is the most important sink for carbon, an enormous amount of research has been emphasized on explorations of options to enrich the soil with abundant carbon (Purakayastha et al., 2019). In the recent few decades, biochar has gained massive recognition in terms of its use as a soil additive which not only ameliorates problematic soils but also augments the carbon hunger of the soils (Ginebra Muñoz, Calvelo-Pereira,

Doussoulin, & Zagal, 2022; Kushwaha and Kumar, 2021; Kushwaha Kumar, & Chaturvedi, 2021; Tamta Kumar, & Kushwaha, 2023). The term biochar has been derived from two words: "bio" from biomass and "char" from the charcoal. So, the term "biochar" in essence refers to the production of char-like substance by utilizing a biomass feedstock. The biomass when subjected to pyrolysis treatment (heating at elevated temperature but in the presence of limited oxygen) yields a solid product in the form of biochar (Lehmann, 2009). The heating of the biomass chiefly produces the three main constituents: (a) *Gas phase*—which comprises mainly volatile gases like carbon dioxide, carbon monoxide, methane, hydrogen, etc., (b) *Liquid phase*—which comprises various oils and tars depending upon their density, and (c) *Solid phase*—which is chiefly constituted by a carbon-concentrated residue.

The proportion of the production of these constituents essentially depends on the supply of the temperature and the duration for which the heat is supplied. As in the case of fast pyrolysis, where a moderately high temperature of less than 500°C is supplied for a very short period (typically for a few seconds) the final product is in the form of liquid which can be as high as 75% weight of the dry matter of the feedstock. While, on the other hand when a very high temperature of >800°C is applied for a longer time, the production of a high proportion (up to 80% of the dry weight of the feedstock) of the volatile gases can be expected. If the objective is to obtain a dense energy-rich solid carbonaceous compound (up to 35% weight of the feedstock), the biomass is heated at a lower temperature of 200°C–300°C but for a longer duration (up to few weeks) (IEA, 2007). This heating of the biomass at lower temperatures yet for an increased time period is also referred to as torrefaction. Here, the transformation of volatile organic vapors in the form of polymerization further enhances the yield of the final solid fraction (El-Naggar, Mahmoud, & Ibrahim, 2019).

Along with the conditions of the production process involved, the composition of the feedstock also assumes significant importance in the composition of the final product obtained. Biomass is chemically composed of mainly three different organic compounds that are cellulose, hemicelluloses, and lignin. Their thermal stability and behavior on increasing the temperature during the pyrolysis dictate the characteristics of the final product obtained. Hemicellulose is a branched organic polysaccharide which decomposes at a relatively lower temperature of 200°C–300°C and hence, it is the chief governing factor of the biochar properties. Cellulose, being an unbranched polysaccharide molecule thermally degrades at a higher temperature range of 300°C–400°C and has a lesser impact on biochar properties compared to the hemicelluloses. On the other hand, lignin is a multifaceted three-dimensional biomolecule which comprises numerous functional groups bonded by a variety of chemical bonds. Thus, the thermal decomposition range is very wide ranging from lower temperatures of 200°C for breaking the lower energy bonds to the temperatures up to 900°C for completion of the degradation procedure (Yang, Yan, Chen, Lee, & Zheng, 2007).

A typical biomass is composed of mainly three elements that are carbon, oxygen, and hydrogen (Usman et al., 2015). The carbon represents the aromaticity of the compound while the hydrogen and oxygen denote the various functional groups present in the entity like phenolic, carboxylic, enolic, etc. The high-temperature treatment during the production of biochar favors the release of O and H elements from the compounds and increases the relative concentration of the C in the biomass. Therefore, the production of biochar from the raw biomass can also be referred to as carbonization (Chen, Chen, Sun, Zheng, & Fu, 2016).

10.2 Properties of the biochar

Mainly the physical and chemical properties of the biochar depend primarily on the feedstock used and the production process involved, yet a few fundamental characteristics are discussed as follows:

i. Carbon content: Biochar includes both labile and recalcitrant fractions of the carbon. Owing to the concentrated aromatization of biochar during the production process, the proportion of the later fraction exceeds. This accounts for the storage of large quantities of the stable carbon fraction in the soil which cannot be turned over by microbial metabolism (Crombie, Masek, Cross, & Sohi, 2015).

ii. pH: High-temperature treatment is responsible for volatilization of various organic acids along with degradation of various acidic functional groups (carboxylic, phenolic). This detachment of O and H elements from the functional groups renders the surface of the molecule with an unbalanced negative charge. These negative charges can readily accept protons when applied to acidic soil as soil amendment (Mukome, Zhang, Silva, Six, & Parikh, 2013). Along with it, the ash content (which is primarily basic in nature) of the biomass is also enhanced. As these processes are conducive to making the biochar basic in reaction, the pH value of the end product generally lies in the alkaline range (Vassilev, Baxter, Andersen, & Vassilev, 2013).

iii. Cation exchange capacity (CEC): CEC being the surface phenomenon, is significantly dependent on the surface area and the abundance of negative charge on the surface. Biochar has been reported to have increased external area and the de-protonated functional groups are responsible for providing a higher negative charge to the surface. It explains the higher CEC of the biochar compared to the corresponding raw feedstock (Zhang, Wang, & Feng, 2021).

iv. Bulk density and porosity: As the biomass is heated during the process of pyrolysis, the volatile constituents (gases) escape from the solid and thus create voids in the structure. These impregnated voids are responsible for increasing the porosity of the biochar and reducing the bulk density (a measure of the total weight of the solid with respect to the volume occupied) compared to the initial raw biomass (Weber and Quicker, 2018).

v. Surface area and pore size distribution: Loss of the volatile matter in the form of gases also enhances the surface area of the resulting biochar. Among the different size proportions, the impact of the pyrolysis on increasing the micropore fraction (<0.05 μm) is more dominant compared to the macropore fraction (>0.05 μm). Results have revealed an increase of up to 70% in the micropore fraction in biochar compared to raw feedstock (Fu et al., 2012).

vi. Water holding capacity: As the biochar becomes more porous, the ability to retain more water in its structure is expected. Although, considering the nature of the biochar as a whole, it is hydrophobic in nature. This can be inferred from the fact that the loss of the polar functional group and increase of aromaticity upon torrefaction renders the structure hydrophobic. This hydrophobic behavior of the biochar helps to avoid further microbial degradation of the molecule (Gray, Johnson, Dragila, & Kleber, 2014).

10.3 Extent of acid soils: world and India

Out of the total 13.15 billion hectare land of the world (excluding permafrost areas) acid soils have an impact on practically 50% area distributed across various continents (Behera, Shukla, Dwivedi, Cerda, & Lakaria, 2019). Classification of the acid soil based on the pH value can be utilized to characterize soils as slightly acidic (pH 5.5–6.5), moderately acidic (pH 4.5–6.5) highly acidic (pH 3.5–4.5) and extremely acidic (pH <3.5). Analysis of the spatial distribution suggests that out of all acid-affected soil, most of the acid soils in the world fall under the category of moderately acidic and slightly acidic (40.74% and 33.06%, respectively) followed by highly acidic (25.92%) and extremely acidic (3.96%) (Von Uexkull & Mutert, 1995).

The distribution of acid soil in India is also widespread. It occupies a total of 49 Mha area representing nearly one-third of the total arable land of the country (Maji, Obi Reddy, & Sarkar, 2012). The acid-dominated area is prevalent in the humid and subhumid regions of the nation. High rainfall in these regions is responsible for the leaching of the basic cations out of the root zone and leaving behind the soil which is more acidic in the reaction. The exchange sites of these acid soils are thus occupied primarily by acidic cations like H^+ and Al^{+3} while they lack basic cations like Ca^{+2}, Mg^{+2} and K^+ which serve a key role in plant nutrition (Behera & Shukla, 2015). The impact of soil acidity on plant growth depends upon the extent of the severity of the physicochemical conditions of the soil along with the tolerance of the plant to these stresses. But by and large, soil acidity limits the growth of plants and thus necessitates the addition of soil amendment in order to utilize the optimum production capacity of the soil within economic limits.

10.4 Biochar interactions in acidic soils

Acidic soils exist in around 50% of the world's arable land and are mainly classified under the Oxisols, Histosols and Alfisols order of soil taxonomy (Singh et al., 2017). Biochar has distinct effects on soil health restoration under problematic soils (Table 10.1) which largely depend upon the composition and quality of biochar as well as the applied ecosystem. Soil acidity is a natural process which is accelerated by intensive farming to a great extent. One of the main causes of farmlands becoming acidic is the acidic parent materials like granite/tonolite, sulfate and nitrate enriched acid rain, use of acid-forming fertilizers, exhaustive nutrient intake by legume crops, and decomposition of organic matter (Rahman et al., 2018). Recently, biochar has gained immense relevance as a potential soil amendment. Biochar is a substantial carbonaceous component produced by the incomplete combustion of diverse organic precursors, such as agricultural leftovers (Novak et al., 2009). Biochar possesses several key properties that make it an effective tool for acid soil reclamation. Firstly, it's high pH buffering capacity enables biochar to raise the pH levels of acidic soils, thereby neutralizing the excessive acidity and creating a more favorable environment for plant growth. Secondly, biochar's high CEC allows it to attract and retain essential nutrients like calcium, magnesium, and potassium, thereby reducing nutrient leaching and enhancing their availability for plant uptake. Additionally, biochar's

TABLE 10.1 Effect of biochar on soil properties of acidic soils

S. No.	Biochar feedstock	Temperature	Rate of application	Crop	Effect on soil properties	References
1.	Maize straw	550°C	40 t ha^{-1}	Winter wheat	Mean soil temperature increased from 8.5°C to 8.9°C, water content increased from 21.5% to 23.4% while bulk density reduced from 1.34 to 1.26 g cm^{-3} compared to control.	Yan et al. (2019)
2.	Eucalyptus wood	350°C	4 t ha^{-1}	Corn	Soil pH increased from 4.79 to 5.16, CEC from 3.64 to 3.92 cmol kg^{-1}, Soil organic carbon from 4.1 to 25.4 g kg^{-1}, while bulk density reduced from 1.58 to 1.34 g cm^{-3} compared to that of control.	Butnan, Deenik, Toomsan, and Vityakon (2018)
3.	Corn stover and Switchgrass	650°C	0, 52, 104, 156 Mg ha^{-1}	Incubation study	Corn stover biochar had increased the CEC values by 87%, 120%, and 142% and switchgrass biochar by 58%, 89%, and 122% at application rates 52, 104, and 156 Mg ha^{-1}, respectively.	Chintala, Mollinedo, Schumacher, Malo, and Julson (2013)
4.	Maize and Sorghum	350°C	1%, 2%, 4%	Trifoliate orange	Biochar increased soil pH continuously, but the soil pH in lime treatments decreased during initial two years	Wu et al. (2020)
5.	Woodchip biochar	525°C	3%	Mustard	Soil aggregate stability was higher relative to the control in all biochar-amended treatments after three years, with 92% increase	Burrell, Zehetner, Rampazzo, Wimmer, and Soja (2016)
6.	Cacao shell	250°C–350°C (average 300°C)	1%		Soil pH increased by 0.5 units and increased the amount of exchangeable base cation	Martinsen et al. (2015)
7.	Rice husk	250°C–350°C (average 300°C)	1%		Soil pH increased by 0.04 units and increased the amount of exchangeable base cation	Martinsen et al. (2015)
8.	Tobacco stalk	500°C	1 t ha^{-1}		Inhibited N and K leaching in light textured soils	Bindu et al. (2017)
9.	Gram, rice residue and maize stover	400°C	17.6 g kg^{-1}		Increased the pH and CEC of soil along with improving the biological properties as depicted by DHA, B-glucosidase	Nain et al. (2022)

(Continued)

TABLE 10.1 (Continued)

S. No.	Biochar feedstock	Temperature	Rate of application	Crop	Effect on soil properties	References
10.	Lantana camara	–	4 t ha^{-1}	Maize	Increases the pH, EC and availability of N and P of acidic red soil thus the maize yield	Masto, Ansari, George, Selvi, and Ram (2013)
11.	Wheat straw	–	5 t ha^{-1}	Rice	Sandy loam soil	Qin et al. (2016)
12.	Hardwood [primarily oak and hickory]	–	25 g kg^{-1}	Soil column study	The biochar amended soils retained more water at gravitydrained equilibrium (up to 15%), had higher cation exchange capacities (up to 20%), and pH values (up to 1 pH unit) relative to the unamended controls	Laird et al. (2010)

porous structure and large surface area create an ideal habitat for beneficial microorganisms, facilitating improved soil fertility and nutrient cycling. Its adsorption capacity for contaminants, including heavy metals and organic pollutants, helps to reduce their harmful effects on both plants and soil organisms. Moreover, the stability and persistence of biochar in the soil contribute to long-term pH maintenance and nutrient retention, ensuring sustained improvements in the reclamation of acidic soils. Finally, biochar's ability to sequester carbon aids in mitigating climate change and supports overall sustainability in the restoration of acid soils. Because of their alkaline nature and substantial potential to buffer pH, biochars have distinctive features that rectify soil acidity (Table 10.1). Since the pH of biochars is typically 7.0, at least 1.5 units higher than acid soil, it has been determined that the incorporation of biochars slightly raises the pH of acid soil (normally 5.5). Overall, the application of biochar in acidic soils can potentially ameliorate the adverse effects of acidity, improve soil fertility, and enhance overall soil health and productivity. However, the effectiveness of biochar application can vary depending on the type of biochar used, the application rate, and the specific characteristics of the acidic soil in question. Therefore, it is essential to consider these factors when using biochar as a soil amendment in acidic environments.

10.4.1 Biochar and soil pH

For agricultural and environmental uses, biochar has many appealing qualities, including a high CEC, large surface area, alkaline pH, more water-holding capacity, more H/C and O/C ratios, and micro-porosity. These qualities of biochar can change the physicochemical makeup of soil. A significant factor in controlling the pH of biochar is its ash content, which contains plenty of inorganic minerals, particularly K, for plant growth (Chen et al., 2016). The pH of biochar also affects how charges are distributed in the soil, which enhances soil CEC by binding cationic groups to soil organic matter

(Basso, Miguez, Laird, Horton, & Westgate, 2013). Additionally, the high surface charge density of biochar can enhance soil CEC and make it easier for cations to be retained and pH levels to be maintained in soil. The alkaline nature of biochar and the existence of polar and nonpolar surface sites for Al adsorption are crucial for lowering soil Al toxicity and increasing the pH of acidic soils (Dai et al., 2017; Qian, Chen and Hu, 2013; Singh et al., 2023).

In a red ferralitic loam, applying biochar has been shown to boost Ca levels and decrease Al toxicity in addition to helping to raise soil pH (Steiner et al., 2007). There has been an increase in soil pH by 1.0 to 1.5 units by treating the soil with wood bark biochar treatment of $10 \, L \, m^{-2}$ (Yamato, Okimori, Wibowo, Anshori, & Ogawa, 2006). Similarly, Shetty and Prakash (2020), whose application of $20 \, t \, ha^{-1}$ of wood biochar increased the pH of acidic soil by 1.95 units, this result points to the higher alkalinity of biochar generated from nonleguminous feedstock. The biochar made from legume straws frequently caused a bigger increase in soil pH than the biochar made from nonlegume straws because of the former's higher alkalinity (Yuan and Xu, 2011). Due to the significant amount of negative charge on its surface, biochar elevated the pH of acidic soils, which eventually improved the soils' capacity to bind specific nutrients (Gaskin, Steiner, Harris, Das, & Bibens, 2008; Novak et al., 2009).

10.4.2 Biochar and nutrient availability

Biochar can improve the fertility of acid soils by enhancing nutrient availability, which in turn increases crop productivity in addition to reducing soil acidity and Al toxicity (Table 10.2). The large surface area, more porosity, and a greater number of functional groups (i.e., hydroxyl, carboxyl, and alkyl groups) in biochar are the most widely accepted causes for the retention of soil nutrients. These functional groups' interactions with nutrients reduce nutrient leaching and improve plant nutrient uptake (Thies and Rillig, 2009). For instance, Olmo, Villar, Salazar, and Alburquerque (2016) found that lignocellulose-based biochar increased P availability while decreasing N and Mn availability in a Haplic Luvisol soil, which also had a favorable effect on root growth. Although adding biochar does not directly change the amount of nitrogen in the soil, it does stop nitrogen from leaching (Chan, Zwieten, Meszaros, Downie, & Joseph, 2007). The complex-bound organic compounds which are present on the biochar surface are solubilized by microbes and made accessible to plant uptake. Moreover, the functional groups help in the adsorption of associated nutrients (i.e., S, Ca, Mg, K, P, and N) and organic molecules.

Biochar has the capacity to store 12% of the CO_2, CH_4, and N_2O (greenhouse gases) produced by anthropogenic sources in ecologically and economically sustainable systems (Vithanage et al., 2015). Applying biochars boosted the ability of additional fertilizers to retain nutrients and reduced the demand for fertilizer, in contrast to simply absorbing agricultural residues, which had short-term advantages on soil fertility (Al-Wabel et al., 2018). As the majority of nutrients from the biomass feedstock, such as Ca, Mg, K, P, Si, and micronutrients, are maintained in the biochar fraction, biochar can operate as a direct nutrition supply when added to marginal soils (Laird et al., 2010). An increase in soil CEC demonstrated an improvement in a soil's capacity to hold onto cation nutrients like Ca^{2+}, Mg^{2+}, K^+, and NH_4^+ (Cheng, Jones, Hill, Bastami, & Tu, 2018). The level of P in the soil

TABLE 10.2 Effect of biochar on nutrient dynamics.

S. No.	Biochar feedstock	Temperature	Rate of application	Crop	Effect on nutrient dynamics	Reference
1.	*Eupatorium adenophorum*	450°C–500°C	40 t ha^{-1}	Maize-mustard	Increase in the soil carbon content by 175%, total nitrogen by 11%, available phosphorus by 422%, potassium by 80.9%, calcium by 78.2%, magnesium by 60.6% compared to control treatment without biochar application.	Pandit et al. (2018)
2.	Eucalyptus wood	350°C	4 t ha^{-1}	Corn	Soil carbon increased by 708%, Potassium content increased by 145%, increase in secondary nutrients like Ca by 268%, Mg by 106% and micronutrient like Mn increased by 311% compared to control without biochar application.	Butnan et al. (2018)
3.	Hardwood [primarily oak (Quercus spp.) and hickory (Carya spp.)]	–	25 g·kg^{-1} 0, 5, 10, or 20 g kg^{-1} of oven dry soil.	Soil column study	The biochar amendments significantly increased total N (up to 7%), organic C (up to 69%), and Mehlich III extractable P, K, Mg, and Ca.	Laird et al. (2010)
4.	walnut shell	450°C	5 t ha^{-1}	Winter wheat and Lentil	higher bioavailability of nutrient ions like K^+ and Ca^{2+} and partially due to improved soil chemical properties like higher CEC.	Safaei Khorram et al. (2020)
5.	Oryza sativa husk biochar	400°C	5 t ha^{-1}	Rice	Yield + 31.3 (%) increased	Steiner et al. (2007)
6.	Maize straw		10–30 Mg ha^{-1}	Maize	Increased available P, K, total N.	Xiao et al. (2016)

increased from 12.75 ppm without amendment to 18.92 ppm after applying 12 t ha^{-1} biochar (Jemal & Yakob, 2021). This might be brought on by an increase in the pH of the acidic soil, which might result in less fixed P being present.

Brantley, Savin, Brye, and Longer (2016) demonstrated that applying poultry litter biochar at a rate of 10 Mg ha^{-1} enhanced the soil's water-soluble P and Mehlich-3 Mg in a Razort loam soil where maize was grown. According to Tammeorg et al. (2014), spruce biochar produced at temperatures between 550°C and 600°C boosted the levels of ammonium acetate-extractable K in an endogleyic umbrisol. In an incubation experiment, Ippolito, Ducey, Cantrell, Novak, and Lentz (2016) showed that plant-available nutrients viz., Zn, Mn and P increased with biochar application rate. If biochar is utilized as an acid soil amendment, the manufacturing of biochars should be planned for specific agricultural uses considering the wide variation in the nutritional composition of biochar and its impact on soil improvement. The enormous surface area, more pore space and more functional groups in biochar are now considered to be the most effective mechanisms for increasing nutrient availability. Future studies are necessary to show how the characteristics of biochar (namely, surface area and functional groups) and nutrient retention in soil interact. Overall, interactions between nutrients that are readily available from biochar and the retention capacity of biochar may cause a shift in soil nutrient availability.

10.4.3 Biochar and aluminum toxicity alleviation

Biochar produced thermally has a propensity to have a high alkaline pH (Lehmann & Joseph, 2015). It is hypothesized that biochar's adsorption has a longer-lasting impact than its liming action on reducing the toxicity of aluminum (Qian et al., 2013). Consequently, the biochar addition seems to be a cutting-edge method for Al toxicity reduction in acid soils. There are many characteristics of biochar that directly or indirectly helps in reducing Al toxicity like the process of biochar production, temperature, pH, electrical conductivity (EC), CEC, calcium carbonate equivalent, cation content, porosity, ash content, surface area, carboxylic and oxygen-containing functional groups *etc.* (Wang and Wang, 2019). In order to reduce soil Al toxicity, alkalinity of biochar and the availability of polar and nonpolar surface sites for Al adsorption are essential. Multiple methods of biochar are capable of reducing the toxicity of soil Al (Figs. 10.1 and 10.2). Al toxicity is significantly reduced by biochar's surface area and functional groups (Shi et al., 2020).

In addition to aliphatic and aromatic groups, the surface of biochar is mostly composed of hydroxyl, epoxy, carboxyl, acyl, carbonyl, ether, ester, amido, sulfonic, and amyl groups. On the surface of biochar, phenolic-OH, carboxyl, carbonyl, and ester groups are the most prevalent (Xiao, Chen, Chen, Zhu, & Schnoor, 2018). The temperature of the pyrolysis during the synthesis of the biochar greatly influences these functional groups. As the pyrolysis temperature rises, the amount of oxygen-containing functional groups dramatically decreases (Xiao et al., 2018). The adsorption potential of biochar is longer-lasting than its liming potential, according to Qian et al. (2013). $Al(OH)_3$ and $Al(OH)_4$ are less hazardous Al species that can be produced because of the liming effect of biochar. Biochar can also absorb the produced Al monomers. Al adsorption where it complexes with organic −OH and −CHO groups through esterification processes or surface adsorption and co-precipitation of Al with silicate particles to form substances like $KAlSi_3O_8$, has been principally responsible for the success of biochar in reducing soil toxicity (Qian et al., 2013).

10.4.4 Biochar and soil nitrification

Global N cycling depends extensively on nitrification. It is the aerobic conversion of ammonium to nitrate, which results in the production of H^+ ions. Leaching of nitrate speeds up soil acidification since nitrification is an acidification process. The effects of biochar on soil nitrification are still debatable as of this writing. According to some studies, adding biochar to acidic soils can speed up the nitrification process, which would then prevent the correction of the acidity of the soil. For instance, Prommer et al. (2014) discovered during a

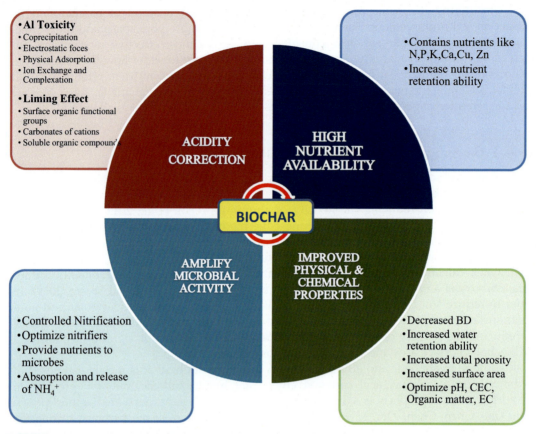

FIGURE 10.1 Mechanism of biochar for ameliorating acidic soil.

field trial that biochar significantly enhanced gross nitrification rates in temperate arable land. In controlled environment investigations, biochars were also discovered to promote the nitrification process (Pereira et al., 2015). Further research is necessary to determine the mechanisms underlying the elevated nitrification caused by biochars. Because acidity can thwart nitrification, the introduction of biochars into acidic soils improved soil pH.

The extent of ammonia-oxidizing genes (amoA) is influenced differently by different biochars, as demonstrated by research by Pereira et al. (2015). For example, walnut shell biochar enhanced amoA whereas hog waste wood lowered it. Increased nitrification can be attributed to biochar's ability to make the environment congenial for nitrifiers in acid soils, which in turn increases the nitrifier population's capacity for nitrification (Prommer et al., 2014). However, the impacts of these key bacteria' activity, abundance, and diversity on the properties of biochar (such as pH, nutrient contents, and adsorption capacity) are not well understood and need further investigation. However, recent research demonstrated that biochars can reduce nitrification and hence avoid acidification in the N-cycle by restricting the amount of NH_4^+ that is available for oxidation reactions (Taghizadeh-Toosi, Clough, Sherlock, & Condron, 2012). According to Yang, Cao, Gao, Zhao, and Li (2015), biochar markedly boosted NH_4^+-N sorption in anthroposols and

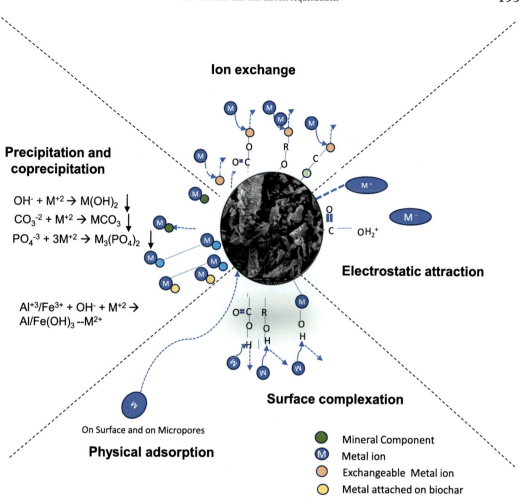

FIGURE 10.2 Mechanism of biochar for reducing Al toxicity in soil.

reduced nitrification. According to Wang et al. (2015), peanut shell biochar lowered nitrification in acid orchard soil by reducing NH_4^+-N concentration and ammonia-oxidizing bacteria (AOB) abundance. As a result, adding biochars is anticipated to slow down nitrification by enhancing NH_4^+ adsorption and reducing microbially-induced ammonia oxidation.

10.5 Biochar and soil carbon sequestration

A promising and practical strategy for storing carbon (C) in soil permanently appears to be the production of biochar and its deposition in the ground (Shaaban et al., 2018). Because biochar is comparatively inert, its deposition in the soil increases the earth's refractory organic C store (Matovic, 2011) thereby, applying biochar is an alternate technique to other standard agricultural practices, which require direct biomass absorption and leads to prompt mineralization and CO_2 release (Crombie et al., 2015). Biochar application can thereby sequester more carbon. Biochar can effectively store carbon for long-term sequestration when put in soil since it can stay there for decades (Bashir et al., 2018). After a year of C mineralization, an incubation study of the biochar-amended Inceptisol soils in India showed that total soil C rose from 410 to 650 g kg^{-1} (Purakayastha et al., 2019).

According to Lehmann, Gaunt, and Rondon (2006) and Sohi, Krull, Lopez-Capel, and Bol (2010), biochar application has the potential to sequester 1 Pg C year^{-1} (the annual mean of the C budget) or even more soil organic carbon. According to Verma, M'hamdi, Dkhili, Brar, and Misra (2014), the C content of biochar may reach 600–800 g kg^{-1}, which is comparable to 2.20 to 2.94 t CO_2 sequestered per ton of biochar. 10% of the world's net primary production turned to charcoal would store 4.8 Gt of carbon per year, or about 20% more than the amount of carbon the atmosphere currently absorbs each year (4.1 Gt C year^{-1}), according to Matovic (2011). A storage capacity of two centuries is provided by the 13.5 t ha^{-1} biochar deposition rate in soil (Sohi et al., 2010). The globe can sequester C effectively with biochar. The total biomass reserves appear to be adequate to meet the need for sequestration and still offer various alternatives to the use of fossil fuels (Yan et al., 2019) (Table 10.3).

TABLE 10.3 A comparative scenario of biochar and other soil amendments in terms of concerns and effectiveness.

Potential amendments	Main effect	Concerns
Lime	1. Increase soil pH, reduce acidity (Schreffler & Sharpe, 2003) 2. Enhance secondary nutrient's concentration (Conyers & Scott, 1989) 3. Minimizes Al toxicity (Ahmad and Tan, 1986)	1. Less effects on subsurface soils (Conyers, Heenan, McGhie, & Poile, 2003) 2. High cost incurred on transportation. 3. Inadvertent addition of heavy metals
Industrial by-products	1. Increase soil pH, reduce acidity (Masud, Li, & Xu, 2015) 2. Reduces Al toxicity (Li, Wang, Xu, & Tiwari, 2010)	1. Risk of Heavy metal.
Plant residues and organic wastes	1. Increase soil pH, reduce acidity (Wang et al., 2013) 2. Increase soil organic carbon content (Martens, 2000) 3. Removes Al toxicity (Haynes and Mokolobate, 2001)	1. Frequent application required (Singh, Kashyap, Kushwaha, & Tamta, 2020) 2. Heavy metal risk (Nicholson, Chambers, Williams, & Unwin, 1999) 3. Eutrophication may occur
Biochar	1. Soil acidity reduction. 2. Nutrient retention and availability (Novak et al., 2009) 3. Enhance microbial population and activity (Gomez, Denef, Stewart, Zheng, & Cotrufo, 2014) 4. Carbon sequestration (Woolf, Amonette, Street-Perrott, Lehmann, & Joseph, 2010)	1. High manufacturing cost 2. More transportation cost. 3. Variability of properties 4. Heavy metal risk

10.6 Future challenges and perspective

Globally, acidic soils are a major issue that lowers agricultural yield (Khalid, Zahir, Arshad, & Asghar, 2009). The use of novel technologies, like biochar, has surfaced as a potential strategy to ameliorate soil acidity even though many measures are being implemented to control soil acidity (Tian et al., 2021). Numerous studies have shown that adding biochar to acidic soils can increase soil pH, the availability of soil nutrients like nitrogen and phosphorus (Spokas, Novak, Venterea, & Cantrell, 2012), soil microbial activity (Chen et al., 2020), water-holding capacity and decrease soil runoff (Busscher, Novak, Ahmedna, & Rehrah, 2011). However, the durability of biochar in soils which is affected by several variables, including soil type, biochar properties, and climatic conditions is crucial to the carbon sequestration potential of biochar (Ameloot, Graber, Verheijen, & Neve, 2015). Additionally, it is important to consider any possible environmental effects of applying biochar, such as modifications to greenhouse gas pollution and the dynamics of soil nutrients (Mendez, Paz-Ferreiro, Gascó, Gutiérrez, & Blanco, 2013) as well as the high cost of production and transportation (Lehmann and Joseph, 2015). The effect of biochar on soil acidity or on soil properties is studied mostly short-term basis under laboratory conditions, the aging effect of biochar in detail in field conditions is needed to develop valid recommendations under various soil and agro-climatic conditions. The effect of biochar on soil physico-chemical properties as well as the microbial communities needs to be further investigated especially in accordance with changes in the biogeochemical cycle (Ding et al., 2016). Researchers are looking forward to biochar as a solution to reduce greenhouse emissions. The most common technique is pyrolysis, which can be categorized into slow, medium, and fart pyrolysis depending on the rate of heating and temperature (Jeffery, Verheijen, van der Velde, & Bastos, 2011; Sohi et al., 2010). Most of the biochar created in India is still using conventional techniques like pit burning and open field burning because the technology is still in its infancy (Sarkar, Singh, Biswas, & Singh, 2019; Singh, Mathur, & Bhatia, 2019). There are many technologies and techniques for producing biochar around the globe, including conventional kilns, retort kilns, gasifiers, and microwave-assisted pyrolysis (Brewer et al., 2016). However, modern methods such as gasifiers, pyrolysis ovens, and microwave-assisted pyrolysis are yet to be introduced on a larger scale, especially Indian context. Sustainability assessment of biochar as a soil amendment with the objective of increasing soil fertility, lowering greenhouse gas pollution, and storing carbon in the soil, in the long run is crucial.

Challenges in adoption and implementation:

- Farmers and officials' ignorance and lack of understanding (Mukherjee, Zimmerman, & Harris, 2017).
- High expenses of manufacturing and poor financial results (Dutta, Sarkar, & Mandal, 2020).
- Insufficient governmental backing and encouragement (Bridle & Sohi, 2019).
- Limited access to money and technology (Bridle & Sohi, 2019).

Challenges particular to India:

- The absence of uniform testing procedures and quality guidelines (Sohi, Krull, Lopez-Capel, & Bol, 2015).
- Ineffective stakeholder cooperation and organization (Sohi et al., 2015).

- The scarcity of manufacturing equipment and feedstocks (Sharma, Wunder, & Ramanathan, 2018).
- Insufficient distribution and transit procedures and facilities for biochar (Sharma et al., 2018).

10.7 Conclusions

The acid soils affect over 50% area of the world and about 33% of arable land in India. The acidity of soil causes nutrient deficiency in several ways and reduces crop productivity. The use of novel technologies, like crop residue-based biochar, has surfaced as a potential strategy to ameliorate soil acidity even though many measures are being implemented for controlling soil acidity. Carbonates, silicates, and functional groups with oxygen are the primary components that make biochar useful for amending acidic soils. Biochar also plays a very important role in enhancing soil physicochemical properties, nutrient availability, soil nitrification, carbon sequestration, and microbial population and activity and in reducing soil acidity, and toxicity of aluminum and other metals. Some issues are also associated with biochar production and its use such as high manufacturing costs, more transportation costs, variability of properties, heavy metal risk, and economic viability. Such issues may be reduced by awareness regarding production technique, use of suitable and easily available feedstock, selection of optimal pyrolysis temperature, and application rate.

References

Ahmad, F., & Tan, K. H. (1986). Effect of lime and organic matter on soybean seedlings grown in aluminum-toxic soil. *Soil Science Society of America Journal*, 50(3), 656–661.

Al-Wabel, M. I., Hussain, Q., Usman, A. R., Ahmad, M., Abduljabbar, A., Sallam, A. S., & Ok, Y. S. (2018). Impact of biochar properties on soil conditions and agricultural sustainability: A review. *Land Degradation and Development*, 29, 2124–2161.

Ameloot, N., Graber, E. R., Verheijen, F. G. A., & Neve, S. (2015). Interactions between biochar stability and soil organisms: Review and research needs. *European Journal of Soil Science*, 66(3), 557–580. Available from https://doi.org/10.1111/ejss.12252.

Bashir, S., Shaaban, M., Mehmood, S., Zhu, J., Fu, Q., & Hu, H. (2018). Efficiency of C3 and C4 plant derived-biochar for Cd mobility, nutrient cycling and microbial biomass in contaminated soil. *Bulletin of Environmental Contamination and Toxicology*, 100, 834–838.

Basso, A. S., Miguez, F. E., Laird, D. A., Horton, R., & Westgate, M. (2013). Assessing potential of biochar for increasing water-holding capacity of sandy soils. *Gcb Bioenergy*, 5(2), 132–143.

Behera, S. K., & Shukla, A. K. (2015). Spatial distribution of surface soil acidity, electrical conductivity, soil organic carbon content and exchangeable potassium, calcium and magnesium in some cropped acid soils of India. *Land Degradion and Development*, 26, 71–79.

Behera, S. K., Shukla, A. K., Dwivedi, B. S., Cerda, A., & Lakaria, B. L. (2019). Alleviating soil acidity: Optimization of lime and zinc use in maize (*Zea Mays* L.) grown on alfisols. *Communications in Soil Science and Plant Analysis*. Available from https://doi.org/10.1080/00103624.2019.1705322.

Bindu, J. P., Reddy, D. D., Santhy, P., Sellamuthu, K., Yassin, M. M., & Naik, R. (2017). Nutrient leaching behaviour of an Alfisol as affected by Tobacco stalk biochar and synthetic Zeolite. *ISTS*.

Brantley, K. E., Savin, M. C., Brye, K. R., & Longer, D. E. (2016). Nutrient availability and corn growth in a poultry litter biochar-amended loam soil in a greenhouse experiment. *Soil Use and Management*, 32(3), 279–288.

Brewer, C. E., Hu, Y., Schmidt-Rohr, K., Loynachan, T. E., Laird, D. A., Brown, R. C., & Novak, J. M. (2016). Influence of production conditions on biochar characteristics and potential environmental applications. *Reviews in Environmental Science and BioTechnology, 15*(2), 243–273. Available from https://doi.org/10.1007/s11157-015-9373-z.

Bridle, T. R., & Sohi, S. P. (2019). The role of biochar in alleviating the challenges of soil degradation and climate change. *European Journal of Soil Science, 70*(4), 779–790. Available from https://doi.org/10.1111/ejss.12803.

Burrell, L. D., Zehetner, F., Rampazzo, N., Wimmer, B., & Soja, G. (2016). Long-term effects of biochar on soil physical properties. *Geoderma, 282,* 96–102.

Busscher, W. J., Novak, J. M., Ahmedna, M. A., & Rehrah, D. (2011). Influence of pecan biochar on physical properties of a Norfolk loamy sand. *Soil Science, 176*(5), 272–280. Available from https://doi.org/10.1097/SS.0b013e318217ab03.

Butnan, S., Deenik, J. L., Toomsan, B., & Vityakon, P. (2018). Biochar properties affecting carbon stability in soils contrasting in texture and mineralogy. *Agriculture and Natural Resources, 51*(6), 492–498.

Chan, K. Y., Zwieten, L. V., Meszaros, I., Downie, A., & Joseph, S. D. (2007). Agronomic values of green waste biochar as a soil amendment. *AJSR, 45*(8), 629–634.

Chen, D., Chen, X., Sun, J., Zheng, Z., & Fu, K. (2016). Pyrolysis polygeneration of pine nut shell: Quality of pyrolysis products and study on the preparation of activated carbon from biochar. *Bioresource Technology, 216,* 629–636.

Chen, D., Zhou, J., Wu, L., Shu, L., Jin, X., Cai, Z., & Li, H. (2020). Influence of biochar on soil properties and bacterial community structure in acidic soil. *Environmental Science and Pollution Research, 27*(26), 32724–32735. Available from https://doi.org/10.1007/s11356-020-09085-7.

Cheng, H., Jones, D. L., Hill, P., Bastami, M. S., & Tu, C. L. (2018). Influence of biochar produced from different pyrolysis temperature on nutrient retention and leaching. *Archives of Agronomy and Soil Science, 64,* 850–859.

Chintala, R., Mollinedo, J., Schumacher, T. E., Malo, D. D., & Julson, J. L. (2013). Effect of biochar on chemical properties of acidic soil. *Archives of Agronomy and Soil Science, 60*(3), 393–404.

Conyers, M. K., Heenan, D. P., McGhie, W. J., & Poile, G. P. (2003). Amelioration of acidity with time by limestone under contrasting tillage. *Soil and Tillage Research, 72*(1), 85–94. Available from https://doi.org/10.1016/S0167-1987(03)00064-3, 2003.

Conyers, M. K., & Scott, B. J. (1989). The influence of surface incorporated lime on subsurface soil acidity. *Australian Journal of Experimental Agriculture, 29*(2), 201–207.

Crombie, K., Masek, O., Cross, A., & Sohi, S. (2015). Biochar synergies and trade-offs between soil enhancing properties and C sequestration potential. *Global Change Biology Bioenergy, 7,* 1161–1175.

Dai, Z., Zhang, X., Tang, C., Muhammad, N., Wu, J., Brookes, P. C., & Xu, J. (2017). Potential role of biochars in decreasing soil acidification - A critical review. *Science of The Total Environment 581–582,* 601–611. Available from https://doi.org/10.1016/j.scitotenv.2016.12.169.

Ding, Y., Liu, Y., Liu, S., Li, Z., Tan, X., Huang, X., . . . Zheng, B. (2016). Biochar to improve soil fertility. A review. *Agronomy for sustainable development, 36*(2), 36.

Dutta, T., Sarkar, A., & Mandal, A. (2020). *Biochar: An alternative soil amendment for agriculture with reduced environmental impact and improved crop productivity. Advances in Soil Science* (pp. 139–155). Springer. Available from https://doi.org/10.1007/978-981-15-2982-1_7.

El-Naggar, A. H., Mahmoud, E. R., & Ibrahim, W. M. (2019). Biochar and compost: A review of remediation techniques for soil improvement. *Journal of Environmental Chemical Engineering, 7*(2), 102899. Available from https://doi.org/10.1016/j.jece.2018.102899.

Fu, P., Hu, S., Xiang, J., Sun, L., Su, S., & Wang, J. (2012). Evaluation of the porous structure development of chars from pyrolysis of rice straw: Effects of pyrolysis temperature and heating rate. *Journal of Analytical and Applied Pyrolysis, 98,* 177–183. Available from https://doi.org/10.1016/j.jaap.2012.08.005.

Gaskin, J., Steiner, C., Harris, K., Das, K., & Bibens, B. (2008). Effect of low temperature pyrolysis conditions on biochar for agricultural use. *Transactions of the ASABE, 51,* 2061–2069.

Ginebra, M., Muñoz, C., Calvelo-Pereira, R., Doussoulin, M., & Zagal, E. (2022). Biochar impacts on soil chemical properties, greenhouse gas emissions and forage productivity: A field experiment. *Science of The Total Environment, 806*150465. Available from https://doi.org/10.1016/j.scitotenv.2021.150465.

Gomez, J. D., Denef, K., Stewart, C. E., Zheng, J., & Cotrufo, M. F. (2014). Biochar addition rate influences soil microbial abundance and activity in temperate soils. *European Journal of Soil Science, 65*(1), 28–39.

Gray, M., Johnson, M. G., Dragila, M. I., & Kleber, M. (2014). Water uptake in biochars: The roles of porosity and hydrophobicity. *Biomass Bioenergy*, 61, 196−205. Available from https://doi.org/10.1016/j.biombioe.2013.12.010.

Haynes, R. J., & Mokolobate, M. S. (2001). Amelioration of Al toxicity and P deficiency in acid soils by additions of organic residues: A critical review of the phenomenon and the mechanisms involved. *Nutrient cycling in agroecosystems*, 59, 47−63.

IEA. (2007). *IEA bioenergy annual report 2006*. International Energy Agency, Paris.

Ippolito, J. A., Ducey, T. F., Cantrell, K. B., Novak, J. M., & Lentz, R. D. (2016). Designer, acidic biochar influences calcareous soil characteristics. *Chemosphere*, 142, 184−191.

Jeffery, S., Verheijen, F. G., van der Velde, M., & Bastos, A. C. (2011). A quantitative review of the effects of biochar application to soils on crop productivity using meta-analysis. *Agriculture, Ecosystems & Environment*, 144(1), 175−187.

Jemal, K., & Yakob, A. (2021). Role of biochar on the amelioration of soil acidity. *Agrotechnology*, 10, 212.

Khalid, M., Zahir, Z. A., Arshad, M., & Asghar, H. N. (2009). Biochar as a potential soil amendment for improving soil properties. *Archives of Agronomy and Soil Science*, 55(5), 565−576. Available from https://doi.org/10.1080/03650340902806422.

Kushwaha, D. P., & Kumar, A. (2021). Modeling of sediment yield and nutrient loss after application of pre-determined dose of top soil amendments. *The Pharma Innovation Journal*, 10(4), 1199−1206. Available from https://www.thepharmajournal.com/archives/2021/vol10issue4/PartQ/10-6-23-477.pdf.

Kushwaha, D. P., Kumar, A., & Chaturvedi, S. (2021). Determining the effectiveness of carbon-based stabilizers blends in arresting soil erosion and elevating properties of Mollisols soils of North Western Himalayas. *Environmental Technology and Innovation*, 23101768. Available from https://doi.org/10.1016/j.eti.2021.101768.

Laird, D. A., Fleming, P., Davis, D. D., Horton, R., Wang, B., & Karlen, D. L. (2010). Impact of biochar amendments on the quality of a typical Midwestern agricultural soil. *Geoderma*, 158, 443−449. Available from https://doi.org/10.1016/j.geoderma.2010.05.013.

Lehmann, J. (2009). Biochar for environmental management: An introduction. *Biochar for environmental management: science and technology*, 25, 15801−15811.

Lehmann, J., Gaunt, J., & Rondon, M. (2006). Bio-char sequestration in terrestrial ecosystems— a review. *Mitigation and Adaptation Strategies for Global Change*, 11, 395−419. Available from https://doi.org/10.1007/s11027-005-9006-5.

Lehmann, J., & Joseph, S. (2015). Biochar for environmental management: An introduction. In J. Lehmann, & S. Joseph (Eds.), *Biochar for environmental management: science, technology and implementation* (pp. 1−12). Routledge.

Li, J. Y., Wang, N., Xu, R. K., & Tiwari, D. (2010). Potential of industrial byproducts in ameliorating soil acidity and aluminum toxicity. *Pedosphere. issas. ac. cn*.

Maji, A. K., Obi Reddy, G. P., & Sarkar, D. (2012). Acid soils of India − their extent and spatial distribution, NBSSLUP Bull. 145, *NBSS&LUP, Nagpur, India*, 138.

Martens, D. A. (2000). Plant residue biochemistry regulates soil carbon cycling and carbon sequestration. *Soil Biology and Biochemistry*, 32(3), 361−369.

Martinsen, V., Alling, V., Nurida, N. L., Mulder, J., Hale, S. E., Ritz, C., & Cornelissen, G. (2015). pH effects of the addition of three biochars to acidic Indonesian mineral soils. *Soil Science and Plant Nutrition*, 61(5), 821−834.

Masto, R. E., Ansari, M. A., George, J., Selvi, V. A., & Ram, L. C. (2013). Co-application of biochar and lignite fly ash on soil nutrients and biological parameters at different crop growth stages of Zea mays. *Ecological Engineering*, 58, 314−322.

Masud, M. M., Li, J. Y., & Xu, R. K. (2015). Application of alkaline slag and phosphogypsum for alleviating soil acidity in an Ultisol profile: A short-term leaching experiment. *Journal of Soils and Sediments*, 15, 365−373.

Matovic, D. (2011). Biochar as a viable carbon sequestration option: Global and Canadian perspective. *Energy*, 36(4).

Mendez, A., Paz-Ferreiro, J., Gascó, G., Gutiérrez, B., & Blanco, F. (2013). Effects of biochar addition on greenhouse gas emissions and soil nutrient dynamics in a laboratory experiment. *Soil Science and Plant Nutrition*, 59(4), 582−590. Available from https://doi.org/10.1080/00380768.2013.801030.

Mukherjee, A., Zimmerman, A. R., & Harris, W. (2017). Surface chemistry variations among a series of laboratory-produced biochars. *Geoderma*, 308, 154−163. Available from https://doi.org/10.1016/j.geoderma.2017.07.029.

Mukome, F. N. D., Zhang, X. M., Silva, L. C. R., Six, J., & Parikh, S. J. (2013). Use of chemical and physical characteristics to investigate trends in biochar feedstocks. *Journal of Agricultural and Food Chemistry*, 61, 2196−2204.

Nain, P., Purakayastha, T. J., Sarkar, B., Bhowmik, A., Biswas, S., Kumar, S., ... Saha, N. D. (2022). Nitrogen-enriched biochar co-compost for the amelioration of degraded tropical soil. *Environmental Technology*, 1–16.

Nicholson, F. A., Chambers, B. J., Williams, J. R., & Unwin, R. J. (1999). Heavy metal contents of livestock feeds and animal manures in England and Wales. *Bioresource Technology, 70*(1), 23–31.

Novak, J. M., Busscher, W. J., Laird, D. L., Ahmedna, M., Watts, D. W., & Niandou, M. A. S. (2009). Impact of biochar amendment on fertility of a southeastern coastal plain soil. *Soil Science, 174*, 105.

Olmo, M., Villar, R., Salazar, P., & Alburquerque, J. A. (2016). Changes in soil nutrient availability explain biochar's impact on wheat root development. *Plant and Soil, 399*, 333–343.

Pandit, N. R., Mulder, J., Hale, S. E., Zimmerman, A. R., Pandit, B. H., & Cornelissen, G. (2018). Multi-year double cropping biochar field trials in Nepal: Finding the optimal biochar dose through agronomic trials and cost-benefit analysis. *Science of the Total Environment, 637*, 1333–1341.

Pereira, E. I. P., Suddick, E. C., Mukome, F. N., Parikh, S. J., Scow, K., & Six, J. (2015). Biochar alters nitrogen transformations but has minimal effects on nitrous oxide emissions in an organically managed lettuce mesocosm. *Biology and Fertility of Soils, 51*, 573–582.

Prommer, J., Wanek, W., Hofhansl, F., Trojan, D., Offre, P., Urich, T., ... Soja, G. (2014). Biochar decelerates soil organic nitrogen cycling but stimulates soil nitrification in a temperate arable field trial. *PLoS One, 9*, e86388.

Purakayastha, T. J., Bera, T., Bhaduri, D., Sarkar, B., Mandal, S., Wade, P., ... Tsang, D. C. (2019). A review on biochar modulated soil condition improvements and nutrient dynamics concerning crop yields: Pathways to climate change mitigation and global food security. *Chemosphere, 227*, 345–365. Available from https://doi.org/10.1016/j.chemosphere.2019.03.170.

Qian, L., Chen, B., & Hu, D. (2013). Effective alleviation of aluminum phytotoxicity by manurederived biochar. *Environmental Science and Technology, 47*, 2737–2745. Available from https://doi.org/10.1021/es3047872.

Qin, X., Wang, H., Liu, C., Li, J., Wan, Y., Gao, Q., & Liao, Y. (2016). Long-term effect of biochar application on yield-scaled greenhouse gas emissions in a rice paddy cropping system: A four-year case study in south China. *Science of the Total Environment, 569*, 1390–1401.

Rahman, Md, Lee, S.-H., Ji, H., Kabir, A., Jones, C., & Lee, K. W. (2018). Importance of mineral nutrition for mitigating aluminum toxicity in plants on acidic soils: Current status and opportunities. *International Journal of Molecular Sciences, 19*, 3073. Available from https://doi.org/10.3390/ijms19103073.

Safaei Khorram, M., Zhang, G., Fatemi, A., Kiefer, R., Mahmood, A., Jafarnia, S., & Li, G. (2020). Effect of walnut shell biochars on soil quality, crop yields, and weed dynamics in a 4-year field experiment. *Environmental Science and Pollution Research, 27*, 18510–18520.

Sarkar, S., Singh, S. K., Biswas, S., & Singh, R. P. (2019). Biochar in India: Current status and prospects. *Energy. Ecology and Environment, 4*(3), 212–222. Available from https://doi.org/10.1007/s40974-019-00138-2.

Schreffler, A. M., & Sharpe, W. E. (2003). Effects of lime, fertilizer, and herbicide on forest soil and soil solution chemistry, hardwood regeneration, and hardwood growth following shelterwood harvest. *Forest Ecology and Management, 177*(1–3), 471–484.

Shaaban, M., Van Zwieten, L., Bashir, S., Younas, A., Núñez-Delgado, A., Chhajro, M. A., ... Hu, R. (2018). A concise review of biochar application to agricultural soils to improve soil conditions and fight pollution. *Journal of Environmental Management, 228*, 429–440.

Sharma, R. K., Wunder, S., & Ramanathan, A. (2018). Sustainable biomass supply chain management: A review on biochar production technology, feedstocks, and their impact on economic and environmental values. *Journal of Cleaner Production, 181*, 335–350. Available from https://doi.org/10.1016/j.jclepro.2018.01.049.

Shetty, R., & Prakash, N. B. (2020). Effect of different biochars on acid soil and growth parameters of rice plants under aluminium toxicity. *Scientific Reports, 10*. Available from https://doi.org/10.1038/s41598-020-69262-x.

Shi, R. Y., Ni, N., Nkoh, J. N., Dong, Y., Zhao, W. R., Pan, X. Y., ... Qian, W. (2020). Biochar retards Al toxicity to maize (*Zea mays* L.) during soil acidification: The effects and mechanisms. *Science of The Total Environment, 719*137448. Available from https://doi.org/10.1016/j.scitotenv.2020.137448.

Singh, N., Mathur, A., & Bhatia, A. (2019). Biochar production technologies: An analysis of their potential towards climate change mitigation and socio-economic impacts. *Journal of Cleaner Production, 234*, 132–146. Available from https://doi.org/10.1016/j.jclepro.2019.06.058.

Singh, S., Luthra, N., Mandal, S., Kushwaha, D. P., Pathak, S. O., Datta, D., ... Pramanick, B. (2023). Distinct behavior of biochar modulating biogeochemistry of salt-affected and acidic soil: A review. *Journal of Soil Science and Plant Nutrition*. Available from https://doi.org/10.1007/s42729-023-01370-9.

Singh, S., Tripathi, D. K., Singh, Swati, Sharma, S., Dubey, N. K., Chauhan, D. K., & Vaculík, M. (2017). Toxicity of aluminium on various levels of plant cells and organism: A review. *Environmental and Experimental Botany, 137*, 177–193. Available from https://doi.org/10.1016/j.envexpbot.2017.01.005.

Singh, S. K., Kashyap, P. S., Kushwaha, D. P., & Tamta, S. (2020). Runoff and sediment reduction using hay mulch treatment at varying land slope and rainfall intensity under simulated rainfall condition. *International Archive of Applied Sciences and Technology, 11*(3), 144–155. Available from https://soeagra.com/iaast/iaast_sept2020/21.pdf.

Sohi, S. P., Krull, E., Lopez-Capel, E., & Bol, R. (2010). A review of biochar and its use and function in soil. *Advances in Agronomy, 105*, 47–82. Available from https://doi.org/10.1016/S0065-2113(10)05002-9.

Sohi, S. P., Krull, E., Lopez-Capel, E., & Bol, R. (2015). A review of biochar and its use and function in soil. *Advances in Agronomy, 130*, 1–76. Available from https://doi.org/10.1016/bs.agron.2014.10.005.

Spokas, K. A., Novak, J. M., Venterea, R. T., & Cantrell, K. B. (2012). Biochar's role as an alternative N-fertilizer: Ammonia capture. *Plant and Soil, 350*(1–2), 35–42. Available from https://doi.org/10.1007/s11104-011-0903-x.

Steiner, C., Teixeira, W. G., Lehmann, J., Nehls, T., De Macêdo, J. L. V., Blum, W. E. H., & Zech, W. (2007). Long term effects of manure, charcoal and mineral fertilization on crop production and fertility on a highly weathered Central Amazonian upland soil. *Plant and Soil, 291*, 275–290.

Taghizadeh-Toosi, A., Clough, T. J., Sherlock, R. R., & Condron, L. M. (2012). Biochar adsorbed ammonia is bioavailable. *Plant and Soil, 350*, 57–69.

Tammeorg, P., Parviainen, T., Nuutinen, V., Simojoki, A., Vaara, E., & Helenius, J. (2014). Effects of biochar on earthworms in arable soil: Avoidance test and field trial in boreal loamy sand. *Agriculture, Ecosystems and Environment, 191*, 150–157.

Tamta, S., Kumar, A., & Kushwaha, D. P. (2023). Potential of roots and shoots of Napier grass for arresting soil erosion and runoff of mollisols of Himalayas. *International Soil and Water Conservation Research*. Available from https://doi.org/10.1016/j.iswcr.2023.02.001.

Thies, J. E., & Rillig, M. C. (2009). Characteristics of biochar: Biological properties. *Biochar for Environmental Management: Science and Technology, 1*, 85–105.

Tian, Y., Sun, X., Yan, X., Wu, W., Fan, J., Wang, L., & Xie, D. (2021). Biochar as an amendment for acid soils: A review of the mechanisms, performance, and challenges. *Environmental Pollution, 271*, 116329. Available from https://doi.org/10.1016/j.envpol.2020.116329.

Von Uexkull, H. R., & Mutert, E. (1995). Global extent, development and economic impact of acid soils. *Plant and Soil, 171*(1), 1–15. Available from https://doi.org/10.1007/BF00009558.

Usman, A. R., Abduljabbar, A., Vithanage, M., Ok, Y. S., Ahmad, M., Ahmad, M., ... Al-Wabel, M. I. (2015). Biochar production from date palm waste: Charring temperature induced changes in composition and surface chemistry. *Journal of Analytical and Applied Pyrolysis, 115*, 392–400. Available from https://doi.org/10.1016/j.jaap.2015.08.016.

Vassilev, S. V., Baxter, D., Andersen, L. K., & Vassilev, C. G. (2013). An overview of the composition and application of biomass ash. Part 1. Phase-mineral and chemical composition and classification. *Fuel, 105*, 40–76.

Verma, M., M'hamdi, N., Dkhili, Z., Brar, K. S., & Misra, K. (2014). Thermochemical transformation of agro-biomass into biochar: Simultaneous carbon sequestration and soil amendment. In S. Brar, G. Dhillon, & C. Soccol (Eds.), *Biotransformation of Waste Biomass into High Value Biochemicals* (pp. 51–70). New York: Springer.

Vithanage, M., Rajapaksha, A. U., Zhang, M., Thielebruhn, S., Lee, S. S., & Ok, Y. S. (2015). Acid activated biochar increased sulfamethazine retention in soils. *Environmental Science and Pollution Research, 22*, 2175–2186.

Wang, J., & Wang, S. (2019). Preparation, modification and environmental application of biochar: A review. *Journal of Cleaner Production, 227*, 1002–1022.

Wang, Z., Zheng, H., Luo, Y., Deng, X., Herbert, S., & Xing, B. (2013). Characterization and influence of biochars on nitrous oxide emission from agricultural soil. *Environmental pollution, 174*, 289–296.

Wang, Z., Zong, H., Zheng, H., Liu, G., Chen, L., & Xing, B. (2015). Reduced nitrification and abundance of ammonia-oxidizing bacteria in acidic soil amended with biochar. *Chemosphere, 138*, 576–583.

Weber, K., & Quicker, P. (2018). Properties of biochar. *Fuel, 217*, 240–261.

Woolf, D., Amonette, J. E., Street-Perrott, F. A., Lehmann, J., & Joseph, S. (2010). Sustainable biochar to mitigate global climate change. *Nature Communications, 1*(1), 1–9. Available from https://doi.org/10.1038/ncomms1053.

Wu, S., Zhang, Y., Tan, Q., Sun, X., Wei, W., & Hu, C. (2020). Biochar is superior to lime in improving acidic soil properties and fruit quality of Satsuma mandarin. *Science of The Total Environment, 714*, 136722.

Xiao, Q., Zhu, L. X., Zhang, H. P., Li, X. Y., Shen, Y. F., & Li, S. Q. (2016). Soil amendment with biochar increases maize yields in a semi-arid region by improving soil quality and root growth. *Crop and Pasture Science (New York, N.Y.), 67*(5), 495–507.

Xiao, X., Chen, B., Chen, Z., Zhu, L., & Schnoor, J. L. (2018). Insight into multiple and multilevel structures of biochars and their potential environmental applications: A critical review. *Environmental Science & Technology, 52*(9), 5027–5047.

Yamato, M., Okimori, Y., Wibowo, I. F., Anshori, S., & Ogawa, M. (2006). Effects of the application of charred bark of *Acacia mangium* on the yield of maize, cowpea and peanut, and soil chemical properties in South Sumatra, Indonesia. *Soil Science and Plant Nutrition, 52*, 489–495. Available from https://doi.org/10.1111/j.1747-0765.2006.00065.x.

Yan, Q., Dong, F., Li, J., Duan, Z., Yang, F., Li, X., ... Li, F. (2019). Effects of maize straw-derived biochar application on soil temperature, water conditions and growth of winter wheat. *European Journal of Soil Science, 70*(6), 1280–1289. Available from https://doi.org/10.1111/ejss.12863.

Yang, F., Cao, X., Gao, B., Zhao, L., & Li, F. (2015). Short-term effects of rice straw biochar on sorption, emission, and transformation of soil NH_4^+-N. *Environmental Science and Pollution Research, 22*, 9184–9192.

Yang, H., Yan, R., Chen, H., Lee, D. H., & Zheng, C. (2007). Characteristics of hemicellulose, cellulose and lignin pyrolysis. *Fuel, 86*, 1781–1788. Available from https://doi.org/10.1016/j.fuel.2006.12.013.

Yuan, J. H., & Xu, R. K. (2011). The amelioration effects of low temperature biochar generated from nine crop residues on an acidic Ultisol. *Soil Use and Management, 27*, 110–115.

Zhang, Y., Wang, J., & Feng, Y. (2021). The effects of biochar addition on soil physicochemical properties: A review. *Catena, 202*105284. Available from https://doi.org/10.1016/j.catena.2021.105284.

11

Biochar as a soil amendment for saline soils reclamation: mechanisms and efficacy

Rashida Hameed[1], Adeel Abbas[1], Guanlin Li[1], Aitezaz A.A. Shahani[2], Beenish Roha[3] and Daolin Du[1]

[1]School of the Environment and Safety Engineering, Institute of Environment and Ecology, Jiangsu University, Zhenjiang, China [2]Key Laboratory of Crop Sciences and Plant Breeding Genetics, College of Agriculture, Yanbian University, Yanji, Jilin, China [3]School of Water Resources and Environment, China University of Geosciences (Beijing), Beijing, China

11.1 Introduction

Salinity refers to the condition wherein a solution (such as water or soil) contains an elevated concentration of soluble salts, reaching levels that exert detrimental effects on the growth and viability of cultivated plant species (Hailu & Mehari, 2021). A soil is categorized as saline when distinct alterations become evident in the upper layer of the earth, characterized by the existence of a white layer of salt crust on the surface of the land, along with a noticeable rise in saline patches dispersed throughout the soil. Additional indicators of saline soil include prolonged waterlogging following precipitation, the deterioration and disintegration of roads, the exacerbation of groundwater and surface water quality rendering them unsuitable for consumption by animals and humans, and the occurrence of salt crystals, predominantly observed in highly saline soils (Kumawat et al., 2023). Soils affected by salt can be classified into three main categories: saline, saline–sodic, or sodic soils. In sodic soils, the predominant cation is sodium, accompanied by carbonate and bicarbonate ions as the dominant anions. Hence, soils containing a high level of soluble salts are termed saline soils, whereas soils containing a substantial amount of exchangeable sodium are categorized as sodic soils. On the other hand, saline–sodic soils showcase a notable abundance of both salts and exchangeable sodium. In contradistinction, regular soils generally exhibit an electrical conductivity (EC) lower than $4\,dS\,m^{-1}$ (Liu et al., 2023a). Based on EC, water or soil solutions can be effectively characterized (Table 11.1).

Salinity stands as a significant menace to the global food security landscape. Areas marked by arid and semiarid conditions in Asia, Australia, Africa, and South America routinely grapple with the emergence of soils adversely impacted by salt. The collective expanse of salt-affected soils encompasses roughly one billion hectares and is projected to amplify on a global scale in the coming years, primarily due to inadequate handling of water and land resources, alongside irregular shifts in regional and global climatic patterns (Irfan et al., 2023). Ramifications stemming from soils impacted by salt encompass a range of issues, including disrupted nutrient equilibrium, unfavorable physical soil traits, osmotic stress, decreased crop yields, toxicity considerations, constrained agricultural output, and diminished economic returns. The excessive accumulation of soluble salts within the soil obstructs the steady and unimpeded growth of crops, resulting in lowered yields (Huang et al., 2023). Moreover, salinity gives rise to imbalances in nutrients, impeding the absorption of vital plant nutrients such as potassium (K), calcium (Ca), magnesium (Mg), phosphorus (P), and more. This ultimately leads to diminished crop yields. Addressing the challenges posed by global food security necessitates the farming of barren salinity-affected areas (Ashilenje et al., 2023).

By employing techniques designed to alleviate the prevalence of soluble salts and/or exchangeable sodium (such as the reclamation of salt-affected soils) or counteracting the detrimental effects of salts on plants (through the management of salt-affected soils), it becomes feasible to harness salt-affected soils for productive crop cultivation. Water-soluble salts inherently exist in all soils, and plants obtain vital nutrients in this soluble state. Nevertheless, the excessive buildup of soluble salts can substantially hinder the growth of plants (Omar et al., 2023). Salinity doesn't just negatively impact the agricultural yield of different crops; it also influences the physical and chemical characteristics of the soil and disrupts the ecological balance in the area (Abbas et al., 2022). The consequences of salinity are mainly evident in reduced crop output, diminished economic gains, and soil erosion (Qian et al., 2023).

TABLE 11.1 Different salinity levels and effects on crops.

Salinity levels	EC (dS m^{-1})	Type	Effect on crops
Non-saline	0–2	Drinking and irrigation	Negligible
Low saline	2–4	Irrigation water	Reduction in the yield of some sensitive crops
Mild saline	4–8	Primary drainage water and groundwater	Growth of many crops is satisfactory and Reduction in the yield is about 50%–60%
High saline	8–16	Secondary drainage water and groundwater	salt-tolerant crop plants growth and yield occur up to a satisfactory level: reduction in the yield is more than 70%
Very high saline	>16	Very saline groundwater	Only salt resistance crops exist

The effects of salinity arise from intricate interactions among diverse processes encompassing morphological, physiological, molecular, and biochemical mechanisms. These interactions profoundly influence seed germination, plant growth, as well as the commitment of water and nutrients (Wahab et al., 2023). Soil salinity poses various hurdles for plants, encompassing ion toxicity, osmotic stress, and scarcity of vital nutrients such as nitrogen, calcium, potassium, phosphorus, iron, and zinc, alongside oxidative stress (Abbas et al., 2021). Consequently, it hampers the plants' capacity to uptake water from the soil. Moreover, soil salinity significantly curtails the plants' uptake of phosphorus, given that phosphate ions combine with calcium ions to form precipitates. Specific elements like sodium, chlorine, and boron can have toxic repercussions on plants, as they tend to be particularly vulnerable to their adverse effects (Rahimzadeh & Ghassemi-Golezani, 2023).

The rapid buildup of sodium within cellular walls can swiftly trigger osmotic stress, ultimately culminating in cell death. Salinity also exerts significant impacts on photosynthesis, mainly by reducing leaf size, chlorophyll levels, and stomatal activity. Additionally, salinity negatively affects the reproductive advancement of seeds, hindering microsporogenesis and stamen filament elongation. This promotes controlled cell death in particular tissue types, leads to the abortion of ovules, and triggers the aging process in fertilized embryos (Piao et al., 2023).

The utilization of biochar additions in saline-sodic and/or sodic soils can positively impact plant growth by improving soil characteristics, particularly the physicochemical attributes. Organic amendments can foster the dissolution of natural calcite ($CaCO_3$) reserves by promoting the generation of carbonic acid within the soil structure. This process of dissolution can subsequently liberate calcium into the soil solution, consequently assisting in the removal of sodium from cation exchange sites (Alghamdi et al., 2023). Furthermore, organic additives play a role in improving particle aggregation within saline-sodic and sodic soils, resulting in the creation of more substantial water-stable clusters. Moreover, the relatively larger particles inherent in organic substances can establish pathways within saline-sodic and/or sodic soils, which often exhibit inadequate structural arrangement. This outcome contributes to enhanced soil permeability, aiding in the efficient removal of sodium through leaching from the soil profile (Cavalcanti-Lima et al., 2023).

11.2 Biochar can serve as a viable amendment for soil with high salinity

Biochar is a carbon-rich material produced by subjecting biomass to pyrolysis at temperatures ranging from 300°C to 1000°C in a low-oxygen environment. Its application in agriculture is increasingly garnering attention due to its capacity to improve soil fertility, enhance crop yields, sequester carbon in the soil, and reduce greenhouse gas emissions. The use of biochar has emerged as a promising approach to addressing soil pollution by immobilizing heavy metals and organic pollutants, as indicated by the research of (Hameed et al., 2020, 2021). Moreover, this strategy holds promise for improving the overall health and quality of the soil, as suggested by the findings (Zhang et al., 2020). Beyond its advantageous impacts, biochar has demonstrated the potential to optimize water and fertilizer utilization efficiency, while also diminishing soil salinity, especially in arid and semiarid areas, as emphasized by (Kong et al., 2021). Different varieties of biochar exhibit a range of effects on distinct soil types, and likewise, the impact of a single type of biochar may vary across various soil compositions (Mishra et al., 2023). The characteristics of biochar, both physical and chemical, are shaped by the particular composition of biomass and the conditions under which pyrolysis is carried out during its manufacturing process (Hameed et al., 2019). Biochar derived from plant residues and manures (non-woody biomass) tends to possess less stable carbon, but higher nutrient content and elevated pH levels. Conversely, biochar obtained from lignocellulosic biomasses (woody biomass) has a more stable carbon content. The use of high-temperature pyrolysis brings about significant changes in the physical and chemical attributes as well as the composition of biochar. On the other hand, low-temperature pyrolysis leads to relatively minor deviations from the original biomass (Osman et al., 2023).

Biochar generated through high-temperature pyrolysis typically presents a greater proportion of fixed carbon and ash, accompanied by lower volatile content when contrasted with biochar produced under low pyrolysis temperatures. This type of biochar tends to possess a high degree of recalcitrance due to its elevated aromaticity. Furthermore, attributes like surface area, pH, elemental composition (including K, P, Ca, etc.), and ratios of C:O and C:N typically experience an upsurge with an increase in pyrolysis temperature (Anand et al., 2023; Nair et al., 2023). Based on empirical observations, the C:O ratio of biochar appears to be relatively unchanged in the context of fast pyrolysis, as opposed to slow pyrolysis. Nevertheless, it has been noted that biochar's surface area experiences a notable augmentation when generated through the process of slow pyrolysis (Zhang et al., 2023b).

The extent of improvements seen in soil and plant quality is shaped by the quantity of biochar applied, the specific soil type (with disparities in organic carbon, nutrient levels, and overall quality), and the duration over which biochar is integrated into the soil (whether applied freshly or showing residual effects). Numerous research findings have pointed out that the influence of biochar is more pronounced in soils that are severely weathered, degraded, and lacking in nutrients, as opposed to soils that are well-structured, nutrient-abundant, and of higher quality (Abiven et al., 2014). Moreover, studies have showcased that biochar has the capacity to enhance plant growth through mechanisms that operate both directly and indirectly. When biochar is applied for direct growth promotion, it supplies plants with crucial mineral nutrients like calcium (Ca), magnesium (Mg), phosphorus (P), potassium (K), and sulfur (S). Conversely, the indirect mechanism revolves around the enhancement of the soil's physical, chemical, and biological properties (Enders et al., 2012; Xu et al., 2012). Hence, the effects of biochar on soil and plant quality can demonstrate considerable variability (Goldschmidt & Buffam, 2023; Qian et al., 2023). For example, a study showcased that utilizing biochar sourced from a blend of bull manure, dairy manure, and pine wood led to a noteworthy rise in total nitrogen levels in a sandy loam Spodosol. However, in a clay loam Alfisol, no substantial increase was noted. Additionally, introducing biochar as an amendment has demonstrated the ability to improve the porous arrangement and water-holding capacity of saline soils. Nevertheless, the degree of this impact depends on factors such as pyrolysis conditions, the type of biomass used, and the quantity of biochar added to the soil (Ponomarev et al., 2023).

Likewise, in another investigation, the utilization of wood-derived biochar was found to stimulate the absorption of phosphorus (P) and nitrogen (N) by corn plants cultivated in sandy loam soil. Nonetheless, in silt loam soil, the assimilation of these nutrients was observed to be impeded (Liu et al., 2023b). Aged biochar displays a higher accumulation of oxygen functional groups, along with an increased anion and cation exchange capacity (CEC), in contrast to fresh biochar. As a result, soils manifest differing reactions when exposed to these two varieties of biochar (Cui et al., 2023; Xu et al., 2023). Much like in soils unaffected by salinity, the introduction of biochar can also bring benefits to soils impacted by high salt levels. This is achieved by increasing surface area, enriching soil organic matter, and enhancing the presence of vital nutrients like calcium (Ca), magnesium (Mg), zinc (Zn), potassium (K), and manganese (Mn). Furthermore, biochar aids in stabilizing soil structure and regulating moisture levels. It also aids in retaining polyvalent cations and supports the removal of sodium (Na) from exchange sites by offering calcium (Ca) in the soil solution (Aziz et al., 2023; Qian et al., 2023).

Biochar further establishes a favorable habitat for soil microorganisms, fostering their role in the restoration of salt-affected soils. Moreover, biochar holds the potential to accelerate salt leaching, thereby shortening the duration necessary to achieve a conventional salt concentration level appropriate for plant growth. Furthermore, the integration of biochar into soil enhances the content and durability of organic compounds. These compounds play a pivotal role in binding soil aggregates for a prolonged period, in contrast to the easily fragmented particles stemming from other carbon-based amendments (Tang et al., 2023; Wang et al., 2023b).

11.3 Role of biochar on soil properties

11.3.1 Role of biochar on nutrient status of salt-affected soils

Soils affected by salinity experience disturbances in their physical, chemical, and biological characteristics, resulting in imbalances of nutrients and the emergence of ion toxicities within the soil structure. The abundance of soluble salts in the soil obstructs the accessibility and uptake of vital nutrients, thus causing adverse repercussions (Etesami et al., 2023). This situation can occur due to the competition among ions or an increase in osmotic pressure, both of which hinder the movement of mineral nutrients to the plant roots. In soils affected by salinity, deficiencies in crucial elements like nitrogen (N), phosphorus (P), and potassium (K) are commonly observed due to various factors. These factors include inadequate organic matter input from plant residues and higher rates of organic matter loss, especially in specific sodic soils. The wetting of sodic soils can trigger the breakdown of aggregates, resulting in improved access to and availability of previously shielded organic matter, thereby accelerating its decomposition (Lu & Fricke, 2023; Saygin et al., 2023). Within sodic soils, the solubility of organic matter is increased, leading to greater leaching losses of mineralized nutrients. Furthermore, heightened salinity levels negatively influence the development and function of soil microbial communities, thus disrupting nutrient transformation mechanisms and indirectly hindering nutrient availability to plants. Additionally, excessive salinity can lead to nitrogen losses to the atmosphere, thereby exacerbating the repercussions on the nutritional state of both soil and plants (Gupta et al., 2023; Khangura et al., 2023).

11.3.2 Biochar and micronutrients and macronutrients in salt-affected soils

The behavior and accessibility of micronutrients in soils affected by salinity show significant deviations from those in regular soils. The introduction of alkaline biochar as a soil amendment into salt-affected soils can exacerbate the issue of restricted access to essential micronutrients like iron (Fe), zinc (Zn), copper (Cu), and manganese (Mn), while simultaneously intensifying boron (B) toxicity. Furthermore, the scarcity of zinc (Zn) is particularly common in calcareous saline-sodic and sodic soils (Bedadi et al., 2023). Molybdenum (Mo) in soils predominantly exists as anions, and its solubility and ease of use are heightened in alkaline conditions, frequently found in saline-sodic and sodic soils. Similarly, the solubility and accessibility of boron (B) also increase with rising pH levels. However, when the pH exceeds 9.0, boron (B) becomes toxic to a variety of crop plants (Havlin, 2020). However, the available pool of research concerning the effects of applying biochar on the solubility and accessibility of micronutrients in salt-affected soils is currently restricted. It holds significant importance to thoroughly investigate the consequences of different types of biochar, including both acidic and alkaline variations, on the dynamics and alterations of pH-sensitive trace elements within these soil conditions (Anawar & Hossain, 2023).

Considering the prevalent reliance on inorganic fertilizers to provide crucial minerals for plant growth, the adoption of organic amendments like farm and poultry manure and compost emerges as a beneficial alternative nutrient source. Biochar, produced from organic waste materials, can potentially encompass a range of plant nutrients, each with differing rates of release (Agbede & Oyewumi, 2023). As a result, integrating biochar into soils impacted by salinity holds the promise of elevating soil fertility and enhancing plant mineral nourishment within these demanding conditions (Fig. 11.1). Biochar encompasses an array of minerals, the composition of which hinges on pyrolysis variables and the type of biomass used. By introducing biochar, the accessibility of phosphorus can be heightened, along with the encouragement of dissolved organic carbon production, ultimately fostering plant growth in salt-affected soils (Mousavi et al., 2023; Wang et al., 2023a).

The accessibility of phosphorus (P) in soil is greatly influenced by the soil's pH, and an optimal pH range for the release of P falls between 5.5 and 7. Studies have indicated that the availability of P decreases in soils with a pH surpassing 7 (Naidu et al., 1991). In soils impacted by elevated salinity, where pH levels exceed the ideal range and organic matter content is minimal, the accessibility of phosphorus (P) becomes a restricting element for the growth of plants (Elgharably, 2008). To tackle this issue, biochar can ameliorate the availability and absorption of phosphorus (P) in salt-affected soils through both direct and indirect means. Biochar acts as a direct supplier of P to plants while concurrently enhancing the state of the growth environment, particularly by increasing the organic carbon content in the soil. This enhancement leads to greater accessibility of P, improved uptake, and enhanced movement of P from plant shoots to grains (Lashari et al., 2013). Moreover, biochar has the potential to enhance the prevalence and dispersion of phosphate-solubilizing bacteria such as Thiobacillus, Pseudomonas, and Flavobacterium in soils. This action, in turn, aids in elevating the accessibility of phosphorus (P) (Liu et al., 2017). The application of biochar was shown to result in an increase of over twofold in the Olsen P levels in soils impacted by salinity (Lashari et al., 2013). This result was attributed to two

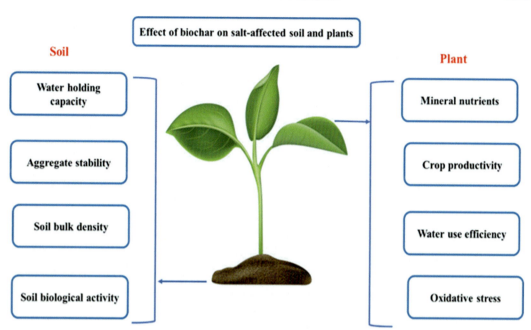

FIGURE 11.1 Effect of biochar on soil and plants physiological, biochemical and microbiological activities.

primary factors linked with the utilization of biochar. Firstly, biochar contains a significant reservoir of accessible phosphorus (P), which contributes to this improvement. Secondly, the application of biochar leads to a reduction in soil pH by approximately 0.3 units likewise (Taghavimehr, 2015) there have been noticeable increases in the concentration of phosphorus (P) observed in salt-affected soils due to the impact of biochar. This effect primarily stems from two mechanisms: firstly, the biochar-induced generation of dissolved organic carbon (DOC) that occupies sorption sites on clay particles (Hameed et al., 2023), hindering P adsorption by soil colloids; and secondly, the release of humic substances that influence P availability by reducing the formation of calcium phosphate crystal phases. However, it's worth noting that contrasting findings exist. Some studies have indicated significant unfavorable interactions between biochar application and P fertilization in relation to plant growth and the phosphorus content in plant tissues within saline-sodic soils (Xu et al., 2016). This adverse effect was linked to the increase in pH caused by biochar application, resulting in the precipitation of the applied phosphorus within the soil. As a result, the accessibility of phosphorus for plant uptake was compromised. The presence of both positive and negative impacts of biochar on the availability and uptake of phosphorus and other nutrients can be attributed to several factors. These include the source material used for producing the biochar, the duration and temperature of the pyrolysis process involved in its creation, and the specific characteristics of the soil where it is applied.

Maintaining an optimal level of potassium (K) is pivotal for fostering plant growth, yet excessive amounts can have detrimental effects on yield and development. The utilization of biochar offers a feasible method for sustaining appropriate K concentrations in soils affected by salinity. This mechanism is recognized as a key contributor to the improved growth of plants under conditions of salt stress following the application of biochar (Omar et al., 2023). Certainly, the nature of both biochar and soil attributes holds substantial influence over the availability of potassium (K). Nevertheless, the indirect consequences of biochar properties on the mineral status of both soil and plants are more critical than its direct effects. The influence of biochar is indirect, impacting the mineral status of soil and plants by modifying mineral accessibility through decreased leaching and heightened retention capacity. This is accomplished by enhancing the physicochemical attributes of the soil (Dreyfus et al., 2023; Nagarajan, 2023).

Soil salinity's presence can indirectly impact the nitrogen (N) condition of the soil by influencing the activities and composition of microorganisms within the rhizosphere. Increased levels of salt in the soil exchange complex or the soil solution can disrupt the formation of nodules by reducing the population of rhizobia in the soil or impeding their capability to infect root hairs. Additionally, the volatilization of NH_3 and N_2O gases can result in significant losses of nitrogen derived from fertilizers within the soil. These losses can be particularly intensified by salinity, as indicated by various studies (Ghosh et al., 2017; Li et al., 2017; Sangwan et al., 2004). The heightened pH levels characteristic of these soils, (Mandal et al., 2016), as emphasized earlier, contribute to and expedite such gas emissions.

Furthermore, an excess of soil salts can hinder plant growth, leading to a reduction in the supply of photosynthates to the root nodules. This, in turn, has a detrimental effect on nitrogen fixation and the overall nitrogen status of the soil (Jini et al., 2023; Lu, & Zhang, Guo, et al., 2023). Through boosting the population and activities of nitrogen-transforming bacteria, the addition of biochar can bolster nutrient processes, particularly nitrogen dynamics, within saline soils. As a result, this enhancement leads to improved availability and effective utilization of nutrients by plants. Additionally, the implementation of biochar can counteract the escape of nitrogen into the atmosphere in the form of NH_3 and N_2O. This advantageous outcome is attributed to the increased adsorption of ammonium onto biochar particles, as well as the lowered rates of nitrification and denitrification (Chen et al., 2023; Zhang et al., 2023a).

Introducing biochar into soils impacted by salinity can amplify plants' ability to acquire nutrients by improving interactions between roots and biochar, as well as by fostering heightened activity of mycorrhizal fungi. This constructive impact on root growth and exploration of the soil can lead to increased plant productivity, particularly when coupled with the inoculation of mycorrhizal fungi (Venugopalan et al., 2023). The joint utilization of biochar and mycorrhizal fungi in salt-affected conditions demonstrated a more substantial boost in plant yield compared to applying biochar or mycorrhizal fungi separately. This advancement can be attributed to heightened phosphorus and manganese uptake, along with the enhancement of the Na/K ratio in salt-affected plants treated with the combined application of mycorrhizal fungi and biochar (Hou et al., 2023).

Despite the substantial volume of research on biochar, none of the studies have specifically focused on investigating the enduring consequences of biochar application on salt-affected soils within the framework of continuous tillage and low-tillage farming systems. As a result, there is a compelling need for extensive field studies conducted over an extended period to delve into the wide spectrum of effects that biochar may have on the fertility status of salt-affected soils distinguished by varying textures and mineralogical attributes (Bhattacharyya et al., 2023). Furthermore, conducting comparative evaluations is crucial to understand the effects of biochar derived from different source materials and applied at various rates on different types of salt-affected soils. This includes studying saline, saline-sodic, and sodic soils as distinct categories (Omar et al., 2023).

11.3.3 Effect of biochar on physicochemical properties of saline soil

Numerous studies have shed light on the favorable effects of biochar on the physicochemical characteristics of saline soil. Introducing biochar to soils impacted by salinity can significantly enhance properties such as total porosity, water retention capacity, and bulk density, primarily owing to its inherently porous configuration (Fig. 11.2). However, the extent of these improvements hinges on several factors connected to the biochar itself, including its source material, pyrolysis conditions, and the rate at which it is applied to the soil (Yue et al., 2023). Furthermore, biochar holds the potential to bolster the structural robustness of salt-affected soils by affecting the soil processes linked to aggregation. It can also produce a favorable influence on both above- and below-ground plant growth, thereby impacting the activities of soil microorganisms within the rhizosphere. Additionally, incorporating biochar into the soil can heighten the levels of calcium, which is acknowledged for its role in encouraging aggregation and aiding in the removal of sodium from the soil profile through leaching. As a result, this contributes to a decrease in the exchangeable sodium percentage (ESP) and an overall enhancement in the physical attributes of deteriorated salt-affected soils (Naorem et al., 2023). A study was undertaken, employing laboratory incubation and column leaching experiments, to explore the influence of biochar on the aggregation and saturated hydraulic conductivity of saline-sodic

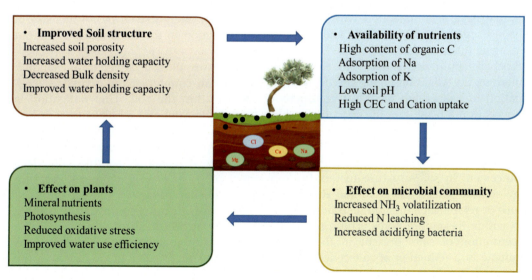

FIGURE 11.2 Possible mechanisms for the effects of biochar on physical and biological properties of salt-affected soils.

soil. The outcomes unveiled that the application of biochar elevated the calcium content in the soil, which, in turn, led to improved aggregation and enhanced saturated hydraulic conductivity. In line with these findings, other investigations have similarly documented a substantial increase in calcium content within saline-sodic soils that were amended with biochar. These observations have translated into noteworthy enhancements in metrics such as the percentage of water-stable aggregates, hydraulic conductivity, and water retention capacity (Fouladidorhani et al.). Nevertheless, the effectiveness of biochar in improving the physical attributes of soils dominated by sodium relies on the calcium content inherent in the particular feedstock utilized. This implies that the results might exhibit variations based on the specific type of biochar employed (Doulgeris et al., 2023).

The impact of biochar on the physical properties of salt-affected soils can be influenced by a range of factors, encompassing soil type, feedstock type, rate of biochar application, type of biochar, and the duration of interaction between biochar and soil. For instance, in a non-calcareous loamy sand soil, the introduction of biochar led to a significant decrease in clay dispersion and the disintegration of aggregates, while concurrently improving the rate of infiltration (Azadi & Raiesi, 2023). However, in a calcareous loam soil, the introduction of biochar did not result in significant alterations to the final infiltration rate. Although the authors did not offer an explanation for the divergent responses of different soils to biochar application, it is worth highlighting that many saline and salt-affected soils around the world are also calcareous. As a result, further research is necessary to uncover the underlying mechanisms that contribute to these variations in observed outcomes (Zonayet et al., 2023). Applying biochar holds promise for ameliorating the physical attributes of degraded salt-affected soils by facilitating the binding of polyvalent cations and clay particles, thereby leading to enhanced aggregation. This advantageous outcome can be attributed to the existence of organic molecules within biochar. A number of studies have showcased an elevation in the proportion of water-stable aggregates in salt-affected soils that have been amended with biochar. This effect is likely driven by an increase in the soil's organic carbon content and a concurrent reduction in the exchangeable sodium percentage (Qian et al., 2023).

Biochar's influence extends to the chemical attributes of soil affected by salinity. Numerous investigations have unveiled that biochars can counteract the excessive accumulation of salts (high electrical conductivity, ECe) within the soil. This mitigation is achieved through the amplification of leaching for soluble salts and a subsequent reduction in the soil's EC. The decrease in ECe can be attributed to two principal mechanisms: *Augmentation of Soil Porosity and Hydraulic Conductivity*: Incorporating biochar enhances soil porosity and hydraulic conductivity, facilitating the drainage of soluble salts. This enhancement in soil structure fosters the movement of water through the soil, carrying away soluble salts. *Adsorption and Retention of Salts*: Biochar possesses the capability to adsorb or retain salts, including sodium (Na), on its surfaces. It can also physically confine salts within its minute pores. These mechanisms collaboratively lead to a decrease in the concentration of salts present in the soil solution. Taken together, these mechanisms culminate in a reduction of salt concentration within the soil, effectively ameliorating the repercussions of excessive salinity (Wang et al., 2023b).

Additionally, the utilization of biochar has demonstrated its efficacy in alleviating the stress induced by salinity by curbing the vertical movement of saline water within the soil profile. This is accomplished by deploying a biochar cover that effectively curtails evaporation, resulting in reduced salt buildup in the upper layers of soil. Moreover, multiple investigations have consistently documented an elevation in soil pH following the introduction of biochar amendments. It's worth noting that a majority of these studies were conducted on soils with initial pH levels lower than that of the applied biochar (initial pH < 5.5 compared to biochar pH > 7.0) (Sudratt & Faiyue, 2023). However, it's important to note that the results might differ when introducing low-pH biochar to high-pH soils, particularly in cases of saline-sodic and sodic soils. Recent studies have unveiled that the introduction of biochar can lead to a noteworthy decrease in pH values within salt-affected soils, although the exact mechanisms driving this pH reduction are not yet fully understood. The substantial decline in pH, observed in both the upper and lower layers of salt-affected soils following biochar addition, indicates the presence of potential factors playing a role in this pH adjustment (Rengel, 2023). This decrease in pH was attributed to the release of $H+$ ions from the ion exchange complex, driven by the influx of Ca^{2+} or Mg^{2+} facilitated by the presence of biochar. Another plausible explanation for the pH reduction in soil amended with biochar is the heightened CEC of biochar, which could potentially enhance the uptake of cations such as K^+, Ca^{2+}, and Mg^{2+} by plants. As a result, the release of H^+ ions from plant roots to maintain charge equilibrium could take place, ultimately leading to a decrease in pH (Sun et al., 2023). Moreover, the expansion of acidogenic soil microorganisms in soils enriched with biochar might also play a role in the decrease of soil pH. In another study, a notable pH decrease was observed in saline soil subsequent to biochar application, and this was attributed to an increase in the release of acidic functional groups during the process of biochar oxidation (Chandarana & Amaresan, 2023).

11.3.4 Carbon stability and its association with soil health restoration

The stability of carbon and its connection to soil health are pivotal factors that underscore the importance of integrating carbon-rich biochar into initiatives aimed at restoring soil health. Biochar, a form of charcoal generated through the pyrolysis process using organic waste materials, is renowned for its capacity to augment soil fertility and overall soil well-being. An essential facet of biochar is its carbon stability, characterized by a substantial carbon content and distinctive chemical attributes that grant it resilience against decomposition over extended time frames. When biochar is introduced into soil, it functions as a durable carbon reservoir, effectively capturing carbon from the atmosphere. This process of carbon sequestration contributes to mitigating climate change by curbing greenhouse gas emissions and promoting the long-term storage of carbon.

Incorporating biochar into soil holds notable ramifications for the restoration of soil health. Biochar contributes to the enhancement of soil structure and stability through the facilitation of aggregation and erosion reduction. Additionally, it heightens the soil's capacity to retain water, fostering improved moisture retention and subsequently decreasing the vulnerability to salt stress (Omondi et al., 2016). Moreover, biochar functions as a habitat for beneficial microorganisms, fostering a diverse soil microbiome that facilitates nutrient cycling and supports plant growth. The integration of carbon-rich biochar into strategies aimed at restoring soil health yields numerous positive outcomes. It amplifies the levels of soil organic carbon, which play a critical role in nutrient retention, soil fertility, and overall soil productivity. Furthermore, biochar has the potential to curtail nutrient leaching, thereby mitigating the environmental impact of agricultural practices and contributing to the safeguarding of water quality. Additionally, soils enriched with biochar have demonstrated enhanced CEC, thereby augmenting nutrient accessibility to plants. Notably, biochar also serves to address contaminated soils by immobilizing heavy metals and reducing their bioavailability, effectively contributing to soil remediation efforts (He et al., 2019). In conclusion, the incorporation of carbon-rich biochar into soil health restoration initiatives is supported by its capacity to fortify carbon stability, sequester atmospheric carbon, and enhance diverse facets of soil health. As a resilient carbon repository with the power to elevate soil fertility, biochar fosters sustainable agriculture, aids in mitigating climate change, and contributes to safeguarding the environment.

11.3.5 Effect of biochar on microorganism in salt-effected soil

Soil microorganisms assume a pivotal role in bolstering soil health through their involvement in essential soil processes, including the decomposition of organic matter, transformation of nutrients, formation of soil aggregates, and establishment of soil structure. However, the activities of these microbial communities and the enzymes they produce, crucial for sustaining plant growth, often decline in saline soils (Yan & Marschner, 2013). Abundant research underscores the detrimental effects of soil salinity on microbial activity, along with its capacity to reshape the composition of the soil's microbial community, particularly impacting salt-sensitive microorganisms. The increase in soil salinity levels elevates the osmotic potential within the soil solution, necessitating microorganisms to invest additional energy for water uptake. Even in conditions of heightened salinity, this response can result in the removal of water from microbial cells (plasmolysis), potentially leading to the mortality of these vital soil microorganisms (Wichern et al., 2006). Research findings consistently highlight the inverse relationship between elevated soil salinity and soil microbial respiration. Notably, (Setia et al., 2010) a significant 50% decrease in soil microbial respiration has been documented when the soil's EC (EC1:5) surpasses 5 dS m^{-1}.

Nevertheless, it's important to acknowledge that soil salinity and sodicity can have adverse impacts on microbial growth, activity, and diversity through various intricate mechanisms (Masmoudi et al., 2023). These mechanisms encompass salt-induced toxicity, reduced water availability, disruption of soil aggregates, and degradation of soil structure, all of which collectively create unfavorable conditions for microbial growth and activity. Additionally, the decrease in organic matter content can limit the availability of crucial energy sources necessary for microbial metabolism (Zeng et al., 2023). Biochar demonstrates the potential to enhance the proliferation and development of soil organisms in salt-affected soils through several underlying mechanisms. Firstly, it facilitates the formation of soil aggregates, providing stable niches for the growth and activity of microorganisms. Secondly, biochar improves the soil's water retention capacity, ensuring a consistent water supply crucial for the sustenance of soil microorganisms. Thirdly, it releases nutrients into the soil matrix, serving as valuable resources for microbial utilization. Additionally, biochar enhances soil carbon biomass and boosts microbial enzymatic activities. Increasing the availability of energy derived from carbon-based compounds, it stimulates soil microorganisms to generate osmotic antistress molecules or expel salt from their cells through heightened metabolic activities (Wichern et al., 2006). The incorporation of biochar into the soil enhances the ability of soil microorganisms to adapt to osmotic stress in saline environments. This practice also promotes the storage of energy within the microbial community, sustaining ongoing metabolic processes and fostering increased biomass production. Furthermore, biochar stimulates the release of dissolved organic carbon and nitrogen from plant roots, acting as essential energy sources for microbial metabolism and growth (Kumar et al., 2023).

11.3.6 Role of biochar in plant growth in salt-effected soil

A significant aspect of plants undergoing salt stress is the accumulation of sodium (Na) and the impaired availability of potassium (K) minerals. Consequently, enhancing the accessibility of potassium (K) and improving the K:Na ratio is deemed a valuable strategy to promote plant growth and yield in saline and saline-sodic soils. The primary factors contributing to the inhibitory impact of salt on plants encompass osmotic and ionic stress. Osmotic stress arises swiftly once roots are exposed to elevated salinity levels, resulting in diminished plant growth and performance. Conversely, ionic stress occurs due to the accumulation of Na+ ions to toxic levels (Munns & Tester, 2008). The excessive accumulation of Na^+ disrupts the usual operations of physiological and biochemical processes within plant cells, ultimately leading to a substantial decrease in yield.

Findings from both laboratory experiments and field investigations have consistently indicated that the incorporation of biochar into saline-affected soils can successfully mitigate salt-induced stress and directly promote plant growth. This enhancement is primarily ascribed to the liberation of essential macro- and micronutrients, such as calcium (Ca), potassium (K), nitrogen (N), phosphorus (P), and zinc (Zn), into the soil. These nutrients play a crucial role in counteracting the adverse consequences of elevated salt concentrations (Drake et al., 2016; Kim et al., 2016).

Given the prominent features of salt-stressed plants involving sodium (Na) accumulation and potassium (K) deficiency, the elevation of the K:Na ratio by increasing potassium availability emerges as a valuable tactic to bolster plant growth and enhance yield in soils affected by sodicity and saline-sodic conditions (Chakraborty et al., 2016). Biochar's ability to raise K concentrations in the soil depends on the feedstock employed. This augmentation of soil K content in salt-affected environments offers a means to counteract the adverse impacts of sodium, thus highlighting a significant benefit linked with the utilization of biochar. As an illustration, (Lin et al., 2015) For instance, a study documented notable increases in wheat and soybean (Glycine max) yields following the introduction of biochar into coastal saline soil. Interestingly, this addition of biochar did not induce alterations in the levels of exchangeable Na, Ca, and Mg. However, it significantly boosted the concentration of exchangeable K by 44% compared to the control. This increase led to an enhanced K:Na ratio within the plants, which played a pivotal role in improving salt tolerance and ultimately promoting better plant growth. In a similar vein (Lashari et al., 2015), researchers observed a significant increase in both the potassium (K) content and the K/Na ratio within the leaf sap of corn plants subjected to salt stress. This augmentation in potassium supply was identified as a central mechanism contributing to the alleviation of salt stress effects on plants.

Another study has been conducted and findings have demonstrated that the presence of salt stress results in a reduction in crop yield and hampers plant growth. This decline in growth and yield metrics observed under saline conditions can be attributed to two primary factors: a decrease in water accessibility and the detrimental impact of sodium ions ($Na+$) on plant health. In previous studies involving potatoes, the application of biochar was shown to alleviate the negative effects of salt stress by adsorbing Na^+ ions from the soil solution (Akhtar et al., 2015b).

The decrease in growth and yield parameters experienced during salinity stress can be linked to two key mechanisms. Firstly, salinity stress reduces water availability by elevating the osmotic potential of the soil solution, impeding the plant's capacity to take up water. This leads to dehydration and stunted growth as a result of limited water uptake. Secondly, the toxicity of $Na+$ ions further contributes to the adverse effects on plants. The excessive accumulation of $Na+$ disrupts essential physiological and biochemical processes, including nutrient absorption and transport, enzyme functions, and ion equilibrium within plant cells. These disruptions cause imbalances in vital nutrients, obstructing overall plant growth and development (Akhtar et al., 2015a).

Organic amendments play a crucial role in promoting plant growth in saline soils. Biochar, depending on its source material, can increase the potassium (K) content in soils, which is a notable advantage of using biochar in salt-affected environments. The effects of biochar on plant growth and yield in salt-affected soils are summarized in Table 11.2. The capacity of biochar to mitigate the adverse impacts of sodium (Na) stands out as one of the key benefits associated with its application (Rengel, 2023). As an example, a study showcased noteworthy improvements in the yields of wheat and soybean when biochar was introduced to coastal saline soil. Intriguingly, this biochar addition didn't affect the levels of exchangeable sodium (Na), calcium (Ca), and magnesium (Mg). Nevertheless, it notably raised the concentration of exchangeable potassium (K) by a remarkable 44% compared to the control. Consequently, this increase in the K:Na ratio within plants enhanced their ability to tolerate salt, thereby fostering amplified plant growth (Zonayet et al., 2023). Biochar exhibits the potential to enhance plant growth in soils affected by salt through a range of mechanisms. These mechanisms include reducing oxidative stress by lowering $O_2\bullet-$ and H_2O_2 levels, alleviating osmotic stress by increasing water retention capacity and availability, decreasing phytohormone production, boosting stomatal density and conductance, and facilitating seed germination (Alharbi et al., 2023).

TABLE 11.2 Effect of biochar on growth/yield of plants grown in salt-affected soils/saline conditions.

Biomass	Temperature	Salt treatment growth medium	Biochar application rate	Experimental crop	Response	Reference
Corn stalk	300°C					
Peanut shell	300°C	Salts 1.5% pH$_{1:25}$ 9.0	10% w/w	Wheat (*Triticum aestivum* L.)/grain yield	Field experiment, increase K/Na ration in plants	Lin et al. (2015)
Corn stalks	400°C					
Peanut shell	350°C	ESP 59 EC 1.0	1. 1.5 t ha^{-1} 2. 5 t ha^{-1} 3. 10 t ha^{-1}	Sesbania (*Sesbania grandiflora* L.)/shoot biomass	Pot experiment, BC-compost is superior to BC, higher rates depressed growth	Luo et al. (2017)
Fagus grandifolia sawdust	378°C	30 g NaCl m^{-2}	1. 5 t ha^{-1} 2. 50 t ha^{-1}	Indian mallow (*Abutilon theophrasti* L.)/plant survival till 60 days Common Self-heal (*Prunella vulgaris*)	Pot experiment, no change in EC/pH of soil, CEC not measured	Thomas et al. (2013)
Hardwood and softwood	500°C	25 mM NaCl water	5% w/w of soil	Wheat (*Zea mays* L.)/grain yield	Pot experiment without leaching	Akhtar et al. (2015b)
Coniferous wood chip	500°C	EC$_{1:5}$ 1.3 dS m^{-1}		Garden lettuce (*Lactuca sativa* L.)/shoot dry matter	Field experiment	Hammer et al. (2015)
		Irrigation 50 mM NaCl water plants irrigation			provision, reduced sodium uptake by plants	
Rice hull,		EC1:5 1.3 dS m^{-1}	5% w/w	Maize (*Zea mays* L.)/dry matter	Pot experiment, 52% increase in P, increased WSA, decreased ESP	Kim et al. (2016)
Peanut shell,	350°C	EC1:5 1.0 dS m^{-1} pH1:2.5 7.98 ESP 59%	1. 1.5% w/w 2. 5.0% w/w 3. 10% w/w	Sesbania pea (*Sesbania cannabina*)	Improved soil health, enhanced nutrient availability, elevated bacterial activities and abundances related to nutrient transformations	Zheng et al. (2018)
Wheat straw, softwood	550°C	200 mM NaCl water plants irrigation	2% (w/w)	Cotton plants	Salt stress inhibited growth and yield but improved water use efficiency	Hou et al. (2023)
Acacia pycnantha	450°C–480°C	ECe 4·75 dS m^{-1}, ESP 6·9	5 Mg ha^{-1}	Eucalyptus (*Eucalyptus viminalis* L.)	Salt stress inhibited the root and shoot length.	Drake et al. (2016)

11.4 Conclusion

The agricultural productivity of lands faces a pressing threat from soil salinization, a prevalent challenge in arid and semiarid regions across the globe. To meet the growing global food demand, the reclamation of unproductive salt-affected soils becomes imperative. Various amendments, both organic and inorganic, can be employed for this purpose. However, the choice of the most suitable and sustainable amendment relies on the specific geographical and physicochemical attributes of the salt-affected land.

Amid these considerations, biochar has emerged as a noteworthy organic and natural amendment, gaining recognition for its efficacy in addressing this predicament. A wealth of research indicates that biochar enhances the physical, chemical, and biological aspects of soil, facilitating the leaching of salts from the root zone and bolstering soil quality metrics such as pH, water retention capacity, and infiltration. Moreover, biochar amplifies the efficiency of plant fertilization, resulting in heightened plant growth and increased yields in salt-affected soils. This amendment also leads to enhancements in factors like available phosphorus, nitrogen, dissolved organic carbon, microbial biomass carbon, microbial respiration, enzyme activities, and concurrently reduces soil salinity.

However, it is worth noting that certain studies have suggested that excessive biochar application may contribute to elevated soil salinity. Consequently, comprehensive scientific investigations are essential to unravel the intricate mechanisms or processes through which biochar impacts plant growth in salt-affected soils. These studies hold immense importance, not only for harnessing the full potential of biochar incorporation but also for minimizing any potential downsides.

Looking ahead, the utilization of biochar as an organic amendment for reclaiming salt-affected lands shows substantial promise. Advances in research and technology are likely to unveil a deeper comprehension of the underlying mechanisms governing biochar's influence on plant growth in such challenging soils. This enhanced understanding will enable the optimization of biochar application rates and methodologies to maximize its benefits while mitigating any potential adverse effects. Thus, biochar's sustainable incorporation into agricultural practices stands as a prospect shaped by evolving insights and innovative strategies.

References

Abbas, A., Huang, P., Hussain, S., Shen, F., Wang, H., & Du, D. (2022). Mild evidence for local adaptation of solidago canadensis under different salinity, drought, and abscisic acid conditions. *Polish Journal of Environmental Studies*, 31(4).

Abbas, A., Yu, H., Cui, H., & Li, X. (2021). Genetic diversity and synergistic modulation of salinity tolerance genes in *Aegilops tauschii* coss. *Plants*, 10(7), 1393.

Abiven, S., Schmidt, M. W., & Lehmann, J. (2014). Biochar by design. *Nature Geoscience*, 7(5), 326–327.

Agbede, T. M., & Oyewumi, A. (2023). The ameliorating effects of biochar and poultry manure on the properties of two degraded soils and sweet potato yield in sub-humid Nigeria. *Communications in Soil Science and Plant Analysis*, 1–15.

Akhtar, S., Andersen, M., & Liu, F. (2015a). Biochar mitigates salinity stress in potato. *Journal of Agronomy and Crop Science*, 201(5), 368–378.

Akhtar, S. S., Andersen, M. N., & Liu, F. (2015b). Residual effects of biochar on improving growth, physiology and yield of wheat under salt stress. *Agricultural Water Management*, 158, 61–68.

Alghamdi, S. A., Alharby, H. F., Abdelfattah, M. A., Mohamed, I. A., Hakeem, K. R., Rady, M. M., & Shaaban, A. (2023). Spirulina platensis-inoculated humified compost boosts rhizosphere soil hydro-physico-chemical properties and atriplex nummularia forage yield and quality in an arid saline calcareous soil. *Journal of Soil Science and Plant Nutrition*, 1–22.

Alharbi, K., Hafez, E. M., Omara, A. E.-D., & Osman, H. S. (2023). Mitigating osmotic stress and enhancing developmental productivity processes in cotton through integrative use of vermicompost and cyanobacteria. *Plants*, *12*(9), 1872.

Anand, A., Gautam, S., & Ram, L. C. (2023). Feedstock and pyrolysis conditions affect suitability of biochar for various sustainable energy and environmental applications. *Journal of Analytical and Applied Pyrolysis*, *170*, 105881.

Anawar, H. M., & Hossain, M. Z. (2023). Sustainable amelioration options and strategies for salinity-impacted agricultural soils. *Climate Change and Legumes: Stress Mitigation for Sustainability and Food Security*, 9.

Ashilenje, D. S., Amombo, E., Hirich, A., Devkota, K. P., Kouisni, L., & Nilahyane, A. (2023). Irrigated barley-grass pea crop mixtures can revive soil microbial activities and alleviate salinity: Evidence from desertic conditions of southern Morocco.

Azadi, N., & Raiesi, F. (2023). Minimizing salinity-induced Pb toxicity to microbial N cycling processes in saline Pb-polluted soils amended with biochar. *Pedobiologia*, *96*, 150861.

Aziz, M. A., Wattoo, F. M., Khan, F., Hassan, Z., Mahmood, I., Anwar, A., Karim, M. F., Akram, M. T., Manzoor, R., & Khan, K. S. (2023). Biochar and polyhalite fertilizers improve soil's biochemical characteristics and sunflower (*Helianthus annuus* L.) yield. *Agronomy*, *13*(2), 483.

Bedadi, B., Beyene, S., Erkossa, T., & Fekadu, E. (2023). *The soils of Ethiopia* (pp. 193–234). Springer.

Bhattacharyya, R., Ghosh, A., Nath, C., Datta, A., & Roy, P. (2023). Carbon management in irrigated arable lands of India. *Indian Journal of Fertilisers*, *19*(5), 460–483.

Cavalcanti-Lima, L. F., Cutrim, M. V., Feitosa, F. Ad. N., Flores-Montes, Md. J., Dias, F. J., Sá, A. Kd. S., Santos, T. P., da Cruz, Q. S., & Lourenço, C. B. (2023). Effects of climate, spatial and hydrological processes on shaping phytoplankton community structure and β-diversity in an estuary-ocean continuum (Amazon continental shelf, Brazil). *Journal of Sea Research* 102384.

Chakraborty, K., Bhaduri, D., Meena, H. N., & Kalariya, K. (2016). External potassium (K+) application improves salinity tolerance by promoting Na+-exclusion, K+-accumulation and osmotic adjustment in contrasting peanut cultivars. *Plant Physiology and Biochemistry*, *103*, 143–153.

Chandarana, K. A., & Amaresan, N. (2023). Predation pressure regulates plant growth promoting (PGP) attributes of bacterial species. *Journal of Applied Microbiology*, lxad083.

Chen, Y., Xu, M., Yang, L., Jing, H., Mao, W., Liu, J., Zou, Y., Wu, Y., Zhou, H., & Yang, W. (2023). A critical review of biochar application for the remediation of greenhouse gas emissions and nutrient loss in rice paddies: Characteristics, mechanisms, and future recommendations. *Agronomy*, *13*(3), 893.

Cui, H., Cheng, J., Shen, L., Zheng, X., Zhou, J., & Zhou, J. (2023). Activation of endogenous cadmium from biochar under simulated acid rain enhances the accumulation risk of lettuce (*Lactuca sativa* L.). *Ecotoxicology and Environmental Safety*, *255*, 114820.

Doulgeris, C., Kypritidou, Z., Kinigopoulou, V., & Hatzigiannakis, E. (2023). Simulation of potassium availability in the application of biochar in agricultural soil. *Agronomy*, *13*(3), 784.

Drake, J. A., Cavagnaro, T. R., Cunningham, S. C., Jackson, W. R., & Patti, A. F. (2016). Does biochar improve establishment of tree seedlings in saline sodic soils? *Land Degradation & Development*, *27*(1), 52–59.

Dreyfus, G., Frederick, C., Larkin, E., & Powers, Y. (2023). Climate change mitigation through organic carbon strategies.

Elgharably, A. G. (2008). Nutrient availability and wheat growth as affected by plant residues and inorganic fertilizers in saline soils.

Enders, A., Hanley, K., Whitman, T., Joseph, S., & Lehmann, J. (2012). Characterization of biochars to evaluate recalcitrance and agronomic performance. *Bioresource Technology*, *114*, 644–653.

Etesami, H., Jeong, B. R., & Glick, B. R. (2023). Potential use of Bacillus spp. as an effective biostimulant against abiotic stresses in crops—A review. *Current Research in Biotechnology* 100128.

Fouladidorhani, M., Shayannejad, M., Mosaddeghi, M.R., Shariatmadari, H., & Arthur, E. Biochar, manure and superabsorbent improve the physical quality of saline-sodic soil under greenhouse conditions. *Soil Science Society of America Journal*.

Ghosh, U., Thapa, R., Desutter, T., Yangbo, H., & Chatterjee, A. (2017). Saline–sodic soils: Potential sources of nitrous oxide and carbon dioxide emissions? *Pedosphere*, *27*(1), 65–75.

Goldschmidt, A., & Buffam, I. (2023). Biochar-amended substrate improves nutrient retention in green roof plots. *Nature-Based Solutions* 100066.

Gupta, S. R., Sileshi, G. W., Chaturvedi, R. K., & Dagar, J. C. (2023). *Agroforestry for sustainable intensification of agriculture in Asia and Africa* (pp. 515–568). Springer.

Hailu, B., & Mehari, H. (2021). Impacts of soil salinity/sodicity on soil-water relations and plant growth in dry land areas: A review. *Journal of Natural Science Research*, *12*(3), 1–10.

Hameed, R., Cheng, L., Yang, K., Fang, J., & Lin, D. (2019). Endogenous release of metals with dissolved organic carbon from biochar: Effects of pyrolysis temperature, particle size, and solution chemistry. *Environmental Pollution*, *255*, 113253.

Hameed, R., Lei, C., & Lin, D. (2020). Adsorption of organic contaminants on biochar colloids: Effects of pyrolysis temperature and particle size. *Environmental Science and Pollution Research*, *27*, 18412–18422.

Hameed, R., Lei, C., Fang, J., & Lin, D. (2021). Co-transport of biochar colloids with organic contaminants in soil column. *Environmental Science and Pollution Research*, *28*, 1574–1586.

Hameed, R., Li, G., Son, Y., Fang, H., Kim, T., Zhu, C., Feng, Y., Zhang, L., Abbas, A., & Zhao, X. (2023). Structural characteristics of dissolved black carbon and its interactions with organic and inorganic contaminants: A critical review. *Science of The Total Environment*, *872*, 162210.

Hammer, E. C., Forstreuter, M., Rillig, M. C., & Kohler, J. (2015). Biochar increases arbuscular mycorrhizal plant growth enhancement and ameliorates salinity stress. *Applied Soil Ecology*, *96*, 114–121.

Havlin, J. L. (2020). *Landscape and land capacity* (pp. 251–265). CRC Press.

He, L., Zhong, H., Liu, G., Dai, Z., Brookes, P. C., & Xu, J. (2019). Remediation of heavy metal contaminated soils by biochar: Mechanisms, potential risks and applications in China. *Environmental Pollution*, *252*, 846–855.

Hou, J., Zhang, J., Liu, X., Ma, Y., Wei, Z., Wan, H., & Liu, F. (2023). Effect of biochar addition and reduced irrigation regimes on growth, physiology and water use efficiency of cotton plants under salt stress. *Industrial Crops and Products*, *198*, 116702.

Huang, P., Hameed, R., Abbas, M., Balooch, S., Alharthi, B., Du, Y., Abbas, A., Younas, A., & Du, D. (2023). Integrated omic techniques and their genomic features for invasive weeds. *Functional & Integrative Genomics*, *23*(1), 44.

Irfan, M., Aslam, H., Maqsood, A., Tazeen, S. K., Mahmood, F., & Shahid, M. (2023). *Plant microbiome for plant productivity and sustainable agriculture* (pp. 99–119). Springer.

Jini, D., Ganga, V., Greeshma, M., Sivashankar, R., & Thirunavukkarasu, A. (2023). Sustainable agricultural practices using potassium-solubilizing microorganisms (KSMs) in coastal regions: A critical review on the challenges and opportunities. *Environment, Development and Sustainability*, 1–24.

Khangura, R., Ferris, D., Wagg, C., & Bowyer, J. (2023). Regenerative agriculture—A literature review on the practices and mechanisms used to improve soil health. *Sustainability*, *15*(3), 2338.

Kim, H.-S., Kim, K.-R., Yang, J. E., Ok, Y. S., Owens, G., Nehls, T., Wessolek, G., & Kim, K.-H. (2016). Effect of biochar on reclaimed tidal land soil properties and maize (*Zea mays* L.) response. *Chemosphere*, *142*, 153–159.

Kong, C., Camps-Arbestain, M., Clothier, B., Bishop, P., & Vázquez, F. M. (2021). Use of either pumice or willow-based biochar amendments to decrease soil salinity under arid conditions. *Environmental Technology & Innovation*, *24*, 101849.

Kumar, A., Bhattacharya, T., Shaikh, W. A., Roy, A., Chakraborty, S., Vithanage, M., & Biswas, J. K. (2023). Multifaceted applications of biochar in environmental management: A bibliometric profile. *Biochar*, *5*(1)11.

Kumawat, K. C., Sharma, B., Nagpal, S., Kumar, A., Tiwari, S., & Nair, R. M. (2023). Plant growth-promoting rhizobacteria: Salt stress alleviators to improve crop productivity for sustainable agriculture development. *Frontiers Plant Science*, *13*, 1101862.

Lashari, M. S., Liu, Y., Li, L., Pan, W., Fu, J., Pan, G., Zheng, J., Zheng, J., Zhang, X., & Yu, X. (2013). Effects of amendment of biochar-manure compost in conjunction with pyroligneous solution on soil quality and wheat yield of a salt-stressed cropland from Central China Great Plain. *Field Crops Research*, *144*, 113–118.

Lashari, M. S., Ye, Y., Ji, H., Li, L., Kibue, G. W., Lu, H., Zheng, J., & Pan, G. (2015). Biochar–manure compost in conjunction with pyroligneous solution alleviated salt stress and improved leaf bioactivity of maize in a saline soil from central China: A 2-year field experiment. *Journal of the Science of Food and Agriculture*, *95*(6), 1321–1327.

Li, Y., Huang, L., Zhang, H., Wang, M., & Liang, Z. (2017). Assessment of ammonia volatilization losses and nitrogen utilization during the rice growing season in alkaline salt-affected soils. *Sustainability*, 9(1), 132.

Lin, X., Xie, Z., Zheng, J., Liu, Q., Bei, Q., & Zhu, J. (2015). Effects of biochar application on greenhouse gas emissions, carbon sequestration and crop growth in coastal saline soil. *European Journal of Soil Science*, 66(2), 329–338.

Liu, H., Todd, J. L., & Luo, H. (2023a). Turfgrass salinity stress and tolerance—A review. *Plants*, 12(4), 925.

Liu, J., Fimognari, L., de Almeida, J., Jensen, C.N.G., Compant, S., Oliveira, T., Baelum, J., Pastar, M., Sessitsch, A., & Moelbak, L. (2023b). Effect of *Bacillus paralicheniformis* on soybean (Glycine max) roots colonization, nutrient uptake and water use efficiency under drought stress. *Journal of Agronomy and Crop Science*.

Liu, S., Meng, J., Jiang, L., Yang, X., Lan, Y., Cheng, X., & Chen, W. (2017). Rice husk biochar impacts soil phosphorous availability, phosphatase activities and bacterial community characteristics in three different soil types. *Applied Soil Ecology*, 116, 12–22.

Lu, Y., & Fricke, W. (2023). Salt stress—Regulation of root water uptake in a whole-plant and diurnal context. *International Journal of Molecular Sciences*, 24(9), 8070.

Lu, C., Zhang, Z., Guo, P., Wang, R., Liu, T., Luo, J., Hao, B., Wang, Y., & Guo, W. (2023). Synergistic mechanisms of bioorganic fertilizer and AMF driving rhizosphere bacterial community to improve phytoremediation efficiency of multiple HMs-contaminated saline soil. *Science of The Total Environment* 163708.

Luo, X., Liu, G., Xia, Y., Chen, L., Jiang, Z., Zheng, H., & Wang, Z. (2017). Use of biochar-compost to improve properties and productivity of the degraded coastal soil in the Yellow River Delta, China. *Journal of Soils and Sediments*, 17, 780–789.

Mandal, S., Thangarajan, R., Bolan, N. S., Sarkar, B., Khan, N., Ok, Y. S., & Naidu, R. (2016). Biochar-induced concomitant decrease in ammonia volatilization and increase in nitrogen use efficiency by wheat. *Chemosphere*, 142, 120–127.

Masmoudi, F., Alsafran, M., Jabri, H. A., Hosseini, H., Trigui, M., Sayadi, S., Tounsi, S., & Saadaoui, I. (2023). Halobacteria-based biofertilizers: A promising alternative for enhancing soil fertility and crop productivity under biotic and abiotic stresses—A review. *Microorganisms*, 11(5), 1248.

Mishra, R. K., Kumar, D. J. P., Narula, A., Chistie, S. M., & Naik, S. U. (2023). Production and beneficial impact of biochar for environmental application: A review on types of feedstocks, chemical compositions, operating parameters, techno-economic study, and life cycle assessment. *Fuel*, 343, 127968.

Mousavi, R., Rasouli-Sadaghiani, M., Sepehr, E., Barin, M., & Vetukuri, R. R. (2023). Improving phosphorus availability and wheat yield in saline soil of the Lake Urmia Basin through enriched biochar and microbial inoculation. *Agriculture*, 13(4), 805.

Munns, R., & Tester, M. (2008). Mechanisms of salinity tolerance. *Annual Review of Plant Biology*, 59, 651–681.

Nagarajan, R. (2023). *Process intensification: Faster, better, cheaper*. CRC Press.

Naidu, R., Syers, J., Tillman, R., & Kirkman, J. (1991). Assessment of plant-available phosphate in limed, acid soils using several soil-testing procedures. *Fertilizer Research*, 30, 47–53.

Nair, R. R., Kißling, P. A., Marchanka, A., Lecinski, J., Turcios, A. E., Shamsuyeva, M., Rajendiran, N., Ganesan, S., Srinivasan, S. V., & Papenbrock, J. (2023). Biochar synthesis from mineral and ash-rich waste biomass, part 2: Characterization of biochar and co-pyrolysis mechanism for carbon sequestration. *Sustainable Environment Research*, 33(1), 1–17.

Naorem, A., Jayaraman, S., Dang, Y. P., Dalal, R. C., Sinha, N. K., Rao, C. S., & Patra, A. K. (2023). Soil constraints in an arid environment—Challenges, prospects, and implications. *Agronomy*, 13(1), 220.

Omar, M., Shitindi, M., Massawe, B. J., Fue, K., Meliyo, J., & Pedersen, O. (2023). Salt-affected soils in Tanzanian agricultural lands: Type of soils and extent of the problem. *Sustainable Environment*, 9(1), 2205731.

Omondi, M. O., Xia, X., Nahayo, A., Liu, X., Korai, P. K., & Pan, G. (2016). Quantification of biochar effects on soil hydrological properties using meta-analysis of literature data. *Geoderma*, 274, 28–34.

Osman, A. I., Farghali, M., Ihara, I., Elgarahy, A. M., Ayyad, A., Mehta, N., Ng, K. H., Abd El-Monaem, E. M., Eltaweil, A. S., & Hosny, M. (2023). Materials, fuels, upgrading, economy, and life cycle assessment of the pyrolysis of algal and lignocellulosic biomass: A review. *Environmental Chemistry Letters*, 1–58.

Piao, J., Che, W., Li, X., Li, X., Zhang, C., Wang, Q., Jin, F., & Hua, S. (2023). Application of peanut shell biochar increases rice yield in saline-alkali paddy fields by regulating leaf ion concentrations and photosynthesis rate. *Plant and Soil*, 483(1), 589–606.

Ponomarev, K., Pervushina, A., Korotaeva, K., Yurtaev, A., Petukhov, A., Tabakaev, R., & Shanenkov, I. (2023). Influence of biochar amendment obtained from organic wastes typical for Western Siberia on morphometric characteristics of plants and soil properties. *Biomass Conversion and Biorefinery*, 1–12.

Qian, S., Zhou, X., Fu, Y., Song, B., Yan, H., Chen, Z., Sun, Q., Ye, H., Qin, L., & Lai, C. (2023). Biochar-compost as a new option for soil improvement: Application in various problem soils. *Science of The Total Environment, 870*, 162024.

Rahimzadeh, S., & Ghassemi-Golezani, K. (2023). The biochar-based nanocomposites improve seedling emergence and growth of dill by changing phytohormones and sugar signaling under salinity. *Environmental Science and Pollution Research*, 1–14.

Rengel, Z. (2023). *Marschner's mineral nutrition of plants* (pp. 665–722). Elsevier.

Sangwan, P., Kumar, V., Singh, J., & Dahiya, S. (2004). Ammonia volatilization losses from surface applied urea in saline soils. *Annals of Biology* (India).

Saygin, S. D., Ozturk, H. S., Akca, M. O., Copty, N. K., Erpul, G., Demirel, B., Saysel, A. K., & Babaei, M. (2023). Solute transport through undisturbed carbonatic clay soils in dry regions under differing water quality and irrigation patterns. *Geoderma, 434*, 116489.

Setia, R., Marschner, P., Baldock, J., & Chittleborough, D. (2010). Is CO_2 evolution in saline soils affected by an osmotic effect and calcium carbonate? *Biology and Fertility of Soils, 46*, 781–792.

Sudratt, N., & Faiyue, B. (2023). Biochar mitigates combined effects of soil salinity and saltwater intrusion on rice (*Oryza sativa* L.) by regulating ion uptake. *Agronomy, 13*(3), 815.

Sun, X., Wang, J., Zhang, M., Liu, Z., E, Y., Lan, Y., He, T., & Meng, J. (2023). Effects of biochar on the Cd uptake by rice and the Cd fractions in paddy soil: A 3-year field experiment. *Agronomy, 13*(5), 1335.

Taghavimehr, J. (2015). Effect of biochar on soil microbial communities, nutrient availability, and greenhouse gases in short rotation coppice systems of Central Alberta.

Tang, C., Yang, J., Xie, W., Yao, R., & Wang, X. (2023). Effect of biochar application on soil fertility, nitrogen use efficiency and balance in coastal salt-affected soil under barley–maize rotation. *Sustainability, 15*(4), 2893.

Thomas, S. C., Frye, S., Gale, N., Garmon, M., Launchbury, R., Machado, N., Melamed, S., Murray, J., Petroff, A., & Winsborough, C. (2013). Biochar mitigates negative effects of salt additions on two herbaceous plant species. *Journal of Environmental Management, 129*, 62–68.

Venugopalan, V., Challabathula, D., & Bakka, K. (2023). *Microbial symbionts and plant health: Trends and applications for changing climate* (pp. 397–437). Springer.

Wahab, A., Munir, A., Saleem, M. H., AbdulRaheem, M. I., Aziz, H., Mfarrej, M. F. B., & Abdi, G. (2023). Interactions of metal-based engineered nanoparticles with plants: An overview of the state of current knowledge, research progress, and prospects. *Journal of Plant Growth Regulation*, 1–21.

Wang, Y., Lin, Q., Liu, Z., Liu, K., Wang, X., & Shang, J. (2023a). Salt-affected marginal lands: A solution for biochar production. *Biochar, 5*(1), 21.

Wang, Y., Wang, S., Zhao, Z., Zhang, K., Tian, C., & Mai, W. (2023b). Progress of euhalophyte adaptation to arid areas to remediate salinized soil. *Agriculture, 13*(3), 704.

Wichern, J., Wichern, F., & Joergensen, R. G. (2006). Impact of salinity on soil microbial communities and the decomposition of maize in acidic soils. *Geoderma, 137*(1–2), 100–108.

Xu, D., Zhang, G., Ni, X., Wang, B., Sun, H., Yu, Y., Mosa, A. A., & Yin, X. (2023). Effect of different aging treatments on the transport of nano-biochar in saturated porous media. *Chemosphere, 323*, 138272.

Xu, G., Lv, Y., Sun, J., Shao, H., & Wei, L. (2012). Recent advances in biochar applications in agricultural soils: Benefits and environmental implications. *CLEAN–Soil, Air, Water, 40*(10), 1093–1098.

Xu, G., Zhang, Y., Sun, J., & Shao, H. (2016). Negative interactive effects between biochar and phosphorus fertilization on phosphorus availability and plant yield in saline sodic soil. *Science of the Total Environment, 568*, 910–915.

Yan, N., & Marschner, P. (2013). Microbial activity and biomass recover rapidly after leaching of saline soils. *Biology and Fertility of Soils, 49*, 367–371.

Yue, Y., Lin, Q., Li, G., Zhao, X., & Chen, H. (2023). Biochar amends saline soil and enhances maize growth: Three-year field experiment findings. *Agronomy, 13*(4), 1111.

Zeng, J., Ma, S., Liu, J., Qin, S., Liu, X., Li, T., Liao, Y., Shi, Y., & Zhang, J. (2023). Organic materials and AMF addition promote growth of taxodium 'zhongshanshan' by improving soil structure. *Forests, 14*(4), 731.

Zhang, C., Zhao, X., Liang, A., Li, Y., Song, Q., Li, X., Li, D., & Hou, N. (2023a). Insight into the soil aggregate-mediated restoration mechanism of degraded black soil via biochar addition: Emphasizing the driving role of core microbial communities and nutrient cycling. *Environmental Research* 115895.

Zhang, N., Reguyal, F., Praneeth, S., & Sarmah, A. K. (2023b). A green approach of biochar-supported magnetic nanocomposites from white tea waste: Production, characterization and plausible synthesis mechanisms. *Science of The Total Environment* 163923.

Zhang, X., Qu, J., Li, H., La, S., Tian, Y., & Gao, L. (2020). Biochar addition combined with daily fertigation improves overall soil quality and enhances water-fertilizer productivity of cucumber in alkaline soils of a semi-arid region. *Geoderma, 363*, 114170.

Zheng, H., Wang, X., Chen, L., Wang, Z., Xia, Y., Zhang, Y., Wang, H., Luo, X., & Xing, B. (2018). Enhanced growth of halophyte plants in biochar-amended coastal soil: Roles of nutrient availability and rhizosphere microbial modulation. *Plant, Cell & Environment, 41*(3), 517–532.

Zonayet, M., Paul, A. K., Faisal-E-Alam, M., Syfullah, K., Castanho, R. A., & Meyer, D. (2023). Impact of biochar as a soil conditioner to improve the soil properties of saline soil and productivity of tomato. *Sustainability, 15*(6), 4832.

CHAPTER 12

Biochar imparting abiotic stress resilience

Debarati Datta[1], Sourav Ghosh[1], Kajal Das[2], Shiv Vendra Singh[3], Sonali Paul Mazumdar[1], Sandip Mandal[4] and Yogeshwar Singh[3]

[1]Division of Crop Production, ICAR—Central Research Institute for Jute and Allied Fibres, Barrackpore, West Bengal, India [2]Division of Germplasm Evaluation, ICAR—National Bureau of Plant Genetic Resources, New Delhi, Delhi, India [3]Department of Agronomy, College of Agriculture, Rani Lakshmi Bai Central Agricultural University, Jhansi, Uttar Pradesh, India [4]Agricultural Energy and Power Division (AEPD), ICAR—Central Institute of Agricultural Engineering, Bhopal, Madhya Pradesh, India

12.1 Introduction

Climate change has a considerable negative impact on catastrophic weather occurrences due to altered atmospheric temperature, humidity, precipitation, and incoming solar radiation (US EPA, 2021). Consequently, the frequency of wet and dry spells has altered groundwater reserves have been overexploited, which eventually has resulted in decreased plant growth and agricultural yield. By the end of the 21st century, the global temperature may have risen by 1.8°C–4°C above preindustrial levels (IPCC, 2018). Additionally, the world population is projected to reach 9.7 billion people by 2064, which will double the amount of food that is consumed, necessitating the need for surplus crop production and placing pressure on the planet's diminishing natural resources (Kumar, Nagar, & Anand, 2021; Vollset et al., 2020).

Abiotic stresses, such as low or high temperature, deficient or excessive water, high salinity, and heavy metals, are hostile to plant growth and development, leading to great crop yield penalty worldwide and posing a severe threat to agriculture and the ecosystem (Wang, Vinocur, & Altman, 2003). It is imperative to develop multistress tolerant crops to satisfy the demands of population increase and to relieve the burden of various abiotic stresses. Extreme weather conditions like flooding, desertification, drought, and storm intensification may be linked to temperature rise. Global yields of rice, wheat, maize, and soybean are reduced by 3%–7% for every 1°C rise in temperature (Gornall et al., 2010). Desertification damages the environment by eroding soil, changing the vegetation's composition, pattern, and structure, and lowering nitrogen levels, phosphorus, and organic matter (OM) in soil. A significant environmental issue, desertification has the potential to affect 35% of the earth's land surface and 32% of the global population (Ghrefat, 2011). Droughts slow down plant growth by reducing leaf water content, food uptake, and photosynthetic activity while increasing the generation of reactive oxygen species (ROS) and harmful metal uptake (Abbas et al., 2018). According to research by India Today, the drought from 2020 to 2022 affected roughly two-thirds of the nation. 64 lakh hectares of agricultural land in India were damaged by floods in just two months, primarily in June and July. About one-third of irrigated areas have been damaged by salinization, mostly as a result of rising sea levels, evaporation, and frequent storm surges. Salinization injures soil properties and minimizes biomass production by decreasing nutrient uptake, damaging photosynthesis, and initiating oxidative stress (Gunarathne et al., 2020).

As a result, harmful metal concentrations in grains and vegetables cultivated in polluted soils have been rising at alarming rates. Overall soil quality and plant growth are inhibited by harsh weather, which negatively affects agriculture (which accounts for around one-fourth of all economic losses) and human livelihoods. Agricultural wastes, waste-derived resources, or industrial goods may be used to improve the quality of problematic and degraded soils (Datta et al., 2022). Biochar is becoming more and more popular due to its potential agronomic advantages, including its improved ability to increase soil fertility and carbon sequestration. Biochar is a carbonaceous product obtained via pyrolysis, with the potential to improve pH, cation exchange capacity (CEC), OM, porosity, surface area, nutrient retention, microbial community, and hydraulic properties of soil (Hossain et al., 2020; Kumar, Bhattacharya, Hasnain, Nayak, & Hasnain, 2020).

Additionally, using agricultural wastes for biochar synthesis (such as straw, food waste, leaves, animal waste, and poultry waste) promotes trash recycling and economic waste management (Chaturvedi, Singh, Dhyani, Govindaraju, & Mandal, 2021; Singh, Chaturvedi, & Datta, 2019). Several studies have reported improved root and shoot biomass, leaf surface area, and root and shoot length under various abiotic stress conditions due to biochar incorporation (Fig. 12.1). Additionally, biochar alters nutrient availability and root exudates, which have an impact on the growth of antagonistic microorganisms (Singh et al., 2023). It also activates plant systemic defenses, releases chemicals that are poisonous to pests, and limits pathogen motility and colonization (Poveda, Martínez-Gómez, Fenoll, & Escobar, 2021). In the past ten years, a sizable amount of research has been conducted on severe environmental stresses and how they affect soil and plant systems (Mansoor et al., 2021).

A comprehensive holistic review that considers the effects of climate change-related risks on soil plants and their reduction by biochar amendment is, however, lacking. Keeping such a scenario into perspective, this chapter is intended to discuss the effects of high temperatures, droughts, floods, and salinization on soil–plant systems, as well as the mechanisms at play and the sustainable function of biochar amendment in improving degraded soils and promoting plant development. Gaining insights into the damages

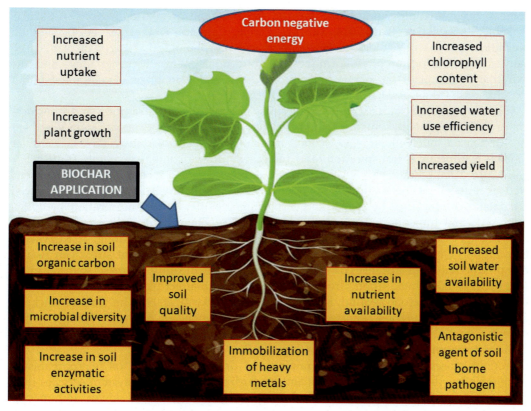

FIGURE 12.1 Impact of biochar for sustainable agriculture: improving soil health and crop productivity.

caused by extreme weather events on the soil—plant systems along with the potential of biochar in repairing them would guide successful and sustainable global functioning and possibly delay the ongoing climate change-associated damages. Moreover, the innovative, novel, and extensive coverage of climate change-triggered soil—plant damage with its potential solution, that is, biochar amendment would be a giant leap for mankind, especially in extreme weather-affected areas.

12.2 Drought

12.2.1 Impact of drought

Globally, drought is a significant barrier to sustained crop productivity. Drought stress is commonly characterized by inconsistent monsoons, decreased soil water retention, impaired leaf water potential, loss of turgor and stomatal closure, suppression of leaf expansion and stomatal conductance and reduced cell enlargement and growth (Mannan, Halder, Karim, & Ahmed, 2016). It has a detrimental impact on plant development and biomass accumulation worldwide, mostly in arid and semiarid regions (Osakabe, Osakabe, Shinozaki, & Tran, 2014; Tardieu, Parent, Caldeira, & Welcker, 2014). Drought can initiate various physiological, biochemical and morphological changes in plants. Too many factors, including phytohormones, osmotic adjustment, ROS signaling, and the hydraulic condition of the plant, contribute to the decline in plant development under drought stress (Khan, Asgher, Fatma, Per, & Khan, 2015).

According to Liu et al. (2017); drought stress affected the redistribution and translocation of Cd in peanuts by causing cellular dryness, decreased cell division, stem lengthening, decreased intercellular CO_2 concentration, increased oxidative stress, irregular stomatal conductance, transpiration, and increased proline content in plants (Kaushal & Wani, 2016; Khalil, Rauf, Monneveux, Anwar, & Iqbal, 2016). Soybean leaf photosynthetic rate was reduced by 26.3% and 37.9% under low and high-intensity drought stress, as well as stomatal conductance by 38.9% and 55.0%, intercellular CO_2 concentration by 15.8% and 17.1%, and transpiration rate by 49.6 and 71.2% (Zhang, Ding, Wang, Su, & Zhao, 2020; Zhang, Xiao, Xue, & Zhang, 2020). Previous research (Wang, Xiong, & Kuzyakov, 2016; Zipper, Qiu, & Kucharik, 2016) have clearly shown that drought stress can reduce crop productivity. For instance, drought stress could significantly decrease soybean grain yield by 24%–50%. In addition, drought stress could also affect plant phenology and thus crop productivity (Farooq, Wahid, Kobayashi, Fujita, & Basra, 2009). The yield of chickpea (*Cicer arietinum* Linn.) was reported to be negatively impacted by water scarcity at the flowering stage (Fang, Turner, Yan, Li, & Siddique, 2010), although soybeans (*Glycine max* L.) did not respond to drought stress in terms of flowering time.

Drought causes cyclical periods of raining and drying, depriving soil microorganisms of the moisture they require to survive and jeopardizing the vital ecosystem services they provide. Droughts can accelerate the decomposition of SOM and lower soil macro-micronutrient levels, which increases CO_2 emissions. Additionally, nitrates, which are largely released as mineral N in soils, have an impact on the SOM's general stability. Drought hinders the productivity of terrestrial ecosystems in its entirety (Ohashi et al., 2015). In order to achieve food security, it is a challenging effort to combat drought stress.

12.2.2 Biochar effect on mitigating drought stress

The application of biochar has significantly improved soil fertility, carbon sequestration, bioenergy production, and the immobilization of organic and inorganic contaminants (Fiaz et al., 2014; Rajapaksha et al., 2016) (Table 12.1). Alkaline pH, adequate EC, good WHC, high surface area, refined physicochemical properties of soil, CEC, and an abundance of micro- and macronutrients in biochar were the main factors favoring the positive effects of biochar on plants grown in drought-stressed soils and limited water conditions (Kumar & Bhattacharya, 2021; Singh, Chaturvedi, Dhyani, & Govindaraju, 2020; Zhang, Ding et al., 2020; Zhang, Xiao et al., 2020). In soils damaged by drought, biochar improved the biometric parameters of maize, okra, and tomato output (Akhtar, Li, Andersen, & Liu, 2014; Haider et al., 2015). Additionally, tomato plants grown in sandy soils with biochar applied boosted their tolerance to wilting and water stress (Mulcahy, Mulcahy, & Dietz, 2013). According to Abideen et al. (2020), there has been a considerable improvement in soil

TABLE 12.1 Effect of biochar on plants facing drought and flood stress.

Sl. No.	Biochar type	Crop	Stress	Effect on plants	References
1.	Rice straw	Wheat	Drought	Increased plant growth (35%–52%), chlorophyll content (58%–63%), gas exchange (40%–85%); decreased metal concentrations (37%–42%), oxidative stress (14%–36%)	Abbas et al. (2018)
2.	Poultry litter	Soybean	Drought	Drought tolerance; increased plant height (3.3%–4.03%), relative water content (4.35%–4.92%), chlorophyll content (7.25%–17%), proline accumulation (22.58%–38.7%)	Mannan et al. (2016)
3.	Cotton sticks	Maize	Drought	Improved relative water content ($\sim$25%), photosynthetic pigments (20%–60%), antioxidant activity (15%–59%); increased root length ($\sim$50%), root dry weight ($>$100%), shoot length ($\sim$25%), shoot dry weight ($>$50%)	Sattar et al. (2019)
4.	Wood chips	Maize	Drought	Increased plant growth (6.5%–7.9%), WUE ($\sim$20%); Improved WHC (soil moisture enhanced from 2.2 to 6.2%)	Haider et al. (2015)
5.	Wheat straw	Wheat	Drought	Improved spike length (6.52%), thousand-grain weight (6.42%), grains per spike (3.07%), biological (9.43%) and economic yield (13.92%); increase WUE ($\sim$20%), chlorophyll content (75%–100%)	Haider et al. (2020)
6.	Hardwood feedstocks	Maize	Flood	Improved water retention in flooded sandy soil, increased soil P and K concentration	Jahromi, Lee, and Fulcher (2020), Jahromi, Lee, Fulcher, Walker et al. (2020)
7.	Corn stalk biochar	Maize	Flood	Decreased the pH and bulk density and increased soil organic carbon (SOC), available nitrogen (AN), and available phosphorus (AP), relative abundance of bacteria, along with the activities of soil enzymes, such as dehydrogenase, β-glucosidase, and alkaline phosphomonoester	Wang, Yin, Wang, Zhao, and Li (2020)
8.	Wheat straw + biochar	Paddy field	Flood	Decrease the nitrogen, phosphorus concentration and organic pollutants, inhibit the diffusion of nonpoint source pollutant, and reduce the risk of water pollution caused by straw returning.	Liu et al. (2020)
9.	Rice husk	–	Flood, Inceptisol	Highest available Si in Inceptisol (8.66 mg kg^{-1}) under flooded condition	Widijanto and Syamsiyah (2022)

fertility, water content, and plant performance during drought in terms of photosynthetic rate, biomass, chlorophyll content, antioxidant activity, and increased WUE. Similarly, the use of biochar from rice husks increased plant-water relationships, fruit and cob yield, and reduced flowering time and proline content in maize plants that were suffering from dryness (Mannan & Shashi, 2020). According to Li and Tan (2021), rice straw-based biochar could increase soil water retention due to its high porosity, surface area, and hydrophilic surface functional groups, primarily —OH and —COOH under drought stress. Application of biochar made from *Lantana camara* increased stomatal conductance, WUE, and total photosynthesis (Batool et al., 2015). Biochar application improved physico-biochemical characteristics in wheat plants under drought stress (Haider et al., 2020). The application of biochar increased grape yield despite water shortages without changing anthocyanin and total acidity levels in the grapes (Genesio, Miglietta, Baronti, & Vaccari, 2015). In a different study, biochar made from poultry litter increased water and proline accumulation and decreased chlorophyll degradation in drought-stressed soybean plants (Mannan, Halder, Karim, & Ahmed, 2017). When rice straw-derived biochar was combined with water-saving irrigation, methane emissions were reduced, irrigation water productivity increased, and rice yield increased, according to Xiao et al. (2018). Additionally, applying biochar to farms that were receiving water-saving irrigation reduced CO_2 emissions by 2.22%, raised WUE by 15.1%—42.5%, and boosted rice production by 9.35%—36.30% (Yang, Jiang, Sun, Ding, & Xu, 2018).

The addition of biochar to soils under drought stress promotes the growth of biotic life, such as microbial and fungal communities, which ultimately boosts drought stress resilience (Fig. 12.2). Biochar's porous pores offer specialized environments for microbial colonization and growth, shielding them from desiccation and other harmful circumstances (unsuitable pH, toxicity, or salinization). As a result, the rhizospheric microorganisms start the solubilization of minerals, which gives plants the nitrogen, phosphorus, potassium, and magnesium they need for healthy growth and lessens the negative impacts of drought. Despite water stress, the application of biochar with *Bradyrhizobium* during droughts increased the biomass, nitrogen uptake, nodulation, and growth in *Lupinus angustifolius* seedlings (Egamberdieva, Reckling, & Wirth, 2017). However, application of birchwood-based biochar to drought-affected potato plants with *Rhizophagus irregularis* (made at 500°C) lowers nitrogen uptake, WUE, and leaf area (Liu et al., 2017). In drought-stressed soils, Liang et al. (2014) found that adding biochar enhanced fungal resilience more than bacterial populations. The study also shows that biochar with a high C/N ratio makes soils more drought-resistant (Liang et al., 2014). Again, repairing drought-stressed soils requires increasing soil water holding capacity, and this is impressively supported by an increase in aggregate stability of soils even with a limited water supply (Baiamonte et al., 2015). Water-holding capacity changes depending on the type of biomass, the method of production, the biochar's thermal treatment parameters, and the application dose (Brantley, Brye, Savin, & Longer, 2015). High CEC, porosity, and surface area of biochar increase the soil's porosity, surface area, and soil—water relationship, enhancing water-holding capacity (Carvalho et al., 2014). The application of biochar increased the water content of sandy loamy silt and vineyard field soils (Baronti et al., 2014). According to Hardie, Clothier, Bound, Oliver, and Close (2014), adding biochar could increase soil porosity and aggregation promoting water infiltration and retention.

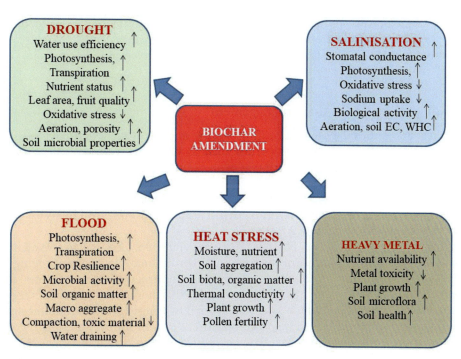

FIGURE 12.2 Use of biochar amendment for abiotic stress mitigation.

12.3 Flood

12.3.1 Impact of flood

The temperature rise due to climate change enables warmer air to hold more moisture. As a result, there are more frequent and severe floods since the strength and length of the precipitation are increased. Floods erode rich topsoils, removing vital nutrients and organic materials that are present in them. Floods also cause anaerobic conditions and reduced mineralization in the soil, which changes the soil's nitrogen and carbon contents (Muhammad, Aziz, Brookes, & Xu, 2017). Floods induce reducing conditions in soil, which alters dissolved organic carbon, alkaline metals, and redox-sensitive elements. (De-Campos, Mamedov, and Huang (2009)) discovered a considerable soil disaggregation and a 20% loss in aggregate stability following two weeks of inundation. The buildup of toxic gases like CO_2 and methane caused by waterlogged circumstances suffocates the roots.

Additionally, waterlogged conditions are linked to oxidation and reduction, which could lower the amount of phosphorus available and the rate at which fungus colonize soils. Floods carry hazardous pathogens, heavy metals, and trace organic contaminants like pesticides and plasticizers into productive soils, which can survive anaerobic conditions and cause plant root damage (Burant, Selbig, Furlong, & Higgins, 2018; Jiang, Lim, Huang, McCarthy, & Hamilton, 2015). The majority of young and old plant species experience stomatal closure, chlorosis, reduced photosynthesis, decreased translocation of food, diminished mineral absorption, altered hormone balance, senescence of leaves, root decay, and death even though well-established healthy plant species are tolerant to flood stress.

12.3.2 Role of biochar in flood management

Biochar has a significant potential to improve the quality of soils that have been damaged by flooding by maneuvering properties such as CEC, pH, nutrient retention, water retention, hydraulic conductivity, available water, infiltration rate, aggregate stability, porosity, and surface area (Obia, Mulder, Martinsen, Cornelissen, & Børresen, 2016). Biochar addition to flood-stressed soil helps in draining standing water faster, supported mainly by altered evaporation rates, bulk density and hydraulic conductivity (Barnes, Gallagher, Masiello, Liu, & Dugan, 2014).

According to Jahromi, Lee, and Fulcher (2020) and Jahromi, Lee, Fulcher, Walker et al. (2020), biochar application assisted in the growth of corn plants (without improving it significantly) and the quality of flooded sandy soils in greenhouse conditions due to increased retention of potassium and phosphorus. Furthermore, biochar immobilizes hazardous substances from soils, which is essential for lowering soil pollution and simultaneously enhancing soil quality. High soil compaction brought on by waterlogging can be improved with the addition of biochar. Biochar reduces tensile strength (42%−242% drop helped by modifying interparticle bonding, friction, clay mineralogy, and SOM), increases consistency, improves SOM content and porosity, enhances aggregation and reduces soil packing (Blanco-Canqui, 2017).

According to Yoo, Kim, and Yoo (2020), even in conditions of extreme waterlogging, biochar application increases aeration and the proportion of macroaggregates (>250 m size). Here, biochar might either encourage the creation of stable aggregates or function as one. Additionally, biochar particles may fill the spaces between the soil aggregates, improving the soil's stability, compaction resistance, and flood resistance (Table 12.1). Moreover, biochar addition alleviates excess water stress by improving drainage assisted by increased hydraulic conductivity of the soil. Reutilization of floodwaters has been proposed to tackle the global problem of water scarcity issues. Keeping in mind the benefits and risks that floodwater reutilization poses, biochar-augmented biofilters have been advocated to improve floodwater quality and remove pollutants from floodwaters via mechanisms involving physisorption and chemisorption (Boehm et al., 2020; Jiang et al., 2015). According to Boehm et al. (2020), installing flood/stormwater biofilters reduces flooding, enhances the quality of the water, and creates green spaces in metropolitan settings. Adding biochar to these biofilters escalates their pollutant removal capability, enhances drinking water quality, improves aquatic life, and avoids surface water pollution. Cow urine-enriched biochar was proposed by Shrestha and Pandit (2017) as an inexpensive, efficient (high pH, CEC, WHC, nutrient retention, surface area, and porosity increase its efficacy), and locally accessible input (could be produced easily using waste/undesirable biomass in barrels, drums, or kilns) method for sustainable farming, food security and improvement of the quality.

12.4 Salinization

12.4.1 Impact of salinization

Salinity is another vital limiting factor for sustainable agriculture. Salinization is one of the major factors for soil degradation, especially in arid and semiarid regions. Globally, over 70 countries have salinity-affected lands, and salinity stress affects more than 6% of the world's total land and about 23% of farmland (Amini, Ghadiri, Chen, & Marschner, 2016). In water-stressed regions in western China, and California, USA, the infliction is as high as 40% (Wang, Xiao, Li, & Li, 2008) and 50% (Letey, 2000), respectively. Salt accumulation in soils has grown in response to extreme weather events including heat waves and droughts. Sea level rise brought on by climate change has caused saltwater intrusion in coastal regions and surrounding aquifers. The precipitation-mediated reduction of salinization has been hindered by the erratic rainfall distribution and timing. As a result, the arable land becomes more salt-stressed, which lowers agricultural productivity and weakens the world economy.

Salinization negatively affects the germination of seeds, the intake of nutrients, and the growth of plants (Zia-Ur-Rehman, Rizwan, Sabir, Ali, & Ahmed, 2016). In addition to disrupting the physiological development of leaves, salinity stress may also make it more difficult for plant roots to absorb water and nutrients (e.g., N) from the soil, which may disturb carbon metabolism. Salinization depletes the water content of leaves, reduces nutrient intake, causes oxidative stress, and hampers the effectiveness of photosynthetic processes (Gunarathne et al., 2020). When a plant is at the initial growth stage, exposure to salt stress results in stomata closure and a decrease in leaf expansion to conserve water in plants (Tardieu et al., 2014). Later on, the salt builds up in the shoots, closing stomata, preventing leaf expansion, and starting premature leaf senescence (Zia-Ur-Rehman et al., 2016). Additionally, by producing ROS, salinization increases oxidative stress in plants and lowers antioxidant enzyme activity. In line with this, stress hormones (especially abscisic acid), a sign of osmotic stress in roots, also accumulate more frequently (Farhangi-Abriz and Torabian, 2017).

Dispersal of saline soil due to salinization decreases hydraulic conductivity and water infiltration of soils (Suarez, Wood, & Lesch, 2006). Moisture content, OM, the carbon/nitrogen ratio, microbial activity, and microbial biomass are all negatively impacted by salinization (Yan, Marschner, Cao, Zuo, & Qin, 2015). Furthermore, salinization inhibits soil respiration by damaging the microbial community structure and enzymatic activity. These deleterious changes deter the generation of plant biomass and ultimately reduce crop yield. Effective solutions are needed for the long-term management of salinization-exposed soils to increase crop yield and farm profitability.

12.4.2 Role of biochar in management of soil salinity

Current practice to remove the salts involves overwatering the land, which causes the salts to leach down out of the rhizosphere under appropriate drainage. The leaching method has gained widespread adoption since it is simple to use. However, it uses substantial water resources. Agriculture water resources have substantially decreased over the past few decades as a result of global warming, population growth, urbanization, and industrialization, endangering the sustainability of the current approach. To prevent soil salinization considering the diminished water availability, this necessitates innovative technology. Biochar, a form of charcoal produced from pyrolysis of biomass waste under limited or no oxygen availability for soil amendment purposes, has the potential to alleviate salinization due reportedly to adsorption of salts, replacement of Na^+ from the exchangeable site of soil particles, reduction of the sodium adsorption ratio, mitigation of the oxidative stress of NaCl, and reduction of salts in plant seedlings. It also improves soil water holding capacity substantially. Biochar is an effective soil amendment to enhance the tolerance of plants to salinization (Table 12.2). Salinization-tolerance and plant growth are enabled by alkaline pH, adequate EC, enhanced WHC, high surface area, porosity, and an abundance of micro- and macronutrients. The strong salt adsorption capacity of biochar, an efficient bioadsorbent, reduces salt contamination from salinized soils (Parkash & Singh, 2020).

Biochar supports photosynthesis, biomass generation, and growth in plants thriving in salinization-stressed soils. There was an increase in photosynthetic rate, root length, shoot biomass, and crop yield (e.g., maize, tomato, and potato), due to biochar application derived from feedstocks like rice hull, wood, and mangrove shrub (Usman et al., 2016). After adding biochar to salinized soils, an improvement in the overall growth of wheat, maize, and halophytes was reported (Lashari et al., 2013; Lashari et al., 2015). The application of biochar could decrease salinization-stress in plants by minimizing phytohormone production; improving photosynthetic parameters; reducing lipid peroxidation in plants; decreasing oxidative stress in plants; augmenting moisture retention in soil; and increasing sodium binding efficiency in soil (Fig. 12.2). These result from inhibited or reduced lipid peroxidation, superoxide ions, and hydrogen peroxide radicals in plant tissues (Farhangi-Abriz & Torabian, 2017). The addition of biochar poultry-manure compost to salinized-soils reduced abscisic acid concentrations in the sap of maize and potato leaves (Lashari et al., 2015). Studies on tomato, wheat, and herbaceous plants found that applying biochar to salinity-stressed plants increased stomatal conductance and stomatal density, which later lowered osmotic stress and increased leaf growth (Thomas et al., 2013).

Glutathione reductase and ascorbate peroxidase levels were found to be reduced when biochar was applied to salinization-affected maize (Kim et al., 2016). The concentration of malondialdehyde in the sap of maize leaves was reduced when salinized soils were amended with biochar made from poultry manure (Lashari et al., 2015). In seedlings of bean plants under salt stress, the addition of biochar reduced the oxidative stress and activities of antioxidant enzymes (Farhangi-Abriz & Torabian, 2017). It's interesting that using biochar in combination with suitable microbial inoculants, like rhizobacteria, promotes the growth and biomass of plants that are affected by salinity (Nadeem, Zahir, Naveed, & Nawaz, 2013). Application of biochar poultry-manure compost to saline soils increased microbial biomass and activities of enzymes (phosphatase, invertase and urease) in the rhizosphere, which propelled maize cultivation (Lu et al., 2015). In salinity stressed-maize and wheat plants, the addition of endophytic bacteria with biochar enhanced photosynthetic rate, leaf area, and plant biomass (Akhtar, Andersen, & Liu, 2015). Application of arbuscular mycorrhizal fungi with biochar improved biomass in salinity-stressed-lettuce plants (Hammer, Forstreuter, Rillig, & Kohler, 2015). According to research on maize,

TABLE 12.2 Effect of biochar on plant facing salinity and heavy metal stress.

Sl. No.	Biochar type	Crop	Stress	Effect on plants	References
1.	Peanut shells	*Sesbania*	Salinization	Improved root length (>160%), leaf surface area, nutrient availability (34.9%–55.3%) nitrogen increase; Increased soil CEC (~15%), OM (90%); promoted microbial activities; reclaimed degraded-coastal soils (exchangeable sodium percentage decrease by 10%)	Zheng et al. (2018)
2.	Wheat straw + poultry manure	Wheat Maize	Salinization	Improved plant height (~40%), root length (>50%), plant density (>80%), decreased salt stress (reduction of 33%–48% Na+ and 41%–48% Cl in leaf sap)	Lashari et al. (2015)
3.	Rice hulls	Maize	Salinization	Improved soil organic carbon (~40%), pH (from 6.9 to 7.3), CEC (~30%); decreased salt stress (decrease exchangeable sodium by 50%)	Kim et al. (2016)
4.	Shrimp waste	Maize	Salinization	Decreased plant height (~11%), shoot dry weigh (~22%), root dry weight (~35%), root colonization (~40%); increased nitrogen uptake (~30%), phosphorus uptake (~30%–50%); increased antioxidant enzyme activity (superoxide dismutase, peroxidase, and ascorbate peroxidase)	Kazemi, Ronaghi, Yasrebi, Ghasemi-Fasaei, and Zarei (2019)
5.	Maple residues	Mung bean	Salinization	Improved root length (~15%), root dry weight (~10%), total root area (~20%),; decreased stress hormones and plant senescence	Nikpour-Rashidabad, Tavasolee, Torabian, and Farhangi-Abriz (2019)
6.	*Lantana* (LBC), and *Parthenium* (PBC) biomass	Wheat	Heavy metals	Heavy metal uptake rates were 63.08% (Cr), 78.07% (Cd), 74.61% (Cu), 78.11% (Pb), 75.73% (Ni), 69.71% (Zn), 28.78% (Mg), and 49.26% (Fe) lower in biochar amendments	Pandey, Suthar, and Chand (2022)
7.	*Cotton straw*	Cotton	Heavy metals	Increased cotton dry weight by an average of 16.82%. Reduced the heavy metal-induced oxidative stress in cotton	Zhu, Wang, Lv, Zhang, and Wang (2020)
8.	*Casuarina*	Summer squash	Heavy metals	Decreased the concentrations of Cd, Co, Cr, Cu, Ni, Pb, and Zn by 25.7%, 52.1%, 12.1%, 32.3%, 31.0%, 85.0%, and 25.2% in the root, and by 37.2%, 66.9%, 24.3%, 40.2%, 42.0%, 89.2%, and 35.5% in the shoot, respectively	Ibrahim, El-Sherbini, and Selim (2022)

(Continued)

TABLE 12.2 (Continued)

Sl. No.	Biochar type	Crop	Stress	Effect on plants	References
9.	Chicken manure and green waste	Indian mustard (Brassica juncea)	Heavy metals	Reduced NH_4NO_3 extractable Cd, Cu, and Pb; increased plant dry biomass by 353% for shoot; reduced Cd, Cu, and Pb accumulation by plants	Park, Choppala, Bolan, Chung, and Chuasavathi (2011)
10.	Sewage sludge	Rice	Heavy metals	Decreased pore water As, Cr, Co, Ni, and Pb due to increase soil pH; mobilization of Cu, Zn, and Cd due to high available concentrations in biochar	Khan, Chao, Waqas, Arp, and Zhu (2013)
11.	Hardwood	Solanum lycopersicum	Heavy metals	Increased pore water As and Cu; immobilization of Cd and Zn due to enhanced pH and DOC	Beesley et al. (2013)

lettuce, and potato plants exposed to salinity, biochar application decreases sodium ion concentrations and increases potassium ion concentrations in xylem sap (Akhtar et al., 2015; Hammer et al., 2015). Additionally, biochar application increases the uptake of nutrients (NPK) and minerals in salinization-exposed plants, which escalates salinization-tolerance (Lashari et al., 2015). An increase in zinc, potassium, manganese, copper, iron, and phosphorus concentrations was observed when biochar was applied to salinity exposed-tomato. The amendment of saline soils with biochar helps minimize salinization stress by improving properties associated with EC, sodium ion adsorption, and sodium ion leaching (Usman et al., 2016). Additionally, high CEC of biochar enables it to switch out sodium ions for calcium or magnesium ions, stabilizing the soil structure (Amini et al., 2016). To conclude, adding biochar to salinization-exposed soils promotes biological activity and plant growth by improving the physicochemical properties of the soil that are directly involved in sodium removal.

12.5 Heat stress

12.5.1 Impact of heat stress

As a result of climate change and global warming, there are more frequent and severe heat waves, severe water shortages, desertification, and long droughts (Warter et al., 2021). Soil temperatures have an impact on soil microclimatic factors, including microbial activity, CO_2 emission, OM decomposition, nutrient mineralization, agrochemical breakdown, soil–water interaction, and plant growth (Wu, Jansson, & Kolari, 2012; Xu, Luo, & Zhou, 2012). An increase in the temperature of the soil increases evaporation and restricts water movement into the soil profile triggering a scarcity of water to plants. The rate of water absorption in the soil and the transport of nutrients by roots simultaneously increase as water viscosity decreases, resulting in a significant impact on photosynthetic activity

(Lahti et al., 2002). Temperature rise boosts metabolic activity in root cells, nutrient uptake, and photosynthesis, which improves root growth and development of lateral roots (Repo, Leinonen, Ryyppö, & Finér, 2004). However, extremely high soil temperature dehydrates soil clay minerals, initiates loss of clay and augments silt content, decreases CEC, transforms aluminum and iron oxides and alters the aggregate stability of soil (Inbar, Lado, Sternberg, Tenau, & Ben-Hur, 2014).

12.5.2 Impact of biochar on heat stress

By acting as a binding agent and boosting the favorable interactions with mycorrhizal fungi and organo-mineral complexes, the addition of biochar helps to increase soil aggregation. Surface hydrophobic-hydrophilic interactions between clay minerals and biochar facilitate the formation of the aggregate (Du, Zhao, Wang, & Zhang, 2017). Additionally, high carbon content and volatile compounds of biochar escalate soil levels of β-glucosidase, boost enzymatic activity, and catalyze the breakdown of cellulose to release glucose, providing energy for microbial growth. High acid phosphatase, urease, and arylsulfatase activity are favored by biochar amendment, which increases the availability of phosphates, nitrogen, and sulfur essential for plant growth (Lopes et al., 2021). Biochar could contain silica, metal oxides, and phenolic compounds, which might signal the induction of the formation of exopolysaccharides biofilm in bacteria in response to drought, pH change, or root desiccation, consequently augmenting microbial colonization. Application of biochar improved the root-zone environment of rice plants by reducing soil bulk density, increasing soil OM content, and altering soil bacterial community structure by increasing the ratio of Proteobacteria to Acidobacteria. as a result, rice plants grown with biochar exhibited improved or up-modulated root morphology, architecture, and physiological traits, such as N assimilation and transport proteins, as well as shoot N uptake and utilization, while the heat-shock and related proteins in roots and leaves were down-modulated. These findings suggest that optimizing management practices can improve the root-zone environment and reduce the negative effects of heat-stress on rice production. They also deepen our understanding of the fundamental eco-physiological mechanisms regulating increased heat-stress tolerance in rice plants.

Soils absorb energy (more than that is lost) with a rise in temperature, which could change the thermal characteristics of the soil (thermal conductivity, diffusivity, reflectance, and temperature). Applying biochar to soils could enhance their thermal characteristics. In order to accomplish this, biochar amendment regulates the accumulation and movement of heat and water in the soil (Fig. 12.2), as well as surface-energy partitioning and heat distribution throughout the soil profile (Usowicz et al., 2020). In a field study, Zhang et al. (2013) revealed that biochar amendment decreased the thermal conductivity of Fluvic Cambisols (sandy loams) susceptible to floods and droughts by 3.48%–7.49%. Further, there was a reduction in diurnal soil temperature fluctuations and moderation of soil temperature extremes (moderation within ±0.4°C to ±0.8°C). In a different study, Zhao, Ren, Zhang, Du, and Wang (2016) found that adding biochar to sandy loams reduced their thermal diffusivity (0.6%–21.5%) and thermal conductivity (0.3%–32.2%). Additionally, the soil's bulk density (24.7%–34.6%) and thermal diffusivity (10.4%–50.8%) decreased, while its moisture content increased, following the addition of biochar. These modifications reduce soil heat conductivity by 24.7%-59.8%, which eventually modifies soil temperatures, influences plant growth, and changes soil biochemical processes. According to Xiong et al. (2020), soil's thermal characteristics are decreased by the addition of biochar and are directly associated with soil moisture and inversely related to soil bulk density. They came to the conclusion that adding biochar to agricultural soil reduces temperature fluctuations.

In medium sandy soils, mango wood-based biochar increased water retention by 137%, while in sandy loamy soils, it increased water retention by 11% (Villagra-Mendoza & Horn, 2018). Similarly, Baiamonte, Crescimanno, Parrino, & De Pasquale, 2019 observed that applying biochar made from forest waste enhanced the plant available water of aridisols by 226%. However, the available water capacity of silty loam soil decreased by 10% when amended with an oak-derived biochar (Mukherjee, Lal, & Zimmerman, 2014). Pine and oak-derived biochars increased the amount of water that sandy soils could hold while lowering the amount in clayey soils. Loamy soil was unaffected by the amendment (Tryon, 1948). Nevertheless, biochar application enhances water retention and water use efficiency (WUE) in plants facing heat-stress (Batool et al., 2015). Liu, Yang, Lu, and Wang (2018) reported biochar amendment reduces the difference in soil temperatures between day and night by about 1°C, with paddy soils experiencing a decrease of 0.66°C–1.39 °C.

High temperatures denature organic acids, consequently increasing their pH (Certini, 2005). The addition of biochar (rich in OM) helps increase the OM content of the soil. High OM increases soil aggregation, which enhances soil structure. It also increases soil stability by providing resistance to mineralization and decomposition. High OM also improves water holding capacity (WHC), nutrient supply and availability, and infiltration rate by enhancing soil permeability (Chang et al., 2021; Janus et al., 2015).

12.6 Heavy metals

12.6.1 Impact of heavy metal on agriculture

Heavy metal(loid) and persistent organic pollutant poisoning of soil, including Cd, Cr, Hg, Pb, Cu, Zn, As, Co, Ni, and Se, is a global issue that endangers the sustainability of the ecosystem, food safety, and human health (Sun et al., 2022). These heavy metals can bioaccumulate in soil, water, and air, affecting the entire ecosystem and posing a threat to the health of all living things. Higher concentrations of metals and chemicals impair plant germination, growth, and production, which are primarily linked to the physiological, biochemical, and genetic components of the plant system. The primary effects of heavy metals on seeds include general abnormalities and a reduction in germination, as well as reduced root and shoot elongation, dry weight, total soluble protein level, oxidative damage, membrane alteration, altered sugar and protein metabolisms, and nutrient loss, contributing to seed toxicity and productivity loss. Ni stress has been reported to reduce yield, disrupt photosynthetic pigments, and create a buildup of Na^+, K^+, and Ca^{2+} in mung beans. By increasing electrolyte leakage, lipid peroxidation, H_2O_2 concentration, antioxidative enzyme activity, and proline levels, the combination of Ni and NaCl in germinating seeds of *Brassica nigra* significantly reduces growth, leaf water potential, pigments, and the photosynthetic apparatus.

Cadmium (Cd) has been shown to induce membrane damage, impair food reserve mobilization by increased cotyledon/embryo ratios of total soluble sugars, glucose, fructose and amino acids, mineral leakage leading to nutrient loss, accumulation in seeds and over-accumulation of lipid peroxidation products in seeds. According to studies, copper (Cu) is hazardous to sunflower seedlings because it causes oxidative stress by producing ROS and lowers catalase activity via oxidation of protein structures (Genchi, Sinicropi, Lauria, Carocci, & Catalano, 2020). Studies have shown that cadmium (Cd) can cause damage to membranes, hinder the mobilization of food reserves by increasing the ratios of total soluble sugars, glucose, fructose, and amino acids in cotyledons to embryos, cause mineral leakage that results in nutrient loss, cause accumulation in seeds, and cause an excessive buildup of lipid peroxidation products in seeds. A variety of in situ and ex situ remediation techniques, such as soil flushing, soil washing, chemical degradation, stabilization, solidification, bioremediation, and phytoremediation, use of lime and compost have been developed to mitigate the potentially harmful effects of contaminated soils and, ideally, restore the ecosystem services (Liu et al., 2018; Zhu, Ye, Ran, Zhang, & Wang, 2022).

12.6.2 Biochar and heavy metal interaction

Recently, biochar amendment has been investigated as a way to lessen soil pollution and play a significant part in cleaning up polluted soils. Existing literature reviews suggest that adding biochar to contaminated soils immobilizes heavy metals and lowers their bioavailability primarily through precipitation, electrostatic interaction, surface adsorption, structural sequestration, and accelerated breakdown (Zama et al., 2018). By raising soil pH and lowering the bioavailability of heavy metals in soils, biochars have also been used to reduce pollution, reducing the health risks to people and animals. Biochar is particularly efficient at immobilizing soil heavy metals due to their stable, porous, and carbonaceous structure. This is accomplished through raising the pH of the soil, cation exchange, direct adsorption, functional group complexation, and metal-hydroxide precipitation. Moreover, heavy metals in biochar are generally considered to be more stable, with low bioavailability after thermal processes (Li et al., 2020).

Biochar amendment has recently been researched for its ability to lessen soil pollution and demonstrate a significant role in remediating damaged soils (Table 12.2). Biochar generated from a variety of feedstocks, such as agricultural manure, animal body waste, or woody plant material, can immobilize heavy metals (Meng & Yuan, 2014). Among the potential nonmediated mechanisms are ion exchange, electrostatic attraction, precipitation, and complexation. Through the ion exchange mechanism, functional groups that include oxygen in general and carboxyl (—COOH) fix metal ions (Ho, Zhu, & Chang, 2017). The increased adsorption capability of biochar for heavy metals can be attributed to high CEC. Numerous studies demonstrate that biochar's high electronegativity makes it possible for positively charged objects to attract one another electrostatically (Ahmad et al., 2018). Cu (II) adsorption increases as pH rises (Tong, Li, Yuan, & Xu, 2011). Initial concentrations of heavy metals are also linked with electrostatic adsorption. Crop waste biochar mostly removes heavy metals through surface complexation. Heavy metals are immobilized via surface complexation on the surface functional groups of biochars with low mineral contents. Functional groups like (OH, COOH) in biochar allow heavy metal binding sites to form complexes, which increase the specific adsorption of metals (Tan, Wang, Zhang, & Huang, 2017). As a result, inorganic ions like Si, S, and Cl in biochar can combine with heavy metals (Tan et al., 2017). Insoluble precipitates are formed due to minerals inside biochar which are fixed with metals.

12.7 Conclusion and future perspective

Due to the negative consequences of climate change, there has been a more than 75% increase in the number of extreme weather events, such as drought and flood, between 2000 and 2019 compared to 1980 to 1999 causing significant deterioration of the soil and water quality. Due to excessive or unforeseen use, the expanding human population has been putting strain on the soil and water resources. While greenhouse gas emissions have intensified, the fertility of agricultural soils has declined globally due to the exposure of soils to frequent flooding, desertification, and salinization. Beneficial properties of biochar such as alkaline pH, highCEC, abundant surface functional groups, remarkable surface area, adequate porosity, excellent water holding capacity, and sufficient nutrient retention capacity can help repair the adverse effects of extreme weather events in the soil–plant system. Applying biochar improves the soil's ability to store water by altering its physio-chemical properties, which increases the amount of water available to plants, improves mineral uptake and controls stomatal conductance. In addition to retaining moisture and nutrients, biochar also inhibits dangerous bacteria, absorbs heavy metals and pesticides, reduces soil erosion, raises the pH of the soil, enhances cationic exchange, and increases soil fertility. To evaluate the long-term effects of applying biochar to soils experiencing real-time temperature rise, drought, flood, and salinization, large-scale field tests should be conducted. The effects of a warmer climate, swampy conditions, water shortages, and extremely salty settings on the mineralization and stability of biochar in various agro-ecological zones need to be investigated. Moreover, further research is necessary for development of modified or engineered biochar with better physicochemical qualities for mitigating the negative effects of temperature rise, drought, flooding, and salinization on soil–plant systems.

References

Abbas, T., Rizwan, M., Ali, S., Adrees, M., Mahmood, A., Zia-ur-Rehman, M., ... Qayyum, M. F. (2018). Biochar application increased the growth and yield and reduced cadmium in drought stressed wheat grown in an aged, contaminated soil. *Ecotoxicology and Environmental Safety, 148*, 825–833.

Abideen, Z., Koyro, H. W., Huchzermeyer, B., Ansari, R., Zulfiqar, F., & Gul, B. J. P. B. (2020). Ameliorating effects of biochar on photosynthetic efficiency and antioxidant defence of *Phragmites karka* under drought stress. *Plant Biology, 22*(2), 259–266.

Ahmad, Z., Gao, B., Mosa, A., Yu, H., Yin, X., Bashir, A., ... Wang, S. (2018). Removal of Cu(II), Cd(II) and Pb(II) ions from aqueous solutions by biochars derived from potassium-rich biomass. *Journal of Cleaner Production, 180*, 437e449.

Akhtar, S. S., Andersen, M. N., & Liu, F. (2015). Biochar mitigates salinity stress in potato. *Journal of agronomy and crop science, 201*(5), 368–378.

Akhtar, S. S., Li, G., Andersen, M. N., & Liu, F. (2014). Biochar enhances yield and quality of tomato under reduced irrigation. *Agricultural Water Management, 138*, 37–44.

Amini, S., Ghadiri, H., Chen, C., & Marschner, P. (2016). Salt-affected soils, reclamation, carbon dynamics, and biochar: A review. *Journal of Soils and Sediments, 16*, 939–953.

Baiamonte, G., Crescimanno, G., Parrino, F., & De Pasquale, C. (2019). Effect of biochar on the physical and structural properties of sandy soil. *Catena, 175*, 294–303.

Baiamonte, G., De Pasquale, C., Marsala, V., Cimò, G., Alonzo, G., Crescimanno, G., & Conte, P. (2015). Structure alteration of a sandy-clay soil by biochar amendments. *Journal of Soils and Sediments, 15*, 816–824.

Barnes, R. T., Gallagher, M. E., Masiello, C. A., Liu, Z., & Dugan, B. (2014). Biochar-induced changes in soil hydraulic conductivity and dissolved nutrient fluxes constrained by laboratory experiments. *PLoS One, 9*(9), 108340.

Baronti, S., Vaccari, F. P., Miglietta, F., Calzolari, C., Lugato, E., Orlandini, S., ... Genesio, L. (2014). Impact of biochar application on plant water relations in *Vitis vinifera* (L.). *European Journal of Agronomy, 53*, 38–44.

Batool, A., Taj, S., Rashid, A., Khalid, A., Qadeer, S., Saleem, A. R., & Ghufran, M. A. (2015). Potential of soil amendments (biochar and gypsum) in increasing water use efficiency of *Abelmoschus esculentus* L. Moench. *Frontiers in plant science, 6*, 733.

Beesley, L., Marmiroli, M., Pagano, L., Pigoni, V., Fellet, G., Fresno, T., ... Marmiroli, N. (2013). Biochar addition to an arsenic contaminated soil increases arsenic concentrations in the pore water but reduces uptake to tomato plants (*Solanum lycopersicum* L.). *Science of the Total Environment, 454*, 598–603.

Blanco-Canqui, H. (2017). Biochar and soil physical properties. *Soil Science Society of America Journal, 81*, 687–711.

Boehm, A. B., Bell, C. D., Fitzgerald, N. J., Gallo, E., Higgins, C. P., Hogue, T. S., ... Wolfand, J. M. (2020). Biochar-augmented biofilters to improve pollutant removal from stormwater–can they improve receiving water quality? *Environmental Science: Water Research & Technology, 6*(6), 1520–1537.

Brantley, K. E., Brye, K. R., Savin, M. C., & Longer, D. E. (2015). Biochar source and application rate effects on soil water retention determined using wetting curves. *Open Journal of Soil Science, 5*(01), 1.

Burant, A., Selbig, W., Furlong, E. T., & Higgins, C. P. (2018). Trace organiccontaminants in urban runoff: Associations with urban land-use. *Environmental Pollution, 242*, 2068–2077.

Carvalho, M. D. M., Maia, A. D. H. N., Madari, B. E., Bastiaans, L., Van Oort, P. A., Heinemann, A. B., ... Meinke, H. (2014). Biochar increases plant-available water in a sandy loam soil under an aerobic rice crop system. *Solid Earth, 5*, 939–952.

Certini, G. (2005). Effects of fire on properties of forest soils: A review. *Oecologia, 143*, 1–10.

Chang, Y., Rossi, L., Zotarelli, L., Gao, B., Shahid, M. A., & Sarkhosh, A. (2021). Biochar improves soil physical characteristics and strengthens root architecture in Muscadine grape (*Vitis rotundifolia* L.). *Chemical and Biological Technologies in Agriculture, 8*(1), 11.

Chaturvedi, S., Singh, S. V., Dhyani, V. C., Govindaraju, K., & Mandal, S. (2021). Characterization, bioenergy value and thermal stability of biochar derived from diverse agriculture and forestry lignocellulosic waste. *Biomass Conversion and Biorefinery*. Available from https://doi.org/10.1007/s13399-020-01239-2.

Datta, D., Chandra, S., Nath, C. P., Kar, G., Ghosh, S., Chaturvedi, S., ... Singh, V. (2022). Soil-plant water dynamics, yield, quality and profitability of spring sweet corn under variable irrigation scheduling, crop establishment and moisture conservation practices. *Field Crops Research, 279*108450.

De-Campos, A. B., Mamedov, A. I., & Huang, C. H. (2009). Short-term reducing conditions decrease soil aggregation. *Soil Science Society of America Journal, 73*(2), 550–559.

Du, Z. L., Zhao, J. K., Wang, Y. D., & Zhang, Q. Z. (2017). Biochar addition drives soil aggregation and carbon sequestration in aggregate fractions from an intensive agricultural system. *Journal of Soils and Sediments, 17*, 581–589.

Egamberdieva, D., Reckling, M., & Wirth, S. (2017). Biochar-based *Bradyrhizobium* inoculum improves growth of lupin (*Lupinus angustifolius* L.) under drought stress. *European Journal of Soil Biology, 78*, 38–42.

Fang, X., Turner, N. C., Yan, G., Li, F., & Siddique, K. H. M. (2010). Flower numbers, pod production, pollen viability, and pistil function are reduced and flower and pod abortion increased in chickpea (*Cicer arietinum* L.) under terminal drought. *Journal of Experimental Botany, 61*, 335–345.

Farhangi-Abriz, S., & Torabian, S. (2017). Antioxidant enzyme and osmoticadjustment changes in bean seedlings as affected by biocharunder salt stress. *Ecotoxicology and Environmental Safety, 137*, 64–70.

Farooq, M., Wahid, A., Kobayashi, N., Fujita, D., & Basra, S. M. A. (2009). Plant drought stress: Effects, mechanisms and management. *Agronomy for Sustainable Development, 29*, 185e212.

Fiaz, K., Danish, S., Younis, U., Malik, S. A., Raza Shah, M. H., & Niaz, S. (2014). Drought impact on Pb/Cd toxicity remediated by biochar in *Brassica campestris*. *Journal of Soil Science and Plant Nutrition, 14*(4), 845e854.

Genchi, G., Sinicropi, M. S., Lauria, G., Carocci, A., & Catalano, A. (2020). The effects of cadmium toxicity. *International Journal of Environmental Research and Public Health, 17*(11), 3782.

Genesio, L., Miglietta, F., Baronti, S., & Vaccari, F. P. (2015). Biochar increases vineyard productivity without affecting grape quality: Results from a four years field experiment in Tuscany. *Agriculture, Ecosystems & Environment, 201*, 20—25.

Ghrefat, A. (2011). Causes, impacts, extent, and control of desertification. In Jessica A. Murphy (Ed.), *Sand Dunes*. Nova Science Publishers, ISBN 978-1-61324-108-0.

Gornall, J., Betts, R., Burke, E., Clark, R., Camp, J., Willett, K., & Wiltshire, A. (2010). Implications of climate change for agricultural productivity in the early twenty-first century. *Philosophical Transactions of the Royal Society B: Biological Sciences, 365*(1554), 2973—2989.

Gunarathne, V., Senadeera, A., Gunarathne, U., Biswas, J. K., Almaroai, Y. A., & Vithanage, M. (2020). Potential of biochar and organic amendments for reclamation of coastal acidic-salt affected soil. *Biochar, 2*, 107—120.

Haider, G., Koyro, H. W., Azam, F., Steffens, D., Müller, C., & Kammann, C. (2015). Biochar but not humic acid product amendment affected maize yields via improving plant-soil moisture relations. *Plant and Soil, 395*, 141—157.

Haider, I., Raza, M. A. S., Iqbal, R., Aslam, M. U., Habib-ur-Rahman, M., Raja, S., ... Ahmad, S. (2020). Potential effects of biochar application on mitigating the drought stress implications on wheat (*Triticum aestivum* L.) under various growth stages. *Journal of Saudi Chemical Society, 24*(12), 974—981.

Hammer, E. C., Forstreuter, M., Rillig, M. C., & Kohler, J. (2015). Biochar increases arbuscular mycorrhizal plant growth enhancement and ameliorates salinity stress. *Applied Soil Ecology, 96*, 114e121.

Hardie, M., Clothier, B., Bound, S., Oliver, G., & Close, D. (2014). Does biochar influence soil physical properties and soil water availability? *Plant and Soil, 376*, 347—361.

Ho, S.-H., Zhu, S., & Chang, J.-S. (2017). Recent advances in nanoscale-metal assisted biochar derived from waste biomass used for heavy metals removal. *Bioresource Technology, 246*, 123e134.

Hossain, M. Z., Bahar, M. M., Sarkar, B., Donne, S. W., Ok, Y. S., Palansooriya, K. N., ... Bolan, N. (2020). Biochar and its importance on nutrient dynamics in soil and plant. *Biochar, 2*, 379—420.

Ibrahim, E. A., El-Sherbini, M. A., & Selim, E. M. M. (2022). Effects of biochar on soil properties, heavy metal availability and uptake, and growth of summer squash grown in metal-contaminated soil. *Scientia Horticulturae, 301*111097.

Inbar, A., Lado, M., Sternberg, M., Tenau, H., & Ben-Hur, M. (2014). Forest fire effects on soil chemical and physicochemical properties, infiltration, runoff, and erosion in a semiarid Mediterranean region. *Geoderma, 221*, 131—138.

IPCC. (2018). Global warming of 1.5°C. An IPCC special report on the impacts of global warming of 1.5°C above pre-industriallevels and related global greenhouse gas emission pathways, in the context of strengthening the global response to the threat of climate change.

Jahromi, N. B., Lee, J., & Fulcher, A. (2020). Effect of biochar application on quality of flooded sandy soils and corn growth under greenhouse conditions. *Agrosystems, Geosci Environ, 3*e20028.

Jahromi, N. B., Lee, J., Fulcher, A., Walker, F., Jagadamma, S., & Arelli, P. (2020). Effect of biochar application on quality of flooded sandy soils and corn growth under greenhouse conditions. *Agrosystems, Geosciences & Environment, 3*(1), e20028.

Janus, A., Pelfrêne, A., Heymans, S., Deboffe, C., Douay, F., & Waterlot, C. (2015). Elaboration, characteristics and advantages of biochars for the management of contaminated soils with a specific overview on *Miscanthus* biochars. *Journal of Environmental Management, 162*, 275—289.

Jiang, S. C., Lim, K. Y., Huang, X., McCarthy, D., & Hamilton, A. J. (2015). Human and environmental health risks and benefits associated with use of urban stormwater. *Wiley Interdisciplinary Reviews: Water, 2*(6), 683—699.

Kaushal, M., & Wani, S. P. (2016). Rhizobacterial-plant interactions: Strategies ensuring plant growth promotion under drought and salinity stress. *Agriculture, Ecosystems & Environment, 231*, 68—78.

Kazemi, R., Ronaghi, A., Yasrebi, J., Ghasemi-Fasaei, R., & Zarei, M. (2019). Effect of shrimp waste—derived biochar and arbuscular mycorrhizal fungus on yield, antioxidant enzymes, and chemical composition of corn under salinity stress. *Journal of Soil Science and Plant Nutrition, 19*, 758—770.

Khalil, F., Rauf, S., Monneveux, P., Anwar, S., & Iqbal, Z. (2016). Genetic analysis of proline concentration under osmotic stress in sunflower (*Helianthus annuus* L.). *Breeding Science, 66*, 463e470.

Khan, S., Chao, C., Waqas, M., Arp, H. P. H., & Zhu, Y. G. (2013). Sewage sludge biochar influence upon rice (*Oryza sativa* L.) yield, metal bioaccumulation and greenhouse gas emissions from acidic paddy soil. *Environmental Science & Technology, 47*, 8624–8632.

Khan, M. I. R., Asgher, M., Fatma, M., Per, T. S., & Khan, N. A. (2015). Drought stress vis a vis plant functions in the era of climate change. *Climate Change and Environmental Sustainability, 3*, 13.

Kim, H. S., Kim, K. R., Yang, J. E., Ok, Y. S., Owens, G., Nehls, T., ... Kim, K. H. (2016). Effect of biochar on reclaimed tidal land soil properties and maize (*Zea mays* L.) response. *Chemosphere, 142*, 153–159.

Kumar, A., & Bhattacharya, T. (2021). Removal of arsenic by wheat straw biochar from soil. *Bulletin of Environmental Contamination and Toxicology*, 1–8.

Kumar, A., Bhattacharya, T., Hasnain, S. M., Nayak, A. K., & Hasnain, M. S. (2020). Applications of biomass-derived materials for energy production, conversion, and storage. *Materials Science for Energy Technologies, 3*, 905–920.

Kumar, A., Nagar, S., & Anand, S. (2021). Nanotechnology for sustainable crop production: Recent development and strategies. In P. Singh, et al. (Eds.), *Plant-microbes-engineered nano-particles (PM-ENPs) Nexus in Agro-Ecosystems, Advances in Science, Technology & Innovation* (pp. 31–47). Cham: Springer.

Lahti, M., Aphalo, P. J., Finér, L., Lehto, T., Leinonen, I., Mannerkoski, H., & Ryyppö, A. (2002). Soil temperature, gas exchange and nitrogen status of 5-year-old Norway spruce seedlings. *Tree Physiology, 22*(18), 1311–1316.

Lashari, M. S., Liu, Y., Li, L., Pan, W., Fu, J., Pan, G., ... Yu, X. (2013). Effects of amendment of biochar-manure compost in conjunction with pyroligneous solution on soil quality and wheat yield of a salt-stressed cropland from Central China Great Plain. *Field Crops Research, 144*, 113–118.

Lashari, M. S., Ye, Y., Ji, H., Li, L., Kibue, G. W., Lu, H., ... Pan, G. (2015). Biochar–manure compost in conjunction with pyroligneous solution alleviated salt stress and improved leaf bioactivity of maize in a saline soil from central China: A 2year field experiment. *Journal of the Science of Food and Agriculture, 95*(6), 1321–1327.

Letey, J. (2000). Soil salinity poses challenges for sustainable agriculture and wildlife. *California Agriculture, 54*(2), 43–48.

Li, H., & Tan, Z. (2021). Preparation of high water-retaining biochar and its mechanism of alleviating drought stress in the soil and plant system. *Biochar, 3*, 579–590.

Li, S., Zou, D., Li, L., Wu, L., Liu, F., Zeng, X., ... Xiao, Z. (2020). Evolution of heavy metals during thermal treatment of manure: A critical review and outlooks. *Chemosphere, 247*125962.

Liang, C., Zhu, X., Fu, S., Méndez, A., Gascó, G., & Paz-Ferreiro, J. (2014). Biochar alters the resistance and resilience to drought in a tropical soil. *Environmental Research Letters, 9*(6), 064013.

Liu, C., Liu, F., Ravnskov, S., Rubæk, G. H., Sun, Z., & Andersen, M. N. (2017). Impact of wood biochar and its interactions with mycorrhizal fungi, phosphorus fertilization and irrigation strategies on potato growth. *Journal of Agronomy and Crop Science, 203*(2), 131–145.

Liu, Y., Li, J., Jiao, X., Li, H., An, Y., & Liu, K. (2020). Effects of straw returning combine with biochar on water quality under flooded condition. *Water, 12*(6), 1633. Available from https://doi.org/10.3390/w12061633.

Liu, Y., Yang, S., Lu, H., & Wang, Y. (2018). Effects of biochar on spatial and temporal changes in soil temperature in cold waterlogged rice paddies. *Soil and Tillage Research, 181*, 102–109.

Lopes, É. M. G., Reis, M. M., Frazão, L. A., da Mata Terra, L. E., Lopes, E. F., dos Santos, M. M., & Fernandes, L. A. (2021). Biochar increases enzyme activity and total microbial quality of soil grown with sugarcane. *Environmental Technology & Innovation, 21*, 101270.

Lu, H., Lashari, M. S., Liu, X., Ji, H., Li, L., Zheng, J., ... Pan, G. (2015). Changes in soil microbial community structure and enzyme activity with amendment of biochar-manure compost and pyroligneous solution in a saline soil from Central China. *European Journal of Soil Biology, 70*, 67–76.

Mannan, M., Halder, E., Karim, M., & Ahmed, J. (2017). Alleviation of adverse effect of drought stress on soybean (*Glycine max.* L.) by using poultry litter biochar. *Bangladesh Agronomy Journal, 19*, 61–69.

Mannan, M. A., Halder, E., Karim, M. A., & Ahmed, J. U. (2016). Alleviation of adverse effect of drought stress on soybean (*Glycine max.* L.) by using poultry litter biochar. *Bangladesh Agronomy Journal, 19*, 61ᵉ69.

Mannan, M.A., & Shashi, M.A. (2020). Amelioration of drought tolerance in maize using rice husk biochar. Maize Prod Use.

Mansoor, S., Kour, N., Manhas, S., Zahid, S., Wani, O. A., Sharma, V., ... Paray, B. A. (2021). Biochar as a tool for effective management of drought and heavy metal toxicity. *Chemosphere, 271*129458.

Meng, X., & Yuan, W. (2014). Can biochar couple with algae to deal with desertification? *Journal of Sustainable Bioenergy Systems, 04*, 194–198.

Muhammad, N., Aziz, R., Brookes, P. C., & Xu, J. (2017). Impact of wheat straw biochar on yield of rice and some properties of *Psammaquent* and *Plinthudult*. *Journal of Soil Science and Plant Nutrition, 17*, 808–823.

Mukherjee, A., Lal, R., & Zimmerman, A. R. (2014). Effects of biochar and other amendments on the physical properties and greenhouse gas emissions of an artificially degraded soil. *Science of The Total Environment, 487*, 26–36.

Mulcahy, D. N., Mulcahy, D. L., & Dietz, D. (2013). Biochar soil amendment increases tomato seedling resistance to drought in sandy soils. *Journal of Arid Environments, 88*, 222–225.

Nadeem, S. M., Zahir, Z. A., Naveed, M., & Nawaz, S. (2013). Mitigation of salinity-induced negative impact on the growth and yield of wheat by plant growth-promoting rhizobacteria in naturally saline conditions. *Annals of Microbiology, 63*(1), 225–232.

Nikpour-Rashidabad, N., Tavasolee, A., Torabian, S., & Farhangi-Abriz, S. (2019). The effect of biochar on the physiological, morphological and anatomical characteristics of mung bean roots after exposure to salt stress. *Archives of Biological Sciences, 71*(2), 321–327.

Obia, A., Mulder, J., Martinsen, V., Cornelissen, G., & Børresen, T. (2016). In situ effects of biochar on aggregation, water retention and porosity in light-textured tropical soils. *Soil and Tillage Research, 155*, 35–44.

Ohashi, M., Kume, T., Yoshifuji, N., Kho, L. K., Nakagawa, M., & Nakashizuka, T. (2015). The effects of an induced short-term drought period on the spatial variations in soil respiration measured around emergent trees in a typical bornean tropical forest, Malaysia. *Plant and Soil, 387*, 337–349.

Osakabe, Y., Osakabe, K., Shinozaki, K., & Tran, L.-S. P. (2014). Response of plants to water stress. *Frontiers in Plant Science, 5*, 86.

Pandey, B., Suthar, S., & Chand, N. (2022). Effect of biochar amendment on metal mobility, phytotoxicity, soil enzymes, and metal-uptakes by wheat (*Triticum aestivum*) in contaminated soils. *Chemosphere, 307*135889.

Park, J. H., Choppala, G. K., Bolan, N. S., Chung, J. W., & Chuasavathi, T. (2011). Biochar reduces the bioavailability and phytotoxicity of heavy metals. *Plant and Soil, 348*, 439–451.

Parkash, V., & Singh, S. (2020). Potential of biochar application to mitigate salinity stress in eggplant. *HortScience: A Publication of the American Society for Horticultural Science, 55*, 1946–1955.

Poveda, J., Martínez-Gómez, Á., Fenoll, C., & Escobar, C. (2021). The use of biochar for plant pathogen control. *Phytopathology, 111*(9), 1490–1499.

Rajapaksha, A. U., Chen, S. S., Tsang, D. C. W., Zhang, M., Vithanage, M., Mandal, S., ... Ok, Y. S. (2016). Engineered/designer biochar for contaminant removal/immobilization from soil and water: Potential and implication of biochar modification. *Chemosphere, 148*, 276e291.

Repo, T., Leinonen, I., Ryyppö, A., & Finér, L. (2004). The effect of soil temperature on the bud phenology, chlorophyll fluorescence, carbohydrate content and cold hardiness of Norway spruce seedlings. *Physiologia Plantarum, 121*, 93–100.

Sattar, A., Sher, A., Ijaz, M., Irfan, M., Butt, M., Abbas, T., ... Cheema, M. A. (2019). Biochar application improves the drought tolerance in maize seedlings. *Phyton; Annales rei Botanicae, 88*(4), 379.

Shrestha, A. J., & Pandit, B. H. (2017). Action research into a flood resilient value chain—biochar-based organic fertilizer doubles productivity of pea in Udayapur, Nepal. *KnE Life Sciences*, 1–19.

Singh, S., Luthra, N., Mandal, S., Kushwaha, D. P., Pathak, S. O., Datta, D., ... Pramanik, B. (2023). Distinct behavior of biochar modulating biogeochemistry of salt-affected and acidic soil: A review. *Journal of Soil Science and plant Nutrition*. Available from https://doi.org/10.1007/s42729-023-01370-9.

Singh, S. V., Chaturvedi, S., & Datta, D. (2019). Biochar: An eco-friendly residue management approach. *Indian Farming, 69*(08), 27–29.

Singh, S. V., Chaturvedi, S., Dhyani, V. C., & Govindaraju, K. (2020). Pyrolysis temperature influences the characteristics of rice straw and husk biochar and sorption/desorption behavior of their biourea composite. *Bioresource Technology, 314*123674. Available from https://doi.org/10.1016/j.biortech.2020.123674.

Suarez, D. L., Wood, J. D., & Lesch, S. M. (2006). Effect of SAR on water infiltration under a sequential rain—irrigation management system. *Agricultural Water Management, 86*(1–2), 150–164.

Sun, W., Yang, K., Li, R., Chen, T., Xia, L., Sun, X., & Wang, Z. (2022). Distribution characteristics and ecological risk assessment of heavy metals in sediments of Shahe reservoir. *Scientific Reports, 12*(1)16239.

Tan, Z., Wang, Y., Zhang, L., & Huang, Q. (2017). Study of the mechanism of remediation of Cd-contaminated soil by novel biochars. *Environmental Science and Pollution Research, 24*, 24844e24855.

Tardieu, F., Parent, B., Caldeira, C. F., & Welcker, C. (2014). Genetic and physiological controls of growth under water deficit. *Plant Physiology, 164*, 1628–1635.

Thomas, S. C., Frye, S., Gale, N., Garmon, M., Launchbury, R., Machado, N., ... Winsborough, C. (2013). Biochar mitigates negative effects of salt additions on two herbaceous plant species. *Journal of Environmental Management, 129*, 62–68.

Tong, X. J., Li, J. Y., Yuan, J. H., & Xu, R. K. (2011). Adsorption of Cu (II) by biochars generated from three crop straws. *Chemical Engineering Journal, 172*(2–3), 828–834.

Tryon, E. H. (1948). Effect of charcoal on certain physical, chemical, and biological properties of forest soils. *Ecological Monographs, 18*, 81–115.

US EPA. (2021). *Climate change indicators: Weather and climate*. pp. 1–11.

Usman, A. R. A., Al-Wabel, M. I., Abdulaziz, A. H., Mahmoud, W. A., EL-Naggar, A. H., Ahmad, M., ... Abdulrasoul, A. O. (2016). Conocarpus biochar induces changes in soil nutrient availability and tomato growth under saline irrigation. *Pedosphere, 26*(1), 27–38.

Usowicz, B., Lipiec, J., Łukowski, M., Bis, Z., Usowicz, J., & Latawiec, A. E. (2020). Impact of biochar addition on soil thermal properties: Modelling approach. *Geoderma, 376*114574.

Villagra-Mendoza, K., & Horn, R. (2018). Effect of biochar addition on hydraulic functions of two textural soils. *Geoderma, 326*, 88–95.

Vollset, S. E., Goren, E., Yuan, C. W., Cao, J., Smith, A. E., Hsiao, T., ... Dolgert, A. J. (2020). Fertility, mortality, migration, and population scenarios for 195 countries and territories from 2017 to 2100: A forecasting analysis for the Global Burden of Disease Study. *The Lancet, 396*(10258), 1285–1306.

Wang, W., Vinocur, B., & Altman, A. (2003). Plant responses to drought, salinity and extreme temperatures: Towards genetic engineering for stress tolerance. *Planta, 28*, 1–14. Available from https://doi.org/10.1007/s00425-003-1105-5.

Wang, J., Xiong, Z., & Kuzyakov, Y. (2016). Biochar stability in soil: Metaanalysis of decomposition and priming effects. *GCB Bioenergy, 8*, 512–523.

Wang, Y., Xiao, D., Li, Y., & Li, X. (2008). Soil salinity evolution and its relationship with dynamics of groundwater in the oasis of inland river basins: Case study from the Fubei region of Xinjiang Province, China. *Environmental Monitoring and Assessment, 140*, 291–302.

Wang, Z., Yin, D., Wang, H., Zhao, C., & Li, Z. (2020). Effects of biochar on waterlogging and the associated change in micro-ecological environment of maize rhizosphere soil in saline-alkali land. *BioResources, 15*(4), 9303.

Warter, M. M., Singer, M. B., Cuthbert, M. O., Roberts, D., Caylor, K. K., Sabathier, R., & Stella, J. (2021). Drought onset and propagation into soil moisture and grassland vegetation responses during the 2012–2019 major drought in Southern California. *Hydrology and Earth System Sciences, 25*(6), 3713–3729.

Widijanto, H., & Syamsiyah, J., 2022. The effect of agricultural waste biochar on the availability of silicon in ultisol and inceptisol under flooded conditions. In *IOP conference series: Earth and environmental science*, 1114 (1), 012035. IOP Publishing.

Wu, S. H., Jansson, P. E., & Kolari, P. (2012). The role of air and soil temperature in the seasonality of photosynthesis and transpiration in a boreal Scots pine ecosystem. *Agricultural and Forest Meteorology, 156*, 85–103.

Xiao, Y., Yang, S., Xu, J., Ding, J., Sun, X., & Jiang, Z. (2018). Effect of biochar amendment on methane emissions from paddy field under water-saving irrigation. *Sustainability, 10*(5), 1371.

Xiong, J., Yu, R., Islam, E., Zhu, F., Zha, J., & Sohail, M. I. (2020). Effect of biochar on soil temperature under high soil surface temperature in coal mined arid and semiarid regions. *Sustainability, 12*(19), 8238.

Xu, X., Luo, Y., & Zhou, J. (2012). Carbon quality and the temperature sensitivity of soil organic carbon decomposition in a tallgrass prairie. *Soil Biology and Biochemistry, 50*, 142–148.

Yan, N., Marschner, P., Cao, W., Zuo, C., & Qin, W. (2015). Influence of salinity and water content on soil microorganisms. *International Soil and Water Conservation Research, 3*(4), 316–323.

Yang, S., Jiang, Z., Sun, X., Ding, J., & Xu, J. (2018). Effects of biochar amendment on CO2 emissions from paddy fields under water-saving irrigation. *International Journal of Environmental Research and Public Health, 15*(11), 2580.

Yoo, S. Y., Kim, Y. J., & Yoo, G. (2020). Understanding the role of biochar in mitigating soil water stress in simulated urban roadside soil. *Science of The Total Environment, 738*139798.

Zama, E. F., Reid, B. J., Arp, H. P. H., Sun, G. X., Yuan, H. Y., & Zhu, Y. G. (2018). Advances in research on the use of biochar in soil for remediation: A review. *Journal of Soils and Sediments, 18*, 2433–2450.

Zhang, Q., Wang, Y., Wu, Y., Wang, X., Du, Z., Liu, X., & Song, J. (2013). Effects of biochar amendment on soil thermal conductivity, reflectance, and temperature. *Soil Science Society of America Journal, 77*(5), 1478–1487.

Zhang, Q., Xiao, J., Xue, J., & Zhang, L. (2020). Quantifying the effects of biochar application on greenhouse gas emissions from agricultural soils: A global meta-analysis. *Sustainability, 12*, 3436. Available from https://doi.org/10.3390/SU12083436.

Zhang, Y., Ding, J., Wang, H., Su, L., & Zhao, C. (2020). Biochar addition alleviate the negative effects of drought and salinity stress on soybean productivity and water use efficiency. *BMC Plant Biology, 20*(1), 11.

Zhao, J., Ren, T., Zhang, Q., Du, Z., & Wang, Y. (2016). Effects of biochar amendment on soil thermal properties in the North China Plain. *Soil Science Society of America Journal, 80*(5), 1157–1166. Available from https://doi.org/10.2136/sssaj2016.01.0020.

Zheng, H., Wang, X., Chen, L., Wang, Z., Xia, Y., Zhang, Y., … Xing, B. (2018). Enhanced growth of halophyte plants in biochar-amended coastal soil: Roles of nutrient availability and rhizosphere microbial modulation. *Plant, Cell & Environment, 41*(3), 517–532.

Zhu, K., Ye, X., Ran, H., Zhang, P., & Wang, G. (2022). Contrasting effects of straw and biochar on microscale heterogeneity of soil O_2 and pH: Implication for N_2O emissions. *Soil Biology and Biochemistry, 166*108564.

Zhu, Y., Wang, H., Lv, X., Zhang, Y., & Wang, W. (2020). Effects of biochar and biofertilizer on cadmium-contaminated cotton growth and the antioxidative defense system. *Scientific Reports, 10*(1)20112.

Zia-Ur-Rehman, M., Rizwan, M., Sabir, M., Ali, S. S., & Ahmed, H. R. (2016). Comparative effects of different soil conditioners on wheat growth and yield grown in saline-sodic soils. *Sains Malaysiana, 45*, 339–346.

Zipper, S. C., Qiu, J., & Kucharik, C. J. (2016). Drought effects on US maize and soybean production: Spatiotemporal patterns and historical changes. *Environmental Research Letters, 11*(9), 094021.

CHAPTER 13

Biochar application in sustainable production of horticultural crops in the new era of soilless cultivation

Ashish Kaushal[1], Rajeev Kumar Yadav[1] and Neeraj Singh[2]

[1]Department of Horticulture, School of Agriculture, Graphic Era Hill University, Dehradun, Uttarakhand, India [2]Department of Horticulture, Ch. Shivnath Singh Shandilya PG College Meerut, Meerut, Uttar Pradesh, India

13.1 Introduction

Horticulture is a significant multidisciplinary enterprise that differs from traditional agriculture like its crops, more intense production methods, and overall profitability. Horticulture enterprise embraces the production, processing, shipping and marketing of fruits, vegetables, ornamental and cut-flower, plantation, spices, medicinal and aromatic crops in domestic and international markets. These crops are grown in soil in open fields or under protected cultivation, pot/containers on substrates, or in growing media or soilless culture in greenhouses. The container's substrates mostly comprise organic matter, including coco peat, peat moss, vermicompost, and other composted materials. Due to rapid urbanization, climate change and reduction in soil fertility, the soil-based production system is facing challenges which ultimately compromise crop productivity (Sharma, Acharya, Kumar, Singh, & Chaurasia, 2018). An alternative approach to address these problems and sustain food production to feed the globally expanding population, growing plants without soil has been developed (Gwynn-Jones et al., 2018). Production of diverse horticultural crops by soilless growing media with biochar has great potential for overcoming the challenge and producing horticultural products in an eco-friendly manner.

Biochar is a solid charcoal-like carbonaceous material produced through a thermochemical conversion process of organic biomass under the quasi-absence of oxygen at temperatures ranging from 350°C to 900°C, a process referred to as pyrolysis (Fan et al., 2021; Lehmann & Joseph, 2009; Singh, Chaturvedi, & Datta, 2019). This process converts the agricultural, industrial, or urban organic matter into different fractions of solid biochar, gas (syngas), and liquid (bio-oil) products. The physical and chemical properties of biochar vary substantially depending on feedstocks, temperatures, pyrolysis methods, and feedstock modifications (Chaturvedi, Singh, Dhyani, Govindaraju, & Mandal, 2023; Singh, Chaturvedi, Dhyani, & Govindaraju, 2020; Yu et al., 2019). The physical properties of biochar include surface area, density, porosity, pore size and volume, hydrophobicity, and water-holding capacity. In general, surface area, porosity, and pore volume increase with increased pyrolysis temperature (Singh et al., 2023). Biochar surface area ranges from 100 to 800 $m^2\,g^{-1}$, and porosity ranges from 50% to 70%. Bulk density is rather low, <0.6 g cm^{-3} (Yu et al., 2019).

Biochar serves a variety of purposes in sustainable horticultural production including, rooting soil-less media by improving media properties, seed germination, crop yield and quality of horticulture produce, and increasing crop tolerance to abiotic and biotic stresses. This chapter summarizes the current application of biochar to improve substrate properties, enhance seed germination, improve root growth, and crop yield, and in mitigation of biotic and abiotic stresses on horticultural crops.

13.2 Biochar for horticultural rooting media

Successful production of horticultural crops requires several cultural inputs, among them the most important is growing substrate (medium). Growing media is a substrate that provides anchorage to the root system and supplies essential nutrients for the growth and development of plants. These substrates may include, soil, sand, compost, peat moss, coco peat, vermiculite, perlite, rock wool, etc. Among them, peat is a widely used rooting media in the protected cultivation of horticultural crops. Peat is the carbon-rich substrate that is formed when plant material, usually in a bogy area is inhibited from decaying fully by acidic and anaerobic conditions.

Mining of peat releases large amounts of stored carbon dioxide, which adds enormous greenhouse gases to the environment which is a serious environmental concern. Biochar is a sustainable alternative to peat, reducing the carbon dioxide emissions brought on by peat use. A sustainable substitute for using peat is biochar, which will cut the carbon dioxide emissions caused by peat usage. Biochar in potting soil combinations is said to boost water storage, nutrient supply, microbial life, and disease control. Biochar in potting soil mixtures allegedly increases water storage, nutrient supply, microbial life, and disease suppression but this depends on feedstock and the production process.

There is interest in substituting some of the peat used in rooting medium in greenhouse horticulture with biochar from renewable organic residual streams (Nemati, Simard, Fortin, & Beaudoin, 2015; Nieto et al., 2016). Peat bogs serve as significant carbon reservoirs and control the local water regime and quality (Steiner & Harttung, 2014). The horticultural interest in biochar, aside from peat substitution, is the use and manipulation of bacterial communities for the protection of plants against diseases, either by direct protection or by inducing plant resilience (Graber et al., 2010; Jaiswal, Frenkel, Elad, Lew, & Graber, 2015). Peat substitution by biochar will preserve peat bogs and lower global carbon dioxide emissions linked with the use of peat extraction and use (Verhagen et al., 2009). The addition of biochar to certain plant growth mediums causes changes in the rhizosphere microbiota as well as chemical reactions in the plant (De Tender et al., 2016). The use of high-input fertigation systems in greenhouse horticulture renders the economic impact of biochar-related improvements in water storage and nutrient delivery less significant than in agricultural applications (Roy et al., 2022). The enhanced control over climatic influences, such as variations in nutrient concentration and water content due to rain, is a benefit of greenhouse testing.

13.2.1 Biochar as a soil-less substrate

Biochar as a growth substrate or amending biochar with various substrates has been utilized in various horticultural crops under a soil-less production system. The effect of biochar on soilless substrates used in hydroponic production systems and as a nursery container substrate has been studied in several horticultural crops. Biochar has excellent properties regarding structural stability of growing media, water, and air capacity, important for use in growing media (Lehmann & Joseph, 2009). Amending biochar into a soil-less substrate affects the physical, chemical, and biological properties of growing media (Huang & Gu, 2019). Amending biochar with soil-less substrates reduces the macronutrient runoff and nutrient retention (Altland & Locke, 2012; Haraz, Bowtell, & Al-Juboori, 2020) and increases water retention in soil-less substrates (Beck, Johnson, & Spolek, 2011) (Table 13.1). Additionally, the use of biochar improved the resistance against biotic and biotic stress, decreased heavy metal toxicity, and influenced growth and development in crop plants (Rosli, Abdullah, Yaacob, & Razali, 2023).

TABLE 13.1 Effect of biochar on growth and development in different horticultural crops.

Sl. no.	Crop	Feedstock and concentration used	Pyrolysis temp.	Effect	Reference
1	Geranium (*Pelargonium zonale* L.)	Peat: biochar, (whole tree *Abies alba* (Mill.) of 100:0, 70:30, 30:70 two fertilizer rates (FERT$_1$, FERT$_2$), and arbuscular mycorrhizal fungi (AMF).	1000°C −1100°C	Higher dry biomass, greater floral clusters, larger and more abundant leaves) and quality resulting in unstressed (lower electrolyte leakage and higher relative water content values) and greener leaves (low L and C, high CHL content.	Conversa, Bonasia, Lazzizera, and Elia (2015)
2	Poinsettia (*Euphorbia pulcherrima*)	Peat-based potting mix amended with biochar at 0%, 20%, 40%, 60%, 80%, or 100% (by volume) at four different fertigation regimes.	450°C	A fertigation rate of 100 mg L^{-1} N to 400 mg L^{-1} N, up to 80% biochar could be used as an amendment to peat-based root substrate with acceptable growth reduction and no changes in quality.	Guo et al. (2018)

(Continued)

TABLE 13.1 (Continued)

Sl. no.	Crop	Feedstock and concentration used	Pyrolysis temp.	Effect	Reference
3	Red lettuce (*Lactuca sativa* L.)	Palm kernel biochar and hydroton used in hydroponic system.	250°C	The combination of palm kernel biochar and hydroton as a substrate enhanced the growth performance, color, and nutritional content.	Rosli et al. (2023)
4	*Mentha* species	20%, 40%, 60%, 80%, or 100%; by vol.) commercial bichar.	–	A 80% (by vol.) of the biochar mixed with 20% of peat moss-based commercial substrate could be used for mint plant production.	Yan, Yu, Liu, Li, and Gu (2020)
5	Tomato (*Solanum lycopersicum*)	Different combination of tomato crop waste biochar used as soilless substrate in hydroponic.	550°C	Biochar helped in a sustainable means to support the growth of tomato under soilless substrate in hydroponic system.	Dunlop, Arbestain, Bishop, and Wargent (2015)
6	Tomato (*Solanum lycopersicum*)	Bamboo biochar.	300°C, 450°C, and 600°C	Biochar produced at 300°C and amended at 3% or pyrolyzed at 450°C and amended at 1% increased plant growth index and fruit quality.	Suthar et al. (2018)
7	Leafy vegetables (cabbage, dill, mallow, red lettuce, and tatsoi)	Rice husk biochar + perlite used as a soilless substrate in hydroponic system.	500°C	Biochar + perlite substrate led to approximately twofold increases in shoot length, number of leaves, and fresh/dry masses of leafy vegetable plants compared with those grown in only perlite substrate.	Awad et al. (2017)
8	Tomato (*Solanum lycopersicum*)	Olive stone biochar.	500°C–600°C	Taller tomato seedling growth in olive stone biochar used in hydroponic system.	Karakaş, Özçimen, and İnan (2017)
9	Cherry tomato (*Solanum lycopersicum*)	Pruned biomass of fruit orchard.	500°C	Application of biochar increased flower and fruit numbers (6 times and 2 times, respectively), acidity (75%), lycopene (28%), and solid soluble content (16%).	**Simiele** et al. (2022)

B. Biochar for soil improvements

13.2.2 Biochar as a container substrate

Different substrates are now being utilized as soil-less growth substrates in container or grow bag production systems including vegetables, herbs, leafy greens, and many fruit crops including strawberries. Peat is the most extensively used substrate in crop production, nursery media, and rooting media along with others like vermiculite, pine bark, rice hull, etc. The physical and chemical properties of these substrates affect crop growth and quality.

The incorporation of biochar affects the chemical and physical properties of container substrates, including nutrient availability, pH, cation exchange capacity (CEC), electrical conductivity (EC), bulk density, total porosity, and container capacity. The inclusion of biochar could also affect microbial activities. Plant growth and development are greatly affected by the properties of biochar incorporated in container substrates viz., the origin of biochar, percentage of biochar applied, and other container substrates components mixed with biochar. Particle size and distribution of a substrate is an important factor that determines the physical properties of the substrate and container capacity (Noguera, Abad, Puchades, Maquieira, & Noguera, 2003). Fonteno, Hardin, and Brewster (1995) described container capacity as the maximum volume of water a substrate can hold after gravity drainage.

It has been reported that the incorporation of biochar with peat and compost made from green waste has increased the container capacity significantly (Méndez, Paz-Ferreiro, Gil, & Gascó, 2015). Whereas other studies reported that adding biochar to container substrate did not affect container capacity (Tian et al., 2012; Vaughn, Kenar, Thompson, & Peterson, 2013). These contradictory results of biochar incorporation into container substrates could be due to the variation in particle sizes of the different biochars used in the substrate components.

Besides container capacity, it has been observed that the inclusion of biochar in the substrate regulates the air space of the container substrate. Air space is the proportion of air-filled large pores (macropores) after drainage (Landis, 1990). Several studies suggested that adding biochar with different substrates increased the air space by increasing the number of macro pores (Méndez et al., 2015; Vaughn et al., 2013). Additionally, the incorporation of biochar facilitates creating good aeration, and drainage and also helps in adsorbing toxic contaminants (Northup, 2013).

The use of biochar has been a standard practice in the orchid industry for many years. In the orchid industry, it is commonly known as horticultural charcoal (Northup, 2013). Northup (2013), demonstrated that biochar can replace the perlite and can avoid the limestone amendment needed for commercial greenhouse soilless substrates based on sphagnum peat. Álvarez et al. (2018) demonstrated that it is possible to grow geranium and petunia as container plants with commercial quality with enhanced physiological responses, using a peat-based substrate mixed with biochar and/or vermicompost (up to 30% vermicompost and 12% biochar). Yang and Zhang (2022) demonstrated that biochar and cow manure addition improve the quality and increase the agronomic value of cornflower (*Centaurea cyanus* L.) when compost green waste was amended with 15% biochar and 10% cow manure which showed an ideal growth media for the cornflower plant.

13.2.3 Biochar as an alternative to peat media

Container plant production in the floriculture industry primarily exploits substrates such as peat moss. Draining peat bogs is an important environmental concern that has enhanced interest in research on complementary products that can be added to peat (Álvarez, Pasian, Lal, López Núñez, & Fernández Martínez, 2017). Several studies have also reported that reduction in the emission of greenhouse gases when biochar was used as a peat substitute for growing plants (Steiner & Harttung, 2014). The decomposition of biochar is comparatively slow (Kuzyakov, Subbotina, Chen, Bogomolova, & Xu, 2009) and can be stored for relatively long periods. Schmidt et al. (2014) demonstrated the synergistic effects of biochar when it is combined with compost in the growth substrate. In a study, Álvarez et al. (2017) reported that when vermicompost (10%–30%) combined with biochar (8%–12%) used as a growing media, geranium (*Pelargonium peltatum*) and petunia (*Petunia hybrida*) induced more growth and flower production than that of the control. These results are of interest in reducing peat base substrate for ornamental plant production in containers and reducing carbon footprint. Guo, Niu, Starman, Volder, and Gu (2018) studied the effect of biochar and different nitrogen fertigation regimes in container substrate on poinsettia's growth and development and found that at a fertigation rate of 100 mg L^{-1} N to 400 mg L^{-1} N, up to 80% biochar could be used as an amendment to peat-based root substrate with acceptable growth reduction with no changes in quality.

13.2.4 Biochar as a nursery substrate

Utilization of biochar in plant nursery enterprises offers interesting possibilities not only in increasing the effectiveness of low-quality irrigation water and decreasing peat usage but also by enhancing nutrient retention and their availability for plant uptake. Several studies revealed that biochar plays a major role in the amelioration of saline-affected soils (Ekinci, Turan, & Yildirim, 2022), especially when poor-quality water is used for irrigation (Saleh, Mahmoud, & Rashad, 2012). Relatively few studies have assessed the potential utilization of biochar in nursery substrates for the urban sector (Gravel, Dorais, & Ménard, 2013; Ruamrungsri et al., 2011; Schulz & Glaser, 2012; Tian et al., 2012). In a study, Dumroese, Heiskanen, Englund, and Tervahauta (2011) found that amending peat media with various rates of pelleted biochar (75% peat and 25% pelleted biochar) improved the hydraulic conductivity and yielded more water at high matric potentials (>10 kPa) and retained a desired 40% air-filled porosity and these results advise that it is appropriate for use in nurseries using small volume containers. Di Lonardo et al. (2017) studied the effect of biochar on salinity reduction by the addition of biochar to the potting medium. The result showed that biochar had no direct effect on plant growth but crucially reduced the cherry laurel salinity damage. It is concluded that tolerance against salinity was due to lower Na + retention in the biochar-containing substrates (Di Lonardo et al., 2017) compared to substrates containing no biochar.

13.3 Biochar-mediated suppression of abiotic stresses

13.3.1 Improving drought tolerance

Production and productivity of horticultural crops are majorly constrained by abiotic stresses such as low or high temperature, drought, saline and acidic stress, and contamination of heavy metals in growth substrate which overall affect sustainability and threaten global food security. Salinity, drought, and heavy metal stresses are three of the most significant primary abiotic pressures that contribute to low growth and agricultural production globally (Yadav et al., 2020). Amending growth substrate with biochar is an efficient and cost-effective management tool for improving crop productivity in horticultural crops. Several investigations have reported a beneficial effect of amending biochar on plants grown under stress conditions, by regulating the metabolic processes involved in growth promotion (Table 13.2).

TABLE 13.2 Effect of biochar on abiotic stresses in horticultural crops.

Stress	Crop	Biochar type and application rate	Pyrolysis temp.	Response	Reference
Drought	Tomato (*Solanum lycopersicum*)	Mixture of rice husk and shell of cotton seed biochar; 0% and 5% by weight.	400°C	Enhanced relative water content and water use efficiency. Improved fruit yield and quality. Decreased membrane stability index. Improved photosynthesis	Akhtar et al. (2014)
Drought	Tomato (*Solanum lycopersicum*)	Rice and maize straw biochar nanoparticles.	350°C	Improve salt tolerance in tomato seedlings treated with nanoparticle biochar	Tao et al. (2023)
Drought	Faba beans (*Vicia faba*)	A 3:100 (w/w) combination of citric acid, and citrus wood biochar applied at a dose of 0, 5, and 10 t ha^{-1}.	–	Improved plant growth and physiological responses. Enhanced contents of N, P, K, and Ca.	Abd El-Mageed et al. (2021)
Drought	Pumpkin (*Cucurbita pepo* L.)	Maize-straw biochar applied at 0, 5, 10 and 20 t ha^{-1}.	350°C	Increased chlorophyll. Enhanced uptake of nutrients. Reduced reactive oxygen.	Langeroodi et al. (2019)
Drought	Cow pea (*Vigna unguiculata* (L.) Walp.)	Mango wood biochar; plastic pots (150 mm × 90 mm) filled with peat moss having biochar (25 g per pot).	500°C	Improved shoot length, root length, plant biomass and leaf area. Improved chlorophyll contents, amylase activity, total soluble sugars. Decreased Na + uptake, MDA, and	Farooq et al. (2020)

(Continued)

TABLE 13.2 (Continued)

Stress	Crop	Biochar type and application rate	Pyrolysis temp.	Response	Reference
Salt	Cabbage (*Brassica oleracea* var. *capitata*)	A mixture of 60% sewage sludge and 40% domestic wastes; 0%, 2.5%, and 5% by soil weight.	550°C	Improved plant growth and biomass, Enhanced chlorophyll, Decreased MDA, H_2O_2, sucrose and proline, Enhanced leaf nutrient elements, decreased Na and Cl content in plant.	Ekinci et al. (2022)
Salt	Brinjal (*Solanum melongena*)	An applied dose of 0% 5% and 10% (w/w) commercial biochar.	–	Biochar treated plants showed significant reduction in Na and Cl uptake, MDA, H_2O_2 and EL and an increase in LRWC, chlorophyll content, antioxidant activity and plant nutrient element uptake thus induced salinity tolerance	Kul (2022)
Salt and drought	Brinjal (*Solanum melongena*)	Maize straw 6% (w/w) biochar.	550°C	Under salt stress, with alternate root-zone drying irrigation treatment, intrinsic water use efficiency (WUE_i) and yield traits were higher than those in deficient irrigation.	Hannachi et al. (2023)
Salt	Common bean (*Phaseolus vulgaris*)	Maple residue biochar 0% 10% and 20% (w/w).	560°C	Enhancing tolerance to salt by decreasing of Na concentration, PAO activity, polyamines, ABA, ACC, and JA contents in plants treated with biochar.	Farhangi-Abriz and Torabian (2018)
Salt	Okra (*Abelmoschus esculentus*)	Wheat straw 0% 5%, and 10% (w/w).	350°C–550°C	The 10% biochar application rate also resulted in the greatest okra plant growth and increased yield, indicating that the effects of salt stress were ameliorated.	Elshaikh et al. (2018)
Copper tocxicity	Lettuce (*Lactuca sativa* L.)	Orchard pruning feedstock-based commercial biochar; 1% biochar (w/w).	550°C	Restored antioxidant activity and flavonoids. Increased total phenols, phenolic acids and anthocyanins	Quartacci et al. (2017)

(*Continued*)

13.3 Biochar-mediated suppression of abiotic stresses

TABLE 13.2 (Continued)

Stress	Crop	Biochar type and application rate	Pyrolysis temp.	Response	Reference
Heavy metal toxicity	Tomato (*Solanum lycopersicum*)	Maize stalks biochar; 0, 5, and 10 t ha^{-1}.	450°C	Improved yield and fruit quality. Decreased uptake of heavy metals in both plant parts and fruits. Improved chlorophyll contents.	Almaroai and Eissa (2020)
Lead (Pb) toxicity	Chicory (*Cichorium intybus* L. var. *intybus*)	Raw bagasse biochar.	600°C	Biochar reduced the concentration of Pb in chicory under hydroponic cultivation.	El-Banna et al. (2019)
Lead (Pb) toxicity	Lettuce (*Lactuca sativa* L.)	Commercial biochar	–	Biochar reduced 80% amount of Pb in lettuce under hydroponic cultivation.	Vannini et al. (2021)
Heavy metal toxicity	Five green leafy vegetables (lettuce—*Lactuca sativa* L., spinach—*Spinacia oleracea* L., corn salad—*Valerianella locusta* L., kale—*Brassica oleracea* L., mustard greens—*Brassica juncea* L.)	Mixture of compost and rabbit-manure-derived biochar 30% (v/w).	550°C	Addition of biochar decreased Cu, Cr, Cd, and Pb availability in the tested substrates, reducing the uptake of Cd in spinach by 61% and Pb in mustard greens by 73% while uptake of Cu and Cr was enhanced in spinach and lettuce by 20%.	Medyńska-Juraszek, Marcinkowska, Gruszka, and Kluczek (2022)

In an experiment, Akhtar, Li, Andersen, and Liu (2014) reported that the addition of 5% (w/w) cotton seed shell and rice husk biochar increased the fruit yield and quality of potted tomato plants subjected to 30% deficit and partial root-zone drying irrigation by 6% and 13%, respectively, compared to fully irrigated control plants under greenhouse. Mulcahy, Mulcahy, and Dietz (2013) reported that adding wood pellets biochar (30% v/v) into a growing substrate enhanced tomato seedling tolerance to drought stress. In pumpkin (*Cucurbita pepo*), Langeroodi, Campiglia, Mancinelli, and Radicetti (2019) demonstrated that adding maize-straw biochar increased drought tolerance in pumpkin plants by decreasing the generation of reactive oxygen species (ROS) and an elevated antioxidant defense system in biochar-treated plants compared to control pumpkin plants. Jabborova, Annapurna, et al. (2021) studied the influence of biochar and arbuscular mycorrhizal fungi (AMF) on the growth of Okra (*Abelmoschus esculentus*) in pot experiments in a net house under drought conditions. The result showed that biochar and AMF together are potentially beneficial for the cultivation of okra in drought-stress conditions. Egamberdieva, Zoghi, Nazarov, Wirth, and Bellingrath-Kimura (2020) reported that amending maize straw biochar improved growth, yield, and nutrient uptake on drought-stressed faba beans (*Vicia faba* L.) compared to unamended plants. Abd El-Mageed, Abdelkhalik, Abd El-Mageed, and Semida (2021) studied the effect of acidified biochar on drought-stressed

Vicia faba in a two-year experiment and reported that biochar addition improved plant growth and drought tolerance in fava bean. In cowpeas, Farooq, Rehman, Al-Alawi, Al-Busaidi, and Lee (2020) reported that amending biochar with growth substrate improved drought tolerance in cowpeas. In a pot experiment in the greenhouse, Yildirim, Ekinci, and Turan (2021) studied the effect of biochar in drought-stressed cabbage plants and reported that amending biochar improved the plant growth, photosynthetic activity, nutrient uptake, and modified physiological and biochemical characteristics in cabbage seedlings under drought conditions (Table 13.2). Increased porosity and pore volume in growing substrate amended with biochar, increases the soil's ability to hold available water under drought stress, which may be responsible for improved drought tolerance (Zulfiqar et al., 2022).

13.3.2 Improving salt tolerance

More than 750 million hectares of land have been harmed by soil salinization worldwide, which amounts to more than 9% of the planet's land surface (Zhongming, Linong, Xiaona, Wangqiang, & Wei, 2021). In commercial production, closed-cycle soilless methods have been recommended because they can reduce water and fertilizer losses (Montesano, Parente, & Santamaria, 2010). However, the major challenging issue with recirculating systems is salt build-up in the substrate in the closed-cycle soilless method. Long-term cycling of nutrient solution in closed soilless systems leads to the build-up of ions like sodium (Na) and chlorine (Cl). In closed soilless systems, especially when irrigation water is of poor quality, the accumulation of ballast ions, Na and Cl, in the recycled nutrient solution is a key limiting issue. Open cycle systems are usually employed to lower the substrate's salt level. However, closed-cycle systems offer better water use efficiency.

The use of salt-resistant varieties and amending substrate with suitable ameliorating agents can mitigate the substrate salinity in the production system. Biochar plays a potential role in reducing substrate salinity in soilless cultures. Several studies have reported the beneficial effect of adding biochar into the soil and soilless substrate. For example, In a pot experiment, Ekinci et al. (2022) investigated the growth, physiology, and biochemistry of cabbage seedlings in response to salinity stress (0 and 150 mM NaCl level) with biochar (weighed at the rate of 0%, 2.5%, and 5% by soil weight), the result demonstrated that biochar mitigated the negative impact of salinity stress on cabbage seedlings by reduction of Cl and Na concentration, and reactive oxygen species (ROS) production, regulating abscisic acid (ABA) content, antioxidant enzyme activity and plant nutrient element content (Table 13.2).

Several other studies have also demonstrated that adding biochar to the growing substrate has a mitigating effect on salinity stress in common bean (*Phaseolus vulgaris*); Farhangi-Abriz and Torabian (2018), cowpea (*Vigna unguiculata* L.); Mehdizadeh, Moghaddam, and Lakzian (2020), spinach (*Spinacea oleracia*); (Sofy, Mohamed, Dawood, Abu-Elsaoud, & Soliman, 2022), dill (*Anethum graveolens*); Ghassemi-Golezani and Rahimzadeh (2022), brinjal (*Solanum melongena* L); Kul (2022), Hannachi, Signore, and Mechi (2023), okra (*Abelmoschus esculentus*); Elshaikh, Zhipeng, Dongli, and Timm (2018), and tomato (*Solanum lycopersicum*); Tao et al. (2023) (Table 13.2). Biochar-mediated mitigation of salt stress in plants could be due to increasing the available water-holding capacity of soil and improving soil chemical properties by limiting Na availability to plant roots, biochar generally contains high K; and both factors lead to increased K uptake and decreased Na uptake, ameliorating the adverse effects of Na (Zulfiqar et al., 2022).

13.3.3 Improving heavy metals tolerance

Currently, a widespread concern is the accumulation of heavy metals and metalloids in food crops causing risks to human health and the environment with the rising urbanization and quick advancements in agricultural and industrial technologies. Potentially harmful toxic metals can be introduced to soil and growing substrates from a variety of sources. The quality of crops and food items might suffer as a result of soil acidity and the excessive accumulation of heavy metals in agricultural and urban soils. Urban and peri-urban fruit and vegetable farming could enhance nutritional quality, food access, and public health. Limiting the uptake of metals by horticultural crops is essential for maintaining production and the quality of horticultural produce as well as for human consumption (Table 13.2).

Biochar has been suggested as a potential de-toxicant of heavy metal contaminants (Zulfiqar et al., 2022) and its utilization in soil, and soilless substrate. In this context, many studies have established that biochar helps in reducing heavy metal stress in horticulture crops. For instance, in lettuce, Quartacci, Sgherri, and Frisenda (2017) reported that biochar amelioration to Cu-contaminated soil restored flavonoids and antioxidant activity and increased phenolic acids, total phenols, and anthocyanins. In another study, Vannini et al. (2021) reported that adding biochar to hydroponic substrate significantly reduces Pb uptake (80%) by lettuce under a hydroponic production system.

Almaroai and Eissa (2020) studied the effect of adding biochar to the growing substrate decreasing the accumulation of heavy metals (Pb and Cd) in tomatoes grown in metal-polluted soil and increasing the tomato yield by 20%–30%. Ibrahim, El-Sherbini, and Selim (2022) added three different biochar (*Casuarina*, mango, and Salix as feedstocks) for their effect on summer squash (*C. pepo* L.) grown in contaminated soil and found that adding biochar helped in reducing the metal uptake up to 4%. Biochar-mediated detoxification of heavy metals such as Pb in celery has also been reported by El-Banna et al. (2019). According to Ahmad et al. (2018), biochar having a large surface area, enough pore volume, and plenty of functional groups are essential for heavy metal adsorption. According to Xiao, Chen, Chen, Zhu, and Schnoor (2018), biochars made at moderate and low temperatures typically have the maximum capacity for adsorbing metal cations.

13.4 Biochar-mediated suppression of biotic stresses

Recently, several studies have shown that the application of biochar can improve resistance against a wide range of plant pathogens. Substrate amendments with biochar enhance resistance to diseases therefore minimize yield reduction in horticultural plants because it contains several compounds viz., benzoic acid, butyric acid, 2-phenoxyethanol, ethylene glycol, cresol, quinines, o-, hydroxyl-propionic acid, and propylene glycol that impact substrate microbial growth (Frenkel et al., 2017; Graber et al., 2010). The amendment of different substrates with biochar can lead to disease resistance and vulnerability in horticultural crops due to changes in plant metabolic pathways (Ji et al., 2022). Generally, two major pathways lead to the system-wide induced resistance (IR) in plants have been described viz., systemic acquired resistance, and induced systemic resistance. Suppression of diseases in plants depends on the amount and origin of amended and some of the relevant findings are documented in Table 13.3. Auxins, brassinosteroids, cytokinins, jasmonic acid, and the synthesis of cell walls, flavonoids, and phenylpropanoids were all shown to be increased by biochar treatment, according to the study (Jaiswal et al., 2020).

In the first report of its kind, it was found that the severity of the fungal foliar diseases caused by *Botrytis cinerea* and *Oidiopsis sicula* Scalia in tomato (*Solanum lycopersicum* L.) and pepper (*Capsicum annuum* L.) was significantly reduced in biochar-amended potting medium (Elad et al., 2010). In another study, it has been demonstrated in *F. ananassa*'s

TABLE 13.3 Effect of biochar on foliar and soil borne disease suppression.

Crop plant	Pathogen	Biochar type and application rate	Response and mechanism	Reference
Tomato	*Fusarium oxysporum f.sp lycopersici*	Wood & Green waste biochar (3% v/v)	Induction of resistance in tomato plants	Akhter, Hage-Ahmed, Soja, and Steinkellner (2015)
	Botrytis cinerea	Eucalyptus wood (0.5%, 1%, 3%)	Induction of resistance in tomato plants	Elad, Cytryn, Harel, Lew, and Graber (2011)
	Phytium aphanidermatum	Pig bone biochar	Biochar acts as a carrier for biological control agents	Postma, Clematis, Nijhuis, and Someus (2013)
Pepper	*Leveilulla taurica*	Citrus wood (1, 3%)	Induction of resistance in pepper plants	Elad et al. (2010)
Basil	*Pythium ultimum*	Spruce bark (50% v/v)	Induction of resistance in basil plants	Gravel et al. (2013)
Carrot	*Pratylenchus penetrans*	Pine wood (0.8%) Pine bark (0.92) Wood (1.24%) Spelt husks (0.64%)	Induction of resistance in carrot plants	George, Kohler, and Rillig (2016)
Pea	*Rhizoctonia solani*	Maple bark (1, 3, 5%)	Induction of resistance in pea plants	Copley, Aliferis, and Jabaji (2015)
Strawberry	*Botrytis cinerea, Colletotrichum acutatum* and *Podosphaera apahanis*	Citrus wood & pepper plant greenhouse wastes (1%, 3%)	Induction of resistance in strawberry plants	Yael, Liel, Hana, Ran, and Shmuel (2012)

resistance to *B. cinerea*, *Podosphaera apahanis*, and *Colletotrichum acutatum* was enhanced when biochar was applied exogenously (Meller Harel et al., 2012). Pepper plants were additionally protected against damages from the broad mite pest (*Polyphagotarsonemus latus* Banks). According to Mehari, Elad, Rav-David, Graber, and Meller Harel (2015), the addition of biochar to soil reduced *B. cinerea*-induced fungal infections of *C. annuum* and *S. lycopersicum*.

Jaiswal, Graber, Elad, and Frenkel (2019) demonstrated preemergence damping off disease control in nursery plants. According to Wang et al. (2019), enhanced soil bacterial populations are frequently linked to improved disease resistance. The disease index of the bacterial wilt produced by *Ralstonia solanacearum* was dramatically reduced by 28.6% and 65.7%, respectively, when peanut shell and wheat straw biochars were applied to the soil (Lu et al., 2016). Chinese ginseng (*Panax ginseng*) root rot disease was reduced by biochar because it is a potent organic compound absorbent from soil (Liu et al., 2022).

Rasool, Akhter, Soja, and Haider (2021) reported that biochar induced the expression of defense-associated genes, including ethylene-responsive Pti4, phenolics, CAT, and POX, jasmonic acid-related PI2, TomloxD, salicylic acid-related PR1a, PR2, and jasmonic acid-related PR2 to reduce Alternaria solani infection-induced damage in tomatoes. According to Liu et al. (2022), a biochar amendment boosted apple seedling growth by absorbing the abiotic component (phloridzin) that causes the apple replant disease from the soil. Biochar helps prevent viral diseases as well as bacterial and fungal illnesses. For instance, Luigi et al. (2022), utilizing the RT-qPCR method and DDCt analysis, found that adding biochar to the soil decreased tomato virus infection rates and potato spindle tuber viroid (PSTVd) replication.

13.5 Enhanced seed germination using biochar

Seed germination is a vital step in the horticulture production system to attain optimal crop development, vigor, and plant population and to improve production and profitability (Zulfiqar et al., 2021). The accessibility of food to the increasing population, fodder for animals, and the supply of medicinal plants, and seed germination determines the continual plant production necessary for the survival of life of humans and animals. The multiplication of the majority of horticultural plants is possible by seed, and if they fail to germinate, this may threaten their continuous existence. Depending on the feedstock, pyrolysis temperature, quantity used, and crop species, studies on the impact of biochar on seed germination showed inhibitory to stimulatory effects. Incorporating biochar into the soil or soilless substrate improves substrate porosity, water-holding capacity, and water availability, thus improving seed germination. The incorporation of biochar enhances seed germination by releasing karrikins, or seed germination hormones (Kochanek, Long, Lisle, & Flematti, 2016).

Additionally, biochar assimilation influences the physio-chemical characteristics of media, facilitating seed germination. For instance, Jabborova, Annapurna, et al. (2021) observed that the incorporation of 1% biochar doubled the seed germination of sweet basil (*Ocimum basilicum*) and increased germination by 28%–30% when applied at 2% and 3%. In addition to biochar alone, biochar coupled with other nanoparticles also aids in seed germination. In another study, Taqdees, Khan, Kausar, Afzaal, and Akhtar (2022), demonstrated that biochar enriched with silicon and zinc nanoparticles improved the germination of salt-stressed radish seeds. Thus, under both normal and stressful circumstances, biochar can be an important factor in the germination of seeds in a variety of horticultural crops. However, the pace of biochar treatment, biochar type, plant species, and environmental factors affect seed germination. Thus, as a noble material, biochar may be expected to enhance seed germination and seedling establishment, providing food safety, environmental protection, and agricultural sustainability all at once. Additionally, based on recent reports, applying biochar seed coating to seed has shown enhanced germination and seedling establishment for many crops including tomato, beans, radish, eggplant, lettuce, etc.

13.6 Biochar-mediated enhanced root growth of horticultural crops

Water availability and uptake as well as the amount of soluble solutes are factors that affect plant water relations such as water potential, osmotic potential, and turgor potential. Additionally, the application of organic or inorganic solutes may change the water relations in plants. For instance, Akhtar et al. (2014) demonstrated that tomato plants under deficit and partial root-zone drying irrigation supplemented with biochar increased relative water content and water use efficiency.

For many horticultural crops which are susceptible to nutritional inadequacies, biochar inclusion into the soilless substrate can be a noble technique to enhance plant nutrition in both nonstressed and stressed environments. In another study, improved root hydraulic conductance and leaf water potential in water-stressed tomato plants grown in sandy loam soil amended with miscanthus straw biochar were reported by Guo, Bornø, Niu, and Liu (2021). In another study, Langeroodi et al. (2019) reported that the use of biochar during water deficit irrigation enhanced the relative water content of pumpkin leaves. Similarly, several studies reported improved root growth and water use efficiency in horticultural crops, for instance, in cowpea (*Vigna unguiculata* (L.) Walp.) Mehdizadeh et al. (2020) in apple, Wang and Wang (2019), in summer savory (*Satureja hortensis* L.), Jabborova, Ma, Bellingrath-Kimura, and Wirth (2021), in strawberry (*Fragaria* × *ananassa* Duch.), Chiomento et al. (2021), in tea plant (*Camellia sinensis* Kuntze) (Yan et al., 2021), in cabbage (*Brassica oleracea* var. *capitata*), Yildirim et al. (2021), in faba bean (*Vicia faba* L.), Abd El-Mageed et al. (2021) and in sweet basil, Jabborova, Ma, et al. (2021) (Table 13.3).

13.7 Conclusion and future perspectives

Diverse sources of biochar have been shown a pivotal role in the sustainable soilless production of horticultural crops. Biochar comes in a wide variety and can affect substrate characteristics, crop growth, and production in various ways. Biochar helps in the mitigation of problems caused by soil-based production systems such as crop growth and development, tolerance to biotic and abiotic stresses, reducing the toxicity of heavy metals, and altering various physiological processes. The feedstock used to produce biochar and the pyrolysis conditions majorly influence the characteristics of biochar and its effects on horticultural ecosystems. Diverse doses of biochar formulations have been applied at an optimal rate in many studies, further standardization of dose-effects needs to be emphasized in many other horticultural crops under soilless production systems which possibly can significantly increase productivity under adverse conditions. The selection of suitable feedstock for biochar production, its optimum application rates, and compatibility with different crops should be taken in mind before biochar application. Additionally, the problem of heavy metals has been worse due to increased industrialization, thus biochar techniques should be developed to reduce the absorption of dangerous substances in fruits and vegetables. For additional opportunities to boost horticulture production, studies combining the use of BC with microbial and nonmicrobial bio-stimulants should be conducted. There is a need to evaluate biochar in nanoform under a soilless horticultural production system. Because using raw biochar in soils or substrates would be financially unfeasible due to the high initial costs, the horticulture sector must develop the necessary technology to produce biochar economically.

References

Abd El-Mageed, T. A., Abdelkhalik, A., Abd El-Mageed, S. A., & Semida, W. M. (2021). Co-composted poultry litter biochar enhanced soil quality and eggplant productivity under different irrigation regimes. *Journal of Soil Science and Plant Nutrition, 21*, 1917–1933.

Ahmad, Z., Gao, B., Mosa, A., Yu, H., Yin, X., Bashir, A., ... Wang, S. (2018). Removal of Cu (II), Cd (II) and Pb (II) ions from aqueous solutions by biochars derived from potassium-rich biomass. *Journal of Cleaner Production, 180*, 437–449.

Akhtar, S. S., Li, G., Andersen, M. N., & Liu, F. (2014). Biochar enhances yield and quality of tomato under reduced irrigation. *Agricultural Water Management, 138*, 37–44.

Akhter, A., Hage-Ahmed, K., Soja, G., & Steinkellner, S. (2015). Compost and biochar alter mycorrhization, tomato root exudation, and development of *Fusarium oxysporum* f. sp. *lycopersici*. *Frontiers in Plant Science, 6*, 529.

Almaroai, Y. A., & Eissa, M. A. (2020). Effect of biochar on yield and quality of tomato grown on a metal-contaminated soil. *Scientia Horticulturae, 265*109210.

Altland, J. E., & Locke, J. C. (2012). Biochar affects macronutrient leaching from a soilless substrate. *HortScience: A Publication of the American Society for Horticultural Science, 47*, 1136–1140.

Álvarez, J. M., Pasian, C., Lal, R., López Núñez, R., & Fernández Martínez, M. (2017). Vermicompost and biochar as substitutes of growing media in ornamental-plant production. *Journal of Applied Horticulture, 19*, 205–214.

Álvarez, J. M., Pasian, C., Lal, R., López, R., Díaz, M. J., & Fernández, M. (2018). Morpho-physiological plant quality when biochar and vermicompost are used as growing media replacement in urban horticulture. *Urban Forestry & Urban Greening, 34*, 175–180.

Awad, Y. M., Lee, S. E., Ahmed, M. B. M., Vu, N. T., Farooq, M., Kim, I. S., ... Meers, E. (2017). Biochar, a potential hydroponic growth substrate, enhances the nutritional status and growth of leafy vegetables. *Journal of Cleaner Production, 156*, 581–588.

Beck, D. A., Johnson, G. R., & Spolek, G. A. (2011). Amending greenroof soil with biochar to affect runoff water quantity and quality. *Environmental Pollution, 159*, 2111–2118.

Chaturvedi, S., Singh, S., Dhyani, V. C., Govindaraju, K., & Mandal, S. (2023). Characterization, bioenergy value and thermal stability of biochar derived from diverse agriculture and forestry lignocellulosic waste. *Biomass Conversion and Biorefinery, 13*, 879–892.

Chiomento, J. L. T., De Nardi, F. S., Filippi, D., dos Santos Trentin, T., Dornelles, A. G., Fornari, M., ... Calvete, E. O. (2021). Morpho-horticultural performance of strawberry cultivated on substrate with arbuscular mycorrhizal fungi and biochar. *Scientia Horticulturae, 282*110053.

Conversa, G., Bonasia, A., Lazzizera, C., & Elia, A. (2015). Influence of biochar, mycorrhizal inoculation, and fertilizer rate on growth and flowering of Pelargonium (*Pelargonium zonale* L.) plants. *Frontiers in Plant Science, 6*, 429.

Copley, T. R., Aliferis, K. A., & Jabaji, S. (2015). Maple bark biochar affects *Rhizoctonia solani* metabolism and increases damping-off severity. *Phytopathology, 105*(10), 1334–1346.

De Tender, C. A., Debode, J., Vandecasteele, B., D'Hose, T., Cremelie, P., Haegeman, A., ... Maes, M. (2016). Biological, physicochemical and plant health responses in lettuce and strawberry in soil or peat amended with biochar. *Applied Soil Ecology, 107*, 1–12.

Di Lonardo, S., Baronti, S., Vaccari, F. P., Albanese, L., Battista, P., Miglietta, F., & Bacci, L. (2017). Biochar-based nursery substrates: The effect of peat substitution on reduced salinity. *Urban Forestry & Urban Greening, 23*, 27–34.

Dumroese, R. K., Heiskanen, J., Englund, K., & Tervahauta, A. (2011). Pelleted biochar: Chemical and physical properties show potential use as a substrate in container nurseries. *Biomass and Bioenergy, 35*, 2018–2027.

Dunlop, S. J., Arbestain, M. C., Bishop, P. A., & Wargent, J. J. (2015). Closing the loop: Use of biochar produced from tomato crop green waste as a substrate for soilless, hydroponic tomato production. *HortScience: A Publication of the American Society for Horticultural Science, 50*, 1572–1581.

Egamberdieva, D., Zoghi, Z., Nazarov, K., Wirth, S., & Bellingrath-Kimura, S. D. (2020). Plant growth response of broad bean (*Vicia faba* L.) to biochar amendment of loamy sand soil under irrigated and drought conditions. environ. *Sustainability, 3*, 319–324.

Ekinci, M., Turan, M., & Yildirim, E. (2022). Biochar mitigates salt stress by regulating nutrient uptake and antioxidant activity, alleviating the oxidative stress and abscisic acid content in cabbage seedlings. *Turkish Journal of Agriculture and Forestry, 46*, 28–37.

Elad, Y., David, D. R., Harel, Y. M., Borenshtein, M., Kalifa, H. B., Silber, A., & Graber, E. R. (2010). Induction of systemic resistance in plants by biochar, a soil-applied carbon sequestering agent. *Phytopathology, 100*, 913–921.

Elad, Y., Cytryn, E., Harel, Y. M., Lew, B., & Graber, E. R. (2011). The biochar effect: Plant resistance to biotic stresses. *Phytopathologia Mediterranea, 50*, 335–349.

El-Banna, M. F., Mosa, A., Gao, B., Yin, X., Wang, H., & Ahmad, Z. (2019). Scavenging effect of oxidized biochar against the phytotoxicity of lead ions on hydroponically grown chicory: An anatomical and ultrastructural investigation. *Ecotoxicology and Environmental Safety, 170*, 363–374.

Elshaikh, N. A., Zhipeng, L., Dongli, S., & Timm, L. C. (2018). Increasing the okra salt threshold value with biochar amendments. *Journal of Plant Interactions, 13*, 51–63.

Fan, M., Li, C., Sun, Y., Zhang, L., Zhang, S., & Hu, X. (2021). In situ characterization of functional groups of biochar in pyrolysis of cellulose. *Science of The Total Environment, 799*149354.

Farhangi-Abriz, S., & Torabian, S. (2018). Biochar increased plant growth-promoting hormones and helped to alleviates salt stress in common bean seedlings. *Journal of Plant Growth Regulation, 37*, 591–601.

Farooq, M., Rehman, A., Al-Alawi, A. K., Al-Busaidi, W. M., & Lee, D. J. (2020). Integrated use of seed priming and biochar improves salt tolerance in cowpea. *Scientia Horticulturae, 272*109507.

Fonteno, W., Hardin, C., & Brewster, J. (1995). *Procedures for determining physical properties of horticultural substrates using the NCSU porometer* (p. 1995) Raleigh, NC, USA: Horticultural Substrates Laboratory, North Carolina State University.

Frenkel, O., Jaiswal, A. K., Elad, Y., Lew, B., Kammann, C., & Graber, E. R. (2017). The effect of biochar on plant diseases: What should we learn while designing biochar substrates? *Journal of Environmental Engineering and Landscape Management, 25*, 105–113.

George, C., Kohler, J., & Rillig, M. C. (2016). Biochars reduce infection rates of the root-lesion nematode *Pratylenchus penetrans* and associated biomass loss in carrot. *Soil Biology and Biochemistry, 95*, 11–18.

Ghassemi-Golezani, K., & Rahimzadeh, S. (2022). Biochar modification and application to improve soil fertility and crop productivity. *Agriculture (Pol'nohospodárstvo), 68*, 45–61.

Graber, E. R., Harel, Y. M., Kolton, M., Cytryn, E., Silber, A., David, D. R., ... Elad, Y. (2010). Biochar impact on development and productivity of pepper and tomato grown in fertigated soilless media. *Plant and Soil, 337*, 481–496.

Gravel, V., Dorais, M., & Ménard, C. (2013). Organic potted plants amended with biochar: Its effect on growth and Pythium colonization. *Canadian Journal of Plant Science, 93*, 1217–1227.

Guo, L., Bornø, M. L., Niu, W., & Liu, F. (2021). Biochar amendment improves shoot biomass of tomato seedlings and sustains water relations and leaf gas exchange rates under different irrigation and nitrogen regimes. *Agricultural Water Management, 245*106580.

Guo, Y., Niu, G., Starman, T., Volder, A., & Gu, M. (2018). Poinsettia growth and development response to container root substrate with biochar. *Horticulturae, 4*(1).

Gwynn-Jones, D., Dunne, H., Donnison, I., Robson, P., Sanfratello, G. M., Schlarb-Ridley, B., & Convey, P. (2018). Can the optimisation of pop-up agriculture in remote communities help feed the world? *Global Food Security, 18*, 35–43.

Hannachi, S., Signore, A., & Mechi, L. (2023). Alleviation of associated drought and salinity stress' detrimental impacts on an eggplant cultivar ('Bonica F_1') by adding biochar. *Plants, 12*, 1399.

Haraz, M. T., Bowtell, L., & Al-Juboori, R. (2020). Biochar effects on nutrients retention and release of hydroponics growth media. *Journal of Agricultural Science, 12*.

Huang, L., & Gu, M. M. (2019). Effects of biochar on container substrate properties and growth of plants-A review. *Horticulturae, 5*, 14.

Ibrahim, E. A., El-Sherbini, M. A., & Selim, E. M. M. (2022). Effects of biochar on soil properties, heavy metal availability and uptake, and growth of summer squash grown in metal-contaminated soil. *Scientia Horticulturae, 301*111097.

Jabborova, D., Annapurna, K., Al-Sadi, A. M., Alharbi, S. A., Datta, R., & Zuan, A. T. K. (2021). Biochar and Arbuscular mycorrhizal fungi mediated enhanced drought tolerance in Okra (*Abelmoschus esculentus*) plant growth, root morphological traits and physiological properties. *Saudi Journal of Biological Sciences, 28*, 5490–5499.

Jabborova, D., Ma, H., Bellingrath-Kimura, S. D., & Wirth, S. (2021). Impacts of biochar on basil (*Ocimum basilicum*) growth, root morphological traits, plant biochemical and physiological properties and soil enzymatic activities. *Scientia Horticulturae, 290*110518.

Jaiswal, A. K., Alkan, N., Elad, Y., Sela, N., Graber, E. R., & Frenkel, O. (2020). Molecular insights into biochar-mediated plant growth promotion and systemic resistance in tomato against Fusarium crown and root rot disease. *Scientific reports, 10*13934.

Jaiswal, A. K., Graber, E. R., Elad, Y., & Frenkel, O. (2019). Biochar as a management tool for soilborne diseases affecting early stage nursery seedling production. *Crop Protection, 120*, 34–42.

Jaiswal, A., Frenkel, O., Elad, Y., Lew, B., & Graber, E. (2015). Non-monotonic influence of biochar dose on bean seedling growth and susceptibility to *Rhizoctonia solani*: The "shifted rmax-effect. *Plant and Soil, 395*, 125–140.

Ji, M., Wang, X., Usman, M., Liu, F., Dan, Y., Zhou, L., ... Sang, W. (2022). Effects of different feedstocks-based biochar on soil remediation: A review. *Environmental Pollution, 294*118655.

Karakaş, C., Özçimen, D., & İnan, B. (2017). Potential use of olive stone biochar as a hydroponic growing medium. *Journal of Analytical and Applied Pyrolysis, 125*, 17–23.

Kochanek, J., Long, R. L., Lisle, A. T., & Flematti, G. R. (2016). Karrikins identified in biochars indicate post-fire chemical cues can influence community diversity and plant development. *PLoS One, 11*, 0161234.

Kul, R. (2022). Integrated application of plant growth promoting rhizobacteria and biochar improves salt tolerance in eggplant seedlings. *Turkish Journal of Agriculture and Forestry, 46*, 677–702.

Kuzyakov, Y., Subbotina, I., Chen, H., Bogomolova, I., & Xu, X. (2009). Black carbon decomposition and incorporation into soil microbial biomass estimated by C labeling. *Soil Biology & Biochemistry, 41*, 210–219.

Landis, T. D. (1990). Growing media. *Containers Growing Media, 2*, 41–85.

Langeroodi, A. R. S., Campiglia, E., Mancinelli, R., & Radicetti, E. (2019). Can biochar improve pumpkin productivity and its physiological characteristics under reduced irrigation regimes? *Scientia Horticulturae, 247*, 195–204.

Lehmann, J., & Joseph, S. (2009). *Biochar for environmental management: Science, technology and implementation*. London: Earthscan.

Liu, C., Xia, R., Tang, M., Chen, X., Zhong, B., Liu, X., ... Zhang, X. (2022). Improved ginseng production under continuous cropping through soil health reinforcement and rhizosphere microbial manipulation with biochar: A field study of Panax ginseng from Northeast China. *Horticulture Research, 9*, 108.

Lu, Y., Rao, S., Huang, F., Cai, Y., Wang, G., & Cai, K. (2016). Effects of biochar amendment on tomato bacterial wilt resistance and soil microbial amount and activity. *International Journal of Agronomy, 2016*, 110.

Luigi, M., Manglli, A., Dragone, I., Antonelli, M. G., Contarini, M., Speranza, S., ... Tomassoli, L. (2022). Effects of biochar on the growth and development of tomato seedlings and on the response of tomato plants to the infection of systemic viral agents. *Frontiers in Microbiology, 13*862075.

Medyńska-Juraszek, A., Marcinkowska, K., Gruszka, D., & Kluczek, K. (2022). The effects of rabbit-manure-derived biochar Co-application with compost on the availability and heavy metal uptake by green leafy vegetables. *Agronomy, 12*, 2552.

Mehari, Z. H., Elad, Y., Rav-David, D., Graber, E. R., & Meller Harel, Y. (2015). Induced systemic resistance in tomato (*Solanum lycopersicum*) against Botrytis cinerea by biochar amendment involves jasmonic acid signaling. *Plant and Soil, 395*, 31–44.

Mehdizadeh, L., Moghaddam, M., & Lakzian, A. (2020). Amelioration of soil properties, growth and leaf mineral elements of summer savory under salt stress and biochar application in alkaline soil. *Scientia Horticulturae, 267*109319.

Meller Harel, Y., Elad, Y., Rav-David, D., Borenstein, M., Shulchani, R., Lew, B., & Graber, E. R. (2012). Biochar mediates systemic response of strawberry to foliar fungal pathogens. *Plant and Soil, 357*, 245–257.

Méndez, A., Paz-Ferreiro, J., Gil, E., & Gascó, G. (2015). The effect of paper sludge and biochar addition on brown peat and coir based growing media properties. *Scientia Horticulturae, 193*, 225–230.

Montesano, F., Parente, A., & Santamaria, P. (2010). Closed cycle subirrigation with low con- centration nutrient solution can be used for soil- less tomato production in saline conditions. *Scientia Horticulturae, 124*, 338–344.

Mulcahy, D. N., Mulcahy, D. L., & Dietz, D. (2013). Biochar soil amendment increases tomato seedling resistance to drought in sandy soils. *Journal of arid environments, 88*, 222–225.

Nemati, M. R., Simard, F., Fortin, J. P., & Beaudoin, J. (2015). Potential use of biochar in growing media. *Vadose Zone Journal, 14*, 2014-06.

Nieto, A., Gascó, G., Paz-Ferreiro, J., Fernández, J. M., Plaza, C., & Méndez, A. (2016). The effect of pruning waste and biochar addition on brown peat based growing media properties. *Scientia Horticulturae, 199*, 142–148.

Noguera, P., Abad, M., Puchades, R., Maquieira, A., & Noguera, V. (2003). Influence of particle size on physical and chemical properties of coconut coir dust as container medium. *Communications in Soil Science and Plant Analysis, 34*, 593–605.

Northup, J. (2013). Biochar as a replacement for perlite in greenhouse soilless substrates. Iowa State University, Graduate Thesis and Dissertations Paper, 13399.

Postma, J., Clematis, F., Nijhuis, E. H., & Someus, E. (2013). Efficacy of four phosphate-mobilizing bacteria applied with an animal bone charcoal formulation in controlling *Pythium aphanidermatum* and *Fusarium oxysporum* f. sp. *radicis lycopersici* in tomato. *Biological Control, 67*(2), 284–291.

Quartacci, M. F., Sgherri, C., & Frisenda, S. (2017). Biochar amendment affects phenolic composition and antioxidant capacity restoring the nutraceutical value of lettuce grown in a copper-contaminated soil. *Scientia Horticulturae, 215*, 9–14.

Rasool, M., Akhter, A., Soja, G., & Haider, M. S. (2021). Role of biochar, compost and plant growth promoting rhizobacteria in the management of tomato early blight disease. *Scientific Reports*, 116092.

Rosli, N. S. M., Abdullah, R., Yaacob, J. S., & Razali, R. B. R. (2023). Effect of biochar as a hydroponic substrate on growth, colour and nutritional content of red lettuce (*Lactuca sativa* L.). *Bragantia, 82*, 20220177.

Roy, A., Chaturvedi, S., Singh, S., Govindaraju, K., Dhyani, V. C., & Pyne, S. (2022). Preparation and evaluation of two enriched biochar-based fertilizers for nutrient release kinetics and agronomic effectiveness in direct-seeded rice. *Biomass Conversion and Biorefinery*. Available from https://doi.org/10.1007/s13399-022-02488-z.

Ruamrungsri, S., Bundithya, W., Potapohn, N., Ohtake, N., Sueyoshi, K., & Ohyama, T. (2011). Effect of NPK levels on growth and bulb quality of some geophytes in substrate culture. *Acta Horticulturae* (886), 213–218.

Saleh, M. E., Mahmoud, A. H., & Rashad, M. (2012). Peanut biochar as a stable adsorbent for removing N [H. sub. 4]-N from wastewater: A preliminary study. *Advances in Environmental Biology, 6*, 2170–2177.

Schmidt, H. P., Kammann, C., Niggli, C., Evangelou, M. W., Mackie, K. A., & Abiven, S. (2014). Biochar and biochar-compost as soil amendments to a vineyard soil: Influences on plant growth, nutrient uptake, plant health and grape quality. *Agriculture, Ecosystems & Environment, 191*, 117–123.

Schulz, H., & Glaser, B. (2012). Effects of biochar compared to organic and inorganic fertilizers on soil quality and plant growth in a greenhouse experiment. *Journal of Plant Nutrition and Soil Science, 175*, 410–422.

Sharma, N., Acharya, S., Kumar, K., Singh, N., & Chaurasia, O. (2018). Hydroponics as an advanced technique for vegetable production: An overview. *Journal of Soil and Water Conservation, 17*, 364–371.

Simiele, M., Argentino, O., Baronti, S., Scippa, G. S., Chiatante, D., Terzaghi, M., & Montagnoli, A. (2022). Biochar enhances plant growth, fruit yield, and antioxidant content of cherry tomato (*Solanum lycopersicum* L.) in a soilless substrate. *Agriculture, 12*, 1135.

Singh, S., Chaturvedi, S., & Datta, D. (2019). Biochar: An eco-friendly residue management approach. *Indian Farming, 69*, 27–29.

Singh, S., Chaturvedi, S., Dhyani, V. C., & Govindaraju, K. (2020). Pyrolysis temperature influences the characteristics of rice straw and husk biochar and sorption/desorption behavior of their biourea composite. *Bioresource Technology, 314*, 123674.

Singh, S., Luthra, N., Mandal, S., Kushwaha, D. P., Pathak, S. O., Datta, D., ... Pramanik, B. (2023). Distinct behavior of biochar modulating biogeochemistry of salt-affected and acidic soil: A review. *Journal of Soil Science and plant Nutrition*. Available from https://doi.org/10.1007/s42729-023-01370-9.

Sofy, M., Mohamed, H., Dawood, M., Abu-Elsaoud, A., & Soliman, M. (2022). Integrated usage of *Trichoderma harzianum* and biochar to ameliorate salt stress on spinach plants. *Archives of Agronomy and Soil Science, 68*, 2005–2026.

Steiner, C., & Harttung, T. (2014). Biochar as growing media additive and peat substitute. *Solid Earth Discuss, 6*, 1023–1035.

Suthar, R. G., Wang, C., Nunes, M. C. N., Chen, J., Sargent, S. A., Bucklin, R. A., & Gao, B. (2018). Bamboo biochar pyrolyzed at low temperature improves tomato plant growth and fruit quality. *Agriculture, 8*, 153.

Tao, R., Zhang, Y., Yang, J., Yang, T., White, J. C., & Shen, Y. (2023). Biochar nanoparticles alleviate salt stress in tomato (*Solanum lycopersicum*) seedlings. *Environmental Science: Nano, 10*, 1800–1811.

Taqdees, Z., Khan, J., Kausar, S., Afzaal, M., & Akhtar, I. (2022). Silicon and zinc nanoparticles-enriched miscanthus biochar enhanced seed germination, antioxidant defense system, and nutrient status of radish under NaCl stress. *Crop and Pasture Science, 73*, 556–572.

Tian, Y., Sun, X., Li, S., Wang, H., Wang, L., Cao, J., & Zhang, L. (2012). Biochar made from green waste as peat substitute in growth media for *Calathea rotundifola* cv. Fasciata. *Scientia Horticulturae, 143*, 15–18.

Vannini, A., Bianchi, E., Avi, D., Damaggio, N., Di Lella, L. A., Nannoni, F., ... Loppi, S. (2021). Biochar amendment reduces the availability of Pb in the soil and its uptake in lettuce. *Toxics, 9*, 268.

Vaughn, S. F., Kenar, J. A., Thompson, A. R., & Peterson, S. C. (2013). Comparison of biochars derived from wood pellets and pelletized wheat straw as replacements for peat in potting substrates. *Industrial Crops and Products, 51*, 437–443.

Verhagen, J., van den Akker, J., Blok, C., Diemont, H., Joosten, H., Schouten, M., ... Wösten, H. (2009). *Climate change. scientific assessment and policy analysis: Peatlands and carbon flows: Outlook and importance for the Netherlands* (p. 2009) Bilthoven, The Netherlands: Netherlands Environmental Assessment Agency PBL, Wab 500102 027.

Wang, J., & Wang, S. (2019). Preparation, modification and environmental application of biochar: A review. *Journal of Cleaner Production, 227*, 1002–1022.

Wang, N., Wang, L., Zhu, K., Hou, S., Chen, L., Mi, D., ... Guo, J. H. (2019). Plant root exudates are involved in *Bacillus cereus* AR156 mediated biocontrol against *Ralstonia solanacearum*. *Frontiers in Microbiology*, 1098.

Xiao, X., Chen, B., Chen, Z., Zhu, L., & Schnoor, J. L. (2018). Insight into multiple and multilevel structures of biochars and their potential environmental applications: A critical review. *Environmental Science & Technology, 52*, 5027–5047.

Yadav, S., Modi, P., Dave, A., Vijapura, A., Patel, D., & Patel, M. (2020). Effect of abiotic stress on crops. In M. Hasanuzzaman, M. C. M. T. Filho, M. Fujita, & T. A. R. Nogueira (Eds.), *Sustainable crop production*. London: IntechOpen.

Yael, B., Liel, G., Hana, B., Ran, H., & Shmuel, G. (2012). Total phenolic content and antioxidant activity of red and yellow quinoa (*Chenopodium quinoa* Willd.) seeds as affected by baking and cooking conditions. *Food and Nutrition Sciences, 03*, 1150–1155.

Yan, J., Yu, P., Liu, C., Li, Q., & Gu, M. (2020). Replacing peat moss with mixed hardwood biochar as container substrates to produce five types of mint (*Mentha* spp.). *Industrial Crops and Products*, 155112820.

Yan, P., Shen, C., Zou, Z., Fu, J., Li, X., Zhang, L., ... Fan, L. (2021). Biochar stimulates tea growth by improving nutrients in acidic soil. *Scientia Horticulturae*, 283110078.

Yang, W., & Zhang, L. (2022). Biochar and cow manure organic fertilizer amendments improve the quality of composted green waste as a growth medium for the ornamental plant *Centaurea Cyanus* L. *Environmental Science and Pollution Research, 29*, 45474–45486.

Yildirim, E., Ekinci, M., & Turan, M. (2021). Impact of biochar in mitigating the negative effect of drought stress on cabbage seedlings. *Journal of Soil Science and Plant Nutrition, 21*, 2297–2309.

Yu, H., Zou, W., Chen, J., Chen, H., Yu, Z., Huang, J., ... Gao, B. (2019). Biochar amendment improves crop production in problem soils: A review. *Journal of Environmental Management, 232*, 8–21.

Zhongming, Z., Linong, L., Xiaona, Y., Wangqiang, Z., & Wei, L. (2021). *Excess salt in soils puts food security at risk*. Rome: FAO.

Zulfiqar, F., Moosa, A., Nazir, M. M., Ferrante, A., Ashraf, M., Nafees, M., ... Siddique, K. H. (2022). Biochar: An emerging recipe for designing sustainable horticulture under climate change scenarios. *Frontiers in Plant Science*, 131018646.

Zulfiqar, F., Wei, X., Shaukat, N., Chen, J., Raza, A., Younis, A., ... Naveed, M. (2021). Effects of biochar and biocharcompost mix on growth, performance and physiological responses of potted *Alpinia zerumbet*. *Sustainability, 13*, 11226.

CHAPTER 14

Biochar-based slow-release fertilizers toward sustainable nutrition supply

Xiuxiu Zhang, Dan Luo and Chongqing Wang

School of Chemical Engineering, Zhengzhou University, Zhengzhou, P.R. China

14.1 Introduction

With the rapid growth of the human population, food security is one of the world's greatest concerns. It has been projected that the number of food yields in 2050 will be about 70% higher than in 2005 (Kopittke, Lombi, Wang, Schjoerring, & Husted, 2019). To address the issue, sustainable agriculture needs to be urgently developed to meet the demand for food and agricultural products. In recent years, fertilizers have played a vital role in supplying plant nutrients, meeting targeted yields and crop requirements, and developing sustainable agriculture. However, some research has indicated that most of the nutrients in conventional fertilizers, such as phosphorus (P), nitrogen (N), and potassium (K) are unable to be utilized by plants due to leaching, photo-degradation, chemical hydrolysis, and microbial decomposition (Tarafder et al., 2020; Xin, Judy, Sumerlin, & He, 2020). Those not only reduce the utilization efficiency of conventional fertilizers, but also cause serious environmental pollution and human health. Generally, the low fertilizer use efficiency is thought to be a result of the release rate of conventional fertilizer being faster than the plant's uptake (Diatta et al., 2020). Therefore, it is essential to design new fertilizers to solve the drawbacks of traditional fertilizers.

To enhance the service life of conventional fertilizers and alleviate environmental issues, a number of novels enhanced efficiency fertilizers, such as slow-release fertilizers (SRFs) and controlled-release fertilizers, have been rapidly developed in recent years (Luo et al., 2023; Marcińczyk & Oleszczuk, 2022). Several studies have reported that SRFs could release nutrients for longer periods than that of traditional fertilizers so that the released nutrients could be adsorbed absolutely by crops and meet the nutrient requirements for plant growth. In addition, the application of SRFs could reduce the emission of nitrous oxide (N_2O) and ammonia in soil, and alleviate the environmental crisis (Ramli, 2019; Saha, Rose, Van Zwieten, Wong, & Patti, 2021). As a result, many studies have already been conducted on the development of different SRFs by several researchers (Chen et al., 2020; Cui et al., 2020; El Sharkawi, Tojo, Chosa, Malhat, & Youssef, 2018; Lyu et al., 2021; Nardi et al., 2018). For instance, Elhassani et al. (2019) developed novel nitrogen-loaded SRFs based on coated urea with hydroxyapatite-encapsulated woodchip species. Results showed that the coated fertilizers released urea at a slower rate than the uncoated urea fertilizers. Importantly, the nutrient release process was found to be significantly controlled by the type of coating materials. Similarly, Giroto, Guimarães, Foschini, and Ribeiro (2017) prepared nanocomposites as SRFs using urea as the substrate to improve the fertilization of urea. Release tests revealed that the nanocomposites displayed a slow release rate of urea. Moreover, the interaction between the morphological changes in the nanocomposite and the fertilizer composition reduced the NH_3 volatilization and increased the phosphorus release, thus improving the utilization efficiency of the fertilizers.

Currently available SRFs are always high-cost and complex production technology. Consequently, new materials and strategies should be devised to fabricate new types of SRFs that are low-cost, efficient, and environmentally friendly. In this context, biochar is gaining great attention as an efficient alternative for SRFs preparation due to its low price, wide source of raw materials, high organic carbon content, and large surface area. Moreover, the low bulk densities of biochar could significantly increase soil water-holding capacity and decrease soil bulk density (Li et al., 2017; Rashid et al., 2021). Recently, several studies of biochar-based SRFs have been gradually reported. Therefore, the objective of this chapter is to give a summarization of various techniques for the preparation of biochar-based SRFs. Moreover, the applications of biochar-based SRFs are also discussed in detail. And the prospects of biochar-based SRFs in sustainable agriculture are provided in conjunction with existing research.

14.2 Biochar-based slow-release fertilizers

Biochar is commonly produced by pyrolysis of natural and waste materials under an inert atmosphere at a wide range of temperatures (Nan et al., 2023; Wang, 2021). Due to its unique physicochemical properties, biochar has received much attention over the past two decades for its ability to modify soil properties, manage waste, and mitigate climate change (Liu et al., 2023; Luo et al., 2023; Wang, Huang, Sun, Yang, & Sillanpää, 2021a). Previous studies have demonstrated that biochar can not only promote nutrient supply and crop yields, but also immobilizes heavy metals and organic contaminants, and thus has promising applications in soil remediation (Hossain et al., 2020; Marcińczyk & Oleszczuk, 2022; Samoraj et al., 2022). For instance, Wang et al. (2021b) prepared a rice hull-derived multiple-modified biochar to stabilize the contamination of cupric ion (Cu (II)) and cadmium ion (Cd (II)) in farmland soil and vegetable soil. It was found from the incubation experiments that the stabilization efficiency of Cu (II) and Cd (II) could reach 100.00% and 92.02% for farmland soil and 100.00% and 90.27% for vegetable soil, respectively. Moreover, the application of multiple-modified biochar increased crop yields by increasing soil cation exchange capacity, available P and K, and organic matter content. In addition, Zhang, Chen, & You, 2016 also investigated the important role of biochar in water retention in soil. Results displayed that the sandy soil with biochar added had a high water adsorption capacity. Jin et al. (2019) revealed that the application of biochar could significantly increase the soil pH, available P, organic carbon, and water retention through a 5-year field trial.

The application of conventional fertilizers is mostly limited by low nutrient use efficiency, high costs, and environmental pollution. To address these issues, biochar-based SRFs have been proposed by combining conventional fertilizers or nutrients with biochar through a variety of methods. To date, a growing number of studies have indicated that biochar-based SRFs are more efficient than conventional fertilizers in retaining nutrients in the soil for longer periods (Chan, Van Zwieten, Meszaros, Downie, & Joseph, 2007; Wang et al., 2022). For instance, An et al. (2020b) developed a novel biochar-based embedded-semiinterpenetrating polymer networks SRFs by grafting biochar into semiinterpenetrating polymer networks. It could be observed that the resulting novel biochar-based SRFs exhibited slower nutrient release, higher water retention, and better degradation ratio than the semiinterpenetrating polymer networks-based SRFs without biochar. Similarly, to reduce the loss of nitrogen and increase nitrogen use efficiency, Bakshi, Banik, Laird, Smith, and Brown (2021) developed biochar-based SRFs with high urea loading. Results indicated that the developed biochar-based SRFs could effectively reduce soil nitrogen loss by reducing NH_3 volatilization and NO_3-N leaching, and additionally provide a large amount of NH_4^+-N to crops for a long period.

In addition, previous studies have reported that the biochar with high contention of P has the possibility to be used as SRFs to enhance crop production (Frišták, Pipíška, & Soja, 2018; Luo et al., 2023; Nam et al., 2018). Therefore, a "P-enriched biochar" fertilizer was fabricated from bacterial biomass waste and its possibility as an SRF was evaluated (Kim, Vijayaraghavan, Reddy, & Yun, 2018). It was found that the water-soluble P content of the biochar was 84.7 mg g^{-1}, which was higher than that of the dried feedstock. Moreover, within five days, P-release from the biochar reached 61% and 52% of the total water-soluble P content in 2% citric acid and water, respectively. Therefore, the results clearly indicated that the "p-enriched biochar" has the feasibility to be used as a source of P recovery and P-SRFs. Similarly, Luo et al. (2021) successfully synthesized a potential "Mg-enriched biochar" fertilizer by pyrolysis of corn straw and biogas effluent. Results showed that the release ratio of P and N from the "Mg-enriched biochar" fertilizers were 6 times and 7 times lower than those of chemical fertilizers, respectively. Furthermore, it was also revealed that the mechanism of P release from the "Mg-enriched biochar" fertilizers was controlled by the "Mg-P" precipitates and the "P-trap" effect, which could significantly promote the growth of corn.

Based on the above mentioned, it can be presumed that biochar-based SRFs will widely apply in agriculture in the future due to the several agronomic benefits, such as providing nutrients (N, P, and K), enhancing nutrient use efficiency, and improving soil water-retention capacity and crop production. The properties of biochar-based SRFs and their functions on soil and plants can vary depending on the raw materials used for preparation. Table 14.1: lists some biochar-based SRFs with different feedstocks and their special functions on plants and soil.

14.3 Preparation methods of biochar-based slow-release fertilizers

Biochar-based SRFs are essential for the development of modern agriculture to ensure food security. Therefore, it is vital to investigate efficient, low-cost and environmentally

TABLE 14.1 Biochar-based slow-release fertilizers with different feedstocks and their special functions on plants and soil.

Fertilizers	Feedstocks	Functions	References
Biochar-based NPK fertilizers	Ca-bentonite and chicken manure	Reduce N, P, and K release rate in coarser sand soil.	Piash, Iwabuchi, and Itoh (2022)
Biochar-based P fertilizers	Coffee husk, poultry litter, H_3PO_4 and MgO	Supply P and Mg to plants, regulate soil pH of acidic.	Carneiro et al. (2018)
P-loaded biochar-based fertilizers	Cotton straw and H_3PO_4	Improve lambda-cyhalothrin adsorption and reduce P release rate.	An et al. (2021c)
Biochar-based slow-release NPK fertilizers	Rice straw, compound fertilizers (15% N, 15% P_2O_5, 15% K_2O), bentonite starch and humic acid	Decrease N leaching and run-off losses, and supply more nutrients to rice.	(Dong et al., 2020)
Biochar-based N/P fertilizers	Oil palm kernel shell, NH_4NO_3 and KH_2PO_4	Reduce nutrient release (55.2% for PO_4^{-1}, 77.4% for NH_3, and NO_3^- was 52.9% for 1 h).	Dominguez et al. (2020)
Biochar-based NPK fertilizers	Sawdust, super phosphate, NH_4NO_3 and KH_2PO_4	Reduce nutrient leaching, enhance water retention and nutrient use efficiency.	Gwenzi, Nyambishi, Chaukura, and Mapope (2018)
Biochar-based N SRFs	Corn straw and $MgCl_2$	Mitigate nitrogen release, increase plant height, shoot and root dry weight, and enhance chlorophyll content and leaf area.	Khajavi-Shojaei, Moezzi, Norouzi Masir, and Taghavi (2020)
Biochar-based waterborne copolymers fertilizers	Polyvinyl alcohol, rice, polyvinylpyrrolidone, and urea	Reduce nutrient release (nutrient leaching of 65.28% on the 22nd day).	Chen et al. (2018b)
Binder-free biochar-based fertilizers	Biogas residue and commercial compound fertilizers	Reduce P and N release (P release rate of 23.93% and N release rate of 41.71% after 413 h).	Yu et al. (2021)
Biochar-based slow N fertilizers	Pine, urea and H_3PO_4	Reduce N loss and provide larger quantities of NH_4^+-N to plants for a longer time.	Bakshi et al. (2021)
Urea loaded Biochar	Rice straw and commercial urea	Depicted >90% urea sorption from aqueous solution followed by sustained urea release pattern	Singh, Chaturvedi, Dhyani, and Govindaraju (2020)
Enriched biochar with N-P-K	Rice husk, NPK and seaweed	Sustained release pattern of NH_4^+, P, and K^+ under batch experiments	Roy et al. (2022)
Biochar-based nitrogen composites developed through urea intercalation	Rice straw and commercial urea	Prolonged the nitrogen release for >25 days compared to commercial urea (9 days) under soil column experiment	Singh et al. (2023)

friendly methods to prepare biochar-based SRFs. To date, various methods, including impregnation, pyrolysis (in situ pyrolysis and co-pyrolysis), granulation, and encapsulation, have been developed to synthesise biochar-based SRFs. Fig. 14.1 shows the preparation methods and some applications of biochar-based SRFs in plants and soil.

14.3.1 Impregnation

The impregnation method for the synthesis of biochar-based SRFs has the advantage of being simple to operate and lowcost. Previous research has reported that biochar impregnated with different materials like urea (Xie, Liu, Ni, Zhang, & Wang, 2011), ammonium salts (Spokas, Novak, & Venterea, 2012), phosphates (Xu, Sun, Shao, & Chang, 2014), and nitrates (Kammann et al., 2015) were used to synthesis biochar-based SRFs, which could effectively increase nutrient content and satisfy the nutritional needs of plants grown. Das and Ghosh (2021) prepared biochar-based slow-release N-P-K fertilizers with four different biomass by impregnation method. The experiments depicted that the release ratio of K_2O, NO_3^-, NH_4^+, and PO_4^{3-} was 74.33%–77.27%, 55.47%–50.84%, 55.47%–50.84%, and 65.31%–68.52%, respectively, after 90 days, which was lower than the fertilizer alone. In addition, the as-prepared biochar-based N-P-K SRFs can also significantly improve the quality of sandy soil by increasing oxidizable carbon, available nutrients, and decreasing sandy soil pH.

The possibility of environmentally friendly SRFs prepared from biochar impregnated with anaerobically digested slurry was investigated (Oh, Shinogi, Lee, & Choi, 2014). It was found that the release behavior of water-soluble K^+, Ca^{2+}, and Mg^{2+} in the as-prepared SRFs was similar to that of commercial fertilizers. However, the water-holding capacity of the soil was greatly enhanced by the utilization of biochar-based SRFs as compared to commercial fertilizers. Lateef et al. (2019) synthesized biochar-based nano-composite by simple impregnation of micro (Mg, N, P, Ca, and K) and macro (Fe, Na, and Zn) nutrients in corncob biochar. The findings displayed that corncob biochar has a rich

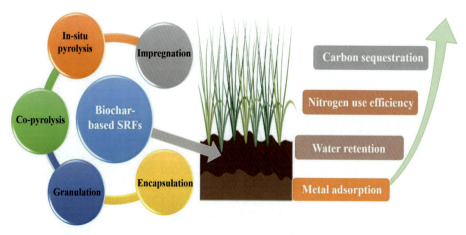

FIGURE 14.1 Preparation methods and applications of biochar-based slow-release fertilizers.

porosity structure, which facilitates and supports nutrients impregnation. Slow-release studies have demonstrated the patterns of micro and macro nutrients released from biochar-based nano-composite over long periods. Moreover, the results of salt index, water absorbance and retention indicated that biochar-based nano-composite was an excellent SRF with great potential to improve soil fertility, plant growth and yields. However, the current problem with the impregnation technology is the low nutrient adsorption capacity of biochar, which largely limits its practical application.

14.3.2 Pyrolysis

14.3.2.1 *In situ pyrolysis*

Due to the benefits in terms of recycling waste biomass, reducing environmental pollution and creating valuable products, pyrolysis has found extensive applications in the production of biochar-based SRFs. Biochar-based SRFs could be synthesized by pyrolysis from many raw materials, such as rice straw (Peng, Ye, Wang, Zhou, & Sun, 2011), manure (Hossain et al., 2020), corncob (Lateef et al., 2019), and sewage sludge (Tomczyk et al., 2020). In some cases, biochar-based fertilizers can be obtained directly from biomass feedstock with high nutrient content. For instance, Lou et al. (2017) reported the preparation of biochar from spent mushroom substrate at different pyrolysis temperatures and its application as a biochar-based SRF for soil remediation. Compared to chemical fertilizers, the spent mushroom substrate biochar prepared at pyrolysis temperatures of 750°C exhibited higher nutrient retention efficiency in total nitrogen and excessive organic matter (represented as COD_{Cr}) leaching.

Similarly, K-rich biochar fertilizers, produced from the feedstock of banana peduncle, were reported (Karim, Kumar, Singh, Panda, & Mishra, 2017). Due to their special physicochemical properties, K-rich biochar fertilizers have the potential to significantly increase soil productivity, including water-holding capacity, soil carbon sequestration, and cation exchange capacity. Roberts, Paul, Dworjanyn, Bird, and de Nys (2015) reported a nutrient-rich seaweed biochar, which was used as a fertilizer for soil amendment, produced from seaweed biomass by the in situ pyrolysis method. In addition, by pyrolyzing cow dung, a high-efficiency engineered biochar was prepared (Chen, Qin, Sun, Cheng, & Shen, 2018a). Experiments confirmed that the application of cow dung-engineered biochar in soil could significantly improve the seed germination, crop growth, and yield of lettuce.

14.3.2.2 *Co-pyrolysis*

For the development of biochar-based SRFs, co-pyrolysis, where nutrients are incorporated into the biomass, is considered as a potential strategy. Moreover, the availability of biochar-based SRFs and the absorption efficiency of plants can also be significantly enhanced by co-pyrolysis. Lustosa Filho et al. (2020) prepared biochar-based SRFs that could supply plants with available P via co-pyrolysis of phosphoric acid (H_3PO_4), poultry litter, triple superphosphate, and magnesium oxide (MgO). Similarly, Chen et al. (2017) also synthesized environmentally friendly montmorillonite-biochar SRFs by co-pyrolysis of low-cost bamboo and montmorillonite. The release results showed that the montmorillonite-biochar SRFs had the slow-release characteristic of P and N. In addition, Piash et al. (2022) prepared the chicken manure biochar-based SRFs through co-pyrolysis of chicken manure with Ca-bentonite, which could satisfy the nutritional requirements of plants.

Due to the merit of low cost and high efficiency, microwave pyrolysis has attracted great interest in the preparation of biochar-based SRFs in recent years. It has been reported that certain chemical reactions may be facilitated by selective heating from microwave irradiation (Lam, Liew, Cheng, & Chase, 2015). Recently, An et al. (2020a) reported a new method for preparing biochar-based SRFs via co-pyrolysis of bentonite, biomass (cotton straw), and K_3PO_4 by microwave irradiation. Compared to the biochar-based SRFs without bentonite, the obtained biochar-based SRFs exhibited better slow-release performance, more positive effects on pepper seedling growth, and lower cost. Similarly, another biochar-based fertilizer was also prepared via microwave pyrolysis using K_3PO_4 and cotton stalk premixed MgO as raw materials, which had a better slow-release performance for P (Liu et al., 2021b).

Apart from fertilizers, water is another crucial factor affecting plant growth. Hence, it is significant to develop novel biochar-based SRFs with high water-holding capacity. Wang, Li, Parikh, and Scow (2019) reported that the water-holding capacity of coarse-textured soils could be improved by the application of biochar particles because of their high porosity structure. An et al. (2021a) developed biochar-based slow-release phosphorus fertilizers by co-pyrolysis technology. Results showed that the maximum swelling capacity of the obtained biochar-based slow-release phosphorus fertilizers could reach 94.2 g g^{-1}, and the water-retention capacity was higher than that of other SRFs, which was positively related to the crosslinking content.

14.3.3 Granulation

To reduce the loss of biochar-based fertilizers in the soil and transportation costs, granulation has been proposed and proved to be a feasible method. In this approach, biochar is typically mixed with chemical fertilizers and binders, and subsequently granulated by disc granulation or mechanical extrusion, which can effectively reduce nutrient release. Consequently, granulation is the most efficient primary preparation method for biochar-based SRFs at present. Generally, the size and quality of the produced granule depend on the type of raw materials, operating pressure, temperature, moisture content, binder materials, and so on (Arshadi, Gref, Geladi, Dahlqvist, & Lestander, 2008; Gilbert, Ryu, Sharifi, & Swithenbank, 2009). For instance, Shi et al. (2020) synthesized the biochar-mineral urea composites by granulation method using woody biochar as the base carrier and bentonite as the mineral binder. Compared to conventional urea fertilizers, the release ratio of dissolved organic carbon and N from the biochar-mineral urea composites was distinctly lower. In addition, the application of biochar-mineral urea composites improved nitrogen use efficiency and maize growth by enhancing N uptake. Similarly, another N−P−K SRFs was prepared by using starch−polyvinyl alcohol as a binder and biochar as a nutrient carrier (Gwenzi et al., 2018). Results displayed that the release rate of NO_3^-, PO_4^{3-}, and K^+ from the obtained N−P−K SRFs was obviously lower than that of conventional fertilizers. Moreover, the water-holding capacity of the N−P−K SRFs-modified soils was also higher compared to other treatments, demonstrating the potential superiority of the N−P−K SRFs over conventional chemical fertilizers.

Some binders are frequently used in granulation methods to produce biochar-based SRFs. However, in some cases, the binder may cause serious damage to the environment due to its inability to degrade (Mostafa et al., 2019). Therefore, it is necessary to develop new strategies for the synthesis of binder-free biochar-based SRFs. In this context, the binder-free biochar-based SRFs from biogas biochar and chemical fertilizers were synthesized (Yu et al., 2021). Results showed that the optimal operating conditions for pelletizing were 7.84% moisture content of the microspheres, 7 mm diameter, 49.54 mm min^{-1} compression speed, and 7.5 MPa molding pressure. Compared with conventional fertilizers, the release time of nutrients in the binder-free biochar-based SRFs was longer, suggesting that the binder-free biochar-based SRFs had excellent sustained release properties. However, the complex preparation and high cost of granular biochar-based SRFs will limit its wide application.

14.3.4 Encapsulation

More recently, encapsulation technology is a novel approach to enhance the slow-release properties of fertilizers, which is primarily aimed at improving nutrient utilization. As a physical barrier, the encapsulated/coated materials on the surface of the fertilizers can impede the transport of nutrients. Hence, several encapsulated biochar-based SRFs have been studied successively. For instance, An et al. (2021b) prepared new encapsulated biochar-based SRFs by co-pyrolysis and encapsulation methods. Compared with unencapsulated biochar-based fertilizers, the as-prepared encapsulated biochar-based SRFs exhibited lower P-release rate constant and superior degradability. Moreover, the pot trials revealed that the pepper seedlings treated with the encapsulated biochar-based SRFs grew much better than those treated with unencapsulated biochar-based fertilizers.

In recent years, many waterborne polymer materials have been reported to be used as coatings for SRFs encapsulation, such as polysulfone, polyvinyl chloride, polyvinylpyrrolidone (PVP), and polystyrene (Sim, Tan, Lim, & Hameed, 2021; Zhou et al., 2015). For some waterborne polymers, nutrients are generally released from the porous pores formed by the expansion of the coating polymer (Abd El-Aziz, Salama, Morsi, Youssef, & El-Sakhawy, 2022). For instance, Chen et al. (2018b) developed a novel polymer-coated urea SRFs using PVP, polyvinyl alcohol, and biochar as coating materials. Obviously, the water absorption of urea was improved by coating biochar and waterborne polymers. Moreover, the soil leaching test revealed that the nutrient release rate was 65.28% after leaching on the 22nd day, indicating that the polymer-coated urea SRFs had an excellent slow-release performance.

In addition, bio-oil, a liquid product in biochar production, can also be deemed as a coating material. For instance, the bio-oil encapsulated biochar-based SRFs by using bio-oil as coating materials were reported (Ye, Zhang, Huang, & Tan, 2019). The results showed that the nutrients (N, P, and K) dissolve rate in water was decreased by approximately 15% after oil-coating treatment. Moreover, the soil leaching rate was also reduced by 3% to 5%, indicating that the sustained release capacity of pristine biochar-based fertilizers could be significantly improved by oil-coated treatment. However, due to the poor degradability of some coating materials, they may remain in the soil after nutrient release, resulting in potential environmental threats. Thus, cheap and bio-degradable coating materials should be developed in the future.

14.4 Applications of biochar-based slow-release fertilizers

14.4.1 Nitrogen use efficiency

Nitrogen use efficiency is broadly defined as the amount of nitrogen used by plants during the current and subsequent seasons. Measures to enhance nitrogen use efficiency are to reduce N losses, including NO_3^- leaching, N_2O emissions, and NH_3 volatilization, which is of great significance to ensure food security and mitigate environmental crises (Zhang et al., 2015). Studies have proved that the utilization of biochar-based SRFs can increase the nitrogen use efficiency value of wheat, maize, and rice by about 21%, 44%, and 74%, respectively (Chen et al., 2023). In addition, Dong et al.(2020) also demonstrated that rice straw biochar-based SRFs, prepared by granulation method, could significantly reduce N losses and provide more nutrients for rice plant growth. Similarly, Wen et al. (2017) synthesized biochar-based slow-release nitrogen fertilizers using a polymer matrix and NH_4^+-loaded biochar as raw materials via microwave irradiation. Results indicated that the biochar-based slow-release nitrogen fertilizers could significantly reduce the nitrogen release rate, which was about 69.8% after 30 days. Compared with NH_4Cl and NH_4^+-loaded biochar, the as-prepared biochar-based slow-release nitrogen fertilizers possessed lower nitrogen leaching loss (10.3%), higher nitrogen use efficiency (64.27%), and higher water-holding capacity, which could efficiently promote the growth of cotton plant. Moreover, the degradation tests showed that the obtained biochar-based slow-release nitrogen fertilizers possessed good degradability, indicating that they might have a wide range of applications in modern sustainable agriculture.

14.4.2 Water retention

The amount of water in the soil depends on the soil texture and the rate of precipitation. Reports have shown that the application of biochar-based SRFs to sandy soils improves the water-holding and retention capacity of the soil more than conventional fertilizers (Rashid et al., 2021). For instance, Shang et al. (2022) designed KH_2PO_4-itaconic acid-modified cotton stalk biochar-based SRFs. Results showed that the water absorption capacity of the obtained biochar-based SRFs was 1.91 g g^{-1}, and the water retention was 28.73% after 30 days, which was related to its rich porous structure. Moreover, the N release rate was 23.64% after 24 days in water, suggesting that the biochar-based SRFs also possessed excellent long-lasting slow-release properties that significantly promote stem and leaf growth. Furthermore, Liu et al. (2021a) reported the fabrication of black liquor-based hydrogel as SRFs via grafting copolymerization induced by redox free radicals. The cultivating plant test results revealed that the water-retention ratio and the maximum water swelling ratio of the black liquor-based hydrogel were 45.25% and 359 g g^{-1}, respectively, which obviously improved the water-retention capacity of the soil.

14.4.3 Metal adsorption

Some metals, such as Zn, Fe, and Cu in the soil, can provide a certain number of micronutrients for crop growth. However, the accumulation of metals in the soil cannot be effectively absorbed by crops, thus affecting the entire soil ecology and potentially causing water pollution through run-off (Zhang et al., 2022a). Researches have demonstrated that some heavy metals could be adsorbed by biochar from soil via cation ion exchange, chelation, complexation, and fixation (Qiu et al., 2015; Xiao, Hu, & Chen, 2020). For instance, to solve Cd (II) pollution, a novel rice husk biochar-impregnated urea fertilizer was produced via impregnation method (Xiang et al., 2021). Results exhibited that the adsorption capacity and cation exchange capacity value of the obtained fertilizers was 14.24 mg g^{-1} and 5.70 cmol kg^{-1}, respectively, which was higher than that of the biochar. It was found that the primary mechanism of Cd (II) adsorption was the ion exchange reactions that occurred between Cd (II) and NH_4^+. Moreover, the O and N functional groups also played a crucial role in enhancing the Cd (II) adsorption capacity of the rice husk biochar-impregnated urea fertilizers.

14.4.4 Carbon sequestration

Over the past few decades, large amounts of oxycarbide, such as carbon dioxide (CO_2) and carbonic oxide (CO), have been released into the atmosphere, leading to global warming as the atmospheric carbon cycle was disrupted. Therefore, increasing soil carbon sequestration is necessary to capture CO_2 from the atmosphere and tackle climate change. Research has proven that biochar has good fixation of NOx and CO_2 (Darby et al., 2016). Subsequently, Sun, Lu, and Feng (2019) reported that the application of biochar as fertilizers on rice paddy soils could suppress CH_4 and N_2O emissions by up to 21.7%–62.3%. Similarly, Zhang, Cheng, Wang, Tahir, and Wang (2022b) found that biochar prepared by co-pyrolysis of peanut shell with a certain amount of potassium dihydrogen phosphate (KH_2PO_4) could significantly improve the carbon sequestration capacity of biochar. Therefore, the application of biochar-based SRFs in soil can increase soil carbon sequestration and reduce atmospheric CO_2 concentrations, thus alleviating environmental crises.

14.5 Challenges and future perspectives

Although the preparation and application of biochar-based SRFs could enhance sustainable agricultural farming practices, increase productivity to ensure a healthy ecosystem, and promote sustainable economic development, the field is still in its infancy. Currently, the development and application of biochar-based SRFs still face a number of challenges, including (i) optimization of the preparation methods of biochar-based SRFs; (ii) clearer explanation of the specific slow-release mechanism of biochar-based SRFs; (iii) large-scale field studies of biochar-based SRFs are still scarce; and (iv) limited tracking studies on the fate and long-term effects of biochar-based SRFs in soil. Therefore, further research on biochar-based SRFs should be considered in the following aspects:

1. Currently, various preparation methods for biochar-based SRFs have been developed, but they cannot meet all the requirements well. In order to promote the concept of using biochar-based SRFs in the world, a simpler and more economical production method should be proposed to meet the demand for biochar-based SRFs in soils and regions with different characteristics. Therefore, future research should be focused on the development of new methods for the production of biochar-based SRFs with specific slow-release patterns of nutrients and also on the use of biochar modifications to control the extent of nutrient release.

2. An in-depth understanding of the mechanisms involved in the slow release of nutrients from biochar-based SRFs is essential to effectively modify their efficacy in the desired direction and to apply biochar-based SRFs in particular ecological tasks. At present, research on the nutrient-loading mechanism of biochar has made some progress, but the understanding of the nutrient slow-release mechanism of biochar-based SRFs is still not deep enough. Accordingly, future research should focus on using advanced technological tools to reveal the mechanism of nutrients slow-release from biochar-based SRFs.

3. According to the research, it is known that most studies on the applications of biochar-based SRFs have been conducted at the laboratory and pilot-scales, and there are few studies on large-scale field applications. Thus, future studies are recommended to transfer from laboratory and pilot-scales to large-scale field applications to assess the reliability, feasibility, reproducibility, and consistency of the results obtained in the laboratories.

4. Studying the fate of biochar-based SRFs in soil, tracking their long-term impacts, and conducting a life-cycle assessment of the entire process is vital for preventing related environmental problems. In addition, this could also ensure the sustainability of biochar-based SRFs in the soil. In this connection, it is urgent to develop and standardize analytical methods suitable for investigating the fate of biochar-based SRFs in the soil and environmental.

5. The characteristics of biochar and biochar-based SRFs depend greatly on the raw materials used to prepare biochar. It has been reported that biochar produced from some feedstocks at lower temperatures may contain toxic compounds, which might be harmful to plants and soil. In order to better utilize biochar for the development of biochar-based SRFs, a relatively complete database should be established to provide information on the raw materials used in the preparation of biochar-based SRFs, the properties of biochar and its effects on the soil system. In addition, the potential risk associated with the application of biochar-based SRFs are also required to be further studied, especially the possibility of releasing toxic substances from biochar under environmental conditions.

14.6 Conclusions

Biochar-based SRFs provide an available way to enhance nutrient use efficiency by different mechanisms to minimize nutrient losses, and their development effectively solves the shortcomings of nutrient loss, environmental pollution and harm to human health that exist in traditional fertilizers. In this chapter, various strategies for the production of biochar-based SRFs were discussed, such as impregnation, pyrolysis (in situ pyrolysis and co-pyrolysis), granulation, and encapsulation. When biochar-based SRFs were applied to soils, they could improve nitrogen use efficiency, water retention, metal adsorption, and carbon sequestration in the soil, thus playing a vital role in improving soil productivity and mitigating climate change. However, there are still some challenges in the development and practical large-scale application of biochar-based SRFs, such as the optimized preparation method, the study of the slow-release mechanism, and the economic and environmental assessment of its application in the process. In addition, the potential risks and environmental issues associated with the application of biochar-based SRFs are also factors that need to be considered. Therefore, the successful development and application of biochar-based SRFs in the future requires more efforts from researchers.

References

Abd El-Aziz, M. E., Salama, D. M., Morsi, S. M., Youssef, A. M., & El-Sakhawy, M. (2022). Development of polymer composites and encapsulation technology for slow-release fertilizers. *Reviews in Chemical Engineering, 38*(5), 603–616.

An, X., Wu, Z., Liu, X., Shi, W., Tian, F., & Yu, B. (2021a). A new class of biochar-based slow-release phosphorus fertilizers with high water retention based on integrated co-pyrolysis and co-polymerization. *Chemosphere, 285*, 131481.

An, X., Wu, Z., Qin, H., Liu, X., He, Y., Xu, X., & Yu, B. (2021b). Integrated co-pyrolysis and coating for the synthesis of a new coated biochar-based fertilizer with enhanced slow-release performance. *Journal of Cleaner Production, 283*, 124642.

An, X., Wu, Z., Shi, W., Qi, H., Zhang, L., Xu, X., & Yu, B. (2021c). Biochar for simultaneously enhancing the slow-release performance of fertilizers and minimizing the pollution of pesticides. *Journal of Hazardous Materials, 407*, 124865.

An, X., Wu, Z., Yu, J., Cravotto, G., Liu, X., Li, Q., & Yu, B. (2020a). Copyrolysis of biomass, bentonite, and nutrients as a new strategy for the synthesis of improved biochar-based slow-release fertilizers. *ACS Sustainable Chemistry & Engineering, 8*(8), 3181–3190.

An, X., Yu, J., Yu, J., Tahmasebi, A., Wu, Z., Liu, X., & Yu, B. (2020b). Incorporation of biochar into semi-interpenetrating polymer networks through graft co-polymerization for the synthesis of new slow-release fertilizers. *Journal of Cleaner Production, 272*, 122731.

Arshadi, M., Gref, R., Geladi, P., Dahlqvist, S. A., & Lestander, T. (2008). The influence of raw material characteristics on the industrial pelletizing process and pellet quality. *Fuel Processing Technology, 89*(12), 1442–1447.

Bakshi, S., Banik, C., Laird, D. A., Smith, R., & Brown, R. C. (2021). Enhancing biochar as scaffolding for slow release of nitrogen fertilizer. *ACS Sustainable Chemistry & Engineering, 9*(24), 8222–8231.

Carneiro, J. S. D. S., Lustosa Filho, J. F., Nardis, B. O., Ribeiro-Soares, J., Zinn, Y. L., & Melo, L. C. A. (2018). Carbon stability of engineered biochar-based phosphate fertilizers. *ACS Sustainable Chemistry & Engineering, 6*(11), 14203–14212.

Chan, K. Y., Van Zwieten, L., Meszaros, I., Downie, A., & Joseph, S. (2007). Agronomic values of greenwaste biochar as a soil amendment. *Soil Research, 45*(8), 629–634.

Chen, L., Chen, X. L., Zhou, C. H., Yang, H. M., Ji, S. F., Tong, D. S., & Chu, M. Q. (2017). Environmental-friendly montmorillonite-biochar composites: Facile production and tunable adsorption-release of ammonium and phosphate. *Journal of Cleaner Production, 156*, 648–659.

Chen, Q., Qin, J., Sun, P., Cheng, Z., & Shen, G. (2018a). Cow dung-derived engineered biochar for reclaiming phosphate from aqueous solution and its validation as slow-release fertilizer in soil-crop system. *Journal of Cleaner Production, 172*, 2009–2018.

Chen, S., Han, Y., Yang, M., Zhu, X., Liu, C., Liu, H., & Zou, H. (2020). Hydrophobically modified water-based polymer for slow-release urea formulation. *Progress in Organic Coatings, 149*, 105964.

Chen, S., Yang, M., Ba, C., Yu, S., Jiang, Y., Zou, H., & Zhang, Y. (2018b). Preparation and characterization of slow-release fertilizer encapsulated by biochar-based waterborne copolymers. *Science of the Total Environment, 615*, 431–437.

Chen, X., Zhou, Y., Li, J., Pillai, S. C., Bolan, N., He, J., & Wang, H. (2023). Activated peroxydisulfate by sorghum straw-based biochar for enhanced tartrazine degradation: Roles of adsorption and radical/nonradical processes. *Environmental Pollution, 316*, 120665.

Cui, Y., Xiang, Y., Xu, Y., Wei, J., Zhang, Z., Li, L., & Li, J. (2020). Poly-acrylic acid grafted natural rubber for multi-coated slow release compound fertilizer: Preparation, properties and slow-release characteristics. *International Journal of Biological Macromolecules, 146*, 540–548.

Darby, I., Xu, C. Y., Wallace, H. M., Joseph, S., Pace, B., & Bai, S. H. (2016). Short-term dynamics of carbon and nitrogen using compost, compost-biochar mixture and organo-mineral biochar. *Environmental Science and Pollution Research, 23*, 11267–11278.

Das, S. K., & Ghosh, G. K. (2021). Developing biochar-based slow-release NPK fertilizer for controlled nutrient release and its impact on soil health and yield. *Biomass Conversion and Biorefinery*, 1–13.

Diatta, A. A., Thomason, W. E., Abaye, O., Thompson, T. L., Battaglia, M. L., Vaughan, L. J., & Filho, J. F. (2020). Assessment of nitrogen fixation by mungbean genotypes in different soil textures using ^{15}N natural abundance method. *Journal of Soil Science and Plant Nutrition, 20*, 2230–2240.

Dominguez, E. L., Uttran, A., Loh, S. K., Manero, M. H., Upperton, R., Tanimu, M. I., & Bachmann, R. T. (2020). Characterisation of industrially produced oil palm kernel shell biochar and its potential as slow release nitrogen-phosphate fertilizer and carbon sink. *Materials Today: Proceedings, 31*, 221–227.

Dong, D., Wang, C., Van Zwieten, L., Wang, H., Jiang, P., Zhou, M., & Wu, W. (2020). An effective biochar-based slow-release fertilizer for reducing nitrogen loss in paddy fields. *Journal of Soils and Sediments, 20*, 3027–3040.

El Sharkawi, H. M., Tojo, S., Chosa, T., Malhat, F. M., & Youssef, A. M. (2018). Biochar-ammonium phosphate as an uncoated-slow release fertilizer in sandy soil. *Biomass and Bioenergy, 117*, 154–160.

Elhassani, C. E., Essamlali, Y., Aqlil, M., Nzenguet, A. M., Ganetri, I., & Zahouily, M. (2019). Urea-impregnated HAP encapsulated by lignocellulosic biomass-extruded composites: A novel slow-release fertilizer. *Environmental Technology & Innovation, 15*, 100403.

Frišták, V., Pipíška, M., & Soja, G. (2018). Pyrolysis treatment of sewage sludge: A promising way to produce phosphorus fertilizer. *Journal of Cleaner Production, 172*, 1772–1778.

Gilbert, P., Ryu, C., Sharifi, V., & Swithenbank, J. (2009). Effect of process parameters on pelletisation of herbaceous crops. *Fuel, 88*(8), 1491–1497.

Giroto, A. S., Guimarães, G. G., Foschini, M., & Ribeiro, C. (2017). Role of slow-release nanocomposite fertilizers on nitrogen and phosphate availability in soil. *Scientific Reports, 7*(1), 46032.

Gwenzi, W., Nyambishi, T. J., Chaukura, N., & Mapope, N. (2018). Synthesis and nutrient release patterns of a biochar-based N–P–K slow-release fertilizer. *International Journal of Environmental Science and Technology, 15*, 405–414.

Hossain, M. Z., Bahar, M. M., Sarkar, B., Donne, S. W., Ok, Y. S., Palansooriya, K. N., & Bolan, N. (2020). Biochar and its importance on nutrient dynamics in soil and plant. *Biochar, 2*, 379–420.

Jin, Z., Chen, C., Chen, X., Hopkins, I., Zhang, X., Han, Z., & Billy, G. (2019). The crucial factors of soil fertility and rapeseed yield-A five year field trial with biochar addition in upland red soil, China. *Science of the Total Environment, 649*, 1467–1480.

Kammann, C. I., Schmidt, H. P., Messerschmidt, N., Linsel, S., Steffens, D., Müller, C., & Joseph, S. (2015). Plant growth improvement mediated by nitrate capture in co-composted biochar. *Scientific Reports, 5*(1), 11080.

Karim, A. A., Kumar, M., Singh, S. K., Panda, C. R., & Mishra, B. K. (2017). Potassium enriched biochar production by thermal plasma processing of banana peduncle for soil application. *Journal of Analytical and Applied Pyrolysis, 123*, 165–172.

Khajavi-Shojaei, S., Moezzi, A., Norouzi Masir, M., & Taghavi, M. (2020). Synthesis modified biochar-based slow-release nitrogen fertilizer increases nitrogen use efficiency and corn (*Zea mays* L.) growth. *Biomass Conversion and Biorefinery*, 1–9.

Kim, J. A., Vijayaraghavan, K., Reddy, D. H. K., & Yun, Y. S. (2018). A phosphorus-enriched biochar fertilizer from bio-fermentation waste: A potential alternative source for phosphorus fertilizers. *Journal of Cleaner Production, 196*, 163–171.

Kopittke, P. M., Lombi, E., Wang, P., Schjoerring, J. K., & Husted, S. (2019). Nanomaterials as fertilizers for improving plant mineral nutrition and environmental outcomes. *Environmental Science: Nano, 6*(12), 3513–3524.

Lam, S. S., Liew, R. K., Cheng, C. K., & Chase, H. A. (2015). Catalytic microwave pyrolysis of waste engine oil using metallic pyrolysis char. *Applied Catalysis B: Environmental, 176*, 601–617.

Lateef, A., Nazir, R., Jamil, N., Alam, S., Shah, R., Khan, M. N., & Saleem, M. (2019). Synthesis and characterization of environmental friendly corncob biochar based nano-composite—A potential slow release nano-fertilizer for sustainable agriculture. *Environmental Nanotechnology, Monitoring & Management, 11*, 100212.

Li, Y., Sun, Y., Liao, S., Zou, G., Zhao, T., Chen, Y., & Zhang, L. (2017). Effects of two slow-release nitrogen fertilizers and irrigation on yield, quality, and water-fertilizer productivity of greenhouse tomato. *Agricultural Water Management, 186*, 139—146.

Liu, G., Zhang, X., Liu, H., He, Z., Show, P. L., Vasseghian, Y., & Wang, C. (2023). Biochar/layered double hydroxides composites as catalysts for treatment of organic wastewater by advanced oxidation processes: A review. *Environmental Research*, 116534.

Liu, X., Li, Y., Meng, Y., Lu, J., Cheng, Y., Tao, Y., & Wang, H. (2021a). Pulping black liquor-based polymer hydrogel as water retention material and slow-release fertilizer. *Industrial Crops and Products, 165*, 113445.

Liu, Z., Tian, F., An, X., Wu, Z., Li, T., & Yang, Q. (2021b). Microwave co-pyrolysis of biomass, phosphorus, and magnesium for the preparation of biochar-based fertilizer: Fast synthesis, regulable structure, and graded-release. *Journal of Environmental Chemical Engineering, 9*(6), 106456.

Lou, Z., Sun, Y., Bian, S., Baig, S. A., Hu, B., & Xu, X. (2017). Nutrient conservation during spent mushroom compost application using spent mushroom substrate derived biochar. *Chemosphere, 169*, 23—31.

Luo, D., Wang, L., Nan, H., Cao, Y., Wang, H., Kumar, T. V., & Wang, C. (2023). Phosphorus adsorption by functionalized biochar: A review. *Environmental Chemistry Letters, 21*(1), 497—524.

Luo, W., Qian, L., Liu, W., Zhang, X., Wang, Q., Jiang, H., & Wu, Z. (2021). A potential Mg-enriched biochar fertilizer: Excellent slow-release performance and release mechanism of nutrients. *Science of the Total Environment, 768*, 144454.

Lustosa Filho, J. F., da Silva Carneiro, J. S., Barbosa, C. F., de Lima, K. P., do Amaral Leite, A., & Melo, L. C. A. (2020). Aging of biochar-based fertilizers in soil: Effects on phosphorus pools and availability to *Urochloa brizantha* grass. *Science of the Total Environment, 709*, 136028.

Lyu, T., Shen, J., Ma, J., Ma, P., Yang, Z., Dai, Z., & Li, M. (2021). Hybrid rice yield response to potted-seedling machine transplanting and slow-release nitrogen fertilizer application combined with urea topdressing. *The Crop Journal, 9*(4), 915—923.

Marcińczyk, M., & Oleszczuk, P. (2022). Biochar and engineered biochar as slow-and controlled-release fertilizers. *Journal of Cleaner Production, 339*, 130685.

Mostafa, M. E., Hu, S., Wang, Y., Su, S., Hu, X., Elsayed, S. A., & Xiang, J. (2019). The significance of pelletization operating conditions: An analysis of physical and mechanical characteristics as well as energy consumption of biomass pellets. *Renewable and Sustainable Energy Reviews, 105*, 332—348.

Nam, W. L., Phang, X. Y., Su, M. H., Liew, R. K., Ma, N. L., Rosli, M. H. N. B., & Lam, S. S. (2018). Production of bio-fertilizer from microwave vacuum pyrolysis of palm kernel shell for cultivation of Oyster mushroom (*Pleurotus ostreatus*). *Science of the Total Environment, 624*, 9—16.

Nan, H., Wang, L., Luo, D., Zhang, Y., Liu, G., & Wang, C. (2023). A bibliometric analysis of biochar application in wastewater treatment from 2000 to 2021. *International Journal of Environmental Science and Technology*, 1—18.

Nardi, P., Ulderico, N. E. R. I., Di Matteo, G., Trinchera, A., Napoli, R., Farina, R., & Benedetti, A. (2018). Nitrogen release from slow-release fertilizers in soils with different microbial activities. *Pedosphere, 28*(2), 332—340.

Oh, T. K., Shinogi, Y., Lee, S. J., & Choi, B. (2014). Utilization of biochar impregnated with anaerobically digested slurry as slow-release fertilizer. *Journal of Plant Nutrition and Soil Science, 177*(1), 97—103.

Peng, X. Y. L. L., Ye, L. L., Wang, C. H., Zhou, H., & Sun, B. (2011). Temperature-and duration-dependent rice straw-derived biochar: Characteristics and its effects on soil properties of an Ultisol in southern China. *Soil and Tillage Research, 112*(2), 159—166.

Piash, M. I., Iwabuchi, K., & Itoh, T. (2022). Synthesizing biochar-based fertilizer with sustained phosphorus and potassium release: Co-pyrolysis of nutrient-rich chicken manure and Ca-bentonite. *Science of the Total Environment, 822*, 153509.

Qiu, M., Sun, K., Jin, J., Han, L., Sun, H., Zhao, Y., & Xing, B. (2015). Metal/metalloid elements and polycyclic aromatic hydrocarbon in various biochars: The effect of feedstock, temperature, minerals, and properties. *Environmental Pollution, 206*, 298—305.

Ramli, R. A. (2019). Slow release fertilizer hydrogels: A review. *Polymer Chemistry, 10*(45), 6073—6090.

Rashid, M., Hussain, Q., Khan, K. S., Alwabel, M. I., Hayat, R., Akmal, M., & Alvi, S. (2021). Carbon-based slow-release fertilizers for efficient nutrient management: Synthesis, applications, and future research needs. *Journal of Soil Science and Plant Nutrition, 21*, 1144—1169.

Roberts, D. A., Paul, N. A., Dworjanyn, S. A., Bird, M. I., & de Nys, R. (2015). Biochar from commercially cultivated seaweed for soil amelioration. *Scientific Reports*, *5*(1), 9665.

Roy, A., Chaturvedi, S., Singh, S., Govindaraju, K., Dhyani, V. C., & Pyne, S. (2022). Preparation and evaluation of two enriched biochar-based fertilizers for nutrient release kinetics and agronomic effectiveness in direct-seeded rice. *Biomass Conversion and Biorefinery*, *14*, 2007–2018. Available from https://doi.org/10.1016/j.scitotenv.2023.165176.

Saha, B. K., Rose, M. T., Van Zwieten, L., Wong, V. N., & Patti, A. F. (2021). Slow release brown coal-urea fertilizer potentially influences greenhouse gas emissions, nitrogen use efficiency, and sweet corn yield in oxisol. *ACS Agricultural Science & Technology*, *1*(5), 469–478.

Samoraj, M., Mironiuk, M., Witek-Krowiak, A., Izydorczyk, G., Skrzypczak, D., Mikula, K., & Chojnacka, K. (2022). Biochar in environmental friendly fertilizers-Prospects of development products and technologies. *Chemosphere*, *296*, 133975.

Shang, A., Yang, K., Lu, Y., Jia, Q., Li, Z., Ma, G., & Mu, J. (2022). A novel slow-release fertilizer derived from itaconic acid–modified biochar: Synthesis, characteristics, and applications in cucumber seedlings. *Journal of Soil Science and Plant Nutrition*, *22*(4), 4616–4626.

Shi, W., Ju, Y., Bian, R., Li, L., Joseph, S., Mitchell, D. R., & Pan, G. (2020). Biochar bound urea boosts plant growth and reduces nitrogen leaching. *Science of the Total Environment*, *701*, 134424.

Sim, D. H. H., Tan, I. A. W., Lim, L. L. P., & Hameed, B. H. (2021). Encapsulated biochar-based sustained release fertilizer for precision agriculture: A review. *Journal of Cleaner Production*, *303*, 127018.

Singh, S., Chaturvedi, S., Dhyani, V. C., & Govindaraju, K. (2020). Pyrolysis temperature influences the characteristics of rice straw and husk biochar and sorption/desorption behavior of their biourea composite. *Bioresource Technology*, *314*, 123674.

Singh, S., Chaturvedi, S., Nayak, P., Dhyani, V. C., Nandipamu, T. M., Singh, D. K., ... Govindaraju, K. (2023). Carbon offset potential of biochar based straw management under rice – wheat system along Indo-Gangetic Plains of India. *Science of the Total Environment*, *897*, 165176.

Spokas, K. A., Novak, J. M., & Venterea, R. T. (2012). Biochar's role as an alternative N-fertilizer: Ammonia capture. *Plant and Soil*, *350*, 35–42.

Sun, H., Lu, H., & Feng, Y. (2019). Greenhouse gas emissions vary in response to different biochar amendments: An assessment based on two consecutive rice growth cycles. *Environmental Science and Pollution Research*, *26*, 749–758.

Tarafder, C., Daizy, M., Alam, M. M., Ali, M. R., Islam, M. J., Islam, R., & Khan, M. Z. H. (2020). Formulation of a hybrid nanofertilizer for slow and sustainable release of micronutrients. *ACS Omega*, *5*(37), 23960–23966.

Tomczyk, B., Siatecka, A., Gao, Y., Ok, Y. S., Bogusz, A., & Oleszczuk, P. (2020). The convertion of sewage sludge to biochar as a sustainable tool of PAHs exposure reduction during agricultural utilization of sewage sludges. *Journal of Hazardous Materials*, *392*, 122416.

Wang, C. (2021). *Production of biochar from renewable resources. Advanced technology for the conversion of waste into fuels and chemicals* (pp. 273–287). Woodhead Publishing.

Wang, C., Huang, R., Sun, R., Yang, J., & Sillanpää, M. (2021a). A review on persulfates activation by functional biochar for organic contaminants removal: Synthesis, characterizations, radical determination, and mechanism. *Journal of Environmental Chemical Engineering*, *9*(5), 106267.

Wang, C., Luo, D., Zhang, X., Huang, R., Cao, Y., Liu, G., & Wang, H. (2022). Biochar-based slow-release of fertilizers for sustainable agriculture: A mini review. *Environmental Science and Ecotechnology*, *10*, 100167.

Wang, D., Li, C., Parikh, S. J., & Scow, K. M. (2019). Impact of biochar on water retention of two agricultural soils–A multi-scale analysis. *Geoderma*, *340*, 185–191.

Wang, Y., Zheng, K., Zhan, W., Huang, L., Liu, Y., Li, T., & Wang, Z. (2021b). Highly effective stabilization of Cd and Cu in two different soils and improvement of soil properties by multiple-modified biochar. *Ecotoxicology and Environmental Safety*, *207*, 111294.

Wen, P., Wu, Z., Han, Y., Cravotto, G., Wang, J., & Ye, B. C. (2017). Microwave-assisted synthesis of a novel biochar-based slow-release nitrogen fertilizer with enhanced water-retention capacity. *ACS Sustainable Chemistry & Engineering*, *5*(8), 7374–7382.

Xiang, A., Gao, Z., Zhang, K., Jiang, E., Ren, Y., & Wang, M. (2021). Study on the Cd (II) adsorption of biochar based carbon fertilizer. *Industrial Crops and Products*, *174*, 114213.

Xiao, J., Hu, R., & Chen, G. (2020). Micro-nano-engineered nitrogenous bone biochar developed with a ball-milling technique for high-efficiency removal of aquatic Cd (II), Cu (II) and Pb (II). *Journal of Hazardous Materials*, *387*, 121980.

Xie, L., Liu, M., Ni, B., Zhang, X., & Wang, Y. (2011). Slow-release nitrogen and boron fertilizer from a functional superabsorbent formulation based on wheat straw and attapulgite. *Chemical Engineering Journal, 167*(1), 342–348.

Xin, X., Judy, J. D., Sumerlin, B. B., & He, Z. (2020). Nano-enabled agriculture: From nanoparticles to smart nano-delivery systems. *Environmental Chemistry, 17*(6), 413–425.

Xu, G., Sun, J., Shao, H., & Chang, S. X. (2014). Biochar had effects on phosphorus sorption and desorption in three soils with differing acidity. *Ecological Engineering, 62*, 54–60.

Ye, Z., Zhang, L., Huang, Q., & Tan, Z. (2019). Development of a carbon-based slow release fertilizer treated by bio-oil coating and study on its feedback effect on farmland application. *Journal of Cleaner Production, 239*, 118085.

Yu, Z., Zhao, J., Hua, Y., Li, X., Chen, Q., & Shen, G. (2021). Optimization of granulation process for binder-free biochar-based fertilizer from digestate and its slow-release performance. *Sustainability, 13*(15), 8573.

Zhang, J., Chen, Q., & You, C. (2016). Biochar effect on water evaporation and hydraulic conductivity in sandy soil. *Pedosphere, 26*(2), 265–272.

Zhang, S., Yang, M., Meng, S., Yang, Y., Li, Y. C., & Tong, Z. (2022a). Biowaste-derived, nanohybrid-reinforced double-function slow-release fertilizer with metal-adsorptive function. *Chemical Engineering Journal, 450*, 138084.

Zhang, X., Davidson, E. A., Mauzerall, D. L., Searchinger, T. D., Dumas, P., & Shen, Y. (2015). Managing nitrogen for sustainable development. *Nature, 528*(7580), 51–59.

Zhang, Y., Cheng, X., Wang, Z., Tahir, M. H., & Wang, M. (2022b). Co-pyrolysis of peanut shell with phosphate fertilizer to improve carbon sequestration and emission reduction potential of biochar. *Fuel Processing Technology, 236*, 107435.

Zhou, Z., Du, C., Li, T., Shen, Y., Zeng, Y., Du, J., & Zhou, J. (2015). Biodegradation of a biochar-modified waterborne polyacrylate membrane coating for controlled-release fertilizer and its effects on soil bacterial community profiles. *Environmental Science and Pollution Research, 22*, 8672–8682.

CHAPTER 15

Microbial dynamics and carbon stability under biochar-amended soils

Shreyas Bagrecha[1], Kadagonda Nithinkumar[2], Nilutpal Saikia[2], Ram Swaroop Meena[3], Artika Singh[4] and Shiv Vendra Singh[4]

[1]Agronomy Section, ICAR—National Dairy Research Institute, Karnal, Haryana, India
[2]Division of Agronomy, ICAR—Indian Agricultural Research Institute, Pusa, New Delhi, India [3]Department of Agronomy, Institute of Agricultural Sciences, Banaras Hindu University, Varanasi, Uttar Pradesh, India [4]Department of Agronomy, College of Agriculture, Rani Lakshmi Bai Central Agricultural University, Jhansi, Uttar Pradesh, India

15.1 Introduction

The pyrolysis process involves converting various carbon-based biowastes, such as woody biomass, agricultural leftovers, animal carcasses, and biosolids, which produce biochar-based soil supplements. During pyrolysis, organic wastes undergo thermal decomposition, forming syngas, biooil, and biochar. Among these products, biochar serves as a reliable means of carbon sequestration and can be utilized as a soil amendment to enhance its properties (Bolan et al., 2023). Soil organic carbon is vital to soil fertility and health through enhancing soil structure, nutrient cycling and water-holding capacity. Biochar increases soil organic carbon content, lowers greenhouse gas emissions, and increases crop yields when applied to the soil. Since the present carbon dioxide (CO_2) concentration is over 400 ppm and is expected to grow to up to 1500 ppm, the loss in soil organic carbon will be directly impacted by the rising CO_2 concentration. According to a study, adding biochar would increase annual soil organic carbon content by 369.8–556.6 kg ha^{-1} (31.8%–47.8%) (Kuzyakov, Subbotina, Chen, Bogomolova, & Xu, 2009). The enhancing effect on soil health is unsurprising given that biochar has been used for millennia to improve agricultural soils. Terra Preta, commonly known as *Terra Preta de Indio*, is a Brazilian *oxisol* transformed anthropogenically by Amazonian tribes. In contrast to the incidental occurrence of charred residues from forest cutting and burning (Novak et al., 2009), it contains a large concentration of charcoal that was applied purposefully by pre-Columbian and Amerindian people (Kim, Sparovek, Longo, De Melo, & Crowley, 2007; Pradhan, Meena, Kumar, & Lal, 2023). Soil health and fertility are also dependent on the microbial community. Bacteria, fungi and other microbes play critical roles in nutrient cycling, organic matter breakdown and soil structure formation. Delgado-Baquerizo et al. (2016) highlight the significance of the correlation between microbial diversity and ecosystem multifunctionality. The study reveals that higher levels of microbial diversity in soil are linked to enhanced ecosystem functioning, specifically in terms of nutrient cycling and organic matter decomposition. The study highlights the importance of preserving and promoting microbial diversity in soils to maintain healthy and productive ecosystems. Singh, Chaturvedi, and Datta (2019) reported that biochar has the ability to provide a favorable environment for soil microbes. This is due to its large surface area and hydrophobic properties, which are known to promote increased microbial activity, improved nutrient retention, and enhanced microbial diversity. Similarly, another study highlights that by enhancing soil aeration, cation exchange capacity, and nutrient bioavailability, enzyme activity enhances microbial diversity that also thermochemically decomposed biomass-based biochar under problematic soils (Meena, Pradhan, Kumar, & Lal, 2023; Singh, Luthra et al., 2023).

Since investigations of plant communities led to the hypothesis that resource availability is a significant determinant of plant productivity or biomass, the link between biodiversity and biomass has been a topic of ongoing controversy in ecology (Warren, Topping, & James, 2009). The diversity and biomass of soil microbial communities play a crucial role in regulating essential ecosystem processes, including organic matter breakdown, the cycling of nitrogen, and gaseous fluxes (Delgado-Baquerizo et al., 2020). However, our knowledge of the biotic and abiotic variables influencing the variety and biomass of soil microorganisms is constantly expanding (Bastida et al., 2021; Pradhan & Meena, 2023a, 2023b). Despite the intricate advantages of biochar application, much remains to discover about the processes underpinning biochar's impacts on soil microbial dynamics and carbon stability. However, research has increased in recent years; there are still several gaps in our understanding of this field. This presents a significant obstacle in the comparison of findings across various studies and impedes the advancement of a holistic comprehension of the underlying mechanisms governing the impacts of biochar on soil characteristics. This limitation arises from the scarcity of data regarding the influence of biochar on diverse soil types and under varying climatic circumstances. Most research endeavors have been carried out over brief timeframes, leaving the long-term persistence of biochar effects still to be fully ascertained. At the same time, understanding biochar's impacts on the microbial population and the interactions between soil microorganisms and carbon cycling is thus critical for optimizing biochar application to enhance soil health (Lehmann & Joseph, 2015; Meena & Pradhan, 2023). Understanding the past research and future needs in this chapter is of utmost importance, which dwells into the complex interaction of biochar, microbial dynamics and carbon stability. The chapter begins with an overview of the microbial communities in soil, including their significance in various biogeochemical processes, influence on the microbial environment, change in microbial abundance, and diversity and potential pathways. Additionally, it covers the function of biochar in carbon sequestration, effects and modifications to carbon stability, the interaction between carbon stability and microbial dynamics, and in the end, a conclusion and perspective.

15.2 Microbial dynamics in biochar-amended soils

15.2.1 Overview of the microbial community in soils

The soil is an extensive ecosystem that harbors a diverse array of microorganisms, and the proper functioning of several soil processes is reliant upon the activities of these microbes. The soil microbial community encompasses a diverse assemblage of microorganisms, including bacteria, fungi, archaea, protozoa, and viruses. The most numerous and varied category of soil microorganisms is bacteria. Many soil bacteria are advantageous to plants, assisting nutrient absorption and protecting them from pathogens (Babalola, 2010). Fungi are another critical group of soil microorganisms with diverse functions. Fungi can form mutualistic associations with plant roots, known as mycorrhizae, which enhance plant nutrient uptake and water absorption (Pradhan & Meena, 2023a, 2023b; Van Der Heijden, Bardgett, & Van Straalen, 2008). Archaea may be found in soil but with less research than bacteria and fungi. Certain archaea are involved in many soil activities such as nitrogen cycling and methane generation (Ishii, Seishi, Kiwamu, & Keishi, 2011). Protozoa, single-celled organisms are essential predators in soil ecosystems, feeding on bacteria and other microbes. Viruses are also present in soils, although they are much less understood than other microorganisms. Overall, the soil microbial community is incredibly diverse and complex and understanding the interactions between microorganisms and their environment is essential for maintaining healthy soil ecosystems. Several factors, including biotic, abiotic, soil treatments and management factors, play deterministic roles in the diversity of soil microbes, and they perform various vital functions in the soil (Fig. 15.1). Hale, Luth, Kenney, and Crowley (2014) identified several techniques employed for the evaluation of microbial community and structure, including ergosterol removal, real-time quantitative polymerase chain reaction, fluorescent in situ hybridization, phospholipid fatty acid quantification, molecular fingerprinting of 16S rRNA gene fragments (specifically final fragment length polymorphism), high-throughput sequencing (also referred to as next-generation sequencing), and soil microscopic analysis.

15.2.2 Biochar impact on soil microbial communities

The impact of biochar on soil microbial populations may vary depending on many parameters, such as the kind of biochar feedstock, application rate, soil type, and management practices. These variables can contribute to biochar's beneficial and detrimental impacts on soil microorganisms. Delgado-Baquerizo et al. (2020) have shown a noteworthy correlation between soil organic carbon content and the diversity of microorganisms. There have been reports indicating that the application of biochar triggers an increase in microbial biomass and activity. Additionally, it has been shown that biochar can enhance microbial diversity and induce changes in the composition of microbial communities, favoring the proliferation of more favorable microbial groups. The possible influence of biochar on soil biotic communities may be attributed to its capacity to safeguard microorganisms from predation and desiccation, as highlighted by Lehmann, Gaunt, and Rondon (2006). Regarding soil pH, McCormack, Ostle, Bardgett, Hopkins, and Vanbergen (2013) conducted a study that investigated the potential of biochar to enhance the abundance of

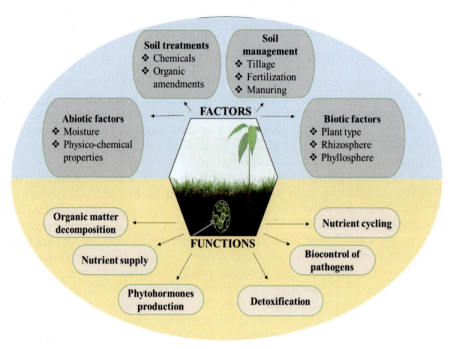

FIGURE 15.1 Factors influencing and functions of soil microbes.

bacteria that thrive in alkaline conditions, such as *Actinobacteria* and *Firmicutes* (Van Zwieten et al., 2010). Multiple studies have shown that biochar can potentially limit the proliferation of basidiomycetes and other fungal species that exhibit a preference for acidic conditions (Wu et al., 2023). The use of biochar as an energy source or nutrient may influence changes in community composition. The introduction of biochar into soil ecosystems has the potential to alter the dynamics of communication among soil organisms. This is primarily attributed to the ability of biochar to absorb signaling molecules, hence affecting the composition and abundance of soil microbial populations. Additionally, biochar may release various bioactive compounds and nutrients, further impacting the interactions within the soil microbial community (Masiello et al., 2013; Pradhan & Meena, 2022). The pathways above are not mutually exclusive and are expected to result in simultaneous alterations in soil microbial communities.

15.2.3 Biochar induced changes in microbial activity and diversity

The use of biochar in soil has been shown to have a wide range of effects on microbial activity and diversity. Different feedstocks, their pyrolysis temperature, period, soil type, and their influence are provided in Table 15.1 (Roy, Meena, Kumar, Jhariya, & Pradhan, 2021; Weralupitiya et al., 2022). It increases soil microbial activity and diversity by providing a habitat for microbes and enhancing nutrient retention in the soil. Soil biological and chemical processes rely mainly on soil microorganisms to maintain terrestrial ecosystems'

TABLE 15.1 The impact of various biochar additives on the diversity of soil microorganisms.

Feedstock	Pyrolysis temp. (°C)	Application rate	Period	Soil	Effect on microbial diversity	References
Swine manure	300	30 g/kg	240 days	Psammaquent soil and Argiustoll soil	The *Actinobacteria* phylum exhibited dominance with a prevalence of 50.1%, while the *Proteobacteria* phylum followed with a proportion of 18.2%. Additionally, the *Chloroflexi* phylum accounted for 13.0% of the observed population, while the *Acidobacteria* phylum constituted 8.8% of the total composition.	Dai, Barberán, Li, Brookes, and Xu (2017)
Corn stem	500–600	0%, 2%, 3% and 4%	36 months	Black soil	The presence of *Fusarium* and *Ustilago* exhibited a reduction subsequent to the introduction of biochar amendment. The addition of biochar resulted in alterations to the beta diversity of the fungus.	Yao et al. (2017)
Rice husk biochar	250–300	10 t ha^{-1}	130 days	Sandy soil	The soil microbial biomass of carbon, nitrogen, and phosphorus exhibited a notable increase.	Singh, Tiwari, Gupta, and Singh (2018)
Bamboo	350–400	5–20 t ha^{-1}	28 months	Sandy soil	In all biochar combinations, the bacterial community exhibited a dominance of *Proteobacteria* (30%), with *Acidobacteria* (20%), *Actinobacteria* (14%), and *Firmicutes* (10%) following suit.	Herrmann et al. (2019)
Cassava stem	400	30 t ha^{-1}	12 months	Ultisol	The addition of biochar to soils resulted in a considerable increase in microbial biomass carbon and nitrogen.	Ullah et al. (2020)
Cunninghamia lanceolate leaf and wood chips	300	1% and 3%	80 days	Red loam soil	The quantity of phosphorus-solubilizing bacteria increased and was promoted by *Acidobacteria*, *Chloroflexi*, *Gemmatimonadetes*, *Bacteroidetes*, *Firmicutes*, and *Ascomycota*.	Zhou et al. (2020)
Swine manure	700	3%	14 days	Sandy clay loam	The quantities of actinomycetes, gram-positive bacteria, and gram-negative bacteria have been seen to exhibit concurrent increases of 33.9%, 28.8%, and 69.7% respectively.	Chen et al. (2021)
Forest logging residue	500	–	24 months	Sandy loam calcareous soil	The microbial diversity in soil was shown to be much more significant in the presence of biochar compared to soil without biochar amendment. The most prevalent bacterial phylum seen was *Actinobacteria*, with *Proteobacteria* being the subsequent most abundant.	Yan, Xue, Zhou, and Wu (2021)
Wheat straw	–	20 t ha^{-1}	24 months	Planosols	Biochar amendment increased bacterial communities' richness and Shannon index	Han et al. (2022)

(*Continued*)

TABLE 15.1 (Continued)

Feedstock	Pyrolysis temp. (°C)	Application rate	Period	Soil	Effect on microbial diversity	References
					by an average of 33% and 10%, respectively.	
Pine and spruce	550	3 t ha^{-1}	36 months	Sandy loam	Microbial biomass carbon was highest in biochar added with manure and fertilizer.	Jiang, Mechler, and Oelbermann (2023)
Sewage sludge	600	1%	5 weeks	Luvisol	The bacterial communities exhibited alterations in response to the rich in nutrients sewage sludge biochar at many levels of assessment, involving alpha diversity, beta diversity, and proportions of high taxonomic ranks.	Efthymiou, Nunes, Jensen, and Jakobsen (2023)
Corn stover	550	1%	–	–	*Actinobacteria, Chloroflexbacteria, Patellobacteria, Saccharimonadales, Arthrobacter*, and *Bacillus* abundance enhanced in biochar-added soils.	Ma, Shao, Ai, Chen, and Zhang (2023)
Date palm residue	350	5% and 10%	1 year	–	The use of biochar resulted in a reduction in fungal abundance and analysis of differentially abundant taxa indicated that the changes in the microbial community composition mainly were attributed to the presence of a select few microorganisms known for their ability to promote plant growth.	Khan, Salman, and Khan (2023)

stability and ecological functions. It enhances the number and vibrancy of microorganisms in the soil, promoting soil fertility and facilitating the cycling of nutrients. Furthermore, biochar has been shown to improve soil structure and increase water retention capacity, producing conducive conditions for the growth and activity of microorganisms. Certain biochar-amended soils have also been proven to enhance the rate of microbial reproduction. For development and metabolism, chemoheterotrophic soil organisms need nutrients and carbon substrates. Energy and cell carbon are both provided by carbon-based substrates for biosynthesis. Biochar adsorbent properties may influence the availability of critical nutrient components and soil carbon substrates' quality and quantity (Swagathnath, Rangabhashiyam, Murugan, & Balasubramanian, 2019).

15.2.3.1 Bacteria

Its effect on bacterial populations in soil is a significant part of its effects. Biochar amendments have been demonstrated to considerably impact bacterial diversity, abundance and community structure. According to several researchers, biochar increases the bacterial population. Biederman, Harpole, and Hartman (2013) conducted a study whereby they observed that introducing biochar resulted in an augmentation of beneficial microorganisms involved in the nutrient cycling process. These microorganisms included nitrogen-fixing bacteria and phosphate-dissolving bacteria. A similar increase in ammonia-oxidizing bacteria and archaea abundance was seen by Chen, Chen, and Grace (2018), indicating improved soil nitrification processes. In a study conducted by Herrmann et al. (2019), bamboo-based biochar was utilized, and the findings indicated that the bacterial community exhibited a predominance of *Proteobacteria*, followed by *Acidobacteria*, *Actinobacteria*, and *Firmicutes* with 30%, 20%, 14% and 10% respectively across all biochar combinations. This observation suggests a positive correlation between these microbial abundances and favorable soil health and crop yield outcomes. The good impacts of biochar on soil microbial diversity were unveiled in a worldwide metaanalysis conducted by Xu et al. (2023). The results of the study indicated that *Acidobacteria* and *Gemmatimonadetes* exhibited distinct responses to the application of biochar. Specifically, *Acidobacteria* had a mostly adverse reaction, whereas *Gemmatimonadetes* displayed a favorable response. The alterations seen in *Acidobacteria* and *Gemmatimonadetes* were shown to be impacted by the characteristics of biochar as well as the prevailing soil conditions. It is worth mentioning that *Acidobacteria* exhibited a drop in abundance as the biochar loading rose, but *Gemmatimonadetes* showed an increase in abundance in response to more excellent soil carbon−nitrogen ratios. Furthermore, there was a progressive rise in bacterial and carbon richness in response to variations in soil pH and biochar load.

Biochar has also been demonstrated to influence the composition and organization of bacterial communities. Recent research has shown changes in the organization of the soil microbial community and movements toward bacterial dominated populations. Zhang et al. (2018) demonstrated that the application of biochar in drip irrigated desert soil led to alterations in the composition of the bacterial community. These changes were shown to promote the proliferation of certain beneficial bacterial taxa that are associated with processes such as nutrient cycling and interactions between plants and microbes. Biochar was shown to enhance microbial diversity and cause changes in the complexity of bacterial networks in clayey soil, examined by Zhang et al. (2020), indicating differences in microbial interactions and possible functional dynamics. Anderson et al. (2011) investigated the effects of biochar-amended soils compared to control soils. The researchers observed time based fluctuations in bacterial family abundances, focusing on those that exceeded 5%. These included *Bradyrhizobiaceae* (approximately 8%), *Hyphomicrobiaceae* (about 14%), *Streptosporangineae* (approximately 6%), and *Thermomonosporaceae* (about 8%). The presence of biochar positively impacted these bacterial families, either by encouraging an increase in abundance or mitigating the extent of their decline. On the other hand, the bacterial families *Streptomycetaceae* (approximately 11%) and *Micromonosporaceae* (about 7%) experienced an adverse effect from the biochar, reducing their abundance.

15.2.3.2 Fungi

The addition of biochar has been shown to have diverse effects on soil microbial populations. Research findings indicate that the use of biochar has been shown to enhance both the diversity and quantity of fungi. In a study conducted by Sun, Smith, and Johnson (2019), it was revealed that the application of biochar to agricultural soils resulted in an increase in fungal biomass and alterations in the composition of fungal communities. The physical attributes of biochar, including its porosity and elevated carbon content, provide a favorable environment conducive to the colonization and growth of fungus. The use of biochar has been shown to enhance the growth rate of mycorrhizal fungi, which establish symbiotic associations with plant roots and improve nutrient uptake. Moreover, previous studies have shown the influence of biochar on the dynamics of fungal competition, perhaps facilitating the proliferation of advantageous fungal species that contribute to the efficient cycling of nutrients and breakdown of organic materials. Zhou et al. (2020) used biochar derived from *Cunninghamia lanceolate* leaf and wood chips. Their research findings demonstrated a notable augmentation in the population of *Firmicutes* and *Ascomycota*, leading to an enhanced abundance of phosphorus-solubilizing bacteria. According to Lehmann et al. (2011), using biochar improved the diversity and balance of the fungal community in soil. The porous structure and extensive surface area of biochar provide an optimal habitat for the colonization and proliferation of fungi. Additionally, it has been discovered that biochar additions change the makeup of microbial communities, favoring the development of specific helpful fungal taxa. The beneficial effects of biochar on fungi populations demonstrate the treasure of this material to enhance the diversity and efficiency of soil dwelling fungi, eventually improving soil health and plant yield. Adding biochar to soil can positively impact fungal populations, promoting their abundance, diversity and functional roles in ecosystem processes.

15.2.3.3 Actinomycetes

Actinomycetes, a family of filamentous bacteria, play a crucial role in decomposing organic matter, producing bioactive compounds, and recycling nutrients within soil ecosystems. The presence of biochar has a notable impact on the actinomycetes population in soil. Multiple studies have shown that the company and variety of actinomycetes significantly increase after the application of biochar amendments. An example of this may be seen in a study conducted by Smith, Santangelo, and Gupta (2014), where it was shown that the introduction of biochar into agricultural soils resulted in a notable augmentation of actinomycetes populations. This finding suggests that biochar fosters a favorable milieu for the growth and proliferation of actinomycetes. In a study conducted by Wu et al. (2017), it was shown that the introduction of biochar to soils contaminated with pollutants resulted in an increase in the population of actinomycetes. These microorganisms play a crucial role in the degradation of recalcitrant organic compounds present in contaminated soils. The findings suggest that the presence of biochar might have a beneficial effect on actinomycetes populations, perhaps enhancing their functional role within soil ecosystems. In order to effectively implement sustainable soil management practices, it is essential to comprehend the impact of biochar on actinomycetes, therefore harnessing the potential benefits offered by these microorganisms.

15.2.3.4 Others

Extensive research has been conducted on the impact of biochar on fungi, bacteria, and actinomycetes. However, it is essential to note that several other microorganisms, including archaea, protozoa, algae, and viruses, may also experience alterations in their behavior and functions due to the presence of biochar. Several archaeal species, namely those involved in the cycling of nutrients and methane production, exhibit changes in their activity and abundance as a result of biochar exposure. The influence of biochar additions on the distribution and composition of archaeal populations in soil and aquatic environments has been elucidated by recent research conducted by Zhang et al. (2019). Biochar has been shown to significantly impact the variety and abundance of protozoan species within the soil ecosystem by providing them with a suitable habitat and a source of nutrition. Based on network analysis, it was shown that phagotrophs served as the keystone group within the studied ecosystem. Furthermore, it was observed that these organisms exhibited sensitivity to the introduction of biochar. According to Asiloglu et al. (2021), the primary taxa that play a crucial role in biochar derived from rice husk and chicken litter are *Conosa* (*Amoebozoa*) and *Discoba* (*Excavata*), respectively. Moreover, the use of biochar has the potential to influence algae, especially in aquatic environments. The introduction of biochar has the potential to influence the growth and composition of algal communities within marine ecosystems. The surface in question has the potential to facilitate the adhesion of algae, or alternatively, it may release nutrients that promote the proliferation of algae. Liu et al. (2020) reported a significant augmentation in the total abundance of soil nematodes as a result of biochar application. The application of biochar resulted in alterations to the community organization of nematodes, increasing the abundance of bacterivores. Furthermore, the presence of biochar positively influenced the nematode population, enhancing the soil's overall quality.

15.2.4 Mechanisms behind biochar impacts on soil microorganisms

The mechanisms behind the effects of biochar on soil microbial communities remain incompletely understood and complex. Possible tools include modifications to soil chemical and physical characteristics, the liberation of vital nutrients and bioactive compounds, and changing microbial interactions within the soil. The application of biochar has been shown to significantly impact many physical features of soil, including its ability to hold water and porosity (Pradhan et al., 2022; Singh, Chaturvedi, Dhyani, & Govindaraju, 2020). These changes in soil properties may notably affect the composition and dynamics of soil microbial communities. The use of biochar has been shown to positively impact water retention in soil and aeration, hence facilitating the growth and activity of microorganisms. Additionally, biochar can impact many soil chemical properties, including pH levels, nutrient accessibility, and the ability to exchange cations. The alterations in question possess the potential to have both direct and indirect effects on soil microorganisms. Alterations in nutrient availability have the potential to induce modifications in microbial growth and activity, whereas fluctuations in soil pH may have an influence on the composition of the microbial community (Kumawat et al., 2022; Li & Tasnady, 2023). Plant-based feedstocks' transportation of anatomical structures occurs via the xylem and phloem vessels. However, during the pyrolysis process, volatile compounds that might occupy micropores are lost, leading to a significant increase in the pore space (Mukherjee, Zimmerman, & Harris, 2011). Since the inception of biochar research, it has been widely acknowledged that micropores serve as habitats for microorganisms, shielding them from potential threats such as grazers and desiccation (Sheoran et al., 2022; Steinbeiss, Gleixner, & Antonietti, 2009). Similar to Mycorrhiza, the assessment of the host plant's reaction is often conducted by quantifying root colonization or the extent of fungal tissue inside the host. Biochar has received considerable attention in the scientific community because of its positive effects on soil enhancement. These effects include promoting carbon storage, improving soil fertility and quality, and immobilizing and transforming various pollutants, including organic compounds and heavy metals (Jeffery et al., 2015; Meena et al., 2022; Zhu, Chen, Zhu, & Xing, 2017). The impacts above are attained by the alteration of soil microbial habitats or the direct manipulation of microbial metabolisms, leading to alterations in the activity of microbes and community compositions. Research has shown that the application of biochar has an impact on the action and biomass of soil microorganisms, resulting in a modification of the ratio between bacteria and fungi, as well as alterations in the activity of soil enzymes and the structure of microbial communities (Mackie, Marhan, Ditterich, Schmidt, & Kandeler, 2015).

The study conducted by Zhu et al. (2017) examined several methods via which biochar influences microbial activity. Fig. 15.2 visually represents likely mechanisms that might help understand these effects. The methods above may be classified into two distinct categories: direct and indirect impacts. Direct influence encompasses several factors. Firstly, it acts as a habitat for soil microbes, providing them with shelter through its pore structures and surfaces. Additionally, it serves as a nutrient source for these microbes, supplying them with essential elements and ions that are adsorbed on their particles, thereby

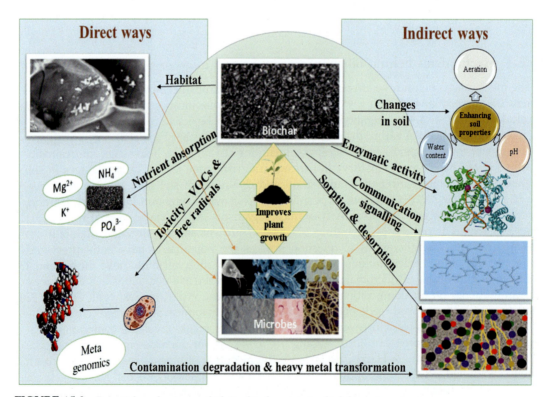

FIGURE 15.2 Potential mechanism underlying biochar on microbial dynamics.

supporting their growth. Lastly, it can potentially reduce toxicity by producing volatile organic compounds and environmentally relentless free radicals. Indirect influence encompasses various mechanisms that impact microbial habitats. These mechanisms involve the modification of soil properties essential for the development of microbial populations, such as aeration conditions, water content, and pH. Additionally, indirect influence affects enzyme activities that play a role in soil elemental cycles associated with microbial processes. It also disrupts microbial communication within and between species through the sorption and hydrolysis of signaling molecules. Furthermore, indirect influence boosts soil contaminants' sorption and degeneration, lowering their accessibility and toxicity to microbial communities. It is crucial to underscore that more experimental validation is necessary for these postulated pathways pertaining to biochar-microbe interactions. It is essential to prioritize the establishment of links between these interaction mechanisms and their corresponding environmental impacts.

15.3 Biochar and carbon stability

15.3.1 Biochar role in soil carbon sequestration

Introducing biochar into the soil matrix leads to changes in the soil's chemical, physical, and biological characteristics. Consequently, this may have an impact on the dynamics of soil organic carbon (Van Zwieten et al., 2014). The loss of soil organic carbon is a significant issue in African and Asian countries since it is closely associated with inadequate input response and poor crop yield (Raj et al., 2021; Yadav et al., 2017). Biochar has been identified as a highly effective technique for sequestering soil organic carbon in cropping soils. This is primarily attributed to its recalcitrant nature, which significantly enhances the stability of carbon in the soil over a prolonged period of time (Chan, Van Zwieten, Meszaros, Downie, & Joseph, 2007). Consequently, biochar holds great promise as a tool for sequestering more significant amounts of carbon when compared to conventional agricultural practices. The impacts of biochar on soil might vary in terms of their scope and amplitude, influenced by several variables such as the precise features of the biochar and the prevailing climatic circumstances under which it is applied (Lehmann & Joseph, 2009). For several reasons, biochar's high carbon concentration is crucial to its ability to sequester carbon. The first benefit is that it makes biochar a reliable source of carbon. When added to soil, biochar functions as a sink by efficiently absorbing and holding carbon over time, preventing its release as CO_2 again (Kumar, Meena et al., 2020; Li & Chan, 2022). During the pyrolysis process, a portion of the carbon in biochar undergoes a transformation, resulting in the formation of an insoluble aromatic structure. This transformation enhances the biochar's resistance to both abiotic and biotic degradation, according to research carried out by Cheng, Lehmann, Thies, Burton, and Engelhard (2006). Carbon in biochar is projected to have a half-life between hundreds and thousands of years (Bruun, Ambus, Egsgaard, & Hauggaard-Nielsen, 2012). The properties of biochar, such as its lability and stability, are mostly influenced by the circumstances under which it is produced by pyrolysis. These parameters include factors such as temperature, the duration of feedstock residency, and the source of the feedstock (Rehrah et al., 2014). According to Singh, Cowie, and Smernik (2012), there exists an inverse relationship between the pyrolysis temperature and the labile fraction, whereas a direct connection is seen between the pyrolysis temperature and the carbon stabilized by soil amendment. The feedstock's chemical and structural composition influences biochar's features, including its three-dimensional organization, porosity, area of surface, nutrient content, availability, and the proportion of labile and recalcitrant carbon content (Tag, Duman, Ucar, & Yanik, 2016). According to Singh et al. (2012), wood-based biochar exhibits greater resistance compared to biochar derived from plants or manure. This can be attributed to its elevated lignin concentration and reduced ash, nitrogen, and volatile matter levels. These characteristics contribute to a more significant amount of condensed and higher molecular weight carbon in wood-based biochar, resulting in an extended mean residence duration. Biochar consists of stable carbon molecules that provide a substantial contribution to the stock of stable organic carbon in the soil, as shown in Fig. 15.3. Which depicts different inflow and outflow of CO_2, uptake of CO_2 by plants and capturing it for a shorter period as a portion of it goes back into the atmosphere via biomass degradation and microbial respiration. At the same time, the anthropogenic emission of CO_2 is continually increasing and the burning of agricultural wastes in open fields directly contributes to it too, creating conditions unfavorable for all life. One alternative is biochar as discussed above, which remains an essential tool for biomass conversion, which is the stable pool. Consequently, a substantial quantity of carbon is retained inside the soil for an extended period, resulting in the sequestration of carbon.

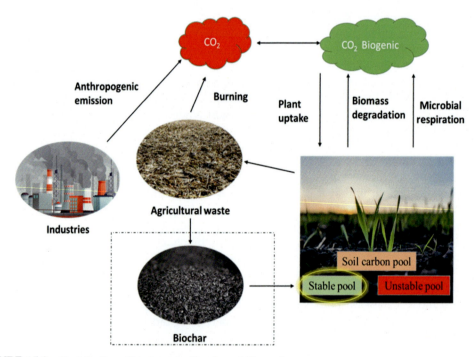

FIGURE 15.3   Contribution of biochar to soil carbon stable pool.

15.3.2 Factors affecting soil carbon stability

The concept of soil carbon stability pertains to the capacity of organic carbon in the soil to withstand degradation and subsequent depletion. Several factors, such as the texture of the soil, humidity, temperature, pH, microbial activity, and the chemical composition of soil organic matter, influence the stability of soil carbon. The significance of soil texture in determining soil carbon stability is of utmost importance, as variations in the sizes of soil particles may give rise to distinct physical and chemical characteristics that influence the process of carbon stabilization. Soils with a fine texture, such as clay soils, often have a greater capacity for carbon stabilization compared to soils with a coarse texture, such as sandy soils. This phenomenon may be attributed to the improved aeration and increased surface area relationship between biochar and soil fragments in uncultivated soils, facilitating the decomposition of soil organic carbon in the presence of sufficient oxygen (Kumar et al., 2020; Rogovska et al., 2011; Wardle, Nilsson, & Zackrisson, 2008). The study revealed that soils with a significant amount of clay, such as *Oxisol* and *Vertisol*, exhibited the lowest rate of Biochar-C mineralization. Sandy clay loam soils, such as *Entisol*, showed a somewhat lower rate, while sandy soils like *Inceptisol* had the most significant rate of biochar-C mineralization. Fang, Singh, Singh, and Krull (2014) found that the stability of biochar-C was significantly influenced by the interactions between biochar and clay in variable charge soil types, such as *Oxisol*. In contrast, soils that included permanently charged minerals, such as *Vertisol* and *Entisol*, or sandy soil types, such as *Inceptisol*, had less impact on biochar-C stability. According to the findings of Brodowski, Amelung, Haumaier, Abetz, and Zech (2005), the chemical interaction between the outermost layer of oxidized biochar and the functional groups of mineral clay and organic matter in the soil has been observed. The present exchange, in conjunction with the enduring nature of biochar-C and indigenous soil organic carbon inside organo-mineral fractions, as well as their confinement within the soil mineral matrix for an extended period, has the potential to diminish the phenomenon of carbon mineralization (Glaser, Lehmann, & Zech, 2002). Many variables contribute to the stability of carbon (Table 15.2).

In a similar vein, the study conducted by Butnan, Deenik, Toomsan, and Vityakon (2017) revealed that the characteristics of biochar exhibited interactions with soil texture and mineral composition, resulting in distinct impacts on soil parameters and the stability of soil organic carbon in both *Ultisol* and Oxisol. The coarse-textured *Ultisol* showed increased soil organic carbon stability due to recalcitrant biochar constituents, and physicochemical and biological protection of native soil organic carbon. However, *Oxisol*, which had a fine texture, exhibited a decline in the stability of soil organic carbon caused by the destruction of soil aggregates, the dispersion of clay particles and the promotion of microbial activity due to biochar-derived labile carbon and the native manganese present in the soil. According to Lehmann et al. (2021), a significant proportion of biochar samples, characterized by an H/Corg ratio below 0.5, had a carbon persistence of over 50% over a century-long period. The average carbon persistence value was found to be 82%.

The data above suggest that both biochar and soil characteristics influence the efficacy of biochar in improving soil organic carbon stability. Multiple processes may be responsible for this phenomenon. Singh et al. (2012) revealed that the material used to produce biochar affects its soil stability. According to their results, biochar generated from manure

TABLE 15.2 Factors affecting carbon stability.

Factor	Description	Influence on carbon stability
Soil type	Composition, texture, mineralogy of soil	In comparison to sandy soils, clay-rich soils often have better carbon stability. Smaller particle sizes and increased surface area can protect carbon more easily.
Organic inputs	Amount and quality of organic materials added.	Better inputs like plant residue can result in more stable carbon. While resistant materials contribute to long-term stability, easily decomposable materials may aid in the storage of carbon for the short term.
Activity of microbes	Decomposition and turnover rates by microorganisms	A faster carbon cycle may result from more excellent microbial activity, reducing stability. Microbial communities are essential for degrading organic waste, and their activity can influence the rate of carbon locked.
Temperature	Soil temperature and its effect on microbial activity.	Warmer conditions often hasten decomposition and reduce carbon stability. A temperature rise can speed up microbial metabolism, which lowers carbon stability and speeds up breakdown.
Moisture	Soil moisture levels affecting microbial processes	Moisture controls microbial activity; too dry or wet soils might be unstable. While severe dryness or saturation can impair microbial function and carbon preservation, optimal moisture levels support microbial activity.
Soil pH	Acidity/alkalinity affecting microbial community	Compared to high pH levels, neutral pH frequently supports more stable carbon. Extreme pH levels can modify microbial populations and enzymatic activity, affecting the carbon decomposition rate.
Land management	Agricultural practices, land use changes	Intensive agriculture and deforestation may reduce the stability of carbon while conservation practices strengthen it. Human activities have the potential to disturb carbon-rich soil strata, increasing carbon loss; conservation practices assist in maintaining soil carbon.
Vegetation cover	Type and density of vegetation cover	Diverse plant cover often promotes more excellent carbon stability through ongoing organic input. Plant litter and root exudates are an ongoing supply of organic material that adds to a stable carbon pool.
Soil disturbance	Physical disruption due to excavation, construction, etc.	The total stability can be decreased by disturbances that release previously stabilized carbon. Mechanical alterations can make previously shielded carbon susceptible to microbial decay, reducing strength.
Depth	Soil profile depth affecting carbon distribution	Reduced microbial activity and shelter from the elements in deeper soils may result in more stable carbon levels. Lower temperatures and increased microbial activity are frequently seen in deeper strata, favorably affecting carbon retention.
Mineral protection	Association of carbon with minerals in the soil	Interactions between carbon and minerals can improve stability by physically preventing carbon from deteriorating. Minerals and organic substances can combine to generate complexes that protect the mineral from microbial assault and lengthen its life.

and heated to 400°C had a higher mineralization level than biochar from plant materials such as leaves and wood when heated to 550°C. The former biochar had more significant non-aromatic carbon and a lower concentration of aromatic condensation. According to Chaturvedi et al. (2021), it was shown that biochar derived from eucalyptus, lantana, and pine needles exhibited a more significant proportion of fixed carbon compared to biochar obtained from straw and husk of rice, maize stover, and sugarcane trash. According to Chen et al. (2022), an increased lignin concentration in biomass feedstock leads to elevated levels of aromaticity and enhanced carbon stability.

On top of that, it is worth noting that the elevated concentrations of mineral nutrients found in manure-derived biochar could potentially lead to impairments in the fundamental carbon structures within the biochar. Consequently, this may lead to a reduction in the quantity of directly linked organic forms and subsequently result in a decline in the carbon stability of the biochar when compared to biochar derived from plants. Alterations in soil pH have the potential to induce modifications in the composition of microbial communities and the chemical composition of soil organic matter, hence exerting an impact on the stability of soil carbon. The process of carbon stabilization tends to exhibit lower levels in acidic soils compared to alkaline soils. The importance of carbon's resistance to breakdown in soil organic carbon turnover is somewhat less significant compared to the capacity of soil microbes to acquire and assimilate organic carbon (Dungait, Hopkins, Gregory, & Whitmore, 2012). According to Sulman, Phillips, Oishi, Shevliakova, and Pacala (2014), the organic carbon produced by soil microorganisms readily adheres to mineral particles within the soil, resulting in enhanced stability and reduced susceptibility to degradation. Therefore, it is regarded as a constituent of stable soil organic carbon (Liang, Schimel, & Jastrow, 2017). Leng and Huang (2018) have posited that the soil stability of biochar is primarily influenced by its composition and structure, including characteristics such as the quantity of organic matter dissolved, level of aromaticity, and level of aromatic condensation. The attributes above are mostly influenced by the biochar pyrolysis temperature used during biochar production. In general, biochar with elevated degrees of aromaticity and aromatic condensing exhibits enhanced resistance to both biotic and abiotic oxidation, hence leading to heightened stability within soil environments.

15.3.3 Biochar induced changes in soil carbon stability

The impact of biochar on enhancing the stability of soil carbon has been discussed, however, a comprehensive understanding of the underlying processes remains incomplete. One proposed mechanism is that biochar provides a physical matrix for soil organic matter to bind to, which can protect it from decomposition. Adding biochar to soil helps to stabilize carbon and reduce carbon dioxide emissions compared to adding raw materials. The presence of biochar has been seen to impact the content of dissolved organic matter (DOM) and the dispersion of organic carbon within soil microaggregates. These effects have implications for carbon sequestration in soil, various biological activities, and the overall quality of soil. The research demonstrates that biochar-amended soil retains similar carbon mineralization rates to unamended soil while dramatically lowering carbon dioxide emissions relative to raw materials. In the Indo-Gangetic rice-wheat system, the biochar-based route reduces the carbon footprint by 176% and 112% over no residue and residue incorporation, respectively (Singh et al., 2023).

The inclusion of different substances into soil yields diverse outcomes in terms of carbon processing. The investigation revealed that adding maize residue to soil resulted in an augmentation of aromatic metabolic products' accumulation within the soluble phase of DOM. The impact of biochar on soil carbon stability may be attributed to alterations in soil microbial populations. As previously mentioned, the use of biochar has been shown to facilitate the proliferation of advantageous microorganisms while inhibiting the development of detrimental microorganisms. This phenomenon has the potential to enhance the process of carbon stabilization. The research also revealed that introducing biochar can alter the soil environment, creating novel microhabitats that support microbial activity and enhance the availability of nutrients (Hernandez-Soriano, Kerré, Kopittke, Horemans, & Smolders, 2016). The priming effect is a significant determinant of the impact of biochar on soil characteristics. The practice of incorporating fresh organic amendments into soil has been well recognized as a means of modifying the pace at which native organic carbon undergoes mineralization. Positive priming is a phenomenon in which the process of carbon mineralization is enhanced, whereas negative priming refers to a phenomenon in which the rate of mineralization is reduced, resulting in a decrease in CO_2 emissions (He et al., 2023).

Various soil amendments impact carbon and nitrogen stability. Mean residence time (MRT), the time it will take for half of C or N to breakdown, is used to assess the strength. The MRT is calculated by monitoring both leaching and gaseous carbon and nitrogen losses. The findings indicate that biochar amendment raised the MRT of carbon considerably but had no impact on the MRT of nitrogen. However, including leaching losses in the MRT calculation significantly decreased the estimated MRT of both carbon and nitrogen compared to calculations based on gaseous emissions alone. The authors have observed that prior investigations have failed to include the impact of leaching losses, resulting in an overestimation of the stability of carbon from the soil and nitrogen. Moreover, the available data on the MRT of soil nitrogen in the presence of biochar or additional amendments is few, mostly because a substantial portion of the nitrogen is lost via the process of leaching (Mukherjee et al., 2011).

Liu et al. (2013) conducted a comprehensive meta analysis of various research. The findings of this study indicate that the introduction of biochar into soil did not have any statistically significant impact on CO_2 emissions across upland agriculture systems, grasslands, and forests. These results align with previous research findings on the subject (Kuzyakov et al., 2009; Singh & Cowie, 2014). However, a significant positive impact on CO_2 emissions was seen in rice cultivation. This effect is believed to be associated with the mineralization of easily decomposable carbon fractions in biochar or its stimulation of the decomposition of naturally occurring soil carbon (Kolb, Fermanich, & Dornbush, 2009; Zimmerman, Gao, & Ahn, 2011). The favorable influence was also connected with high soil organic carbon content and substantial synthetic nitrogen fertilizer fertilizer inputs in rice fields (Kimetu & Lehmann, 2010). It is essential to emphasize the importance of not allowing short-term carbon release in soils treated with biochar to diminish its long-term capacity for trapping carbon (Jones et al., 2011). The application of biochar has consistently shown a positive impact on organic carbon in the soil levels across several land-use categories. Notably, the most substantial benefit has been seen in rice fields, where the presence of waterlogging conditions contributes to the preservation of soil organic carbon by impeding its decomposition (Biederman & Harpole, 2013; Xie et al., 2013). In upland cropping systems, soil organic carbon response to biochar was lower, likely because of the accelerated soil organic carbon turnover rate associated with agricultural practices such as soil tillage or harvest (Liu et al., 2006).

Measuring CO_2 release and total mineralization coefficient reveals that increasing the quantity of swine manure compost applied to *Oxisols* may be stabilized by increasing the biochar application rate, hence limiting rapid mineralization. Adding biochar to Inceptisols with moderate alkalinity stabilizes the compost's organic matter to some extent but has a minor influence on its mineralization. In moderately acidic *Inceptisol* higher biochar treatment rate may maintain compost organic matter, but a higher concentration may significantly accelerate the mineralization of the compost (Tsai & Chang, 2019).

15.4 Interactions between microbial dynamics and carbon stability

Microbial diversity and function are critical factors that impact soil carbon stability under biochar-amended soil. Biochar can influence microbial diversity and function by providing a habitat for microorganisms to thrive in, as discussed earlier, and carbon in biochar can fuel up the below-ground microbial dynamics (Crowther et al., 2019). A fraction of the organic carbon undergoes conversion into more enduring forms that exhibit reduced susceptibility to rapid decomposition. According to Six, Conant, Paul, and Paustian (2002), the stable soil organic matter that is formed exhibits increased resistance to microbial destruction, hence playing a role in the extended sequestration of carbon within the soil. Soil bacteria can degrade both labile carbon found in low-temperature biochar and aromatic carbon compounds common in high-temperature biochar. The involvement of soil microorganisms in these processes has been shown in recent studies (Kuzyakov et al., 2009).

As biochar matures, it undergoes a natural degradation process that may lead to the formation of functional groups that contain oxygen and modifications to its charge on the surface. The alterations described have the potential to impact the interfacial interactions that occur among the soil matrices and microorganisms, as noted by Cheng et al. (2006). The regulation of soil carbon stability may be influenced more significantly by microbial activity throughout the soil rather than the chemical makeup of soil organic matter. The biochar's porous characteristics provide an environment suitable for microbial colonization, while its extensive surface area promotes microbial adhesion and the production of biofilms.

In addition, biochar supplies microorganisms with essential nutrients and energy, encouraging their development and activity. Microbial enzyme activity is another critical factor that affects soil carbon stability. Microbes uses enzymes to breakdown complex organic molecules into simpler compounds that can be used for energy or nutrient acquisition. The activity of microbial enzymes can impact carbon stabilization by either increasing or decreasing carbon degradation rates. Biochar can influence microbial enzyme activity by providing a surface for enzyme adsorption and reaction. It also alters the chemical structure of soil organic matter, making it accessible to microbial enzymes. Furthermore, biochar can stimulate the production of specific enzymes that promote carbon stabilization, such as phenol oxidase (Sinsabaugh et al., 2008). Microbial metabolites, such as organic acids and exopolysaccharides, can potentially influence soil carbon's stability. The presence of these metabolites has the potential to induce the formation of aggregates of soil particles, enhance the soil's ability to retain water, and inhibit the degradation of organic matter in the soil. Furthermore, incorporating biochar into sandy-textured soils significantly increased microbial biomass carbon, which ultimately led to the retention of soil nutrients in the soil, soil moisture, enhancing crop productivity and soil aggregation (Gaskin et al., 2010; Jeffery, Verheijen, van der Velde, & Bastos, 2011). Furthermore, the relationship between the diversity of microbes and carbon stability in soils amended with biochar encompasses several beneficial effects. These include the promotion of microbial activity and metabolic capacity, the enhancement of nutrient cycling and availability, and the bolstering of soil resilience against environmental pressures. Simultaneously, this interaction reduces carbon dioxide emissions, mitigates carbon loss through leaching, stabilizes soil organic matter, improves water retention and soil moisture regulation, and protects against soil erosion (Fig. 15.4).

Moreover, the use of biochar in sandy soils, characterized by inherent low fertility and vulnerability to nutrient leaching and runoff, has the potential to enhance soil fertility and stimulate soil microbial activity in comparison to loam or silt soils (Lehmann, 2007; Woolf, Amonette, Street-Perrott, Lehmann, & Joseph, 2010). The use of biochar has been seen to

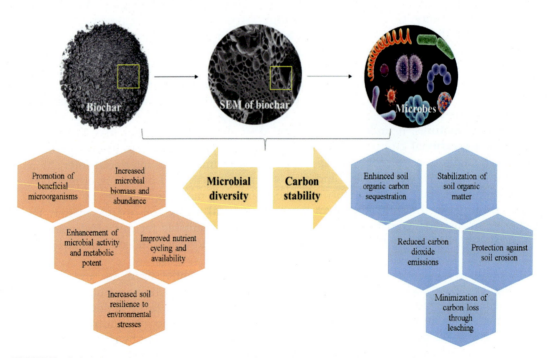

FIGURE 15.4 Interactive influence of biochar on microbial diversity and carbon stability.

result in a reduction in soil CO_2 fluxes and a beneficial effect on soil carbon content in soils with moderate acidity levels. Simultaneously, no notable enhancement was detected in soils with neutral or alkaline pH levels. The capacity of biochar to regulate soil pH through enhanced microbial consumption (Bruun, Jensen, & Jensen, 2008), leads to a more pronounced priming effect on the mineralization of native soil organic carbon (Foereid, Lehmann, & Major, 2011; Jones et al., 2011), and promoting increased crop or biomass productivity, thereby providing additional organic carbon substrate for biochar induced soil organic carbon mineralization, could explain the variations observed in acidic soil conditions (Liu et al., 2013).

Nevertheless, adding biochar hinders the process of soil carbon mineralization in neutral or alkaline soils, mostly due to the increased soil pH. Furthermore, the application of biochar to acidic soils resulted in a significant increase in the carbon content of soil microbial biomass, in comparison to soil conditions that were either neutral or alkaline. This effect may be attributed to the alkaline buffering properties of biochar, as shown by Steiner et al. (2008). The correlation between microbial dynamics and soil carbon stability is intricate and diverse, necessitating more investigation to enhance comprehension of the fundamental processes and ramifications for ecosystem operation and carbon sequestration.

15.5 Conclusion and future perspectives

The implications of these findings are substantial as numerous factors are to be considered in this interaction. Biochar has significant potential to improve soil health and mitigate climate change because it can boost soil carbon sequestration. However, further research is necessary to develop a comprehensive understanding of the mechanisms behind the effects of biochar on soil properties and identify the most effective approaches for its application and maintenance. The findings of this chapter suggest that the application of biochar may be tailored to specific soil types and climatic conditions in order to optimize its effectiveness as a means of improving soil health and sequestering carbon and can be summarized as

- Biochar considerably affects soil microbial populations, boosting the development of certain microorganism groups and increasing their metabolic activity.
- Biochar can preserve soil carbon by encouraging the synthesis of resistant carbon molecules and decreasing the rates of organic matter degradation.
- The interactions between microbial dynamics and carbon stability in biochar-amended soils are intricate and varied, with the microbial community playing an essential role in soil carbon sequestration.

Despite various knowledge gaps, there are several prospects for more study in this sector. Some possible research areas include standardized approaches for investigating microbial dynamics and carbon stability in biochar-amended soils and standardized methodology would allow for more reliable comparisons of findings across research and would aid in the development of a thorough knowledge of the processes behind biochar's impacts on soil characteristics. Conducting a comprehensive, extended investigation would enable us to evaluate biochar's enduring effects on soil carbon sequestration over an extended period. Additionally, it would allow the identification of optimal application rates and suitable feedstock choices tailored to specific geographical regions. Research investigating the impacts of biochar with respect to context-specific assessments will definitely aid in optimizing the use of biochar as a tool for enhancing microbial diversity, carbon stability, addressing climate change, optimizing and advance sustainable land management techniques with the right policies, information sharing, and capacity building.

References

Anderson, C. R., Condron, L. M., Clough, T. J., Fiers, M., Stewart, A., Hill, R. A., & Sherlock, R. R. (2011). Biochar induced soil microbial community change: Implications for biogeochemical cycling of carbon, nitrogen and phosphorus. *Pedobiologia, 54*(5–6), 309–320.

Asiloglu, R., Samuel, S. O., Sevilir, B., Akca, M. O., Acar Bozkurt, P., Suzuki, K., & Harada, N. (2021). Biochar affects taxonomic and functional community composition of protists. *Biology and Fertility of Soils, 57*, 15–29.

Babalola, O. O. (2010). Beneficial bacteria of agricultural importance. *Biotechnology Letters, 32*, 1559–1570.

Bastida, F., Eldridge, D. J., García, C., Kenny Png, G., Bardgett, R. D., & Delgado-Baquerizo, M. (2021). Soil microbial diversity–Biomass relationships are driven by soil carbon content across global biomes. *The ISME Journal, 15*(7), 2081–2091.

Biederman, L. A., & Harpole, W. S. (2013). Biochar and its effects on plant productivity and nutrient cycling: A meta-analysis. *GCB Bioenergy, 5*(2), 202–214.

Biederman, L. A., Harpole, W. S., & Hartman, K. (2013). Nutrient retention in biochar-amended soil. *Soil Biology and Biochemistry, 57*, 896–906.

Bolan, S., Hou, D., Wang, L., Hale, L., Egamberdieva, D., Tammeorg, P., & Bolan, N. (2023). The potential of biochar as a microbial carrier for agricultural and environmental applications. *Science of the Total Environment*, 163968.

Brodowski, S., Amelung, W., Haumaier, L., Abetz, C., & Zech, W. (2005). Morphological and chemical properties of black carbon in physical soil fractions as revealed by scanning electron microscopy and energy-dispersive X-ray spectroscopy. *Geoderma, 128*(1–2), 116–129.

Bruun, S., Jensen, E. S., & Jensen, L. S. (2008). Microbial mineralization and assimilation of black carbon: Dependency on degree of thermal alteration. *Organic Geochemistry, 39*, 839–845.

Bruun, E. W., Ambus, P., Egsgaard, H., & Hauggaard-Nielsen, H. (2012). Effects of slow and fast pyrolysis biochar on soil C and N turnover dynamics. *Soil Biology and Biochemistry, 46*, 73–79.

Butnan, S., Deenik, J. L., Toomsan, B., & Vityakon, P. (2017). Biochar properties affecting carbon stability in soils contrasting in texture and mineralogy. *Agriculture and Natural Resources, 51*(6), 492–498.

Chan, K. Y., Van Zwieten, L., Meszaros, I., Downie, A., & Joseph, S. (2007). Agronomic values of green waste biochar as a soil amendment. *Australian Journal of Soil Resources, 45*, 629–634.

Chaturvedi, S., Singh, S. V., Dhyani, V. C., Govindaraju, K., Vinu, R., & Mandal, S. (2021). Characterization, bioenergy value, and thermal stability of biochars derived from diverse agriculture and forestry lignocellulosic wastes. *Biomass Conversion and Biorefinery*, 1–14.

Chen, B., Koziel, J. A., Białowiec, A., Lee, M., Ma, H., O'Brien, S., & Brown, R. C. (2021). Mitigation of acute ammonia emissions with biochar during swine manure agitation before pump-out: Proof-of-the-concept. *Frontiers in Environmental Science, 9*, 613614.

Chen, X., Wu, W., Han, L., Gu, M., Li, J., & Chen, M. (2022). Carbon stability and mobility of ball milled lignin- and cellulose-rich biochar colloids. *Science of The Total Environment, 802*, 149759.

Chen, Z., Chen, B., & Grace, J. K. (2018). Effects of biochar on the abundance and community structure of ammonia-oxidizing bacteria and archaea in soil. *Chemosphere, 199*, 287–294.

Cheng, C. H., Lehmann, J., Thies, J. E., Burton, S. D., & Engelhard, M. H. (2006). Oxidation of black carbon by biotic and abiotic processes. *Organic Geochemistry, 37*, 1477–1488.

Crowther, T. W., Van den Hoogen, J., Wan, J., Mayes, M. A., Keiser, A. D., Mo, L., & Maynard, D. S. (2019). The global soil community and its influence on biogeochemistry. *Science (New York, N.Y.), 365*(6455), eaav0550.

Dai, Z., Barberán, A., Li, Y., Brookes, P. C., & Xu, J. (2017). Bacterial community composition associated with pyrogenic organic matter (biochar) varies with pyrolysis temperature and colonization environment. *Msphere*, 2(2), 10−1128.

Delgado-Baquerizo, M., Maestre, F. T., Reich, P. B., Jeffries, T. C., Gaitan, J. J., Encinar, D., & Singh, B. K. (2016). Microbial diversity drives multifunctionality in terrestrial ecosystems. *Nature Communications*, 7(1), 10541.

Delgado-Baquerizo, M., Reich, P. B., Trivedi, C., Eldridge, D. J., Abades, S., Alfaro, F. D., & Singh, B. K. (2020). Multiple elements of soil biodiversity drive ecosystem functions across biomes. *Nature Ecology & Evolution*, 4(2), 210−220.

Dungait, J. A., Hopkins, D. W., Gregory, A. S., & Whitmore, A. P. (2012). Soil organic matter turnover is governed by accessibility not recalcitrance. *Global Change Biology*, 18(6), 1781−1796.

Efthymiou, A., Nunes, I., Jensen, B., & Jakobsen, I. (2023). Response of bacterial communities to the application of sewage sludge biochar and *Penicillium aculeatum* in rhizosphere and bulk soil of wheat. *Applied Soil Ecology*, 190, 104986.

Fang, Y., Singh, B., Singh, B. P., & Krull, E. (2014). Biochar carbon stability in four contrasting soils. *European Journal of Soil Science*, 65(1), 60−71.

Foereid, B., Lehmann, J., & Major, J. (2011). Modeling black carbon degradation and movement in soil. *Plant and Soil*, 345, 223−236.

Gaskin, J. W., Speir, R. A., Harris, K., Das, K. C., Lee, R. D., Morris, L. A., & Fisher, D. S. (2010). Effect of peanut hull and pine chip biochar on soil nutrients, corn nutrient status, and yield. *Agronomy Journal*, 102(2), 623−633.

Glaser, B., Lehmann, J., & Zech, W. (2002). Ameliorating physical and chemical properties of highly weathered soils in the tropics with charcoal − A review. *Biology and Fertility of Soils*, 35(4), 219−230.

Hale, L., Luth, M., Kenney, R., & Crowley, D. (2014). Evaluation of pinewood biochar as a carrier of bacterial strain Enterobacter cloacae UW5 for soil inoculation. *Applied Soil Ecology*, 84, 192−199.

Han, Z., Xu, P., Li, Z., Lin, H., Zhu, C., Wang, J., & Zou, J. (2022). Microbial diversity and the abundance of keystone species drive the response of soil multifunctionality to organic substitution and biochar amendment in a tea plantation. *GCB Bioenergy*, 14(4), 481−495.

He, Y., Wang, Y., Jiang, Y., Yin, G., Cao, S., Liu, X., & Chen, F. (2023). Drivers of soil respiration and nitrogen mineralization change after litter management at a subtropical Chinese sweetgum tree plantation. *Soil Use and Management*, 39(1), 92−103.

Hernandez-Soriano, M. C., Kerré, B., Kopittke, P. M., Horemans, B., & Smolders, E. (2016). Biochar affects carbon composition and stability in soil: A combined spectroscopy-microscopy study. *Scientific Reports*, 6(1), 25127.

Herrmann, L., Lesueur, D., Robin, A., Robain, H., Wiriyakitnateekul, W., & Bräu, L. (2019). Impact of biochar application dose on soil microbial communities associated with rubber trees in North East Thailand. *Science of the Total Environment*, 689, 970−979.

Ishii, S., Seishi, I., Kiwamu, M., & Keishi, S. (2011). Nitrogen cycling in rice paddy environments: Past achievements and future challenges. *Microbes and Environments*, 26(4), 282−292.

Jeffery, S., Bezemer, T. M., Cornelissen, G., Kuyper, T. W., Lehmann, J., Mommer, L., ... van Groenigen, J. W. (2015). The way forward in biochar research: Targeting trade-offs between the potential wins. *Globl Change Biology and Bioenergy*, 7, 1−13.

Jeffery, S., Verheijen, F. G., van der Velde, M., & Bastos, A. C. (2011). A quantitative review of the effects of biochar application to soils on crop productivity using meta-analysis. *Agriculture, Ecosystems & Environment*, 144(1), 175−187.

Jiang, R. W., Mechler, M. A., & Oelbermann, M. (2023). Exploring the effects of one-time biochar application with low dosage on soil health in temperate climates. *Soil Security*, 100101.

Jones, D. L., Murphy, D. V., Khalid, M., Ahmad, W., Edwards-Jones, G., & DeLuca, T. H. (2011). Short-term biochar-induced increase in soil CO_2 release is both biotically and abiotically mediated. *Soil Biology and Biochemistry*, 43(8), 1723−1731.

Khan, M. A., Salman, A. Z., & Khan, S. T. (2023). Indigenously produced biochar retains fertility in sandy soil through unique microbial diversity sustenance: A step toward the circular economy. *Frontiers in Microbiology*, 14.

Kim, J. S., Sparovek, G., Longo, R. M., De Melo, W. J., & Crowley, D. (2007). Bacterial diversity of terra preta and pristine forest soil from the Western Amazon. *Soil Biology and Biochemistry*, 39, 684−690.

Kimetu, J. M., & Lehmann, J. (2010). Stability and stabilisation of biochar and green manure in soil with different organic carbon contents. *Soil Research*, 48(7), 577−585.

Kolb, S. E., Fermanich, K. J., & Dornbush, M. E. (2009). Effect of charcoal quantity on microbial biomass and activity in temperate soils. *Soil Science Society of America Journal*, 73(4), 1173–1181.

Kumar, R., Sharma, P., Gupta, R. K., Kumar, S., Sharma, M. M. M., Singh, S., & Pradhan, G. (2020). Earthworms for eco-friendly resource efficient agriculture. In Kumar, et al. (Eds.), *Resources use efficiency in agriculture* (pp. 47–84). Singapore: Springer Nature. Available from https://doi.org/10.1007/978-981-15-6953-1_2.

Kumar, S., Meena, R. S., Datta, R., Verma, S. K., Yadav, G. S., Pradhan, G., Molaei, A., Rahman, G. K. M. M., & Mashuk, H. A. (2020). Legumes for carbon and nitrogen cycling: An organic approach. carbon and nitrogen cycling in soil. In R. Datta, et al. (Eds.), *Carbon and nitrogen cycling in soil* (pp. 337–375). Singapore: Springer Nature. Available from https://doi.org/10.1007/978-981-13-7264-3_10, 337–375.

Kumawat, A., Bamboriya, S. D., Meena, R. S., Yadav, D., Kumar, A., Kumar, S., ... Pradhan, G. (2022). Legume-based inter-cropping to achieve crop, soil, and environmental health security. In R. S. Meena, & S. Kumar (Eds.), *Advances in legumes for sustainable intensification* (pp. 307–328). Elsevier. Available from https://doi.org/10.1016/B978-0-323-85797-0.00005-7.

Kuzyakov, Y., Subbotina, I., Chen, H., Bogomolova, I., & Xu, X. (2009). Black carbon decomposition and incorporation into soil microbial biomass estimated by 14C labeling. *Soil Biology and Biochemistry*, 41(2), 210–219.

Lehmann, J. (2007). Bioenergy in the black. *Frontiers in Ecology*, 5, 381–387.

Lehmann, J., & Joseph, S. (2009). *Biochar for environmental management: Science and technology* (p. 416) London & Sterling: Earthscan.

Lehmann, J., & Joseph, S. (2015). Biochar for environmental management: An introduction. *In Biochar for environmental management: Science, technology and implementation*, 1–12.

Lehmann, J., Cowie, A., Masiello, C. A., Kammann, C., Woolf, D., Amonette, J. E., & Whitman, T. (2021). Biochar in climate change mitigation. *Nature Geoscience*, 14(12), 883–892.

Lehmann, J., Gaunt, J., & Rondon, M. (2006). Bio-char sequestration in terrestrial ecosystems – A review. *Mitigation and Adaptation Strategies for Global Change*, 11(2), 403–427.

Lehmann, J., Rillig, M. C., Thies, J., Masiello, C. A., Hockaday, W. C., & Crowley, D. (2011). Biochar effects on soil biota—a review. *Soil Biology and Biochemistry*, 43(9), 1812–1836.

Leng, L., & Huang, H. (2018). An overview of the effect of pyrolysis process parameters on biochar stability. *Bioresource Technology*, 270, 627–642.

Li, S., & Chan, C. Y. (2022). Will Biochar suppress or stimulate greenhouse gas emissions in agricultural fields? Unveiling the dice game through data syntheses. *Soil Systems*, 6, 73.

Li, S., & Tasnady, D. (2023). Biochar for soil carbon sequestration: Current knowledge, mechanisms, and future perspectives. *Journal of Carbon Research*, 9(3), 67.

Liang, C., Schimel, J. P., & Jastrow, J. D. (2017). The importance of anabolism in microbial control over soil carbon storage. *Nature Microbiology*, 2(8), 1–6.

Liu, Q. H., Shi, X. Z., Weindorf, D. C., Yu, D. S., Zhao, Y. C., Sun, W. X., & Wang, H. J. (2006). Soil organic carbon storage of paddy soils in China using the 1: 1,000,000 soil database and their implications for C sequestration. *Global Biogeochemical Cycles*, 20(3).

Liu, X., Zhang, A., Ji, C., Joseph, S., Bian, R., Li, L., & Paz-Ferreiro, J. (2013). Biochar's effect on crop productivity and the dependence on experimental conditions—A meta-analysis of literature data. *Plant and Soil*, 373, 583–594.

Liu, X., Zhang, D., Li, H., Qi, X., Gao, Y., Zhang, Y., & Li, H. (2020). Soil nematode community and crop productivity in response to 5-year biochar and manure addition to yellow cinnamon soil. *BMC Ecology*, 20(1), 1–13.

Ma, B., Shao, S., Ai, L., Chen, S., & Zhang, L. (2023). Influences of biochar with selenite on bacterial community in soil and Cd in peanut. *Ecotoxicology and Environmental Safety*, 255, 114742.

Mackie, K. A., Marhan, S., Ditterich, F., Schmidt, H. P., & Kandeler, E. (2015). The effects of biochar and compost amendments on copper immobilization and soil microorganisms in a temperate vineyard. *Agriculture, Ecosystems & Environment*, 201, 58.

Masiello, C. A., Chen, Y., Gao, X., Liu, S., Cheng, H. Y., Bennett, M. R., & Silberg, J. J. (2013). Biochar and microbial signaling: Production conditions determine effects on microbial communication. *Environmental Science & Technology*, 47(20), 11496–11503.

McCormack, S. A., Ostle, N., Bardgett, R. D., Hopkins, D. W., & Vanbergen, A. J. (2013). Biochar in bioenergy cropping systems: Impacts on soil faunal communities and linked ecosystem processes. *Gcb Bioenergy*, 5(2), 81–95.

Meena, R. S., & Pradhan, G. (2023). Industrial garbage-derived biocompost enhances soil organic carbon fractions, CO_2 biosequestration, potential carbon credits, and sustainability index in a rice-wheat ecosystem. *Environmental Research*, 116525. Available from https://doi.org/10.1016/j.envres.2023.116525.

Meena, R.S., Kumawat, A., Kumar, S., Prasad, S.K., Pradhan, G., Jhariya, M.K., ... Raj, A. (2022). Effect of legumes on nitrogen economy and budgeting in South Asia. In Meena, R.S., & Kumar, S. (eds.), *Advances in legumes for sustainable intensification*, pp. 619–638. https://doi.org/10.1016/B978-0-323-85797-0.00001-X.

Meena, R. S., Pradhan, G., Kumar, S., & Lal, R. (2023). Using industrial wastes for rice-wheat cropping and food-energy-carbon-water-economic nexus to the sustainable food system. *Renewable and Sustainable Energy Reviews, 187*, 113756. Available from https://doi.org/10.1016/j.rser.2023.113756.

Mukherjee, A., Zimmerman, A. R., & Harris, W. (2011). Surface chemistry variations among a series of laboratory-produced biochars. *Geoderma, 163*, 155–247.

Novak, J. M., Busscher, W. J., Laird, D. L., Ahmedna, M., Watts, D. W., & Niandou, M. A. S. (2009). Impact of biochar amendment on fertility of a southeastern coastal plain soil. *Soil Science, 174*, 105–112.

Pradhan, G., & Meena, R. S. (2022). Diversity in the rice–wheat system with genetically modified zinc and iron-enriched varieties to achieve nutritional security. *Sustainability, 14*, 9334. Available from https://doi.org/10.3390/su14159334.

Pradhan, G., & Meena, R. S. (2023). Interaction impacts of biocompost on nutrient dynamics and relations with soil biota, carbon fractions index, societal value of CO_2 equivalent, and ecosystem services in the wheat-rice farming. *Chemosphere*, 139695. Available from https://doi.org/10.1016/j.chemosphere.2023.139695.

Pradhan, G., & Meena, R. S. (2023). Utilizing waste compost to improve the atmospheric CO_2 capturing in the rice-wheat cropping system and energy-cum-carbon credit audit for a circular economy. *Science of the Total Environment*, 164572. Available from https://doi.org/10.1016/j.scitotenv.2023.164572.

Pradhan, G., Meena, R. S., Kumar, S., & Lal, R. (2023). Utilizing industrial wastes as compost in wheat-rice production to improve the above and below-ground ecosystem services. *Agriculture, Ecosystems and Environment, 358*, 108704. Available from https://doi.org/10.1016/j.agee.2023.108704.

Pradhan, G., Meena, R. S., Kumar, S., Jhariya, M. K., Khan, N., Shukla, U. N., ... Kumar, S. (2022). Legumes for eco-friendly weed management in an agroecosystem. In Meena, R.S., & Kumar, S. (eds.), *Advances in legumes for sustainable intensification*, pp. 133–154. https://doi.org/10.1016/B978-0-323-85797-0.00033-1.

Raj, A., Jhariya, M. K., Banerjee, A., Meena, R. S., Nema, S., Khan, N., ... Pradhan, G. (2021). Agroforestry a model for ecological sustainability. In M. K. Jhariya, R. S. Meena, A. Raj, & S. N. Meena (Eds.), *Natural resources conservation and advances for sustainability* (pp. 289–308). Elsevier. Available from https://doi.org/10.1016/B978-0-12-822976-7.00002-8.

Rehrah, D., Reddy, M. R., Novak, J. M., Bansode, R. R., Schimmel, K. A., Yu, J., ... Ahmedna, M. (2014). Production and characterization of biochars from agricultural by-products for use in soil quality enhancement. *Journal of analytical and applied pyrolysis, 108*, 301–309.

Rogovska, N., Laird, D., Cruse, R., Fleming, P., Parkin, T., & Meek, D. (2011). Impact of biochar on manure carbon stabilization and greenhouse gas emissions. *Soil Science Society of America Journal, 75*(3), 871–879.

Roy, O., Meena, R. S., Kumar, S., Jhariya, M. K., & Pradhan, G. (2021). Assessment of land use systems for CO_2 sequestration, carbon credit potential, and income security in Vindhyan Region, India. *Land Degradation and Development, 33*(4), 670–682. Available from https://doi.org/10.1002/ldr.4181.

Sheoran, S., Ramtekey, V., Kumar, D., Kumar, S., Meena, R. S., Kumawat, A., ... Shukla, U. N. (2022). Grain legumes: Recent advances and technological interventions. In R. S. Meena, & S. Kumar (Eds.), *Advances in legumes for sustainable intensification* (pp. 507–532). Elsevier. Available from https://doi.org/10.1016/B978-0-323-85797-0.00025-2.

Singh, B. P., & Cowie, A. L. (2014). Long-term influence of biochar on native organic carbon mineralisation in a low-carbon clayey soil. *Scientific Reports, 4*(1), 3687.

Singh, B. P., Cowie, A. L., & Smernik, R. J. (2012). Biochar carbon stability in a clayey soil as a function of feedstock and pyrolysis temperature. *Environmental Science & Technology, 46*(21), 11770–11778.

Singh, C., Tiwari, S., Gupta, V. K., & Singh, J. S. (2018). The effect of rice husk biochar on soil nutrient status, microbial biomass and paddy productivity of nutrient poor agriculture soils. *Catena, 171*, 485–493.

Singh, S., Chaturvedi, S., Nayak, P., Dhyani, V. C., Nandipamu, T. M. K., Singh, D. K., ... Govindaraju, K. (2023). Carbon offset potential of biochar based straw management under rice-wheat system along Indo-Gangetic Plains of India. *Science of The Total Environment*, 165176.

Singh, S., Luthra, N., Mandal, S., Kushwaha, D. P., Pathak, S. O., Datta, D., ... Pramanick, B. (2023). Distinct behavior of biochar modulating biogeochemistry of salt-affected and acidic soil: A review. *Journal of Soil Science and Plant Nutrition*, 1–17.

Singh, S. V., Chaturvedi, S., & Datta, D. (2019). Biochar: An eco-friendly residue management approach. *Indian Farming*, 69(08), 27–29.

Singh, S. V., Chaturvedi, S., Dhyani, V. C., & Govindaraju, K. (2020). Pyrolysis temperature influences the characteristics of rice straw and husk biochar and sorption/desorption behavior of their biourea composite. *Bioresource Technology*, 314, 123674.

Sinsabaugh, R. L., Lauber, C. L., Weintraub, M. N., Ahmed, B., Allison, S. D., Crenshaw, C., ... Zeglin, L. H. (2008). Soil enzyme activities and biodiversity measurements as integrative microbiological indicators. In *Microbial Ecology in Sustainable Agroecosystems CRC Press*, 221–236.

Six, J., Conant, R. T., Paul, E. A., & Paustian, K. (2002). Stabilization mechanisms of soil organic matter: Implications for C-saturation of soils. *Plant and Soil*, 241(2), 155–176.

Smith, A. P., Santangelo, J. D., & Gupta, V. V. S. R. (2014). Effect of biochar amendment on soil microbial community and soil properties in temperate soils. *Journal of Soil Science and Plant Nutrition*, 14(4), 991–1002.

Steinbeiss, S., Gleixner, G., & Antonietti, M. (2009). Effect of biochar amendment on soil carbon balance and soil microbial activity. *Soil Biology and Biochemistry*, 41, 1301–1310.

Steiner, C., Teixeira, W. G., Lehmann, J., Nehls, T., de Macêdo, J. L. V., Blum, W. E. H., & Zech, W. (2008). Long term effects of manure, charcoal and mineral fertilization on crop production and fertility on a highly weathered Central Amazonian upland soil. *Plant and Soil*, 291(1–2), 275–290.

Sulman, B. N., Phillips, R. P., Oishi, A. C., Shevliakova, E., & Pacala, S. W. (2014). Microbe-driven turnover offsets mineral-mediated storage of soil carbon under elevated CO_2. *Nature Climate Change*, 4(12), 1099–1102.

Sun, T., Smith, J., & Johnson, A. (2019). Effects of biochar on fungal biomass and community composition in agricultural soils. *Journal of Soil Science*, 50(3), 123–135.

Swagathnath, G., Rangabhashiyam, S., Murugan, S., & Balasubramanian, P. (2019). Influence of biochar application on growth of Oryza sativa and its associated soil microbial ecology. *Biomass Conversion and Biorefinery*, 9, 341–352.

Tag, A. T., Duman, G., Ucar, S., & Yanik, J. (2016). Effects of feedstock type and pyrolysis temperature on potential applications of biochar. *Journal of Analytical and Applied Pyrolysis*, 120, 200–206.

Tsai, C. C., & Chang, Y. F. (2019). Carbon dynamics and fertility in biochar-amended soils with excessive compost application. *Agronomy*, 9(9), 511.

Ullah, S., Liang, H., Ali, I., Zhao, Q., Iqbal, A., Wei, S., & Jiang, L. (2020). Biochar coupled with contrasting nitrogen sources mediated changes in carbon and nitrogen pools, microbial and enzymatic activity in paddy soil. *Journal of Saudi Chemical Society*, 24(11), 835–849.

Van Der Heijden, M. G., Bardgett, R. D., & Van Straalen, N. M. (2008). The unseen majority: Soil microbes as drivers of plant diversity and productivity in terrestrial ecosystems. *Ecology Letters*, 11(3), 296–310.

Van Zwieten, L., Kimber, S., Morris, S., Chan, K. Y., Downie, A., Rust, J., & Cowie, A. (2010). Effects of biochar from slow pyrolysis of paper mill waste on agronomic performance and soil fertility. *Plant and Soil*, 327, 235–246.

Van Zwieten, L., Singh, B. P., Kimber, S. W. L., Murphy, D. V., Macdonald, L. M., Rust, J., & Morris, S. (2014). An incubation study investigating the mechanisms that impact N_2O flux from soil following biochar application. *Agricultural Ecosystem and Environment*, 191, 53–62.

Wardle, D. A., Nilsson, M. C., & Zackrisson, O. (2008). Fire-derived charcoal causes loss of forest humus. *Science (New York, N.Y.)*, 320(5876), 629, 629.

Warren, J., Topping, C. J., & James, P. (2009). A unifying evolutionary theory for the biomass–diversity–fertility relationship. *Theoretical Ecology*, 2, 119–126.

Weralupitiya, C., Gunarathne, V., Keerthanan, S., Rinklebe, J., Biswas, J. K., Jayasanka, J., & Vithanage, M. (2022). Influence of biochar on soil biology in the charosphere. In *Biochar in Agriculture for Achieving Sustainable Development Goals*, 273–291.

Woolf, D., Amonette, J. E., Street-Perrott, F. A., Lehmann, J., & Joseph, S. (2010). Sustainable biochar to mitigate global climate change. *Nature Communications*, 1(1), 56.

Wu, D., Ren, C., Ren, D., Tian, Y., Li, Y., Wu, C., & Li, Q. (2023). New insights into carbon mineralization in tropical paddy soil under land use conversion: Coupled roles of soil microbial community, metabolism, and dissolved organic matter chemodiversity. *Geoderma*, 432, 116393.

Wu, H., Zeng, G., Liang, J., Chen, J., Xu, P., Lai, C., & Huang, D. (2017). Effects of biochar on the abundance and community composition of denitrifiers in a reclaimed soil from a wastewater irrigation area. *Environmental Science and Pollution Research, 24*(6), 5694–5704.

Xie, Z., Xu, Y., Liu, G., Liu, Q., Zhu, J., Tu, C., & Hu, S. (2013). Impact of biochar application on nitrogen nutrition of rice, greenhouse-gas emissions and soil organic carbon dynamics in two paddy soils of China. *Plant and Soil, 370*, 527–540.

Xu, W., Xu, H., Delgado-Baquerizo, M., Gundale, M. J., Zou, X., & Ruan, H. (2023). Global meta-analysis reveals positive effects of biochar on soil microbial diversity. *Geoderma, 436*, 116528.

Yadav, R. K., Yadav, M. R., Kumar, R., Parihar, C. M., Yadav, N., Bajiya, R., … Yadav, B. (2017). Role of biochar in mitigation of climate change through carbon sequestration. *International Journal of Current Microbiology and Applied Science, 6*(4), 859–866.

Yan, T., Xue, J., Zhou, Z., & Wu, Y. (2021). Biochar-based fertilizer amendments improve the soil microbial community structure in a karst mountainous area. *Science of the Total Environment, 794*, 148757.

Yao, Q., Liu, J., Yu, Z., Li, Y., Jin, J., Liu, X., & Wang, G. (2017). Three years of biochar amendment alters soil physiochemical properties and fungal community composition in a black soil of northeast China. *Soil Biology and Biochemistry, 110*, 56–67.

Zhang, G., Guo, X., Zhu, Y., Liu, X., Han, Z., Sun, K., & Han, L. (2018). The effects of different biochars on microbial quantity, microbial community shift, enzyme activity, and biodegradation of polycyclic aromatic hydrocarbons in soil. *Geoderma, 328*, 100–108.

Zhang, H., Sun, H., Zhou, S., Bai, N., Zheng, X., Li, S., & Lv, W. (2019). Effect of straw and straw biochar on the community structure and diversity of ammonia-oxidizing bacteria and archaea in rice-wheat rotation ecosystems. *Scientific Reports, 9*(1), 9367.

Zhang, S., Song, Y., Hu, F., Qiu, L., Feng, Y., Cai, X., & Wang, H. (2020). Biochar improves soil microbial community diversity and alters microbial network complexity in a clayey soil. *Science of the Total Environment, 707*, 136066.

Zhou, C., Heal, K., Tigabu, M., Xia, L., Hu, H., Yin, D., & Ma, X. (2020). Biochar addition to forest plantation soil enhances phosphorus availability and soil bacterial community diversity. *Forest Ecology and Management, 455*, 117635.

Zhu, X., Chen, B., Zhu, L., & Xing, B. (2017). Effects and mechanisms of biochar-microbe interactions in soil improvement and pollution remediation: A review. *Environmental Pollution, 227*, 98–115.

Zimmerman, A. R., Gao, B., & Ahn, M. Y. (2011). Positive and negative carbon mineralization priming effects among a variety of biochar-amended soils. *Soil biology and Biochemistry, 43*(6), 1169–1179.

CHAPTER 16

Nutrient enriched and co-composted biochar: system productivity and environmental sustainability

Cícero Célio de Figueiredo[1], Leônidas Carrijo Azevedo Melo[2], Carlos Alberto Silva[2], Joisman Fachini[1], Jefferson Santana da Silva Carneiro[3], Everton Geraldo de Morais[2], Ornelle Christiane Ngo Ndoung[1], Shiv Vendra Singh[4] and Tony Manoj Kumar Nandipamu[5]

[1]Department of Agronomy, School of Agronomy and Veterinary Medicine, University of Brasilia, Brasilia, Brazil [2]Department of Soil Science, School of Agricultural Sciences, Federal University of Lavras, Lavras, Minas Gerais, Brazil [3]School of Technological Education, Xinguara, Pará, Brazil [4]Department of Agronomy, College of Agriculture, Rani Lakshmi Bai Central Agricultural University, Jhansi, Uttar Pradesh, India [5]Department of Agronomy, G. B. Pant University of Agriculture & Technology, Pantnagar, Uttarakhand, India

16.1 Introduction

Biochar is a solid material, rich in carbon, obtained by thermochemical processes from various organic residues. Biochar has several interesting physicochemical characteristics that make it a multifunctional material that can be used for different purposes, such as nutrient and microbial carrier, soil amendment to improve soil health, immobilizing agent for remediation of toxic metals and organic contaminants in soil and water, catalyst for industrial applications, porous material for mitigating greenhouse gas (GHG) emissions and odorous compounds, and feed supplement to improve animal health and nutrient intake efficiency and, thus, productivity (Bolan et al., 2022). Biochar as a soil amendment has been widely studied and the results indicate that it can partially or fully replace chemical fertilizers in agricultural production (Faria, Figueiredo, de, Coser, Vale, & Schneider, 2018; Singh, Chaturvedi, Dhyani, & Govindaraju, 2020), increase soil organic carbon stocks (Chagas, Figueiredo, & Ramos, 2022) and reduce the deleterious effects of heavy metals (HMs) (Guo, Song, & Tian, 2020), among other functions. However, the performance of biochar as a soil amendment is extremely dependent on the feedstock, the pyrolysis conditions, and the application rate, among others (Ippolito et al., 2020). In addition, the low concentration or imbalanced nutrients may limit the use of biochar as a fertilizer. To overcome the low concentration of nutrients in biochar, high doses are normally applied, resulting in prohibitive costs for large-scale adoption. In developed countries, biochar costs range from US$300 to US$700 per ton (Clare et al., 2014).

In view of this, several strategies have been assessed to make the use of biochar as a fertilizer viable. Enriching biochar with one or multiple nutrients has been a promising strategy to increase nutrient use efficiency, reduce final product costs, as well as reduce the environmental damage caused by chemical fertilizers, such as nutrient leaching and GHG emissions (Ndoung, Figueiredo, & Ramos, 2021). In this chapter, we present biochar as a matrix for producing sustainable fertilizers. Techniques for enriching biochar with nutrients and the effects of enriched biochar on crop productivity are also covered throughout the text. Additionally, slow-release mechanisms of biochar-based fertilizers (BBFs) are presented for nitrogen, sulfur, phosphorus, and potassium. Finally, the effects of cocomposted biochar (COMBI) on nutrient supply to plants are also discussed.

16.2 Biochar as a matrix to produce sustainable fertilizers

16.2.1 Enrichment techniques

In addition to having direct effects as a fertilizer, biochar is also considered a matrix for producing sustainable fertilizers. In this case, the presence of nutrients, high cation-exchange capacity (CEC), high porosity and surface properties are the main characteristics of biochar considered in the production of sustainable fertilizers.

There are several techniques to enrich biochar with nutrients in order to produce BBFs. Biochar enrichment can be performed by (1) mixing biochar with a nutrient solution followed by drying; (2) blending the biochar and fertilizer in powder form, followed by granulation/pelleting; and (3) copyrolysis which involves mixing the raw material with nutrients followed by pyrolysis (Sim, Tan, Lim, & Hameed, 2021). Overall, these techniques are grouped into four categories: (1) treatment or direct mixing of nutrient-rich raw materials followed by slow pyrolysis; (2) mixing organic residue with highly water-soluble fertilizers, rock powder or other material (natural or synthetic) rich in a specific nutrient followed by pyrolysis; (3) nutrient enrichment can also be done after pyrolysis (postpyrolysis process); and (4) in some cases the mixture may be re-submitted to pyrolysis (Ndoung et al., 2021). In direct treatment, nutrient-rich feedstocks are mixed and subsequently subjected to pyrolysis to produce BBFs. The main feedstocks used include sludges, composts, manures, algae, bone waste, mushroom substrate compost, municipal and some crop wastes (Melo, Lehmann, Carneiro, & Camps-Arbestain, 2022; Ndoung et al., 2021).

BBFs with good nutrient concentrations were obtained by mixing bone waste and dairy cattle carcasses, with P contents ranging from 12.7% to 15.3% (Zwetsloot et al., 2016).

In the second category, BBF can be obtained by enriching organic residue with nutrients from soluble mineral fertilizers, rock powders and waste from the mineral industry, followed by pyrolysis. In general, this type of enrichment seeks to provide one or more nutrients present in low concentrations in the organic material that will be pyrolyzed. In addition to fertilizer production, this mixture has also been used to functionalize biochar for HM holding capacity and water retention (Ndoung et al., 2021). Nutrient sources commonly used in blends include soluble fertilizers (urea, single superphosphate, triple superphosphate, monoammonium phosphate (MAP), di-ammonium phosphate, potassium chloride, potassium sulfate), minerals (bentonite clay, phosphate rock, phosphogypsum), and chemical solutions ($FeSO_4 \cdot 0.7H_2O$, $MgCl_2$, $CaCl_2$). For instance, maize stalk was enriched with triple superphosphate, diatomite and urea and submitted to pyrolysis at 300°C, 450°C, and 600°C; the BBF produced from this mixture showed higher amounts of N, Si, P, Ca, and higher CEC than the feedstocks (Chen et al., 2021). Poultry litter and coffee husk were mixed with phosphoric acid, magnesium oxide and triple superphosphate and pyrolyzed at 500°C. In this case, the BBF released P slowly and increased P content in the soil and performed similarly to soluble fertilizers regarding plant growth (Carneiro et al., 2021). Similarly, BBF was efficiently produced from a mixture of wheat straw and urea, bentonite clay, rock phosphate, Fe_2O_3, and $FeSO_4 \cdot 0.7H_2O$ (Chew et al., 2020).

Postpyrolysis is the most used technique to produce BBFs. In this case, biochars are treated with nutrient-rich sources, including soluble mineral fertilizer, rock powder, clay, compost, wastewater, and so on (Ndoung et al., 2021). Postpyrolysis BBFs show several advantages compared to their original feedstocks, such as acting as slow-release fertilizer (Fachini, Figueiredo, & Vale, 2022), increase HM adsorption (Ahmad et al., 2018), reduce HM toxicity in plants (Dad et al., 2021), decrease nutrient leaching (Dong et al., 2020), improve water retention and mycorrhizal colonization, increase available nutrients in the soil, and, hence, increase plant growth (Joseph et al., 2015).

According to Ndoung et al. (2021), in general, the increase in soil nutrient content after BBF application could be attributed to:

1. The pyrolysis that results in the concentration of nutrients within the biochar;
2. The reduction of nutrient fixation in the soil, since biochar can alter the adsorption and desorption balance in the soil;
3. The incorporation of nutrient sources into biochar, such as minerals, clays, compost, or mineral fertilizers;
4. The ability of biochar to adsorb nutrients on its surface and release them gradually and, as a result, reduce loss by leaching.

16.2.2 Effects on crop productivity

Biochar addition up to $20\,t\,ha^{-1}$ along with inorganic fertilizers has been shown to increase crop productivity by 15% beyond the application of conventional fertilizers only (Ye, Camps-Arbestain, & Shen, 2020), showing a synergistic effect of biochar and fertilizers. This has motivated several studies, using different approaches, on BBF development aiming to reduce application rates and make biochar viable at large scale. Under field conditions, there are several studies that show an increase in crop productivity for BBF when compared with conventional sources. For instance, a BBF made with urea and triple superphosphate increased, on average, by 10% maize productivity in a Cd-contaminated soil in China when compared with conventional fertilizer (Chen et al., 2021). A N-based BBF made with urea, biochar and additives (bentonite and gelatinized maize flour) was shown to increase maize grain yield up to 19% (average of 9%) when compared with urea only under tropical conditions in Brazil (Puga, Grutzmacher, & Cerri, 2020). A recent metaanalysis comprising 148 pairwise comparisons from 40 articles indicated that, on average, BBF applied at very low application rates (mean of 0.9 t ha^{-1}—only field studies considered) increased crop productivity by 10% when compared with fertilized controls and 186% compared with nonfertilized controls (Melo et al., 2022). The increase in crop productivity promoted by BBF was higher in tropical than in temperate regions; and the most pronounced effect of BBF was verified in highly weathered soils (Oxisols and Ultisols) and weekly developed soils (Entisols and Inceptisols) when compared to other soil types. Another remarkable result was the higher effect of BBFs with high carbon contents ($>30\%$) when compared with those of low carbon contents ($<30\%$) on crop productivity (Melo et al., 2022). At such low application rates, biochar was pointed as a fertilizer enhancer in BBF rather than a soil amendment, especially because biochar can be designed to solve specific problems that conventional fertilizers cannot. Different mechanisms explain the good performance of BBFs to increase crop productivity. Some of these mechanisms involved in the release, absorption and use efficiency of nutrients from BBFs have already been studied in different soil and crop conditions and are discussed below.

16.3 Slow-release mechanisms of biochar-based fertilizers

16.3.1 Nitrogen, phosphorus, and sulfur dynamics

Nitrogen and P have complex dynamics in soil-plant systems. Both nutrients are applied in relatively high amounts in cultivated soils. The losses of N occur mainly through leaching and volatilization, while P can be lost by leaching, runoff or fixation in soil. In tropical climates, the main problem is P fixation in clays of weathered soils. Thus different strategies have been developed for BBF to increase N and P use efficiency and, in general, slow-release dynamics are sought. The functional surface properties and porosity of biochar makes it ideal for nutrient adsorption, such as ammonium (NH_4^+), but a review article reveals the limitation of biochar adsorption properties to generate highly concentrated N fertilizers (Rasse, Weldon, & Joner, 2022). This does not exclude the utility of biochar in regulating the N dynamics in soil by increasing the N adsorption and reducing leaching (Yao, Gao, Zhang, Inyang, & Zimmerman, 2012) or reducing N_2O emissions (Cayuela, Sánchez-Monedero, & Roig, 2013) when biochar is applied as soil amendment. Thus other strategies including granulation of N with biochar and additives (e.g., clays, polymers, chitosan, etc.) are needed to produce highly concentrated fertilizers and control the N release rate satisfactorily.

For P, there are different strategies to produce BBF that will generate different mechanisms of nutrient release dynamics. Biochar impregnated with cations (i.e., Ca^{2+}, Mg^{2+}, Al^{3+}, or Fe^{3+}) are usually used to remove P from aqueous media. Biochar normally has a negative surface charge due to oxygen functional groups and cation impregnation generates positive charges that adsorb from low to high amounts of phosphate from an aqueous solution and the energy of such adsorption will depend on the metal used and the metal content in the feedstock. Nardis, Carneiro, De Souza, De Barros, and Melo (2021) impregnated sewage sludge, poultry litter, and pig manure with magnesium chloride and pyrolyzed these mixtures aiming to adsorb P. It was observed that precipitation and ligand exchange were the main mechanisms of P adsorption, and despite the slow-release behavior of P, maize plants were able to access the P in the short-term from BBFs and they performed equivalently or higher than triple superphosphate. In another study, Al^{3+} and Mg^{2+} were impregnated at up to 20% in pig manure and their biochar was used to remove phosphate from aqueous solution (Nardis et al., 2021). Pig manure biochar doped with Mg^{2+} removed the most P (maximum 231 mg g^{-1}) mainly through precipitation, surface complexation, and electrostatic attraction, and it was recommended for reuse as a BB.

Copyrolysis of feedstock enriched with phosphate turns soluble phosphate compounds into low-solubility phosphates (e.g., pyrophosphate), which reduces P diffusion and release rate. This can be a strategy to reduce fast P fixation in highly weathered soils or reduce P leaching in sandy soils. For instance, poultry litter was enriched with triple superphosphate or phosphoric acid both combined with magnesium oxide aiming to reduce acidity and generate an enhanced efficiency biochar-based P fertilizer (Lustosa Filho, Barbosa, Carneiro, & Melo, 2019). In this study, water-soluble P reduced greatly (to <2%), but P soluble in citric acid or neutral ammonium citrate plus water showed high levels, and a greenhouse experiment with maize confirmed the efficiency of both fertilizers, especially the one made with phosphoric acid and applied in powder form. In another study, similar P BBFs made from poultry litter and coffee husk were studied in sequence crop cultivations [grass (three cycles), maize, and bean] in a greenhouse experiment (Carneiro et al., 2021). It was observed that P from BBFs show a slow and steady release that reduces its fast sorption (fixation) in P-fixing soils. Thus plants were able to absorb P without compromising their growth and biomass production.

The mechanisms of P delivery from BBF are still under investigation, but the slow-release behavior associated with the negative charge surface and buffering capacity caused by the biochar matrix in the BBF seems to play a central role in increasing P use efficiency. Sugarcane straw biochar was activated with KOH and neutralized with H_3PO_3 aiming to produce a biochar fertilizer of enhanced efficiency (Borges, Strauss, Camelo, Sohi, & Franco, 2020). It was observed that this BBF outperformed triple superphosphate, especially in clayey soil, due mainly to the pH buffering at the range of 7–8 for a long time, while TSP caused a pH decrease to 3 and remained lower than the BBF up to 40 d, which made the difference regarding the P availability in the vicinity of the fertilizer granules. BBF can also be produced from copyrolysis of feedstock and rock phosphate. Phosphorus can be solubilized and made available to plants through the incorporation of phosphate-solubilizing bacteria (de Amaral Leite, de Souza Cardoso, & de Almeida Leite, 2020). This can be an inexpensive and sustainable approach to explore insoluble P sources such as rock phosphates or even bone chars. The chemical modification of P-rich feedstock such as poultry manure with $Mg(OH)_2$ followed by pyrolysis is also efficient in increasing P solubility and availability to plants due to the increase in the formation of magnesium phosphates (less crystalline) and lower formation of calcium phosphates (more crystalline) (Leite et al. in preparation).

Nitrogen mineralized from organic matter (OM) only supplies part of the N required by crops. In most crop fields, N must be supplemented by mineral fertilizers, especially in sites in which the use efficiency of N by crops is reduced. Globally, most synthetic N supplied to crops is lost to air and groundwater with serious impacts on the environmental quality (Ayele & Atlabachew, 2021; Gao et al., 2022). Additionally, N greenhouse emissions from rising fertilizer use threaten the efforts to reduce global warming. Conversely, where the supply of N to crops is limited, the reduced availability of N in soil hampers plant growth, and food yield as well (Gao et al., 2022). To be effectively used by crops, N needs to be gradually released by fertilizers. Nitrogen supplied by mineral fertilizers is readily available to plants, therefore it is prone to leach to subsoil layers less exploited by plant roots, especially in sites in which the nitrification rate is high. Overall, the overuse and misleading management of N fertilizers cause ammonia (NH_3) volatilization, nitrate leaching, and nitrous oxide (N_2O) emission, with subsequent pollution of surface water, groundwater and air, besides rising costs of fertilization; conversely, low rates of N added to soil reduce food production (Wang et al., 2022; Wen et al., 2017).

The use of slow-release biochar-based N fertilizers (BBNFs) is a promising strategy to increase N use efficiency (NUE), enhance crop N acquisition and decrease off-site N losses (Gao et al., 2022; Wang et al., 2022). Improved efficiency of N-fertilizer use is achieved when N supply is synchronized with crop growth stages of higher N demand (Fageria & Baligar, 2005; Liao et al., 2020). The methods to formulate biochar N fertilizers include in situ pyrolysis, copyrolysis, impregnation, encapsulation, and granulation (Barbosa, Correa, Carneiro, & Melo, 2022; Wang et al., 2022). In most BBNF synthesis routes, pyrolysis can promote the conversion of minerals into N organic forms or the formation of the novel and chemically more stable N chemical species in biochar, mainly in those produced at high temperatures (Almendros, Knicker, & González-Vila, 2003). Stability, pools, reactivity, interaction with the biochar matrix, and solubility of N chemical forms will, in the last instance, determine the agronomic value of biochar or composites, and, consequently, its agronomic efficiency in nourishing crops, while avoiding N loss and air pollution. Biochar can be an efficient N-carrying matrix for plants because it contains functional groups capable of adsorbing N-ammonium and N-nitrate, increases water retention in the soil, and reduces soil N leaching (Rasse et al., 2022). Nitrogen is more gradually released by BBNFs than soluble N sources such as MAP and ammonium sulfate (Cen, Wei, Muthukumarappan, Sobhan, & McDaniel, 2021), but most studies have focused on the interaction of biochar and urea, the main N source used in agriculture (Barbosa et al., 2022; Liao et al., 2020; Shi et al., 2020). Also, the C present in stable and carbonized matrices persists for a longer time in the soil, interacts with the soil biota and controls the rates of nitrification, denitrification (flow of N_2O from the soil to the atmosphere) and urea hydrolysis. According to Tang, Liu, Li, and Liu (2022), the mixture of biochar and ammonium fertilizer reduced soil N_2O emissions by approximately 31%, while the combination of cotton-derived biochar and nitrate fertilizer decreased soil N_2O emissions by 63%–71%. Urea added to the soil along with biochar and biochar + urea inhibitors reduce soil ammonia emissions by 27% and 69%, respectively, over the exclusive use of urea (Dawar et al., 2021). Upon urea, both biochar and urea inhibitors improved biomass and yield, and NUE by wheat plants, respectively, at 38%, 22%, and 27%, respectively (Dawar et al., 2021). In fact, the intensive use of urea in the world is one of the main causes of intensive losses of N by leaching, volatilization, and N_2O emission from intensive N-fertilized crop fields (Rasse et al., 2022; Shi et al., 2020). Upon conventional urea, Biochar-urea-bentonite-sepiolite composite produced from urban green waste reduced the N leaching in soil, improved the retention and slower N-ammonium conversion into nitrate, while increasing shoot biomass by 14%, maize root growth by 25% and NUE by maize (Shi et al., 2020). However, in Biochar-urea-bentonite composite produced from oilseed rape straws, compared with conventional urea, the nitrification rate was higher in BBNFs-treated soils, but the NUE of N was higher (+59%) due to the reduction of denitrification (Liao et al., 2020). Thus the potential of biochar to affect the dynamic of nitrification depends on the biochar properties, thus, charred matrices are capable of controlling the abundance of ammonium or nitrate forms in soil (Barbosa et al., 2022; Liao et al., 2020; Shi et al., 2020).

Mixing mineral N fertilizers with biochar is justified by the fact that biochar contains inner pores and channels in which mineral N diffusion is supposed to happen or N physical adsorption in charred can take place. The main mechanisms related to biochar influence on the N cycle and its nutrient use efficiency by crops are: (1) adsorption at surface functional groups, (2) protonation of NH_3 and induced conversion into NH_4^+, (3) oxidation of NH_4^+ to NO_3^- (affecting nitrate leaching rate), (4) adsorption of nitrate at biochar exchange sites, (5) adsorption of ammonium at biochar charges, (6) microbial induced N immobilization, (7) adsorption mineral N forms in biochar micropores (Mandal et al., 2016). In addition, biochar has a variable and broad CEC over soil matrices (Lago, Silva, Melo, & de Morais, 2021). In addition, ammonium is adsorbed at biochar negative charges in low-energy electrostatic bonds, although the amounts of NH_4^+ are reduced and rely on feedstock and pyrolysis temperature employed in the biochar synthesis (Gai et al., 2014; Weldon et al., 2022). As the pyrolysis temperature increases, the abundance of organic functional groups in biochar is lower, as well as the potential for ammonium retention (Cai, Qi, Liu, & He, 2016; Phuong, Hoang, Luan, & Tan, 2021). BBNFs release N to crops gradually (Liao et al., 2020). In fact, the pattern of N release relies on biochar properties, mainly pH, elemental composition, organic N content and pools in the charred matrices, N pools and contents in raw feedstocks, and the processes and sources of mineral N used in the synthesis of BBNFs as well (Gao et al., 2022). In addition, types of N precursors can make important contributions to the properties of BBNFs.

In addition to mineral N, organic N pools are present in BBNFs. Organic N in biochar is gradually released, therefore less prone to be leached and supposed to meet the nutritional demands of crops at different growth phases. In the BBNFs synthesis, it is necessary to optimize the pyrolysis conditions (temperature, retention time, feedstock, etc.), so that the N in the carbonized matrix is high and the losses of N by volatilization during pyrolysis are reduced. Even so, there are losses of N during pyrolysis and the N in biochar is less labile than that of the biomass that gave rise to it (Liu, Li, Jiang, & Yu, 2017; Schellekens et al., 2018; Wang, Arbestain, Hedley, & Bishop, 2012). Overall, the N content in crop residues is reduced, thus it is necessary to choose organic residues with a higher N content, such as manures derived from intensive animal production systems and sewage sludge, as well as cakes derived from materials with high content of proteins. At pyrolysis temperatures greater than 450°C, the N in the carbonized matrix has a greater aromatic character, as heterocyclic N is prevalent in biochar, therefore this N pool is less susceptible to decomposition (Almendros et al., 2003; Liu et al., 2017). The reduction of N losses during pyrolysis and the neoformation of more labile N compounds in biochar is the right condition to reduce the amount of mineral N in the formulation of N-biochar-based composites.

It is unlikely that simple pyrolysis of N-rich residues would produce fertilizers that are high in N and capable of nourishing plants. Therefore there are several synthesis routes for BBNFs, but, technically, it is not feasible to formulate fertilizers with high N without the use of mineral N fertilizers. Nitrogen N losses during pyrolysis are variable, being governed by pyrolysis temperature and feedstock. In addition, the remaining N in biochar is less susceptible to mineralization than the N in the pristine feedstock (Li, Barreto, Li, Chen, & Hsieh, 2018; Wang et al., 2012). Therefore high doses of BBFs would be required to nourish the crops. As a result, the alternatives would be physical mixtures of carbonized N matrices with mineral fertilizers, encapsulation of urea with biochar and the use of biochar with manure with subsequent pelletizing or granulation of the mixture. The use of BBNFs leads to improved corn growth over ammonium sulfate, besides reducing corn losses (Ameer, Wei, Wu, & Rubel, 2022). In the formulation of composites, there is a greater chance that sources rich in nitrate produce BBNFs of greater agronomic value than ammonium counterpart sources because ammoniacal N is more prone than nitrate to volatilized when N is mixed with high pH biochar. When mixed with biochar with pH in the alkaline range, N as ammonium is more prone to be converted into ammonia gas form.

In the synthesis of biochar-mineral fertilizer composites, it is necessary to check the content and the chemical nature of the functional groups in the carbonized matrix, the pore distribution and the CEC of the biochar. It is possible to functionalize biochars using acids, H_2O_2, and KOH, so that new, more reactive functional groups (C—O, RCOOH, C = O, etc.) are formed. Biochars with a greater presence of functional groups capable of retaining mineral N in fillers and in pores of the carbonized matrix may be more effective in retaining mineral N. In the case of urea, in addition to encapsulation, there are already reports in the literature of diffusion of this N source in biochar pores, which increases the efficiency of the use of amide N. In encapsulating urea with biochar, it is possible to use biodegradable clays and polymers (sodium alginate, cellulose acetate, and ethyl cellulose), in mixtures with biochar, which delays the hydrolysis and leaching of urea (Gao et al., 2022; Li et al., 2018; Wang et al., 2012). The agronomic efficiency of carbonized sources or biochar composites in supplying N to crops depends on the forms of N present in the N sources, the rate of N release and the ability of the matrices to carbonize to reduce N loss processes and maintain in the soil suitable $NH_4^+:NO_3^-$ ratios for crop nutrition. Regarding the kinetics of N release by carbonized matrices, the pyrolysis temperature is a determining factor, since high temperatures produce carbonized matrices with low N contents, reduced CEC and stable heterocyclic forms of N with high aromatic character, therefore less susceptible to mineralization. The rate of N release by biochars and composites is regulated by the susceptibility of organic forms to decomposition. In a broad and seminal review, Gao et al. (2022) verified that the release of N associated with carbonized matrices is more gradual, with partial mineralization of N in relation to mineral sources. In this direction, the N from urea was hydrolyzed and released into the soil in less than 10 days,

16.3 Slow-release mechanisms of biochar-based fertilizers

whereas Biochar-urea-based composites regulate the ammonium: nitrate ratio available to plants, thus affecting the crop growth. This is because biochar affects urea hydrolysis and nitrification rate (Gao et al., 2022).

The advantages associated with the use of N BBFs occur due to the more efficient use of N by crops, reduction of N losses, and reduction of GHG (N_2O) emissions by replacing mineral sources of N (more polluting) with composites of N-biochar (Gao et al., 2022). Some factors limit the use of biochars as N-carrying matrices, given that there is volatilization of N during pyrolysis, there may be immobilization of mineral N in the carbonized matrix and the speed of N release may not be compatible with the nutritional phases of greater demand for N by crops. N immobilization happens when biochars with a higher C:N ratio are added to soil, mainly those charred matrices with high contents of lignocellulosic, although N initial immobilization in biochar-treated soils contributed to reducing N_2O emissions (Borchard et al., 2019). The use of BBNFs in agriculture is advantageous because it encompasses modern concepts of the circular economy since organic waste is properly disposed of. The stoichiometric basis of the mixture of N mineral sources with biochar has not yet been established and it is necessary to carry out a life cycle analysis of the newly synthesized fertilizers, but, potentially, there is greater energy efficiency, less pollution and a reduction in GHGs emissions when N mixed with carbonized matrices is used to replace mineral N fertilizers.

One of the main mechanisms involved in the increase of BBNFs on the greater efficiency of N use is the control of the speed of N release, with N gradually releasing N to the plants, an effect conditioned by the properties of the formulated BBNFs (Gao et al., 2022). However, if this release is too slow, plants with a short life cycle, or those with high initial N demand, such as annual crops, may be harmed, so the study of release kinetics and/or mineralization must be accompanied by studies of plant growth. Thus the same BBNFs can present different N use efficiencies due to the effect of the crop, time of application and cultivation conditions, and the ammonium: nitrate ratio supplied to crops. Thus the ideal is the development of BBNFs and characterization of N release patterns, N forms and effects on their proportion so that N can be supplied to suit different crop species in different growth stages and production systems.

Therefore compared to mineral and organic fertilizers, BBNFs have greater advantages, as they contain both mineral forms of N that are used in the initial development of plants, as well as organic forms of N, which are mineralized throughout the plant growing season. Regarding the mineral forms, another advantage is that BBNFs, due to their ability to retain N over soluble N sources, cause even this mineral N to be released gradually. In addition, biochars have high hydrophobicity, especially when using higher pyrolysis temperatures, which makes the BBNFs hydration process difficult, consequently reducing N release. Since BBNFs have gradual release due to the different mechanisms mentioned, after cultivation, a greater residual effect is expected, so their use should be evaluated not only for the current crop, but throughout the cultivation of different plant species successively grown. The residual effect of BBNFs formulated with urea shows a longer N residual effect than that produced from the exclusive use of urea, which was explained due to the gradual N release pattern of BBFs (Barbosa et al., 2022).

Sulfur is known as one of the hidden nutrients, thus its use is highly recommended in most cropland areas. Sulfate is the major source of S uptake by crops in the soil. Potential

S sources to crops are elemental S from volcanic areas and S compounds trapped in filters when petroleum is refined, and natural gas and coal are burnt, and their evolved gases are trapped in biochar. Nowadays, more severe legislation restricts the levels of S in the air, which could increase S inputs recycled for agriculture. Sulfur in the form of H_2S, a corrosive and toxic gas, is one of the final products of fermentation. The use of biochar to retain H_2S in fermentation streams is a feasible route to produce S-enriched biochar that could reduce air pollution while benefiting agriculture as an additional S source to crops (Zhang, Voroney, Price, & White, 2016). Accordingly, biochar was already used to adsorb H_2S generated in biodigesters, and the S-enriched biochar (SulfaChar) and synthetic S fertilizer (control) were used to nourish soybean plants grown in potting soil in a 90-day greenhouse study (Zhang et al., 2016). Elemental S and $S-SO_4^{2-}$ were the dominant S forms in Sulfachar (36.5%), confirming biochar, besides retaining H_2S, converted it into nontoxic S forms capable of nourishing soybean plants. Compared with the sulfate, SulfaChar sharply increased (31%–49%) corn biomass, while only slightly improved soybeans growth (+4 to +14%). SulfaChar also improved the uptake of S, N, P, K, Ca, and Mg, and crop acquisition of Zn, Mn, and B as well (Zhang et al., 2016).

It should be highlighted that most crop residues are known by their reduced levels (0.2%–0.5%) of S (Silva, 2008), which is prone to be volatilized during pyrolysis, although in some biochar S content increases, decreases or resemble the levels of pristine feedstocks (Leng et al., 2022). It is a high challenge to produce biochar-based S fertilizers only with in situ pyrolysis, but it is possible to create new routes to formulate S composites with increased contents of S in biochars. With this purpose, exogenous forms of S should be added to biochar, or previously mixed with feedstocks before the carbonization of many S-poor feedstocks. Elemental S is recycled predominantly when S is recovered from the refineries of oil and gas (Eow, 2002), and the low price of this recycled S is a stimulus for its use for agricultural purposes (Mattiello et al., 2017). The addition of S to carbonized samples, or to enrich biochars with gaseous forms of S, such as H_2S, is a suitable process to produce composites that have high levels of S capable of nourishing plants while improving their growth and yield. Elemental S could also be used to acidify biochar with pH in the alkaline range, as well as sulfuric acid, to remove HMs and lower the pH of the final biochar. Leaching of biosolids with a 3% H_2SO_4 solution lowered the HMs content and the ash content of biochar, preserving its physicochemical attributes, while improving the pyrolysis conditions of the treated sewage sludge (Hakeem et al., 2022). The use of sulfuric acid in the carbonization of rice husk produced sulfonated biochar with greater C content, high capacity to sequester C and abundant SO_3, H and OH functional groups capable of efficiently absorbing Cd^{2+} (Zhou et al., 2021). Acidification of biochar due to oxidation of elemental sulfur can solubilize salts in the ash of charred matrices, providing slow-release readily available forms of sulfate salts in biochars. The study carried out by Mattiello et al. (2017) demonstrated that the mixing of elemental S with Zn in granulated fertilizers increased S and Zn contents in an acid soil over time, as well as Zn diffusion, and gradually released available Zn and S in a slightly alkaline soil.

It is worth mentioning that the use of elemental S in the pyrolysis process should be carried out with care due to the elevated risk of elemental S explosion. Thus it is not safe, nor it is recommended to mix elemental S with organic waste, notably those rich in nitrate, again, due to the risk of explosive reactions during pyrolysis, although elemental S could be used as an additive in postpyrolysis processes to enrich the biochar with S and reduce the pH of alkalinized biochars. Sulfur interacts with other nutrients, notably N, as well as optimizes the uptake of phosphate and micronutrients (Mattiello et al., 2017). The use of organic waste rich in S, notably sewage sludge and other wastes enriched in S, is recommended for the production of Biochar-based S fertilizers, although S is largely volatilized during pyrolysis, notably at high-temperature charring processes. Overall, sewage sludge is richer in S (0.3–2.3 wt.%) than crop residues (Silva, 2008), and inadequate disposal and combustion lead to the formation and emission of SO_2, which triggers acid rain problems (Dewil, Baeyens, Roels, & Van De Steene, 2009). Accordingly, pig manure and sewage sludge-derived biochars were investigated regarding their capacity to adsorb and convert hydrogen sulfide into S forms available to crops (Xu, Cao, Zhao, & Sun, 2014). Moisture, alkalinity degree, and contents of inorganic salts in the biochar ruled the biochar's ability to retain H_2S and convert it into SO_4^{2-}, which predominates as $CaSO_4$ in the sewage sludge biochar (SSB), and in mostly soluble $(K, Na)_2SO_4$ forms in pig manure-derived biochars (Xu et al., 2014). Overall, S functional groups found in biochars include organic sulfur (e.g., C–S, –C–S–C–, C=S, thiophene, and sulfone) and inorganic sulfur (e.g., sulfate, sulfide, sulfite, and elemental S), which can be identified and determined through FTIR, X-ray photoelectron spectroscopy (XPS), and XANES (Leng et al., 2022). Sulfur organic groups found in biochar matrices play an important role in removing pollutants, as well as in nourishing crops and acting as a solid acid catalyst to accelerate the formation of polar groups capable of interacting and complexing metals in the charred matrix (Leng et al., 2022).

16.3.2 Potassium release dynamics

Studies on BBFs enriched exclusively with potassium are scarce. Most works involve the enrichment of at least two macronutrients, usually, N and P. SSB is the most commonly used raw material in K-enriched BBF production. SSB belongs to the group of biochars that have the highest concentrations of the main plant nutrients, especially P, Ca, N, and Zn (Yue, Cui, Lin, Li, & Zhao, 2017). Therefore SSB can be used directly as a fertilizer. However, the final stage of sewage treatment promotes the precipitation of N and P, making the sewage sludge (SS) concentrated in these nutrients and poor in K. During sewage treatment, K is not incorporated into the SS biomass, being eliminated in the form of soluble salts along with the liquid effluent (Kirchmann, Börjesson, Kätterer, & Cohen, 2016). As a consequence of the low K concentration in the SS, the content of this nutrient in the SSB has not been sufficient for the adequate supply of K for several crops (Chagas, Figueiredo, & Paz-Ferreiro, 2021; Faria et al., 2018), requiring complementation by mineral fertilization or the need to apply high doses of SSB for an adequate supply of K (Faria et al., 2018).

To overcome this limitation, enriching SSB with K sources is an alternative with great potential to generate new BBFs. Enrichment of SSB with soluble sources of K (KCl and K_2SO_4) was successfully performed (Fachini et al., 2021). These new fertilizers in the form of granules and pellets were evaluated for the K release in pure silica and the results showed that this enriched made it possible to obtain a slow-release K fertilizer

(Fachini et al., 2022). This performance of K-enriched BBF as a slow-release fertilizer has been proven in both substrate/soil and water.

In general, the experimental conditions used in different studies, such as the K release method, incubation time, fertilizer dose, soil moisture or water holding capacity, in addition to the type of substrate, lead to divergences in the results. Soil moisture level has been highlighted as an important factor in the assessment of K release from BBFs. An et al. (2020) found a total K release from K-enriched BBF of about 82% in soil incubated for 20 days and kept at 40% moisture. Piash, Iwabuchi, and Itoh (2022) reported a 22% slower release of K over 90 days in clayey soil (with 60% moisture) when compared to pure biochar. When incubated in pure silica sand for 30 days, BBF reduced the K release by 77% compared to KCl when the moisture level was between 10 and 20% (Fachini et al., 2022). The slow release of K from K-enriched BBF has also been proven in water. Gwenzi, Nyambishi, and Mapope (2018) observed that K release was 1.5 times slower than soluble mineral fertilizer when incubated in water for 68 days. Similarly, total K release was approximately 74.33%–77.27% in water over the 90-day period (Das & Ghosh, 2021). The release of available K was less than 80% when 0.5 g of BBF was submerged in 50 mL of distilled water for 28 days (Wu et al., 2021). Kim, Hensley, and Labbe (2014) observed a total release of 78%–87% of K after 18 days of K-enriched BBF incubated in water.

The dynamic of K release from BBFs is strongly influenced by fertilizer characteristics such as high surface area, porous microstructure, surface charge, and multiple functional groups that allow the retention of K ions through physical and chemical adsorption (Das & Ghosh, 2021; Gwenzi et al., 2018). The mechanisms involved in the slow-release dynamics of K from BBFs are shown in Fig. 16.1. K adsorption mechanisms include surface complexation, inner-sphere complexation, surface ion exchange, coprecipitation, electrostatic attraction, and cation interactions between solute ions and biochar aromatic rings (Ye, Zhang, Huang, & Tan, 2019). Carboxyl is the main functional group responsible for surface complexation with metal cations due to the formation of C–O bonds, while

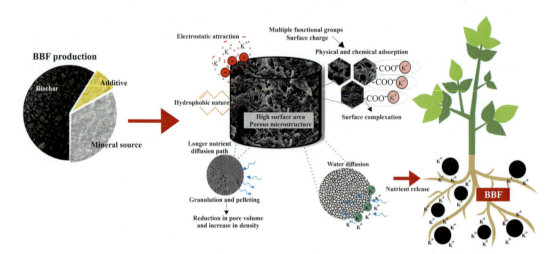

FIGURE 16.1 Mechanisms involved in the slow-release dynamics of K from biochar-based fertilizers.

hydroxyl reacts with metal cations via internal complexation (Sawalha, Peralta-Videa, Saupe, Dokken, & Gardea-Torresdey, 2007). In addition, these functional groups can undergo ionization in an alkaline environment, acquiring negative charges and thus exerting an electrostatic attraction on metallic cations such as K. Therefore the greater the number of carboxyl and hydroxyl in the biochar, the greater the adsorption of K and the slower its release (Ye et al., 2019).

The main characteristics of biochar responsible for the slow release of K are porosity and surface area which confer greater exposure to nutrient adsorption sites (Gong, Tan, Zhang, & Huang, 2019). Furthermore, the copyrolysis of clay and biomass results in the formation of hydrophobic compounds, such as alkene/alkane, and aromatic compounds (Cheng et al., 2022) which, in addition to repelling water, promote the formation of hydrogen bonds and electrostatic attraction (Xiao, Chen, Chen, Zhu, & Schnoor, 2018). K ions can be strongly adsorbed at the crystal exchange sites of bentonite and ash during pyrolysis forming stable K species such as microcline ($KAlSi_3O_8$) and kalsilite ($KAlSiO_4$) (Piash et al., 2022; Wu et al., 2021).

In addition to the physicochemical characteristics of the biochar, the physical form (granule and pellets) also promotes a slower release of K from BBF (Fachini et al., 2022; Gwenzi et al., 2018; Kim et al., 2014; Roy et al., 2022). This is because, during the granulation and pelleting processes, the fertilizer particles are forced against each other leading to reduced pore volume and increased density (Reza, Uddin, Lynam, & Coronella, 2014). In addition, pellet sizes are larger which results in a longer nutrient diffusion path and greater mechanical strength (Jiang et al., 2014). Binders such as lignin and starch (Fachini et al., 2022; Gwenzi et al., 2018), as well as coating materials (Das & Ghosh, 2021), can reduce fertilizer disintegration and therefore reduce K release.

Several mathematical models can be used to describe the kinetics of K release from BBFs. All models suggest that K release is mainly controlled by desorption and diffusion processes. According to Fachini et al. (2022), the release of K from BBF can be divided into three steps: (1) the initial rapid release of the water-soluble fraction present on the biochar surface by the rapid diffusion process; (2) a slower release of the poorly soluble fraction found in biochar pores and channels. This process also occurs by diffusion; however, it is slower than the previous phase because the pore network creates physical protection, limiting the access of water to this fraction located inside the biochar matrix; and (3) the disintegration of the insoluble fraction which is strongly adsorbed to the solid phase. This fraction is generally not released during the short-term leaching experiment. After soil application, the release of nutrients from this insoluble reservoir can occur through BBF microbial degradation (Das & Ghosh, 2021; Fachini et al., 2022; Gwenzi et al., 2018).

16.4 Cocomposted biochar as an enriched nutrient source

The miraculous biochar can be put into use in several ways, one of which is cocomposting with organic amendments viz. FYM, manures, compost, and vermicompost, to enrich the biochar with the organic nutrient sources before the aerobic or anaerobic composting. The deliberations were on the rise in the recent past to substantiate the potential benefits of COMBI viz. reduced emissions, improved soil structure, soil quality, microbial habitat,

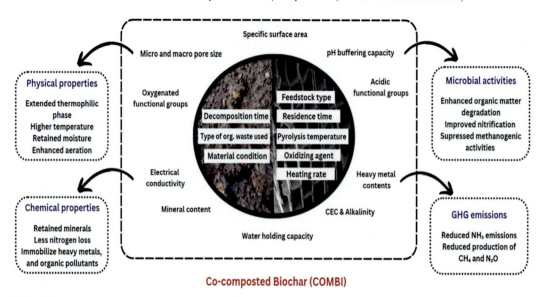

FIGURE 16.2 Schematic representation of cocomposted biochar and its sustainable ecological effects on soil.

and system productivity (Fig. 16.2). Both techniques of composting have distinct effects on the end product; the aerobic process utilizes O_2 to decompose organic wastes coupled with higher temperature and quick decomposition rates, in contrast to slower decomposition and lower temperatures in the absence of oxygen with an anaerobic process (Osman et al., 2022; Zainudin, Singam, Sazili, Shirai, & Hassan, 2022). Better aeration during composting facilitates more microbial abundance and activity resulting in better mineralization and nutrient availability, whereas the entire process devoid of oxygen improves compost quality and OM content (Chiaramonti, Casini, Barsali, Yambanis, & Ruiz, 2019). Either way, the end product of both processes has a positive impact on output and significantly enhances its quality and nutritional composition. The proportion rationale of biochar and organic material affects the final compost quality, and upon its addition to the soil alters various soil physicochemical and biological properties constructively. An optimum ratio of biochar (10%—15%) with the organic materials is recommended (Steiner, Das, Melear, & Lakly, 2010), however, higher doses of biochar loading were also reported by Chowdhury, de Neergaard, and Jensen (2014). The pyrolysis temperature, heating rate, and residence period of biochar, type and quality of organic material influence the composting quality. The incorporation of biochar into composting accelerates the breakdown of organic waste materials through a more organized approach, resulting in the creation of a compost product that is more nourishing and environmentally safe. This cocomposting process significantly enhances the quality of both biochar and compost, subsequently leading to improved soil health and higher crop yields, as noted by Ravindran et al. (2022). The COMBI demonstrates superior physicochemical properties, which prove more advantageous in enhancing soil quality in agricultural areas compared to using biochar and compost separately (Tian, Tang, Ge, & Tsang, 2023).

A significant number of studies on COMBI were on the rise in the last few years and unswerving potential results were being reported on soil nutrients, health, quality, and productivity. An improvement in rice-sorghum system productivity by 75% through better vegetative growth and development after soil application of charcoal and manure cocompost in Brazil was reported by Steiner et al. (2007). When biochar is incorporated into compost, its filtration properties lead to decreased leaching and volatilization of plant nutrients, particularly nitrogen. Biochar's surfaces can absorb nutrients, reducing losses and enhancing the fertilizer value of the compost (Joseph et al., 2018). Studies consistently show that cocomposting with biochar reduces nitrogen losses by over 50%, favoring nitrate over ammonia/ammonium (Kammann et al., 2015). Additionally, biochar's porous structure improves aeration and moisture regulation during the composting process, supporting nutrient-cycling microbes and promoting soil health and plant vitality (Fischer & Glaser, 2012). In a study by Holatko et al. (2022), the nitrogen-enriched biochar cocompost brought a significant increase in maize and cowpea yield by 23.8% and 20.7%, respectively, over unamended soil. On the other hand, the COMBI-biochar liming effect on the soils with pH 4−5 brought a drastic increase in the yield of cereal grasses by 39.7% (Wang, Villamil, Davidson, & Akdeniz, 2019). The after-harvest soil properties with the application of COMBI in maize reportedly increased N content by 45%−56%, and cob yield by 18%−24% (Naeem et al., 2018). Green vegetable waste, *Eupatorium*, and cattle farmyard manure were cocomposted with biochar (80:20 ratio), which reportedly improved the soil CEC, potassium, calcium, and magnesium with a 24.3% increase in corn yield (Pandit et al., 2019). Soil application of *Acacia* green waste and fowl manure (10 t ha^{-1}) improved the barley yield by 30%−49%, increased dissolved organic carbon by 34%, cation-exchange capacity by 24%, and improved soil fertility (Agegnehu, Nelson, & Bird, 2016). The results indicate that incorporating COMBI can enhance crop productivity through better soil fertility, nutrient retention, compost quality, and reduced GHG emissions. Nevertheless, it is essential to recognize that the actual impact of COMBI on crop productivity may differ depending on variables like the type of biochar, composting method, and specific management conditions.

The role of biochar toward GHG emissions and their mitigation is well established through potential carbon sequestration (Singh et al., 2023). Among the different potential areas where biochar can be put into use; the emissions from the decomposition of organic wastes can be mitigated by biochar through a combined composting process. Apart from mitigating potential emissions from agricultural ecosystems, GHG fluxes from composting can be reduced through dynamic changes brought by biochar in the composting process. The mitigation of GHG emissions while composting was construed by the addition of biochar, which is evidence of astounding results on COMBI. The addition of biochar to the pig manure for composting significantly reduced CO_2 emissions by 26.1%−51.0% over control, an apparent moment enlightens the adsorption capacity of biochar (Vandecasteele et al., 2017; Wang et al., 2018). Soil aeration while composting; an important phenomenon was managed efficiently through biochar by reducing anaerobic sites and anaerobe activity while improving methanotroph activity and reducing the CH_4 fluxes (Sonoki et al., 2013; Yin et al., 2021). Composting through the incorporation of wheat straw biochar and sewage sludge (1:1) reported 80% reduction in methane emissions in contrast to biochar-devoid composting (Awasthi et al., 2016). On the other hand, the NO_2−N emissions from composting were drastically reduced by 25.9% over control, when biochar was added to wood chips, pig manure, and sawdust for composting. The biochar surface sites have the potential to adsorb NO_2-N and convert to nitrogen through various bio-geochemical reactions (He et al., 2019). Similarly, the NH_4 emissions from the decomposition of poultry litter were reduced by 47% through the addition of pine chips biochar to obtain ecologically sustainable COMBI (Steiner et al., 2010). Overall, from the findings on COMBI, it is evident that biochar demonstrates remarkable potential in mitigating GHG emissions during composting processes. Its incorporation significantly reduces CO_2, CH_4, NO_2-N, and NH_4 emissions, highlighting its impressive adsorption and conversion capabilities. The incorporation of biochar in composting presents a promising approach for ecologically sustainable management and carbon sequestration.

16.5 Conclusion

In future studies, BBFs should be tested for different crops, soil types and agricultural systems. The effect of successive application of BBFs in increasing the residual levels of P, N, K, S, and mainly, micronutrients in soils should also be investigated, as well as the growth and yield of crops successfully cultivated in both temperate and tropical soils field conditions. Pyrolysis conditions (temperature, retention time, etc.), the influence of feedstock and functionalization and the use of additives to reduce the losses of C, N, and S during pyrolysis should be a priority for future studies. Overall, carbon-based fertilizers are a new technology that must be tested and used in cropland areas, since they combine the principles of circular economy (waste cycling) with the synthesis of fertilized with improved agronomic efficiency for highly weathered soils. In tropical areas, the fertilizer sector must adopt the technologies available for the synthesis of fertilizers that fit better to crops and soil types in order to fully or partially replace conventional soluble mineral fertilizers of low efficiency massively used in tropical soils. Important advancements have been reached, but more could be done with the production of composites, optimization of pyrolysis conditions, coapplication of nutrients with biochar, mixtures of biochar with layered double hydroxides, use of soluble and bioactive compounds of biochar organic matrix, biochar nanoparticles, and so on. The main challenges to developing this new sector of slow-release BBFs include the following challenges (Wang et al., 2022): (1) the synthesis of high agronomic value slow-release fertilizers, (2) the definition of BBF nutrient release patterns and the bioavailability of its nutrient chemical species, and (3) the long-term assessment of BBFs in real crop field conditions. Finally, in the tropical region, the results, technologies generated, BBF synthesis routes and agronomic efficiency and performance in lab conditions signalizes that BBFs could be suitable sources to nourish crops in highly weathered soils.

References

Agegnehu, G., Nelson, P. N., & Bird, M. I. (2016). Crop yield, plant nutrient uptake and soil physicochemical properties under organic soil amendments and nitrogen fertilization on nitisols. *Soil and Tillage Research, 160*, 1–13.

Ahmad, M., Usman, A. R. A., Al-Faraj, A. S., Ahmad, M., Sallam, A., & Al-Wabel, M. I. (2018). Phosphorus-loaded biochar changes soil heavy metals availability and uptake potential of maize (*Zea mays* L.) plants. *Chemosphere, 194*, 327–339.

Almendros, G., Knicker, H., & González-Vila, F. J. (2003). Rearrangement of carbon and nitrogen forms in peat after progressive thermal oxidation as determined by solid-state ^{13}C- and ^{15}N-NMR spectroscopy. *Organic Geochemistry, 34*, 1559–1568.

Ameer, S., Wei, L., Wu, Y., & Rubel, R. (2022). Effects of biochar-based control release nitrogen fertilizers on corn growth in greenhouse trials. In *2022 ASABE annual international meeting* (p. 1). American Society of Agricultural and Biological Engineers.

An, X., Wu, Z., Yu, J., Cravotto, G., Liu, X., Li, Q., ... Yu, B. (2020). Copyrolysis of biomass, bentonite, and nutrients as a new strategy for the synthesis of improved biochar-based slow-release fertilizers. *ACS Sustainable Chemistry & Engineering, 8*(8), 3181–3190.

Awasthi, M. K., Wang, Q., Huang, H., Li, R., Shen, F., Lahori, A. H., ... Zhang, Z. (2016). Effect of biochar amendment on greenhouse gas emission and bio-availability of heavy metals during sewage sludge co-composting. *Journal of Cleaner Production, 135*, 829–835.

Ayele, H. S., & Atlabachew, M. (2021). Review of characterization, factors, impacts, and solutions of Lake eutrophication: Lesson for lake Tana, Ethiopia. *Environ Sci Pollut Res, 28*, 14233–14252. Available from https://doi.org/10.1007/s11356-020-12081-4.

Barbosa, C. F., Correa, D. A., Carneiro, J. S. D. S., & Melo, L. C. A. (2022). Biochar phosphate fertilizer loaded with urea preserves available nitrogen longer than conventional urea. *Sustainability, 14*(2), 686.

Bolan, N., Hoang, S. A., Beiyuan, J., Gupta, S., Hou, D., Karakoti, A., ... Van Zwieten, L. (2022). Multifunctional applications of biochar beyond carbon storage. *International Materials Reviews, 67*(2), 150–200. Available from https://doi.org/10.1080/09506608.2021.1922047.

Borchard, N., Schirrmann, M., Cayuela, M. L., Kammann, C., Wrage-Mönnig, N., Estavillo, J. M., & Novak, J. (2019). Biochar, soil and land-use interactions that reduce nitrate leaching and N_2O emissions: A meta-analysis. *Science of the Total Environment, 651*, 2354–2364.

Borges, B. M. M. N., Strauss, M., Camelo, P. A., Sohi, A. P., & Franco, H. C. J. (2020). Re-use of sugarcane residue as a novel biochar fertilizer – Increased phosphorus use efficiency and plant yield. *Journal of Cleaner Production, 262*, 121406.

Cai, Y., Qi, H., Liu, Y., & He, X. (2016). Sorption/desorption behavior and mechanism of NH_4^+ by biochar as a nitrogen fertilizer sustained-release material. *Journal of Agricultural and Food Chemistry, 64*(24), 4958–4964.

Carneiro, J. S. S., Ribeiro, I. C. A., Nardis, B. O., Barbosa, C. F., Lustosa Filho, J. F., & Melo, L. C. A. (2021). Long-term effect of biochar-based fertilizers application in tropical soil: Agronomic efficiency and phosphorus availability. *Science of the Total Environment, 760*, 143955.

Cayuela, M., Sánchez-Monedero, M., Roig, A., et al. (2013). Biochar and denitrification in soils: When, how much and why does biochar reduce N_2O emissions? *Scientific Reports, 3*, 1732. Available from https://doi.org/10.1038/srep01732.

Cen, Z., Wei, L., Muthukumarappan, K., Sobhan, A., & McDaniel, R. (2021). Assessment of a biochar-based controlled release nitrogen fertilizer coated with polylactic acid. *Journal of Soil Science and Plant Nutrition, 21*(3), 2007–2019.

Chagas, J. K., Figueiredo, C. C., & Paz-Ferreiro, J. (2021). Sewage sludge biochars effects on corn response and nutrition and on soil properties in a 5-yr field experiment. *Geoderma, 401*, 115323.

Chagas, J. K., Figueiredo, C. C., & Ramos, M. L. G. (2022). Biochar increases soil carbon pools: Evidence from a global meta-analysis. *Journal of Environmental Management, 305*, 114403.

Chen, Z., Pei, J., Wei, Z., Ruan, X., Hua, Y., Xu, W., ... Guo, Y. (2021). A novel maize biochar-based compound fertilizer for immobilizing cadmium and improving soil quality and maize growth. *Environmental Pollution, 277*, 116455.

Cheng, J., Liao, Z., Hu, S.-C., Geng, Z.-C., Zhu, M.-Q., & Xu, W.-Z. (2022). Synthesis of an environmentally friendly binding material using pyrolysis byproducts and modified starch binder for slow-release fertilizers. *Science of the Total Environment, 819*, 153146.

Chew, J., Zhu, L., Nielsen, S., Graber, E., Mitchell, D. R. G., Horvat, J., ... Fan, X. (2020). Biochar based fertilizer: Supercharging root membrane potential and biomass yield of rice. *Science of the Total Environment, 713*, 136431.

Chiaramonti, D., Casini, D., Barsali, T., Yambanis, Y.H., & Ruiz, J.P. (2019). Biochar and organic matter co-composting: A critical review. In *Biochar II: Production, characterization and applications*, ECI Symposium Series.

Chowdhury, M. A., de Neergaard, A., & Jensen, L. S. (2014). Potential of aeration flow rate and bio-char addition to reduce greenhouse gas and ammonia emissions during manure composting. *Chemosphere, 97*, 16–25.

Clare, A., Shackley, S., Joseph, S., Hammond, J., Pan, G. X., & Bloom, A. (2014). Competing uses for China's straw: The economic and carbon abatement potential of biochar. *GCB Bioenergy, 7*, 1272–1282.

Dad, F. P., Khan, W., Tanveer, M., Ramzani, P. M. A., Shaukat, R., & Muktadir, A. (2021). Influence of iron-enriched biochar on Cd sorption, its ionic concentration and redox regulation of radish under cadmium toxicity. *Agriculture, 11*, 1.

Das, S. K., & Ghosh, K. G. (2021). Developing biochar-based slow-release N-P-K fertilizer for controlled nutrient release and its impact on soil health and yield. *Biomass Conversion and Biorefinery*. Available from https://doi.org/10.1007/s13399-021-02069-6.

Dawar, K., Fahad, S., Jahangir, M. M. R., Munir, I., Alam, S. S., Khan, S. A., & Danish, S. (2021). Biochar and urease inhibitor mitigate NH_3 and N_2O emissions and improve wheat yield in a urea fertilized alkaline soil. *Scientific Reports, 11*(1), 1–11.

de Amaral Leite, A., de Souza Cardoso, A. A., de Almeida Leite, R., et al. (2020). Selected bacterial strains enhance phosphorus availability from biochar-based rock phosphate fertilizer. *Annals of Microbiology, 70*, 6. Available from https://doi.org/10.1186/s13213-020-01550-3.

Dewil, R., Baeyens, J., Roels, J., & Van De Steene, B. (2009). Evolution of the total sulphur content in full-scale wastewater sludge treatment. *Environmental Engineering Science, 26*(4), 867–872.

Dong, D., Wang, C., Van Zwieten, L., Wang, H., Jiang, P., Zhou, M., & Wu, W. (2020). An effective biochar-based slow-release fertilizer for reducing nitrogen loss in paddy fields. *Journal of Soils and Sediments, 20*, 3027–3040.

Eow, J. S. (2002). Recovery of sulfur from sour acid gas: A review of the technology. *Environmental Progress., 21*, 143–162.

Fachini, J., Figueiredo, C. C., Frazão, J. J., Rosa, S. D., Silva, J., & Vale, A. T. (2021). Novel K-enriched organomineral fertilizer from sewage sludge-biochar: Chemical, physical and mineralogical characterization. *Waste Management, 135*, 98–108.

Fachini, J., Figueiredo, C. C., & Vale, A. T. (2022). Assessing potassium release in natural silica sand from K-enriched sewage sludge biochar fertilizers. *Journal of Environmental Management, 314*, 115080.

Fageria, N. K., & Baligar, V. C. (2005). Enhancing nitrogen use efficiency in crop plants. *Advances in Agronomy, 88*, 97–185.

Faria, W. M., Figueiredo, C. C., de Coser, T. R., Vale, A. T., & Schneider, B. G. (2018). Is sewage biochar capable of replacing inorganic fertilizers for corn production? Evidence from a two - year field experiment. *Archives of Agronomy and Soil Science, 64*, 505–519.

Fischer, D., & Glaser, B. (2012). Synergisms between compost and biochar for sustainable soil amelioration. *Management of Organic Waste, 1*, 167–198.

Gai, X., Wang, H., Liu, J., Zhai, L., Liu, S., Ren, T., & Liu, H. (2014). Effects of feedstock and pyrolysis temperature on biochar adsorption of ammonium and nitrate. *PLoS One, 9*(12), e113888.

Gao, Y., Fang, Z., Van Zwieten, L., Bolan, N., Dong, D., Quin, B. F., & Chen, W. (2022). A critical review of biochar-based nitrogen fertilizers and their effects on crop production and the environment. *Biochar, 4*, 1–19.

Gong, H., Tan, Z., Zhang, L., & Huang, Q. (2019). Preparation of biochar with high absorbability and its nutrient adsorption−desorption behaviour. *Science of the Total Environment, 694*, 133728.

Guo, M., Song, W., & Tian, J. (2020). Biochar-facilitated soil remediation: Mechanisms and efficacy variations. *Frontiers in Environmental Science, 8*, 521512. Available from https://doi.org/10.3389/fenvs.2020.521512.

Gwenzi, W., Nyambishi, T. J., & Mapope, N. (2018). Synthesis and nutrient release patterns of a biochar-based N-P-K slow-release fertilizer. *International Journal of Environmental Science and Technology, 15*, 405–414.

Hakeem, I. G., Halder, P., Marzbali, M. H., Patel, S., Rathnayake, N., Surapaneni, A., & Shah, K. (2022). Mild sulphuric acid pre-treatment for metals removal from biosolids and the fate of metals in the treated biosolids derived biochar. *Journal of Environmental Chemical Engineering, 10*(3), 107378.

He, X., Yin, H., Han, L., Cui, R., Fang, C., & Huang, G. (2019). Effects of biochar size and type on gaseous emissions during pig manure/wheat straw aerobic composting: Insights into multivariate-microscale characterization and microbial mechanism. *Bioresource Technology, 271*, 375–382.

Holatko, J., Hammerschmiedt, T., Kucerik, J., Baltazar, T., Radziemska, M., Havlicek, Z., … Brtnicky, M. (2022). Soil properties and maize yield improvement with biochar-enriched poultry litter-based fertilizer. *Materials, 15*(24), 9003.

Ippolito, J. A., Cui, L., Kammann, C., Wrage-Mönnig, N., Estavillo, J. M., Fuertes-Mendizabal, T., ... Borchard, N. (2020). Feedstock choice, pyrolysis temperature and type influence biochar characteristics: A comprehensive meta-data analysis review. *Biochar, 2*, 421–438.

Jiang, L., Liang, J., Yuan, X., Li, H., Li, C., Xiao, Z., ... Zeng, G. (2014). Co-pelletization of sewage sludge and biomass: The density and hardness of pellet. *Bioresource Technology, 166*, 435–443.

Joseph, S., Anawar, H. M., Storer, P., Blackwell, P., Chia, C., Lin, Y., ... Solaiman, Z. M. (2015). Effects of enriched biochars containing magnetic iron nanoparticles on mycorrhizal colonisation, plant growth, nutrient uptake and soil quality improvement. *Pedosphere, 25*, 749–760.

Joseph, S., Kammann, C. I., Shepherd, J. G., Conte, P., Schmidt, H. P., Hagemann, N., & Graber, E. R. (2018). Microstructural and associated chemical changes during the composting of a high temperature biochar: Mechanisms for nitrate, phosphate and other nutrient retention and release. *Science of the Total Environment, 618*, 1210–1223.

Kammann, C. I., Schmidt, H. P., Messerschmidt, N., Linsel, S., Steffens, D., Müller, C., & Joseph, S. (2015). Plant growth improvement mediated by nitrate capture in co-composted biochar. *Scientific Reports, 5*(1), 11080.

Kim, P., Hensley, D., & Labbé, N. (2014). Nutrient release from switchgrass-derived biochar pellets embedded with fertilizers. *Geoderma, 234*, 341–351.

Kirchmann, H., Börjesson, G., Kätterer, T., & Cohen, Y. (2016). From agricultural use of sewage sludge to nutrient extraction: A soil science outlook. *Ambio, 46*, 143–154.

Lago, B. C., Silva, C. A., Melo, L. C. A., & de Morais, E. G. (2021). Predicting biochar cation exchange capacity using Fourier transform infrared spectroscopy combined with partial least square regression. *Science of The Total Environment, 794*, 148762.

Leng, L., Liu, R., Xu, S., Mohamed, B. A., Yang, Z., Hu, Y., & Li, H. (2022). An overview of sulfur-functional groups in biochar from pyrolysis of biomass. *Journal of Environmental Chemical Engineering*, 107185.

Li, S., Barreto, V., Li, R., Chen, G., & Hsieh, Y. P. (2018). Nitrogen retention of biochar derived from different feedstocks at variable pyrolysis temperatures. *Journal of Analytical and Applied Pyrolysis, 133*, 136–146.

Liao, J., Liu, X., Hu, A., Song, H., Chen, X., & Zhang, Z. (2020). Effects of biochar-based controlled release nitrogen fertilizer on nitrogen-use efficiency of oilseed rape (*Brassica napus* L.). *Scientific Reports, 10*(1), 1–14.

Liu, W. J., Li, W. W., Jiang, H., & Yu, H. Q. (2017). Fates of chemical elements in biomass during its pyrolysis. *Chemical Reviews, 117*(9), 6367–6398.

Lustosa Filho, J. F., Barbosa, C. F., Carneiro, J. S. S., & Melo, L. C. A. (2019). Diffusion and phosphorus solubility of biochar-based fertilizer: Visualization, chemical assessment and availability to plants. *Soil and Tillage Research, 194*, 104298. Available from https://doi.org/10.1016/j.still.2019.104298.

Mandal, S., Thangarajan, R., Bolan, N. S., Sarkar, B., Khan, N., Ok, Y. S., & Naidu, R. (2016). Biochar-induced concomitant decrease in ammonia volatilization and increase in nitrogen use efficiency by wheat. *Chemosphere, 142*, 120–127.

Mattiello, E. M., Silva, R. C., Degryse, F., Baird, R., Gupta, V. V., & McLaughlin, M. J. (2017). Sulfur and zinc availability from co-granulated Zn-enriched elemental sulfur fertilizers. *Journal of Agricultural and Food Chemistry, 65*(6), 1108–1115.

Melo, L. C. A., Lehmann, J., Carneiro, J. S. S., & Camps-Arbestain, M. (2022). Biochar-based fertilizer effects on crop productivity: A meta-analysis. *Plant and Soil, 472*, 45–58.

Naeem, M. A., Khalid, M., Aon, M., Abbas, G., Amjad, M., Murtaza, B., ... Ahmad, N. (2018). Combined application of biochar with compost and fertilizer improves soil properties and grain yield of maize. *Journal of Plant Nutrition, 41*(1), 112–122.

Nardis, B. O., Carneiro, J. S. S., De Souza, I. M. G., De Barros, R. G., & Melo, L. C. A. (2021). Phosphorus recovery using magnesium-enriched biochar and its potential use as fertilizer. *Archives of Agronomy and Soil Science, 67*, 1017–1033. Available from https://doi.org/10.1080/03650340.2020.1771699.

Ndoung, O. C. N., Figueiredo, C. C., & Ramos, M. L. G. (2021). A scoping review on biochar-based fertilizers: Enrichment techniques and agro-environmental application. *Heliyon, 7*, e08473. Available from https://doi.org/10.1016/j.heliyon.2021.e08473.

Osman, A. I., Fawzy, S., Farghali, M., El-Azazy, M., Elgarahy, A. M., Fahim, R. A., ... Rooney, D. W. (2022). Biochar for agronomy, animal farming, anaerobic digestion, composting, water treatment, soil remediation, construction, energy storage, and carbon sequestration: A review. *Environmental Chemistry Letters, 20*(4), 2385–2485.

Pandit, N. R., Schmidt, H. P., Mulder, J., Hale, S. E., Husson, O., & Cornelissen, G. (2019). Nutrient effect of various composting methods with and without biochar on soil fertility and maize growth. *Archives of Agronomy and Soil Science, 66,* 250−265.

Phuong, N. V., Hoang, N. K., Luan, L. V., & Tan, L. V. (2021). Evaluation of adsorption capacity in water of coffee husk-derived biochar at different pyrolysis temperatures. *International Journal of Agronomy,* 2021.

Piash, M. I., Iwabuchi, K., & Itoh, T. (2022). Synthesizing biochar-based fertilizer with sustained phosphorus and potassium release: Co-pyrolysis of nutrient-rich chicken manure and Ca-bentonite. *Science of the Total Environment, 822,* 153509.

Puga, A. P., Grutzmacher, P., Cerri, C. E. P., et al. (2020). Biochar-based nitrogen fertilizers: Greenhouse gas emissions, use efficiency, and maize yield in tropical soils. *Science of the Total Environment, 704,* 135375. Available from https://doi.org/10.1016/j.scitotenv.2019.135375.

Rasse, D. P., Weldon, S., Joner, E. J., et al. (2022). Enhancing plant N uptake with biochar-based fertilizers: Limitation of sorption and prospects. *Plant and Soil, 475,* 213−236. Available from https://doi.org/10.1007/s11104-022-05365-w.

Ravindran, B., Awasthi, M. K., Karmegam, N., Chang, S. W., Chaudhary, D. K., Selvam, A., ... Munuswamy-Ramanujam, G. (2022). Co-composting of food waste and swine manure augmenting biochar and salts: Nutrient dynamics, gaseous emissions and microbial activity. *Bioresource Technology, 344,* 126300.

Reza, M. T., Uddin, M. H., Lynam, J. G., & Coronella, C. J. (2014). Engineered pellets from dry torrefied and HTC biochar blends. *Biomass Bioenergy, 63,* 229−238.

Roy, A., Chaturvedi, S., Singh, S., Govindaraju, K., Dhyani, V. C., & Pyne, S. (2022). Preparation and evaluation of two enriched biochar-based fertilizers for nutrient release kinetics and agronomic effectiveness in direct-seeded rice. *Biomass Conversion and Biorefinery.* Available from https://doi.org/10.1007/s13399-022-02488-z.

Sawalha, M. F., Peralta-Videa, J. R., Saupe, G. B., Dokken, K. M., & Gardea-Torresdey, J. L. (2007). Using ftir to corroborate the identity of functional groups involved in the binding of Cd and Cr to saltbush (*Atriplex canescens*) biomass. *Chemosphere, 66*(8), 1424e1430.

Schellekens, J., Silva, C. A., Buurman, P., Rittl, T. F., Domingues, R. R., Justi, M., ... Trugilho, P. F. (2018). Molecular characterization of biochar from five Brazilian agricultural residues obtained at different charring temperatures. *Journal of Analytical and Applied Pyrolysis, 130,* 106−117.

Shi, W., Ju, Y., Bian, R., Li, L., Joseph, S., Mitchell, D. R., & Pan, G. (2020). Biochar bound urea boosts plant growth and reduces nitrogen leaching. *Science of the Total Environment, 701,* 134424.

Silva, C. A. (2008). Uso de resíduos orgânicos na agricultura. In G. A. Santos, L. S. Silva, L. P. Canellas, & F. O. Camargo (Eds.), *Fundamentos da matéria orgânica do solo; ecossistemas tropicais & subtropicais* (pp. 597−624). Porto Alegre: Metrópole,.

Sim, D. H. H., Tan, I. A. W., Lim, L. L. P., & Hameed, B. H. (2021). Encapsulated biochar-based sustained release fertilizer for precision agriculture: A review. *Journal of Cleaner Production, 303,* 127018. Available from https://doi.org/10.1016/j.jclepro.2021.127018.

Singh, S., Chaturvedi, S., Dhyani, V. C., & Govindaraju, K. (2020). Pyrolysis temperature influences the characteristics of rice straw and husk biochar and sorption/desorption behavior of their biourea composite. *Bioresource Technology, 314,* 123674. Available from https://doi.org/10.1016/j.biortech.2020.123674.

Singh, S., Chaturvedi, S., Nayak, P., Dhyani, V. C., Nandipamu, T. M., Singh, D. K., ... Govindaraju, K. (2023). Carbon offset potential of biochar based straw management under rice − Wheat system along Indo-Gangetic Plains of India. *Science of the Total Environment, 897,* 165176. Available from https://doi.org/10.1016/j.scitotenv.2023.165176.

Sonoki, T., Furukawa, T., Jindo, K., Suto, K., Aoyama, M., & Sánchez-Monedero, M. Á. (2013). Influence of biochar addition on methane metabolism during thermophilic phase of composting. *Journal of Basic Microbiology, 53*(7), 617−621.

Steiner, C., Das, K. C., Melear, N., & Lakly, D. (2010). Reducing nitrogen loss during poultry litter composting using biochar. *Journal of Environmental Quality, 39*(4), 1236−1242.

Steiner, C., Teixeira, W. G., Lehmann, J., Nehls, T., de Macêdo, J. L. V., Blum, W. E., & Zech, W. (2007). Long term effects of manure, charcoal and mineral fertilization on crop production and fertility on a highly weathered Central Amazonian upland soil. *Plant and Soil, 291,* 275−290.

Tang, Z., Liu, X., Li, G., & Liu, X. (2022). Mechanism of biochar on nitrification and denitrification to N_2O emissions based on isotope characteristic values. *Environmental Research, 212*(part A), 113219. Available from https://doi.org/10.1016/j.envres.2022.113219.

Tian, W., Tang, Y., Ge, D., & Tsang, D. C. (2023). Biochar-assisted anaerobic ammonium oxidation. *Biochar Applications for Wastewater Treatment*. Wiley.

Vandecasteele, B., Willekens, K., Steel, H., D'Hose, T., Van Waes, C., & Bert, W. (2017). Feedstock mixture composition as key factor for C/P ratio and phosphorus availability in composts: Role of biodegradation potential, biochar amendment and calcium content. *Waste and Biomass Valorization, 8*(8), 2553–2567.

Wang, C., Luo, D., Zhang, X., Huang, R., Cao, Y., Liu, G., & Wang, H. (2022). Biochar-based slow-release of fertilizers for sustainable agriculture: A mini review. *Environmental Science and Ecotechnology*, 100167.

Wang, Q., Awasthi, M. K., Ren, X., Zhao, J., Li, R., Wang, Z., ... Zhang, Z. (2018). Combining biochar, zeolite and wood vinegar for composting of pig manure: The effect on greenhouse gas emission and nitrogen conservation. *Waste Management, 74*, 221–230.

Wang, T., Arbestain, M. C., Hedley, M., & Bishop, P. (2012). Chemical and bioassay characterisation of nitrogen availability in biochar produced from dairy manure and biosolids. *Organic Geochemistry, v. 51*, 45–54.

Wang, Y., Villamil, M. B., Davidson, P. C., & Akdeniz, N. (2019). A quantitative understanding of the role of co-composted biochar in plant growth using meta-analysis. *Science of The Total Environment, 685*, 741–752.

Weldon, S., van der Veen, B., Farkas, E., Kocatürk-Schumacher, N. P., Dieguez-Alonso, A., Budai, A., & Rasse, D. (2022). A re-analysis of NH_4^+ sorption on biochar: Have expectations been too high? *Chemosphere*, 134662.

Wen, P., Wu, Z., Han, Y., Cravotto, G., Wang, J., & Ye, B. C. (2017). Microwave-assisted synthesis of a novel biochar-based slow-release nitrogen fertilizer with enhanced water-retention capacity. *ACS Sustainable Chem. istryEngineering, 5*(8), 2017.

Wu, W., Yan, B., Zhong, L., Zhang, R., Guo, X., Cui, X., ... Chen, G. (2021). Combustion ash addition promotes the production of K-enriched biochar and K release characteristics. *Journal of Cleaner Production, 311*, 127557.

Xiao, X., Chen, B., Chen, Z., Zhu, L., & Schnoor, J. L. (2018). Insight into multiple and multilevel structures of biochars and their potential environmental applications: A critical review. *Environmental Science & Technology, 52*(9), 5027–5047.

Xu, X., Cao, X., Zhao, L., & Sun, T. (2014). Comparison of sewage sludge-and pig manure-derived biochars for hydrogen sulfide removal. *Chemosphere, 111*, 296–303.

Yao, Y., Gao, B., Zhang, M., Inyang, M., & Zimmerman, A. R. (2012). Effect of biochar amendment on sorption and leaching of nitrate, ammonium, and phosphate in a sandy soil. *Chemosphere, 89*(11), 1467–1471.

Ye, L., Camps-Arbestain, M., Shen, Q., et al. (2020). Biochar effects on crop yields with and without fertilizer: A meta-analysis of field studies using separate controls. *Soil Use and Management, 36*, 2–18.

Ye, Z., Zhang, L., Huang, Q., & Tan, Z. (2019). Development of a carbon-based slow release fertilizer treated by bio-oil coating and study on its feedback effect on farmland application. *Journal of Cleaner Production, 239*, 118085.

Yin, Y., Yang, C., Li, M., Zheng, Y., Ge, C., Gu, J., ... Chen, R. (2021). Research progress and prospects for using biochar to mitigate greenhouse gas emissions during composting: A review. *Science of The Total Environment, 798*, 149294.

Yue, Y., Cui, L., Lin, Q., Li, G., & Zhao, X. (2017). Efficiency of sewage sludge biochar in improving urban soil properties and promoting grass growth. *Chemosphere, 173*, 551–556.

Zainudin, M. H. M., Singam, J. T., Sazili, A. Q., Shirai, Y., & Hassan, M. A. (2022). Indigenous cellulolytic aerobic and facultative anaerobic bacterial community enhanced the composting of rice straw and chicken manure with biochar addition. *Scientific Reports, 12*(1), 5930.

Zhang, H., Voroney, R. P., Price, G. W., & White, A. J. (2016). Sulfur-enriched biochar as a potential soil amendment and fertiliser. *Journal of Soil Research, 55*, 93–99.

Zhou, Z., Yao, D., Li, S., Xu, F., Liu, Y., Liu, R., & Chen, Z. (2021). Sustainable production of value-added sulfonated biochar by sulfuric acid carbonization reduction of rice husks. *Environmental Technology & Innovation, 24*, 102025.

Zwetsloot, M. J., Lehmann, J., Bauerle, T., Vanek, S., Hestrin, R., & Nigussie, A. (2016). Phosphorus availability from bone char in a P-fixing soil influenced by root-mycorrhizae-biochar interactions. *Plant and Soil, 408*, 95–105.

SECTION C

Environmental sustainability and bioremediation

CHAPTER 17

Biochar-led methanogenic and methanotrophic microbial community shift: mitigating methane emissions

Tony Manoj Kumar Nandipamu[1], Prayasi Nayak[1], Sumit Chaturvedi[1], Vipin Chandra Dhyani[1], Rashmi Sharma[2] and Nishanth Tharayil[3]

[1]Department of Agronomy, G. B. Pant University of Agriculture & Technology, Pantnagar, Uttarakhand, India [2]Department of Agronomy, School of Agriculture, Graphic Era Hill University, Dehradun, Uttarakhand, India [3]Plant Ecophysiology, Plant and Environmental Sciences Department, Clemson University, SC, United States

17.1 Introduction

A gradual escalation in greenhouse gas (GHG) accumulation in the atmosphere results in an impending climate change through a rise in earth's surface temperature through the increase in methane (CH_4), carbon dioxide (CO_2), nitrous oxide (N_2O), ozone (O_3), water vapor levels (Al-Ghussain, 2019). In general, the greenhouse effect helps life to sustain on the earth by making it suitable to thrive upon, otherwise, the surface temperature would have dropped beyond where no life form would find it suitable to live (Attenborough, 2020). Meanwhile, rampant global warming as a result of the prolonged accumulation of GHGs ushered calamities and disasters, taking a severe toll on the various cycles of the ecosystem. Of all the GHGs, the carbon-based GHGs CO_2, CH_4 and chlorofluorocarbons pose a severe environmental threat. The latent global warming potential of CH_4 is 21 times more than that of CO_2 and has an average lifespan of 12 ± 3 years in the atmosphere, which is widely being emitted from wetlands, rice fields, ruminants, coal mines, anthropogenic activities to the recent Nord stream pipeline leaks (Leshchenko, Shulzhenko, Kaplin, Maistrenko, & Shcherbyna, 2023). Even though major contributors to GHGs are industries and the transport sector, surprisingly around 18.4% of the world's total GHG emissions were from agriculture and land use systems. Methane is a major driving force of the GHGs, which alone contributes to roughly a 30% increase in the world's temperature rise (IEA, 2023). Wetlands, the major methane emitting source contributed around 194.0 Tg year^{-1} followed by anthropogenic interventions of agriculture (141.4 Tg year^{-1}) according to IEA (2023). Paddy fields, biomass burning and composting are the major drivers contributing towards the significant increase in CH$_4$ around the globe (Fig. 17.1).

Rice cultivation spreads over 165.2 million hectares (Mha) and holds 22.3% of the world's area under cereal production and is the most important crop in South and South-East Asia (FAOSTAT, 2022). Rice-based cropping systems stand atop as the major contributing factor of food production systems globally, with rice as the first and integral food crop of the system. Adopting and practising the high yield output-based intensive cropping systems viz. rice-rice, rice alone, and rice-wheat cropping systems to meet the food demand of the rising population, inadvertently encouraged the methane emissions from agriculture over the past six decades (Fig. 17.2). Overall CH$_4$ emissions from rice fields stand tall among different sources of GHG emissions, which are significantly contributed by rice-grown countries of South and

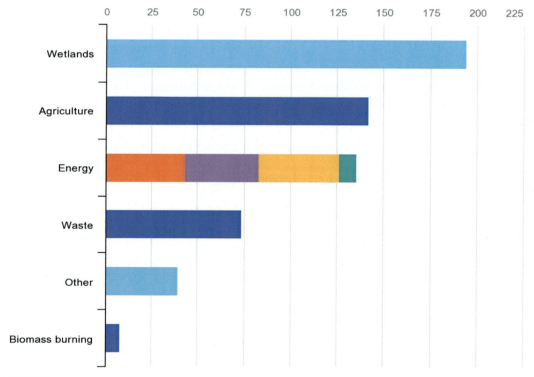

FIGURE 17.1 Methane emissions from different sectors. *Source: Adapted from IEA. (2021). Sources of methane emissions, International Energy Agency, Paris. IEA. Licence: CC BY 4.0. https://www.iea.org/data-and-statistics/charts/sources-of-methane-emissions-2021.*

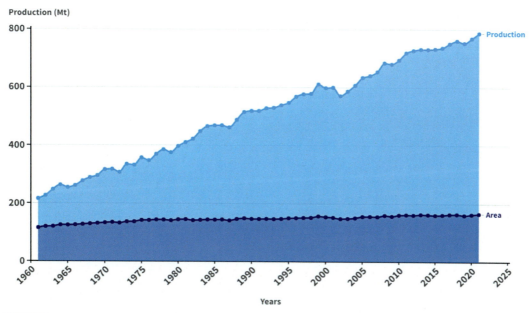

FIGURE 17.2 Global rice cultivated area and production levels over the past decades (FAOSTAT, 2020). *Source: FAOSTAT. (2020). Methane emissions from different countries. Food and Agriculture Organization. <https://www.fao.org/faostat/en/#data/GR> Accessed 15.12.22.*

South-East Asia (Fig. 17.3). Around 30% and 11% of global agricultural CH_4 and N_2O, respectively, are emitted from rice fields, which pose a serious threat towards rapidly increasing climate change (Hussain et al., 2015). Considerable emissions from rice fields are released under the influence of methane-emitting methanogenic fauna when provided with anaerobic conditions. Prominent methanogenic microbes in flooded paddy environments include *Methanosarcina*, *Methanobacterium*, *Methanocella*, *Methanosaeta*, and *Methanomassiliicoccus* (Lu et al., 2022). Dominant microbes under favorable circumstances (viz. optimum temperature, redox potential, soil pH, texture, soil organic carbon and moisture content) generate CH_4 responsible for the inclination of surface temperatures. Mitigating the global warming potential by seeking and applying the solution at the roots would help reduce the emissions, where the potential of biochar offers a viable solution. Biochar is a carbon-rich material produced under anoxic conditions through a thermo-chemical process with a higher surface area, mesopores, surface charges and moisture-holding capacity, which can address the microbial abundance and communal shift of soil methanogenic and methanotrophic bacteria. Few studies documented the role of biochar as a propitious solution in mitigating CH_4 emissions (Awad et al., 2018; Han et al., 2016; Wu, Song, et al., 2019; Wu, Zhang, Dong, Li, & Xiong, 2019) and its potential was extensively explored and ecological benefits were widely demonstrated (Ji, Zhou, Zhang, Luo, & Sang, 2020; Nan et al., 2020a, 2020b; Pratiwi and Shinogi, 2016; Yoo, Kim, Lee, & Ding, 2016). In this chapter, an attempt has been made to establish biochar relations with microbial dynamics and GHG emissions, particularly methane.

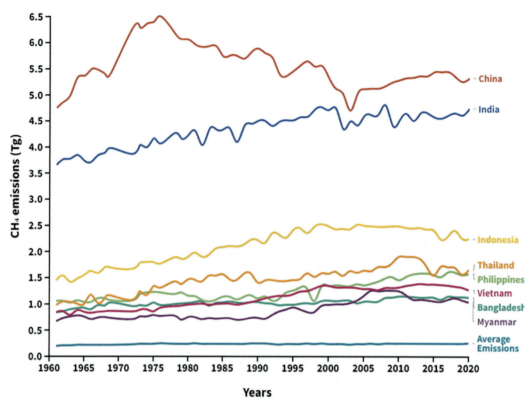

FIGURE 17.3 Long term methane emissions from major rice cultivated countries of the world. *Source:* FAOSTAT. (2022). *Methane emissions from cultivated rice fields. Food and Agriculture Organization.* <https://www.fao.org/faostat/en/#data/GR> Accessed 15.12.22.

17.2 Overview of the biochar's potential to mitigate methane emissions

Biochar, a carbonaceous material produced through the pyrolysis of biomass, has garnered considerable interest for its potential to mitigate methane (CH_4) emissions within various environmental contexts (Li & Tasnady, 2023: Osman et al., 2022). Its unique physicochemical attributes, particularly the intricate interactions with soil matrices and microbial communities, position it as a promising tool in the global endeavor to curtail anthropogenic contributions to climate change, predominantly through methane abatement. Methane, a potent GHG exhibiting significantly higher global warming potential than carbon dioxide (CO_2) over shorter timescales, is predominantly emitted from anthropogenic sources such as agriculture, livestock husbandry, and wetland ecosystems (Yaashikaa, Kumar, Varjani, & Saravanan, 2020). While conventional methods of CH_4 reduction exhibit efficacy, they often exhibit limitations pertaining to feasibility, sustainability, and comprehensive ecological benefits (Lehmann, Gaunt, & Rondon, 2006). In this context, biochar emerges as an innovative avenue for CH_4 mitigation while simultaneously offering improvements in soil fertility and overall environmental quality. The intrinsic structure of biochar plays a pivotal role in its efficacy for CH_4 mitigation. Biochar, characterized by a high surface area and porous morphology, provides a conducive habitat for a diverse range of microbial populations, including methanotrophic bacteria (Steiner, Bayode, & Ralebitso-Senior, 2016). These bacteria possess the unique ability to oxidize CH_4, converting it to carbon dioxide. The colonization of biochar surfaces by methanotrophic bacteria facilitates the oxidation of CH_4 within microsites, resulting in a reduction in net CH_4 emissions from soil, wetlands, and other methane-rich environments (Zhang et al., 2019). Notably, the symbiotic interaction between biochar and methanotrophic bacteria can induce a microbial shift that reinforces CH_4 oxidation dynamics. The provision of a biochar substrate alters the spatial distribution and metabolic activity of these bacteria, leading to accelerated CH_4 consumption rates. Consequently, biochar-amended soils exhibit enhanced CH_4 oxidation capacity, potentially attenuating methane emissions at a system level.

In conjunction with its structural attributes, biochar's physicochemical properties significantly contribute to CH_4 mitigation. The functional groups present on the biochar surface, including hydroxyl, carboxyl, and aromatic fractions, facilitate the adsorption of methane and other volatile organic compounds (Lévesque, Oelbermann, & Ziadi, 2022; Yaashikaa et al., 2020). Adsorption on biochar surfaces immobilizes CH_4, thereby minimizing its release into the atmosphere (Tisserant & Cherubini, 2019). This dual functionality of biochar enabling CH_4 oxidation and adsorption profoundly influences methane dynamics within soil systems (Pan et al., 2021). The integration of biochar into cropping systems garners multifaceted advantages beyond CH_4 reduction. The improvements in soil structure, water-holding capacity, and nutrient retention attributed to biochar amendments collectively augment crop productivity and soil health (Zhang, Zhang, Xu, Dong, & Xiong, 2022; Zhang, Zhao, Liu, Zhao, & Li, 2022). This promotes increased carbon sequestration potential and contributes to global efforts in CO_2 mitigation. Moreover, the enduring nature of carbon sequestration in biochar-amended soils, with residence times ranging from 1000 to more than 10,000 years, underscores its significance as a long-term climate mitigation strategy (Lehmann et al., 2006). However, the efficacy of biochar in CH_4 mitigation is intrinsically tied to contextual parameters. The type of biochar, its physicochemical attributes, application rates, soil characteristics, and prevailing environmental conditions collectively dictate the extent of methane reduction achievable (Kumar et al., 2023). Tailoring biochar applications to site-specific conditions is pivotal for maximizing its methane mitigation potential. Consequently, the successful deployment of biochar as a CH_4 mitigation strategy necessitates a holistic understanding of these factors, underscoring the need for ongoing research to refine best practices. Long-term studies are essential to elucidate the temporal dynamics of biochar-microbial interactions and the durability of CH_4 abatement effects over prolonged periods (Zhang et al., 2019).

The biochar presents a promising avenue for mitigating CH_4 emissions in diverse environmental contexts. Its distinctive attributes, spanning from structural niches to multifaceted physicochemical properties, stimulate a microbial shift favoring CH_4 oxidation and adsorption. This intricate relationship not only addresses CH_4 emissions but also augments soil fertility, crop productivity, and carbon sequestration potential. Nonetheless, the variable nature of biochar efficacy mandates context-specific approaches to harness its CH_4 mitigation potential fully. As global imperatives to combat climate change persevere, biochar's potential as an innovative and sustainable tool in the reduction of methane emissions stands as a beacon of hope and progress towards a more environmentally balanced future.

17.3 Methanogenesis and methane emissions from cropping systems

The anaerobic respiration of methanogens to generate CH_4 as the final metabolic product is known as methanogenesis (Lyu, Shao, Akinyemi, & Whitman, 2018; Serrano-Silva, Sarria-Guzmán, Dendooven, & Luna-Guido, 2014). Methanogens are distinctive and strict anaerobic archaea that respire without oxygen and have a very low energy yield (≤ 1 ATP). The distinguished methanogenic process is carried out by limited obligate archaea, which respire and generate CH_4 without fermentation or alternative electron acceptors under anoxic conditions. Methanogens prefer soil organic matter (dead microbes, plant biomass, root debris, animal waste, and organic fertilizers) to meet carbon requirements and emit methane as a by-product. Methanogens can be classified into three groups based on their substrate usage: hydrogenotrophic, aceticlastic, and methylotrophic. Methanogens are predominantly hydrogenotrophic which utilize H_2, formate, or certain simple alcohols as electron donors, and reduce CO_2 to CH_4 (Table 17.1). Aceticlastic methanogens thrive in H_2-reduced habitats suitable for acetate formation from organic matter, which serve as a substrate to form methane and CO_2 in rice paddies, wetlands and biogas chambers. The acetate substrate-based methanogens contribute around 70%–90% and formate, CO_2/H_2-based archaea back 10%–30% of CH_4 emissions (Norina, 2007; Parmentier et al., 2015). The prolific growth of methanogenic communities, survival and activity depends on mesophilic temperatures (around 25°C), substrate availability, and suitable environment for CH_4 production. The methylotrophic methanogens survive in various geologic habitats

TABLE 17.1 Substrate consumption-based methanogenesis classification.

S. No.	Methanogenesis	Source	End product	Orders	References
1.	Hydrogenotrophic	Hydrogen and CO_2	$CH_4 + 2H_2O$	*Methanococcales, Methanobacteriales, Methanosarcinales, Methanomicrobiales, Methaopyrales,* and *Methanocellales*	Stadtman and Barker (1951) Balch, Fox, Magrum, Woese, and Wolfe (1979) Kurr et al. (1991) Sakai et al. (2008)
2.	Aceticlastic	Acetate	$CH_4 + CO_2$	*Methanosarcinales* only	Fenchel, King, and Blackburn (2012) Malyan et al. (2016) Conrad (2020)
3.	Methylotrophic	Methanol, Methylamines/methylated thiols	$3CH_4 + CO_2 + 2H_2O$	*Methanomassiliicoccales, Methanobacteriales* and *Methanosarcinales*	Söllinger and Urich (2019) Kurth, Op den Camp, and Welte (2020)

viz., hot springs, marine, and hypersaline conditions, although their role in methane emissions is negligible (Lyu et al., 2018; Nazaries, Murrell, Millard, Baggs, & Singh, 2013).

Rice-based systems remain major methane emitting sources over the world especially in developing nations whose contribution remains at 1.3% of the total GHG emissions (IEA, 2021). Flooded rice cultivation, the most popular practice in East and South-East Asian countries remains a major hotspot for CH_4 emissions, where China and India have remained on the top for the last few decades (Fig. 17.4). Major rice-based cropping systems

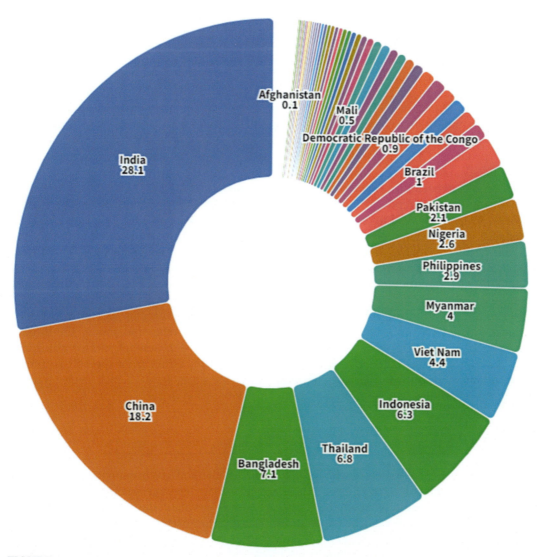

FIGURE 17.4 Percent methane contribution of countries from paddy cultivation. Source: *FAOSTAT. (2020). Methane emissions from different countries. Food and Agriculture Organization. <https://www.fao.org/faostat/en/#data/GR> Accessed 15.12.22.*

of India include; rice-rice, rice-wheat, rice-rapeseed/mustard, rice-groundnut, rice-potato and rice-pulses. Predominant rice-growing belts of India, viz. Indo-Gangetic Plains, Krishna-Godavari delta, belts of Southern Odisha, Gujarat and Assam have been exploited for extensive rice cultivation either through rice-wheat, or rice-rice double cropping upon moisture availability around the year (Panda et al., 2018). Despite being grown over an area of 45.7 Mha CH_4 emissions were quantified at around 4.75 Tg which are on the rise since the green revolution (FAOSTAT, 2022; IndiaStat, 2022). The increased area under rice-rice, rice-wheat systems, and summer rice under irrigated conditions and faulty management practices, that is, nutrient, water and resource management, led to the rise in methane emissions. Rice systems act as an excellent source of soil organic carbon storage, despite their significant contribution towards methane and nitrous oxide emissions (Liu et al., 2021b). Under strict anaerobic conditions, the methanogens reduce the carbonaceous compounds viz., carbohydrates, acetate, formate and CO_2 through catalysis of different enzymes to generate ATP for methanogenesis (Dubey, 2005; Welte and Deppenmeier, 2011). The abundance of methanogenic archaea along with naturally occurring anaerobic microbial communities is evident with proliferated growth and development under the influence of soil organic matter. The hydrogenotrophic and aceticlastic methanogens actively produce CH_4 under the flooded paddy fields where soil E_h drops less than -200 eV (Ali, Oh, & Kim, 2008; Lyu et al., 2018). The relative abundance of methanogenic archaea viz., *Methanosaeta, Methanoregula, Methanocella, Methanospirillum* and *Methanobacterium* was reported in rice fields from transplanting to harvest (Lee, Kim, Kim, Madsen, & Jeon, 2014; Liu et al., 2021b; Pramanik & Kim, 2018). The activity and abundance of the methanogens depend on the soil temperature, moisture and substrate availability during the growing season and have a direct correlation with CH_4 flux in the rice-wheat cropping system (Bhullar, Iravani, Edwards, & Olde Venterink, 2013; Ma, Qiu, & Lu, 2010). The continuous rice-wheat rotation reportedly increased CH_4 efflux from the wheat due to the abundance of anaerobic methane-producing archaea in rice due to soil moisture, temperature, organic carbon and management interventions (Xu et al., 2022). Methane derived through methanogenesis from submerged anaerobic soil to the atmosphere is channeled through ebullition, diffusion, and through aerenchyma conduits of rice plants (Singh, Tiwari, Boudh, & Singh, 2017). The CH_4 efflux from the soil is intervened by methanogen counterparts termed "methane munchers" which were found actively proliferating under aerobic conditions of rice-based systems.

17.4 Methanotrophs and methane oxidation from cropping systems

The methane flux of methanogens under an anaerobic soil environment has a strong relation with bio-chemical reactions at the ionic level and leverages methane-consuming microbes. Methane-oxidizing methanotrophs are more prevalent towards the submerged and oxidized zone of the rice rhizosphere. Methanotrophic bacteria have the potential to consume around $70-300$ Tg CH_4 yr^{-1} globally and reduce atmospheric methane emission by 50% (Lyu et al., 2018). The methanotroph biota under aerobic conditions oxidizes methane from the rhizosphere zone to CO_2 through methanol, formaldehyde, formate and methane monooxygenase (MMO) enzymes. Methane oxidation greatly reduces methane diffusion into the atmosphere; around 30%−60% of CH_4 was oxidized before it reaches the atmosphere (Gupta et al., 2021; Khalil, Rasmussen, & Shearer, 1998; Saha et al., 2022; Sass, Fisher, Wang, Turner, & Jund, 1992). The commonly reported methanotrophs from rice fields include *Methylobacter, Methylocaldum, Methylomonas, Methylococcus, Methylosarcina, Methylomicrobium, Methylocystis, Methylosinus*, etc. (Chen & Murrell, 2010; Lee et al., 2013). Methanotrophs are classified into types I and II, where the former dominates the aerobic environment with poor CH_4 and high O_2 and the latter thrive in methane-rich anaerobic zones (Bhattacharyya et al., 2019; Mayumi, Yoshimoto, Uchiyama, Nomura, & Nakajima-Kambe, 2010). Methanotrophs in the rhizospheric zone oxidize methane to CO_2 by extracting the necessary oxygen from the atmosphere through the aerenchyma tissue (Malyan et al., 2021). Studies revealed an increased abundance of methanotrophs with the addition of organic matter in the soybean-wheat system, as evident around 34.5%−67.5% of CH_4 was consumed over the entire duration (Mohanty, Kollah, Chaudhary, Singh, & Singh, 2015). Methane consumption in the upper 0−5 cm soil layer was higher than below 5−15 cm of soybean-wheat, maize-wheat and maize-gram cropping systems grown under varying tillage practices (Kollah et al., 2020, Mohanty et al., 2017). Findings from the above studies indicate that optimum aerobic conditions and substrate availability result in the abundance and healthy activity of methanotrophs on organic matter availability. Similarly, nitrogen availability in the soil has an inhibitory effect on methanotrophs, and proper nutrient management must be ensured in cropping systems to promote methane oxidation (Peng et al., 2019). The adoption of resource-efficient conservation agriculture practices aims at improved soil health; microbial biodiversity favors methane consumption and mitigates CH_4 emissions through altering soil hydro-thermal regimes of different cropping systems (Smith, Reay, & Smith, 2021). The oxidation potential of methanotrophs in cropping systems can be further enhanced by adopting carbon-negative technologies viz. conservation agriculture and biochar whose profound impact is being enticed in modern-day agriculture.

17.5 Biochar-led methanogenesis and methanotrophy

Biochar is a carbon-enriched material produced through various thermo-chemical reactions during the combustion of biomass in a process known as pyrolysis. Burning organic materials including manure, forest leaves, organic waste, algae, sewage sludge, and wood under ceased oxygen conditions produce biochar, a porous, amorphous, stable, and low-density carbon substance (Atkinson, Fitzgerald, & Hipps, 2010; Laufer and Tomlinson, 2012; McHenry, 2009; Shackley, Sohi, & Haszeldine, 2009). Upon pyrolysis, the end product obtained was a recalcitrant carbon source, which is resistant to weathering and decomposition with a longer shelf-life (around 1000−10,000 years) (Shenbagavalli, Mahimairaja, 2012). When applied to soil, biochar, as a stable source of C, remains for a longer period, imparting long-term soil C for a prolonged period. The potential of biochar is much realized in agriculture through terrestrial carbon sequestration, reduced GHG emissions besides improved soil quality. The mitigation potential of biochar stands tall among all the natural methods of carbon capture, utilization and storage. Biochar with its miraculous physicochemical properties channelize the soil fauna and amplifies microbial diversity towards the reduction of GHGs from the cropping systems. Soil application of biochar influences methanogenesis and methanotrophy and regulates the CH_4 emissions at a scale when coupled with efficient crop management practices. The influence of biochar on soil methane fluxes may depend on water management and soil pH and has been reported to attenuate short-term methane emissions from paddy soil (Jeffery, Verheijen, Kammann, & Abalos, 2016). Potential methane emissions from rice fields can be effectively suppressed through biochar in various ways, that is, (i) through improved soil aeration and promotes methane oxidation in oxygen-rich zone by methanotrophic bacteria, (ii) biochar-led evolution of methanogenic archaeal communities and methanogenesis inhibition, and (iii) methane adsorption on biochar (Davamani, Parameswari, & Arulmani, 2020; Ji et al., 2020; Nan et al., 2020a, 2020b). Some studies have found that applying biochar to the soil instantly reduced methane emissions, perhaps because biochar promoted soil aeration by altering the soil's physicochemical properties (Nan, Xin, Qin, Waqas, & Wu, 2021). Biochar addition can lower CH_4 emissions from paddy fields, conversely enhance the fluxes from wheat and soybean cropping systems, but still the mechanisms involved are not completely understood (Qi et al., 2021) (Table 17.2). Employing biochar helped to cutback CH_4 emissions from a grass stand *Brachiaria humidicola* and soybean (Rondon, Ramirez, & Lehmann, 2005), while an increase in methane flux from the wheat crop stand was reported by Ma et al. (2022) and Karhu, Mattila, Bergström, and Regina (2011). The biochar incorporation might enhance microbial activity and organic matter decomposition under the influence of soil moisture and biochar type, which probably influence the methane flux (Zimmerman, Gao, & Ahn, 2011). Applying biochar in rice fields alters microbial activity by influencing the soil pH, water and nutrient retention, especially NH_4^+, which interferes with methanogenesis by scaling down the abundance of methanogens (Qi et al., 2021) (Fig. 17.5). In contrast, biochar plays a different role in the soils of crops other than paddy, by enhancing the methanotrophic activity by enhancing the oxidation of methane fluxes from the soil.

Soil application of biochar stimulates the CH_4 oxidizing microbes and improves their abundance and biodiversity, especially in crops like paddy (Huang et al., 2019, Wu, Song, et al., 2019; Wu, Zhang, et al., 2019). Biochar modifies the soil condition and promotes conducive situations for methanotrophic archaea, in contrast to methane-generating microbes. The addition of carbon through biochar happens to act as a reserve for nutrient sources through enhanced adsorption and controlling losses, pH-stimulated nutrient release and better aeration which are highly desirable for the methane-oxidizing archaea (Jeffery et al., 2016; Ribas et al., 2019). Biochar application insinuates the CH_4 oxidation and cuts down the methanogenic activity with the type and amount of biomass, and pyrolysis temperature which varies with the inherent properties of the biomass used (Kubaczyński et al., 2022). The soil incorporation of biochar in tobacco-rice rotation by Gao, Liu, Shi, and Lv (2022) proved to reduce 8.47 kg CH_4 uptake ha^{-1} $tonne^{-1}$ of biochar applied. Soil incorporation of biochar at increased rates enhanced the rice biomass yield by 8%–48% and suppressed CH_4 emissions by 7%–42% in rice grown over two consecutive seasons (Nguyen, Van Nguyen, 2023). Most of the studies from the recent past indicate that biochar application instigates the soil-borne methanogenic and methanotrophic microbiotas abundance, development through dynamically loaded and throughput properties which administer methane oxidation and fluxes.

TABLE 17.2 Biochar's impact on methanobacteria in crop systems; mechanism and the fate of CH_4.

S. No.	Cropping system	Biochar used	Experiment type	Duration (years)	Fate of methane	Mechanism	References
1.	Rice-Rice	Rice straw	Greenhouse	2	Reduced by 7%–42%	Decreased methanogen: methanotroph ratio	Nguyen, Van Nguyen (2023)
2.	Rice-Rice	Wheat straw	Incubation	2	Reduced by 19.8%–37.1%	Decreased methanogen: methanotroph ratio	Liu et al. (2021a)
3.	Rice-Rice	Rice straw	Growth chambers	2	Reduced by 39.5%	Decreased methanogen: methanotroph ratio	Han et al. (2016)
4.	Tobacco-Rice	Tobacco straw	Field	2	Reduced	Decreased methanogen: methanotroph ratio	Huang et al. (2019)
5.	Rice	Rice straw	Incubation	1	Reduced	Decreased methanogen: methanotroph ratio	Lu et al. (2022)
6.	Rice-Rice	Rice straw	Greenhouse	2	Reduced by 22.5%–95.7%	Decreased methanogen: methanotroph ratio	Qi et al. (2021)
7.	Corn	Wheat straw	Field	2	Reduced GWP of CH_4 by 9.8%	N fertilization under a high biochar dose slightly reduced methanotroph activity	Zhang et al. (2012)
8.	Rice	Corn straw	Field	1	Reduced 91.2% of cumulative CH_4 emissions	Increased methanotrophs	Feng, Xu, Yu, Xie, and Lin (2012)
9.	Rice	Rice straw	Field	4	Reduced by 41%	Decreased methanogen: methanotroph ratio	Nan, Wang, Wang, Yi, and Wu (2020a)
10.	Rice	Rice straw	Field	4	Reduced by 28%–57%	Abundance of methanogens: methanotrophs decreased by 11%–31%	Wang et al. (2019)
11.	Rice-Rice	Rice straw	Field	3	Reduced by 14%–43%	Decreased methanogen: methanotroph ratio	Nan et al. (2020b)

17.6 Biochar stimulates the methanogenic and methanotrophic microbial shift

The biochar potential to amend the soil properties is vital in mitigating GHG emissions, especially methane by altering the microbial population dynamics under different cropping systems grown over wide ecological regions. Crops grown with biochar application rendered better yields and productivity yet were marred by poor management practices which led to unsought CH_4 emissions and poor methane oxidation. Being a carbon-rich substance with better properties, influences soil carbon cycling especially soil-dissolved

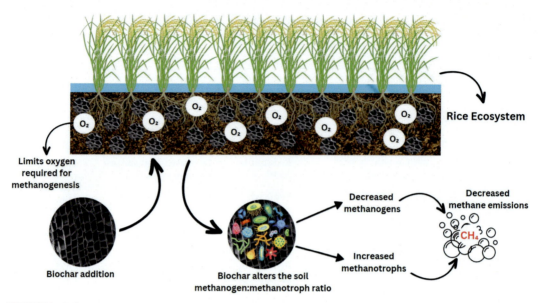

FIGURE 17.5 Biochar derived methano-microbial dynamic alteration to mitigate CH_4 emissions.

organic carbon (DOC) made up of minute molecular organic acids to complex humic substances further transforms to ammonium also indirectly influences methane oxidation (Dong, Singh, Li, Lin, & Zhao, 2019). The methanotrophs are nitrogen-hungry and their requirement for NH_4^+-N restricts ammonium oxidation and oxidizes CH_4 fluxes under flooded rice systems (Megraw & Knowles, 1987). Biochar drives the microbial shift in the existing soil fauna by facilitating favorable conditions for specific microbial genera around the rhizospheric soil (Yan et al., 2022; Zhang & Shen, 2022). However, the increase in soil pH with biochar application resulted in reduced methanogenic microbe diversity, through the addition of more carbon groups and decreased soil CEC (Khan, Chao, Waqas, Arp, & Zhu, 2013; Shen et al., 2014). Few studies also reported the enhanced soil properties (MBC, MBN, pH, and DOC) influenced the abundance of methanogenic microbes, which streamlined the ratio of methanogens to methanotrophs and increased CH_4 emissions from the paddy soils and even hypothesized the CH_4 adsorption on the mesopores of biochar (Gao et al., 2022). Biochar produced through low-temperature pyrolysis biochar decreased soil redox potential (Eh) and increased the abundance of methanogens, while high-temperature pyrolysis biochar increased the soil Eh and reduced the CH_4 emissions in the paddy fields (Cai, Feng, & Zhu, 2018).

17.6.1 Promotes methanogenic microbial shift and inhibits methanogenesis activity

Some of the studies on biochar soil addition showed a positive influence on methanogenic activity in contrast to the inhibition of methanogenic archaea. The major concept behind the enhanced methanogenesis through methanogenic biota was believed that biochar improves soil N through an increase in soil NH_4^+-N (Megraw & Knowles, 1987; Wang et al., 2019). The application of ammonia-based fertilizers indirectly influences the abundance and diversity of methane-producing and oxidizing archaea (Wang et al., 2019). The relative abundance of methane-producing *Merthanosarcina* and *Methanosaeata* was decreased by 39.2% and 13.6%, when different rice straw biochars were applied, respectively. The balance in the activity and proliferation of methanogenic to methanotrophic microbes was mediated through soil pH adjustment, DOC and soil NH_4^+-N upon the addition of biochar. Apart from the adsorbing useful nutrients on the higher specific surface area of biochar; the enhanced porosity enables microbes to colonize in mesopores (Thies & Rillig, 2012). The higher ratio of methanogens to methanotrophs promotes soil CH_4 fluxes, while the reverse is true for methane oxidation (Wang et al., 2019) (Table 17.2). On the other hand, results from the long-term biochar experiments reveal the positive influence of aged biochar by reducing the aceticlastic and hydrogenotrophic methanogens, which ultimately reduced the methanogen to methanotrophic microbe ratio in the rice-based cropping system (Han et al., 2016; Sriphirom et al., 2020). A high throughput sequencing of the biochar applied soils reveals the inhibited dominance of methanogenic archaeal communities in the rice-based cropping system, where the abundance and activity of methanogens were impaired with the increase in soil pH and soil aeration (Fu et al., 2021; Pietikäinen, Kiikkilä, & Fritze, 2000). The CH_4 emissions can be cut down by balancing the methanogens: methanotrophs ratio through biochar application which can be weighed upon the consistence results obtained through biochar application (Chen et al., 2017; Feng et al., 2012). Methane emissions were brought down by a remarkable decrease in methanogenic archaea by 55.5% in soil when biochar was applied at 12.5 t ha^{-1} (Kumputa, Vityakon, Saenjan, & Lawongsa, 2019). Most of the studies on biochar soil treatment revealed a significant impact on methane-producing archaeal diversity without straining the harmony of methanogenic communities and inhibited CH_4 emissions over the crop duration. In contrast to this, some reports revealed an increase in methanogenic activity from biochar application through altered microbial dynamics which is yet due to be established with sound scientific justification (Lorenz & Lal, 2014).

17.6.2 Abundance of methanotrophs and oxidation activity

Biochar has proven its potential to impart beneficial soil properties apart from subsidizing the soil microbial communities' nutrition by adsorption through preventing nutrient leaching and improving the soil labile carbon. The methanotrophic microbial colonies dwell in soil by binding to the surface and in pores of biochar and oxidize the CH_4 fluxes even in the hypoxic conditions around the rhizosphere. Most of the methanotrophic genera prefer aerobic conditions and some species of *Methylosinus* carry out methane oxidation under anaerobic situations thriving aside to methanogens (In 't Zandt et al., 2018). The activity of all the methanotrophs especially type II methanotrophic bacteria (*Methylocystis*, *Methylosinus*, *Methylocella*, and *Methylocapsa*) involves copper catalytical activity in the conversion of methane oxidation to methanol (Knapp, Fowle, Kulczycki, Roberts, & Graham, 2007; Trotsenko, Murrell, 2008). The biochar has the potential to adsorb copper under the acidic environment of continuous rice-rice or continuous rice cropping systems where the soil organic carbon is more which influences the availability of copper (Geaneth, Francisco, Faz, & Acosta, 2022; Noguchi, Hasegawa, Shinmachi, & Yazaki, 1997; Rawat, Saxena, & Sanwal, 2019; Zhu et al., 2017). The biofiltration of methane before breaking out into the atmosphere from flooded soil is efficiently tackled by the methanotrophs under the influence of biochar at the oxic-anoxic soil interface (Feng et al., 2012; Han et al., 2016; Reddy, Yargicoglu, Yue, & Yaghoubi, 2014; Reim, Lüke, Krause, Pratscher, & Frenzel, 2012). Soil biochar addition tends to enhance the diversity of the methanotroph communities and their abundance through elevated labile C and N and optimal pH conditions for their proliferation (Jeffery et al., 2016; Lehmann et al., 2006; Ouyang, Tang, Yu, & Zhang, 2014). The methanotrophic communities' hierarchy and microbial structure were not altered for the sake of oxidizing methane emissions when biochar was applied. Studies conducted to evaluate the methane emissions in response to biochar addition in rice reported a dip in CH_4 fluxes in a range of 20%–51% compared to no soil application (Cai et al., 2018; Qin et al., 2016; Wang et al., 2019; Yang et al., 2019). In the upland regions with poor soil moisture and better oxygen supply, the addition of biochar did increase the indigenous methanotrophic oxidation which is not significant to assess the response (Di et al., 2011). However, biochar addition tends to increase methane uptake by 96% in clover–clover–wheat–bean–oat crop rotation taken up in the polar region (Karhu et al., 2011). Incorporating rice husk biochar into the rice-wheat system effectively brought down CH_4 emissions by 10.9% and 22.8% from no-residue and residue-incorporated tillage systems (Singh et al., 2023). The dynamic response of methanotrophs in terms of diversity and activity indicates the effect of temperature, crop cultivation and soil management methods with biochar amendment which grabs the attention towards future technologies to mitigate methane emissions.

17.7 Physicochemical properties of biochar relevant to methane mitigation

An ever-looming threat of climate change is no longer a myth and has severe implications on diverse sectors, especially agriculture through soil degradation, soil-moisture-nutrient imbalances and an ultimate yield reduction. To combat climate change, there is a dire need for potential tools for mitigation and adaptation; one of which is carbon-negative biochar. Biochar is carbonaceous material produced under anoxic conditions called pyrolysis through thermo-chemical reactions. Biochar has miraculous physical and chemical properties that are relevant to mitigating methane emissions, as supported by several studies (Nan et al., 2021; Sun et al., 2022). The astounding properties of biochar have the tendency to alter the soil physico-chemical state and retain conditions suitable for the growth and development of flora and fauna under a relatively diverse set of environments. Conversely, the properties of biochar were governed by the type of feedstock, pyrolysis temperature, heating rate and residence or cooling period. The black and charred output of high-temperature pyrolysis (600°C–800°C) with a longer residence period have better properties than those prepared at a relatively lower temperature range (300°C–450°C) (Zhang, Zhang, et al., 2022; Zhang, Zhao, et al., 2022). The following properties explain the mechanisms and influence of biochar on the diverse and dynamic soil floral and faunal relations.

17.7.1 High porosity

The biochar upon pyrolysis was charred into fine particles due to the degree of organic matter decomposition (Katyal, Thambimuthu, & Valix, 2003) and micropore formation (Chen & Chen, 2009). The porous structure of biochar helps to alter the oxidation and reduction capacity of soils, meanwhile, acts as a haven for soil microbes and mediates the soil aeration under anaerobic conditions (Hussain et al., 2017). High through-put research on biochar porosity revealed the longitudinal pores of various sizes ranging from micro- to macropores. Where the larger micropores are resultant of biomass vascular bundles, and act as channels to route pyrolytic vapors during internal combustion (Lee, Lee, Gan, Thangalazhy-Gopakumar, & Ng, 2017). These pores have significant importance in improving soil quality and providing habitats for symbiotic micro-organisms (Thies & Rillig, 2012). The methane microbes viz. methanogens and methanotrophs are significantly influenced by the biochar porosity; especially under flooded rice, harbor methanotrophs oxidize the methane emitted under anaerobic conditions in the surface layers (Sriphirom, Towprayoon, Yagi, Rossopa, & Chidthaisong, 2022).

17.7.2 Large specific surface area

The higher specific surface area of biochar provides the added advantage of more surface charges with dynamic adsorption capacities. The biochar prepared from the feedstocks with more lignin content at higher temperatures possesses a more specific surface area with the desired properties to amend the soil properties (Liu et al., 2022). The increase in pyrolysis temperatures improves the surface area either through ward-off of pore-blocking substances or creating thermal to clear open the pores and increase the area (Rafiq et al., 2016). The huge influence of pyrolysis temperature on biochar leads to the destruction of aliphatic alkyls and ester groups and exposes the lignin core to improve the surface area of the biochar (Chen & Chen, 2009). The larger surface area strengthens the biochar's capture potential to adsorb gases and carbon via binding and provides a conducive habitat for microbial consortia and proliferation. The significant influence of biochar's specific surface area on microbial consortia strongly influences the methane release dynamics and oxidation in the soils, besides the beneficial alteration of diverse mechanisms and ecological cycles.

17.7.3 Strong ion exchange capacity

Biochar has a higher ion exchange capacity, which can significantly affect the soil physicochemical properties. The strong ionization capacity of biochar coupled with high porosity and alkalinity affects soil nutrient availability, microbial abundance and diversity (Mukherjee & Lal, 2013). Apart from these, the soil water holding capacity and nutrient retention capacity were governed by the available charges on the biochar's surface and its strong association with the exchange of ions and strongly adheres to the substance surface. Chemical modification or enrichment of biochar also increases the ion-exchange potential of biochar, where ion-exchange potential was increased by 10.6 times when zeolite was used as an additive with biochar (Zheng et al., 2020). Studies also reported that an increase in the cation exchange capacity (CEC) of biochar was observed with an increase in pyrloyzing temperature upon loss of volatile matter and increased surface micropores (Gomez-Eyles, Beesley, Moreno-Jimenez, Ghosh, & Sizmur, 2013; Cely, Tarquis, Paz-Ferreiro, Méndez, & Gascó, 2014). The ion exchange capacity of biochar indirectly influences the methane emissions from the soil by manipulating the physicochemical properties of soil and the mechanism of anaerobic decomposition and oxidation in paddy soils (Nan et al., 2021; Wu, Song, et al., 2019; Wu, Zhang, et al., 2019).

17.7.4 Carbon content

Biochar, a carbon-rich material, whose stability and integrity were deciphered by the nature of carbon structures. These dynamic amorphous carbon structures were formed through cellulose degradation at higher pyrolysis temperatures, which also results in the formation of micropores (Vamvuka & Sfakiotakis, 2011; Zhao, Ta, & Wang, 2017). The typical biochars prepared at higher temperatures are highly stable and have >65% fixed carbon content, which is highly reliant on feedstock type and pyrolysis conditions (Wijitkosum & Jiwnok, 2019). The stable biochar sequesters more carbon into the soil and captures it for a longer duration (Han et al., 2016). Meanwhile, the aged biochar in the soils provides a safe haven in mesopores helps microbial communities to proliferate and alters the microbial abundance in accordance with soil management conditions (Lee et al., 2023).

17.7.5 Electro-chemical properties

The electro-chemical properties or more versatile electron transfer capability of biochar aids in microbial relations and energy transfer under anaerobic conditions. The higher specific surface area and hydrophobicity of biochar facilitate electron transmission among anaerobic microbial communities. Though the pyrolysis temperature and type of feedstock drive the electro-chemical potential of biochar, the soil-applied biochar facilitates congenial microclimate for electron exchange between microbes and carbon substrates under anaerobic conditions (Sun et al., 2022; Wang et al., 2022). By channeling the necessary electron sharing under anoxic soil conditions, and holding off the intermediary metabolites produced by acetotrophic methanogens, biochar reduces the substrate availability for CH_4 generation (Manga, Boutikos, Semiyaga, Olabinjo, & Muoghalu, 2023). These handy assets of biochar promote the methanogenic archaeal activity even under the absence of oxygen and oxidize the methane fluxes produced under submerged or anoxic conditions (Wang et al., 2022).

Overall, biochar has physical and chemical properties that can influence methane mitigation through various mechanisms, including adsorption of methane, modification of microbial communities, and improvement of soil conditions. These properties contribute to the potential of biochar as a promising material for mitigating methane emissions and promoting sustainable soil management practices.

17.8 Conclusion and future perspectives

In conclusion, the use of biochar as a soil amendment has shown great potential for mitigating methane emissions in cropping systems through the promotion of a microbial shift through diversification and altering dynamics. The mechanisms by which biochar reduces methane emissions are complex and involve changes in soil physicochemical properties, such as increased soil pH and nutrient availability, as well as alterations in the microbial community structure and function. The effectiveness of biochar in mitigating methane emissions depends on various factors, including feedstock type, pyrolysis conditions, application rate, and soil type. Nevertheless, the use of biochar holds promise for sustainable agriculture by reducing GHG emissions while simultaneously improving soil health and system productivity. Future research should aim to further elucidate the mechanisms underlying the biochar-microbial interactions and optimize biochar application strategies for specific cropping systems and environmental conditions.

The use of biochar offers a promising strategy for mitigating the effect of climate change in agricultural soils. Nevertheless, further extensive research is necessary to validate these findings across a wide range of agricultural soil types, considering different types and amounts of biochar. To assess the long-term impact of biochar, extended experiments are required to monitor its influence on soil methane emissions and consumption following irrigation, rainfall, and nitrogen fertilization. Future investigations should thoroughly examine the potential effects of applying biochar on the dynamics of methanogenic and methanotrophic populations, taking into account soil type, irrigation, nitrogen sources, and management practices. Additionally, comprehensive analyses should encompass the effects of biochar amendments on soil carbon relations under diverse environmental conditions, considering the role of various microbial consortia and their interaction with methane-producing and methane-consuming bacteria. Lastly, it is essential that future research ensures the stability and sustainability of soil-applied biochar, considering both environmental and economic perspectives.

References

Al-Ghussain, L. (2019). Global warming: Review on driving forces and mitigation. *Environmental Progress & Sustainable Energy, 38*(1), 13–21.

Ali, M. A., Oh, J. H., & Kim, P. J. (2008). Evaluation of silicate iron slag amendment on reducing methane emission from flood water rice farming. *Agriculture, Ecosystems & Environment, 128*(1–2), 21–26.

Atkinson, C. J., Fitzgerald, J. D., & Hipps, N. A. (2010). Potential mechanisms for achieving agricultural benefits from biochar application to temperate soils: A review. *Plant and Soil, 337*, 1–18.

Attenborough, D. (2020). *A life on our planet: My witness statement and a vision for the future.* Random House.

Awad, Y. M., Wang, J., Igalavithana, A. D., Tsang, D. C., Kim, K. H., Lee, S. S., ... Ok, Y. S. (2018). Biochar effects on rice paddy: Meta-analysis. *Advances in Agronomy, 148*, 1–32.

Balch, W. E., Fox, G. E., Magrum, L. J., Woese, C. R., & Wolfe, R. (1979). Methanogens: Re-evaluation of a unique biological group. *Microbiological Reviews, 43*(2), 260–296.

Bhattacharyya, P., Dash, P. K., Swain, C. K., Padhy, S. R., Roy, K. S., Negi, S., ... Mohapatra, T. (2019). Mechanism of plant mediated methane emission in tropical lowland rice. *Science of the Total Environment, 651*, 84–92.

Bhullar, G. S., Iravani, M., Edwards, P. J., & Olde Venterink, H. (2013). Methane transport and emissions from soil as affected by water table and vascular plants. *BMC Ecology, 13*, 1–9.

Cai, F., Feng, Z., & Zhu, L. (2018). Effects of biochar on CH_4 emission with straw application on paddy soil. *Journal of Soils and Sediments, 18*(2), 599–609. Available from https://doi.org/10.1007/s11368-017-1761-x.

Cely, P., Tarquis, A. M., Paz-Ferreiro, J., Méndez, A., & Gascó, G. (2014). Factors driving the carbon mineralization priming effect in a sandy loam soil amended with different types of biochar. *Solid Earth, 5*(1), 585–594. Available from https://doi.org/10.5194/se-5-585-2014.

Chen, B., & Chen, Z. (2009). Sorption of naphthalene and 1-naphthol by biochars of orange peels with different pyrolytic temperatures. *Chemosphere, 76*(1), 127–133.

Chen, W., Liao, X., Wu, Y., Liang, J. B., Mi, J., Huang, J., ... Wang, Y. (2017). Effects of different types of biochar on methane and ammonia mitigation during layer manure composting. *Waste Management, 61*, 506–515. Available from https://doi.org/10.1016/j.wasman.2017.01.014.

Chen, Y., & Murrell, J. C. (2010). *Ecology of aerobic methanotrophs and their role in methane cycling. Handbook of Hydrocarbon and Lipid Microbiology* (pp. 3067–3076). Berlin, Heidelberg: Springer. Available from https://doi.org/10.1007/978-3-540-77587-4_229.

Conrad, R. (2020). Importance of hydrogenotrophic, aceticlastic and methylotrophic methanogenesis for methane production in terrestrial, aquatic and other anoxic environments: A mini review. *Pedosphere, 30*(1), 25–39. Available from https://doi.org/10.1016/s1002-0160(18)60052-9.

Davamani, V., Parameswari, E., & Arulmani, S. (2020). Mitigation of methane gas emissions in flooded paddy soil through the utilization of methanotrophs. *Science of The Total Environment, 726*138570. Available from https://doi.org/10.1016/j.scitotenv.2020.138570.

Di, H. J., Cameron, K. C., Shen, J. P., Winefield, C. S., O'Callaghan, M., Bowatte, S., & He, J. Z. (2011). Methanotroph abundance not affected by applications of animal urine and a nitrification inhibitor, dicyandiamide, in six grazed grassland soils. *Journal of Soils and Sediments, 11*(3), 432–439. Available from https://doi.org/10.1007/s11368-010-0318-z.

Dong, X., Singh, B. P., Li, G., Lin, Q., & Zhao, X. (2019). Biochar has little effect on soil dissolved organic carbon pool 5 years after biochar application under field condition. *Soil Use and Management, 35*(3), 466–477. Available from https://doi.org/10.1111/sum.12474.

Dubey, S. K. (2005). Microbial ecology of methane emission in rice agroecosystem: A review. *Applied Ecology and Environmental Research, 3*(2), 1–27.

FAOSTAT. (2020). Methane emissions from different countries. Food and Agriculture Organization. <https://www.fao.org/faostat/en/#data/GR> Accessed 15.12.22.

FAOSTAT. (2022). Methane emissions from cultivated rice fields. Food and Agriculture Organization. <https://www.fao.org/faostat/en/#data/GR> Accessed 15.12.22.

Fenchel, T., King, G. M., & Blackburn, T. H. (2012). Bacterial metabolism. In T. Fenchel, G. M. King, & T. H. Blackburn (Eds.), *Bacterial Biogeochemistry* (pp. 1–34). Elsevier.

Feng, Y., Xu, Y., Yu, Y., Xie, Z., & Lin, X. (2012). Mechanisms of biochar decreasing methane emission from Chinese paddy soils. *Soil Biology and Biochemistry, 46*, 80–88. Available from https://doi.org/10.1016/j.soilbio.2011.11.016.

Fu, L., Lu, Y., Tang, L., Hu, Y., Xie, Q., Zhong, L., ... Zhang, S. (2021). Dynamics of methane emission and archaeal microbial community in paddy soil amended with different types of biochar. *Applied Soil Ecology, 162*103892. Available from https://doi.org/10.1016/j.apsoil.2021.103892.

Gao, J., Liu, L., Shi, Z., & Lv, J. (2022). Biochar amendments facilitate methane production by regulating the abundances of methanogens and methanotrophs in flooded paddy soil. *Frontiers in Soil Science, 2*. Available from https://doi.org/10.3389/fsoil.2022.801227.

Geaneth, I., Francisco, E., Faz, Á., & Acosta, J. A. (2022). Soil organic carbon dynamics in two rice cultivation systems compared to an agroforestry cultivation system. *Agronomy, 12*(1), 17. Available from https://doi.org/10.3390/agronomy12010017.

Gomez-Eyles, J. L., Beesley, L., Moreno-Jimenez, E., Ghosh, U., & Sizmur, T. (2013). The potential of biochar amendments to remediate contaminated soils. *Biochar and Soil Biota, 4*, 100–133.

Gupta, K., Kumar, R., Baruah, K. K., Hazarika, S., Karmakar, S., & Bordoloi, N. (2021). Greenhouse gas emission from rice fields: A review from Indian context. *Environmental Science and Pollution Research, 28*(24), 30551–30572.

Han, X., Sun, X., Wang, C., Wu, M., Dong, D., Zhong, T., ... Wu, W. (2016). Mitigating methane emission from paddy soil with rice-straw biochar amendment under projected climate change. *Scientific Reports, 6*(1), 1–10.

Huang, Y., Wang, C., Lin, C., Zhang, Y., Chen, X., Tang, L., ... Song, T. (2019). Methane and nitrous oxide flux after biochar application in subtropical acidic paddy soils under tobacco-rice rotation. *Scientific Reports, 9*(1), 1–10.

Hussain, M., Farooq, M., Nawaz, A., Al-Sadi, A. M., Solaiman, Z. M., Alghamdi, S. S., ... Siddique, K. H. M. (2017). Biochar for crop production: Potential benefits and risks. *Journal of Soils and Sediments, 17*(3), 685–716.

Hussain, S., Peng, S., Fahad, S., Khaliq, A., Huang, J., Cui, K., & Nie, L. (2015). Rice management interventions to mitigate greenhouse gas emissions: A review. *Environmental Science and Pollution Research International*, 22(5), 3342–3360. Available from https://doi.org/10.1007/s11356-014-3760-4.

IEA. (2021). Sources of methane emissions, International Energy Agency, Paris. IEA. Licence: CC BY 4.0. https://www.iea.org/data-and-statistics/charts/sources-of-methane-emissions-2021.

IEA. (2023). Methane and climate change. <https://www.iea.org/reports/global-methane-tracker-2022/methane-and-climate-change> Accessed 18.12.22.

In 't Zandt, M. H., van den Bosch, T. J. M., Rijkers, R., van Kessel, M. A. H. J., Jetten, M. S. M., & Welte, C. U. (2018). Co-cultivation of the strictly anaerobic methanogen Methanosarcina barkeri with aerobic methanotrophs in an oxygen-limited membrane bioreactor. *Applied Microbiology and Biotechnology*, 102(13), 5685–5694. Available from https://doi.org/10.1007/s00253-018-9038-x.

IndiaStat. (2022). State/Season-wise area, production and productivity of rice in India (2020–2021). IndiaAgriStat. <https://www.indiastatagri.com/table/agriculture/state-season-wise-area-production-productivity-ric/1423615> Accessed 21.12.22..

Jeffery, S., Verheijen, F. G., Kammann, C., & Abalos, D. (2016). Biochar effects on methane emissions from soils: A meta-analysis. *Soil Biology and Biochemistry*, 101, 251–258. Available from https://doi.org/10.1016/j.soilbio.2016.07.021.

Ji, M., Zhou, L., Zhang, S., Luo, G., & Sang, W. (2020). Effects of biochar on methane emission from paddy soil: Focusing on DOM and microbial communities. *Science of the Total Environment*, 743140725.

Karhu, K., Mattila, T., Bergström, I., & Regina, K. (2011). Biochar addition to agricultural soil increased CH_4 uptake and water holding capacity – Results from a short-term pilot field study. *Agriculture, Ecosystems & Environment*, 140(1–2), 309–313. Available from https://doi.org/10.1016/j.agee.2010.12.005.

Katyal, S., Thambimuthu, K., & Valix, M. (2003). Carbonisation of bagasse in a fixed bed reactor: Influence of process variables on char yield and characteristics. *Renewable Energy*, 28(5), 713–725.

Khalil, K., Rasmussen, M. A., & Shearer, M. J. (1998). Effects of production and oxidation processes on methane emissions from rice fields. *Journal of Geophysical Research: Atmospheres*, 103(D19), 25233–25239. Available from https://doi.org/10.1029/98JD01116.

Khan, S., Chao, C., Waqas, M., Arp, H. P. H., & Zhu, Y. G. (2013). Sewage sludge biochar influence upon rice (*Oryza sativa* L.) yield, metal bioaccumulation and greenhouse gas emissions from acidic paddy soil. *Environmental Science & Technology*, 47(15), 8624–8632.

Knapp, C. W., Fowle, D. A., Kulczycki, E., Roberts, J. A., & Graham, D. W. (2007). Methane monooxygenase gene expression mediated by methanobactin in the presence of mineral copper sources. *Proceedings of the National Academy of Sciences*, 104(29), 12040–12045.

Kollah, B., Bakoriya, M., Dubey, G., Parmar, R., Somasundaram, J., Shirale, A., ... Mohanty, S. R. (2020). Methane consumption potential of soybean-wheat, maize-wheat and maize-gram cropping systems under conventional and no-tillage agriculture in a tropical vertisol. *The Journal of Agricultural Science*, 158(1–2), 38–46.

Kubaczyński, A., Walkiewicz, A., Pytlak, A., Grządziel, J., Gałązka, A., & Brzezińska, M. (2022). Biochar dose determines methane uptake and methanotroph abundance in Haplic Luvisol. *Science of The Total Environment*, 806, 151259.

Kumar, A., Bhattacharya, T., Shaikh, W. A., Roy, A., Chakraborty, S., Vithanage, M., & Biswas, J. K. (2023). Multifaceted applications of biochar in environmental management: A bibliometric profile. *Biochar*, 5(1).

Kumputa, S., Vityakon, P., Saenjan, P., & Lawongsa, P. (2019). Carbonaceous greenhouse gases and microbial abundance in paddy soil under combined biochar and rice straw amendment. *Agronomy*, 9(5), 228.

Kurr, M., Huber, R., König, H., Jannasch, H. W., Fricke, H., Trincone, A., ... Stetter, K. O. (1991). *Methanopyrus kandleri*, gen. and sp. nov. represents a novel group of hyperthermophilic methanogens, growing at 110° C. *Archives of Microbiology*, 156, 239–247.

Kurth, J. M., Op den Camp, H. J. M., & Welte, C. U. (2020). Several ways one goal-methanogenesis from unconventional substrates. *Applied Microbiology and Biotechnology*, 104(16), 6839–6854. Available from https://doi.org/10.1007/s00253-020-10724-7.

Laufer, J.,, & Tomlinson, T. (2012). Biochar field studies: An IBI research summary.

Lee, H. J., Kim, S. Y., Kim, P. J., Madsen, E. L., & Jeon, C. O. (2014). Methane emission and dynamics of methanotrophic and methanogenic communities in a flooded rice field ecosystem. *FEMS Microbiology Ecology*, 88(1), 195–212.

Lee, J., Jeong, H., Gwon, H., Lee, H., Park, H., Kim, G., ... Lee, S. (2023). Effects of biochar on methane emissions and crop yields in east Asian paddy fields: A regional scale meta-analysis. *Sustainability, 15*(12), 9200.

Lee, X. J., Lee, L. Y., Gan, S., Thangalazhy-Gopakumar, S., & Ng, H. K. (2017). Biochar potential evaluation of palm oil wastes through slow pyrolysis: Thermochemical characterization and pyrolytic kinetic studies. *Bioresource Technology, 236*, 155–163. Available from https://doi.org/10.1016/j.biortech.2017.03.105.

Lee, Y., Park, J., Gang, K., Ryu, C., Yang, W., Jung, J., & Hyun, S. (2013). Production and characterization of biochar from various biomass materials by slow pyrolysis. *Technical Bulletin-Food and Fertilizer Technology Center, 197*.

Lehmann, J., Gaunt, J., & Rondon, M. (2006). Bio-char sequestration in terrestrial ecosystems – A review. *Mitigation and Adaptation Strategies for Global Change, 11*(2), 403–427. Available from https://doi.org/10.1007/s11027-005-9006-5.

Leshchenko, I., Shulzhenko, S., Kaplin, M., Maistrenko, N., & Shcherbyna, E. (2023). Assessment of the greenhouse gases reduction by the oil and gas sector of Ukraine to Meet international climate agreements. In A. Zaporozhets, & O. Popov (Eds.), *Systems, decision and control in energy IV. Studies in systems, decision and control* (456). Cham: Springer. Available from https://doi.org/10.1007/978-3-031-22500-0_13.

Lévesque, V., Oelbermann, M., & Ziadi, N. (2022). Biochar in temperate soils: Opportunities and challenges. *Canadian Journal of Soil Science, 102*(1), 1–26.

Li, S., & Tasnady, D. (2023). Biochar for soil carbon sequestration: Current knowledge, mechanisms, and future perspectives. *Journal of Carbon Research, 9*(3), 67.

Liu, J., Qiu, H., Wang, C., Shen, J., Zhang, W., Cai, J., ... Wu, J. (2021a). Effects of biochar amendment on greenhouse gas emission in two paddy soils with different textures. *Paddy and Water Environment, 19*(1), 87–98. Available from https://doi.org/10.1007/s10333-020-00821-8.

Liu, Y., Ge, T., van Groenigen, K. J., Yang, Y., Wang, P., Cheng, K., ... Kuzyakov, Y. (2021b). Rice paddy soils are a quantitatively important carbon store according to a global synthesis. *Communications Earth & Environment, 2*(1), 1–9. Available from https://doi.org/10.1038/s43247-021-00229-0.

Liu, Z., Zhen, F., Zhang, Q., Qian, X., Li, W., Sun, Y., ... Qu, B. (2022). Nanoporous biochar with high specific surface area based on rice straw digestion residue for efficient adsorption of mercury ion from water. *Bioresource Technology, 359*, 127471.

Lorenz, K., & Lal, R. (2014). Biochar application to soil for climate change mitigation by soil organic carbon sequestration. *Journal of Plant Nutrition and Soil Science, 177*(5), 651–670.

Lu, Y., Liu, Q., Fu, L., Hu, Y., Zhong, L., Zhang, S., ... Xie, Q. (2022). The effect of modified biochar on methane emission and succession of methanogenic archaeal community in paddy soil. *Chemosphere, 304*135288. Available from https://doi.org/10.1016/j.chemosphere.2022.135288.

Lyu, Z., Shao, N., Akinyemi, T., & Whitman, W. B. (2018). Methanogenesis. *Current Biology, 28*(13), R727–R732. Available from https://doi.org/10.1016/j.cub.2018.05.021.

Ma, K. E., Qiu, Q., & Lu, Y. (2010). Microbial mechanism for rice variety control on methane emission from rice field soil. *Global Change Biology, 16*(11), 3085–3095.

Ma, X., Lv, M., Huang, F., Zhang, P., Cai, T., & Jia, Z. (2022). Effects of biochar application on soil hydrothermal environment, carbon emissions, and crop yield in wheat fields under ridge & furrow rainwater harvesting planting mode. *Agriculture, 12*(10), 1704. Available from https://doi.org/10.3390/agriculture12101704.

Malyan, S. K., Bhatia, A., Kumar, A., Gupta, D. K., Singh, R., Kumar, S. S., ... Jain, N. (2016). Methane production, oxidation and mitigation: A mechanistic understanding and comprehensive evaluation of influencing factors. *Science of The Total Environment, 572*, 874–896. Available from https://doi.org/10.1016/j.scitotenv.2016.07.182.

Malyan, S. K., Kumar, S. S., Singh, A., Kumar, O., Gupta, D. K., Yadav, A. N., ... Kumar, A. (2021). Understanding methanogens, methanotrophs, and methane emission in rice ecosystem. *Microbiomes and the Global Climate Change*, 205–224.

Manga, M., Boutikos, P., Semiyaga, S., Olabinjo, O., & Muoghalu, C. C. (2023). Biochar and its potential application for the improvement of the anaerobic digestion process: A critical review. *Energies, 16*(10), 4051.

Mayumi, D., Yoshimoto, T., Uchiyama, H., Nomura, N., & Nakajima-Kambe, T. (2010). Seasonal change in methanotrophic diversity and populations in a rice field soil assessed by DNA-stable isotope probing and quantitative real-time PCR. *Microbes and environments, 25*(3), 156–163.

McHenry, M. P. (2009). Carbon-based stock feed additives: A research methodology that explores ecologically delivered C biosequestration, alongside live weights, feed use efficiency, soil nutrient retention, and perennial fodder plantations. *Journal of the Science of Food and Agriculture, 90*(2), 183–187.

Megraw, S. R., & Knowles, R. (1987). Methane production and consumption in a cultivated humisol. *Biology and Fertility of Soils, 5*, 56–60.

Mohanty, S. R., Bandeppa, G. S., Dubey, G., Ahirwar, U., Patra, A. K., & Bharati, K. (2017). Methane oxidation in response to iron reduction-oxidation metabolism in tropical soils. *European Journal of Soil Biology, 78*, 75–81.

Mohanty, S., Kollah, B., Chaudhary, R. S., Singh, A. B., & Singh, M. (2015). Methane uptake in tropical soybean–wheat agroecosystem under different fertilizer regimes. *Environmental Earth Sciences, 74*, 5049–5061.

Mukherjee, A., & Lal, R. (2013). Biochar impacts on soil physical properties and greenhouse gas emissions. *Agronomy, 3*(2), 313–339.

Nan, Q., Wang, C., Wang, H., Yi, Q., & Wu, W. (2020a). Mitigating methane emission via annual biochar amendment pyrolyzed with rice straw from the same paddy field. *Science of The Total Environment, 746*141351. Available from https://doi.org/10.1016/j.scitotenv.2020.141351.

Nan, Q., Wang, C., Yi, Q., Zhang, L., Ping, F., Thies, J. E., & Wu, W. (2020b). Biochar amendment pyrolysed with rice straw increases rice production and mitigates methane emission over successive three years. *Waste Management, 118*, 1–8. Available from https://doi.org/10.1016/j.wasman.2020.08.013.

Nan, Q., Xin, L., Qin, Y., Waqas, M., & Wu, W. (2021). Exploring long-term effects of biochar on mitigating methane emissions from paddy soil: A review. *Biochar, 3*(2), 125–134. Available from https://doi.org/10.1007/s42773-021-00096-0.

Nazaries, L., Murrell, J. C., Millard, P., Baggs, L., & Singh, B. K. (2013). Methane, microbes and models: Fundamental understanding of the soil methane cycle for future predictions. *Environmental Microbiology, 15*(9), 2395–2417.

Nguyen, B. T., & Van Nguyen, N. (2023). Biochar addition balanced methane emissions and rice growth by enhancing the quality of paddy soil. *Journal of Soil Science and Plant Nutrition*, 1–12. Available from https://doi.org/10.1007/s42729-023-01249-9.

Noguchi, A., Hasegawa, I., Shinmachi, F., & Yazaki, J. (1997). *Possibility of copper deficiency and impediment in ripening in rice from application of crop residues in submerged soil. Plant nutrition for sustainable food production and environment* (pp. 799–800). Netherlands: Springer.

Norina, E. (2007). Methane emission from northern wetlands. Term paper. HS 2007. pp. 1–20.

Osman, A. I., Fawzy, S., Farghali, M., El-Azazy, M., Elgarahy, A. M., Fahim, R. A., ... Rooney, D. W. (2022). Biochar for agronomy, animal farming, anaerobic digestion, composting, water treatment, soil remediation, construction, energy storage, and carbon sequestration: A review. *Environmental Chemistry Letters, 20*(4), 2385–2485.

Ouyang, L., Tang, Q., Yu, L., & Zhang, R. (2014). Effects of amendment of different biochars on soil enzyme activities related to carbon mineralisation. *Soil Research, 52*(7), 706–716.

Pan, S., Dong, C., Su, J., Wang, P., Chen, C., Chang, J., ... Hung, C. (2021). The role of biochar in regulating the carbon, phosphorus, and nitrogen cycles exemplified by soil systems. *Sustainability, 13*(10), 5612.

Panda, B. B., Satpathy, B. S., Nayak, A. K., Tripathi, R., Shahid, M., Mohanty, S., ... Nayak, P. K. (2018). *Agroecological Intensification of Rice based cropping system*. India: ICAR.

Parmentier, J. W., Zhang, W., Mi, Y., Zhu, X., Hayes, D. J., Zhuang, Q., ... McGuire, A. D. (2015). Rising methane emissions from northern wetlands associated with sea ice decline. *Geophysical Research Letters, 42*(17), 7214–7222. Available from https://doi.org/10.1002/2015GL065013.

Peng, Y., Wang, G., Li, F., Yang, G., Fang, K., Liu, L., ... Yang, Y. (2019). Unimodal response of soil methane consumption to increasing nitrogen additions. *Environmental Science & Technology, 53*(8), 4150–4160.

Pietikäinen, J., Kiikkilä, O., & Fritze, H. (2000). Charcoal as a habitat for microbes and its effect on the microbial community of the underlying humus. *Oikos, 89*(2), 231–242. Available from https://doi.org/10.1034/j.1600-0706.2000.890203.x.

Pramanik, P., & Kim, P. J. (2018). *Methanogens harboring in rice rhizosphere reduce labile organic carbon compounds to produce methane gas. Rice Crop-Current Developments*. IntechOpen.

Pratiwi, E. P. A., & Shinogi, Y. (2016). Rice husk biochar application to paddy soil and its effects on soil physical properties, plant growth, and methane emission. *Paddy and Water Environment, 14*, 521–532.

Qi, L., Ma, Z., Chang, S. X., Zhou, P., Huang, R., Wang, Y., ... Gao, M. (2021). Biochar decreases methanogenic archaea abundance and methane emissions in a flooded paddy soil. *Science of The Total Environment, 752*141958. Available from https://doi.org/10.1016/j.scitotenv.2020.141958.

Qin, X., Li, Y., Wang, H., Liu, C., Li, J., Wan, Y., ... Liao, Y. (2016). Long-term effect of biochar application on yield-scaled greenhouse gas emissions in a rice paddy cropping system: A four-year case study in south

China. *Science of The Total Environment*, 569-570, 1390–1401. Available from https://doi.org/10.1016/j.scitotenv.2016.06.222.

Rafiq, M. K., Bachmann, R. T., Rafiq, M. T., Shang, Z., Joseph, S., & Long, R. (2016). Influence of pyrolysis temperature on physico-chemical properties of corn stover (*Zea mays* L.) biochar and feasibility for carbon capture and energy balance. *PLoS One*, 11(6), e0156894.

Rawat, J., Saxena, J., & Sanwal, P. (2019). Biochar: A sustainable approach for improving plant growth and soil properties. In V. Abrol, & P. Sharma (Eds.), *Biochar - An Imperative Amendment for Soil and the Environment*. IntechOpen.

Reddy, K. R., Yargicoglu, E. N., Yue, D., & Yaghoubi, P. (2014). Enhanced microbial methane oxidation in landfill cover soil amended with biochar. *Journal of Geotechnical and Geoenvironmental Engineering*, 140(9), 04014047.

Reim, A., Lüke, C., Krause, S., Pratscher, J., & Frenzel, P. (2012). One millimetre makes the difference: High-resolution analysis of methane-oxidizing bacteria and their specific activity at the oxic–anoxic interface in a flooded paddy soil. *The ISME Journal*, 6(11), 2128–2139.

Ribas, A., Mattana, S., Llurba, R., Debouk, H., Sebastià, M. T., & Domene, X. (2019). Biochar application and summer temperatures reduce N_2O and enhance CH_4 emissions in a Mediterranean agroecosystem: Role of biologically-induced anoxic microsites. *Science of the Total Environment*, 685, 1075–1086.

Rondon, M.A., Ramirez, J.A., & Lehmann, J. (2005). Greenhouse gas emissions decrease with charcoal additions to tropical soils. In *Proceedings of the 3rd USDA Symposium on Greenhouse Gases and Carbon Sequestration*, Baltimore, USA, March 21–24, 2005, p. 208.

Saha, M. K., Mia, S., Biswas, A. A. A., Sattar, M. A., Kader, M. A., & Jiang, Z. (2022). Potential methane emission reduction strategies from rice cultivation systems in Bangladesh: A critical synthesis with global meta-data. *Journal of Environmental Management*, 310, 114755.

Sakai, S., Imachi, H., Hanada, S., Ohashi, A., Harada, H., & Kamagata, Y. (2008). *Methanocella paludicola* gen. nov., sp. nov., a methane-producing archaeon, the first isolate of the lineage 'Rice Cluster I', and proposal of the new archaeal order *Methanocellales* ord. nov. *International Journal of Systematic and Evolutionary Microbiology*, 58(4), 929–936.

Sass, R. L., Fisher, F. M., Wang, Y. B., Turner, F. T., & Jund, M. F. (1992). Methane emission from rice fields: The effect of floodwater management. *Global Biogeochemical Cycles*, 6(3), 249–262.

Serrano-Silva, N., Sarria-Guzmán, Y., Dendooven, L., & Luna-Guido, M. (2014). Methanogenesis and methanotrophy in soil: A review. *Pedosphere*, 24(3), 291–307.

Shackley, S., Sohi, S., & Haszeldine, S. (2009). Biochar, reducing and removing CO_2 while improving soils: A significant and sustainable response to climate change. *UK Biochar Research*, 1–12.

Shen, J., Tang, H., Liu, J., Wang, C., Li, Y., Ge, T., ... Wu, J. (2014). Contrasting effects of straw and straw-derived biochar amendments on greenhouse gas emissions within double rice cropping systems. *Agriculture, Ecosystems & Environment*, 188, 264–274.

Shenbagavalli, S., & Mahimairaja, S. (2012). Production and characterization of biochar from different biological wastes. *International Journal of Plant, Animal and Environmental Sciences*, 2(1), 197–201.

Singh, C., Tiwari, S., Boudh, S., & Singh, J. S. (2017). Biochar application in management of paddy crop production and methane mitigation. Agro-Environmental Sustainability. *Managing Environmental Pollution*, 123–145.

Singh, S., Chaturvedi, S., Nayak, P., Dhyani, V. C., Nandipamu, T. M. K., Singh, D. K., ... Govindaraju, K. (2023). Carbon offset potential of biochar based straw management under rice-wheat system along Indo-Gangetic Plains of India. *Science of The Total Environment*, 165176.

Smith, P., Reay, D., & Smith, J. (2021). Agricultural methane emissions and the potential formitigation. *Philosophical Transactions of the Royal Society A*, 379(2210), 20200451.

Söllinger, A., & Urich, T. (2019). Methylotrophic methanogens everywhere—Physiology and ecology of novel players in global methane cycling. *Biochemical Society Transactions*, 47(6), 1895–1907.

Sriphirom, P., Chidthaisong, A., Yagi, K., Boonapatcharoen, N., Tripetchkul, S., & Towprayoon, S. (2020). Evaluating the effect of different biochar application sizes on methane emission reduction from rice cultivation. *IOP Conference Series. Earth and Environmental Science*, 463(1), 012170. Available from https://doi.org/10.1088/1755-1315/463/1/012170.

Sriphirom, P., Towprayoon, S., Yagi, K., Rossopa, B., & Chidthaisong, A. (2022). Changes in methane production and oxidation in rice paddy soils induced by biochar addition. *Applied Soil Ecology*, 179, 1.

Stadtman, T. C., & Barker, H. A. (1951). Studies on the methane fermentation IX: The origin of methane in the acetate and methanol fermentations by methanosarcina. *Journal of Bacteriology*, 61(1), 81–86.

Steiner, C., Bayode, A., & Ralebitso-Senior, T. K. (2016). Feedstock and production parameters: Effects on biochar properties and microbial communities. *Biochar Application*, 41–54.

Sun, Z., Feng, L., Li, Y., Han, Y., Zhou, H., & Pan, J. (2022). The role of electrochemical properties of biochar to promote methane production in anaerobic digestion. *Journal of Cleaner Production*, 362, 132296.

Thies, J. E., & Rillig, M. C. (2012). *Characteristics of biochar: Biological properties. Biochar for Environmental Management* (pp. 117–138). Routledge.

Tisserant, A., & Cherubini, F. (2019). Potentials, limitations, co-benefits, and trade-offs of biochar applications to soils for climate change mitigation. *Land*, 8(12), 179.

Trotsenko, Y. A., & Murrell, J. C. (2008). Metabolic aspects of aerobic obligate methanotrophy. *Advances in Applied Microbiology*, 63, 183–229.

Vamvuka, D., & Sfakiotakis, S. (2011). Effects of heating rate and water leaching of perennial energy crops on pyrolysis characteristics and kinetics. *Renewable Energy*, 36(9), 2433–2439.

Wang, C., Shen, J., Liu, J., Qin, H., Yuan, Q., Fan, F., ... Wu, J. (2019). Microbial mechanisms in the reduction of CH_4 emission from double rice cropping system amended by biochar: A four-year study. *Soil Biology and Biochemistry*, 135, 251–263. Available from https://doi.org/10.1016/j.soilbio.2019.05.012.

Wang, S., Shi, F., Li, P., Yang, F., Pei, Z., Yu, Q., ... Liu, J. (2022). Effects of rice straw biochar on methanogenic bacteria and metabolic function in anaerobic digestion. *Scientific Reports*, 12(1), 1–14.

Welte, C., & Deppenmeier, U. (2011). Proton translocation in methanogens. *Methods in Enzymology*, 494, 257–280. Available from https://doi.org/10.1016/B978-0-12-385112-3.00013-5.

Wijitkosum, S., & Jiwnok, P. (2019). Elemental composition of biochar obtained from agricultural waste for soil amendment and carbon sequestration. *Applied Sciences*, 9(19), 3980.

Wu, Z., Song, Y., Shen, H., Jiang, X., Li, B., & Xiong, Z. (2019). Biochar can mitigate methane emissions by improving methanotrophs for prolonged period in fertilized paddy soils. *Environmental Pollution*, 253, 1038–1046.

Wu, Z., Zhang, X., Dong, Y., Li, B., & Xiong, Z. (2019). Biochar amendment reduced greenhouse gas intensities in the rice-wheat rotation system: Six-year field observation and meta-analysis. *Agricultural and Forest Meteorology*, 278, 107625.

Xu, P., Zhou, W., Jiang, M., Khan, I., Wu, T., Zhou, M., ... Hu, R. (2022). Methane emission from rice cultivation regulated by soil hydrothermal condition and available carbon and nitrogen under a rice–wheat rotation system. *Plant and Soil*, 480(1-2), 283–294.

Yaashikaa, P. R., Kumar, P. S., Varjani, S., & Saravanan, A. (2020). A critical review on the biochar production techniques, characterization, stability and applications for circular bioeconomy. *Biotechnology Reports*, 28.

Yan, H., Cong, M., Hu, Y., Qiu, C., Yang, Z., Tang, G., ... Jia, H. (2022). Biochar-mediated changes in the microbial communities of rhizosphere soil alter the architecture of maize roots. *Frontiers in Microbiology*, 131023444. Available from https://doi.org/10.3389/fmicb.2022.1023444.

Yang, S., Xiao, Y., Sun, X., Ding, J., Jiang, Z., & Xu, J. (2019). Biochar improved rice yield and mitigated CH_4 and N_2O emissions from paddy field under controlled irrigation in the Taihu Lake Region of China. *Atmospheric Environment*, 200, 69–77. Available from https://doi.org/10.1016/j.atmosenv.2018.12.003.

Yoo, G., Kim, Y. J., Lee, Y. O., & Ding, W. (2016). Investigation of greenhouse gas emissions from the soil amended with rice straw biochar. *KSCE Journal of Civil Engineering*, 20, 2197–2207.

Zhang, A., Liu, Y., Pan, G., Hussain, Q., Li, L., Zheng, J., & Zhang, X. (2012). Effect of biochar amendment on maize yield and greenhouse gas emissions from a soil organic carbon poor calcareous loamy soil from Central China Plain. *Plant and Soil*, 351(1–2), 263–275. Available from https://doi.org/10.1007/s11104-011-0957-x.

Zhang, C., Zeng, G., Huang, D., Lai, C., Chen, M., Cheng, M., ... Wang, R. (2019). Biochar for environmental management: Mitigating greenhouse gas emissions, contaminant treatment, and potential negative impacts. *Chemical Engineering Journal*, 373, 902–922.

Zhang, J., & Shen, J.-L. (2022). Effects of biochar on soil microbial diversity and community structure in clay soil. *Annals of Microbiology*, 72(1). Available from https://doi.org/10.1186/s13213-022-01689-1.

Zhang, X., Zhang, Q., Xu, X., Dong, Y., & Xiong, Z. (2022). Biochar mitigated yield-scaled N_2O and NO emissions and ensured vegetable quality and soil fertility: A 3-year greenhouse field observation. *Agronomy*, 12(7), 1560.

Zhang, X., Zhao, B., Liu, H., Zhao, Y., & Li, L. (2022). Effects of pyrolysis temperature on biochar's characteristics and speciation and environmental risks of heavy metals in sewage sludge biochars. *Environmental Technology & Innovation, 26*, 102288.

Zhao, S., Ta, N., & Wang, X. (2017). Effect of temperature on the structural and physicochemical properties of biochar with apple tree branches as feedstock material. *Energies, 10*(9), 1293.

Zheng, X. J., Chen, M., Wang, J. F., Liu, Y., Liao, Y. Q., & Liu, Y. C. (2020). Assessment of zeolite, biochar, and their combination for stabilization of multimetal-contaminated soil. *ACS Omega, 5*(42), 27374–27382.

Zhu, Z., Ge, T., Xiao, M., Yuan, H., Wang, T., Liu, S., ... Kuzyakov, Y. (2017). Belowground carbon allocation and dynamics under rice cultivation depends on soil organic matter content. *Plant and Soil, 410*(1–2), 247–258. Available from https://doi.org/10.1007/s11104-016-3005-z.

Zimmerman, A. R., Gao, B., & Ahn, M. (2011). Positive and negative carbon mineralization priming effects among a variety of biochar-amended soils. *Soil Biology and Biochemistry, 43*(6), 1169–1179. Available from https://doi.org/10.1016/j.soilbio.2011.02.005.

CHAPTER 18

Biochar enhances carbon stability and regulates greenhouse gas flux under crop production systems

Anamika Barman[1], Anurag Bera[2], Priyanka Saha[1], Saptaparnee Dey[3], Suman Sen[1], Ram Swaroop Meena[2], Shiv Vendra Singh[4] and Amit Kumar Singh[4]

[1]Division of Agronomy, ICAR—Indian Agricultural Research Institute, Pusa, New Delhi, India [2]Department of Agronomy, Institute of Agricultural Sciences, Banaras Hindu University, Varanasi, Uttar Pradesh, India [3]Division of Soil Science and Agricultural Chemistry, ICAR—Indian Agricultural Research Institute, Pusa, New Delhi, India [4]Department of Agronomy, College of Agriculture, Rani Lakshmi Bai Central Agriculture University, Jhansi, Uttar Pradesh, India

18.1 Introduction

Global warming and consequent climate change a significant environmental issues currently having an alarming impact on the world. Considering the irreversible effects of climate change, many people believe it to be a far greater threat to mankind than COVID-19. Increased concentrations of greenhouse gases (GHGs) due to anthropogenic activities and their persistence in the atmosphere will continue to have long-lasting impacts on global climate for many years as already indicated by sea level rise, altered patterns of precipitation, frequent droughts, heat extremes and melting of glaciers. In this challenging circumstance, over the coming 20 years (by 2040), the Earth's average surface temperature will rise by 1.5°C above preindustrial levels, and by the middle of the century, it may even reach 2°C. without stern emission reduction strategies (IPCC, 2021). Furthermore, according to an Intergovernmental Panel on Climate Change (IPCC), "special assessment on the implications of global warming of 1.5°C above preindustrial level" published in 2018, two-fifths of the world's population dwelt in places where the temperature was above 1.5°C (IPCC, 2018). Given this, The IPCC stated that to limit global temperature rise to 1.5°C, a global net zero by 2050 was the minimum required. All anthropogenic GHG emissions must be reduced to zero through adaptation and mitigation strategies, after removal via natural and artificial sinks. The most critical contributor to global warming is the continual increase in concentrations of GHGs such as carbon dioxide (CO_2), methane (CH_4), nitrous oxide (N_2O), and nitrogen trifluoride (N_2F_3) in the atmosphere. Net anthropogenic emissions of GHGs totaled 596.6 GtCO2-eq in 2019. This was 12% (6.5 $GtCO_2$-eq) more than in 2010 and 54% (21 $GtCO_2$-eq) more than in 1990. Compared to the period from 2000 to 2009, the annual average from 2010 to 2019 was 566.0 $GtCO_2$-eq, or 9.1 $GtCO_2$-eq year^{-1} higher. Over the past ten years, this average emissions growth has been the greatest. The average annual growth rate decreased from 2.1% per year in 2000−09 to 1.3% per year in 2010−19 (IPCC, 2022). Among all of these GHGs, CO_2 is the main one, accounting for around 72% of the increase in global temperature, followed by CH_4 (20%), N_2O (5%), and F-gases (3%). Therefore, a global need exists to stop the global warming and climate change through a reduction in the concentration of GHG$_s$. Studies show that carbon emissions cause climate change, and burning crop residues is a more significant factor (Jacobson, 2014; Pradhan, Meena, Kumar, & Lal, 2023).

Burning of crop residue has grown to be a significant environmental issue that affects human health and accelerates global warming (Bhuvaneshwari, Hettiarachchi, & Meegoda, 2019). Through the practical application of sustainable management approaches coupled with the efforts and policies of the government, the issue of crop residue burning can be addressed. The surplus biomass waste with a high organic content can be appropriately repurposed or diverted for destruction to energy choices rather than burning, which helps to reduce climate change by lowering the amount of GHG emissions (Meena, Pradhan, Kumar, & Lal, 2023; Palanivelu, Ramachandran, & Raghavan, 2021; Singh, Chaturvedi, & Datta, 2019). By absorbing CO_2 that has previously been released into the atmosphere and storing it in a long-lived pool, together with lowering CH_4, N_2O, and F-gas emissions at the source, global warming can be mitigated. To counteract climate change, it is therefore essential to identify technologies that absorb and store already released GHGs and those that minimize anthropogenic GHG emisssion at the source. Although the traditional approaches for soil carbon sequestration are being encouraged, such as agronomic interventions, soil management, agroforestry, and afforestation, it require a lot of time (Menon et al., 2007). In recent years, the conversion of crop residue biomass to "biochar" has received substantial attention in climate change mitigation by offering disposal of crop biomass waste, which facilitates carbon sequestration and reduction of GHG emissions coupled with food security and environmental benefits (Majumder, Neogi, Dutta, Powel, & Banik, 2019). For preserving agriculture's sustainability and long-term ecological balance, it is essential to have depth knowledge of biochar production from various biomass wastes, the carbon storage potential of biochar, the impact of biochar on soil microbial biomass carbon (SMBC) and carbon fractions, and the role of biochar in GHG emissions reduction. Therefore, this chapter will go through the functions of

18.2 Surplus crop residue vis-a-vis biochar production

Agricultural wastes are products that originate from different farming operations. The term "harvest trash," more often known as "crop residue," refers to the field residues as well as the residues that remain after a crop has been harvested and turned into a helpful resource in an agricultural field (Bhuvaneshwari et al., 2019). Field residues may include leaves, stalks, stubbles and seed pods. The term "gross residue potential" describes the total quantity of residue that can be produced from any crop. In contrast, the term "surplus residue potential" describes the crop residue that remains after any competing applications, such as animal bedding, composting, or cattle feeding. According to estimates, India has a gross amount of 686 million tonnes (MT) of crop residues annually produced by 26 crops (Hiloidhari, Das, & Baruah, 2014). For cereal crops, the ratios of residue to economic yield range from 1.4 for rice, 1.3 for wheat, 2.0 for maize and 1.5 for oats and barley. Cereal crops contribute the most to the total residue produced in India, with 368 MT (54%) followed by 111 MT (16%) of sugarcane. When gross residues are considered, rice (154 MT) contributes the most, followed by wheat (131 MT) (Jat & Gerard, 2014). Cereal contributes the highest amount of surplus residue (89 MT), followed by sugarcane (69 MT) (Jat & Gerard, 2014). A substantial amount of crop residue is left in the fields due to the use of mechanical farming in rice-based cropping systems due to the increasing labor and time restrictions of intensive agriculture. According to IPCC, more than 25% of the crop residues were burned on farms. Jain, Bhatia, and Pathak (2014) additionally noted that across all states, the percentage of burned paddy trash ranged from 8% to 80%. In terms of crop residue, rice contributed the most (43%), followed by wheat (21%), sugarcane (19%), and oilseed crops (5%). Burning agricultural residue produces 1515 kg of CO_2, 92 kg of CO, 3.83 kg of NO_x, 0.4 kg of SO_2, 2.7 kg of CH_4, and 15.7 kg of nonmethane volatile organic chemicals per tonne (Andreae & Merlet, 2001). Crop residue burning in-situ has been associated with a slew of issues, including air pollution, human and animal health hazards, destruction of beneficial microflora, declining soil fertility, global warming, and others. Biochar synthesis using surplus crop residues might be a viable solution to dispose of crop wastes in this challenging situation safely. Soil biochar can affect many biogeochemical processes due to its resistance to aromatic structure and reactive surfaces. In addition, it can also serve as a sink for atmospheric CO_2. The application of biochar offers a sustainable management practice because of its potential to boost soil organic carbon (SOC), promote soil fertility, mitigate GHG emissions, and increase agricultural productivity (Kaur & Dhundhara, 2022). In India, 517.82 MT of crop residues may create 212.04 ± 44.27 MT of biochar. Approximately 376.11 ± 78.52 MT of CO_2-eq carbon is sequestered, and 1.66 ± 0.46 MT of soil nutrients are retained when biochar is incorporated into the soil (Anand, Kumar, & Kaushal, 2022). A further estimate calculates India's market for biochar generated from crop residues is worth roughly $500 billion. Therefore, it is the need of the hour to utilize surplus crop residues for biochar production to achieve the twin benefits of climate change mitigation and adaptation. The availability of gross crop residues in India and their utilization is depicted in Fig. 18.1.

18.3 Biochar and its significant properties

Biochar is a carbonaceous solid produced by thermochemical conversion of biomass at temperatures between 300°C and 1000°C in a partial or anaerobic situation (Xie, Reddy, Wang, Yargicoglu, & Spokas, 2015). It has a higher surface area, negative functional group, recalcitrant aromatic structure, and highly porous structure (Purakayastha et al., 2019). Due to its aromatic structure, biochar is more stable in terms of physical and chemical properties compared to the organic matter from which it was formed. The surface area of biochar varies from 0.5 to 450 $m^2\,g^{-1}$ (Brassard, Godbout, & Raghavan, 2016). The characteristics of biochar differ greatly depending on the source and type of crop residue used and the pyrolysis condition (Singh, Chaturvedi, Dhyani, & Govindaraju, 2020). Its properties also depend on temperature, lignin content, nitrogen content and reactor design (Palanivelu et al., 2021; Singh et al., 2020). Among several biomass conversion technologies, the thermochemical conversion process is advantageous. The thermochemical process

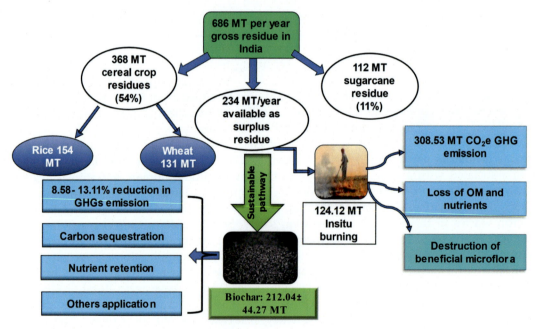

FIGURE 18.1 Crop residue availability in India. Data Source: *Hiloidhari, M., Das, D., & Baruah, D. C. (2014). Bioenergy potential from crop residue biomass in India. Renewable and* Sustainable Energy Reviews, 32, 504–512; *Jat, M. L., & Gerard, B. (2014). Nutrient management and use efficiency in wheat systems of South Asia. Advances in Agronomy, 125, 171–259.*

includes pyrolysis, gasification, combustion, torrefaction and hydrothermal carbonization. Low, moderate temperature (300°C–700°C) with longer residence time yields solid carbonaceous substance biochar, and moderate temperature (425°C–600°C) with short residence time yields bio-oil. Longer residence time with higher temperatures (>800°C) produces syngas. It has been found that biochar with characteristics like C/N ratio 30, H/C_{org} <0.7 and O/C_{org} <0.2 can store carbon in soil and help in the mitigation of climate change (Brassard et al., 2017; Chaturvedi, Singh, Dhyani, Govindaraju, & Mandal, 2021). Biochar's physical, chemical and morphological properties vary with the kind of crop residue from which it is prepared. Purakayastha, Kumari, and Pathak (2015) assessed the properties of biochar prepared from four different crop residues, that is, rice, wheat, maize, and pearl millet. According to their findings, biochar had alkaline pH values, with maize stover exhibiting the highest value (10.7). In recent years, biochar has become increasingly relevant for use in agriculture for crop production. In many instances, adding biochar and biochar with compost to the soil enhances its biophysical and chemical characteristics and the availability of nutrients to plants. Biochar is an important eco-friendly biostimulant that primarily boosts agricultural production and minimizes the negative effects of various abiotic stresses. As a fertilizer, it increases crop output, and as a soil conditioner, it enhances soil quality. Additionally, after applying biochar to soil, significant improvements in root growth, crop growth and development, and yield were seen (Abiven, Hund, Martinsen, & Cornelissen, 2015). Application of biochar has been found to boost crop yields by enhancing soil water holding capacity in light or sandy soils or drainage in clayey soils, as well as nutrient uptake and usage efficiency through increasing soil CEC for a given fertilizer application rate (Agegnehu et al. 2015). In addition, biochar helps to boost the soil microbial population by offering soil microorganisms a habitat and releasing crucial nutrients from its matrix (Pradhan & Meena, 2023; Tauqeer, Turan, Farhad, & Iqbal, 2022). The seed germination, growth, dry matter, nutritional quality and yield of the crop are enhanced with biochar application. Using biochar with other amendments, such as microbes and plant nutrients, will increase its ability to remove trace metals from soil while boosting crop yields. Furthermore, under contaminated condition, biochar can maintain Osmo protectant and osmotic potential of plant by synthesizing stressed protein and proline content (Haider et al., 2022).

18.3.1 Fortified biochar

Nutrient-enriched biochar fertilizers are gaining more and more attention in agricultural production, due to their economic and environmental co-benefits in sustainable crop production. Minerals contained in biochar during pyrolysis play an important role in releasing nutrients to the plant and further carbon stabilization (Giudicianni et al., 2021). Pyrolysis of biomass to produce biochar is affected by the addition of nutrients externally and inherent nutrient content. Before pyrolysis, the amount of inorganic nutrients in biochar can be increased through the addition of refined minerals to the feedstock. Appropriate selection of minerals is very important for the overall increase of stable carbon content in biochar and stable carbon yield. Recent studies have investigated several techniques for producing fertilizers with macronutrients (nitrogen, phosphorous, and potassium) enriched biochar. In this regard, N is an important plant nutrient because N volatilizes from the feedstock material during pyrolysis, and most of the remaining N in the biochar is inaccessible to plants (Ye et al., 2020). Therefore, doping biochar feedstock with minerals such as CaO can enhance the retention of N and produce more stable nitrogen compounds (Buss et al., 2022). Additionally, they observed that before pyrolysis, doping biomass with two minerals P and K could increase the availability of one or both nutrients by the formation of highly soluble potassium phosphate. When P was added in the forms of ammonium phosphate, phosphoric acid, and calcium dihydrogen phosphate, the ability of biochar to sequester carbon substantially increased (Fuentes et al., 2008). Furthermore, enrichment of biochar with vermiculite can increase biochar's carbon retention capacity and stability. Enrichment of biochar with nutrients can increase the nutrient content of biochar, reduce fertilizer consumption, improve the biochar's carbon sequestration potential, and optimize nutrient release from biochar (Meena & Pradhan, 2023; Roy et al., 2022). Additionally, biochar enriched with nutrients increases the efficiency of nutrient use by the plant, improves soil physical, chemical and biological properties, ameliorates the problematic soil and ultimately augments the crop yield. Although nutrient-rich biochar is a potential technology for sustainable crop production still there is lack of systematic research regarding the large variety of feedstock and nutrients, and their potential to remove carbon dioxide from the environment.

18.4 Carbon storage potential of biochar

With climate change as a potential outcome in recent times, reducing atmospheric CO_2 and its storage in terrestrial ecosystems is a top priority for us. Due to its tiny particle size and high carbonaceous content, biochar has been identified as a valuable resource for maintaining the Earth's existing carbon storage (Nanda, Dalai, Berruti, & Kozinski, 2016). Biochar has much potential for long-term carbon sequestration due to its aromaticity, existence of amorphous structure and decreased accessibility to decomposers. Biochar improves soil quality, generates energy, and has industrial use, it also shows qualities that could mitigate environmental changes caused by rising atmospheric CO_2 levels (Palanivelu et al., 2021). Most biochar comes from the pyrolysis of residual plant biomass that would have otherwise been wasted. Therefore, it is possible to obtain the product at a reasonable cost. The mechanism of biochar-mediated CO_2 emissions reduction is depicted in Fig. 18.2. Here, it is shown that plants use atmospheric CO_2 and by converting solar energy into chemical energy, it forms carbohydrates from CO_2. When the plant withers, it's dry biomass either burnt in the field or it gets decomposed. In this method, almost 99% of carbon reverts into the atmosphere as CO_2. Waving out this large proportion of carbon and sequestering the carbon in the form of soil organic carbon (SOC) is necessary to reduce the contribution of GHG emissions from agricultural practices (Singh, Chaturvedi, et al., 2023; Singh, Luthra, et al., 2023). For this purpose, biomass from plant bodies is utilized as a feedstock here, and thermal conversion, that is, pyrolysis, is employed to produce a charcoal-like product, namely "biochar" and syn-gas or biofuel for power generation. With this strategy, carbon dioxide emissions can be reduced by as much as 50% (Lehmann, Gaunt, & Rondon, 2006). In this section, we are more focused on discussing the potential for biochar to store carbon. The relative inertness of biochar makes it an

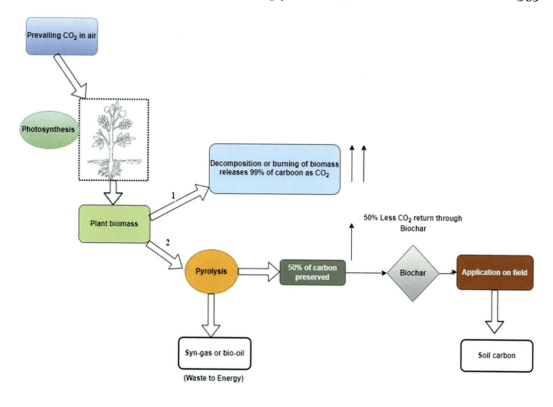

FIGURE 18.2 Biochar-mediated carbon cycle.

important input to enhance the SOC pool (Glaser, Haumaier, Guggenberger, & Zech, 2001). Compared to conventional agricultural approaches such as the direct incorporation of biomass, which leads to quick mineralization and subsequent CO_2 release, Thus, to sequester more carbon in the form of SOC, applying biochar is a potential approach (Bruun et al., 2011). Biochar's ability to store carbon is mostly the result of its long-term stability and inaccessibility to decomposers.

18.4.1 Long-term stability of biochar

Biochar must be stable to sequester C in soil for a long time. Biochar is so stable because its composition shifts when cellulose and lignin are completely degraded, giving way to the emergence of aromatic structures with furan-like chemicals. These alterations to the organic bond composition significantly impact biochar's durability. The thermal conversion process, that is, pyrolysis by which biochar is prepared, is also responsible for enhancing its stability. In the pyrolysis process, it has been found that the volatile and nitrogen content of biochar is decreased when the temperature is raised from 400°C to 600°C. Still, the ash and amount of fixed carbon are simultaneously enhanced (Purakayastha et al., 2016). Since the biochar created at 600°C would have a higher C:N ratio, it would be more stable in soil. The pyrolysis of organic materials to produce biochar significantly increases the carbon's recalcitrance in the biomass, contributing to the compound's long-term stabilization. The feedstocks used for biochar preparation also determine how long biochar remains stable in soil. In a CO_2 efflux study, it was found that maize biochar was the most stable and effective at preserving the original soil organic C (Kumar, Sharma, et al., 2020; Pradhan & Meena, 2023a; Zhang et al., 2022). Additionally, it was revealed that biochar made from wheat and pearl millet had decreased C mineralization. However, C mineralization was greater in the rice biochar. Increased mineralization is correlated with decreased stability. Hence, the characteristics of feedstocks and their potential in carbon storage must be assessed before their utilization as biochar.

18.4.2 Inaccessibility of biochar to decomposers

Biochar is exceptionally resistant to microbial breakdown. Intrinsic resistance, the separation of the substrate from the decomposers, and the creation of contacts between mineral surfaces are the primary mechanisms at work in soils that and stabilize biochar upon entry and considerably enhance its residence period in soil. In contrast to free organic matter, biochar has primarily been detected in fractions of SOM that are contained within aggregates, thus the accessibility to decomposers gets reduced (Brodowski, John, Flessa, & Amelung, 2006; Liang et al., 2008). The effects of biochar on soil microorganisms are both immediate and indirect due to its provision of carbon (C) and other nutrients and its ability to alter soil conditions and provide extra room for colonization (e.g., moisture and pH). The mineralization results, however, did not differ between biochar-rich soils with 27%, 10%, and 0.3% clay, indicating that increased aggregation in the finer-textured soils had no consequence on biochar mineralization (Liang et al., 2008). Biochar's ability to withstand microorganisms depends on several factors, including the chemical composition of the biochar that resulted from the heat treatment, the carbonization temperature attained during pyrolysis, the types and nature of the initial crop residues (Conti, Rombolà, Modelli, Torri, & Fabbri, 2014; Enders, Hanley, Whitman, Joseph, & Lehmann, 2012; Kumar, Meena, et al., 2020). Thus, biomass used for biochar needs to be accessed along with the temperature threshold to know the degradability of the biochar in the soil environment.

There has been discussion of biochar-based carbon sequestration in soil for almost a decade and it has shown considerable promise as a carbon-negative technique (Lehmann, 2007; Raj et al., 2021). Although there are still some unknowns when trying to estimate how long carbon in biochar stays in the soil based on its structural, chemical, and soil properties, it is generally accepted that the biochar's carbon stays in the soil for a much longer period than the biomass it was made from. Since quantifying how much stable carbon is present in biochar is difficult, many researchers resort to proxies. Importantly, how much carbon will still be present in soil after 100 years is of the utmost importance? In various research papers, the carbon sequestration potential of biochar has been estimated to be in the range of 0.7–1.8 Gt CO_2-C (eq) per year^{-1} (Smith, 2016).

18.4.3 Impact of biochar on soil microbial biomass carbon and carbon fractions

SMBC is an important component of the soil organic matter (SOM) that regulates the transformation and storage of nutrients under different soil systems (Meena et al., 2022; Singh, Chaturvedi, et al., 2023; Singh, Luthra, et al., 2023). It is a labile SOC fraction component. It is a key factor in the functioning of ecosystems and is crucial to biological systems. Furthermore, it is a significant indicator of nutrient dynamics and soil organic C stability changes. In general, the impacts of biochar on SMBC rely on the properties of the biochar, the temperature at which it is pyrolyzed, the feedstock used, and the diversity of the microbial biomass, which is reliant on the soil's characteristics (Li, Wang, Chang, Jiang, & Song, 2020; Pradhan & Meena, 2022). Adding biochar may enhance the SMBC and the soil microbial community, which will ultimately affect nutrient cycling, plant growth and development, mineralization of SOC, reduction in GHG emissions, and carbon sequestration. Many evidence are there that increase in SMBC with biochar addition. A 4-year study showed that the application of biochar with 9 t/ha/year increases the SMBC by 45% to 294% at soil depths of 10–20 and 20–30 cm compared to no biochar addition (Zhang et al., 2014). Biochar application 15 t ha^{-1}, 30 t ha^{-1}, and 30 t ha^{-1} + 50 kg P$_2$O$_5$ ha^{-1} resulted in a significant increase of MBC by 4.5, 8.2, and 8.3 folds, respectively, in comparison to the control (Amoakwah et al., 2022). Compared to soil total organic carbon, certain labile fractions of SOC are more important to sustain soil fertility. The SMBC and light fractions of SOC are essential fractions of SOM and for the short-term cycling of nutrients in the soil because they quickly respond to environmental changes. Microorganisms mineralized the organic matter and serve as bonding agents for macroaggregation. Furthermore, SMBC positively correlated with all labile fractions of organic C and macroaggregates (Demisie, Liu, & Zhang, 2014). When biochar was applied, the contents of soil organic C and recalcitrant organic C fraction increased by 11.0%–22.1% and 18.4%–32.3%, respectively, at 6 and 12 t ha^{-1} of biochar application. The percentage of recalcitrant organic C has increased than that for SOC. Aryl C content increased by 26.5%–95.4% at all rates, whereas phenolic C increased by 23.8%–56.7% at 6 and 12 t/ha, both larger than those for SOC (Zheng et al., 2022).

18.4.4 Decomposition rate of biochar

The rate at which biochar decomposes in the soil is crucial because it impacts how long biochar can provide advantages to the soil, how long it can be used to add carbon to the soil, and how it can minimize the effects of climate change. Most of the constituents in biochar are aromatic and mostly resistant to microbial breakdown and chemical transformation. Therefore, the biochar decomposition rate is very slow compared to the direct incorporation of crop residue, which is readily mineralized and releases CO$_2$. Similarly, compared to other compounds, including sugars, lipids, lignin, etc, the decomposition of biochar significantly reduced over time. As a result of its relative inertness, applying biochar helps to replenish the soil's recalcitrant organic carbon pool. The decomposition rate of biochar mainly depends on the state of the biochar, soil characteristics, and climatic conditions (Roy & Dias, 2017; Sheoran et al., 2022). Even when temperature is constant, pyrolysis time affects the subsequent decomposition rate. In comparison to biochar generated at higher temperatures (550°C–650°C), it has been reported that biochar produced at lower temperatures (400°C–450°C) mineralizes more quickly (Fang, Singh, Singh, & Krull, 2014). The cause of this is due to the biochar's increased aromaticity and degree of aromatic condensation and its inherent chemical resistance to biological and chemical breakdown. According to Purakayastha et al. (2015), soil with rice straw biochar had higher levels of C mineralization than the maize, wheat, and pearl millet biochar throughout the incubation period. Additionally, they reported less soil C enrichment with rice straw biochar primarily caused by the lower amount of C content and a higher rate of C mineralization of the biochar. Furthermore, they investigated in an incubation study, the application of biochar could increase total soil C (TSC) at the end of one year of C mineralization. Kuzyakov, Bogomolova, and Glaser (2014) used ^{14}C-labeled biochar to trace its decomposition to CO_2 for 8.5 years and between the fifth and eighth years, the degradation rates of biochar calculated by $^{14}CO_2$ efflux were $7 \times 10^{4\%}$ day^{-1}. Due to methodological limitations and the fact that the changes are too modest for any meaningful experimental period, it is difficult to estimate the decomposition and modifications of biochar directly.

18.5 Biochar mitigating greenhouse gases emission

Reducing GHGs to mitigate climate change is necessary to maintain the sustainability of the global environment. Biochar can sequestrate carbon into the soil and provide adaptation and mitigation co-benefits. As part of the UN Framework Convention on Climate Change, the Paris Agreement and the Kyoto Protocol were created to reduce GHG emissions. Nitrous oxide (N_2O), carbon dioxide (CO_2), methane (CH_4), as the primary GHGs that had to be reduced. Among those GHG_s, the role of CO_2 is highest. Numerous scientists have suggested that biochar could be a sustainable option to offset the GHG emissions which are presented in Tables 18.1 and 18.2.

18.5.1 Biochar and CO_2 sequestration in carbon pools

Biochar is a potential material that could reduce the atmospheric CO_2 release into the environment. Biochar is produced at high temperatures so has more furan-like compounds with aromaticity and turbotrain structure. So, it is recalcitrance in nature and stable and thus has less decomposition rate. The particle size of biochar is small, so remains within aggregate, not freely as SOM. They also form complexes with clay mineral surfaces. For these two reasons, biochar has reduced accessibility to decomposers. Thus, their decomposition rate is very slow, which helps in carbon sequestration. Purakayastha et al. (2015) studied the effect of different biochars on sandy loam soil in India and found total C mineralization over 1-year period was observed lowest in maize stover biochar, followed by wheat straw biochar and pearl millet stalk biochar showed comparable C mineralization, which was at par with the control treatment. This translated into the enrichment of organic C contents in soil. So, maize biochar also contributed to TSC, which indicates carbon sequestration. There are more strong structural surface functional groupings in the maize stover biochar, including aromatic C = C, which make it more stable than other biochar. El

TABLE 18.1 Effect of biochar application on soil organic carbon and carbon sequestration.

S. No.	Applied ecosystem. (soil type and crop)	Biochar application Feedstock	Pyrolysis temp.	Application rate	Effect (% reduction/increment) and possible mechanism	Reference
1.	Sandy clay loam soil of Northeast China, typic Argialboll (USDA system), biochar applied to corn crop.	Corn straw	700°C.	10, 20, and 30 t ha^{-1}	The total organic carbon (TOC) concentration in macro aggregates was considerably higher after biochar application at 10, 20, and 30 t ha^{-1} (28.6%, 22.1%, and 23.2%, respectively) compared to control. Incorporating biochar into soils may help preserve soil organic carbon in macro-aggregate size fractions, resulting in the improvement of SOC.	Joseph et al. (2020)
2.	Clay loam soil of, biochar was applied to maize crop.	Straw biochar (particle size of 0.002–2.0 mm)	500°C–600°C.	10, 20, 30, and 50 t ha^{-1}	Application of biochar enhanced the organic matter content by 2.15 to 5.88 g kg^{-1}. Due to an extensive specific surface area, biochar can not only promote the formation of soil aggregates but also form stable organic-inorganic complexes through adsorption, thereby facilitating soil biochemistry and soil organic matter formation. Application of biochar at 30 and 50 t ha^{-1} increases OM content by 30.9% and 96.1%, respectively.	Cen et al. (2021)

(Continued)

TABLE 18.1 (Continued)

S. No.	Applied ecosystem. (soil type and crop)	Biochar application Feedstock	Pyrolysis temp.	Application rate	Effect (% reduction/increment) and possible mechanism	Reference
3.	Sandy clay loam, biochar was applied to rice crop	Rice husk	350°C–400°C	Biochar 12 t ha^{-1}, biochar 12 t ha^{-1} and N fertilizer 90 kg ha^{-1}, N fertilizer 90 kg ha^{-1} alone.	In the first and second years, the POC (Particulate Organic Matter) levels varied between 1.5 and 3 g kg^{-1}. POC levels were dramatically improved when biochar and nitrogen were mixed, recording increases of 50%, 33%, and 16% over the control, N and biochar alone, at 12 t ha^{-1} in 2016. Additionally, combined biochar and N treatment increased equivalent amounts by 50%, 50%, and 16% in 2017 over to control, N, and biochar, respectively.	Oladele and Adetunji (2021)
4.	Sandy loam soil. Biochar was applied in winter wheat–summer maize rotation.	Peanut shells	500°C	22.5 t ha^{-1}.	The biochar altered treatments increased the TOC level in all fractions by 12.99%–126.40% as compared to control and NPK treatments. Furthermore, the BC recalcitrant C molecules significantly restrict C mineralization, increasing the ability of soil to sequester C.	Zhang et al. (2021)

5.	Sub-tropical mid hill location of Sikkim, biochar amendment was applied in maize (Zea mays L.)-black gram (Vigna mungo var. viridis L.).	Maize stalk biochar	700°C	0, 2.5, and 5.0 t ha^{-1}	As the rate of biochar application increased, the microbial biomass carbon significantly increased. The goat dung with biochar treatment (5 t ha^{-1}) had the highest microbial biomass carbon (476.58 mg kg^{-1} of soil) among the manures. The availability of substrates and habitats for soil microbiological communities could be improved by biochar, enhancing microbial diversity.	Das et al. (2021)
6.	Ferralsol (FAO soil classification system), biochar was applied to Moso bamboo plantation.	Bamboo leaf biochar	500°C	5 and 10 t ha^{-1}	Application of biochar at 5 and 10 t ha^{-1} increased the concentrations of water soluble organic carbon and MBC by an average of 13.9% and 27.0% and 12.4% and 15.1%, respectively, in comparison to the control. Regardless of the treatment or sampling date, O-alkyl C (48.3%–55.4%) predominated the SOC, followed by alkyl C (24.2%–27.2%). This is because adding biochar to the soil increased the amount of lignin and aromatic compounds, which are resistant to biological deterioration and have a direct impact on the makeup and activity of the soil microbial population.	Li et al. (2018)

(Continued)

TABLE 18.1 (Continued)

S. No.	Applied ecosystem. (soil type and crop)	Biochar application Feedstock	Pyrolysis temp.	Application rate	Effect (% reduction/ increment) and possible mechanism	Reference
7.	Pot culture, where rice seedlings are transplanted.	Rice, wheat and maize straw	500°C	The proportion of biochar in the soil was 2%. The experiment included four treatments viz., soil only, soil + wheat straw biochar, soil + rice straw biochar, and soil + maize straw biochar	The SOC content of each treatment increased by 34.5%–38.0% compared to the no-biochar treatment after 5 days of applying biochar. After applying straw biochar, the amount of SOC increased, probably due to the biochar's high organic carbon content.	Jing et al. (2020)
8.	Silt loam soil, biochar was applied to winter wheat crop.	Corn cobs	360°C	4.5 and 9.0 Mg ha^{-1} yr^{-1}	When compared to the no-biochar control, biochar application significantly increased the contents of the SIC (3.2%–24.3%), >53 m particulate organic carbon (POC, 38.2%–166.2%), and total SOC (15.8%–82.2%). The increased SOC accumulation was mostly attributable to the high carbon content of biochar, which is a natural feature, as well as the potential negative stimulating effect of biochar on the native SOC. Furthermore, biochar's aromatic structure significantly enhances its capacity to resist biodegradation, which significantly raises SOC.	Shi et al. (2021)

9.	Two soil condition *viz.*, Sandy loam and clayey soil, biochar applied to maize crop.	Maize residue	200°C, 400°C, and 600°C	5 and 10 g kg^{-1} soil	The carbon pool index values for biochar in both soils were greater than those for uncharred feedstock, and they tended to rise consistently with treatments that used higher temperatures for pyrolysis at both addition rates. Because of their high aromatic C content (lower atomic H/C and O/C ratios), high-temperature biochars have a resistant and persistent character that is a necessary component for boosting soil C sequestration ability.	Khadem, Raiesi, Besharati, and Khalaj (2021)
10.	Sandy loam, biochar was applied to corn crop.	Corn residue	400°C–500°C	15, 30, and 45 t ha^{-1}	SOC sequestration in the top 15-cm depth was dramatically improved by biochar application at 30 and 45 t ha^{-1} by 19% and 37%, respectively, in the first growing season and by 12% and 15% in the second.	Yang et al. (2020)
11.	Pot trials were conducted. 500 g of soil samples were put in plastic containers and thoroughly mixed with the rice and maize biochar. After that, seeds for eggplant (Solanum melongena) were planted.	Rice husk and corn stover.	Rice husk: 550°C&&&Corn stover: 650°C	0.5%, 1.5%, or 3.0% (wt./wt.) of rice husk and maize biochar were added to the soil.	By the end of the 107th day of the growth period, the addition of biochar (3.0 wt.%) results in an overall increase of SOC content of 328% for corn stover biochar addition and 417% for rice husk biochar addition.	Mohan et al. (2018)

(*Continued*)

TABLE 18.1 (Continued)

S. No.	Applied ecosystem. (soil type and crop)	Biochar application Feedstock	Pyrolysis temp.	Application rate	Effect (% reduction/ increment) and possible mechanism	Reference
12.	A loamy soil sample was used in an incubation investigation for 120 days.	Rice husk (RH), pecan shell (PS), and bamboo (BB).	550°C	0%, 1%, 2%, and 5% of biochar were added to the soil, as well as 100% pure biochar.	After 120 days of incubation, 5% RH, PS, and BB treatments reduced the cumulative mineralized quantities of SOC by 7.95%–10.7%, while 1% or 2% biochar treatments had no apparent differences from the control.	Liu et al. (2018)
13.	Fluvisol, biochar amendment was applied to winter wheat-summer maize rotation.	Rice husks and cotton seed hulls	400°C	0, 30, 60, and 90 t ha^{-1}	Increases in the biochar application rate were followed by an upward trend in TOC content, which was considerably higher at high biochar rates than at lower biochar rates. the impact of biochar production on native-SOC content, for instance, 30 t ha^{-1} of biochar produced lowered native-SOC content while 60 and 90 t ha^{-1} of biochar produced increased n-SOC content. The soil in this study may have different microbial biomass or diversity due to the presence of biochar, which could alter n-SOC turnover.	Sun et al. (2020)

14.	Soil was calcareous. It was a 5 years long term experiment. Wheat and maize crops were grown every year.	Rice husks and cotton seed hulls.	400°C	30, 60, and 90 t ha^{-1}	According to the findings, long-term biochar application at 30, 60, and 90 t ha^{-1} enhanced soil total inorganic carbon by 18.8%, 42.4%, and 62.3%, as well as native soil inorganic carbon by 7.8%, 20.2%, and 28.3%, respectively, in the 0–20 cm soil layer. Adding biochar at similar rates also led to increases in total inorganic carbon of 13.4%, 22.8%, and 30.5%, respectively, in the soil's top 20 to 40 cm. Due to its porosity, biochar has the potential to improve soil microbial activity and porosity. In order for SIC to precipitate, soil microbes are important because they produce CO_2 when SOC is broken down.	Dong, Singh, Li, Lin, and Zhao (2019)
15.	Two paddy soils from northern and southern China were used. The experiment was performed under a rice/wheat annual rotation system	Rice straw	500°C	11.3 Mg ha^{-1}	Following the application of biochar (BC), the increase in aromatic and phenolic C led to a significant increase in SOC sequestration. The high concentration of aromatic C in soil explained how BC helped to contribute to long-term C sequestration since condensed aromatic moieties are a stable component that are extremely resistant to chemical and microbial destruction. A higher effect was seen in southern China than northern China, despite the four-year BC amendment encouraging SOC sequestration in two soils.	Bi et al. (2020)

TABLE 18.2 Effect of biochar application on greenhouse gas emissions.

S. No.	Applied ecosystem. (soil type and crop)	Biochar application			Effect (% reduction/increment) and possible mechanism	References
		Feedstock	Pyrolysis temp	Application rate		
1.	Soil type: Anthraquic Ustorthent, Grossarenic Kandiustalf, and Ustic Quartzipsamment according to USDA Soil Survey Staff (2014) guidelines.	Rice husk	350°C	5, 15, and 25 t ha^{-1}	A cumulative emission of 3.06 g m^{-2} CO$_2$-C was generated by applying biochar at a rate of 25 t ha^{-1}, which was significantly greater than the emissions of 2.82 and 2.78 g m^{-2} that were produced by applying biochar at a rate of 15 and 5 t ha^{-1}, respectively.	Olaniyan et al. (2020)
2.	Silty loam soil texture, biochar applied to maize crop.	Maize straw	400°C–450°C	0, 10, 20, and 30 t ha^{-1}	There were no appreciable variations between the biochar application rates of 10, 20, and 30 t ha^{-1}, ranging from 3.79 × 10^3 to 3.59 × 10^3 kg CO$_2$-C ha^{-1}, and the cumulative CO$_2$ emissions in the biochar treatments were much lower than in the "no amendment" control. This may be attributed to the sorption of labile C onto the surface or into the pores of the biochar as the amount of biochar was added, which led a progressive reduction in CO$_2$ emissions.	Shen, Zhu, Cheng, Yue, and Li (2017)
3.	Black Chernozem, silty clay loam texture.	Wheat straw	450°C	Wheat straw (St): 30 t ha^{-1}, biochar (Bc): 20 t ha^{-1}, and wheat straw plus its biochar (StBc): 30 t ha^{-1} of wheat straw and 20 t ha^{-1} of biochar application.	In the control, St, and StBc treatments, soil N$_2$O fluxes were around 60.0 g N$_2$O-N ha^{-1} day^{-1}, which was almost 25.6% greater than that in the Bc treatment. The average soil CO$_2$ fluxes during the growth season were roughly twice as high in the St and StBc treatments as they were in the CK and Bc treatments.	Duan et al. (2020)

4.	Submerged paddy soil with a pH value of 6.31.	Rice straw	500°C	7.5 t ha^{-1}	Conventional fertilization (CF), CF with straw, and optimal fertilization + 15% less fertilizer all resulted in substantial increases in cumulative CH4 emissions of 11.80% and 2.35%, respectively, while CF + Biochar (B) significantly reduced cumulative CH4 emissions by 27.80% and 28.46%, respectively. Optimal fertilization (OF) boosted the ability of biochar to reduce N2O emissions, with CF + B and OF + B obtaining the greatest N$_2$O reductions of 30.56% and 32.21%, respectively.	Li, He, Zhou, He, and Yang (2023)
5.	Littoral clay salt soil. Biochar was applied to paddy.	Wheat straw	450°C	N fertilizer: 0 and 300 kg ha^{-1}. Biochar: 0, 20 (B_1), and 40 (B_2) t ha^{-1}. (N_0B_0, N_0B_1, N_0B_2, N_1B_0, N_1B_1, N_1B_2)	In comparison to the N0B0 value of the treatment, biochar amendment raised the total N$_2$O emissions from the N0B1 and N0B2 treatments by 23.4% and 30.5%, respectively. Regardless of the addition of N fertilizer, biochar amendment raised the cumulative CH4 emissions by an average of 14.4% and 14.6% in the B1 and B2 treatments, respectively, in comparison to values from sites without biochar treatments.	Sun, Deng, Fan, Li, and Liu (2020)

(*Continued*)

TABLE 18.2 (Continued)

S. No.	Applied ecosystem. (soil type and crop)	Biochar application			Effect (% reduction/increment) and possible mechanism	References
		Feedstock	Pyrolysis temp	Application rate		
6.	The soil was loamy texture slightly with acidic (pH = 6.0) in nature.	Wheat straw	350°C–550°C	Biochar: 0 (B_0), 20 (B_1) and 40 (B_2) t ha^{-1} Nitrogen: 0 (N_0) and 150 (N_1) kg ha^{-1}	Compared to B0N1, total CH_4 emissions were reduced by 40.4% under B2N1. In the first two rice seasons, the biochar-amended soils had much lower N_2O emissions than the unamended ones. Since the additional carbon may provide more substrates for the soil microbes, the addition of organic materials (such as crop straw) could enhance the emission of soil N_2O.	Liu et al. (2019)
7.	Silty clay loam, Flanagan silt loam association (Soil Survey Staff, NRCS, USDA).	Yellow pine	550°C	Biochar applied at 100 Mg ha^{-1}, N fertilizer at 269 kg N ha^{-1}, and biochar + N fertilizer addition at 100 Mg ha^{-1} and 269 kg N ha^{-1}.	Comparing fertilized and unfertilized treatments with and without biochar application, there was a trend towards lower total annual N2O emissions, but these changes were not statistically significant. Potential N_2O yield ($N_2O/(N_2 + N_2O)$) in the biochar + N and biochar treatments was much lower than in the nitrogen treatment.	Edwards et al. (2018)
8.	4 sites for paddy rice production and 3 sites for maize production of China used for this study purpose.	Wheat straw	350°C–500°C	20 t ha^{-1}	When compared to no biochar application, biochar application decreased the CFs by 20.37 to 41.29 t carbon dioxide equivalent ha^{-1} (CO_2-eq ha^{-1}) for paddy rice production and 28.58 to 39.49 t CO_2-eq ha^{-1} for maize cultivation, respectively.	Xu et al. (2019)

9.	Ultisol (USDA taxonomy), biochar was applied to typical double rice paddy system.	Wheat straw	500°C	0, 24, and 48 t ha^{-1}	It was found that adding biochar consistently reduced annual total CH4 emissions over the course of four years by 20% to 51%. Adding biochar considerably decreased annual net greenhouse gas emissions (NGHGE) and GHG intensity (GHGI) by 156% to 264% and 159% to 278%, respectively, on a 4-year average.	Wang et al. (2018)
10.	Clay loam soil, biochar applied to rice crop.	Activated rice hull biochar pellets with 40% of N (ARHBP40%), and activated palm biochar pellets with 40% of N (APBP-40%)	850°C	ARHBP-40%, 36 kg ha^{-1} based on total nitrogen requirement, APBP-40%, 36 kg ha^{-1} based on total nitrogen requirement.	During rice cultivation, N_2O emissions occurred in the following order: Control > APBP-40% > ARHBP-40%. In comparison to the control, N_2O emissions in the ARHBP-40% were 27% lower. This decrease might be caused by the fact that activated palm biochar can only adsorb PO4-P in aqueous solutions, but activated rice hull biochar can also adsorb NH4-N.	Shin et al. (2021)

naggar and his co-workers El-Naggar et al. (2019) suggested that the labile pool of biochar consists of only 3% with a mean residue time of 108 days and the nonlabile part consists of 97% with a mean residue time of 556 days. So, after the addition of biochar to soil, initially, CO_2 production is there due to the labile portion of biochar. Then it gradually decreases because of the complex structure. CO_2 also gets adsorbed on pore space. The ability of biochar to increase soil carbon sequestration is widely recognized. However, several studies have revealed the detrimental effects of biochar on GHG emissions. In a study of agricultural soil in China, it was reported that biochar had no impact on soil respiration. According to a study, the addition of biochar did not affect soil bacteria' ability to use carbon or their ability to release CO_2 into the atmosphere (Liu et al., 2016; Roy, Meena, Kumar, Jhariya, & Pradhan, 2021). There is evidence that using biochar increases CO_2 emissions. The abiotic release of inorganic carbon brings this on, the breakdown of biochar's labile components, and the breakdown of its humus or organic components (Spokas, Koskinen, Baker, & Reicosky, 2009).

18.5.2 Role of biochar in mitigation of CH_4 emission

A substantial contributor to global warming, methane (CH_4) is a significant GHG in the atmosphere. CH_4 is emitted from natural sources, that is, paddy fields and human activities. The global CH_4 emission from rice paddy fields was estimated to be 14.8–41.7 Tg year^{-1} based on the 2006 IPCC Guidelines for National Greenhouse Gas Inventories and country-specific activity data (Kumawat et al., 2022; Yan, Akiyama, Yagi, & Akimoto, 2009). Therefore, it is the need of the hour to adopt an efficient method to reduce CH_4 emissions from the environment. Numerous researchers claimed that using biochar can lower net CH_4 emissions in both short-term and incubation tests (Nan, Wang, Wang, Yi, & Wu, 2020). The effect of biochar application on methanogenic bacterial activity is depicted in Fig. 18.3. When biochar is applied to the paddy soil it can inhibit the methanogenic bacteria activity by the mechanism of decreasing dissolved organic carbon, increasing oxygen input and promoting methanogenic activity (Nan et al., 2020). Compared to the conventional rate of nitrogen and the optimum rate of nitrogen application in rice fields, biochar application combined with various nitrogen fertilizer doses reduced CH_4 emission by 13.2%–27.1% (He et al., 2020). Applying biochar at 20 and 40 t ha^{-1} decreased CH_4 emission by 11.2% and 17.5%, respectively, on average during the 6-year study (Wu, Zhang, Dong, Li, & Xiong, 2019). Biochar had a small and significant effect on methane emission and ultimately reduced GHG intensity. However, adding biochar can also increase GHG emissions. A field study with wheat straw biochar amended soil over two consecutive rice growing cycles found that the emission of CH_4 enhanced by 49% and 31% at the application rates of 40 and 10 Mg ha^{-1}, respectively (Pradhan et al., 2022; Zhang et al., 2012). For evaluating these emissions data, considerable vigilance is required when applying biochar on a large scale.

18.5.3 Role of biochar in mitigation of N_2O emission

Biochar can help bring down N_2O emissions, according to numerous research (Xiang, Liu, Ding, Yuan, & Lin, 2015). In terms of yearly global anthropogenic N_2O emissions,

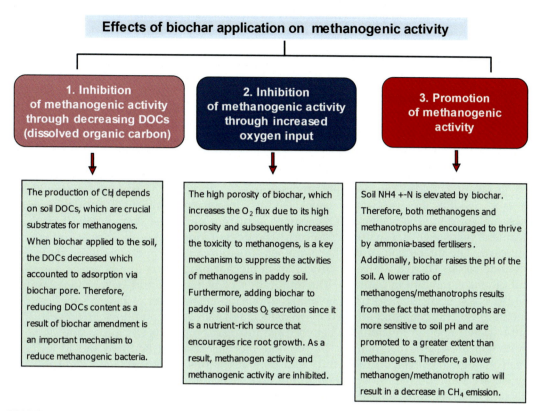

FIGURE 18.3 Effect of biochar application on methanogenic activity.

agricultural activities account for around 59% (4.1 Tg N_2O-N yearr^{-1}) of them. The main contributors to this growth are increased animal production and the application of N fertilizers. In the case of N_2O, biochar by its liming effect, shifts pH and acts as a source of electrons. Studies have shown that it forms N_2O to N_2 thus reducing N_2O emission. Though the effect of biochar on GHG emission is variable and depends on soil type, feedstock and pyrolysis condition, Zheng et al. (2019) studied the effect of wheat straw (pyrolysis at 350°C–550°C) biochar on GHG emission in calcareous fluvo-aquic loamy aquic Fluvent (China) soil with and without N fertilizer. They found in no fertilizer treatment biochar application increased CO_2 emission at high application dose for liable C. Fertilizer addition reduced microbial respiration. However, N_2O emission reduced significantly with biochar application rate in N fertilized soil as biochar increased pH, favoured activity of N_2O reductase from denitrifier and reduced N_2O emission. Nelissen, Saha, Ruysschaert, and Boeckx (2014) conducted a study on silt loam soil using willow, pine maize and wood mixture feedstock. They found that nitrate concentrations decreased by 6%–16% on biochar addition by nonelectrostatic sorption of NH_4^+ or NO_3^- through biochar's micropores. Lee et al. (2022) evaluate the effect of biochar application by using different feedstock (barley straw and poultry manure) and biochar application rates (0, 5, 10, and 20 t ha^{-1}).

Application of biochar 20 t ha^{-1} considerably reduced N$_2$O emissions by 33.2% compared to the control. The N$_2$O emissions from all soils were greatly reduced by barley straw biochar. Compared to biochar made from poultry manure, barley straw reduced emissions of N$_2$O by 74.5%. To reduce N2O emissions and halt climate change, biochar application could be a viable option.

18.6 Challenges and future prospects

Although biochar provides several benefits, its application has not become quite common in farming fields due to many productions and application-associated restrictions. The crop residues are used for livestock feeding, manuring to the field, mulching purposes, fuel for domestic needs, and sometimes industrial purposes. Preparation of biochar at farmers' fields also is a matter of concern. The degradation rate of biochar in soil is not quite clear. The interaction of biochar with soil, microorganisms and climate needs to be well understood. There are several difficulties in biochar production due to processing difficulties and variability in biomass. The application of biochar should be location-specific thus there is a need for field trials before its application. The effect of biochar application sometimes is conflicting and affected by feedstock and pyrolysis conditions, making it difficult to determine the relationship between the intrinsic pollutants in biochar and its toxicity to soil meso-and macrofauna. To promote biochar as a soil additive and to reduce climate change; however, further study and demonstrations of its production and use in the field are still required. Most of the studies had been conducted at individual field level or laboratory level; thus, long-term field experiments need to be conducted. There are currently relatively few long-term studies with GHG mitigation. To achieve carbon neutrality and net zero by 2050, reducing GHG emissions is of utmost importance. Thus, there is a need for location and feedstock-specific biochar production and its largescale application in farmers' fields to mitigate GHG emissions and long-term carbon storage.

References

Abiven, S., Hund, A., Martinsen, V., & Cornelissen, G. (2015). Biochar amendment increases maize root surface areas and branching: A shovelomics study in Zambia. *Plant and Soil, 395*(1), 45–55.

Agegnehu, G., Bass, A. M., Nelson, P. N., Muirhead, B., Wright, G., & Bird, M. I. (2015). Biochar and biochar-compost as soil amendments: Effects on peanut yield, soil properties and greenhouse gas emissions in tropical North Queensland, Australia. *Agriculture, Ecosystems & Environment, 213*, 72–85.

Amoakwah, E., Arthur, E., Frimpong, K. A., Lorenz, N., Rahman, M. A., Nziguheba, G., & Islam, K. R. (2022). Biochar amendment impacts on microbial community structures and biological and enzyme activities in a weathered tropical sandy loam. *Applied Soil Ecology, 172*, 104364.

Anand, A., Kumar, V., & Kaushal, P. (2022). Biochar and its twin benefits: Crop residue management and climate change mitigation in India. *Renewable and Sustainable Energy Reviews, 156*, 111959.

Andreae, M. O., & Merlet, P. (2001). Emission of trace gases and aerosols from biomass burning. *Global Biogeochemical Cycles, 15*(4), 955–966.

Bhuvaneshwari, S., Hettiarachchi, H., & Meegoda, J. N. (2019). Crop residue burning in India: Policy challenges and potential solutions. *International Journal of Environmental Research and Public Health, 16*(5), 832.

Bi, Y., Cai, S., Wang, Y., Zhao, X., Wang, S., Xing, G., & Zhu, Z. (2020). Structural and microbial evidence for different soil carbon sequestration after four-year successive biochar application in two different paddy soils. *Chemosphere, 254*, 126881.

Brassard, P., Godbout, S., & Raghavan, V. (2016). Soil biochar amendment as a climate change mitigation tool: Key parameters and mechanisms involved. *Journal of Environmental Management, 181*, 484−497.

Brassard, P., Godbout, S., Raghavan, V., Palacios, J. H., Grenier, M., & Zegan, D. (2017). The production of engineered biochars in a vertical auger pyrolysis reactor for carbon sequestration. *Energies, 10*(3), 288.

Brodowski, S., John, B., Flessa, H., & Amelung, W. (2006). Aggregate-occluded black carbon in soil. *European Journal of Soil Science, 57*(4), 539−546.

Bruun, E. W., Hauggaard-Nielsen, H., Ibrahim, N., Egsgaard, H., Ambus, P., Jensen, P. A., & Dam-Johansen, K. (2011). Influence of fast pyrolysis temperature on biochar labile fraction and short-term carbon loss in a loamy soil. *Biomass and Bioenergy, 35*(3), 1182−1189.

Buss, W., Wurzer, C., Manning, D. A., Rohling, E. J., Borevitz, J., & Mašek, O. (2022). Mineral-enriched biochar delivers enhanced nutrient recovery and carbon dioxide removal. *Communications Earth & Environment, 3*(1), 1−11.

Cen, R., Feng, W., Yang, F., Wu, W., Liao, H., & Qu, Z. (2021). Effect mechanism of biochar application on soil structure and organic matter in semi-arid areas. *Journal of Environmental Management, 286*, 112198.

Chaturvedi, S., Singh, S. V., Dhyani, V. C., Govindaraju, K., & Mandal, S. (2021). Characterization, bioenergy value and thermal stability of biochar derived from diverse agriculture and forestry lignocellulosic waste. *Biomass Conversion and Biorefinery*. Available from https://doi.org/10.1007/s13399-020-01239-2.

Conti, R., Rombolà, A. G., Modelli, A., Torri, C., & Fabbri, D. (2014). Evaluation of the thermal and environmental stability of switchgrass biochars by Py−GC−MS. *Journal of Analytical and Applied Pyrolysis, 110*, 239−247.

Das, S. K., Ghosh, G. K., Avasthe, R., Choudhury, B. U., Mishra, V. K., Kundu, M. C., ... Dhakre, D. S. (2021). Organic nutrient sources and biochar technology on microbial biomass carbon and soil enzyme activity in maize-black gram cropping system. *Biomass Conversion and Biorefinery*, 1−11.

Demisie, W., Liu, Z., & Zhang, M. (2014). Effect of biochar on carbon fractions and enzyme activity of red soil. *Catena, 121*, 214−221.

Dong, X., Singh, B. P., Li, G., Lin, Q., & Zhao, X. (2019). Biochar increased field soil inorganic carbon content five years after application. *Soil and Tillage Research, 186*, 36−41.

Duan, M., Wu, F., Jia, Z., Wang, S., Cai, Y., & Chang, S. X. (2020). Wheat straw and its biochar differently affect soil properties and field-based greenhouse gas emission in a Chernozemic soil. *Biology and Fertility of Soils, 56*, 1023−1036.

Edwards, J. D., Pittelkow, C. M., Kent, A. D., & Yang, W. H. (2018). Dynamic biochar effects on soil nitrous oxide emissions and underlying microbial processes during the maize growing season. *Soil Biology and Biochemistry, 122*, 81−90.

El-Naggar, A., Lee, S. S., Rinklebe, J., Farooq, M., Song, H., Sarmah, A. K., ... Ok, Y. S. (2019). Biochar application to low fertility soils: A review of current status, and future prospects. *Geoderma, 337*, 536−554.

Enders, A., Hanley, K., Whitman, T., Joseph, S., & Lehmann, J. (2012). Characterization of biochars to evaluate recalcitrance and agronomic performance. *Bioresource Technology, 114*, 644−653.

Fang, Y., Singh, B., Singh, B. P., & Krull, E. (2014). Biochar carbon stability in four contrasting soils. *European Journal of Soil Science, 65*(1), 60−71.

Fuentes, M. E., Nowakowski, D. J., Kubacki, M. L., Cove, J. M., Bridgeman, T. G., & Jones, J. M. (2008). Survey of influence of biomass mineral matter in thermochemical conversion of short rotation willow coppice. *Journal of the Energy Institute, 81*(4), 234−241.

Giudicianni, P., Gargiulo, V., Grottola, C. M., Alfè, M., Ferreiro, A. I., Mendes, M. A. A., ... Ragucci, R. (2021). Inherent metal elements in biomass pyrolysis: A review. *Energy & Fuels, 35*(7), 5407−5478.

Glaser, B., Haumaier, L., Guggenberger, G., & Zech, W. (2001). The 'Terra Preta' phenomenon: A model for sustainable agriculture in the humid tropics. *The Science of Nature, 88*(1), 37−41.

Haider, F. U., Wang, X., Farooq, M., Hussain, S., Cheema, S. A., Ul Ain, N., ... Liqun, C. (2022). Biochar application for the remediation of trace metals in contaminated soils: Implications for stress tolerance and crop production. *Ecotoxicology and Environmental Safety, 230*, 113165.

He, T., Yuan, J., Luo, J., Lindsey, S., Xiang, J., Lin, Y., ... Ding, W. (2020). Combined application of biochar with urease and nitrification inhibitors have synergistic effects on mitigating CH_4 emissions in rice field: A three-year study. *Science of The Total Environment, 743*, 140500.

Hiloidhari, M., Das, D., & Baruah, D. C. (2014). Bioenergy potential from crop residue biomass in India. *Renewable and sustainable energy reviews*, 32, 504−512.

IPCC. (2018). Summary for policymakers. In *Global Warming of 1.5°C. An IPCC Special Report on the impacts of global warming of 1.5°C above pre-industrial levels and related global greenhouse gas emission pathways, in the context of strengthening the global response to the threat of climate change, sustainable development, and efforts to eradicate poverty* [Masson-Delmotte, V., P. Zhai, H.-O. Pörtner, D. Roberts, J. Skea, P.R. Shukla, A. Pirani, W. Moufouma-Okia, C. Péan, R. Pidcock, S. Connors, J.B.R. Matthews, Y. Chen, X. Zhou, M.I. Gomis, E. Lonnoy, T. Maycock, M. Tignor, and T. Waterfield (eds.)].

IPCC. (2021). Climate change 2021: The physical science basis. in *The Working Group I contribution to the Sixth Assessment Report addresses the most up-to-date physical understanding of the climate system and climate change, bringing together the latest advances in climate science.*

IPCC (2022). The IPCC Sixth Assessment Report WGIII climate assessment of mitigation pathways: from emissions to global temperatures. <https://www.ipcc.ch/report/ar6/wg3/downloads/report/IPCC_AR6_WGIII_SPM.pdf>

Jacobson, M. Z. (2014). Effects of biomass burning on climate, accounting for heat and moisture fluxes, black and brown carbon, and cloud absorption effects. *Journal of Geophysical Research: Atmospheres*, 119(14), 8980−9002.

Jain, N., Bhatia, A., & Pathak, H. (2014). Emission of air pollutants from crop residue burning in India. *Aerosol and Air Quality Research*, 14(1), 422−430.

Jat, M. L., & Gerard, B. (2014). Nutrient management and use efficiency in wheat systems of South Asia. *Advances in Agronomy*, 125, 171−259.

Jing, Y., Zhang, Y., Han, I., Wang, P., Mei, Q., & Huang, Y. (2020). Effects of different straw biochars on soil organic carbon, nitrogen, available phosphorus, and enzyme activity in paddy soil. *Scientific Reports*, 10(1), 8837.

Joseph, U. E., Toluwase, A. O., Kehinde, E. O., Omasan, E. E., Tolulope, A. Y., George, O. O., … Hongyan, W. (2020). Effect of biochar on soil structure and storage of soil organic carbon and nitrogen in the aggregate fractions of an Albic soil. *Archives of Agronomy and Soil Science*, 66(1), 1−12.

Kaur, M., & Dhundhara, S. (2022). Crop residues: A potential bioenergy resource. *International conference on chemical, bio and environmental engineering* (pp. 359−378). Cham: Springer.

Khadem, A., Raiesi, F., Besharati, H., & Khalaj, M. A. (2021). The effects of biochar on soil nutrients status, microbial activity and carbon sequestration potential in two calcareous soils. *Biochar*, 3, 105−116.

Kumar, R., Sharma, P., Gupta, R. K., Kumar, S., Sharma, M. M. M., Singh, S., & Pradhan, G. (2020). Earthworms for eco-friendly resource efficient agriculture. In Kumar, et al. (Eds.), *Resources use efficiency in agriculture* (pp. 47−84). Singapore: Springer Nature. Available from https://doi.org/10.1007/978-981-15-6953-1_2.

Kumar, S., Meena, R. S., Datta, R., Verma, S. K., Yadav, G. S., Pradhan, G., Molaei, A., Rahman, G. K. M. M., & Mashuk, H. A. (2020). Legumes for carbon and nitrogen cycling: An organic approach. In R. Datta, et al. (Eds.), *Carbon and nitrogen cycling in soil* (pp. 337−375). Singapore: Springer Nature, Carbon and Nitrogen Cycling in Soil.. Available from https://doi.org/10.1007/978-981-13-7264-3_10.

Kumawat, A., Bamboriya, S. D., Meena, R. S., Yadav, D., Kumar, A., Kumar, S., Raj, A., & Pradhan, G. (2022). Legume-based inter-cropping to achieve crop, soil, and environmental health security. In R. S. Meena, & S. Kumar (Eds.), *Advances in legumes for sustainable intensification* (pp. 307−328). Elsevier. Available from https://doi.org/10.1016/B978-0-323-85797-0.00005-7.

Kuzyakov, Y., Bogomolova, I., & Glaser, B. (2014). Biochar stability in soil: Decomposition during eight years and transformation as assessed by compound-specific 14C analysis. *Soil Biology and Biochemistry*, 70, 229−236.

Lee, J. M., Park, D. G., Kang, S. S., Choi, E. J., Gwon, H. S., Lee, H. S., & Lee, S. I. (2022). Short-term effect of biochar on soil organic carbon improvement and nitrous oxide emission reduction according to different soil characteristics in agricultural land: A laboratory experiment. *Agronomy*, 12(8), 1879.

Lehmann, J. (2007). A handful of carbon locking. *Nature*, 447, 10−11.

Lehmann, J., Gaunt, J., & Rondon, M. (2006). Bio-char sequestration in terrestrial ecosystems—A review. *Mitigation and Adaptation Strategies for Global Change*, 11(2), 403−427.

Li, D., He, H., Zhou, G., He, Q., & Yang, S. (2023). Rice yield and greenhouse gas emissions due to biochar and straw application under optimal reduced N fertilizers in a double season rice cropping system. *Agronomy*, 13(4), 1023.

Li, X., Wang, T., Chang, S. X., Jiang, X., & Song, Y. (2020). Biochar increases soil microbial biomass but has variable effects on microbial diversity: A meta-analysis. *Science of the Total Environment*, 749, 141593.

Li, Y., Li, Y., Chang, S. X., Yang, Y., Fu, S., Jiang, P., ... Zhou, J. (2018). Biochar reduces soil heterotrophic respiration in a subtropical plantation through increasing soil organic carbon recalcitrancy and decreasing carbon-degrading microbial activity. *Soil Biology and Biochemistry, 122*, 173–185.

Liang, B., Lehmann, J., Solomon, D., Sohi, S., Thies, J. E., Skjemstad, J. O., ... Wirick, S. (2008). Stability of biomass-derived black carbon in soils. *Geochimica et Cosmochimica Acta, 72*(24), 6069–6078.

Liu, X., Zheng, J., Zhang, D., Cheng, K., Zhou, H., Zhang, A., ... Pan, G. (2016). Biochar has no effect on soil respiration across Chinese agricultural soils. *Science of the Total Environment, 554*, 259–265.

Liu, X., Zhou, J., Chi, Z., Zheng, J., Li, L., Zhang, X., ... Pan, G. (2019). Biochar provided limited benefits for rice yield and greenhouse gas mitigation six years following an amendment in a fertile rice paddy. *Catena, 179*, 20–28.

Liu, Y., Chen, Y., Wang, Y., Lu, H., He, L., & Yang, S. (2018). Negative priming effect of three kinds of biochar on the mineralization of native soil organic carbon. *Land Degradation & Development, 29*(11), 3985–3994.

Majumder, S., Neogi, S., Dutta, T., Powel, M. A., & Banik, P. (2019). The impact of biochar on soil carbon sequestration: meta-analytical approach to evaluating environmental and economic advantages. *Journal of Environmental Management, 250*, 109466.

Meena, R. S., Pradhan, G., Kumar, S., & Lal, R. (2023). Using industrial wastes for rice-wheat cropping and food-energy-carbon-water-economic nexus to the sustainable food system. *Renewable and Sustainable Energy Reviews, 187*113756. Available from https://doi.org/10.1016/j.rser.2023.113756.

Meena, R. S., & Pradhan, G. (2023). Industrial garbage-derived biocompost enhances soil organic carbon fractions, CO_2 biosequestration, potential carbon credits, and sustainability index in a rice-wheat ecosystem. *Environmental Research*116525. Available from https://doi.org/10.1016/j.envres.2023.116525.

Meena, R.S., Kumawat, A., Kumar, S., Prasad, S.K., Pradhan, G., Jhariya, M.K., Banerjee, A., & Raj, A. (2022). Effect of legumes on nitrogen economy and budgeting in South Asia. In Meena, R.S., Kumar, S. (eds.), *Advances in legumes for sustainable intensification* (pp. 619–638). <https://doi.org/10.1016/B978-0-323-85797-0.00001-X>.

Menon, S., Denman, K.L., Brasseur, G., Chidthaisong, A., Ciais, P., Cox, P.M., ... Zhang, X. (2007). Couplings between changes in the climate system and biogeochemistry (No. LBNL-464E). Lawrence Berkeley National Lab. (LBNL), Berkeley, CA (United States).

Mohan, D., Abhishek, K., Sarswat, A., Patel, M., Singh, P., & Pittman, C. U. (2018). Biochar production and applications in soil fertility and carbon sequestration–a sustainable solution to crop-residue burning in India. *RSC Advances, 8*(1), 508–520.

Nan, Q., Wang, C., Wang, H., Yi, Q., & Wu, W. (2020). Mitigating methane emission via annual biochar amendment pyrolyzed with rice straw from the same paddy field. *Science of the Total Environment, 746*, 141351.

Nanda, S., Dalai, A. K., Berruti, F., & Kozinski, J. A. (2016). Biochar as an exceptional bioresource for energy, agronomy, carbon sequestration, activated carbon and specialty materials. *Waste and Biomass Valorization, 7*(2), 201–235.

Nelissen, V., Saha, B. K., Ruysschaert, G., & Boeckx, P. (2014). Effect of different biochar and fertilizer types on N_2O and NO emissions. *Soil Biology and Biochemistry, 70*, 244–255.

Oladele, S. O., & Adetunji, A. T. (2021). Agro-residue biochar and N fertilizer addition mitigates CO_2-C emission and stabilized soil organic carbon pools in a rain-fed agricultural cropland. *International Soil and Water Conservation Research, 9*(1), 76–86.

Olaniyan, J. O., Isimikalu, T. O., Raji, B. A., Affinnih, K. O., Alasinrin, S. Y., & Ajala, O. N. (2020). An investigation of the effect of biochar application rates on CO_2 emissions in soils under upland rice production in southern Guinea Savannah of Nigeria. *Heliyon, 6*(11).

Palanivelu, K., Ramachandran, A., & Raghavan, V. (2021). Biochar from biomass waste as a renewable carbon material for climate change mitigation in reducing greenhouse gas emissions—A review. *Biomass Conversion and Biorefinery, 11*(5), 2247–2267.

Pradhan, G., & Meena, R. S. (2023). Interaction impacts of biocompost on nutrient dynamics and relations with soil biota, carbon fractions index, societal value of CO_2 equivalent, and ecosystem services in the wheat-rice farming. *Chemosphere*139695. Available from https://doi.org/10.1016/j.chemosphere.2023.139695.

Pradhan, G., & Meena, R. S. (2022). Diversity in the rice–wheat system with genetically modified zinc and iron-enriched varieties to achieve nutritional security. *Sustainability, 14*, 9334. Available from https://doi.org/10.3390/su14159334.

Pradhan, G., & Meena, R. S. (2023a). Utilizing waste compost to improve the atmospheric CO_2 capturing in the rice-wheat cropping system and energy-cum-carbon credit audit for a circular economy. *Science of the Total Environment*164572. Available from https://doi.org/10.1016/j.scitotenv.2023.164572.

Pradhan, G., Meena, R. S., Kumar, S., & Lal, R. (2023). Utilizing industrial wastes as compost in wheat-rice production to improve the above and below-ground ecosystem services. *Agriculture, Ecosystems and Environment*, 358108704. Available from https://doi.org/10.1016/j.agee.2023.108704.

Pradhan, G., Meena, R.S., Kumar, S., Jhariya, M.K., Khan, N., Shukla, U.N., Singh, A.K., Sheoran, S. & Kumar, S. (2022). Legumes for eco-friendly weed management in an agroecosystem. In: Meena, R.S., Kumar, S. (eds.). *Advances in legumes for sustainable intensification* (pp. 133–154). <https://doi.org/10.1016/B978-0-323-85797-0.00033-1>

Purakayastha, T. J., Bera, T., Bhaduri, D., Sarkar, B., Mandal, S., Wade, P., ... Tsang, D. C. (2019). A review on biochar modulated soil condition improvements and nutrient dynamics concerning crop yields: Pathways to climate change mitigation and global food security. *Chemosphere*, 227, 345–365.

Purakayastha, T. J., Das, K. C., Gaskin, J., Harris, K., Smith, J. L., & Kumari, S. (2016). Effect of pyrolysis temperatures on stability and priming effects of C3 and C4 biochars applied to two different soils. *Soil and Tillage Research*, 155, 107–115.

Purakayastha, T. J., Kumari, S., & Pathak, H. (2015). Characterisation, stability, and microbial effects of four biochars produced from crop residues. *Geoderma*, 239, 293–303.

Raj, A., Jhariya, M. K., Banerjee, A., Meena, R. S., Nema, S., Khan, N., Yadav, S. K., & Pradhan, G. (2021). Agroforestry a model for ecological sustainability. In M. K. Jhariya, R. S. Meena, A. Raj, & S. N. Meena (Eds.), *Natural resources conservation and advances for sustainability* (pp. 289–308). Elsevier. Available from https://doi.org/10.1016/B978-0-12-822976-7.00002-8.

Roy, A., Chaturvedi, S., Singh, S. V., Govindaraju, K., Dhyani, V. C., & Pyne, S. (2022). Preparation and evaluation of two enriched biochar-based fertilizers for nutrient release kinetics and agronomic effectiveness in direct-seeded rice. *Biomass Conversion and Biorefinery*. Available from https://doi.org/10.1007/s13399-022-02488-z.

Roy, O., Meena, R. S., Kumar, S., Jhariya, M. K., & Pradhan, G. (2021). Assessment of land use systems for CO_2 sequestration, carbon credit potential, and income security in Vindhyan Region, India. *Land Degradation and Development*, 33(4), 670–682. Available from https://doi.org/10.1002/ldr.4181.

Roy, P., & Dias, G. (2017). Prospects for pyrolysis technologies in the bioenergy sector: A review. *Renewable and Sustainable Energy Reviews*, 77, 59–69.

Shen, Y., Zhu, L., Cheng, H., Yue, S., & Li, S. (2017). Effects of biochar application on CO_2 emissions from a cultivated soil under semiarid climate conditions in Northwest China. *Sustainability*, 9(8), 1482.

Sheoran, S., Ramtekey, V., Kumar, D., Kumar, S., Meena, R. S., Kumawat, A., Pradhan, G., & Shukla, U. N. (2022). Grain legumes: Recent advances and technological interventions. In R. S. Meena, & S. Kumar (Eds.), *Advances in legumes for sustainable intensification* (pp. 507–532). Elsevier. Available from https://doi.org/10.1016/B978-0-323-85797-0.00025-2.

Shi, S., Zhang, Q., Lou, Y., Du, Z., Wang, Q., Hu, N., ... Song, J. (2021). Soil organic and inorganic carbon sequestration by consecutive biochar application: Results from a decade field experiment. *Soil Use and Management*, 37(1), 95–103.

Shin, J., Park, D., Hong, S., Jeong, C., Kim, H., & Chung, W. (2021). Influence of activated biochar pellet fertilizer application on greenhouse gas emissions and carbon sequestration in rice (*Oryza sativa* L.) production. *Environmental Pollution*, 285, 1174.

Singh, S., Chaturvedi, S., Nayak, P., Dhyani, V. C., Nandipamu, T. M., Singh, D. K., Pratibha, G., Mathyam, P., Srinivasrao, K., & Govindaraju, K. (2023). Carbon offset potential of biochar based straw management under rice–wheat system along Indo-Gangetic Plains of India. *Science of the Total Environment*, 897165176.

Singh, S., Luthra, N., Mandal, S., Kushwaha, D. P., Pathak, S. O., Datta, D., Sharma, R., & Pramanik, B. (2023). Distinct behavior of biochar modulating biogeochemistry of salt-affected and acidic soil: A review. *Journal of Soil Science and plant Nutrition*. Available from https://doi.org/10.1007/s42729-023-01370-9.

Singh, S. V., Chaturvedi, S., & Datta, D. (2019). Biochar: An eco-friendly residue management approach. *Indian Farming*, 69(08), 27–29.

Singh, S. V., Chaturvedi, S., Dhyani, V. C., & Govindaraju, K. (2020). Pyrolysis temperature influences the characteristics of rice straw and husk biochar and sorption/desorption behavior of their biourea composite. *Bioresource Technology*, 314123674. Available from https://doi.org/10.1016/j.biortech.2020.123674.

Smith, P. (2016). Soil carbon sequestration and biochar as negative emission technologies. *Global change biology, 22* (3), 1315–1324.

Soil Survey Staff. (2014). *Keys to Soil Taxonomy* (12th Edn). Washington DC: USDA-Natural Resources Conservation Service.

Spokas, K. A., Koskinen, W. C., Baker, J. M., & Reicosky, D. C. (2009). Impacts of woodchip biochar additions on greenhouse gas production and sorption/degradation of two herbicides in a Minnesota soil. *Chemosphere, 77* (4), 574–581.

Sun, L., Deng, J., Fan, C., Li, J., & Liu, Y. (2020). Combined effects of nitrogen fertilizer and biochar on greenhouse gas emissions and net ecosystem economic budget from a coastal saline rice field in southeastern China. *Environmental Science and Pollution Research, 27*, 17013–17022.

Sun, Z., Zhang, Z., Zhu, K., Wang, Z., Zhao, X., Lin, Q., & Li, G. (2020). Biochar altered native soil organic carbon by changing soil aggregate size distribution and native SOC in aggregates based on an 8-year field experiment. *Science of the Total Environment, 708*, 134829.

Tauqeer, H. M., Turan, V., Farhad, M., & Iqbal, M. (2022). Sustainable agriculture and plant production by virtue of biochar in the era of climate change. *Managing plant production under changing environment* (pp. 21–42). Singapore: Springer.

Wang, C., Liu, J., Shen, J., Chen, D., Li, Y., Jiang, B., & Wu, J. (2018). Effects of biochar amendment on net greenhouse gas emissions and soil fertility in a double rice cropping system: A 4-year field experiment. *Agriculture, Ecosystems & Environment, 262*, 83–96.

Wu, Z., Zhang, X., Dong, Y., Li, B., & Xiong, Z. (2019). Biochar amendment reduced greenhouse gas intensities in the rice-wheat rotation system: Six-year field observation and meta-analysis. *Agricultural and Forest Meteorology, 278*, 107625.

Xiang, J., Liu, D., Ding, W., Yuan, J., & Lin, Y. (2015). Effects of biochar on nitrous oxide and nitric oxide emissions from paddy field during the wheat growth season. *Journal of Cleaner Production, 104*, 52–58.

Xie, T., Reddy, K. R., Wang, C., Yargicoglu, E., & Spokas, K. (2015). Characteristics and applications of biochar for environmental remediation: A review. *Critical Review Environmental Science & Technology, 45*(9), 939–969.

Xu, X., Cheng, K., Wu, H., Sun, J., Yue, Q., & Pan, G. (2019). Greenhouse gas mitigation potential in crop production with biochar soil amendment—A carbon footprint assessment for cross-site field experiments from China. *Gcb Bioenergy, 11*(4), 592–605.

Yan, X., Akiyama, H., Yagi, K., & Akimoto, H. (2009). Global estimations of the inventory and mitigation potential of methane emissions from rice cultivation conducted using the 2006 Intergovernmental Panel on Climate Change Guidelines. *Global Biogeochemical Cycles, 23*(2).

Yang, W., Feng, G., Miles, D., Gao, L., Jia, Y., Li, C., & Qu, Z. (2020). Impact of biochar on greenhouse gas emissions and soil carbon sequestration in corn grown under drip irrigation with mulching. *Science of the Total Environment, 729*, 138752.

Ye, L., Camps-Arbestain, M., Shen, Q., Lehmann, J., Singh, B., & Sabir, M. (2020). Biochar effects on crop yields with and without fertilizer: A meta-analysis of field studies using separate controls. *Soil Use and Management, 36*(1), 2–18.

Zhang, A., Bian, R., Pan, G., Cui, L., Hussain, Q., Li, L., ... Yu, X. (2012). Effects of biochar amendment on soil quality, crop yield and greenhouse gas emission in a Chinese rice paddy: A field study of 2 consecutive rice growing cycles. *Field Crops Research, 127*, 153–160.

Zhang, Q. Z., Dijkstra, F. A., Liu, X. R., Wang, Y. D., Huang, J., & Lu, N. (2014). Effects of biochar on soil microbial biomass after four years of consecutive application in the north China plain. *PLoS One, 9*(7), e102062.

Zhang, R., Qu, Z., Liu, L., Yang, W., Wang, L., Li, J., & Zhang, D. (2022). Soil respiration and organic carbon response to biochar and their influencing factors. *Atmosphere, 13*(12), 2038.

Zhang, S., Cui, J., Wu, H., Zheng, Q., Song, D., Wang, X., & Zhang, S. (2021). Organic carbon, total nitrogen, and microbial community distributions within aggregates of calcareous soil treated with biochar. *Agriculture, Ecosystems & Environment, 314*, 107408.

Zheng, H., Liu, D., Liao, X., Miao, Y., Li, Y., Li, J., ... Ding, W. (2022). Field-aged biochar enhances soil organic carbon by increasing recalcitrant organic carbon fractions and making microbial communities more conducive to carbon sequestration. *Agriculture, Ecosystems & Environment, 340*, 108.

CHAPTER 19

The use of biochar to reduce carbon footprint: toward net zero emission from agriculture

Anurag Bera[1,3], Ram Swaroop Meena[1], Anamika Barman[2] and Priyanka Saha[2]

[1]Department of Agronomy, Institute of Agricultural Sciences, Banaras Hindu University, Varanasi, Uttar Pradesh, India [2]Division of Agronomy, ICAR—Indian Agricultural Research Institute, Pusa, New Delhi, India [3]Department of Agronomy, Tea Research Association, North Bengal Regional R & D Center, Jalpaiguri, West Bengal, India

19.1 Introduction

Climate change is no longer a concern for the distant future; instead, it has become an urgent crisis that can be tackled only by combining efforts worldwide. The average global surface temperature is rising due to the build-up of greenhouse gases (GHGs) caused by anthropogenic activities such as the generation of electricity, manufacturing goods, clearing of forests, use of automobiles, and the intensive crop production system. From 1901 to 2020, the Earth's average temperature increased by around 1.1°C (NASA Earth Observatory, 2015), whereas the last decade of 2011—20 was the warmest in recorded history. Since the 1980s, each successive decade has set a record for warmth (IPCC, 2007). Not only the increase in temperature, but climate change also encompasses a series of adverse consequences such as rising sea levels and altered weather patterns like drought, flood, heat waves, cyclones and many more. Now, more than ever, it is critical to cut GHG emissions to head off the worst effects of climate change. Several international treaties and agreements are adopted globally, which reflect the commitment and strategies for combating climate change. The Paris Agreement's (2015) target of "holding the increase in the global average temperature to well below 2°C above preindustrial levels and pursuing efforts to limit the temperature increase to 1.5°C above preindustrial levels" has raised the bar for international climate policy and brought greater attention to the role all sectors can play in mitigating climate change. Agriculture is one of the most critical sectors of our living, contributing significantly to global warming due to the GHGs it releases. Agriculture and food production are associated with emitting three GHGs, that is, CO_2, CH_4 and N_2O. During 2018, 9.3 billion metric tonnes of CO_2 equivalent (CO_2-e) was emitted worldwide because of agriculture, which includes emissions from both agricultural operations inside and outside of farms as well as those caused by changes in land use and land cover (Tubiello et al., 2013; FAO). After dragging its feet on climate change, the world is finally mobilizing toward a goal of lessening GHG emissions, which has been named a mission called "net zero emission." It refers to a carbon-neutral state in which the total amount of GHGs emitted into the atmosphere is equal to the amount removed from the atmosphere by that sector.

As such, there has been an effort to identify the factors that can significantly cut emissions of GHGs. After long years of research, scientists worldwide have recommended a C-rich, pyrolysed porous product, namely "biochar" as one of the potent amendments to capture C in the form of soil carbon (SC), thereby lessening releases of CO_2 into the atmosphere. Several prior studies have documented biochar's distinctive physicochemical features, such as its large specific surface area (SSA), abundant hydrophilic surface functional groups, high liming capability, and substantial cation exchange capacity (CEC) (Li et al., 2014; Pradhan, Meena, Kumar, & Lal, 2023; Singh, Chaturvedi, & Datta, 2019). All of the biochar's qualities work toward its ultimate goal of increasing SC stock, which is necessary for achieving a carbon and noncarbon GHG emissions equilibrium in agroecosystems. When talking about agricultural ecosystems, soil becomes an essential sink for C which can prevent the release of CO_2 into the atmosphere. Around 2500 gigatons (GT) of the 3170 GT total C in terrestrial ecosystems can be found in the soil. Both organic and inorganic forms of C (1550 GT) are present in soil (950 GT). A further $10-60$ mg C hm^{-2} (1 $hm^2 = 104$ m^2) can be stored in soil (Song, Pan, Zhang, Zhang, & Wang, 2016). Sequestering C in the soil can be an excellent technique to reduce CO_2 concentrations in the atmosphere (Meena, Pradhan, Kumar, & Lal, 2023); side-by-side it can increase crop growth and soil productivity (Baronti et al., 2010). In underdeveloped countries, increasing the soil organic pool by 1 Mg ha^{-1} $year^{-1}$ might increase crop yield by 32 Mt $year^{-1}$ (Lal, 2006). Therefore, improving the soil C pool should be our main objective to attain net zero emissions from the agricultural production system. Adding biochar to the farm ecosystem can improve soil organic carbon (SOC) storage and reduce GHG emissions. This chapter will highlight the contribution of the agricultural production system to GHG emissions, the potential and mechanism of biochar-led GHG emissions mitigation and ultimately, the path to achieving net-zero emission goals from the agricultural sector.

19.2 Carbon footprint and its significance in the changing global environment

Global warming refers to the increase in the earth's average surface temperature and its consequences, which have become a massive threat to the entire human race (Pradhan & Meena, 2023; Sivakumar, 2011; Smith & Leiserowitz, 2012). Human activities, such as the excessive burning of fossil fuels and coals, have raised atmospheric concentrations of

GHGs, which in turn has contributed to global warming. These GHGs serve as a blanket or cap, preventing part of the heat from escaping from Earth into space (Hansen, 2004; Kweku et al., 2018; Meena & Pradhan, 2023). In response, the planet's average surface temperature is rising, and natural disasters like storms, heat waves, floods, and droughts are becoming increasingly prevalent (Singh & Singh, 2012). Over the past few decades, the scientific communities have expressed grave concerns regarding global warming and climate change. Numerous international treaties, including the United Nations Framework Convention on Climate Change (1992), the Kyoto Protocol (1997), the Bali Roadmap (2007), the Copenhagen Agreement (2009), and the Paris Agreement (2016) have been ratified already, which illustrate the resolve and actions of the government to combat climatic changes. Countries have made emissions reduction pledges and new initiatives by consensus. As a result, the novel ideas of a low-C economy, low-C city, C trade, C pricing, C tax, and other measures to cut C emissions have emerged as a crucial aspect of global development strategy (Hertwich & Peters, 2009; Larsen & Hertwich, 2009; Lenzen, Wood, & Wiedmann, 2010; Pradhan & Meena, 2023a). Recent studies on low-C issues have centered on accounting for and reducing emissions, trading platforms for C emissions, levies on C emissions, and emission regulations. One of the most fundamental and important areas of low-C study is the carbon footprint (CF). CFs are the sum of all GHG emissions produced by any human activity, institution, service, location, or product, measured in terms of carbon dioxide equivalent (CO_2-e). A growing interest in assessing the CF has been observed among researchers and policymakers. In this section, we will explore the fundamentals of the CF and see how it has the potential to become an effective monitor of CO_2 emissions.

19.3 Carbon footprint

The concept of the CF developed from the idea of the ecological footprint, which quantifies the impact of human activity on the planet's natural resources. Similarly, The CF is a measurement of the entire quantity of CO_2 emissions generated by an action or accumulated over the life stages of a product, both directly and indirectly (Pradhan & Meena, 2022; Weidema, Thrane, Christensen, Schmidt, & Løkke, 2008; Wiedmann, 2007). The CF is a type of footprint indicator that includes the ecological, water and land footprint (Fang, Heijungs, & de Snoo, 2014). The Industrial Revolution in the 1820s marked the beginning of rapid climate change (Foley et al., 2013). Human activities like burning fossil fuels for energy and repeatedly cutting down trees have increased atmospheric GHG levels, making it more challenging to establish a C-neutral lifestyle. Many lifestyle changes can help reduce a person's CF, including devouring less meat and dairy and wasting less food, upgrading to more energy-efficient home appliances, buying fewer goods significantly (especially disposable items, like fast fashion), and taking fewer trips (especially plane trips). Individuals, businesses, organizations, governments, and nations leave a CF (Damert, Morris, & Guenther, 2020; Finkbeiner, 2009). Once the CF is calculated, steps can be taken to lessen it, through measures like greener technology, increased energy efficiency, improved process and product management, retooled Green Public or Private Procurement, altered consumption habits, and C offsetting, among others (Roy, Meena, Kumar, Jhariya, & Pradhan, 2021; Sundarakani, Goh, de Souza, & Shun, 2008; Singh et al., 2023a).

Calculating a product or organizational CF involves a multistep process known as life cycle assessment, during which numerous aspects of the product or organization's operational activities are evaluated. The functional boundaries must be understood to quantify emission data and derive the CF. CO_2-e is calculated by multiplying emission data from various actions by a standard emission factor. While the idea of a person's CF has been around since the 1990s as an indicator to quantify the impact, it took off in 2003 when oil and gas firm BP ran an ad campaign asking random individuals on the street what their CF was. That drive excited individuals about figuring out their CF and enlightened them on the importance of such studies. Since then, CF analysis and monitoring have become mainstream and are typically considered when making projections. To compare the relative impact of different GHGs on global warming, their global warming potential is sometimes expressed as the warming impact of a fixed amount of carbon dioxide equivalent (CO_2-e). Compared to the 298 CO_2-e produced by one unit of N_2O, only 23 CO_2-e are made by one team of CH_4 (Li et al., 2014; Pradhan et al., 2022). The CF per unit area-kg CO_2-e. ha^{-1} number is sometimes used to describe the global warming potential of all gases.

19.3.1 Trend and analysis of current CO_2 emission

The percentage of greenhouse effect attributable to CO_2 is exceptionally high, hovering around 75%. The majority of CO_2 emissions (76% of the total), including those from electricity and heat generation, transportation, industrial processes, and residential use, all come from the combustion of fossil fuels. Another source (24% of the total) of man-made CO_2 emissions includes agriculture, forestry, and other land uses, largely attributable to deforestation (Fig. 19.1). In this scenario, the trend analysis approach becomes essential for modeling CO_2 and non-CO_2 GHG emissions from various sources and developing effective policies and strategies to lower their proportions.

Analysis results demonstrated that the models are suitable for making long-term estimates of CO_2 emissions for strategic planning. Historically, CO_2 has made up around 0.03% of the total volume of the earth's atmosphere; but, because of human actions like burning fossil fuels and cutting down trees, that number has climbed by about 25% since

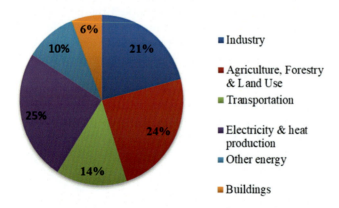

FIGURE 19.1 Global CO_2 emissions by different economic sectors.

the beginning of the industrial era. In 2021, the average atmospheric concentration of CO_2 reached 414.72 parts per million, an all-time high. CO_2 levels in the atmosphere have steadily risen over the past 60 years, at a rate around 100 times faster than in the past due to natural causes, such as the end of the last ice age 11,000−17,000 years ago. According to research presented at Conferences of the Parties (COP27) of the United Nations Framework Convention on Climate Change (UNFCC) in Sharm El-Sheikh, C emissions are expected to climb slightly in 2022 compared to 2021, with the most significant increase expected in India (+6%). If projections hold, the United States (+1.5%) will see the second-highest rise in emissions. However, China (−0.9%) and the European Union (−0.8%) are setting an encouraging example by lowering their emissions (Kumawat et al., 2022; Moosmann et al., 2021) (Fig. 19.2). In 2019, global emissions reached 36.3 GT, decreasing to 34.5 GT in 2020 due to the prolonged economic drag caused by COVID-19, and rising to 36.3 GT in 2021. Increased oil consumption and fossil fuel burning are the primary drivers behind the anticipated 1% increase in global fossil CO_2 emissions in 2022 (range 0.1%−1.9%), reaching 36.6 GT (Hoang et al., 2021; Newell, Raimi, Villanueva, & Prest, 2021; Sheoran et al., 2022). More alarmingly, the COP27 report notes that there is now a 50% probability that global warming will reach 1.5°C (the lower limit of temperature rise compared to preindustrial levels specified by the Paris Agreement) over the next nine years if emission trends are continued (Arora & Arora, 2023; Meena et al., 2022; Siva, 2022). Economic factors and international treaties, such as the Kyoto Protocol, the Paris Agreement, and the UNFCC COPs, all of which aim to reduce GHG emissions collectively, inform the analysis of the relationships between energy use and CO_2 emissions. Energy consumption trends across countries and progress toward emission reduction goals are investigated annually. Following the direction, decisions and policies are developed. Thus, it is crucial to assess the data on C emissions regularly and to adopt measures accordingly.

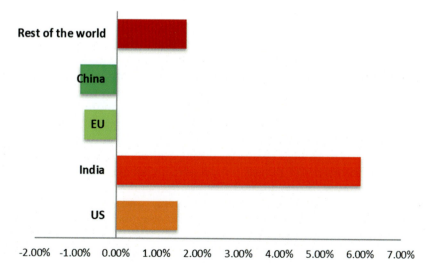

FIGURE 19.2 Status of CO_2 emission of top emitter country/Union (COP 27).

19.3.2 Agricultural practices for C footprint management

The agricultural sector is responsible for 18% of global emissions of GHGs, primarily CO_2, CH_4, and N_2O. Because of the widespread conversion of forested and peat-land ecosystems to agricultural use, agriculture has become a significant contributor to global warming and climate change. Non- CO_2 emissions, on the other hand, are primarily the result of agricultural and livestock processes. It is therefore, essential to lessen the impacts of climate change by lowering GHG emissions. This can be achieved by thoroughly investigating the CF produced by agricultural practices (Ozlu, Arriaga, Bilen, Gozukara, & Babur, 2022; Raj et al., 2021). Reducing GHG emissions and, consequently, climate change, could be a goal of efforts to regulate agrarian management by evaluating the agricultural CF. Farming activities like driving tractors, pumping water for irrigation, using a power tiller, harvesting, threshing, winnowing, transporting the produce, and so on all use a considerable amount of energy. Management strategies for improving energy efficiency while lowering CF, are particularly promising in this setting (Paris et al., 2022). When dealing with climate change, CF relies on the soil system for inputs and outputs. Understanding the inputs-outputs perspective requires analyzing the mechanisms of C sequestration and stabilization in the face of rising temperatures and other environmental stresses. Implementing efficient management practices in fields, that is, conservation tillage, organic fertilization, biochar-like organic amendments additions, manuring, crop rotation, and improved residue management increases C sequestration. Similarly, when the mechanical processes that contribute to the natural sequestration of soil C are taken into account, factors including crop density, crop type, and photosynthesis become significant in CF assessments (Ozlu et al., 2022). Using inorganic fertilizers without careful management can have unforeseen repercussions in conventional agriculture systems, such as lowering soil pH and moisture, GHG emission, spreading soil colloids, reducing soil aggregate stability, and many more (Khalid, Tuffour, & Bonsu, 2014; Kumar et al., 2020; Lawal & Girei, 2013). The C sequestration/GHG emission ratio becomes quite significant during fertilization in the cropping system. In recent times, low-cost organic fertilization has garnered more attention as of late because of its ability to provide the nutrients progressively as per the requirement of the crop while simultaneously protecting the C stability of the soil (Zhao et al., 2009). Manuring has been shown to improve soil health by influencing soil properties, including biological indicators, soil microbial community compositions, microbial biomass, earthworm populations, enzyme activities, soil aggregation, porosity, soil bulk density and compaction, soil pH and soil water relation (Ozlu, Sandhu, Kumar, & Arriaga, 2019). Biochar, a C-rich pyrolyzed biomass, has been gaining popularity as an organic amendment due to its ability to improve soil fertility, soil structure, nutrient availability, soil-water retention, and SC storage (Bartocci, Bidini, Saputo, & Fantozzi, 2016). C added to soils by biochar treatments can persist for decades or longer. Biochar can be used to counteract climate change by lowering soil degradation and soil-borne GHG emissions, increasing C sequestration and soil nutrient content, and decreasing soil erosion. (Hagemann et al., 2017). SOC increases when organic matter, such as agricultural leftovers, decomposes at the soil's surface. Conventional tillage practices may speed up the mineralization of SOC and the release of CO_2 because they break up the soil aggregates and expose labile organic matter, which makes soil microorganisms work harder to oxidize SOM.

Agricultural CO_2 emissions are strongly correlated with the frequency and severity of soil disturbances (Kumar et al., 2020a; La Scala, Bolonhezi, & Pereira, 2006). Conservation tillage or minimum tillage has come up with potential solutions for climate change mitigation purposes, which suggests reduced soil disturbance residue cover and rotation of crops to maintain sustainability among all components. Herbicide, another potential contributor of GHG emissions can also be restricted for use by pairing it with conservation tillage and applying only at critical stages (Cordeau, 2022). N_2O and CH_4 emissions from the soil are caused by biological and naturally occurring mechanisms in the soil ecosystem, both of which can be controlled by maintaining good water and nutrient status and using appropriate farming techniques. In this way, CF can be reduced by maintaining a balanced soil ecosystem, adopting resilient farming practices, stimulating soil microbial activity, and forming a sizable C pool.

19.4 Greenhouse gas emissions and the role of agriculture's production system

19.4.1 Agricultural sectors and greenhouse gas inventories

Agriculture is one of the most susceptible sectors to climate change and a significant contributor to it (FAO, 2016; Mbow et al., 2019). Land use change and agricultural production account for about 20%—25% of GHGs released by humans (IPCC, 2019). As the world's population rises, so will the need for food, feed, fiber, and fuel, putting a strain on farmers to raise productivity. Expanding crop areas by clearing uncultivated land is the quickest way to increase grain production, but this threatens ecosystem biodiversity (Isbell et al., 2011); and decreases environmental quality (Foley et al., 2011) by using up C supplies in natural soils and vegetation (Liu, Cutforth, Chai, & Gan, 2016). Since most agricultural inputs significantly contribute to GHG emissions (Goglio et al., 2014; Yue, Xu, Hillier, Cheng, & Pan, 2017), farming negatively impacts the environment. In particular, the incautious and unbalanced use of inorganic fertilizers and pesticides in high-yielding farming systems contributes to GHG emissions (Lehmann & Joseph, 2015). In addition to agricultural production, animal husbandry is recognized as a significant contributor to GHG emissions, accounting for around 18% of world emissions (Gerber et al., 2013). CH_4, N_2O, and CO_2 from livestock contribute approximately 44%, 29%, and 27% of the world's total GHG emissions (Forabosco, Chitchyan, & Mantovani, 2017; Gerber et al., 2013; Rojas-Downing, Nejadhashemi, Harrigan, & Woznicki, 2017). Global GHG emissions in 2017 were 51 billion tonnes CO_2-e (Gt CO_2-e year^{-1}); if land use emissions are included, they can reach 56 Gt CO_2-e year^{-1}. In 2017, the percentage of agriculture's global CO_2-e emissions from all human activities was 20%. The emissions from agriculture were 11.1 Gt CO_2-e year^{-1}, made up of 6.1 Gt CO_2-e year^{-1} from crop and livestock operations within the farm gate and 5.0 Gt CO_2-e year^{-1} from agricultural land usage. Within the farm gate, crop and livestock operations contributed 11%, while related land use contributed another 9%. Changes in the overall trend and the relative proportion of CO_2 emissions attributable to crop and livestock activities relative to total CO_2 emissions across all sectors have been observed over time (Fig. 19.3).

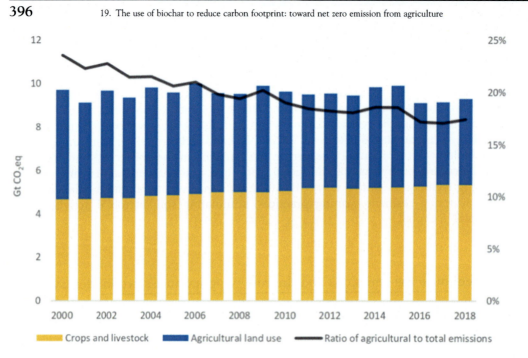

FIGURE 19.3 Annual emissions from crops, livestock, and related land use, and agriculture's percentage share of global greenhouse gas emissions, 2000−18.

19.4.2 Effect of land use and farming practices on soil greenhouse gas emission

Several factors affect crop production's GHG emission footprint. The complete life cycle analysis showed that CO_2 is emitted during the growing of a crop, transportation of numerous input goods to the farm, and manufacturing processes that take place off-farm. The significant factors contributing toward GHG emission from agriculture have been discussed below (Fig. 19.4).

19.4.2.1 In situ crop residue decomposition by microbes

A significant source of nitrogen (N) in the soil is the crop residue left over from no-till systems or the biomass worked into tilled systems after harvest. It transforms such as mineralization, nitrification, and denitrification, all of which contribute to CO_2, CH_4, and N_2O production under aerobic or anaerobic conditions (Wang et al., 2018). In Uttar Pradesh, India, total emissions from burning agricultural waste in the field in 2007 were 251 gigagrams (Gg) of CH_4 and 6.5 Gg of N_2O (Bhatia, Jain, & Pathak, 2013).

19.4.2.2 Injudicious use of N fertilizers

Agriculture is responsible for around 80% of the N_2O that humans emit into the atmosphere each year; more than half of these emissions come from farming practices (Pires, da Cunha, de Matos Carlos, & Costa, 2015). The intensity of the emissions is influenced by

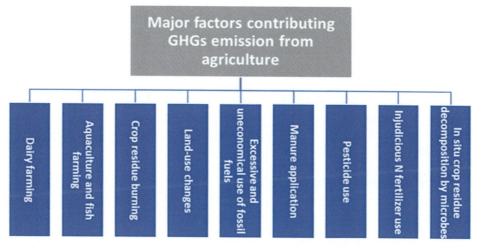

FIGURE 19.4 List of agricultural factors contributing to greenhouse gas emissions.

environmental conditions such as precipitation and potential evapotranspiration during N fertilization (Carlson et al., 2017; Millar et al., 2018). For instance, the direct leaching and emissions of N_2O due to using N fertilizers are proportional to the relationship between precipitation and potential evapotranspiration (Tongwane et al., 2016).

19.4.2.3 Manure application

Over half of agriculture's GHG emissions come from fertilizer and manure on the field's surface (Ren et al., 2017). The amount and scope of manure's impact on GHG emissions depend on factors such as the manure's composition, the methods used to apply it, and the rates at which it is applied. Since solid manures add inert forms of C and N to the soil, their potential GHG emission footprint is lower than that of liquid manures (Aguirre-Villegas & Larson., 2017). Manure incorporation, however, is less susceptible to GHG emissions than surface application.

19.4.2.4 Excessive uneconomical use of fossil fuels

In recent years, fuel-driven machinery has become increasingly common in on-farm operations like planting seeds, applying fertilizer and pesticides, harvesting field crops, and off-farm operations like fertilizer manufacture, shipping, storage, and delivery to the farm gate. In general, manufacturing N fertilizers using fossil fuels before field application produces more emissions than the production and use of pesticides in crop fields (Liu et al., 2016).

19.4.2.5 Land use changes

To mitigate the negative consequences of climate change, especially during drought and humid spells, farmers may use land-use change as an adaptive feedback mechanism (Permpool, Bonnet, & Gheewala, 2016; Lungarska & Chakir., 2018). In tropical climates, primary forest conversion to cropland (about 25%), perennial crops (about 30%), and forest conversion to grassland (about 12%) results in significant soil organic C loss (Don, Schumacher, & Freibauer, 2011).

19.4.2.6 Crop residue burning

The CF of GHG is greatly affected by the burning of agricultural crop residues and contributes to the release of GHGs (CO_2, N_2O, and CH_4). Between 48% and 95% of this residue is burned in Thailand and the Philippines open fields. In India, Thailand, and the Philippines, open-field burning of rice straw results in GHG emissions of respectively 0.05%, 0.18%, and 0.56% (Jain, Bhatia, & Pathak, 2014).

19.4.2.7 Dairy, aquaculture and fish farming

Livestock, especially ruminants, occupy 70%–80% of all anthropogenic land uses worldwide (FAO, 2009; Bellarby et al., 2013). The enteric fermentation of ruminant animals creates CH_4, whereas the primary sources of N_2O are the denitrification and nitrification of soils (which are used to grow feed). Aquaculture refers to using different methods and powers to raise a wide variety of fish species. Recently, experts worldwide have produced data on GHG emissions and their CF for several aquaculture products, particularly marine species (salmon and shrimp) (Robb, MacLeod, Hasan, & Soto, 2017).

Rice cultivation (10%), the use of synthetic fertilizer (12.6%), burning crop residue and pasture (4.7%), leftover crop residues (4.7%), enteric emissions (39.7%), manure left on pastures (16.1%), poor manure management (6.7%), and manure application (3.7%) significantly contribute to total GHG emission, according to a more recent report (Patra & Babu, 2017).

19.4.2.8 Climate innovative farming practices for reducing GHG emission

Although the agricultural sector serves as a contributor to GHG emissions, also, as a mitigation approach, is managed properly. Recent cultivation methods and technologies are being employed related to the mitigation climate-related issues. The main three approaches are reduction of emission, enhancement of removal and avoiding emission (Fig. 19.5). By storing C in the soil and lowering CH_4 and N_2O emissions from the soil through changes in land-use management, GHG emissions from agriculture can be

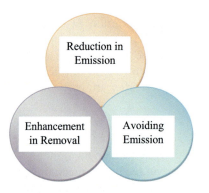

FIGURE 19.5 Approaches for mitigation of climate change.

mitigated. The quantity of C stored in the soil increases when crop mixtures are changed to include more perennial or deep-rooted plants. The accumulation of soil C is promoted by cultivation methods that leave residues and minimize tillage, intense tillage. Methane and nitrous oxide emissions can be decreased through modifications in crop genetics and effective management of irrigation, fertilizer use, and soils. Such choices are crucial for increasing soil fertility and mitigating global warming.

19.4.2.9 Diversifying crop rotations to reduce carbon footprint

Crop diversity is increasingly recognized as a critical strategy for raising agroecosystem production and reducing CF (Gan et al., 2015; Yang, Gao, Zhang, Chen, & Sui, 2014). The previous crop heavily influences the GHG emission footprint of the next crop in a rotation due to its effect on the N_2 cycle and the soil's organic and mineral components. Therefore, it is more important to calculate the net GHG balance and GHG emission footprint of an entire crop sequence than for a single crop in a system.

19.4.2.10 Carbon sequestration

SC sequestration involves enclosing atmospheric carbon dioxide (CO_2) in a stable carbon form within soils, where it will not be quickly remitted. Soil parameters, environmental variables, native plants, and human-made factors strongly influence the capacity for soil C sequestration in an agroecosystem. As a result, some agro-ecosystems have low SOC stocks, whereas others have substantial SOC stocks. It needs a special adoption of best management approaches (such as conservation agriculture, precision farming, integrated nutrient management, etc.) that generate a positive C budget to restore SOC in C-depleted systems. A 1-m soil profile's uppermost 30–100 cm contains roughly half of the world's SOC. Recently, a plan emerged to increase the amount of SOC sequestered globally to 4 mm each year to combat climate change and boost food security (Lal, 2018).

19.4.2.11 Genetic enhancement of crops and animals

The creation of novel varieties and hybrids, genetic selection, enhanced breeding methods, and genetic engineering and modification technology have all contributed to a significant rise in food output. We can now find traits in a genome that boost productivity and confer resilience to pests and drought. Other characteristics, including improved input uptake efficiency or the ability to produce inputs inside the plant and the enhancement of beneficial soil nutrients, will be easier to identify and improve thanks to scientific advancements.

19.4.2.12 Conservation agriculture

Conservation agriculture can improve the use of natural resources, including water, air, fossil fuels, and soil, by adopting resource-saving technologies like zero- or minimum tillage with direct sowing, permanent or semipermanent residue cover, and crop rotations. These innovations can increase agriculture's sustainability by protecting the resource base with greater input efficiency and reducing GHG emissions.

19.4.2.13 Increasing input-use efficiency

The essential components of agricultural production are land, other natural resources, labor, and capital. We have relatively plenty of labor and capital, but our supply of land and other natural resources is limited and even declining. The overall amount of water used for agriculture is anticipated to rise by 13% by 2025. Without improving irrigation efficiency, we may need 50% more water by 2050 to meet the world's food needs. Since natural gas is a primary component in the production of urea, the direct nitrogenous fertilizer used in the nation, the emission of CO_2 in the case of fertilizer application is indirect. Here, using nitrogenous fertilizer has resulted in the direct emission of N_2O. Hence, reducing the natural gas subsidies and replacing outdated factories and technology could mitigate this case.

19.4.2.14 Different agronomic measures

SC storage can reduce GHG emissions by incorporating a more efficient set of crop management practices that boost yields and increase residual C (Sarma et al., 2018). Using methods such as cultivating superior crop types, altering crop rotations with perennial crops, and substituting N fertilizer with manure have all been shown to decrease emissions of GHGs (which encourages more below-ground biomass), and avoid or reduce the intensity and extent of fallow land use (Sarma et al., 2018) (Table 19.1).

TABLE 19.1 Agronomic practices and input details.

Agronomic practices	Details
Improved mechanization	Fuel efficient, multitasking, fast operating, compatible with biodiesel
Manure management	Improved storage, proper handling, anaerobic digestion, efficient use, additive use
Tillage alteration	Reduced tillage, minimum tillage, zero tillage
Irrigation alteration	Intermittent drainage, mid-season drainage, controlled irrigation, alternate wet and dry, timely irrigation
Crop rotation and Intensification	Crop rotation, crop diversification, legume intensification, cover crop, crop intensification, direct seed rice, grazing management, agroforestry
Fertilizer management	Site-specific nutrient management, lime management, biofertilizer, integrated nutrient management
Choice of cultivar	N efficient, short duration, less CH_4 emitting
Residue management	Mulching, removal of straw, biochar, composting, straw incorporation
Bioenergy production	Reducing bioethanol and biodiesel C footprint, energy crops

19.5 Potential of biochar to cut down on greenhouse effects

Biochar has gained significance in modern agricultural and environmental management since the discovery of the so-called terra preta in the Brazilian Amazon by the late Wim

Sombroek. Biochar's proven effectiveness in sequestering C in soil and lowering GHG emissions from soil (CO_2, CH_4, and N_2O) has increased its value for combating climate change and global warming. There has been a recent surge in interest in using biochar as a soil amendment due to the substance's potential to improve soil quality in various ways. These include reducing nutrient losses, increasing moisture retention, water holding capacity, hydraulic conductivity, soil aeration (Schmidt et al., 2014), soil microbial activity (Purakayastha, Kumari, & Pathak, 2015), and promoting agricultural productivity (Jones, Rousk, Edwards-Jones, DeLuca, & Murphy, 2012). Thus, applying biochar in crop fields can be an efficient strategy for active crop management and lessening the CF of specific agricultural land use. This section will mainly emphasize how biochar becomes an effective tool by enhancing SC sequestration and reducing GHG emissions from agricultural systems.

19.5.1 Biochar as a tool for sequestering C in soil

The role of SC in maintaining secure food supplies, functional ecosystems, and a thriving environment is becoming increasingly important in the face of climate change. Agricultural land use restoration, adoption of conservation tillage practices and application of biochar as a soil amendment are all considered to be diverse methods of sequestering C in the soil (Lal, 2008). Biochar is a C-rich porous substance obtained by thermo-chemically converting biomass at temperatures between 350 and 700°C in an oxygen-deficient or no oxygen environment (Amonette & Joseph, 2009; Singh et al., 2020). It has been demonstrated to increase SC levels due to its unique physicochemical properties. Biochar can store C in an inert state for very long periods in both managed and wild ecosystems; the half-life of biochar-C typically ranges from 10^2 to 10^7 years (Zimmerman, 2010; Chaturvedi, Singh, Dhyani, Govindaraju, & Mandal, 2021). Without compromising food security, habitat, or soil conservation, annual biochar production has the potential to reduce net emissions of CO_2, CH_4, and N_2O by up to 1.8 $PgCO_2$-C equivalent (CO_2-C_e) per year (12% of current anthropogenic CO_2-C_e emission), and total net emissions over a century by 130 Pg CO_2-e (Woolf, Amonette, Street-Perrott, Lehmann, & Joseph, 2010; Paustian et al., 2016). In a comparison analysis conducted by Paustian et al. (2016); multiple agriculture-based GHG mitigation practices were evaluated based on their average CO_2 removal rates from the atmosphere. Biochar application was shown to be the most successful technique there (Fig. 19.6).

Biochar's ability to mitigate climate change is primarily attributable to its highly refractory character, as dictated by feedstock and pyrolysis temperature, which slows down the pace at which C fixed by photosynthesis is released back into the atmosphere. Recent studies suggest that shorter pyrolysis time and higher pyrolysis temperature can produce higher biochar recalcitrance (Singh et al., 2020). In another dimension, in a developing country like India, a considerable amount of biomass is produced each year, and most of the surplus biomass residues are subjected to on-farm burning, accounting for 93–141 million tons (MT) each year. Among the cereal residues, 44 million tonnes of rice, followed by 24.5 MT of wheat residues, and 80% of the cotton fiber residues are burnt on-farm. This kind of malpractice releases a large amount of GHG into the atmosphere, damaging the health of the soil,

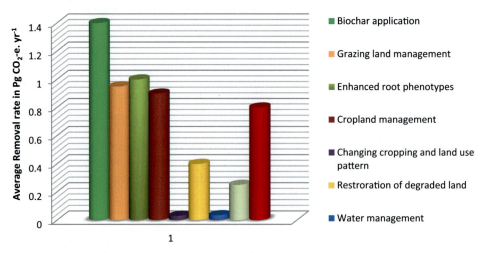

FIGURE 19.6 Average greenhouse gas removal potential of various agriculture practices. *Source: Paustian, K., Lehmann, J., Ogle, S., Reay, D., Robertson, G. P., & Smith, P. (2016). Climate-smart soils.* Nature, 532(7597), 49–57.

and taking away valuable plant nutrients in the crop residues. 70–80 MT of rice and wheat straw are burned annually in the Indian state of Punjab, releasing almost 140 MT of CO_2 into the air along with CH_4, N_2O, and other air pollutants. In this context, biochar presents a multifaceted chance to convert large-scale crop residues into profitable assets like biochar rather than costly liabilities (Punia, Nautiyal, & Kant, 2008).

19.5.2 Meta-analysis of biochar's effects on biogenic greenhouse gas fluxes

GHGs in the atmosphere trap the heat, increasing the global average surface temperature. Human activities contribute to the ongoing increases in the three most important GHGs contributing to global warming: CO_2, CH_4, and N_2O. (IPCC, 2007). There is new data to suggest that biochar can aid in mitigating global warming by lowering emissions of GHGs such as CH_4 and N_2O (Leng & Huang, 2018). It was projected by Woolf et al. (2010) that if biochar were used globally, it might offset as much as 12% of the world's current anthropogenic CO_2-C_e emissions. While biochar has the potential to reduce GHG emissions, many unanswered concerns remain, particularly concerning the nature and extent of the reduction in soil GHG emissions that will occur after its application. When applied to soil, biochar can alter soil microbiome composition and activity, soil pH, and biogeochemical processes (Singh et al., 2023b), all affecting soil GHG fluxes (Chan, Van Zwieten, Meszaros, Downie, & Joseph, 2008; Spokas & Reicosky, 2009). A rough sense of biochar's potential in GHG mitigation can be gleaned from the few metaanalysis studies that have examined the shift in GHG fluxes produced by its application. In the laboratory condition, Liu et al. (2011) found that adding biochar to waterlogged paddy soil lowered CH_4 and CO_2 emissions. This was attributed to a combination of factors, including the biochar amendment reduced N_2O emissions from pastureland and soybean soil by 80% and 50%, respectively, by reducing microbial conversion and denitrification, according to findings by Rondon, Ramirez, and Lehmann (2005), limitation of methanogen activity, an increase in pH, and a decrease in microbial biomass C. Biochar's impact on soil GHG emissions has been replicated across multiple investigations (Van Zwieten et al., 2010; Feng, Xu, Yu, Xie, & Lin, 2012). However, variability is also observed in biochar-related studies. For example, Zhang et al. (2010) reported applying biochar at a rate of 40 t ha^{-1} reduced N$_2$O emissions by 21%–28% and 10.7%–41.8% in paddy and maize fields, respectively, while increasing CH$_4$ emissions by 41% in paddy and CO$_2$ emissions by 12% in maize. In a meta-analysis study, 177 observations from 51 publications reported that significantly ($p < 0.05$), N$_2$O emissions were significantly reduced by biochar application by biochar application by 19% and 15% in both field and laboratory studies, respectively, for an average reduction of 16%. More than 76% of the field trials assessing the influence of biochar on N$_2$O emission were done for less than 0.5 years, and in these experiments, N$_2$O emissions were observed to fall significantly ($p < 0.05$) by 21% (Song et al., 2016). For CO$_2$ flux analysis, 77 observations from 31 publications were considered, which reported that Biochar dramatically reduced CO$_2$ emissions by 5% in rice fields but raised them by 12% in upland areas, according to field trials. Although biochar application shows an effect on CO$_2$ fluxes, not immediately (<0.5 years), it requires a substantial period (up to 1 year) to show its impact (Song et al., 2016). In the case of CH$_4$ flux, 42 observations of 19publications were considered, which implied that compared to nonbiochar treated controls, biochar application considerably ($p < 0.05$) increased CH$_4$ emission by 19% in the field but significantly ($p < 0.05$) reduced CH$_4$ emission by 18% in the laboratory conditions (Song et al., 2016). These results highlight that several parameters, including research locations, experiment duration, biochar application rate, biochar feedstock, and pyrolysis processes, considerably affect the influence of biochar amendment on soil GHG fluxes. Therefore, to determine whether or not the policy will successfully reduce GHG emissions, all three of the most essential GHGs must be taken into account simultaneously and then compared with the conventional situations. Understanding the potential significance of biochar in reducing global climate change is hampered by our lack of knowledge about the overall influence on three GHG emissions, reducing the predictive accuracy of models calculating reduced soil GHG emissions due to biochar application. However, biochar can be recommended for application in crop fields for its climate change mitigation abilities and its numerous positive effects, as realized from many experiments.

19.6 Related favorable impacts of biochar on climate change mitigation

19.6.1 Negative priming effect of biochar on the mineralization of soil organic carbon

The role of biochar in SOC mineralization remains unclear and contentious. Biochar application alters the SOC mineralization through "priming effects," ultimately changing the soil organic C mineralization rates. The priming impact of biochar on SOC mineralization might be either positive or negative, or it may have no effect at all. Biochar's ability to prime soil depends on factors such as the soil type, labile OM present, the biochar's characteristics, and the pyrolysis conditions (Abbruzzini et al., 2017). For instance, A larger concentration of labile OC in manure-based biochar makes it more susceptible to rapid decomposition and an eventual beneficial priming effect than biochar produced from plants (Singh and Cowie, 2014). The negative priming effect is defined as any retardation in the SOC mineralization due to adding a new substrate or inhibiting microbial activity due to specific changes in the soil environment (Zimmerman et al., 2011). The presence of the substrate is a crucial factor in native SOC decomposition. Biochar application affects substrate availability by releasing soil-labile organic C for use by soil microorganisms or by adsorption to the biochar surface. Based on the pyrolysis conditions, biochar with similar feedstock can exhibit various priming effects. A rapid increase in soil organic C due to the breakdown of labile organic C has positive priming effect of biochar application. The negative priming effect of biochar applications is caused by the material's high C content and aromatic structure, which impedes biodegradation and disintegration (Rasul et al., 2021). Another significant mechanism for the negative priming effect of biochar is the increase in soil aggregate formation caused by its application, which protects the soil's organic C from microbial decomposition (Zheng et al., 2018). In addition, the adsorption of dissolved organic carbon (DOC) into the surface of biochar reduces the decomposition. It increases the stability of soil organic C. Biochar can increase the stability of SOC through two mechanisms: encapsulating organic matter on its surface and protecting it from further degradation through sorption to the biochar's surface (Rasul et al., 2022). Another factor in biochar-induced negative priming is the release of toxic substances during biomass combustion. These include dioxins, furans, benzene, phenol, methoxy-phenol, carboxylic acid, ketone, ethane, and polyaromatic hydrocarbons. These harmful substances are microbial inhibitors that stop microbial activity and proliferation (Zhu et al., 2017). Li et al. (2014) conducted a study in North China plain to a sandy loam soil with four treatments, that is, control, biochar (0.5% of soil mass), inorganic nitrogen (100 mg kg^{-1}), and combined biochar and nitrogen. They reported that the CO_2 emissions from native SOC were reduced by 64.9%—68.8% when biochar and N amendment were combined. This shows that biochar application had a negative priming effect, inhibiting both native SOC degradation and the stimulation impact of inorganic N on native SOC degradation. Yang et al. (2022) noted the negative priming effect of biochar application in sandy clay loam soil and sandy soil. At the end of the incubation period, adding biochar that had been pyrolyzed at 300°C, 450°C, or 600°C reduced the quantity of native SOC mineralized in sandy clay loam soil by 12%, 15%, and 13%, respectively (180 days).

19.6.2 Increasing soil C stock can reduce soil surface albedo

Surface albedo is the fraction of incoming solar radiation reflected into space. The surface energy balance relies significantly on the albedo of the surface. On the Earth's surface, the net radiation flux is the total soil heat flux, sensible energy and latent energy. A disparity in the world's energy budget is the primary cause of climate change. Applying biochar sequestrate C removes CO_2 from the atmosphere, increases the SC stock, and mitigates climate change. Although there is no significant direct relationship between soil organic C and albedo, certain practices that aid in SC sequestration can affect the albedo of the soil surface. Biochar application affects the energy budget in several ways. By collecting C and keeping it out of the atmosphere, they reduce the absorption of solar irradiance. However, they also influence sun irradiance absorption through their albedo. Biochar is black organic matter, which darkens the soil color and affects the reflectance and soil temperature. Dark soil absorbs light and reflects less light. Biochar shows low reflectivity to radiation, thus reducing the surface soil albedo. Furthermore, the decrease in surface albedo is attributed to lower surface temperature. However, the knowledge and field-level experiments about the influence of biochar on soil surface albedo have not been quantified yet. Zhang et al. (2017) reported that applying biochar 45 and 5 t ha^{-1} year^{-1} decreased the surface albedo significantly compared to the control treatment in maize and wheat crops. Additionally, they noted that as the crop canopy increases the amount of surface albedo decreases and vanishes. Applying 30–60 t ha^{-1} biochar to the surface soil decreases the soil surface albedo by 40% compared to the control. Further, they reported that at early and late growing season the albedo value was recorded 0.2 and 0.3 for control treatments, whereas 30–60 t ha^{-1} biochar application recorded 0.08–0.12, that is, 40% reduction. Meyer et al. (2012) reported that biochar application can change the surface albedo and counteract climate change. They reported an annual average albedo reduction value of 0.05 due to application of 30–32 t ha^{-1} of biochar in Germany. Usowicz et al. (2016) reported the effect of biochar from wood off in grassland and fallow in the temperate climate of Poland. They found that increasing biochar application reduces the albedo value. Biochar applications at the world scale equivalent to 120 t ha^{-1} reduce negative radiative forcings by 5% for croplands, 11% for grasslands, and 23% overall (Verheijen et al., 2013).

19.6.3 Alteration in plant growth rates after biochar application

Biochar promotes plant growth and advancement in numerous ways. The recalcitrance C sources of biochar alter the plant performance, competition and growth. Biochar-amended soil can change the plant community composition by affecting seed germination rate and plant establishment (Voorde et al., 2014). Its unique properties—including its increased surface area, presence of an oxygen-containing functional group, increased CEC, and increased porosity—make it an effective tool for enhancing soil health and crop yields. However, large variation is observed in crop productivity due to biochar application as there is so much heterogeneity in biochar properties, soil, and environmental conditions. Simiele et al. (2022) conducted a study with *Solanum lycopersicum var. cerasiforme* to know the effect of biochar application (20% application rates) on plant growth, fruit yield and quality. They reported that biochar increased leaf area by 26% and 36% compared with biochar untreated plants. Additionally, the biochar increased root length, surface area, and root, stem, and leaf biomasses twofold compared to untreated plants. Biochar application also increases the fruit and flower numbers, acidity, lycopene, and solid soluble content. After pyrolysis, biochar modified with HNO_3 or $HNO_3 + H_3PO_4$ increased the concentrations of water-soluble macro and micronutrients and promoted plant growth by improving plant nutrient uptake (Sahin et al., 2017). Application of biochar to agricultural soils can greatly increase plant productivity. In a metaanalysis study, the overall yield response was estimated to be 16.0% ± 1.3% regardless of the biochar or soil circumstances. While this happened, a significant variance in plant productivity response was seen under various biochar or soil conditions, ranging from −31.8% to 974% (Dai et al., 2020). Compared to the control, the biochar amendment greatly increased the biomass in maize plant (Qiao-Hong et al., 2014). Applying biochar increased the rice and wheat root, straw and grain biomasses by 3%−19%, 10%−19%, and 10%−16%, respectively compared with nitrogen fertilizer. Further, it improves grain nutrient and physiological use efficiency, by 20%−53% and 38%−230%, respectively (Zhang et al., 2010).

19.7 Conclusion

Agricultural and food production systems support the livelihoods of millions of people around the globe. The world has achieved self-sufficiency in food production following the green revolution in the mid-1970s, while keeping this momentum has been challenging due to the increasing scarcity of natural resources, depleting the land's inherent fertility over time. To overcome the issue, the application of inorganic fertilizers and synthetic chemicals has grown rapidly in last few decades which made our food production system more GHG-intensive and as well as more responsible for global warming. In this scenario, it is necessary to adopt climate-change-adaptive farming practices that, if effective, will lessen the contribution of agriculture to climate change. According to several research studies, biochar application in agricultural fields has been demonstrated to have mitigating effects in field conditions and laboratory incubations. With the addition of biochar, both in the field and the lab, N_2O emissions were drastically reduced. Similarly, CO_2 emissions from paddy fields were reduced by a substantial quantity after biochar application, but it didn't show mitigating effects while used in upland conditions and laboratory incubations. However, biochar treatment reduced CH_4 emissions in laboratory incubations, indicating that the effect of biochar on rice fields was not completely predictable. Considering the potential for an increase or decrease in one of the three GHG fluxes as a result of biochar amendments, we find that CO_2-C equivalent outputs from rice cultivation are reduced. At the same time, emissions from upland areas are elevated. While there is solid support for promoting the use of biochar in paddy fields as a means to mitigate climate change, upland areas should approach the practice with prudence. These outcomes, nevertheless, are very context-dependent and dependent on variables such as the feedstock used to generate biochar, the pyrolysis temperature, and the application rate. Moreover, our current understanding of the impact of biochar additions on GHG Fluxes is mostly based on brief laboratory incubation, which presents a hurdle because it is unclear what would happen in the long run. Therefore, more extensive, multiregional, multibiome field investigations are needed to advocate using biochar in agricultural land to mitigate climate change and lead us toward net zero emission from agricultural production system.

References

Abbruzzini, T. F., Moreira, M. Z., de Camargo, P. B., Conz, R. F., & Cerri, C. E. P. (2017). Increasing rates of biochar application to soil induce stronger negative priming effect on soil organic carbon decomposition. *Agricultural Research, 6*, 389−398.

Aguirre-Villegas, H. A., & Larson, R. A. (2017). Evaluating greenhouse gas emissions from dairy manure management practices using survey data and lifecycle tools. *Journal of Cleaner Production, 143,* 169–179.

Amonette, J. E., & Joseph, S. (2009). Characteristics of biochar: Microchemical properties. In S. Lehmann Jand Joseph (Ed.), Biochar for environmental management. (pp. 33–52). London: Earthscan.

Arora, P., & Arora, N. K. (2023). COP27: A summit of more misses than hits. *Environmental Sustainability,,* 1–7.

Baronti, S., Alberti, G., Camin, F., Criscuoli, I., Genesio, L., Mass, R., . . . Miglietta, F. (2017). Hydrochar enhances growth of poplar for bioenergy while marginally contributing to direct soil carbon sequestration. *GCB Bioenergy, 9*(11), 1618–1626.

Bartocci, P., Bidini, G., Saputo, P., & Fantozzi, F. (2016). Biochar pellet carbon footprint. *Chemical. Engineering Transactions, 50,* 217–222.

Bellarby, J., Tirado, R., Leip, A., Weiss, F., Lesschen, J. P., & Smith, P. (2013). Livestock greenhouse gas emissions and mitigation potential in Europe. *Global Change Biology, 19*(1), 3–18.

Bhatia, A., Jain, N., & Pathak, H. (2013). Methane and nitrous oxide emissions from Indian rice paddies, agricultural soils and crop residue burning. *GHGs: Science and Technology, 3*(3), 196–211.

Carlson, K. M., Gerber, J. S., Mueller, N. D., Herrero, M., MacDonald, G. K., Brauman, K. A., ... West, P. C. (2017). Greenhouse gas emissions intensity of global croplands. *Nature Climate Change, 7*(1), 63–68.

Chan, K. Y., Van Zwieten, L., Meszaros, I., Downie, A., & Joseph, S. (2008). Using poultry litter biochars as soil amendments. *Soil Research, 46*(5), 437–444.

Chaturvedi, S., Singh, S. V., Dhyani, V. C., Govindaraju, K., & Mandal, S. (2021). Characterization, bioenergy value and thermal stability of biochar derived from diverse agriculture and forestry lignocellulosic waste. *Biomass Conversion and Biorefinery*. Available from https://doi.org/10.1007/s13399-020-01239-2.

Cordeau, S. (2022). Conservation agriculture and agroecological weed management. *Agronomy, 12*(4), 867.

Dai, Y., Zheng, H., Jiang, Z., & Xing, B. (2020). Combined effects of biochar properties and soil conditions on plant growth: A meta-analysis. *Science of the Total Environment, 713,* 136635.

Damert, M., Morris, J., & Guenther, E. (2020). Carbon footprints of organizations and products. *Responsible consumption and production,* 59–72.

Don, A., Schumacher, J., & Freibauer, A. (2011). Impact of tropical land-use change on soil organic carbon stocks–a meta-analysis. *Global Change Biology, 17*(4), 1658–1670.

Fang, K., Heijungs, R., & de Snoo, G. R. (2014). Theoretical exploration for the combination of the ecological, energy, carbon, and water footprints: Overview of a footprint family. *Ecological Indicators, 36,* 508–518.

FAO. (2009). *The state of food and agriculture. Livestock in balance.* Rome: FAO.

FAO. (2016). Climate change, agriculture and food security The State of Food and Agriculture 2106 (SOFA), Rome: FAO. <http://www.fao.org/3/i6030e/I6030E.pdf>.

Feng, Y., Xu, Y., Yu, Y., Xie, Z., & Lin, X. (2012). Mechanisms of biochar decreasing methane emission from Chinese paddy soils. *Soil Biology and Biochemistry, 46,* 80–88.

Finkbeiner, M. (2009). Carbon footprinting—opportunities and threats. *The International Journal of Life Cycle Assessment, 14,* 91–94.

Foley, S. F., Gronenborn, D., Andreae, M. O., Kadereit, J. W., Esper, J., Scholz, D., ... Crutzen, P. J. (2013). The Palaeoanthropocene–The beginnings of anthropogenic environmental change. *Anthropocene, 3,* 83–88.

Foley, J. A., Ramankutty, N., Brauman, K. A., Cassidy, E. S., Gerber, J. S., Johnston, M., ... Zaks, D. P. (2011). Solutions for a cultivated planet. *Nature, 478*(7369), 337–342.

Forabosco, F., Chitchyan, Z., & Mantovani, R. (2017). Methane, nitrous oxide emissions and mitigation strategies for livestock in developing countries: A review. *South African Journal Of Animal Science, 47*(3), 268–280.

Gan, Y., Hamel, C., O'Donovan, J. T., Cutforth, H., Zentner, R. P., Campbell, C. A., ... Poppy, L. (2015). Diversifying crop rotations with pulses enhances system productivity. *Scientific Reports, 5*(1), 14625.

Gerber, P. J., Hristov, A. N., Henderson, B., Makkar, H., Oh, J., Lee, C., ... Oosting, S. (2013). Technical options for the mitigation of direct methane and nitrous oxide emissions from livestock: A review. *Animal, 7*(s2), 220–234.

Goglio, P., Grant, B. B., Smith, W. N., Desjardins, R. L., Worth, D. E., Zentner, R., & Malhi, S. S. (2014). Impact of management strategies on the global warming potential at the cropping system level. *Science of the Total Environment, 490,* 921–933.

Hagemann, N., Joseph, S., Schmidt, H. P., Kammann, C. I., Harter, J., Borch, T., ... Kappler, A. (2017). Organic coating on biochar explains its nutrient retention and stimulation of soil fertility. *Nature Communications, 8*(1), 1089.

Hansen, J. (2004). Defusing the global warming time bomb. *Scientific American, 290*(3), 68−77.

Hertwich, E. G., & Peters, G. P. (2009). Carbon footprint of nations: A global, trade-linked analysis. *Environmental Science & Technology, 43*(16), 6414−6420.

Hoang, A. T., Nižetić, S., Olcer, A. I., Ong, H. C., Chen, W. H., Chong, C. T., ... Nguyen, X. P. (2021). Impacts of COVID-19 pandemic on the global energy system and the shift progress to renewable energy: Opportunities, challenges, and policy implications. *Energy Policy, 154*, 112322.

IPCC (2019). Special report on climate change, desertification, land degradation, sustainable land management, food security, and greenhouse gas fluxes in terrestrial ecosystems. *Summary for Policy Makers*. Cambridge: IPCC.

IPCC. (2007).In S. Solomon, D. Qin, M. Manning, Z. Chen, M. Marquis, K. B. Averyt, M. Tignor, and H. L. Miller (Eds.), *Climate change 2007: The physical science basis* (pp. 100−101).

Isbell, F., Calcagno, V., Hector, A., Connolly, J., Harpole, W. S., Reich, P. B., ... Loreau, M. (2011). High plant diversity is needed to maintain ecosystem services. *Nature, 477*(7363), 199−202.

Jain, N., Bhatia, A., & Pathak, H. (2014). Emission of air pollutants from crop residue burning in India. *Aerosol and Air Quality Research, 14*(1), 422−430.

Jones, D. L., Rousk, J., Edwards-Jones, G., DeLuca, T. H., & Murphy, D. V. (2012). Biochar-mediated changes in soil quality and plant growth in a three year field trial. *Soil biology and Biochemistry, 45*, 113−124.

Khalid, A. A., Tuffour, H. O., & Bonsu, M. (2014). Influence of poultry manure and NPK fertilizer on hydraulic properties of a sandy soil in Ghana. *International Journal of Scientific Research in Agricultural Sciences, 1*(2), 16−22.

Kumar, S., Meena, R. S., Datta, R., Verma, S. K., Yadav, G. S., Pradhan, G., Molaei, A., Rahman, G. K. M. M., & Mashuk, H. A. (2020). Legumes for carbon and nitrogen cycling: An organic approach. carbon and nitrogen cycling in soil. In R. Datta, et al. (Eds.), *Carbon and Nitrogen Cycling in Soil* (pp. 337−375). Singapore: Springer Nature, P: 337−375. Available from https://doi.org/10.1007/978-981-13-7264-3_10.

Kumar, R., Sharma, P., Gupta, R. K., Kumar, S., Sharma, M. M. M., Singh, S., & Pradhan, G. (2020a). Earthworms for eco-friendly resource efficient agriculture. In Kumar, et al. (Eds.), *Resources use efficiency in agriculture* (pp. 47−84). Singapore: Springer Nature, P. Available from https://doi.org/10.1007/978-981-15-6953-1_2.

Kumawat, A., Bamboriya, S. D., Meena, R. S., Yadav, D., Kumar, A., Kumar, S., ... Pradhan, G. (2022). Legume-based inter-cropping to achieve crop, soil, and environmental health security. In R. S. Meena, & S. Kumar (Eds.), *Advances in legumes for sustainable intensification* (pp. 307−328). Elsevier, P-. Available from https://doi.org/10.1016/B978-0-323-85797-0.00005-7.

Kweku, D. W., Bismark, O., Maxwell, A., Desmond, K. A., Danso, K. B., Oti-Mensah, E. A., ... Adormaa, B. B. (2018). Greenhouse effect: Greenhouse gases and their impact on global warming. *Journal of Scientific Research and Reports, 17*(6), 1−9.

Lal, R. (2006). Enhancing crop yields in the developing countries through restoration of the soil organic carbon pool in agricultural lands. *Land Degradation & Development, 17*(2), 197−209.

Lal, R. (2008). Carbon sequestration in soil. *CABI Reviews*, 20. (2008).

Lal, R. (2018). Digging deeper: A holistic perspective of factors affecting soil organic carbon sequestration in agroecosystems. *Global Change Biology, 24*(8), 3285−3301.

Larsen, H. N., & Hertwich, E. G. (2009). The case for consumption-based accounting of greenhouse gas emissions to promote local climate action. *Environmental Science & Policy, 12*(7), 791−798.

Lawal, H. M., & Girei, H. A. (2013). Infiltration and organic carbon pools under the long term use of farm yard manure and mineral fertilizer.

Leng, L., & Huang, H. (2018). An overview of the effect of pyrolysis process parameters on biochar stability. *Bioresource Technology, 270*, 627−642.

Lehmann, J., & Joseph, S. (Eds.) (2015). *Biochar for environmental management: Science, technology and implementation*. Routledge.

Lenzen, M., Wood, R., & Wiedmann, T. (2010). Uncertainty analysis for multi-region input−output models−a case study of the UK's carbon footprint. *Economic Systems Research, 22*(1), 43−63.

Liu, C., Cutforth, H., Chai, Q., & Gan, Y. (2016). Farming tactics to reduce the carbon footprint of crop cultivation in semiarid areas. A review. *Agronomy for Sustainable Development, 36*, 1−16.

Liu, Y., Yang, M., Wu, Y., Wang, H., Chen, Y., & Wu, W. (2011). Reducing CH4 and CO2 emissions from waterlogged paddy soil with biochar. *Journal of Soils and Sediments, 11*, 930−939.

Li, Z., Zhang, R., Wang, X., Chen, F., Lai, D., & Tian, C. (2014). Effects of plastic film mulching with drip irrigation on N_2O and CH_4 emissions from cotton fields in arid land. *The Journal of Agricultural Science, 152*(4), 534−542.

Lungarska, A., & Chakir, R. (2018). Climate-induced land use change in France: Impacts of agricultural adaptation and climate change mitigation. *Ecological Economics*, 147, 134–154.

La Scala, N., Bolonhezi, D., & Pereira, G. T. (2006). Short-term soil CO2 emission after conventional and reduced tillage of a no-till sugar cane area in southern Brazil. *Soil and Tillage Research*, 91(1–2), 244–248.

Mbow, C., Rosenzweig, C., Barioni, L.G., Benton, T.G., Herrero, M., Krishnapillai, M., ... & Xu, Y. (2019). *Food security*.

Meena, R. S., Kumawat, A., Kumar, S., Prasad, S. K., Pradhan, G., Jhariya, M. K., ... Raj, A. (2022). Effect of legumes on nitrogen economy and budgeting in South Asia. In Meena, R.S., Kumar, S. (Eds.), *Advances in legumes for sustainable intensification* (pp. 619–638). <https://doi.org/10.1016/B978-0-323-85797-0.00001-X>.

Meena, R. S., Pradhan, G., Kumar, S., & Lal, R. (2023). Using industrial wastes for rice-wheat cropping and food-energy-carbon-water-economic nexus to the sustainable food system. *Renewable and Sustainable Energy Reviews*, 187113756. Available from https://doi.org/10.1016/j.rser.2023.113756.

Meena, R. S., & Pradhan, G. (2023). Industrial garbage-derived biocompost enhances soil organic carbon fractions, CO_2 biosequestration, potential carbon credits, and sustainability index in a rice-wheat ecosystem. *Environmental Research*116525. Available from https://doi.org/10.1016/j.envres.2023.116525.

Meyer, S., Bright, R. M., Fischer, D., Schulz, H., & Glaser, B. (2012). Albedo impact on the suitability of biochar systems to mitigate global warming. *Environmental Science & Technology*, 46(22), 12726–12734.

Millar, N., Urrea, A., Kahmark, K., Shcherbak, I., Robertson, G. P., & Ortiz-Monasterio, I. (2018). Nitrous oxide (N_2O) flux responds exponentially to nitrogen fertilizer in irrigated wheat in the Yaqui Valley, Mexico. *Agriculture, Ecosystems & Environment*, 261, 125–132.

Moosmann, L., SiemonS, A., Fallasch, F., Schneider, L., Urrutia, C., Wissner, N., & Oppelt, D. (2021). The COP26 climate change conference. In *Glasgow climate change conference*, October–November.

NASA Earth Observatory (2015). *World of change: Global temperatures*.

Newell, R., Raimi, D., Villanueva, S., & Prest, B. (2021). *Global energy outlook 2021: Pathways from Paris* (p. 8) Resources for the Future.

Ozlu, E., Arriaga, F. J., Bilen, S., Gozukara, G., & Babur, E. (2022). Carbon footprint management by agricultural practices. *Biology*, 11(10), 1453.

Ozlu, E., Sandhu, S. S., Kumar, S., & Arriaga, F. J. (2019). Soil health indicators impacted by long-term cattle manure and inorganic fertilizer application in a corn-soybean rotation of South Dakota. *Scientific Reports*, 9(1), 11776.

Paris, B., Vandorou, F., Balafoutis, A. T., Vaiopoulos, K., Kyriakarakos, G., Manolakos, D., & Papadakis, G. (2022). Energy use in open-field agriculture in the EU: A critical review recommending energy efficiency measures and renewable energy sources adoption. *Renewable and Sustainable Energy Reviews*, 158, 112098.

Patra N. K., Babu S. C. (2017). *Mapping Indian agricultural emissions*. <http://www.indiaenvironmentportal.org.in/files/file/>.

Paustian, K., Lehmann, J., Ogle, S., Reay, D., Robertson, G. P., & Smith, P. (2016). Climate-smart soils. *Nature*, 532(7597), 49–57.

Permpool, N., Bonnet, S., & Gheewala, S. H. (2016). Greenhouse gas emissions from land use change due to oil palm expansion in Thailand for biodiesel production. *Journal of Cleaner Production*, 134, 532–538.

Pires, M. V., da Cunha, D. A., de Matos Carlos, S., & Costa, M. H. (2015). Nitrogen-use efficiency, nitrous oxide emissions, and cereal production in Brazil: Current trends and forecasts. *PLoS One*, 10(8), e0135234.

Pradhan, G., & Meena, R. S. (2023). Interaction impacts of biocompost on nutrient dynamics and relations with soil biota, carbon fractions index, societal value of CO_2 equivalent, and ecosystem services in the wheat-rice farming. *Chemosphere*139695. Available from https://doi.org/10.1016/j.chemosphere.2023.139695.

Pradhan, G., & Meena, R. S. (2022). Diversity in the rice–wheat system with genetically modified zinc and iron-enriched varieties to achieve nutritional Security. *Sustainability*, 14, 9334. Available from https://doi.org/10.3390/su14159334.

Pradhan, G., & Meena, R. S. (2023a). Utilizing waste compost to improve the atmospheric CO_2 capturing in the rice-wheat cropping system and energy-cum-carbon credit audit for a circular economy. *Science of the Total Environment*164572. Available from https://doi.org/10.1016/j.scitotenv.2023.164572.

Pradhan, G., Meena, R.S., Kumar, S., Jhariya, M.K., Khan, N., Shukla, U.N., ... Kumar, S. (2022). Legumes for eco-friendly weed management in an agroecosystem. In Meena, R.S., Kumar, S. (Eds.), *Advances in legumes for sustainable intensification* (pp. 133–154). https://doi.org/10.1016/B978-0-323-85797-0.00033-1.

Pradhan, G., Meena, R. S., Kumar, S., & Lal, R. (2023). Utilizing industrial wastes as compost in wheat-rice production to improve the above and below-ground ecosystem services. *Agriculture, Ecosystems and Environment, 358*108704. Available from https://doi.org/10.1016/j.agee.2023.108704.

Punia, M., Nautiyal, V. P., & Kant, Y. (2008). Identifying biomass burned patches of agriculture residue using satellite remote sensing data. *Current Science*, 1185–1190.

Purakayastha, T. J., Kumari, S., & Pathak, H. (2015). Characterisation, stability, and microbial effects of four biochars produced from crop residues. *Geoderma, 239*, 293–303.

Qiao-Hong, Z. H. U., Xin-Hua, P. E. N. G., Huang, T. Q., Zu-Bin, X. I. E., & Holden, N. M. (2014). Effect of biochar addition on maize growth and nitrogen use efficiency in acidic red soils. *Pedosphere, 24*(6), 699–708.

Raj, A., Jhariya, M. K., Banerjee, A., Meena, R. S., Nema, S., Khan, N., ... Pradhan, G. (2021). Agroforestry a model for ecological sustainability. In M. K. Jhariya, R. S. Meena, A. Raj, & S. N. Meena (Eds.), *Natural Resources Conservation and Advances for Sustainability* (pp. 289–308). Elsevier, P. Available from https://doi.org/10.1016/B978-0-12-822976-7.00002-8.

Rasool, M., Akhter, A., & Haider, M. S. (2021). Molecular and biochemical insight into biochar and Bacillus subtilis induced defense in tomatoes against Alternaria solani. *Scientia Horticulturae, 285*, 110203.

Rasul, M., Cho, J., Shin, H. S., & Hur, J. (2022). Biochar-induced priming effects in soil via modifying the status of soil organic matter and microflora: A review. *Science of the Total Environment, 805*, 150304.

Ren, F., Zhang, X., Liu, J., Sun, N., Wu, L., Li, Z., & Xu, M. (2017). A synthetic analysis of greenhouse gas emissions from manure amended agricultural soils in China. *Scientific reports, 7*(1), 8123.

Robb, D. H., MacLeod, M., Hasan, M. R., & Soto, D. (2017). Greenhouse gas emissions from aquaculture: A life cycle assessment of three Asian systems. *FAO Fisheries and Aquaculture Technical Paper* (p. 609).

Rojas-Downing, M. M., Nejadhashemi, A. P., Harrigan, T., & Woznicki, S. A. (2017). Climate change and livestock: Impacts, adaptation, and mitigation. *Climate risk management, 16*, 145–163.

Rondon, M., Ramirez, J. A., & Lehmann, J. (2005). Charcoal additions reduce net emissions of greenhouse gases to the atmosphere. In *Proceedings of the 3rd USDA Symposium on Greenhouse Gases and Carbon Sequestration in Agriculture and Forestry* (Vol. 208, pp. 21–24). Baltimore: USDA.

Roy, O., Meena, R. S., Kumar, S., Jhariya, M. K., & Pradhan, G. (2021). Assessment of land use systems for CO_2 sequestration, carbon credit potential, and income security in Vindhyan Region, India. *Land Degradation and Development, 33*(4), 670–682. Available from https://doi.org/10.1002/ldr.4181.

Sahin, O., Taskin, M. B., Kaya, E. C., Atakol, O. R. H. A. N., Emir, E., Inal, A., & Gunes, A. Y. D. I. N. (2017). Effect of acid modification of biochar on nutrient availability and maize growth in a calcareous soil. *Soil Use and Management, 33*(3), 447–456.

Sarma, B., Farooq, M., Gogoi, N., Borkotoki, B., Kataki, R., & Garg, A. (2018). Soil organic carbon dynamics in wheat-Green gram crop rotation amended with vermicompost and biochar in combination with inorganic fertilizers: A comparative study. *Journal of Cleaner Production, 201*, 471–480.

Schmidt, H. P., Kammann, C., Niggli, C., Evangelou, M. W., Mackie, K. A., & Abiven, S. (2014). Biochar and biochar-compost as soil amendments to a vineyard soil: Influences on plant growth, nutrient uptake, plant health and grape quality. *Agriculture, Ecosystems & Environment, 191*, 117–123.

Sheoran, S., Ramtekey, V., Kumar, D., Kumar, S., Meena, R. S., Kumawat, A., ... Shukla, U. N. (2022). Grain legumes: Recent advances and technological interventions. In R. S. Meena, & S. Kumar (Eds.), *Advances in legumes for sustainable intensification* (pp. 507–532). Elsevier, P-. Available from https://doi.org/10.1016/B978-0-323-85797-0.00025-2.

Simiele, M., Argentino, O., Baronti, S., Scippa, G. S., Chiatante, D., Terzaghi, M., & Montagnoli, A. (2022). Biochar enhances plant growth, fruit yield, and antioxidant content of cherry tomato (Solanum lycopersicum L.) in a soilless substrate. *Agriculture, 12*(8), 1135.

Singh, S., Chaturvedi, S., & Datta, D. (2019). Biochar: An eco-friendly residue management approach. *Indian Farming, 69*(08), 27–29.

Singh, B. P., & Cowie, A. L. (2014). Long-term influence of biochar on native organic carbon mineralisation in a low-carbon clayey soil. *Scientific reports, 4*(1), 3687.

Singh, B.R., & Singh, O. (2012). Study of impacts of global warming on climate change: Rise in sea level and disaster frequency. *Global warming—impacts and future perspective*.

Singh, S. V., Chaturvedi, S., Dhyani, V. C., & Govindaraju, K. (2020). Pyrolysis temperature influences the characteristics of rice straw and husk biochar and sorption/desorption behavior of their biourea composite. *Bioresource Technology, 314*123674. Available from https://doi.org/10.1016/j.biortech.2020.123674.

Singh, S., Chaturvedi, S., Nayak, P., Dhyani, V. C., Nandipamu, T. M., Singh, D. K., ... Govindaraju, K. (2023a). Carbon offset potential of biochar based straw management under rice–wheat system along Indo-Gangetic Plains of India. *Science of the Total Environment, 897*165176. Available from https://doi.org/10.1016/j.scitotenv.2023.165176.

Singh, S., Luthra, N., Mandal, S., Kushwaha, D. P., Pathak, S. O., Datta, D., ... Pramanik, B. (2023b). Distinct behavior of biochar modulating biogeochemistry of salt-affected and acidic soil: A review. *Journal of Soil Science and plant Nutrition*. Available from https://doi.org/10.1007/s42729-023-01370-9.

Siva, N. (2022). COP27: A "collective failure.". *The Lancet, 400*(10366), 1835.

Sivakumar, B. (2011). Global climate change and its impacts on water resources planning and management: Assessment and challenges. *Stochastic Environmental Research and Risk Assessment, 25*, 583–600.

Smith, N., & Leiserowitz, A. (2012). The rise of global warming skepticism: Exploring affective image associations in the United States over time. *Risk Analysis: An International Journal, 32*(6), 1021–1032.

Song, X., Pan, G., Zhang, C., Zhang, L., & Wang, H. (2016). Effects of biochar application on fluxes of three biogenic greenhouse gases: A meta-analysis. *Ecosystem Health and Sustainability, 2*(2), e01202.

Spokas, K. A., & Reicosky, D. C. (2009). *Impacts of sixteen different biochars on soil greenhouse gas production*.

Sundarakani, B., Goh, M., de Souza, R., & Shun, C. (2008). *Measuring carbon footprints across the supply chain*.

Tongwane, M., Mdlambuzi, T., Moeletsi, M., Tsubo, M., Mliswa, V., & Grootboom, L. (2016). Greenhouse gas emissions from different crop production and management practices in South Africa. *Environmental Development, 19*, 23–35.

Tubiello, F. N., Salvatore, M., Rossi, S., Ferrara, A., Fitton, N., & Smith, P. (2013). The FAOSTAT database of greenhouse gas emissions from agriculture. *Environmental Research Letters, 8*(1), 015009.

Usowicz, B., Lipiec, J., Łukowski, M., Marczewski, W., & Usowicz, J. (2016). The effect of biochar application on thermal properties and albedo of loess soil under grassland and fallow. *Soil and Tillage Research, 164*, 45–51.

van de Voorde, T. F., Bezemer, T. M., Van Groenigen, J. W., Jeffery, S., & Mommer, L. (2014). Soil biochar amendment in a nature restoration area: Effects on plant productivity and community composition. *Ecological Applications, 24*(5), 1167–1177.

Van Zwieten, L., Kimber, S., Morris, S., Chan, K. Y., Downie, A., Rust, J., ... Cowie, A. (2010). Effects of biochar from slow pyrolysis of papermill waste on agronomic performance and soil fertility. *Plant and Soil, 327*, 235–246.

Verheijen, F. G., Jeffery, S., van der Velde, M., Penížek, V., Beland, M., Bastos, A. C., & Keizer, J. J. (2013). Reductions in soil surface albedo as a function of biochar application rate: implications for global radiative forcing. *Environmental Research Letters, 8*(4), 044008.

Wang, J., Xi, F., Liu, Z., Bing, L., Alsaedi, A., Hayat, T., ... Guan, D. (2018). The spatiotemporal features of greenhouse gases emissions from biomass burning in China from 2000 to 2012. *Journal of Cleaner Production, 181*, 801–808.

Weidema, B. P., Thrane, M., Christensen, P., Schmidt, J., & Løkke, S. (2008). Carbon footprint: A catalyst for life cycle assessment? *Journal of industrial Ecology, 12*(1), 3–6.

Wiedmann, T. (2007). Minx. J: A definition of 'carbon footprint'. *ISA Research Report*, 07-01.

Woolf, D., Amonette, J. E., Street-Perrott, F. A., Lehmann, J., & Joseph, S. (2010). Sustainable biochar to mitigate global climate change. *Nature communications, 1*(1), 56.

Yang, X., Gao, W., Zhang, M., Chen, Y., & Sui, P. (2014). Reducing agricultural carbon footprint through diversified crop rotation systems in the North China Plain. *Journal of Cleaner Production, 76*, 131–139.

Yue, Q., Xu, X., Hillier, J., Cheng, K., & Pan, G. (2017). Mitigating greenhouse gas emissions in agriculture: From farm production to food consumption. *Journal of Cleaner Production, 149*, 1011–1019.

Zhang, A., Cui, L., Pan, G., Li, L., Hussain, Q., Zhang, X., ... Crowley, D. (2010). Effect of biochar amendment on yield and methane and nitrous oxide emissions from a rice paddy from Tai Lake plain, China. *Agriculture, Ecosystems & Environment, 139*(4), 469–475.

Zhao, Y., Wang, P., Li, J., Chen, Y., Ying, X., & Liu, S. (2009). The effects of two organic manures on soil properties and crop yields on a temperate calcareous soil under a wheat–maize cropping system. *European Journal of Agronomy, 31*(1), 36–42.

Zheng, H., Wang, X., Luo, X., Wang, Z., & Xing, B. (2018). Biochar-induced negative carbon mineralization priming effects in a coastal wetland soil: Roles of soil aggregation and microbial modulation. *Science of the Total Environment, 610*, 951–960.

Zhu, X., Chen, B., Zhu, L., & Xing, B. (2017). Effects and mechanisms of biochar-microbe interactions in soil improvement and pollution remediation: A review. *Environmental Pollution, 227*, 98–115.

Zimmerman, A. R., Gao, B., & Ahn, M. Y. (2011). Positive and negative carbon mineralization priming effects among a variety of biochar-amended soils. *Soil Biology & Biochemistry, 43*(6), 1169–1179. Available from https://doi.org/10.1016/j.soilbio.2011.02.005.

CHAPTER 20

Biochar as climate-smart strategy to address climate change mitigation and adoption in 21st century

Dipita Ghosh[1,2], Subodh Kumar Maiti[1], Sk Asraful Ali[3], Sayantika Bhattacharya[4], Tony Manoj Kumar Nandipamu[5], Biswajit Pramanick[6] and Manpreet Singh Preet[4]

[1]Department of Environmental Science & Engineering, Indian Institute of Technology (Indian School of Mines), Dhanbad, Jharkhand, India [2]Forest Operations and Biomass Utilization, Ecological Restoration Institute, Northern Arizona University, Flagstaff, AZ, United States [3]Division of Agronomy, ICAR—Indian Agricultural Research Institute, Pusa, New Delhi, India [4]School of Agriculture, Graphic Era Hill University, Dehradun, Uttarakhand, India [5]College of Agriculture, Govind Ballabh Pant University of Agriculture & Technology, Pantnagar, Uttarakhand, India [6]Department of Agronomy, College of Agriculture, Dr. Rajendra Prasad Central Agricultural University, Samastipur, Bihar, India

20.1 Introduction

"I want you to act as if your house is on fire, because it is." The famous quote by the teenage activist, Greta Thunberg sheds light on the severity of climate change and the importance of bringing about mitigative changes to the increasing global temperature. One great challenge for mankind in the 21st century is dealing with global climate issues which are causing havoc worldwide. Planet Earth is facing the worst climate issues in the form of droughts, floods, forest fires and storms which are ultimately leading to deaths, famine, loss of livelihood, pan/epidemics, and displacement of populations (Friedlingstein, Allen, Canadell, Peters, & Seneviratne, 2019). These catastrophic changes will be most severe in developing and least developed countries—precisely those countries with the least ability to adapt. This human-induced climate crisis not only affects man but all components of the ecosystem. The continuous increase in greenhouse gas (GHG) emissions, particularly carbon dioxide (CO_2), methane (CH_4), and nitrous oxide (N_2O), is driving unprecedented changes in our climate system. To address this issue, it is crucial to reduce GHG emissions and enhance the capacity of natural carbon (C) sinks which will help mitigate the rise in atmospheric CO_2 levels.

Biochar is a black charcoal-like material produced through the anoxic thermochemical conversion of biomass using pyrolysis processes (Ghosh & Maiti, 2020; Lehmann & Joseph, 2015). These processes not only generate recoverable heat and fuels like gases and condensable volatiles but also produce solid biochar. Biochar is an environmentally persistent material known for its high carbon content and low levels of oxygen and hydrogen. Biochar has garnered increasing attention for its potential to sequester carbon dioxide from the atmosphere while enhancing soil productivity and resilience. The production of biochar involves heating biomass under controlled conditions, resulting in the conversion of volatile organic compounds into stable carbon. This transformation not only captures carbon from the atmosphere but also yields a highly porous material with unique physicochemical properties that can significantly influence soil health and plant growth.

A study conducted by Ghosh and Maiti (2023) reported that the application of invasive weed biochar increased the total C-stock of a coal mine spoil by 13% and 91% at 10 and 20 t ha^{-1}, respectively. Fidel, Laird, and Parkin (2017) reported that biochar has the potential to improve the soil's inorganic carbon by 0.023–0.045 mg C kg^{-1} and organic carbon by 0.001–0.0069 mg C kg^{-1}. According to Woolf et al. (2010), estimated that biochar has the capability to reduce annual net emissions of carbon dioxide, methane, and nitrous oxide by up to 1.8 Pg CO_2-C equivalent per year, which is equivalent to 12% of current anthropogenic CO_2-C equivalent emissions. Over a century, biochar could potentially reduce total net emissions by 130 Pg CO_2-C equivalents. The remarkable aspect is that this potential can be achieved without compromising food security, habitat, or soil conservation. Furthermore, Lehmann et al. (2021) stated that biochar systems have the potential to contribute to emission reductions of 3.4–6.3 PgCO_2e (1 Pg = 1 Gt) globally, with approximately half of this reduction attributed to CO_2 removal. However, trade-offs exist between using biochar for soil sequestration and producing renewable bioenergy. The choice of replacing different energy sources with biochar is crucial. When biochar replaces coal, the emissions of biochar systems increase by 3%. In contrast, when biochar replaces renewable energy sources, emissions decrease significantly by 95%.

This chapter aims to explore the role of biochar as a climate-smart strategy to address climate change mitigation and adaptation in the 21st century. Its role in carbon sequestration, emission reduction, and the improvement of soil resilience in the face of a changing climate has been reported. Through a comprehensive review of existing research and case studies, we will delve into the potential benefits and challenges associated with widespread biochar adoption. By examining the carbon sequestration capabilities, biochar applications, waste management potential, and impact on ecosystems, we aim to provide an in-depth understanding of biochar's role in the fight against climate change. With this understanding, policymakers, scientists, and stakeholders can collaborate to harness the full potential of biochar and drive meaningful change in the pursuit of a more sustainable and climate-resilient planet.

20.2 Factors affecting the ability of biochar to fix carbon in the soil

The ability of biochar to fix carbon in the soil, also known as carbon sequestration, is influenced by various factors. Some of the key factors affecting the carbon fixation capacity of biochar include:

1. Biochar Properties: The characteristics of biochar, such as its surface area, porosity, pH, and chemical composition, can significantly impact its carbon sequestration potential. Biochar with high surface area and porosity promotes soil heath which improves the vegetation growth and microbial activities. This influnces the potential of crabon fixation of the ecosystem.
2. Feedstock Type: The source of biomass used to produce biochar can affect its carbon stability. Different feedstocks have varying amounts of recalcitrant carbon, which influences how long the carbon remains sequestered in the soil. Feedstock is one of the most important factors affecting the biochar properties and in turn, its ability to fix carbon in the soil upon application.
3. Pyrolysis Conditions: The temperature and duration of pyrolysis for biochar production influence its properties and carbon stability. Higher pyrolysis temperatures generally have high recalcitrant carbon which result in biochar with lower reactivity and higher carbon sequestration potential.
4. Soil Properties: The characteristics of the soil, such as soil texture, organic matter content, pH, and microbial activity, can affect the interactions between biochar and soil, influencing carbon sequestration.
5. Climate and Environmental Conditions: Climatic factors, such as temperature, moisture, and vegetation cover, can impact biochar-soil interactions and subsequent carbon sequestration rates.
6. Land Management Practices: The agricultural or land management practices, including the frequency of biochar application, tillage, and crop rotation, can affect the long-term stability of carbon in the soil.
7. Biochar Aging: Over time, biochar can undergo aging processes in the soil, which may alter its carbon stability and sequestration potential.
8. Microbial Activity: Soil microorganisms play a crucial role in the decomposition of organic matter, including biochar. Microbial activity can affect the turnover of carbon in the soil and the stability of biochar-derived carbon.

Understanding these factors is essential for optimizing biochar application and management strategies to enhance carbon sequestration in the soil and contribute to climate change mitigation efforts. Additionally, long-term monitoring and research are essential to assess the effectiveness and sustainability of biochar as a carbon sequestration method in different soil and environmental conditions.

20.3 Biochar and climate change

20.3.1 Biochar application for carbon stabilization and sequestration in soil

The biochar produced from agricultural, industrial biomass, forestry wastes and residues can be put into use in diverse sectors in several ways to mitigate the ill effects of climate change on the environment. The application of biochar has the potential to mitigate the agricultural GHG emissions which stand for around 18.4% of global emissions, besides being used in various other sectors viz. infrastructure and construction, energy storage, aquaculture, as insulation material and additive in asphalt (Osman et al. 2022; Ritchie, Roser, & Rosado, 2020). Agriculture is the potential sector contributing to climate change, through intensive agricultural practices, which have caused considerable soil carbon loss to the atmosphere (Lal, 2004). In the context of global warming and soil degradation, adding biochar (the solid product of organic matter pyrolysis) to soil has been considered a promising strategy to enhance long-term soil carbon sequestration through dynamic carbon stabilization and GHG mitigation (Bamminger, Poll, & Marhan, 2018; Singh et al., 2023a).

The atmospheric CO_2, the major driver for carbon sequestration is channeled through terrestrial photosynthesizing plants, algae and bacteria, upon decaying release carbon into the soil (Fig. 20.1). This way, each year, agriculture fixes 30 Giga tonnes of carbon but when the plant's decay, 30 Giga tonnes of carbon return to the atmosphere, resulting in little net change. The carbon stored in this form would retain unless disturbed, but converting the surface available crop residues into biochar and incorporating them into the soil would retain the carbon for >1000 years (Brassard et al. 2019; Chaturvedi, Singh, Dhyani, Govindaraju, & Mandal, 2021). Carbon sequestration through biochar is considered as "biomass with carbon removal and sequestration" approach, which helps to reduce emissions and lock carbon down permanently (Kuzyakov, Subbotina, Chen, Bogomolova, & Xu, 2009), making it a potential tool against climate change. Intergovernmental Panel on Climate Change (IPCC) has listed biochar as an important

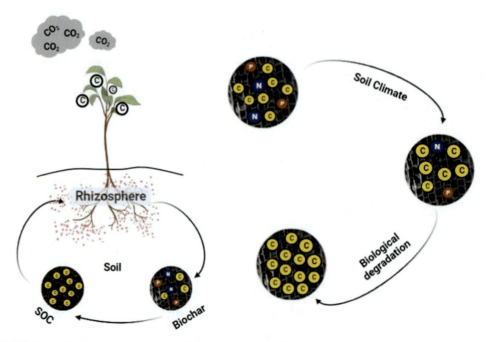

FIGURE 20.1 Carbon transformation and migration in the soil.

carbon-negative option (IPCC, 2021) and a potential driver of carbon capture, utilization, and storage technology to mitigate global climate change (Wang, Deng, Yang, Hou, & Hou, 2023). Carbon negative is the technology which reduces CO_2 instead of just offsetting it, whereas most processes today such as green energy (solar/wind) are carbon neutral, which neither release excess CO_2 nor sequester any carbon. This important factor is one of the reasons biochar stands apart from other climate change initiatives.

The higher carbon content and more stability of the biochar explain its potential to sequester the carbon for a long term in the soil in contrast to the biomass. The primary reason for the higher stability of biochar in soils is their chemical recalcitrance, that is, resistant to microbial decomposition due to the presence of aromatic bonds such as polycyclic aromatic hydrocarbons (PAHs), −OC, C−O−C functional groups (Liang et al. 2008; Singh, Chaturvedi, Dhyani, & Govindaraju, 2020; Tomczyk, Sokołowska, & Boguta, 2020). During pyrolysis, the biomass undergoes devolatilization and the solid portion gets enriched in carbon. The H and O are preferably removed over C and the H/C and O/C ratios tend to decrease as biomass undergoes its transformation into biochar. According to Krull, Baldock, Skjemstad, and Smernik (2009), decreasing the H/C ratio in biochar indicates an increasing aromatic structure in the biochar, providing enhanced stabilization than biomass. Biochar addition to soil actively stabilizes carbon through increased aggregation and physical stabilization of soil organic carbon (SOC). The pyrolysis temperature of biochar significantly influences the physical stability of soil through particle aggregation when applied to the soil (Singh et al., 2023b; Wang, Fonte, Parikh, Six, & Scow, 2017). Besides the type of feedstock, the pyrolyzing temperature, O/C ratios, and soil clay content also influence the mean residence time (MRT) of biochar in the soil. Although feedstocks pyrolyzed at high temperatures ($>800°$ C) recover less biochar with lower O/C ratio (<0.2), this biochar has a higher MRT in the soil (>1000 years). Biochar being more resistant to weathering and degradation possesses a recalcitrant carbon pool approximately 97%, whose MRT last beyond 600 years (Wang, Xiong, & Kuzyakov, 2016). Although there is still a chance that biochar can decompose in soils, it is worth noting that biochar decomposes at a much slower rate with soils that have a clay content of 40%−70%. Microbial decomposition of biochar in soil is limited due to its high stability, but it affects soil microorganisms by altering the soil environment (Zhang & Shen, 2022). Biochar enhances microbial growth, metabolic activities, and community diversity. It changes rhizosphere soil microbial communities by increasing root secretions and providing nutrients. Biochar also affects microbial abundance, activity, and community structure (Yan et al. 2022).

The addition of biochar into the soil positively influences the soil's physicochemical properties and fertility over the long run with improved soil health. Biochar tends to alter the soil's physical properties through increased surface area, and more negative surface charges to hold more nutrients and moisture and help thrive diverse microbial flora. Besides, soil incorporation of biochar improves organic carbon storage through more sorption of organic carbon from added external sources. Altered soil properties hold better moisture and nutrients and help for better crop establishment under stress conditions, ultimately increasing the assimilation of atmospheric CO_2 and influencing the soil C cycle. Converting the crop residues into biochar by preventing the open-air burning of major crop straws would help to trap the carbon for a longer period. The soil carbon sequestration potential of biochar was estimated at 0.5−2 Gt CO_2 year^{-1} by 2050 at a cost of

$30-120$ ton^{-1} of CO_2 (Smith, 2016). The broader academic literature envisions sequestration rates between 1 and 35 Gt CO_2 year^{-1} with estimates of the cumulative potential ranging from 78 to 477 Gt CO_2 by the end of this century (Fuss et al. 2018). The miraculous properties of biochar and its uses in sequestering carbon prove to be more potential to recuperate the damages caused to soil health over the years (Table 20.1). It also promotes better crop production and has a direct impact on higher yields for agriculture which has been its primary use and center of most research surrounding biochar production. And it is highly imperative towards inclusion of biochar as soil amendment for better soil health, besides mitigating the GHG emissions thereby minimizing the negative impact of agriculture on the environment.

TABLE 20.1 Mean residence time of soil-applied biochar of different feedstocks.

S. No.	Feedstock	Production	MRT (years)	References
1	Gamma grass	Pyrolysis for 3 h at 650°C	4419	Zimmerman and Gao (2013)
2	Pecan shells	Pyrolysis for 30 min at 700°C	3857	Novak et al. (2010)
3	Pine wood	Pyrolysis for 3 h at 650°C	3202	Zimmerman (2010)
4	Eucalyptus wood	Pyrolysis for 40 min at 550°C	3080	Singh, Cowie, and Smernik (2012)
5	Oak wood	Pyrolysis for 3 h at 650°C	951	Zimmerman and Gao (2013)
6	Barley roots	Pyrolysis with some air for 24 h at 500°C	893	Bruun, Clauson-Kaas, Boburská, and Thomsen (2014)
7	Cedar wood	Pyrolysis for 3 h at 650°C	605	Zimmerman (2010)
8	Corn stover	Pyrolysis for 3 h at 350°C	311	Herath et al. (2015)
9	Bubinga wood	Pyrolysis for 3 h at 400°C	269	Zimmerman (2010)
10	Poultry litter	Pyrolysis for 40 min at 400°C	239	Singh et al. (2012)
11	Cow manure	Pyrolysis for 40 min at 400°C	170	Singh et al. (2012)
12	Savana grass	Fire-derived char	130	Bird, Moyo, Veenendaal, Lloyd, and Frost (1999)
13	Miscanthus straw	Pyrolysis for 20 min at 575°C	58	Bai et al. (2013)
14	Rye straw	Pyrolysis for 2 h at 350°C	41	Hamer, Marschner, Brodowski, and Amelung (2004)
15	Corn silage	Pyrolysis for 2 h at 600°C	14	Bamminger, Marschner, and Jüschke (2014)
16	Wheat straw	Pyrolysis for 2 h at 525°C	12	Bruun, Ambus, Egsgaard, and Hauggaard-Nielsen (2012)
17	Sugarcane bagasse	Pyrolysis for 40 min at 350°C	11	Cross and Sohi (2011)

20.3.2 Indirect impacts of biochar on CO_2-equivalent emissions

20.3.2.1 Methane

Methane (CH_4) is a significant GHG with a global warming potential of 25 (IPCC, 2007). The agricultural sector, mainly rice production contributes to 50% of anthropogenic CH_4 emissions (Brevik, 2012). The total emission of CH_4 from soil is determined by the balance between CH_4 production through methanogenesis and CH_4 consumption through methanotrophic processes. Under aerobic and well-drained conditions, the soil is a net sink for CH_4. However, conditions that promote anaerobic environments, such as warm temperatures and the presence of labile carbon, favor CH_4 emissions (Lehmann & Joseph, 2015).

Biochar application can have both positive and negative effects on soil methane (CH_4) emissions. In some cases, biochar application has been reported to reduce CH_4 emissions due to its high surface area which can limit the GHG release into the atmosphere (Nguyen & Van Nguyen, 2023). Additionally, biochar can promote the growth of methanotrophic bacteria that consume CH_4, leading to lower emissions. In the study conducted by Feng, Xu, Yu, Xie, and Lin (2012), the application of biochar produced by slow pyrolysis of corn stalks at temperatures of 300°C and 500°C resulted in a significant reduction in cumulative paddy CH_4 emissions compared to control soils. Additionally, the study by Karhu, Mattila, Bergström, and Regina (2011) also reported that biochar amendment increased CH_4 uptake in soil due to improved soil aeration and increased CH_4 diffusion through the soil. According to the study, the high porosity of biochar could increase soil water-holding capacity (WHC), stabilizing fluctuations in the CH_4 flux caused by changes in water content.

Conversely, there are instances where biochar application has been associated with increased CH_4 emissions. The addition of biochar can provide a carbon-rich substrate that enhances microbial activity, including methanogenesis. Moreover, under certain conditions, biochar can alter soil aeration, water dynamics, and temperature, creating favorable conditions for methanogenesis and subsequently increasing methane emissions. In the study by Spokas and Reicosky (2009), the addition of different types of biochar to soils resulted in varied effects on CH_4 emissions. Some types of biochar emitted CH_4 when incubated alone, likely due to off-gassing from pores or surface desorption. When mixed into agricultural and landfill soils, most biochars decreased net CH_4 oxidation rates, leading to increased CH_4 production. The authors speculated that this decrease in methanotrophic activity could be due to potential inhibitors on the biochar surface or altered pH and metal toxicity. Yu, Tang, Zhang, Wu, and Gong (2013) reported that in soils with low moisture content, biochar increased CH_4 emission due to its effect on soil pH and microbial activity. In contrast, in soils with higher moisture content, biochar enhanced CH_4 emissions throughout the incubation period. The addition of organic C from biochar provided an available substrate for methanogens and created anaerobic microsites, favoring CH4 emissions.

In summary, the impact of biochar on CH_4 emissions can be complex and depends on factors such as the rate of biochar application, the nature of the biochar, soil moisture content, and microbial activity. These variables should be carefully considered when evaluating the effect of biochar amendment on CH_4 emissions in different soil and environmental conditions.

20.3.2.2 Nitrous oxide

Soil biochar amendment has been reported to help reduce N_2O emissions which is a powerful GHG with a global warming potential of 298 (IPCC, 2007). Agriculture accounts for approximately 70% of the N_2O atmospheric loading which is sometimes affected by abiotic factors but mostly is influenced by biotic or microbial factors. Soil microbial nitrification, denitrification, and nitrate ammonification are the three main mechanisms of the nitrogen cycle and the major contributors to N_2O emissions from soil (Baggs, 2011). Reports suggest that the application of biochar decreases N_2O emissions (Bruun, Müller-Stöver, Ambus, & Hauggaard-Nielsen, 2011; Kettunen & Saarnio, 2013). The observed reductions in N_2O emissions can be mainly attributed to several factors resulting from the impact of biochar on soil properties and microorganisms. These factors include changes in pH that alter the N_2O-to-N_2 ratio during denitrification, modifications in the abundance of microorganisms, increased adsorption of NH_4^+ or NO_3^-, and improvements in soil aeration and porosity, which affect soil water dynamics and lead to lower denitrification rates. Thus, the influence of biochar on microorganisms' activity is a key driver of these changes.

Biochar can retain nitrogen compounds, such as NH_4^+ and NO_3^-, as indicated by studies conducted by Kettunen and Saarnio (2013). This retention capability can furthermore influence the nitrogen cycle in the soil (Clough et al., 2010). According to Kammann, Ratering, Eckhard, and Müller (2012), the reduced N_2O/N_2 ratio is a result of biochar's adsorption of NH_4^+ or NO_3^- from initial nitrogen fertilization and soil mineralization. Angst et al. (2013) investigated the impact of biochar amendment on N_2O emissions from slurry, manure, or chemical fertilizer. They observed significantly lower N_2O emissions from samples amended with biochar compared to control samples. Their explanation involved the sorptive properties of biochar, which potentially decreased the availability of organic substrate and NH_4^+-N. Sun, Li, Chen, Wang, and Xiong (2014) also reported that biochar adsorbs free NH_4^+ through enhanced physical retention, leading to reduced N_2O emissions. In conclusion, the adsorption of ammonium onto biochar surfaces hinders nitrification and denitrification processes (Sohi, Krull, Lopez-Capel, & Bol, 2010).

The impact of biochar on soil N_2O emissions can vary depending on the specific soil type to which it is added. The results on the potential mitigation of N_2O by biochar can differ due to various factors, including environmental conditions, soil characteristics, and crop management practices (Cayuela et al., 2014). Given these factors, the outcomes of studies evaluating the effects of biochar on N_2O emissions might show variability, and the potential benefits of biochar application in reducing N_2O emissions need to be assessed within the context of specific soil types and local agricultural practices. This highlights the importance of conducting region-specific research to better understand the role of biochar in mitigating GHG emissions and promoting sustainable soil and crop management practices.

20.3.2.3 Improvement of plant yield and improved C-sequestration in plant biomass

Biochar application has been shown to enhance carbon sequestration in plant biomass, leading to increased carbon storage in the form of organic matter within plant tissues. Several mechanisms contribute to this improvement:

- Increased Nutrient Availability: Biochar amendments improve soil nutrient retention and availability. As a result, plants have enhanced access to essential nutrients such as nitrogen, phosphorus, and potassium. This nutrient availability supports more robust plant growth and greater biomass accumulation, which in turn results in higher carbon sequestration in the plant tissues.
- Enhanced Plant Growth and Root Development: Biochar promotes better root growth and development. The porous structure of biochar creates a favorable habitat for beneficial soil microbes, improving soil health and nutrient uptake. With healthier root systems, plants can access more nutrients and water, leading to increased biomass production and carbon sequestration.
- Improved Water Use Efficiency: Biochar-amended soils exhibit improved water retention and distribution. During periods of water scarcity, plants supplied with biochar-amended soil are better equipped to conserve and utilize available water, which sustains plant growth and contributes to increased carbon sequestration.
- Reduced GHG Emissions: Biochar application can indirectly contribute to improved carbon sequestration by mitigating GHG emissions, such as nitrous oxide (N_2O) and methane (CH_4), from the soil. Lower GHG emissions led to fewer carbon losses, allowing more carbon to be retained and sequestered in the plant biomass.
- Longevity of Biochar-Derived Carbon: Biochar itself is a stable form of carbon and has a slow rate of decomposition in the soil. When incorporated into the soil, biochar-derived carbon can persist for several decades or even centuries, contributing to long-term carbon sequestration in plant biomass.
- Promotion of Soil Carbon Storage: Biochar application can enhance SOC storage, creating a more favorable environment for plant growth and carbon sequestration. The increased SOC levels in the soil provide additional organic matter for plants to incorporate into their biomass.
- By improving soil fertility, nutrient availability, water use efficiency, and plant growth, biochar application creates favorable conditions for enhanced carbon sequestration in plant biomass. The combination of direct effects on plant growth and indirect effects on soil carbon dynamics makes biochar a valuable tool for promoting sustainable agriculture and climate change mitigation through increased carbon sequestration in plant biomass. However, the effectiveness of biochar application can be influenced by various factors, including biochar properties, soil type, crop type, and management practices. Thus, careful consideration and optimization are necessary to achieve the desired outcomes in specific agricultural contexts.

20.4 Impact of biochar application on carbon footprint under different landforms

Biochar, a carbon-rich material produced through the pyrolysis of biomass, has gained significant attention as a climate-smart strategy to mitigate climate change and promote sustainable development. In coarse-textured soils, biochar appears to improve crop output by increasing soil fertility and decreasing nutrient leaching (Uzoma et al., 2011; Verheijen et al., 2014). The mineralization rates of biochar are also thought to be slower than those of the original biomass, making it relatively stable in soil (Spokas, Baker, & Reicosky, 2010).

This makes biochar appealing as a method for storing carbon in addition to its ability to improve soil quality and reduce the discharge of pollutants into the environment (Clough & Condron, 2010). This chapter explores the potential of biochar as a versatile tool for climate change mitigation and adaptation in the 21st century. Furthermore, this chapter highlights the importance of biochar in promoting climate-smart agriculture, restoring degraded lands, and fostering climate resilience. Overall, biochar demonstrates great potential as a transformative solution in addressing climate change and advancing sustainable development goals in the 21st century.

20.4.1 Agricultural soil

Biochar is a good carbon sink and is recommended as an efficient countermeasure to rising GHG emissions to the environment as a soil amendment (Mukherjee & Lal, 2013). The addition of biochar can increase the stock of SOC and may be able to slow down global warming. The enormous soil disturbance brought on by human activities, according to Sapkota et al. (2017) and Singh et al. (2020), has created a soil carbon source that is partially to blame for the increase in atmospheric CO_2 concentrations. According to Minasny et al. (2018), the addition of biochar can enhance agricultural SOC by 0.4% year^{-1} on a worldwide scale, compensating for human sources' global GHG emissions. According to research by Downie, Van Zwieten, Smernik, Morris, and Munroe (2011), ancient Australian Aboriginal oven mounds with soil that had been historically infused with charcoal between 650 and 1609 years ago had significantly higher soil carbon stores than the nearby soil. According to Woolf et al. (2010), biochar can be used on a worldwide scale to possibly store up to 12% of anthropogenic GHG emissions in a system that is environmentally sustainable. Straw is a biomass that, when gasified to create biochar, contains aromatic carbon compounds that are more stable and capable of storing carbon than the original feedstock in modified soil (Hansen et al., 2016). In addition to slowing down the degradation of SOC, biochar can significantly adsorb soil-dissolved carbon (Lu et al., 2014). Hailegnaw, Mercl, Pračke, Száková, and Tlustoš (2019) also observed a decrease in nitrate and dissolved organic carbon in soil treated with wood chip biochar.

Biochar is being used more frequently as a cost-efficient method of sequestering carbon in efforts to combat the problem of global climate change. Due to increased physical protection of soil aggregation, Zheng, Wang, Luo, Wang, and Xing (2018) demonstrated that biochar amendment greatly helped SOC stability in coastal soil and discovered a biochar-induced negative priming effect. As a result, the distribution of biochar particles throughout the gradient of soil particle sizes may have a significant impact on how much carbon is stored in the soil. Additionally, using biochar may cause soil to aggregate more, which is anticipated to result in native SOC having lower bioavailability mostly due to the preservation of soil aggregates (Du, Zhao, Wang, & Zhang, 2017). Applying biochar constantly over a 10-year field trial in Shandong Province, China, caused the soil's inorganic carbon to also contribute to soil carbon sequestration. There are three biochar treatments: no biochar (Control), 4.5 Mg ha^{-1} year^{-1} of biochar application (B4.5), and 9.0 Mg ha^{-1} year^{-1} of biochar application (B9.0), respectively. In comparison to the no-biochar control, the results showed that applying biochar considerably increased the contents of soil inorganic carbon (3.2%–24.3%), particulate organic carbon (POC, 38.2%–166.2%), and total SOC (15.8%–82.2%). Following the application of biochar, SIC shows that carbon was sequestered and that the applied biochar was primarily assigned to the POC fraction. Additionally, the field observations of decreased SOC and increased microbial biomass carbon levels imply that the application of biochar may have a beneficial priming effect on native SOC.

20.4.2 Mining spoil

Due to the devastation of flora and soil structure caused by open-cast coal mining, the carbon sink is completely lost. The establishment of plant cover is severely constrained by the poor organic matter and WHC of coal mine debris (Maiti, 2013). Surface mining increases carbon dioxide (CO_2) emissions on a global scale due to deforestation and land degradation (Ahirwal & Maiti, 2017; Shrestha & Lal, 2006). Restoration of postmining sites can increase the amount of time that atmospheric carbon (C) is stored in the soil and biomass. Global implementation of revegetation combined with biochar amelioration as a method of effective climate change mitigation is possible (Majumder, Neogi, Dutta, Powel, & Banik, 2019; Norini et al., 2019). Biochar is particularly resilient in nature due to its high MRT and aromaticity. In contrast to the control (4.92 mol CO_2 m^{-2} s^{-1}), *Lantana camara* biochar reduced mine spoil CO_2 flux to 3% (2.60 mol CO_2 m^{-2} s^{-1}) and 2% (2.85 mol CO_2 m^{-2} s^{-1}), according to a study by Ghosh and Maiti (2021) About 3.92 Mg C ha^{-1} $year^{-1}$ was determined to be the rate of C buildup for mining spoil that has been reclaimed for 8 years. The use of biochar boosted the inorganic and biogenic carbon pool, particularly the recalcitrant pool, indicating that biochar can be a useful method of enhanced carbon fixation in the spoil, coupled with plantation activities. The C-stock grew by 36%–42% in just 6 months of treatment, allowing the mining spoil's stubborn carbon content to be fixed for a longer time. This demonstrates that biochar, together with forestry-based rehabilitation of coal mine debris, has significant potential for fixing carbon. As a result, carbon stock rises with reclamation age, and adding biochar can raise it almost to the level of a reference forest site.

20.4.3 Agroforestry

The capacity of Indian agroforestry systems to sequester carbon ranges from 0.003 to 3.98 Mg C ha^{-1} $year^{-1}$ in soil to 0.25 to 19.14 Mg C ha^{-1} $year^{-1}$ in biomass, according to Dhyani, Ram, and Dev (2016). Agroforestry's ability to store biomass carbon over time is influenced by rotating harvests. Many agroforestry tree species could be harvested after 5-60 years, depending on their use (Chaturvedi et al. 2017). According to the data, the agroforestry system may store carbon for 5–60 years before gradually releasing it to the atmosphere after tree harvesting. How long the C-clocking period lasts, however, is also influenced by the use of materials like pulp, paper, wood for fuel, and lumber.

20.4.4 Forestry

In the cycling of carbon in terrestrial ecosystems and the reduction of global warming, forests are essential. Intensive forest management and global climate change have had a significant negative impact on the quality of forest soils through soil acidity, a decrease in the amount of organic carbon in the soil, a deterioration in the biological features of the soil, and a loss in soil biodiversity. Increasing SOC concentrations by incorporating biochar into the soil may be a valuable tactic in intensively managed ecosystems to prevent SOC depletion. The mechanisms underpinning the effects of putting biochar on soil CO_2 emissions can be summarized by the following four processes: (1) After being applied to soils, the labile organic carbon in the biochar adds to the soil's pool of labile organic carbon, enhancing soil CO_2 emissions (Yoo & Kang, 2012; Mukherjee and Lal, 2013; Spokas, 2013). (2) Biochar can lower soil surface CO_2 emissions by absorbing soil CO_2 molecules due to its high adsorption ability. (3) According to Liang et al. (2010) and Jones et al. (2011), the addition of biochar to soil indirectly changes its water content, porosity, aggregation, pH, and CEC. (4) The variety and activity of microbial taxa involved in CO_2 production can be greatly impacted by the use of biochar (Liu et al. 2011; Liu, Liu, Show, & Tay, 2009; Mitchell, Simpson, Soong, & Simpson, 2015; Wang et al. 2016; Zhang et al. 2012).

However, there are significant differences between research on how applying biochar affects soil CO_2 fluxes in forest ecosystems. CO_2 emissions either increased, decreased, or remained unchanged when biochar was added to forest soils. In temperate forest soil, Mitchell et al. (2015) found that applying sugar maple biochar increased soil CO_2 emissions by 5, 10, and $20 \, t \, ha^{-1}$. In comparison to untreated control soils, soils altered with biochar released higher CO_2 emissions. This was reinforced by the findings of Hawthorne et al. (2017), who discovered that CO_2 fluxes from Douglas-fir Forest soil treated with 10% biochar were much higher than those from the same soil treated with 1% biochar.

Despite these potential benefits, it is important to note that the effectiveness of biochar in carbon sequestration can vary depending on several factors, including the type of biochar, soil type, climate conditions, and management practices (Table 20.2). To maximize its benefits, the appropriate application rate and integration with existing agricultural practices should be carefully considered.

20.5 Biochar-based carbon crediting

Carbon credits are a tradable commodity that represents a certain amount of carbon dioxide equivalent (CO_2e) emissions reduced or sequestered. They are often used as a tool to incentivize actions that mitigate climate change. Organizations or individuals that reduce their carbon emissions or sequester carbon can earn carbon credits, which can then be sold or used to offset their own emissions. The concept of biochar-based carbon crediting involves utilizing biochar as a tool to sequester carbon from the atmosphere and then receiving carbon credits for carbon stored in the biochar. This can provide financial incentives for individuals, farmers, or businesses to adopt biochar practices, which in turn can lead to increased adoption of sustainable land management practices and reduced GHG emissions. It's important to note that the effectiveness of biochar-based carbon crediting depends on factors such as the type of biomass used for pyrolysis, the production process, the characteristics of the soil in which biochar is applied, and the monitoring and verification mechanisms in place to ensure that the carbon remains sequestered over time.

TABLE 20.2 Potential of biochar in enhancing carbon sequestration in different land use systems.

S. No.	Biochar Feedstock	Pyrolysis temp.	Rate of application	System	Response	References
1.	Wheat straw	350°C–550°C	10–40 t ha^{-1}	Rice-based cropping system	Soil organic carbon (SOC) storage @ 2–12 observed in paddy field of China	Zhang et al. (2010)
2.	Rice husk	400°C	41 t ha^{-1}	Rice-based cropping system	Net accumulation of 0.8 g kg^{-1} of SOC observed in the rice growing soils of Phillipines	Haefele et al. (2011)
3.	Forest wood	500°C	11 t ha^{-1}	Sorghum	SOC storage @ 2 g kg^{-1}	Steiner et al. (2007)
4.	Cow manure	550°C	10–20 t ha^{-1}	Maize	SOC storage @ 4.3 g kg^{-1}	Uzoma et al. (2011)
5.	Corn stalks	400°C	16–32 t ha^{-1}	Soybean-wheat crop rotation	SOC accumulation of 8.9–13.8 g kg^{-1}	Lin et al. (2015)
6.	Winter wheat straw	700°C–750°C	5% (w/w), i.e., corresponding to 100 t ha^{-1}	Incubation study	Relatively lower CO_2 emissions observed after incubation study and Total organic carbon remain stabled through-out the incubation period	Hansen et al. (2016)
7.	Bamboo Leaf	500°C	5 t ha^{-1}	Chinese chestnut forest	SOC storage in chestnut plantations increased by 9.4% in 1 year after the application of biochar.	Singh et al. (2020)
8.	Apple branches	550°C	20 t ha^{-1}	Apple Orchard	SOC sequestration rate of apple orchard after two years of experiment was 5.23 t ha^{-1} year^{-1}	Han et al. (2021)
9.	Switch grass	550°C	15 t ha^{-1}	Grassland	Net carbon storage of 7.5 t ha^{-1} has been recorded	Liu et al. (2021)
10.	Wood chips	500°C	10 t ha^{-1}	Wheat cropping system	Biochar-amended soils showed a 50% increase in carbon sequestration over a 10-year period.	Lehmann et al. (2021)
11.	Rice husk	400°C	8 t ha^{-1}	Rice	Enhanced carbon sequestration in paddy soils with a net carbon storage of 6.1 t ha^{-1}	Wang, Pan, Liu, Zhang, and Xiong (2019)

(Continued)

TABLE 20.2 (Continued)

S. No.	Biochar Feedstock	Pyrolysis temp.	Rate of application	System	Response	References
12.	Miscanthus	450°C	7 t ha^{-1}	Energy crop	A net carbon storage of 6.5 t ha^{-1} has been observed	Zhang, Li, Xing, and Xu (2020)
13.	Apple branches	450°C	20 t ha^{-1}	Winter wheat	adding biochar to subsurface soil (10–20 cm) can effectively avoid severe disturbance of the surface soil environment and thus benefit soil carbon sequestration in the long term.	Smith (2016)
14.	Invasive weed	450°C	10–20 t ha^{-1}	Reclaimed mine spoil	the total C-stock increased by 13% and 91% at 10 and 20 t ha^{-1}.	Ghosh and Maiti (2023)
15.	Rice straw	500°C for 3–4 min	0, –8 Mg ha^{-1}	semiarid subtropical agricultural soil	Total organic C stocks after 5 years were found to be almost double the amount of biochar C added.	Mavi et al. (2023)

However, while the concept has potential, there are also challenges to consider, such as accurately quantifying carbon sequestration, addressing potential unintended environmental impacts, and establishing robust methodologies for monitoring and verifying the carbon stored in biochar. Biochar-based carbon crediting is an evolving concept and its implementation and recognition in various carbon credit markets is growing each day.

20.6 Conclusions and future recommendations

Continued research and development are essential for deepening our understanding of biochar's potential as a carbon sequestration tool. Through scientific investigations, we can gain insights into the optimal feedstock selection, pyrolysis conditions, and application techniques to maximize carbon sequestration efficiency. This knowledge helps refine biochar production processes and improves its effectiveness in mitigating climate change. Research and development efforts enable the optimization of biochar properties to enhance its carbon sequestration potential. This includes improving biochar stability, porosity, surface area, and nutrient retention capacity. Understanding the relationships between biochar production parameters and their physical and chemical properties allows for targeted modifications to tailor biochar characteristics for specific soil types and environmental conditions. Disseminating knowledge about biochar and its carbon sequestration potential to diverse stakeholders, including farmers, landowners, policymakers, and the public creates awareness which in turn enhances the understanding and acceptance of biochar as a climate change mitigation strategy, facilitating its wider adoption and implementation. In conclusion, research, development, and implementation efforts are critical for realizing the full potential of biochar as a tool for carbon sequestration. These efforts advance our knowledge, refine biochar production and application techniques, and support the integration of biochar into climate change mitigation and land management strategies.

References

Ahirwal, J., & Maiti, S. K. (2017). Assessment of carbon sequestration potential of revegetated coal mine overburden dumps: A chronosequence study from dry tropical climate. *Journal of Environmental Management*, 201, 369–377.

Angst, T. E., Patterson, C. J., Reay, D. S., Anderson, P., Peshkur, T. A., & Sohi, S. P. (2013). Biochar diminishes nitrous oxide and nitrate leaching from diverse nutrient sources. *Journal of Environmental Quality*, 42(3), 672–682.

Baggs, E. M. (2011). Soil microbial sources of nitrous oxide: Recent advances in knowledge, emerging challenges and future direction. *Current Opinion in Environmental Sustainability*, 3(5), 321–327.

Bai, M., Wilske, B., Buegger, F., Esperschütz, J., Kammann, C. I., Eckhardt, C., ... Breuer, L. (2013). Degradation kinetics of biochar from pyrolysis and hydrothermal carbonization in temperate soils. *Plant and Soil*, 372, 375–387.

Bamminger, C., Poll, C., & Marhan, S. (2018). Offsetting global warming-induced elevated greenhouse gas emissions from an arable soil by biochar application. *Glob Chang Biol*, 24, e318–e334.

Bamminger, C., Marschner, B., & Jüschke, E. (2014). An incubation study on the stability and biological effects of pyrogenic and hydrothermal biochar in two soils. *European Journal of Soil Science*, 65, 72–82.

Bird, M. I., Moyo, C., Veenendaal, E. M., Lloyd, J., & Frost, P. (1999). Stability of elemental carbon in a savanna soil. *Global Biogeochemical Cycles, 13*(4), 923−932.

Brassard, P., Godbout, S., Lévesque, V., Palacios, J. H., Raghavan, V., Ahmed, A., ... Verma, M. (2019). Biochar for soil amendment. *Char and Carbon Materials Derived from Biomass,* 109−146.

Brevik, E. C. (2012). Soils and climate change: Gas fluxes and soil processes. *Soil Horizons, 53*(4), 12−23.

Bruun, E. W., Ambus, P., Egsgaard, H., & Hauggaard-Nielsen, H. (2012). Effects of slow and fast pyrolysis biochar on soil C and N turnover dynamics. *Soil Biology and Biochemistry, 46,* 73−79.

Bruun, E. W., Müller-Stöver, D., Ambus, P., & Hauggaard-Nielsen, H. (2011). Application of biochar to soil and N2O emissions: Potential effects of blending fast-pyrolysis biochar with anaerobically digested slurry. *European Journal of Soil Science, 62*(4), 581−589.

Bruun, S., Clauson-Kaas, S., Bobuľská, L., & Thomsen, I. K. (2014). Carbon dioxide emissions from biochar in soil: Role of clay, microorganisms and carbonates. *European Journal of Soil Science, 65*(1), 52−59.

Cayuela, M. L., Van Zwieten, L., Singh, B. P., Jeffery, S., Roig, A., & Sánchez-Monedero, M. A. (2014). Biochar's role in mitigating soil nitrous oxide emissions: A review and meta-analysis. *Agriculture, Ecosystems & Environment, 191,* 5−16.

Chaturvedi, S., Singh, S. V., Dhyani, V. C., Govindaraju, K., & Mandal, S. (2021). Characterization, bioenergy value and thermal stability of biochar derived from diverse agriculture and forestry lignocellulosic waste. *Biomass Conversion and Biorefinery.* Available from https://doi.org/10.1007/s13399-020-01239-2.

Chaturvedi, O.P., Handa, A.K., Uthappa, A.R., Sridhar, K.B., Kumar, N., Chavan, S.B., & Rizvi, J. (2017). Promising agroforestry tree species in India. Central Agroforestry Research Institute: Jhansi, India; South Asia Regional Programme of the World Agroforestry Research Centre: New Delhi, India, pp. 1−190.

Clough, T. J., & Condron, L. M. (2010). Biochar and the nitrogen cycle: Introduction. *Journal of Environmental Quality, 39*(4), 1218−1223.

Clough, T. J., Bertram, J. E., Ray, J. L., Condron, L. M., O'callaghan, M., Sherlock, R. R., & Wells, N. S. (2010). Unweathered wood biochar impact on nitrous oxide emissions from a bovine-urine-amended pasture soil. *Soil Science Society of America Journal, 74*(3), 852−860.

Cross, A., & Sohi, S. P. (2011). The priming potential of biochar products in relation to labile carbon contents and soil organic matter status. *Soil Biology and Biochemistry, 43*(10), 2127−2134.

Dhyani, S. K., Ram, A., & Dev, I. (2016). Potential of agroforestry systems in carbon sequestration in India. *Indian Journal of Agriculture Science, 86,* 1103−1112.

Downie, A. E., Van Zwieten, L., Smernik, R. J., Morris, S., & Munroe, P. R. (2011). Terra preta Australis: Reassessing the carbon storage capacity of temperate soils. *Agriculture, Ecosystems and Environment, 140,* 137e147.

Du, Z., Zhao, J., Wang, Y., & Zhang, Q. (2017). Biochar addition drives soil aggregation and carbon sequestration in aggregate fractions from an intensive agricultural system. *Journal of Soils and Sediments, 17,* 581−589.

Feng, Y., Xu, Y., Yu, Y., Xie, Z., & Lin, X. (2012). Mechanisms of biochar decreasing methane emission from Chinese paddy soils. *Soil Biology and Biochemistry, 46,* 80−88.

Fidel, R. B., Laird, D. A., & Parkin, T. B. (2017). Impact of biochar organic and inorganic carbon on soil CO_2 and N_2O emissions. *Journal of Environmental Quality, 46*(3), 505−513.

Friedlingstein, P., Allen, M., Canadell, J. G., Peters, G. P., & Seneviratne, S. I. (2019). Comment on "The global tree restoration potential.". *Science (New York, N.Y.), 366*(6463), eaay8060.

Fuss, S., Lamb, W. F., Callaghan, M. W., Hilaire, J., Creutzig, F., Amann, T., ... Minx, J. C. (2018). Negative emissions—Part 2: Costs, potentials and side effects. *Environmental Research Letters, 13*(6), 063002.

Ghosh, D., & Maiti, S. K. (2020). Can biochar reclaim coal mine spoil? *Journal of Environmental Management, 272,* 111097.

Ghosh, D., & Maiti, S. K. (2021). Effect of invasive weed biochar amendment on soil enzymatic activity and respiration of coal mine spoil: A laboratory experiment study. *Biochar, 3*(4), 519−533.

Ghosh, D., & Maiti, S. K. (2023). Invasive weed-based biochar facilitated the restoration of coal mine degraded land by modulating the enzyme activity and carbon sequestration. *Restoration Ecology, 31*(3), e13744.

Haefele, S. M., Konboon, Y., Wongboon, W., Amarante, S., Maarifat, A. A., Pfeiffer, E. M., & Knoblauch, C. J. F. C. R. (2011). Effects and fate of biochar from rice residues in rice-based systems. *Field Crops Research, 121*(3), 430−440.

Hailegnaw, Niguss Solomon, Mercl, Filip, Pračke, Kateřina, Száková, Ji. řina, & Tlustoš, Pavel (2019). High temperature-produced biochar can be efficient in nitrate loss prevention and carbon sequestration. *Geoderma, 338,* 48−55.

Hamer, U., Marschner, B., Brodowski, S., & Amelung, W. (2004). Interactive priming of black carbon and glucose mineralisation. *Organic Geochemistry*, 35(7), 823–830.

Han, J., Zhang, A., Kang, Y., Han, J., Bo, Y., Hussain, Q., . . . Khan, M. A. (2021). Biochar promotes soil organic carbon sequestration and reduces net global warming potential in apple orchard: A two-year study in the Loess Plateau of China. *Science of The Total Environment*, 803, 150035.

Hansen, V., Müller-Stöver, D., Munkholm, L. J., Peltre, C., Hauggaard-Nielsen, H., & Jensen, L. S. (2016). The effect of straw and wood gasification biochar on carbon sequestration, selected soil fertility indicators and functional groups in soil: An incubation study. *Geoderma*, 269, 99–107.

Hawthorne, I., Johnson, M. M. S., Jassal, R. S., Black, T. A., Grant, N. J., & Smukler, S. M. (2017). Application of biochar and nitrogen influences fluxes of CO_2, CH_4 and N_2O in a forest soil. *Journal of Environment Management*, 192, 203–214.

Herath, H. M. S. K., Camps-Arbestain, M., Hedley, M. J., Kirschbaum, M. U. F., Wang, T., & Van Hale, R. (2015). Experimental evidence for sequestering C with biochar by avoidance of CO_2 emissions from original feedstock and protection of native soil organic matter. *Gcb Bioenergy*, 7(3), 512–526.

IPCC. (2007). IPCC fourth assessment report: Climate change 2007. http://www.ipcc.ch/publications_and_data/ar4/wg1/en/ch2s2-10-2.html. Page consulted on October 2014.

IPCC. (2021). Climate change 2021: The physical science basis. IPCC Contribution of Working Group I to the Sixth Assessment Report of the Intergovernmental Panel on Climate Change. https://www.ipcc.ch/report/sixth-assessment-report-working-group-i/

Jones, D. L., Murphy, D. V., Khalid, M., Ahmad, W., Edwards-Jones, G., & DeLuca, T. H. (2011). Short-term biochar-induced increase in soil CO_2 release is both biotically and abiotically mediated. *Soil Biology and Biochemistry*, 43(8), 1723–1731.

Kammann, C., Ratering, S., Eckhard, C., & Müller, C. (2012). Biochar and hydrochar effects on greenhouse gas (carbon dioxide, nitrous oxide, and methane) fluxes from soils. *Journal of Environmental Quality*, 41(4), 1052–1066.

Karhu, K., Mattila, T., Bergström, I., & Regina, K. (2011). Biochar addition to agricultural soil increased CH4 uptake and water holding capacity—Results from a short-term pilot field study. *Agriculture, Ecosystems & Environment*, 140(1–2), 309–313.

Kettunen, R., & Saarnio, S. (2013). Biochar can restrict N_2O emissions and the risk of nitrogen leaching from an agricultural soil during the freeze-thaw period. *Agricultural and Food Science*, 22(4), 373–379.

Krull, E. S., Baldock, J. A., Skjemstad, J. O., & Smernik, R. J. (2009). Characteristics of biochar: Organochemical properties. In J. Lehmann, & S. Joseph (Eds.), *Biochar for environmental management: Science and technology* (pp. 53–66). Sterling/London: Earthscan.

Kuzyakov, Y., Subbotina, I., Chen, H., Bogomolova, I., & Xu, X. (2009). Black carbon decomposition and incorporation into soil microbial biomass estimated by 14C labeling. *Soil Biology and Biochemistry*, 41(2), 210–219.

Lal, R. (2004). Soil carbon sequestration to mitigate climate change. *Geoderma*, 123, 1–22.

Lehmann, J., & Joseph, S. (2015). *Biochar for environmental management: An introduction. Biochar for environmental management* (pp. 1–13). Routledge.

Lehmann, J., Cowie, A., Masiello, C. A., Kammann, C., Woolf, D., Amonette, J. E., . . . Whitman, T. (2021). Biochar in climate change mitigation. *Nature Geoscience*, 14(12), 883–892.

Liang, B., Lehmann, J., Sohi, S. P., Thies, J. E., O'Neill, B., Trujillo, L., . . . Luizão, F. J. (2010). Black carbon affects the cycling of non-black carbon in soil. *Organic Geochemistry*, 41(2), 206–213.

Liang, B., Lehmann, J., Solomon, D., Sohi, S., Thies, J. E., Skjemstad, J. O., . . . Wirick, S. (2008). Stability of biomass-derived black carbon in soils. *Geochimica et Cosmochimica Acta*, 72(24), 6069–6078.

Lin, X. W., Xie, Z. B., Zheng, J. Y., Liu, Q., Bei, Q. C., & Zhu, J. G. (2015). Effects of biochar application on greenhouse gas emissions, carbon sequestration and crop growth in coastal saline soil. *European Journal of Soil Science*, 66(2), 329–338.

Liu, Q. S., Liu, Y., Show, K. Y., & Tay, J. H. (2009). Toxicity effect of phenol on aerobic granules. *Environmental Technology*, 30(1), 69–74.

Liu, X., Wang, Z., Zheng, K., Han, C., Li, L., Sheng, H., & Ma, Z. (2021). Changes in soil carbon and nitrogen stocks following degradation of alpine grasslands on the Qinghai-Tibetan Plateau: A meta-analysis. *Land Degradation & Development*, 32(3), 1262–1273.

Liu, Y., Yang, M., Wu, Y., Wang, H., Chen, Y., & Wu, W. (2011). Reducing CH_4 and CO_2 emissions from waterlogged paddy soil with biochar. *Journal of Soils and Sediments, 11*, 930–939.

Lu, Weiwei, Ding, Weixin, Zhang, Junhua, Li, Yi, Luo, Jiafa, Bolan, Nanthi, & Xie, Zubin (2014). Biochar suppressed the decomposition of organic carbon in a cultivated sandy loam soil: A negative priming effect. *Soil Biology and Biochemistry, 76*, 12–21.

Maiti, S. K. (2013). *Ecorestoration of coal mine degraded lands*. New Delhi: Springer. Available from https://doi.org/10.1007/978-81-322-0851-8.

Majumder, S., Neogi, S., Dutta, T., Powel, M. A., & Banik, P. (2019). The impact of biochar on soil carbon sequestration: Meta-analytical approach to evaluating environmental and economic advantages. *Journal of Environmental Management, 250*, 109466.

Mavi, M. S., Singh, G., Choudhary, O. P., Singh, A., Vashisht, B. B., Sekhon, K. S., ... Singh, B. (2023). Successive addition of rice straw biochar enhances carbon accumulation in soil irrigated with saline or non-saline water. *Environmental Research, 217*, 114733.

Minasny, B., Arrouays, D., McBratney, A.B., Angers, D.A., Chambers, A., Chaplot, V., ... Winowiecki, L. (2018). Rejoinder to comments on Minasny et al., 2017. Soil carbon 4 per mille. *Geoderma, 292*, 59–86; *Geoderma, 309*, 124–129.

Mitchell, P. J., Simpson, A. J., Soong, R., & Simpson, M. J. (2015). Shifts in microbial community and waterextractable organic matter composition with biochar amendment in a temperate forest soil. *Soil Biology & Biochemistry, 81*, 244–254.

Mukherjee, A., & Lal, R. (2013). Biochar impacts on soil physical properties and greenhouse gas emissions. *Agronomy, 3*(2), 313–339.

Nguyen, B. T., & Van Nguyen, N. (2023). Biochar addition balanced methane emissions and rice growth by enhancing the quality of paddy soil. *Journal of Soil Science and Plant Nutrition*, 1–12.

Norini, M. P., Thouin, H., Miard, F., Battaglia-Brunet, F., Gautret, P., Guégan, R., ... Motelica-Heino, M. (2019). Mobility of Pb, Zn, Ba, As and Cd toward soil pore water and plants (willow and ryegrass) from a mine soil amended with biochar. *Journal of Environmental Management, 232*, 117–130.

Novak, J. M., Busscher, W. J., Watts, D. W., Laird, D. A., Ahmedna, M. A., & Niandou, M. A. (2010). Short-term CO_2 mineralization after additions of biochar and switchgrass to a typic Kandiudult. *Geoderma, 154*(3–4), 281–288.

Osman, A. I., Fawzy, S., Farghali, M., El-Azazy, M., Elgarahy, A. M., Fahim, R. A., ... Rooney, D. W. (2022). Biochar for agronomy, animal farming, anaerobic digestion, composting, water treatment, soil remediation, construction, energy storage, and carbon sequestration: A review. *Environmental Chemistry Letters, 20*(4), 2385–2485.

Ritchie, H., Roser, M., & Rosado, P. (2020). CO_2 and greenhouse gas emissions. *Our World in Data*. Available from https://ourworldindata.org/co2-and-greenhouse-gas-emissions.

Sapkota, T. B., Jat, R. K., Singh, R. G., Jat, M. L., Stirling, C. M., Jat, M. K., ... Gupta, R. K. (2017). Soil organic carbon changes after seven yearsof conservation agriculture in a rice-wheat system of the eastern Indo-Gangetic Plains. *Soil Use and Management, 33*, 81–89.

Shrestha, R. K., & Lal, R. (2006). Ecosystem carbon budgeting and soil carbon sequestration in reclaimed mine soil. *Environment International, 32*(6), 781–796.

Singh, S. V., Chaturvedi, S., Dhyani, V. C., & Govindaraju, K. (2020). Pyrolysis temperature influences the characteristics of rice straw and husk biochar and sorption/desorption behavior of their biourea composite. *Bioresource Technology, 314*, 123674.

Singh, S., Chaturvedi, S., Nayak, P., Dhyani, V. C., Nandipamu, T. M., Singh, D. K., ... Govindaraju, K. (2023a). Carbon offset potential of biochar based straw management under rice – wheat system along Indo-Gangetic Plains of India. *Science of the Total Environment, 897*, 165176.

Singh, S., Luthra, N., Mandal, S., Kushwaha, D. P., Pathak, S. O., Datta, D., ... Pramanik, B. (2023b). Distinct behavior of biochar modulating biogeochemistry of salt-affected and acidic soil: A review. *Journal of Soil Science and Plant Nutrition*. Available from https://doi.org/10.1007/s42729-023-01370-9.

Singh, B. P., Cowie, A. L., & Smernik, R. J. (2012). Biochar carbon stability in a clayey soil as a function of feedstock and pyrolysis temperature. *Environmental Science & Technology, 46*(21), 11770–11778.

Smith, P. (2016). Soil carbon sequestration and biochar as negative emission technologies. *Global Change Biology, 22*(3), 1315–1324.

Sohi, S. P., Krull, E., Lopez-Capel, E., & Bol, R. (2010). A review of biochar and its use and function in soil. *Advances in Agronomy, 105*, 47–82.

Spokas, K. A. (2013). Impact of biochar field aging on laboratory greenhouse gas production potentials. *GCB Bioenergy, 5*(2), 165–176.

Spokas, K. A., Baker, J. M., & Reicosky, D. C. (2010). Ethylene: Potential key for biochar amendment impacts. *Plant and Soil, 333*(1–2), 443–452.

Spokas, K. A., & Reicosky, D. C. (2009). Impact of sixteen different biochars on soil greenhouse gas production. *Annals of Environmental Science*, 179–193.

Steiner, C., et al. (2007). Long term effects of manure, charcoal and mineral fertilization on crop production and fertility on a highly weathered central Amazonian upland soil. *Plant and Soil, 291*(1–2), 275–290.

Sun, L., Li, L., Chen, Z., Wang, J., & Xiong, Z. (2014). Combined effects of nitrogen deposition and biochar application on emissions of N_2O, CO_2 and NH_3 from agricultural and forest soils. *Soil Science and Plant Nutrition, 60*(2), 254–265.

Tomczyk, A., Sokołowska, Z., & Boguta, P. (2020). Biochar physicochemical properties: Pyrolysis temperature and feedstock kind effects. *Reviews in Environmental Science and Bio/Technology, 19*, 191–215.

Uzoma, K. C., Inoue, M., Andry, H., Fujimaki, H., Zahoor, A., & Nishihara, E. (2011). Effect of cow manure biochar on maize productivity under sandy soil condition. *Soil use and Management, 27*(2), 205–212.

Verheijen, F. G. A., Graber, E. R., Ameloot, N., Bastos, A. C., Sohi, S., & Knicker, H. (2014). Biochars in soils: New insights and emerging research needs. *European Journal of Soil Science, 65*(1), 22–27.

Wang, J. Y., Xiong, Z. Q., & Kuzyakov, Y. (2016). Biochar stability in soil: Metaanalysis of decomposition and priming effects. *GCB Bioenergy, 8*(3), 512–523.

Wang, D., Fonte, S. J., Parikh, S. J., Six, J., & Scow, K. M. (2017). Biochar additions can enhance soil structure and the physical stabilization of C in aggregates. *Geoderma, 303*, 110–117.

Wang, J., Pan, X., Liu, Y., Zhang, X., & Xiong, Z. (2019). Effects of biochar amendment in two soils on greenhouse gas emissions and crop production. *Plant and Soil, 360*, 287–298.

Wang, L., Deng, J., Yang, X., Hou, R., & Hou, D. (2023). Role of biochar toward carbon neutrality. *Carbon Research, 2*(1). Available from https://doi.org/10.1007/s44246-023-00035-7.

Woolf, D., Amonette, J. E., Street-Perrott, F. A., Lehmann, J., & Joseph, S. (2010). Sustainable biochar to mitigate global climate change. *Nature Communications, 1*(1), 56.

Yan, H., Cong, M., Hu, Y., Qiu, C., Yang, Z., Tang, G., . . . Jia, H. (2022). Biochar-mediated changes in the microbial communities of rhizosphere soil alter the architecture of maize roots. *Frontiers in Microbiology, 13*, 1023444.

Yoo, G., & Kang, H. (2012). Effects of biochar addition on greenhouse gas emissions and microbial responses in a short-term laboratory experiment. *Journal of Environmental Quality, 41*(4), 1193–1202.

Yu, L., Tang, J., Zhang, R., Wu, Q., & Gong, M. (2013). Effects of biochar application on soil methane emission at different soil moisture levels. *Biology and Fertility of Soils, 49*, 119–128.

Zhang, A., Cui, L., Pan, G., Li, L., Hussain, Q., Zhang, X., . . . Crowley, D. (2010). Effect of biochar amendment on yield and methane and nitrous oxide emissions from a rice paddy from Tai Lake plain, China. *Agriculture, Ecosystems & Environment, 139*(4), 469–475.

Zhang, A., Li, X., Xing, J., & Xu, G. (2020). Adsorption of potentially toxic elements in water by modified biochar: A review. *Journal of Environmental Chemical Engineering, 8*(4), 104196.

Zhang, A., Liu, Y., Pan, G., Hussain, Q., Li, L., Zheng, J., & Zhang, X. (2012). Effect of biochar amendment on maize yield and greenhouse gas emissions from a soil organic carbon poor calcareous loamy soil from Central China Plain. *Plant and Soil, 351*, 263–275.

Zhang, J., & Shen, J.-L. (2022). Effects of biochar on soil microbial diversity and community structure in clay soil. *Annals of Microbiology, 72*(1).

Zheng, H., Wang, X., Luo, X., Wang, Z., & Xing, B. (2018). Biochar-induced negative carbon mineralization priming effects in a coastal wetland soil: Roles of soil aggregation and microbial modulation. *Science of the Total Environment, 610–611*, 951–960.

Zimmerman, A. R. (2010). Abiotic and microbial oxidation of laboratory-produced black carbon (biochar). *Environmental Science & Technology, 44*(4), 1295–1301.

Zimmerman, A. R., & Gao, B. (2013). The stability of biochar in the environment. In N. Ladygina, & F. Rineau (Eds.), *Biochar and Soil Biota* (pp. 1–40). Boca Raton, USA: CRC Press.

CHAPTER 21

Biochar for pollutants bioremediation from soil and water ecosystem

Amit K. Dash[1], Saloni Tripathy[2], A. Naveenkumar[2], Tanmaya K. Bhoi[3], Arpna Kumari[4], Divya[5], Ashish M. Latare[6], Tony Manoj Kumar Nandipamu[5], Virendra Singh[5], Md. Basit Raza[7], Anuj Saraswat[8] and Jehangir Bhada[9]

[1]ICAR—Indian Institute of Seed Science, Mau, Uttar Pradesh, India [2]Division of Soil Science and Agricultural Chemistry, ICAR—Indian Agricultural Research Institute, Pusa, New Delhi, India [3]Forest Protection Division, ICFRE—Arid Forest Research Institute (ICFRE—AFRI), Jodhpur, Rajasthan, India [4]Department of Applied Biological Chemistry, The University of Tokyo, Tokyo, Japan [5]Department of Agronomy, College of Agriculture, Govind Ballabh Pant University of Agriculture and Technology, Pantnagar, Uttarakhand, India [6]Institute of Agricultural Sciences, Banaras Hindu University, Varanasi, Uttar Pradesh, India [7]ICAR-Directorate of Floricultural Research, Pune, Maharashtra, India [8]Department of Soil Science, College of Agriculture, Govind Ballabh Pant University of Agriculture and Technology, Pantnagar, Uttarakhand, India [9]Ecosystem Sciences Department, University of Florida, FL, United States

21.1 Introduction

Rapid industrialization has led to the depletion of natural resources and the development of enormous amounts of toxic waste that contaminate water and soil, endangering human health and the environment (Majhi et al., 2022; Priyadarshanee & Das, 2021; Dash, Dwivedi, Dey, & Meena, 2023). Emerging contaminants (ECs) are substances occurring artificially or naturally and are not typically regulated in the natural environment but can potentially harm the environment or public health. Most of these harmful substances were previously unknown, unrecognized, or unidentified, but they have recently come to light as contaminants linked to possible environmental concerns (Singh, Luthra, & Bhattacharya, 2024). Several of these substances are additives that are commonly found in industrial and domestic items, such as plasticizers for toys and food containers, surface-active chemicals for detergents and water- and oil-repellent products, and flame retardants for textiles and other products (Eggen, Heimstad, Stuanes, & Norli, 2013). Their fate in engineered systems, water/wastewater treatment facilities, and our surroundings is still largely unknown. An essential area of considerable interest that requires further research is the possible impact and threat to human health, animals, aquatic life, and ecosystems by the accumulation of ECs. Fig. 21.1. shows a schematic representation of the pathways by which ECs reach the environment. These pollutants can be persistent and widely distributed, frequently bioactive and bioaccumulative. Since most ECs are not currently regulated, it is necessary to constantly monitor and report on any potential occurrences of them in water sources and effluent discharges. ECs are not always completely novel compounds. That term can be used for any substance under the following three broad categories.

1. Compounds recently entered the environment, such as industrial additives.
2. The second group comprises substances that may have been present in the environment for a long time but whose presence was recently discovered and whose relevance has only recently begun to draw attention (such as medications).
3. The third group consists of substances that have been there for a while but whose potential harm to people and the environment has recently come to light (such as hormones) (Stefanakis & Becker, 2016).

As people become more conscious of the negative impacts of developing pollutants, research has been done on various remediation strategies. The existing conventional

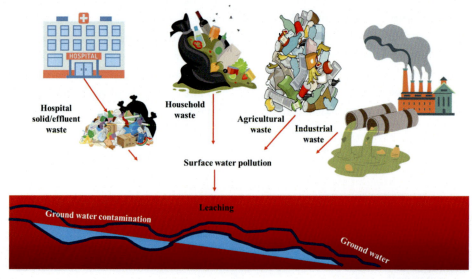

FIGURE 21.1 Diagram showing pathways of emerging pollutants from sources to the environment (Kumar et al., 2022).

21.1 Introduction

wastewater treatment facilities are not designed to effectively address the problem of ECs due to their complex interaction in the environment, such as organo-mineral combinations, thus making it difficult to effectively remove these substances (Morin-Crini et al., 2022). In the past few years, many techniques have been developed to alleviate the problem of ECs in the environment. These methods encompass a range of approaches, such as biodegradation, photocatalytic degradation, nanofiltration, membrane bioreactors, separation, enhanced chemical oxidation, and others (Li et al., 2019). Unfortunately, these approaches exhibit limited efficacy or lack financial feasibility when used on a large scale. For instance, microbial degradation is an important technology now accessible for treating wastewater and remediating soil. However, some ECs are not easily biodegradable thus shifting the focus from microbial degradation to photocatalytic degradation (Majone et al., 2015).

Yet, these methods are frequently ineffective or unprofitable when used on a large scale. For instance, microbial degradation used for the reclamation of soil and treatment of wastewater may be ineffective for a wide range of contaminants (Majone et al., 2015). Unfortunately, applying ultraviolet light to plants on a wide scale is still neither practical nor cost-effective for treating effluent present in water (Mirzaei, Leonardi, & Neri, 2016). Dolar et al. (2012) reported that membranes offer a practical method that may be utilized to lower the concentration of various pollutants. However, they only serve as physical filters in those processes. The concentrations of most drugs or pharmaceuticals in sewage water effluent may be drastically reduced with ozone treatment. Unfortunately, these procedures are typically ineffective for removing recalcitrant pollutants, including pharmaceuticals. The risk of ECs being transmitted from the environment to humans must, therefore be reduced, and it is vital to find effective strategies for eliminating or significantly reducing the concentration of ECs in the ecosystem (Amin et al., 2021).

Of all the widely followed techniques for the remediation of persistent organic and inorganic pollutants (Including ECs), biochar is an efficient and environment-friendly technique (Liao, Zhang, Wang, & Tang, 2017; Rani, Shanker, & Jassal, 2017; Sharma, Khardia, & Saraswat, 2022). The components used to create biochar are easily available. A variety of biomass, such as crops, forest wastes, and animal manure, have been explored as raw materials for the synthesis of biochar (Chaturvedi, Singh, Dhyani, Govindaraju, & Mandal, 2021). Under such scenarios, biochar is becoming more widely acknowledged as a helpful substance for carbon (C) sequestration and also soil and water reclaimant because of its strong sorption abilities and high carbon content (Huang et al., 2017). Biochar is a C-rich by-product of different feedstocks, which is produced at higher temperatures (450–600 °C approximately) in the partial presence of oxygen that is termed as thermal degradation of biomass that enhances the agronomic variables and significantly remediates soil pollutants (Huang, Lee, & Huang, 2021). These C-containing adsorbents have a very high surface area over normal organic substrates, resulting in a strong affinity with a higher potential to bind organic and inorganic pollutants (Singh, Luthra, et al., 2023). Biochar reduces the bioavailability of contaminants in the environment, protecting plants and animals from the dangers of prolonged exposure to these substances. Several factors significantly influence the sorption ability of biochar during the pyrolysis process, including the period of residence, the kind of material employed, the amount of heat supplied, and the pace at which heat is transferred. Similarly, physicochemical attributes such as surface area, surface charge, and chemical functionality substantially influence the extent to which both inorganic and organic contaminants adhere to surfaces (Zhu, Kwon, & Pignatello, 2005).

21.2 Occurrence and fates of emerging contaminants in the ecosystem

Antibiotics, hormones, illegal medicines, endocrine-disrupting substances (EDCs), cosmetics, personal care items, chemical pesticides, surfactants, industrial goods, microplastics, nanoparticles, and nanomaterials are examples of ECs, as shown in Fig. 21.2. They are divided into three categories: those that pose a genuine risk to human health or the environment. They could be of industrial origin or come from wastewater municipal (domestic), agricultural, medical, or research facilities.

Pharmaceutical compounds found in wastewater include sex and steroid hormones, human and livestock antibiotics, and over-the-counter human medicines. They have been found in groundwater, sludge, sediments, natural waterways, and effluents from wastewater treatment plants. Aquatic life is a soft target for accumulating these pollutants since they are continuously exposed to wastewater residues throughout their lives. The effects of medicines on land and aquatic organisms are little understood from an ecotoxicological standpoint, and a thorough investigation of these effects is crucial. Various chemicals such as galaxolide and tonalide, which are often found in various personal care products (PCPs) and consumer items, are quite harmful to the environment. The most likely fates for PCPs and their metabolites include conversion to CO_2 and water, mixing with receiving water bodies as the original or mineralized product, and solubilization by solids like sludge or biosolids, particularly if the material or the biologically controlled transformation product is lipophilic. Due to the androgenic or estrogenic properties, EDCs can cause harm to the endocrine system even in low quantities. Human activities and various industrial processes release natural and artificial EDCs into the environment, which transfer through sewage treatment systems before ending up in the soil, surface water, sediment, and groundwater. The majority of studies have focused only on estrogenic substances. EDCs are found in wastewater at very low concentrations ($ng\ L^{-1}$ or $mg\ L^{-1}$). These substances are quite concerning since it is uncertain how long they will be exposed to people and what harm they may cause. Possible issues from these growing toxins in the environment include anomalous physiological functions and reproductive impairment, an increase in cancer cases, the emergence of antibiotic-resistant microorganisms, and maybe increasing chemical toxicity (Wilkinson, Hooda, Barker, Barton, & Swinden, 2017).

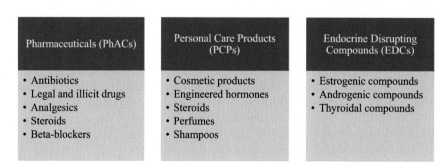

FIGURE 21.2 Occurrence of emerging contaminants in the environment and their sources.

Emerging volatile pollutants may be released into the atmosphere during production or incineration and through surface water or volatilization. Other contaminants, like the herbicide atrazine, have been found in precipitation that is geographically distant from its source regions, suggesting that they may have entered the atmosphere through sprays used on the ground. Concerning their adsorption affinity to suspended solid/sediment particles and resistance to biodegradation ECs appear to decline in rivers at varying rates. Attenuation of ECs can be attributed to adsorption to suspended solids in water from rivers, sediments, and banks; certain substances preferentially bind to suspended solids over staying in a dissolved state. According to Gregg, Prahl, and Simoneit (2015), adsorption on suspended solids may help transit these chemicals in aquatic environments. Environmental destiny is largely determined by the physiochemical features of pollutants as well as the characteristics of the adsorbing substance.

21.3 Properties of biochar affecting removal of emerging contaminants

The distinct physicochemical characteristics of biochar impact the pollutant sorption capacity and are crucial for forming porous soil connections. The primary variables in biochar that may affect the fate of pesticides in environment are its constituents, presence or absence of functional groups, reaction, and surface properties. The basic chemical structure of biochar gives it the capacity to hold cations strongly and to exchange them with other cations, which is thought to play a primary part in the process of pesticide sorption in soil (Kim, Kim, Kim, Alessi, & Baek, 2021). Through various mechanisms, including surface chelation and acidity, biochar can reduce the availability of pollutants. Variations in the feedstock and temperature during synthesis cause changes in the functional chemistry and elemental composition of charred substances, ultimately impacting contaminants' sorption onto it. Furthermore, the increased heat during preparation expands the micropores and surface properties of the biochar, supplying more locations for pesticide sorption (Abbas et al., 2018). The ability of biochar to adsorb pesticide compounds comes from functional groups on biochar surfaces (Baronti et al., 2010). The alcohol (−OH) groups are changed to phenolic hydroxyl groups at a high temperature, enhancing the aromatic structure, the O/C and H/C ratio, and the surface area of the biochar (Yu et al., 2019). More surface specificity is found in high-temperature biochar. However, a reduction in functional groups may decrease the sorption capacity.

Biochar samples are often alkaline. When the temperature rises, the amount of ash O and H decreases, and the pH and aromaticity of the biochar increase (Ronsse, van Hecke, Dickinson, & Prins, 2013). High pH of biochar can promote the breakdown of carbamate and organophosphorus insecticides via an alkali catalysis process (Mukome, Zhang, Silva, Six, & Parikh, 2013). The effectiveness of biochar for adsorbing organic and inorganic pollutants is also greatly influenced by its surface area and porosity through pore filling and surface adsorption (Deng et al., 2017). Biochar will have a stronger capacity for adsorption due to its higher surface area and porous structure. According to reports, plant biochar has the highest surface areas (Wei et al., 2019). Depending on potential applications, biochar's porosity and surface properties may help adsorb contaminants from the environment.

The efficiency of removal of ECs is not only affected by the properties and quality of biochar, rather, it is also governed by the properties of contaminants, thus modulating the overall efficacy of biochar in reclaiming pollutants from the surroundings. The interactions between biochar and organic compounds as electron donors and acceptors may be impacted by the aromaticity of pollutants, which would then change the adsorption process (Dai, Zhang, Xing, Cui, & Sun, 2019). For the adsorption of aromatic compounds, it is advisable to utilize biochar made at high temperatures while undergoing pyrolysis. Adsorption is frequently accelerated by ionizable functional groups like $-NHR$, $-NH_2$, $-NHCOR$, $-CONH_2$, and $-OH$. Acidic soil conditions cause basic pesticides to produce cations, boosting adsorption, while nonionic pesticides may briefly polarize, and promote adsorption on a charged surface. Pesticide ionizability is influenced by soil pH (Rasool, Rasool, & Gani, 2022). The hydrophilicity and lipophilicity of the contaminants also impact the adsorption and desorption behavior of biochar (Khalid et al., 2020). As opposed to lipophilic substances, biochar often interacts less with water-soluble pollutants (Rasool et al., 2022).

21.4 Biochar application for removal of emerging contaminants from aqueous system

The ECs have been found in drinking water, sediments, wastewater, natural waters, groundwater, and sludge. Because of their extremely harmful impacts, even when present at low concentrations, ECs have drawn more attention as the world's water resources are being depleted daily. To lessen the environmental and health problems, it is crucial to establish a treatment method that emphasizes the safe and efficient removal of ECs from wastewater. Biochar has shown effective bioremediation potential under an aqueous ecosystem (Table 21.1).

21.4.1 Heavy metals

The US Geological Survey defined ECs as all chemicals in everyday life, including drugs and medicines, particularly those sold without a prescription, personal hygiene and sanitation products, such as soaps, disinfectants, odors, and food chemical additives (preservatives, colorants, and adulterants) that exist in the environment, water, soil, and atmosphere. Unfortunately, to these contaminants, we can still add the classic, heavy metal ions such as mercury, lead, cadmium, thallium, silver, and others found in our lands and water (Lodeiro, Capelo, Oliveira, & Lodeiro, 2019; Singh, Sharma, & Balan, 2024). These contaminants are constantly released into the environment, putting people and all other life forms on Earth at risk. Because of their nonbiodegradable nature, toxic effects, persistence, accumulation, and biomagnification occurring in the trophic systems, along with negative impacts on human beings and other living organisms, heavy metals in the water ecosystem have raised significant environmental concerns (Kumari, Rajput, & Singh,

TABLE 21.1 Removal of pollutants from the aqueous ecosystem by biochar.

Biochar	Pyrolysis temp (°C)	Heavy metals	Adsorption capacity (mg/g)	Mechanism	References
Alkali lignin	400	Pb	1003.7	Ion exchange, surface complexation and mineral precipitation	Wu et al. (2021)
Pomelo peel	600	Cr	39.3	Ion exchange, surface complexation and reduction, π–π interaction and electron donor–acceptor complex	Dong et al. (2021)
Corn straw	500	Cr	117.0	Electrostatic attraction, complexation, ion exchange and reduction	Qu et al. (2021)
Bamboo	700	U	49.6	Strong complexation, reduction, and co-precipitation	Lyu et al. (2021)
Rice husk	300	Propylparaben (PP)	70.6	Degradation	Nikolaou, Vakros, Diamadopoulos, and Mantzavinos (2020)
Cassava waste residues	500	Sulfamerazine (SMR), oxytetracycline (OTC)	3.86 (SMR) and 8.37 (OTC)	Sorption to modified cassava biochar in OTC > SMR sequence	Luo et al. (2021)
Cornstalk, orange peel, peanut hull	300, 500, and 700	TC	123.4, 99.0, and 109.9	Enhanced adsorption of TC upon biochar $KMnO_4$ treatment	Fan et al. (2021)
Food waste	300, 450, and 600	Fluoride	123.4	Adsorption (91.4%) on Al-modified biochar in a pH range of 5–11	Meilami et al. (2021)
Bagasse	500	Sulfonamides	—	H_2O_2 laden biochar adsorbed sulfonamide antibiotics with ~89% efficiency	Qin et al. (2019)

2024). According to the reports of Tan et al. (2015), biochar is an efficient, green, and ecologically safe adsorbent that is efficient for the removal of a wide range of metals (As, Cd, Pb, Cu, Zn, Mn, Mg, Cr, and Cr) through various mechanisms including electrostatic attraction, ion exchange reactions, physisorption, surface chelation, precipitation, and ligand exchange (Anyakora, Ehianeta, & Umukoro, 2013; Roy, Meena, Kumar, Jhariya, & Pradhan, 2021; Velayatzadeh, 2023). It has been demonstrated that biochar can significantly lower Cu(II) concentrations in aqueous solutions (Meng et al., 2014; Saraswat, Nath, et al., 2023; Saraswat, Ram, Raza, et al., 2023). For instance, biochars generated from canola, soybean, and peanut straws at 400 °C by oxygen-limited pyrolysis were abundant in d-COOH and phenolic hydroxyl groups. They could specifically adsorb Cu(II) through surface complexes (Tong, Li, Yuan, & Xu, 2011). Being a highly effective and competitively priced permeable reactive barrier media, biochar can also be used to effectively remove radionuclides from groundwater, such as cesium and uranium (Dinis & Fiúza, 2021). In a study, biochar was prepared using biomass from *Arachis hypogaea*, *Brassica napus*, and *Glycine max* straw in a muffle furnace for 3.75 hours at 300, 400, and 500 °C, and was further used for removal of Cu (Tong et al., 2011). The outcomes revealed that the adsorption capacity of Cu^{2+} occurred in the following order: peanut straw char > soybean straw char > canola straw char. Synthesis of biochar from *Sargassum hemiphyllum*-derived biochar at 700°C, displayed remarkable Cu (II) electrosorption capability (75–120 mg g^{-1}). The Cu (II) electrosorption process was best explained by the Langmuir model, revealing monolayer electrosorption on the biochar surface through physical sorption, ion exchange, surface complexation, surface precipitation, electrostatic interaction, and metal-π interaction mechanisms (Truong et al., 2023). In another study, cacao pod husk biochar was utilized to produce biochar at a temperature of 500 ± 5 °C, which exhibited a high affinity for adsorbing toxic metals such as Pb^{2+}, Hg^{2+}, and Cd^{2+} from aqueous medium. The results evaluated using Langmuir constant revealed the strong affinity between biochar and heavy metals through metal-π interaction mechanisms (Tokarčíková et al., 2023). A study to assess the impact of magnetically modified maize biochar on Zn(II), Cd(II) and Pb(II) reported >90% adsorption onto the biochar through C–H bonds in the $-CH_2$ and $-CH_3$ groups which also enhanced by the improved biochar surface area (Abbey et al., 2023). Utilizing tea waste biochar to harness the adsorption potential of Pb(II) and Cd(II) from the aqeos media revealed 81.6% and 38.6% reduction from the solution through physical sorption, ion exchange, surface complexation mechanisms (Zhang et al., 2023). The recent studies emphasize the multi-occular role of biochar in stabilizing the heavy metals through intricate mechanisms in the aqueous medium to filter the waste water efficiently through environment-friendly approaches.

21.4.2 Organic pollutants

Due to its unique qualities, bichar has been employed as an inventive, environmentally friendly, and reasonably priced adsorbent in wastewater treatment. Biochar can be applied in a wide range of water and wastewater treatment processes. As a porous, biostable, and readily available substance, biochar has been widely employed as a support or supplement medium during anaerobic digestion and a filter media for removing pathogens, heavy metals, and suspended debris. The effectiveness of biochar as a support-based catalyst for degrading dyes and resistant pollutants has also been studied by Enaime et al. (2020). Various pollutants, such as color pigments, organic chemicals, pesticides, antibiotics, and organic toxins, have a high affinity for biochar. Some examples are methylene blue, trichloroethylene, atrazine, norflurazon, fluridone, sulfamethoxazole, bisphenol-A, ofloxacin, and norfloxacin, phenanthrene and naphthalene, N-nitroso dimethylamine, and 17α-ethinyl estradiol (Chen, Zhou, & Lin, 2015). Organic pollutants present in aqueous systems can undergo decontamination through interactions with biochar, involving physical, chemical, and redox reactions. Physisorption refers to the physical interaction between the surface of biochar and organic pollutants, whereas chemical interactions entail the chemical binding of organic pollutants to biochar. Physisorption is driven by electrostatic forces and intermolecular attraction, predominantly influenced by Van der Waals weak forces, leveraging the surface charge of biochar (Ambaye, Rene, Dupont, Wongrod, & Van Hullebusch, 2020). Numerous experiments were conducted to examine the adsorption potential of biochar for organic pollutants from the aqueous system. Cationic methylene blue dye was removed using biochars derived from eucalyptus, palm, and anaerobic digestion waste as sorbents (Sun, Wan, & Luo, 2013). The removal efficiency of eucalyptus and palm bark biochars was 78.3 and 89.8%, respectively. Ahmad et al. (2013) used biochar made from pine needles to sorb trichloroethylene from water at pyrolysis temperatures ranging from 300 to 700°C. Due to its large surface area, microporosity, and amount of carbonization, the findings indicated that charred substances generated @ 700 °C had the best removal efficacy for trichloroethylene. The most common sorption mechanism reported behind the effective removal of these pollutants was pore filling in the homogenized (packed) soil columns. Biochar produced from the chip of *Pinus* could reduce the total atrazine leaching by 52% when it was pyrolyzed between 300 and 550 °C (Delwiche, Lehmann, & Walter, 2014). In addition to pollutants, biochar can act as a superior sorbent for various ions like nitrate, ammonium, phosphate, and fluoride (Saraswat, Ram, Kouadri, et al., 2023; Wang et al., 2015). The biochar produced from sugarcane bagasse could effectively adsorb nitrate by increasing the contact time (Kameyama, Miyamoto, Shiono, & Shinogi, 2012). Similarly, Yao et al. (2012) evaluated the capacity of biochars produced from Brazilian pepperwood @ 600 °C to sorb PO_4^{3}, NO_3^-, and NH_4^+ in light soil using column leaching studies. Compared to the soil-alone, they found that the biochar significantly decreased the overall amounts of PO_4^{3-}, NO_3^-, and NH_4^+ in the leachates. Fan et al. (2017) investigated a new biochar derived from sewage sludge as an adsorbent for removing methylene blue (MB), a common dyeing wastewater component. They achieved nearly 100% MB removal efficiency within 10 hours, with a maximum adsorption capacity of 29.85 mg/g at pH 7. The effective removal was attributed to the biochar's coarse, porous texture and large surface area (25 m^2/g). In another study, Dastidar and co-workers developed biochar from Spirulina platensis algae biomass residue for Congo red (CR) and MB adsorption. The adsorption performance of these biochars (82.6% for CR, 92.6% for MB) was comparable to commercial activated carbon, indicating their potential for wastewater treatment (Nautiyal, Subramanian, & Dastidar, 2017). However, the low adsorption capacity of pristine biochars may not meet practical demands due to limited functionalities like poor hydrophilicity, surface area, and active sites. Consequently, biochar-based hybrid materials are gaining attention for better removal efficiencies. For instance, Wang, Wang, and Herath (2017) coated biochar with Al_2O_3 nanoparticles, enhancing MB adsorption capacity.

Similarly, Kayan, Kalderis, Kulaksız, and Gözmen (2017) modified biochar with Fe for better removal of malachite green. Zhang, Sun, Min, and Ren (2018) synthesized biochars from sludge pyrolysis, effectively removing phenol via ozonation. Frontistis and co-workers used malt rootlet-derived biochar to activate sodium persulfate for sulfamethoxazole (SMX) degradation (Kemmou, Frontistis, Vakros, Manariotis, & Mantzavinos, 2018), while Zhu et al. (2018) utilized reed-derived biochar for peroxydisulfate (PDS) activation and SMX degradation, achieving significant removal efficiencies.

21.5 Biochar application for removal of emerging contaminants from the soil system

Biochar plays a significant role in removing ECs from the soil through various mechanisms. The adsorption capacity of biochar, along with its unique physio-chemical properties, allows it to effectively bind and immobilize contaminants, preventing their movement and reducing their bioavailability. The most serious issue is the regular contamination of soil ecosystems with polycyclic aromatic hydrocarbons (PAHs), dyes, antibiotics, pesticides, and herbicides as a result of the organic waste discharged from market areas, households, farming activities, and industries (Ahmad et al., 2014). Effective removal efficiency depicted by biochar under the soil ecosystem has been documented in Table 21.2.

TABLE 21.2 Removal of pollutants from the soil ecosystem by biochar.

Biochar	Pyrolysis temperature (°C)	Pollutant	Removal efficiency (%)	References
Wood	450	Metalaxyl	70.1	You et al. (2021)
Sewage sludge	700	PAHs	74.0	Godlewska and Oleszczuk (2022)
Green waste	650	Pb	92.9	Pan et al. (2021)
Kenaf bar	600	Cd	90.1	Qian, Liang, Zhang, Huang, and Diao (2022)
Sawdust	550	Polystyrene	>94.8%	Siipola, Pflugmacher, Romar, Wendling, and Koukkari (2020)
Sugarcane bagasse	750	Polystyrene based latex	>99.9%	Ganie, Khandelwal, Tiwari, Singh, and Darbha (2021)
Food waste	600	Fluoride	91.4%	Dewage, Liyanage, Pittman, Mohan, and Mlsna (2018)
Rice straw	600	PAHs	58.8%	Zhang, Min, Tang, Rafiq, and Sun (2020)
Rice husk	700	PCBs	91%	Silvani et al. (2019)
Pig manure	700	Clothianidin and Imidacloprid	90.5 and 81.4%	Zhang et al. (2020)

Abbreviations: *PCB*:-Polychlorinated biphenyl; *PAH*: polyaromatic hydrocarbon.

21.5.1 Heavy metals

Heavy metal related soil pollution is increasingly a major concern. The primary causes of soil pollution are both anthropogenic and natural (Dash et al., 2021) The natural processes include rock weathering, volcanic eruption, soil erosion, and landslides. Major anthropogenic activities contributing to heavy metal release into the environment are smelting, mining, and agrochemicals (pesticides and fertilizers). Effective use of biochar reduces the biological availability of heavy metals and the stress that toxins cause to the biotic component of soil (Sharma et al., 2020). Chen et al. (2015) examined the Cd(II) adsorption potential of biochar made from municipal wastes. The outcomes demonstrated that the effectiveness of Cd(II) removal increased as the biochar dose was raised, with the highest removal capacity of 42.80 mg g^{-1} being attained at the biochar application rate of 0.2% (w/w). According to Namgay et al. (2010), applying biochar reduced the levels of As, Pb, and Cd in maize shoots. Similarly, the effective removal of As, Cd, and Zn using biochar was also documented by Beesley and Marmiroli (2011). These researchers established that the biochar surface induced metal sorption and that the procedure was not instantly reversible.

21.5.2 Synthetic dyes

The industries of textile materials, wood, pulp, and printing effluents contain synthetic colors, which are then dumped into the environment (Tkaczyk, Mitrowska, & Posyniak, 2020). Such colors can seriously harm the environment and have negative effects on biological oxygen demand, chemical oxygen demand, and total organic carbon (Jadhav et al., 2012; Raj et al., 2021). As they are resistant to light, oxidizing chemicals, and aerobic remediation, they are challenging to cure using standard approaches. Biochar has been used to remove the dye, with several researchers having hypothesized alternative processes. For example, some studies explored a cost-based system for treating the released dyes using straw biochar (Xu, Xiao, Yuan, & Zhao, 2011). Similarly, According to Mui et al. (2010), bamboo biochar has monolayer sorption capacities of 0.0406, 0.0416, and 0.998 mmol g^{-1} on acid blue 25, acid yellow 117, and methylene blue, respectively.

21.5.3 Antibiotics

Antibiotics are medications primarily used to treat serious illnesses brought on by pathogenic microbial infections and save lives by reducing microbial infections and mortality rates (Patel et al., 2022). In addition to having beneficial effects on health, they also endanger the environment by polluting it. A serious global catastrophe is developing due to the emerging multidrug resistance in bacteria and rising antibiotic contamination. Antibiotics are widely utilized, and a significant amount of this usage (30–90%) is primarily eliminated through anthropogenic excretion in urine and feces (Gao et al., 2012; Sabri et al., 2020). These ECs pose a severe threat to society and the environment because they may lead to the spread of antibiotic-resistant genes in livestock farms (Patel et al., 2019). To properly treat resources contaminated with antibiotics, several actions are required. Beacause its antibacterial characteristic has no impact on biochar's catalysis or adsorption, biochar is the most appealing solution for removing antibiotics. Depending on the characteristics of both the antibiotic and the biochar being utilized, the sorption of antibiotics can vary significantly (Zhang et al., 2013). Besides that, some researchers have demonstrated that biochar has an adsorption capacity that is comparable to or even superior to that of commercially available activated carbon. Depending on the kind of antibiotic, the removal rate might even reach 100% in rare circumstances (Ahmed, Zhou, Ngo, & Guo, 2015). Biochars from *Pinus* spp and *Bambusa vulgaris* being produced under various temperatures and chemical circumstances and studied for their ability to remove sulfonamides (sulfamethoxazole and sulfapyridine) through adsorptive means, were shown to have strong adsorption capabilities (Xie, Chen, Xu, Zheng, & Zhu, 2014).

21.5.4 Polycyclic aromatic hydrocarbons

The constant exploitation of crude oil and its derivatives has led to the ubiquity of PAHs, which are components of petroleum hydrocarbons. Because of its sorptive qualities, biochar has been shown to reduce the concentrations of both total and bioavailable PAHs (Beesley, Jimenez, & Eyles, 2010). For instance, it has been demonstrated that pore-filling processes, multilayer adsorption, surface coverage, condensation in capillary pores, and adsorption into the polymeric matrix all contribute to the adsorption of PAHs to wood biochars (Werner & Karapanagioti, 2005). Some prominent examples have included the decrease in phenanthrene exposure to soil biota using biochar and a 50% reduction in the accumulation of PAHs in waters of soil pores (Beesley et al., 2010). Shi et al., (2011) observed a notable decrease in phenanthrene uptake by maize seedlings after adding biochar made from rice straw to phenanthrene-contaminated soil. The denser and more resistant moieties caused a greater drop in total PAH concentrations, even though increased microbial breakdown seems improbable given that the charcoal sorbents decrease molecular bioavailability (Rhodes et al., 2008).

21.5.5 Pesticides

To manage pests and diseases in agriculture, agrochemicals of organic nature are purposefully introduced to various parts of the environment like air, water, and soil. The enhanced adsorption with a reduction in the dispersion of pesticide, when biochar is present, can lessen the danger of accumulation of such xenobiotics in the surroundings and its exposure to life forms from the standpoint of the environment's and people's health. The increased crop production and lower residues in farm products may also result from decreased bioavailability and plant uptake. Adsorption plays a crucial role in determining the fate of pesticides in soil, wherein pesticide molecules adhere to the surface of solid particles through various mechanisms like bonding, hydrogen bonding, ionic bonding, covalent bonding, and van der Waals forces (Dong et al., 2024). This process dictates the accumulation and movement of pesticides in the environment, impacting environmental and food safety. Introducing biochar into soil can notably enhance the adsorption of herbicides like atrazine, nicosulfuron, and diuron (Liu et al., 2018). Multiple studies have highlighted those soils amended with biochar exhibit improved pesticide sorption compared to biochar-free soils, particularly for atrazine, 2,4-D, carbaryl, and thiacloprid (Ogura et al., 2021). For example, Clay, Krack, Bruggeman, Papiernik, and Schumacher (2016) noted a 4.5 to 6-fold increase in atrazine sorption efficiency after amending soil with 10% maize stalk biochar. Additionally, Zhang et al. (2020) found that amending soil with biochar pyrolyzed at low temperatures (300 °C) promoted the biodegradation of imidacloprid and clothianidin by supplying organic carbon and available nitrogen for microorganisms, albeit inhibiting chemical degradation. Conversely, biochars pyrolyzed at high temperatures (500 to 700 °C) favored chemical degradation while inhibiting biodegradation. Optimizing the application rate and frequency of biochar is essential to achieve ideal soil conditions for pesticide remediation (Liu et al., 2018). Different dosages may alter pH and soil electrical conductivity (EC) to varying degrees (Palansooriya et al., 2022). The efficacy of pesticide sorption in biochar-amended soils also involves competition, as soil is a complex system with various organic and inorganic contaminants coexisting. The interaction of biochar with inorganic and organic particles in the soil, as well as other pollutants, may lead to competition and pore blockage, thereby reducing the sorption capacity for targeted pesticides (Liu et al., 2018). Graber, Tsechansky, Khanukov, and Oka (2011) reported that the fumigant 1,3-dichloro propene was less effective in soil modified with biochar. The findings demonstrated that to achieve the full effect on nematode survival in the soil where biochar @ 1% was added, the fumigant application rate was required to be twice as previously. Similar findings were made by Nag et al. (2011), who reported that adding biochar to soil had a significant impact on the effectiveness of herbicide atrazine in controlling weeds like *Lolium perenne* and that to achieve the necessary control of weeds, the rate of application of herbicide may have to be enhanced upto 4 times. They added that the chemistry of the herbicide compound and mechanism of action affected how much biochar affected herbicide efficacy.

21.6 Challenges and future perspective

Depending on the type and properties of ECs, the adsorption mechanisms by biochar vary. As a result, choosing the right biochar is crucial and demands more attention. Most recent studies have been done on a single pollutant or collection of contaminants. Therefore, future research should be done to evaluate the biochar's ability to adsorb multiple contaminants from the soil and water ecosystem. Additionally, only small-scale laboratory applications of biochar have been made. It is crucial to conduct an in-depth analysis of the use of biochar and determine its technical and economic feasibility. The elucidation of mainly unidentified breakdown products and their synthesis during biological degradation should also be the subject of future research. Such breakdown products' potential environmental toxicity needs to be taken into account. Much remains to be clarified regarding the factors and circumstances affecting the attenuation of ECs utilizing biochar in the environment.

21.7 Conclusions

The adverse effects of rapid industrialization on natural resources and the environment are evident through the emergence of emerging contaminants (ECs). While conventional remediation methods often prove inadequate, biochar offers a promising, eco-friendly solution due to its high sorption capabilities and availability from various biomass sources. By reducing the bioavailability of contaminants, biochar plays a crucial role in safeguarding ecosystems and human health from the harmful impacts of ECs, highlighting its potential as a key tool in environmental remediation efforts. However, the efficacy of biochar in EC removal is contingent upon multiple factors, including its composition, surface area, and porosity, as well as the properties of the contaminants themselves. Further research is warranted to optimize biochar-based remediation strategies and enhance their effectiveness in mitigating the impacts of ECs on environmental and human health. Heavy metals such as mercury, lead, cadmium, and copper can be efficiently adsorbed by biochar through various mechanisms such as electrostatic attraction, ion exchange reactions, and surface chelation. Similarly, organic pollutants like dyes, pesticides, antibiotics, and toxins exhibit high affinity for biochar, undergoing physical and chemical interactions leading to their removal from water. Physisorption and chemical binding play significant roles in the adsorption of organic pollutants onto biochar surfaces. Additionally, biochar-based hybrid materials show promise for enhancing the removal efficiency of ECs from aqueous systems. Overall, biochar offers a sustainable and environmentally friendly solution for mitigating the impacts of ECs on water quality and ecosystem health. Further research and development in this area are essential for optimizing biochar-based remediation strategies and addressing emerging environmental challenges effectively.

References

Abbas, Z., Ali, S., Rizwan, M., Zaheer, I. E., Malik, A., Riaz, M. A., et al. (2018). A critical review of mechanisms involved in the adsorption of organic and inorganic contaminants through biochar, Wabel, M.I *Arabian Journal of Geosciences, 11*, 1–23.

Abbey, C. Y. B., Duwiejuah, A. B., & Quianoo, A. K. (2023). Removal of toxic metals from aqueous phase using cacao pod husk biochar in the era of green chemistry. *Applied Water Science, 13*(2). Available from https://doi.org/10.1007/s13201-022-01863-5.

Ahmad, M., Lee, S. S., Rajapaksha, A. U., Vithanage, M., Zhang, M., Cho, J. S., & Ok, Y. S. (2013). Trichloroethylene adsorption by pine needle biochars produced at various pyrolysis temperatures. *Bioresource Technology, 143*(0), 615–622.

Ahmad, M., Rajapaksha, A. U., Lim, J. E., Zhang, M., Bolan, N., Mohan, D., & Ok, Y. S. (2014). Biochar as a sorbent for contaminant management in soil and water: A review. *Chemosphere, 99*, 19–33.

Ahmed, M. B., Zhou, J. L., Ngo, H. H., & Guo, W. (2015). Adsorptive removal of antibiotics from water and wastewater: Progress and challenges. *Science of the Total Environment, 532*, 112–126.

Ambaye, T. G., Rene, E. R., Dupont, C., Wongrod, S., & Van Hullebusch, E. D. (2020). Anaerobic digestion of fruit waste mixed with sewage sludge digestate biochar: Influence on biomethane production. *Frontiers in Energy Research, 8*, 31.

Amin, S., Solangi, A. R., Hassan, D., Hussain, N., Ahmed, J., & Baksh, H. (2021). Recent trends in development of nanomaterials based green analytical methods for environmental remediation. *Current Analytical Chemistry, 17*(4), 438–448.

Anyakora, C., Ehianeta, T., & Umukoro, O. (2013). Heavy metal levels in soil samples from highly industrialized Lagos environment. *African Journal of Environmental Science and Technology, 7*(9), 917–924.

Baronti, S., Alberti, G., Delle Vedove, G., Di Gennaro, F., Fellet, G., & Genesio, L. (2010). The biochar option to improve plant yields: First results from some field and pot experiments in Italy. *Italian Journal of Agronomy, 5*(1), 3–12.

Beesley, L., Jimenez, E. M., & Eyles, J. L. G. (2010). Effects of BC and greenwaste compost amendments on mobility, bioavailability and toxicity of inorganic and organic contaminants in a multi-element polluted soil. *Environment and Pollution, 158*(0), 2282–2287.

Beesley, L., & Marmiroli, M. (2011). The immobilisation and retention of soluble arsenic, cadmium and zinc by biochar. *Environmental Pollution, 159*(2), 474–480.

Chaturvedi, S., Singh, S., Dhyani, V. C., Govindaraju, K., & Mandal, S. (2021). Characterization, bioenergy value and thermal stability of biochar derived from diverse agriculture and forestry lignocellulosic waste. *Biomass Conversion and Biorefinery*. Available from https://doi.org/10.1007/s13399-020-01239-2.

Chen, C., Zhou, W., & Lin, D. (2015). Sorption characteristics of N-nitrosodimethylamine onto biochar from aqueous solution. *Bioresource Technology, 179*(0), 359–366.

Chen, T., Zhou, Z., Han, R., Meng, R., Wang, H., & Lu, W. (2015). Adsorption of cadmium by biochar derived from municipal sewage sludge: Impact factors and adsorption mechanism. *Chemosphere, 134*, 286–293.

Clay, S. A., Krack, K. K., Bruggeman, S. A., Papiernik, S., & Schumacher, T. E. (2016). Maize, switchgrass, and ponderosa pine biochar added to soil increased herbicide sorption and decreased herbicide efficacy. *Journal of Environmental Science and Health, Part B, 51*(8), 497–507.

Dai, Y., Zhang, N., Xing, C., Cui, Q., & Sun, Q. (2019). The adsorption, regeneration and engineering applications of biochar for removal organic pollutants: A review. *Chemosphere, 223*, 12–27.

Dash, A. K., Dwivedi, B. S., Dey, A., & Meena, M. C. (2023). Temperature sensitivity of soil organic carbon as affected by crop residue and nutrient management options under conservation agriculture. *Journal of Soil Science and Plant Nutrition, 23*(3), 4183–4197. Available from https://doi.org/10.1007/s42729-023-01335-y.

Dash, A. K., Singh, C., & Verma, S. (2021). *Chemistry and recent advances in management of acid sulphate soils. Latest trends in soil science*. New Delhi: Integrated Publication.

Delwiche, K. B., Lehmann, J., & Walter, M. T. (2014). Atrazine leaching from biochar-amended soils. *Chemosphere, 95*(0), 346–352.

Deng, H., Feng, D., He, J. X., Li, F. Z., Yu, H. M., & Ge, C. J. (2017). Influence of biochar amendments to soil on the mobility of atrazine using sorption–desorption and soil thin-layer chromatography. *Ecological Engineering, 99*, 381–390.

Dewage, N. B., Liyanage, A. S., Pittman, C. U., Jr, Mohan, D., & Mlsna, T. (2018). Fast nitrate and fluoride adsorption and magnetic separation from water on α-Fe_2O_3 and Fe_3O_4 dispersed on Douglas fir biochar. *Bioresource Technology, 263*, 258–265.

Dinis, M. D. L., & Fiúza, A. (2021). Mitigation of uranium mining impacts—A review on groundwater remediation technologies. *Geosciences, 11*(6), 250.

Dolar, D., Gros, M., Rodriguez-Mozaz, S., Moreno, J., Comas, J., Rodriguez-Roda, I., & Barceló, D. (2012). Removal of emerging contaminants from municipal wastewater with an integrated membrane system, MBR–RO. *Journal of Hazardous Materials, 239–240*, 64–69. Available from https://doi.org/10.1016/j.jhazmat.2012.03.029.

Dong, F. X., Yan, L., Zhou, X. H., Huang, S. T., Liang, J. Y., Zhang, W. X., & Diao, Z. H. (2021). Simultaneous adsorption of Cr(VI) and phenol by biochar-based iron oxide composites in water: Performance, kinetics and mechanism. *Journal of Hazardous Materials, 416* 125930.

Dong, N., Wang, Z., Wang, J., Song, W., Du, L., Gu, X., & Li, S. (2024). Preparation of CPVC-based activated carbon spheres and insight into the adsorption-desorption performance for typical volatile organic compounds. *Environmental Pollution, 343*, 123177.

Eggen, T., Heimstad, E. S., Stuanes, A. O., & Norli, H. R. (2013). Uptake and translocation of organophosphates and other emerging contaminants in food and forage crops. *Environmental Science and Pollution Research International, 20*(7), 4520–4531. Available from https://doi.org/10.1007/s11356-012-1363-5.

Enaime, G., Baçaoui, A., Yaacoubi, A., & Lübken, M. (2020). Biochar for wastewater treatment—Conversion technologies and applications. *Applied Sciences, 10*(10), 3492.

Fan, X., Qian, Z., Liu, J., Geng, N., Hou, J., & Li, D. (2021). Investigation on the adsorption of antibiotics from water by metal loaded sewage sludge biochar. *Water Science and Technology, 83*(3), 739–750.

Ganie, Z. A., Khandelwal, N., Tiwari, E., Singh, N., & Darbha, G. K. (2021). Biochar-facilitated remediation of nanoplastic contaminated water: Effect of pyrolysis temperature induced surface modifications. *Journal of Hazardous Materials, 417*, 126096.

Gao, P., Munir, M., & Xagoraraki, I. (2012). Correlation of tetracycline and sulfonamide antibiotics with corresponding resistance genes and resistant bacteria in a conventional municipal wastewater treatment plant. *Science of the Total Environment, 421–422*, 173–183.

Godlewska, P., & Oleszczuk, P. (2022). Effect of biomass addition before sewage sludge pyrolysis on the persistence and bioavailability of polycyclic aromatic hydrocarbons in biochar-amended soil. *Chemical Engineering Journal, 429*, 132143.

Graber, E. R., Tsechansky, L., Khanukov, J., & Oka, Y. (2011). Sorption, volatilization, and efficacy of the fumigant 1,3-dichloropropene in a biochar-amended soil. *Soil Science Society of America Journal, 75*(4), 1365–1373.

Gregg, T., Prahl, F. G., & Simoneit, B. R. T. (2015). Suspended particulate matter transport of polycyclic aromatic hydrocarbons in the lower C olumbia River and its estuary. *Limnology and Oceanography, 60*(6), 1935–1949.

Huang, D., Liu, L., Zeng, G., Xu, P., Huang, C., Deng, L., & Wan, J. (2017). The effects of rice straw biochar on indigenous microbial community and enzymes activity in heavy metal-contaminated sediment. *Chemosphere, 174*, 545–553.

Huang, W. H., Lee, D. J., & Huang, C. (2021). Modification on biochars for applications: A research update. *Bioresource Technology, 319*, 124100.

Jadhav, S. B., Surwase, S. N., Kalyani, D. C., Gurav, R. G., & Jadhav, J. P. (2012). Biodecolorization of azo dye Remazol orange by *Pseudomonas aeruginosa* BCH and toxicity (oxidative stress) reduction in *Allium cepa* root cells. *Applied Biochemistry and Biotechnology, 168*(0), 1319–1334.

Jia, M., Wang, F., Bian, Y., Stedtfeld, R. D., Liu, G., Yu, J., & Jiang, X. (2018). Sorption of sulfamethazine to biochars as affected by dissolved organic matters of different origin. *Bioresource Technology, 248*, 36–43. Available from https://doi.org/10.1016/j.biortech.2017.08.082.

Kameyama, K., Miyamoto, T., Shiono, T., & Shinogi, Y. (2012). Influence of sugarcane bagasse-derived biochar application on nitrate leaching in Calcaric dark red soil. *Journal of Environmental Quality, 41*(4), 1131–1137.

Kayan, B., Kalderis, D., Kulaksız, E., & Gözmen, B. D. W. T. (2017). Adsorption of Malachite Green on Fe-modified biochar: influencing factors and process optimization. *Desalination and Water Treatment, 74*, 383–394.

Kemmou, L., Frontistis, Z., Vakros, J., Manariotis, I. D., & Mantzavinos, D. (2018). Degradation of antibiotic sulfamethoxazole by biochar-activated persulfate: factors affecting the activation and degradation processes. *Catalysis Today, 313*, 128–133.

Khalid, S., Shahid, M., Murtaza, B., Bibi, I., Natasha, N. M. A., Asif Naeem, M., & Niazi, N. K. (2020). A critical review of different factors governing the fate of pesticides in soil under biochar application. *Science of the Total Environment, 711*, 134645.

Kim, H. B., Kim, J. G., Kim, T., Alessi, D. S., & Baek, K. (2021). Interaction of biochar stability and abiotic aging: Influences of pyrolysis reaction medium and temperature. *Chemical Engineering Journal, 411*, 128441.

Kumar, R., Qureshi, M., Vishwakarma, D. K., Al-Ansari, N., Kuriqi, A., Elbeltagi, A., & Saraswat, A. (2022). A review on emerging water contaminants and the application of sustainable removal technologies. *Case Studies in Chemical and Environmental Engineering, 6*, 100219.

Kumari, A., Rajput, S., Singh, S. V., et al. (2024). New dimensions into the removal of pesticides using an innovative ecofriendly technique: nanoremediation. In *Nano-Bioremediation for Water and Soil Treatment*, Apple Academic Press.

Li, L., Zou, D., Xiao, Z., Zeng, X., Zhang, L., & Jiang, L. (2019). Biochar as a sorbent for emerging contaminants enables improvements in waste management and sustainable resource use. *Journal of Cleaner Production, 210*, 1324–1342.

Liao, X., Zhang, C., Wang, Y., & Tang, M. (2017). The abiotic degradation of methyl parathion in anoxic sulfur-containing system mediated by natural organic matter. *Chemosphere, 176*, 288–295.

Liu, G., Zheng, H., Jiang, Z., Zhao, J., Wang, Z., Pan, B., & Xing, B. (2018). Formation and physicochemical characteristics of nano biochar: insight into chemical and colloidal stability. *Environmental Science & Technology, 52*(18), 10369–10379.

Lodeiro, C., Capelo, J. L., Oliveira, E., & Lodeiro, J. F. (2019). *New toxic emerging contaminants: Beyond the toxicological effects,* . *Environmental science and pollution research* (Vol. 26). Springer Verlag Issue 1. Available from https://doi.org/10.1007/s11356-018-3003-1.

Luo, J., Li, X., Ge, C., Müller, K., Yu, H., Deng, H., ... Wang, H. (2021). Preparation of ammonium-modified cassava waste-derived biochar and its evaluation for synergistic adsorption of ternary antibiotics from aqueous solution. *Journal of Environmental Management, 298*113530. Available from https://doi.org/10.1016/j.jenvman.2021.113530.

Lyu, P., Wang, G., Wang, B., Yin, Q., Li, Y., & Deng, N. (2021). Adsorption and interaction mechanism of uranium (VI) from aqueous solutions on phosphate-impregnation biochar cross-linked Mg Al layered double-hydroxide composite. *Applied Clay Science, 209*, 106146.

Majhi, P. K., Raza, B., Behera, P. P., Singh, S. K., Shiv, A., Mogali, S. C., & Behera, B. (2022). Future-proofing plants against climate change: A path to ensure sustainable food systems. *Biodiversity, functional ecosystems and sustainable food production* (pp. 73–116). Berlin: Springer.

Majone, M., Verdini, R., Aulenta, F., Rossetti, S., Tandoi, V., & Kalogerakis, N. (2015). In situ groundwater and sediment bioremediation: Barriers and perspectives at European contaminated sites. *New Biotechnology, 32*(1), 133−146.

Meilani, V., Lee, J., Kang, J., Lee, C., Jeong, S., & Park, S. (2021). Application of aluminum-modified food waste biochar as adsorbent of fluoride in aqueous solutions and optimization of production using response surface methodology. *Microporous and Mesoporous Materials, 312*110764. Available from https://doi.org/10.1016/j.micromeso.2020.110764.

Meng, J., Feng, X. L., Dai, Z. M., Liu, X. M., Wu, J. J., & Xu, J. M. (2014). Adsorption characteristics of Cu(II) from aqueous solution onto biochar derived from swine manure. *Environmental Science and Pollution Research International, 21*(11), 7035−7046.

Mirzaei, A., Leonardi, S. G., & Neri, G. (2016). Detection of hazardous volatile organic compounds (VOCs) by metal oxide nanostructures-based gas sensors: A review. *Ceramics International, 42*(14), 15119−15141.

Morin-Crini, N., Lichtfouse, E., Fourmentin, M., Ribeiro, A. R. L., Noutsopoulos, C., Mapelli, F., & Crini, G. (2022). Removal of emerging contaminants from wastewater using advanced treatments. A review. *Environmental Chemistry Letters, 20*(2), 1333−1375.

Mui, E. L. K., Cheung, W. H., Valix, M., & McKay, G. (2010). Dye adsorption onto char from bamboo. *Journal of Hazardous Materials, 177*(1−3), 1001−1005.

Mukome, F. N. D., Zhang, X. M., Silva, L. C. R., Six, J., & Parikh, S. J. (2013). Use of chemical and physical characteristics to investigate trends in biochar feedstocks. *Journal of Agricultural and Food Chemistry, 61*(9), 2196−2204.

Nag, S. K., Kookana, R., Smith, L., Krull, E., Macdonald, L. M., & Gill, G. (2011). Poor efficacy of herbicides in biochar-amended soils affected by their chemistry and mode of action. *Chemosphere, 84*(11), 1572−1577.

Namgay, T., Singh, B., & Singh, B. P. (2010). Influence of biochar application to soil on the availability of As, Cd, Cu, Pb, and Zn to maize (*Zea mays* L.). *Soil Research, 48*(7), 638−647.

Nautiyal, P., Subramanian, K. A., & Dastidar, M. G. (2017). Experimental investigation on adsorption properties of biochar derived from algae biomass residue of biodiesel production. *Environmental Processes, 4*, 179−193.

Nikolaou, S., Vakros, J., Diamadopoulos, E., & Mantzavinos, D. (2020). Sonochemical degradation of propylparaben in the presence of agro-industrial biochar. *Journal of Environmental Chemical Engineering, 8*(4), 104010.

Ogura, A. P., Lima, J. Z., Marques, J. P., Sousa, L. M., Rodrigues, V. G. S., & Espíndola, E. L. G. (2021). A review of pesticides sorption in biochar from maize, rice, and wheat residues: Current status and challenges for soil application. *Journal of Environmental Management, 300*, 113753.

Palansooriya, K. N., Sang, M. K., Igalavithana, A. D., Zhang, M., Hou, D., Oleszczuk, P., ... Ok, Y. S. (2022). Biochar alters chemical and microbial properties of microplastic-contaminated soil. *Environmental Research, 209*, 112807. Available from https://doi.org/10.1016/j.envres.2022.112807.

Pan, H., Yang, X., Chen, H., Sarkar, B., Bolan, N., Shaheen, S. M., & Wang, H. (2021). Pristine and iron-engineered animal- and plant-derived biochars enhanced bacterial abundance and immobilized arsenic and lead in a contaminated soil. *Science of the Total Environment, 763*144218.

Patel, A. K., Katiyar, R., Chen, C. W., Singhania, R. R., Awasthi, M. K., Bhatia, S., & Dong, C. D. (2022). Antibiotic bioremediation by new generation biochar: Recent updates. *Bioresource Technology, 358*, 127384.

Patel, M., Kumar, R., Kishor, K., Mlsna, T., Pittman, C. U., Jr, & Mohan, D. (2019). Pharmaceuticals of emerging concern in aquatic systems: Cchemistry, occurrence, effects, and removal methods. *Chemical Reviews, 119*(6), 3510−3673.

Priyadarshanee, M., & Das, S. (2021). Biosorption and removal of toxic heavy metals by metal tolerating bacteria for bioremediation of metal contamination: A comprehensive review. *Journal of Environmental. Chemical Engineering, 9*(1), 104686.

Qian, W., Liang, J. Y., Zhang, W. X., Huang, S. T., & Diao, Z. H. (2022). A porous biochar supported nanoscale zero-valent iron material highly efficient for the simultaneous remediation of cadmium and lead contaminated soil. *Journal of Environmental Sciences, 113*, 231−241.

Qiao, M., Ying, G. G., Singer, A. C., & Zhu, Y. G. (2018). Review of antibiotic resistance in China and its environment. *Environment International, 110*, 160−172.

Qin, P., Huang, D., Tang, R., Gan, F., Guan, Y., & Lv, X. (2019). Enhanced adsorption of sulfonamide antibiotics in water by modified biochar derived from bagasse. *Open Chemistry, 17*(1), 1309−1316. Available from https://doi.org/10.1515/chem-2019-0141.

Qu, J., Wang, Y., Tian, X., Jiang, Z., Deng, F., Tao, Y., & Zhang, Y. (2021). KOH-activated porous biochar with high specific surface area for adsorptive removal of chromium (VI) and naphthalene from water: Affecting factors, mechanisms and reusability exploration. *Journal of Hazardous Materials, 401*123292.

Raj, A., Jhariya, M. K., Banerjee, A., Meena, R. S., Nema, S., Khan, N., ... Pradhan, G. (2021). Agroforestry a model for ecological sustainability. In M. K. Jhariya, R. S. Meena, A. Raj, & S. N. Meena (Eds.), *Natural resources conservation and advances for sustainability* (pp. 289–308). Elsevier. Available from https://doi.org/10.1016/B978-0-12-822976-7.00002-8..

Rani, M., Shanker, U., & Jassal, V. (2017). Recent strategies for removal and degradation of persistent and toxic organochlorine pesticides using nanoparticles: A review. *Journal of Environmental Management, 190*, 208–222.

Rasool, S., Rasool, T., & Gani, K. M. (2022). A review of interactions of pesticides within various interfaces of intrinsic and organic residue amended soil environment. *Chem. Eng. J. Ad.V., 11*, 100301.

Rhodes, A. H., Carlin, A., & Semple, K. T. (2008). Impact of black carbon in the extraction and mineralization of phenanthrene in soil. *Environmental Science and Technology, 42*(3), 740–745.

Ronsse, F., van Hecke, S., Dickinson, D., & Prins, W. (2013). Production and characterization of slow pyrolysis biochar: Influence of feedstock type and pyrolysis conditions. *GCB Bioenergy, 5*(2), 104–115.

Roy, O., Meena, R. S., Kumar, S., Jhariya, M. K., & Pradhan, G. (2021). Assessment of land use systems for CO_2 sequestration, carbon credit potential, and income security in Vindhyan Region, India. *Land Degradation and Development, 33*(4), 670–682. Available from https://doi.org/10.1002/ldr.4181.

Sabri, N. A., Schmitt, H., Van der Zaan, B., Gerritsen, H. W., Zuidema, T., Rijnaarts, H. H. M., & Langenhoff, A. A. M. (2020). Prevalence of antibiotics and antibiotic resistance genes in a wastewater effluent-receiving river in the Netherlands. *Journal of Environmental Chemical Engineering, 8*(1), 102245.

Saraswat, A., Nath, T., Omeka, M. E., Unigwe, C. O., Anyanwu, I. E., & Ugar, S. I. (2023). Irrigation suitability and health risk assessment of groundwater resources in the Firozabad industrial area of north-central India: An integrated indexical, statistical, and geospatial approach. *Frontiers in Environmental Science, 11*, 296.

Saraswat, A., Ram, S., Kouadri, S., Raza, M. B., Hombegowda, H. C., Kumar, R., ... & Jena, R. K. (2023). Groundwater quality, fluoride health risk and geochemical modelling for drinking and irrigation water suitability assessment in Tundla block, Uttar Pradesh, India. Groundwater for Sustainable Development, 100991.

Saraswat, A., Ram, S., Raza, M. B., Islam, S., Sharma, S., Omeka, M. E., & Golui, D. (2023). Potentially toxic metals contamination, health risk, and source apportionment in the agricultural soils around industrial areas, Firozabad, Uttar Pradesh, India: A multivariate statistical approach. *Environmental Monitoring and Assessment, 195*(7), 863.

Sharma, G. K., Jena, R. K., Hota, S., Kumar, A., Ray, P., Fagodiya, R. K., & Ray, S. K. (2020). Recent development in bioremediation of soil pollutants through biochar for environmental sustainability. *Biochar applications in agriculture and environment management*, 123–140.

Sharma, S., Khardia, N., & Saraswat, A. (2022). Nanobiochar: Production and application. *A Monthly Peer Reviewed Magazine for Agriculture and Allied Sciences*, 69.

Sharma, S., Singh, Y. V., Saraswat, A., Prajapat, V., & Kashiwar, S. R. (2021). Groundwater quality assessment of different villages of Sanganer block in Jaipur District of Rajasthan. *Journal of Soil Salinity and Water Quality, 13*(2), 248–254.

Shi, M., Hu, L. C., Huang, Z. Q., & Dai, J. Y. (2011). The influence of bio-char inputting on the adsorption of phenanthrene by soils and by maize seedlings. *Journal of Agro-Environment Science, 30*(0), 912–916.

Siipola, V., Pflugmacher, S., Romar, H., Wendling, L., & Koukkari, P. (2020). Low-cost biochar adsorbents for water purification including microplastics removal. *Applied sciences, 10*(3), 788.

Silvani, L., Hjartardottir, S., Bielská, L., Škulcová, L., Cornelissen, G., Nizzetto, L., & Hale, S. E. (2019). Can polyethylene passive samplers predict polychlorinated biphenyls (PCBs) uptake by earthworms and turnips in a biochar amended soil? *Science of the Total Environment, 662*, 873–880.

Singh, S., Luthra, N., Mandal, S., Kushwaha, D. P., Pathak, S. O., Datta, D., ... Pramanik, B. (2023). Distinct behavior of biochar modulating biogeochemistry of salt-affected and acidic soil: A review. *Journal of Soil Science and plant Nutrition*. Available from https://doi.org/10.1007/s42729-023-01370-9.

Singh, S. V., Luthra, N., Bhattacharya, S., et al. (2024). Occurrence of emerging contaminants in soils and impacts on rhizosphere. In *Emerging Contaminants Sustainable Agriculture and the Environment*, Elsevier Sciences.

Singh, S. V., Sharma, R., Balan, P., et al. (2024). Nanomaterials for inorganic pollutants removal from contaminated water. In *Nano-Bioremediation for Water and Soil Treatment*, Apple Academic Press.

Stefanakis, A. I., & Becker, J. A. (2016). A review of emerging contaminants in water: Classification, sources, and potential risks. *Impact of water pollution on human health and environmental sustainability*, 55–80.

Sun, L., Wan, S., & Luo, W. (2013). Biochars prepared from anaerobic digestion residue, palm bark, and eucalyptus for adsorption of cationic methylene blue dye: Characterization, equilibrium, and kinetic studies. *Bioresource Technology, 140*, 406–413.

Tan, X., Liu, Y., Zeng, G., Wang, X., Hu, X., Gu, Y., & Yang, Z. (2015). Application of biochar for the removal of pollutants from aqueous solutions. *Chemosphere, 125*(0), 70–85.

Tkaczyk, A., Mitrowska, K., & Posyniak, A. (2020). Synthetic organic dyes as contaminants of the aquatic environment and their implications for ecosystems: A review. *Science of The Total Environment, 717*137222. Available from https://doi.org/10.1016/j.scitotenv.2020.137222.

Tokarčíková, M., Peikertová, P., Barabaszová, K. Č., Životský, O., Gabor, R., & Seidlerová, J. (2023). Regeneration possibilities and application of magnetically modified biochar for heavy metals elimination in real conditions. *Water Resources and Industry, 30*, 100219.

Tong, X., Li, J., Yuan, J., & Xu, R. (2011). Adsorption of Cu(II) by biochars generated from three crop straws. *Chemical Engineering Journal, 172*(2–3), 828–834.

Truong, Q. M., Nguyen, T. B., Chen, W. H., Chen, C. W., Patel, A. K., Bui, X. T., ... Dong, C. D. (2023). Removal of heavy metals from aqueous solutions by high performance capacitive deionization process using biochar derived from Sargassum hemiphyllum. *Bioresource Technology, 370*, 128524.

Velayatzadeh, M. (2023). *Heavy metals in surface soils and crops*. IntechOpen. Available from http://doi.org/10.5772/intechopen.108824.

Wang, C., Wang, Y., & Herath, H. (2017). Polycyclic aromatic hydrocarbons (PAHs) in biochar – their formation, occurrence and analysis: A review. *Organic Geochemistry, 114*, 1–11.

Wang, Z. H., Guo, H. Y., Shen, F., Yang, G., Zhang, Y. Z., Zeng, Y. M., & Deng, S. H. (2015). Biochar produced from oak sawdust by lanthanum (La)-involved pyrolysis for adsorption of ammonium (NH_4^+), nitrate (NO_3^-), and phosphate (PO_4^{3-}). *Chemosphere, 119*(0), 646–653.

Wei, S., Zhu, M., Fan, X., Song, J., Peng, P., Li, K., & Song, H. (2019). Influence of pyrolysis temperature and feedstock on carbon fractions of biochar produced from pyrolysis of rice straw, pine wood, pig manure and sewage sludge. *Chemosphere, 218*, 624–631.

Werner, D., & Karapanagioti, H. K. (2005). Comment on modelling-maximum adsorption capacities of soot and soot-like materials for PAHs and PCBs. *Environmental Science and Technology, 39*(0), 381–382.

Wilkinson, J., Hooda, P. S., Barker, J., Barton, S., & Swinden, J. (2017). Occurrence, fate and transformation of emerging contaminants in water: An overarching review of the field. *Environmental Pollution, 231*(1), 954–970.

Wu, F., Chen, L., Hu, P., Wang, Y., Deng, J., & Mi, B. (2021). Industrial alkali lignin-derived biochar as highly efficient and low-cost adsorption material for Pb(II) from aquatic environment. *Bioresource Technology, 322*, 124539.

Xie, M., Chen, W., Xu, Z., Zheng, S., & Zhu, D. (2014). Adsorption of sulfonamides to demineralized pine wood biochars prepared under different thermochemical conditions. *Environmental Pollution, 186*(0), 187–194.

Xu, R. K., Xiao, S. C., Yuan, J. H., & Zhao, A. Z. (2011). Adsorption of methyl violet from aqueous solutions by the biochars derived from crop residues. *Bioresource Technology, 102*(22), 10293–10298.

Yao, Y., Gao, B., Zhang, M., Inyang, M., & Zimmerman, A. R. (2012). Effect of biochar amendment on sorption and leaching of nitrate, ammonium, and phosphate in a sandy soil. *Chemosphere, 89*(11), 1467–1471.

You, X., Suo, F., Yin, S., Wang, X., Zheng, H., Fang, S., & Li, Y. (2021). Biochar decreased enantioselective uptake of chiral pesticide metalaxyl by lettuce and shifted bacterial community in agricultural soil. *Journal of Hazardous Materials, 417*126047.

Yu, H., Zou, W., Chen, J., Chen, H., Yu, Z., & Huang, J. (2019). Biochar amendment improves crop production in problem soils: A review. *Journal of Environmental Management, 232*, 8–21.

Zhang, G., Liu, X., Sun, K., He, Q., Qian, T., & Yan, Y. (2013). Interactions of simazine, metsulfuron-methyl, and tetracycline with biochars and soil as a function of molecular structure. *Journal of Soils and Sediments, 13*(9), 1600–1610.

Zhang, N., Reguyal, F., Praneeth, S., & Sarmah, A. K. (2023). A novel green synthesized magnetic biochar from white tea residue for the removal of Pb (II) and Cd (II) from aqueous solution: Regeneration and sorption mechanism. *Environmental Pollution, 330*, 121806.

Zhang, P., Min, L., Tang, J., Rafiq, M. K., & Sun, H. (2020). Sorption and degradation of imidacloprid and clothianidin in Chinese paddy soil and red soil amended with biochars. *Biochar, 2*, 329–341.

Zhang, P., Sun, H., Min, L., & Ren, C. (2018). Biochars change the sorption and degradation of thiacloprid in soil: Insights into chemical and biological mechanisms. *Environmental Pollution, 236*, 158–167.

Zhu, D., Kwon, S., & Pignatello, J. J. (2005). Adsorption of single-ring organic compounds to wood charcoals prepared under different thermochemical conditions. *Environmental Science and Technology, 39*(11), 3990–3998.

Zhu, L., Tong, L., Zhao, N., Wang, X., Yang, X., & Lv, Y. (2020). Key factors and microscopic mechanisms controlling adsorption of cadmium by surface oxidized and aminated biochars. *Journal of Hazardous Materials, 382*, 121002.

CHAPTER 22

Tailored biochar: a win–win strategy to remove inorganic contaminants from soil and water

Saptaparnee Dey[1], T.J. Purakayastha[1] and Anurag Bera[2]

[1]Division of Soil Science and Agricultural Chemistry, ICAR—Indian Agricultural Research Institute, Pusa, New Delhi, India [2]Department of Agronomy, Institute of Agricultural Sciences, Banaras Hindu University, Varanasi, Uttar Pradesh, India

22.1 Introduction

Inorganic contaminants, viz., heavy metals, nitrate, and phosphate, are the major pollutants which can jeopardize soil health and crop productivity. They can also result in contsmination of food chain, groundwater pollution, and eutrophication. Agricultural soils are contaminated with a wide variety of metal(loid)s and organic contaminants due to widespread anthropogenic activities (Pradhan, Meena, Kumar, & Lal, 2023). With the increase in population, the use of heavy metals and the addition of nitrate phosphatic fertilizers has increased to a great extent. Inorganic contaminants also result from industrial production, mining, sewage sludge, and diffuse sources such as metal piping, traffic, and combustion by-products from coal-burning power stations. Concomitantly excess amounts of these inorganic contaminants, including heavy metals and nutrients, pose a severe issue in our soil and water (Singh, Sharma, & Balan, 2024). Nevertheless, in soil, inorganic contaminants lead to heavy metal accumulation, deteriorate soil health, and cause bioaccumulation of metals, leading to biomagnification. There is an urgent need for the simultaneous removal of several inorganic impurities from wastewater and soil systems. Different methods including chemical precipitation, ion exchange, reverse osmosis, and biological processes are available to remove these inorganic contaminants from water, but they are costly and not effective (Chen et al., 2022a; Kumari, Rajput, & Singh, 2024). Moreover, in soil, inorganic contaminant removal can be done with phytoremediation, bioremediation or other management practices, but they are not easy and cost-effective. Nowadays, one of the most promising technologies is application of biochar.

The discovery of biochar goes back to 1960 by the Late Wim Sombreak who, during a soil survey in the Brazilian Amazon basin, discovered Terra Preta (tera pretado indio in Portuguese) where plant parts voluntarily introduced into the soil by the tribes over thousands of years—a sort of slash and char strategy. Biochar is a stable carbon-rich byproduct produced by pyrolyzing biomass in an oxygen-free environment (Amonette & Joseph, 2009; Singh, Chaturvedi, & Datta, 2019). It is a win—win strategy for carbon sequestration, reducing greenhouse gas emissions, land remediation, contaminant immobilization and soil fertilization (Meena, Pradhan, Kumar, & Lal, 2023; Zhang et al., 2019). However, pristine biochars have low cation exchange capacities (CECs), and many weathering studies have indicated that it will take decades for biochar's surface to naturally oxidize. In this regard, numerous researchers looked at various physical (ball milling, microwave pyrolysis, steam activation), chemical (acid, alkali and other chemical treatments), and biological techniques (Pradhan & Meena, 2023; Rajapaksha et al., 2016) for modification of pristine biochars. Among them, chemical treatment is comparatively easy, efficient, and cost-effective. Pore filling, electrostatic interactions, H-bonding, ion exchange, complexation, precipitation, and redox are suggested mechanisms for the metal(loid) and inorganic contaminants adsorption by biochar (Zhang et al., 2019) as depicted in Fig. 22.2. Myriad studies are available where engineered biochar removed inorganic contaminants from soil and water more efficiently than pristine biochar. This chapter aims to discuss the recent development of tailored biochar (with special reference to chemical modification), their efficiency in removing inorganic contaminants from both soil and water and the probable mechanism behind it.

22.2 Occurrence and fates of inorganic pollutants in the soil and water ecosystem

The intricate interplay of inorganic pollutants within soil and water ecosystems is quite complex, and scientists are very interested in understanding this because it greatly affects environmental dynamics and human health. These pollutants, stemming from anthropogenic activities encompassing industrial effluents, agricultural practices, and urban runoff, traverse complex pathways governed by their physicochemical attributes and prevailing biogeochemical processes (Singh, Luthra, & Bhattacharya, 2024). These pollutants have multifaceted interactions in soil, including adsorption onto mineral surfaces, complexation with organic matter, and precipitation as insoluble compounds. These processes collectively contribute to their immobilization within the soil matrix. At the same time, metals such as cadmium (Cd), lead (Pb), and mercury (Hg) exhibit a propensity for persistent accumulation, potentially compromising soil quality and facilitating bioaccumulation through plant uptake (Hameed et al., 2023; Meena & Pradhan, 2023).

In aquatic environments, the fate of inorganic pollutants is also complex. When released into water, these pollutants disperse due to water movement and often interact at the sediment-water boundary. Here, they stick to sediment particles, staying out of the water. However, they can be stirred back up by forces like water currents and organisms living in the sediment. Furthermore, introducing excessive inorganic compounds, notably nitrates and phosphates, instigates eutrophication processes (El-Sheekh et al., 2021;

Roy et al., 2022). This culminates in accelerated algal growth, depletion of oxygen during decomposition, and potential disruptions in aquatic ecosystems.

Inorganic pollutant removal is paramount, prompting scientists to investigate diverse approaches. A multidisciplinary strategy is necessary, and one such avenue is the utilization of biochar. Additionally, collaborative efforts spanning fields like chemistry, engineering, and environmental science are being employed to develop effective solutions. This collective approach enhances our understanding of pollutant behavior and fosters innovative techniques for a cleaner environment.

22.3 Modification methods for the preparation of engineered biochar

Numerous studies exist where researchers attempt to prepare engineered biochar using various modification techniques which significantly improve the bioremedial efficiency of biochar (Table 22.1). These methods are mainly classified into physical, chemical, and biological modification summarized in Fig. 22.1.

Chemical treatments modify the physical and chemical properties of biochar. Treatment with acid and oxidizing agents add different functional groups, such as carboxyl, ketone, etc., on the biochar surface, which augment surface charge on biochar and help to adsorb cationic contaminants from soil and water by electrostatic attraction (Kharel et al., 2019). One of the most common methods of engineered biochar preparation is acid treatment. In general, hydrochloric (HCl), phosphoric (H_3PO_4) acids, sulfuric (H_2SO_4) and nitric (HNO_3) acids (individually or in combination), as well as organic acids such as citric ($C_5H_8O_7$), and oxalic (C_2H_2O) acid are employed for this purpose (Chen et al., 2022a). Furthermore, alkali treatments (mainly NaOH, KOH) are also reported to enhance porosity and functional groups on biochar and enhance the removal efficiency of contaminants. Various oxidizing agents, including hydrogen peroxide (H_2O_2), ozone (O_3), and potassium permanganate ($KMnO_4$), also enhance oxygen-containing surface functional groups on biochar. Nevertheless, different iron salts [(for instance, goethite (FeOOH), hematite (Fe_2O_3)] and other oxide-loaded biochar, nonmetallic heteroatom–doped biochar are reported to improve inorganic contaminant removal (cationic and anionic) by adsorption, chemical precipitation, complex formation, redox processes, etc. Moreover, treatment with different surfactants, clay, and organic solvents increased surface area and enhanced functional groups of biochar, making tailored biochar a potential candidate for contaminant removal from soil and water (Rajapaksha et al., 2016).

Furthermore, there are different methods for biological modifications, like- aerobic/anaerobic digestion, biofilm production, etc. (Rajapaksha et al., 2016). Different physical modifications are ball milling (Lyu et al., 2018), microwave pyrolysis (Mohamed, Kim, Ellis, & Bi, 2016), steam activation (Chia, Downie, Munroe, 2015), gas purging (Rajapaksha et al., 2016), etc. Steam activation mainly increases the porosity and surface area of biochar, resulting in higher inorganic contaminant removal from soil and water systems (Chen et al. 2022a). Microwave activation is an advanced technique of modification of biochar where high-frequency electromagnetic irradiation is used with frequencies ranging from 0.03 to 300 GHz (Duan, Oleszczuk, Pan, & Xing, 2019). It could heat both the inner and outer surface of biochar simultaneously, resulting in a more functional group and

TABLE 22.1 Recent developments in inorganic contaminant removal by engineered biochar.

Result	Quantity	Mechanism	References
Hydrous zirconium oxide (ZrO) modified biochar reduced As contamination in leaves and stem of indicator plant cowpea (Vigna unguiculata L.)	Concentration was reduced by 42.6% (leaves) and 30.8% (stems)	Sorption	Chen et al. (2022a)
CaSiO$_3$ modified rice husk and wood biochar by ball milling increased Cd(II) sorption from solution	Sorption by wood biochar-57% Sorption by modified rice husk biochar-98%	Cation exchange of Ca^{2+}, precipitation. carbon-Si coupled structure of rice husk derived biochar is very important for the sorption of Cd(II)	Liu et al. (2023)
KOH-modified chicken feather biochar enhanced Cd(II) sorption	Adsorption capacity-62.14 mg g^{-1}	Increased porosity, more functional groups	Chen et al. (2021)
Mg/Fe bimetallic oxide-modified biochar derived from *Eupatorium adenophorum* increased the removal of Pb(II) and Cd(II) from wastewater.	Pb(II) and Cd(II) adsorption are 252.70 and 156.60 mg g^{-1}, respectively	Increased functional groups, change in chemical properties	Cheng et al. (2023)
α-Fe$_2$O$_3$ modified biochar increased sorption of Cu(II), Pb(II), As(V)	Sorption of Cu(II) (258.22 mg · g^{-1}), Pb(II) (390.60 mg · g^{-1}) by α-Fe$_2$O$_3$ modified biochar prepared at 350°C, and sorption of As(V) (5.78 mg g^{-1}) by modified biochar prepared at 250°C	Chemisorption electrostatic adsorption, ion exchange, precipitation and complexation	Li et al. (2023)
Novel MnFe$_2$O$_4$ loaded biochar increased Cd(II) sorption	Sorption by modified and unmodified biochar were 127.37 mg g^{-1} and 25.73 mg g^{-1}, respectively	Net-like porous structure and 18-fold enlarged specific surface area	Bai et al. (2023)
Fe−Mn oxide-modified corn straw biochar enhanced Hg(II) and Cd(II)	The maximum adsorption capacities of Hg(II) and Cd(II) were 86.82 and 131.03 mg g^{-1}, respectively.	Chemical complexation reaction (by forming monodentate or multidentate inner sphere complexes)	Sun, Sun, Xu, Wang, and Liang (2023)
KMnO$_4$ modified walnut shell biochar enhanced Cu(II), Pb(II), Cd(II), Zn(II) sorption from monometal solutions	From Langmuir model the adsorption maxima for Cu(II), Pb(II), Cd(II), Zn(II) were 30.18, 70.37, 44.94 and 58.96 mg · g^{-1}, respectively, in monometal solutions	Mn−O and Mn−OH appeared on biochar surface and metals were adsorbed mainly by replacing the Mn^{2+} in O−Mn and formed complexes with the surface functional groups (−OH, −COOH)	Chen et al. (2023b)
Fe−Mn oxide-biochar increased As sorption from aqueous solutions	Adsorption capacity-8.80 mg g^{-1}	FTIR and XPS analysis showed that As(III) was oxidized to As(V) on the surface of Fe- and Mn-oxide particles and adsorbed on the Fe−Mn oxide surface forming As-O bonds.	Lin, Song, Huang, Khan, and Qiu (2019)

(Continued)

TABLE 22.1 (Continued)

Result	Quantity	Mechanism	References
Nano-Ce_2O functionalized biochar reduced phosphate leaching and enhanced total phosphate content in soil	Enhanced total P content of the surface soil by 7.22%	Chemisorption, electrostatic attraction, surface precipitation and ligand exchange	Feng et al. (2017)
Corn cob biochar treated with H_3PO_4 + 0.3 M NaOH increased NH_4^+ adsorption	Increased sorption by 42%	Increased functional group, CEC, pore space, surface area	Vu et al. (2018)
Rice straw biochar treated with O_3-HCl-$FeCl_3$ increased retention of NH_4^+–N, NO_3^-–N, PO_4^{3-}–P and K^+ by a sandy loam soil	Retention of NH_4^+-N, NO_3^-–N, PO_4^{3-}–P and K^+ enhanced by 33.7, 27.8, 15.0, and 5.74% over a comparable dose of untreated rice straw biochar.	Increased CEC and AEC of biochar, outer and inner sphere complex formation	Dey et al. (2023)
MgO-modified peanut shell biochar increased phosphate sorption	Maximum phosphate adsorption capacity 1809 mg·g^{-1}	Protonation, electrostatic attraction, monodentate complexation and bidentate complexation	Liang et al. (2023)
Fe–Mn oxide-biochar [biochar: $FeSO_4$: $KMnO_4$::18:3:1] (Weight ratio) increased As adsorption from aqueous solutions	Adsorption capacity- 8.80 mg g^{-1}	FTIR and XPS analysis showed that As(III) was oxidized to As(V) on the surface of Fe- and Mn-oxide particles and adsorbed on the Fe–Mn oxide surface forming As-O bonds.	Lin et al. (2019)
MgO-biochar nanocomposites increased removal of phosphate from aqueous solutions	62.9% removal	Increased surface area, phosphate sorption by surface precipitation	Zheng, Wan, Chen, Chen, and Gao (2020)
Ozone oxidation followed by ammonium hydroxide reaction chemically modified *Pisum sativum* peels waste biochar enhanced Cu(II) sorption	23% removal	Amino groups helped in removal of contaminant	El-Nemr et al. (2020)
Impact of physical/chemical/biological/mixed treatment of biochar to remove inorganic contaminants			
Poplar woodchip biochar modified by ball milling improves Hg(II) removal	Maximum adsorption capacity 320 mg g^{-1}	Surface adsorption, ligand exchange, surface complexation	Lyu et al. (2020)
Biochar pyrolyzed at different pyrolysis temperatures (300°C, 500°C, and 700°C) and ball milled with 3-trimethoxysilylpropanethiol	Hg(II) adsorption to the tune of 401.8, 379.6 and 270.6 mg g^{-1} by engineered biochar prepared at 300, 500, and 700 °C respectively and sorption of methyl mercury 108.16, 85.27 and 39.14 mg g^{-1} respectively	Increased pore size, surface area, O-containing functional group which helped in loading of -SH group	Zhao et al. (2022)

(Continued)

TABLE 22.1 (Continued)

Result	Quantity	Mechanism	References
Steam-activated canola straw biochar pyrolyzed at 700°C increased lead(II) adsorption	Highest adsorption capacity 195 mg g^{-1}	Increased surface area	Kwak et al. (2019)
Ball milled cow bone meal pyrolyzed biochar increased Cd(II), Cu(II) and Pb(II) sorption	Maximum adsorption capacity of engineered biochar for Cd(II), Cu(II) and Pb(II) were 165.77, 287.58 and 558.88 mg g^{-1}, respectively which indicated increase of sorption capacity to the tune of 93.91.%, 75.56% and 64.61% compared to unmilled biochar	Surface complexation, cation exchange, chemical precipitation, electrostatic interaction and cation-π bonding	Xiao, Hu, and Chen (2020)
Ball-milled biochar was entrapped in Ca-alginate beads which enhanced Cd(II) adsorption from water	Maximum Cd(II) adsorption capacity 227.1 mg g^{-1}	Surface complexation, attraction, high surface area	Wang et al. (2018)
Ball-milled bamboo biochar increased ammonium adsorption from water	Maximum adsorption capacity 22.9 mg g^{-1}	Increased surface area	Qin et al. (2020)
NaOH and ball milling co-modified biochar increased Cd(II) adsorption	Maximum Cd(II) adsorption capacity 83.59 mg g^{-1}	Increased specific surface area, mineral content and cation exchange capacity of biochar, which improved precipitation, cation exchange and complexation with Cd(II)	Du et al. (2022)
Pseudomonas sp. NT-2 loaded maize biochar increased Cd(II) and Cu(II) sorption	Reduced bioavailability	Surface complexation, cation exchange, chemical precipitation, electrostatic interaction by maize biochar and functional groups (i.e., −NH bond, −COOH, −OH) on the NT-2 cell helped in biosorption and biomineralization	Tu et al. (2020)
Ball-milled bone-derived biochar loaded with Fe$_3$O$_4$ increased Pb(II) sorption	Maximum adsorption capacity 339.39 mg g^{-1}	Electrostatic attraction, cation exchange precipitation, complexation, coprecipitation	Qu et al. (2023)
Biologically modified biochar prepared from digestion residue of corn straw silage increased Cd(II) sorption from water	Adsorption capacity 175.44 mg g^{-1}	Increased surface area, electrostatic attraction	Tao et al. (2019)

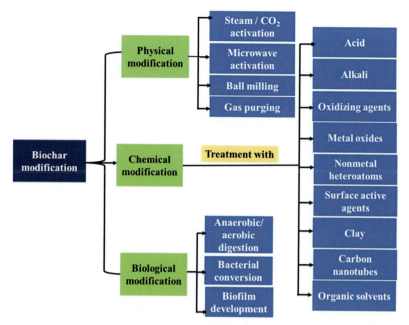

FIGURE 22.1 Different methodologies for the preparation of engineered biochar.

increased surface area of engineered biochar (Zhang et al. 2022). Ball milling is an eco-friendly and low-cost technique which can adjust the particle size to the nanoscale (<1000 nm) under nonequilibrium conditions (Lyu et al. 2020; Pradhan & Meena, 2023a). Researchers have explored a method to enhance biochar's gas adsorption capabilities by subjecting it to treatment with a mixture of high-temperature CO_2 and ammonia. This approach has been studied for its potential to adsorb carbon dioxide (CO_2) gas, as documented by Rajapaksha et al. (2016).

Thus, the modification techniques of biochar make the tailored biochar a promising candidate for the removal of contaminants from soil and water systems by electrostatic attraction, chemical complexation, surface precipitation, pore fillings, H-bonding, surface adsorption, ion exchange, etc. (Dey et al., 2023; Kharel et al., 2019; Rajapaksha et al., 2016). The proposed mechanism of contaminant removal by engineered biochar is shown in Fig. 22.2.

22.4 Chemical modification of biochar and their effect on contaminant bioremediation from soil and water

22.4.1 Acid-treated biochar

Acid modification of biochar results in the incorporation of various oxygen-containing functional groups on its surface, such as carboxylic, phenolic OH, alcoholic OH, ketone, etc. which raise the surface charge and, as a result, its CEC (Pradhan & Meena, 2022;

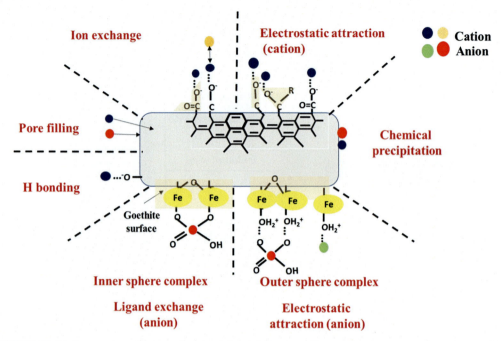

FIGURE 22.2 Proposed mechanism of removal/immobilization of inorganic contaminants by engineered biochar from soil and water. Source: *Reused and modified from Dey, S., Purakayastha, T. J., Sarkar, B., Rinklebe, J., Kumar, S., Chakraborty, R., ... Shivay, Y. S. (2023). Enhancing cation and anion exchange capacity of rice straw biochar by chemical modification for increased plant nutrient retention. Science of The Total Environment, 886, 163681; License number-5594590050510.*

Wang et al. 2019, Wang & Wang, 2019). Wang & Wang (2019) found that when maize stalks biochar was treated with H_3PO_4, there was a dramatic increase in the oxygen-containing groups compared to unmodified biochar. While Uchimiya, Bannon, and Wartelle (2012) reported that being a strong oxidizing agent, acid mixture increased O/C ratios more in biochar as compared to single acid treatment, that is, 30% HNO_3. Furthermore, through ligand- and proton-promoted reactions, oxalic and other organic acids improved pollutant sorption by engineered biochar (Vithanage et al., 2015). Thus, altering biochar's physical and chemical characteristics with high porosity and/or high surface charge may be employed to effectively remove inorganics from soil and water (Roy, Meena, Kumar, Jhariya, & Pradhan, 2021; Singh, Chaturvedi, Dhyani, & Govindaraju, 2020; Uchimiya et al., 2012).

Acid treatment induced laminar carboxylate structure on the biochar surface of *Opuntia ficus indica* cactus fibers, which had a high affinity for Cu(II) ions and helped to remove Cu(II) from aqueous solution by forming outer and inner sphere complexes at near-neutral and acidic pH, respectively (Hadjittofi, Prodromou, & Pashalidis, 2014; Pradhan et al., 2022). Trivalent lanthanides were more strongly bound after 8 M HNO_3 treatments of *Luffa cylindrica* biochar (Liatsou, Pashalidis, Oezaslan, & Dosche, 2017). Because of its superior vascular structure and insertion of acid anhydride groups (hydrolyzed rapidly to carboxylic acid), trivalent lanthanides from aqueous solution were more readily adsorbed on the surface of designed biochar.

To investigate the performance of tailored biochar in the removal of heavy metals from polluted soil, a 63-day incubation experiment was conducted using 5% HCl-treated modified coconut shell biochar. It was reported that biochar reduced acid-soluble Ni, Cd, and Zn levels by 57.2%, 30.1%, and 12.7%, respectively, and effectively immobilized acid-soluble Ni, Cd, and Zn (Liu et al., 2022). Moreover, Hazrati, Farahbakhsh, Cerdà, and Heydarpoor (2021) discovered that citric acid treatment of ultrasound pretreated sewage sludge biochar had a significant ability for pollutant removal. It reduced the availability of Pb, Zn, and Cd in soil by 85.3%, 82.9%, and 30.6%, respectively, and decreased the ecological risk index. The scanning electron microscopy (SEM) and Fourier transform infrared (FTIR) spectroscopy results corroborated that this modified biochar might improve the sorption of certain heavy metals through chelation. Furthermore, citric acid improved the sorption of Cd and Pb from the soil in chickpea biochar (Kumawat et al., 2022; Nazari, Rahimi, & Nezhad, 2019). Li et al. (2016) treated wheat straw biochar with 6 M HCl followed by $FeCl_3$ (maintaining the ratio of iron to biochar at 0.7) to enhance the AEC of biochar, which would increase nitrate and phosphate sorption. It was found that HCl treatment resulted in the protonation of surface hydroxyl groups of biochar and increased nitrate adsorption, indicating a rise in AEC. Potassium-iron rice biochar composite augmented nitrate sorption but showed poor sorption capacity for NH_4^+ signifying higher AEC (Chandra, Medha, & Bhattacharya, 2020). Myriad studies have shown that acid treatment increases the quantity of carboxylic, phenolic OH, alcoholic OH and carbonyl groups. Acid-treated biochar emerges as a feasible option for the removal of cationic contaminants from soil and water systems.

22.4.2 Alkali-treated biochar

Numerous investigations have shown that alkali treatments change the physicochemical characteristics of biochar. Alkali treatment (mainly NaOH, KOH) integrates new O-containing functional groups (mostly lactone, OH groups, etc.) on the surface of biochar, enhances graphitic carbon content (Fan et al., 2010), and improves surface area and porosity (Dey et al., 2023; Fan et al., 2010). These properties make alkali-modified biochar a viable option for inorganic pollutant sorption.

Jin et al. (2014) reported that 2 M KOH treatment of municipal solid waste biochar significantly enhanced arsenic (V) adsorption due to increased surface area, pore volume, phenolic hydroxyl groups (—OH), and carboxylic groups. Moreover, hydrothermal carbonization of switch grass was able to absorb more Cu(II) and Cd(II) after being treated with KOH (Regmi et al., 2012; Sheoran et al., 2022). Surface complexation, electrostatic attraction, and/or surface precipitation increased the Cd(II) sorption capacity after NaOH treatment. A subsequent study also validated the adsorption technique. In this regard, An et al. (2021) reported that pomelo peel biochar developed with NaOH modification recorded higher Mn(II) adsorption. Moreover, attenuated total reflection FTIR (ATR-FTIR) spectroscopy, X-ray photoelectron spectroscopy (XPS) and X-ray diffraction (XRD) evidenced that the modification process introduced $-COO^-$ and CO_3^{2-} on biochar, which could combine with Mn(II) to form $-COOMn^+$ and $MnCO_3$ through cation exchange and surface precipitation. Nevertheless, corncobs, cotton stalks, and peanut shells used as biochar feedstock successfully removed ammonium ions from water (Liu, Xue, Gao, Cheng, & Yang, 2016). The efficacy of heavy metal (Cu, Pb, Ni) removal from wastewater was improved by ultrasound (170 kHz) pretreated softwood biochar after alkali activation, and this designed biochar recorded specific adsorption in the following order: Cu > Pb > Ni (Peter, Chabot, & Loranger, 2021). It was further supported by peaks in FTIR of treated biochar where KOH treatment had broadened the band at $3400\ cm^{-1}$ signifying a higher amount of hydrogen-bonded OH group of alcohols and phenols (Meena et al., 2022; Yakout, 2015). Zhong et al. (2021) prepared millet bran biochar modified with NaOH for Cr (VI) sorption from an aqueous solution. The authors postulated that lactate might have formed hydrogen bonds with oxygen- or nitrogen-containing groups on biochar via other groups and might have formed chelate with Cr (VI) via $-COOH$ (or $-OH$). This facilitated electron transport from Cr (VI) to biochar ($-OH$ and $-NH_2$ group) and thus reduced Cr (VI) in water.

Alkali-modified biochar was also used for the remediation of inorganic contaminants from soil. Rizwan et al. (2020) found that the AEC of NaOH-treated rice straw biochar was 81 cmol (e^-) kg^{-1}, compared to 25.1 cmol (e^-) kg^{-1} for untreated biochar. The N-containing functional groups (amide, imide, lactone, pyrrolic, and pyridinic groups) might have been boosted during pyrolysis by NaOH pretreatment, which raised the positive charge on the surface. Furthermore, Liu et al. (2021b) reported that the application of KOH-modified rice hull biochar decreased the amount of available Zn in the soil, because the KOH treatment increased the "-conjugated aromatic structure," strengthened Zn(II)−π interaction, and facilitated $Zn(OH)_2$ formation in the more alkaline soil. Furthermore, it has been reported that after alkali treatment CEC was augmented by six folds which might be due to the increase in the amount of active pore sites, surface functional groups, and exchangeable cations (Wongrod, Simon, van Hullebusch, Lens, & Guibaud, 2018). These factors might help heavy metals and other contaminants bind on the biochar surface.

22.4.3 Oxidizing agent−treated biochar

Various oxidizing agents, including hydrogen peroxide (H_2O_2), ozone (O_3), potassium permanganate ($KMnO_4$), and others, have been employed in numerous studies to prepare engineered biochar for the removal of inorganic contaminants from soil and water. Oxidizing agents furnish biochar with oxygen-containing functional groups, enhancing surface charge on biochar. Kharel et al. (2019) observed that surface ozonization of biochar markedly increased the CEC of biochar. It was further confirmed by SEM FTIR spectroscopy analysis of biochar, which showed the existence of several functional groups that increase biochar's surface charge enabling it to remove toxins from soil and water systems more effectively.

Recently, H_2O_2-modified biochars have been getting more attention because they are effective and cheap. By oxidizing the carbonized surface of biochar, H_2O_2 modification was also able to raise the carboxylic content from 16.4% to 22.3% and enhanced other O-containing functional groups (Xue et al., 2012) making it better at removing ammonium from water. They also found that this engineered biochar had 20 times more Pb(II) sorption from water over pristine biochar. Furthermore, chemically activating *Miscanthus giganteas* (a perennial warm-season (C4) grass) biochar with H_2O_2 increased the removal of Cu(II) and Zn(II) from water (Cibati, Foereid, Bissessur, & Hapca, 2017). Nevertheless, NaOH + H_2O_2-modified coffee waste biochar was designed with O-containing functional groups which enhanced strontium (Sr) adsorption from water by electrostatic attraction (Shin et al., 2022).

The maximum Pb(II) adsorption from water increased from 2 to 93 mg g^{-1} in the case of saw dust hydrochar modified by H_2O_2 and ultrasonication (Xia et al., 2019) According to the mechanism analysis, the main mechanisms for Pb(II) sorption were complexation with free carboxyl, hydroxyl, and cation interaction, with their corresponding contribution percentages accounting for 62.14%, 27.14%, and 10.74%, respectively (Xia et al., 2019). Biochar is also treated with potassium permanganate ($KMnO_4$). When compared with untreated biochar, the surface area decreased but pore volume and the average pore width increased after $KMnO_4$ treatment (Qian & Chen., 2014). These findings implied that $KMnO_4$ oxidation could enlarge the micropores. The decrease in surface area could be caused by the formation of oxidation products such as Mn oxides, which could block biochar pores or cause pore walls to collapse.

Oxidizing agents are also used with other chemicals for the preparation of engineered biochar. El-Nemr, Abdelmonem, Ismail, Ragab, and El Nemr (2020) reported that ozone oxidation followed by ammonium hydroxide reaction chemically modified *Pisum sativum* peels waste biochar which enhanced Cu(II) adsorption to the tune of 23% over untreated biochar. Energy dispersive X-ray spectroscopy (EDAX) and FTIR spectroscopy confirmed that amino groups were successfully formed on the modified biochar surface. Furthermore, to reduce heavy metal risk (including Cd, Cr, Cu, Pb, and Zn) Fe-rich biochar was prepared from red mud and reed straw and further oxidized with H_2O_2 to improve sludge dewatering and transformation of heavy metals. Dey et al. (2023) reported that rice straw biochar treated with O_3-HCl-$FeCl_3$ markedly decreased the leaching of NH_4^+-N, NO_3^--N, PO_4^{3-}-P and K^+ from sandy loam soil.

22.4.4 Metal oxide-loaded biochar

Recently, metal oxides were added to biochar to improve its properties and, as a result, its absorption capacity (Raj et al., 2021). Peng et al. (2019) reported that biochar could not absorb phosphate due to electrostatic repulsion since its surface is negatively charged. As a result, modifying biochar by treating it with iron salts is becoming increasingly important in the removal of anionic pollutants. Different iron salts [for instance, goethite (FeOOH), hematite (Fe_2O_3)] produce amorphous surface active iron oxyhydroxide that works as an anion-binding site and enhances the positive charge on the surface of biochar. Furthermore, reducing agents such as iron sulfide (FeS) and nano zero-valent iron (nZVI) are used throughout the pollutant removal process (Chen et al., 2022a).

Biochar treated with Fe salts [$Fe(NO_3)_3$] remained mainly amorphous as goethite (FeOOH) and displayed greater phosphate sorption (Wu, Zhang, Wang, & Ding, 2020). HCl-$FeCl_3$ treatment enhanced nitrate and phosphate sorption from polluted water by forming outer and inner sphere complex respectively (Li et al., 2016).

Modifying biochar with nano zero-valent iron (nZVI), FeS, and FeOOH enhanced redox active pollutant species removal from soil and water by redox processes. Zhu et al. (2018) discovered effective Cr(VI) adsorption by nZVI-modified biochar. Furthermore, the change of the lethal Cr(VI) into the less harmful Cr(III) was seen, demonstrating the function of Fe°. Liu, Wang, Han, Hu, and Qiu (2021a) demonstrated the conversion of labile U(VI) species to nonlabile U(IV) species by peanut shell biochar (PBC) treated with FeS and starch, and XPS analysis highlighted the role of Fe° and S(II) in the reduction process.

Different oxides (separately or in combinations) are also used to modify biochar to increase their contaminant removal efficiency. For instance, nano-MnO_2-biochar composite recorded higher Cu(II) sorption from aqueous solution. After adsorption, Cu(II) was transformed into CuO, $Cu(C_2H_3O_2)_2$, and $Cu(OH)_2$ and got precipitated or chelated on the biochar surface (Zhou et al. 2017). Biochar impregnated with ZnO nanoparticles helped to form Zn-Cr oxide complexes and also assisted the reduction of Cr(VI) to Cr(III) and thus augmented Cr(VI) removal from aqueous solution (Yin, Guo, & Peng, 2018; Yu, Jiang, 2018, Yu, Lian, Cui, & Liu, 2018). Moreover, nano-ZnO/ZnS modified biochar increased the sorption of Pb(II), Cu(II), and Cr(VI) from the solution (Li, Zhang, Gao, & Li, 2018). Li et al. (2017) reported that the MnOx-loaded biochar increases the sorption of Cd(II) due to an increase in micropore sizes and the number of oxygen-containing functional groups, cation exchange capacity, pore structure and specific surface areas as compared to the unmodified biochar.

Biochar has a negatively charged surface so it generally cannot adsorb phosphate due to electrostatic repulsion (Peng et al., 2019). So several studies are being conducted on different engineered or modified biochar through impregnation of minerals which increase phosphate adsorption thereby reducing cost and rate of application (Kumar et al., 2020; Peng et al., 2021). Moreover, Peng et al. (2021) revealed that Fe/Al hydroxide modification of corn straw biochar increased oxidation resistance on the biochar surface than the raw biochar. So application of 2% (w/w) modified biochar significantly decreased P leaching by 81.3% from the soil column. This result was further supported by a sorption study which revealed maximum Langmuir P adsorption capacities were greater (528 mg kg^{-1}) for modified biochar than unmodified ones (393 mg kg^{-1}). Fe/Al hydroxides with their higher specific surface area and functional groups have a higher affinity for P and form stable bonds with P (Yang et al., 2019). Thus Fe/Al hydroxides-modified biochar can potentially reduce P leaching in fertile calcareous soil (Peng et al., 2021). Moreover, $MgCl_2$-modified biochar was reported to have maximum adsorption capacity of NH_4-N of 30 mg g^{-1} (Gong, Ni, Xiong, Cheng, & Xu, 2017). Yin et al. (2018) reported that Al-Mg modified soybean straw biochar had maximum adsorption capacity of NH_4^+-N and NO_3^--N of 0.70, 40.63 mg g^{-1}, respectively.

22.4.5 Nonmetallic heteroatom−doped biochar

Nonmetallic heteroatom doping is a novel way of functionalizing biochar and enhancing its ability to bind pollutants by altering its properties (Chen et al. 2022a). It has significant environmental remediation potential. It eliminates pollutants by electrostatic attraction, H-bond formation, and other mechanisms (Cheng, Ji, Liu, Mu, & Zhu, 2021). Corn straw biochar doped with N recorded higher graphitic-N (46%) and higher surface area which resulted in better sorption of cadmium (Cd(II)) (Yin et al. 2018, Yu, Jiang, 2018,

Yu, Lian, Cui, & Liu, 2018) by forming complex with hydroxyl groups, graphitic-N (Yin et al. 2020). Nevertheless, S doping increased the number of functional groups on biochar, such as sulfur rings, C–S, C=S, C–S–C, and C–S–O (Chen et al., 2022a) which enhanced the affinity of novel biochar for inorganic compounds. Yang et al. (2019) prepared a corn straw Fe composite biochar and then added S^{2-} and Mn^{2+} and reported that doping with S enhanced the removal of Pb(II) from water by precipitating PbS. However, Sui et al. (2021) reported that boron can be another excellent potential heteroatom for modifying the electron distribution, influencing the physicochemical properties of biochar, and providing new sorption sites for removing pollutants. Iron was removed by ion exchange, chemical complexation, and coprecipitation with B-doped biochar (Sui et al. 2021). In addition, Wang, Gao, and Wan (2018), Wang et al. (2018) created magnetic biochar that was doped with lanthanum (La), resulting in a La-O-Sb inner sphere complex and a significant increase in Stibium Sb(V) sorption. The mechanism of sorption of Sb(V) was corroborated by FTIR and XPS analysis. Moreover, apart from heavy metals, the maximum sorption capacity of NH_4^+, NO_3^- was reported to be augmented by 1.9, 11.2 folds when pristine oak sawdust biochar was doped with Lanthanum (Wang, Shen, Shen, & Li, 2015a).

22.4.6 Surface active agent—treated biochar

Surfactants are classified into four categories based on their hydrophilic group: cationic, anionic, nonionic, and Gemini (Paria, 2008). Cationic surfactants can easily adhere with negatively charged biochar surfaces through electrostatic attraction and improve positive charge on biochar. So, cationic surfactants are used to improve the removal of anionic contaminants by biochar mostly through the ion exchange process (Murad et al., 2022). In this regard, Murad et al. (2022) treated biochar with cetyltrimethylammonium bromide, a typical cationic surfactant, to considerably improve the adsorption and immobilization of Cr(VI) in its anionic form from water and soil. The adsorption kinetics and the pseudo-second-order model's best fit corroborated that chemisorption was the predominant way of interaction between Cr(VI) and biochar. Bioavailability, leachability, and bioaccessibility of this contaminant were diminished by 92%, 100%, and 97%, respectively, following the application of engineered biochar to soil, indicating that biochar is more efficient at removing pollutants. Moreover, surface-adsorbed sodium dodecyl benzene sulfonate, an anionic surfactant, changed the surface functional groups of the biochar, making it negatively charged. It dramatically enhanced the removal of Cr(III) from wastewater, although the removal of Cr(VI) was significantly lowered (Chai et al., 2018). TX-100, a nonionic surfactant, improved the removal of Cr(III) by engineered biochar by lowering the interfacial tension between the biochar and the solution phase. However, it increased Cr(VI) removal at low concentrations but hindered removal at high concentrations. So, it implied that the kind and concentration of surfactants must be considered while using biochar to treat wastewater containing coexisting pollutants.

22.4.7 Clay-modified biochar

Vermiculite, montmorillonite, kaolinite, zeolite, and illite are readily available natural clay minerals that are used to dope biochar to increase the functional groups (e.g., $-NH_2$,

−COOH, −OH), CEC, porosity, surface area of biochar-clay composites, thereby removal of inorganic contaminants from soil and water (Premarathna et al. 2019). Feedstock is mixed with clay suspension before pyrolysis to prepare clay-impregnated biochar (Sizmur, Fresno, Akgül, Frost, & Moreno-Jiménez, 2017). Fu, Zhang, Xia, Lei, and Wang (2020) reported biochar made from corncobs doped with montmorillonite had a Pb(II) sorption capacity of 140 mg g^{-1} in aqueous solution. Furthermore, attapulgite, zeolite, starch-modified corn straw biochar furnished with hematite adsorbed Cr(VI) from aqueous solution by electrostatic interaction, hydrogen bond formation, redox reaction and coprecipitation (Zhu et al., 2020). So, clay modification methods of biochar composites are becoming an emerging technique to remove inorganic contaminants from soil and wastewater.

22.4.8 Carbon nanotube-coated biochar

Carbon nanotubes (CNTs) can remove pollutants from soil and water due to their large surface area and nanostructure, but it is costly. So, biochar can be used as carrier material for these CNTs to make them cost-effective. Furthermore, CNTs, graphene, and graphene oxides (GOs) are now used to prepare biochar composites that increase the surface area and functional groups of biochar while reducing its cost (Rajapaksha et al. 2016). Sweetgum biochar, when coupled with CNTs and GO by slow pyrolysis, improved the sorption of Pb(II) and Cd(II) from contaminated water due to its higher specific surface area. Huang, Zhang, Wang, Lv, and Kang (2012) discovered that bamboo charcoal coated with CNTs immobilized Pb(II) by generating complexes with acidic functional groups. However, CNTs are mainly used to remove dyes and organic pollutants from wastewater.

22.4.9 Layered double hydroxide−modified biochar

Layered double hydroxides (LDHs) have a layered stacking structure and are effective anion exchange adsorbents (Chen et al. 2022b). According to research, biochar produced by LDH could reclaim contaminated soils and water. For instance, Lyu, Wang, Cao, Wang, and Deng (2021) observed that applying a Mg/Al LDH-biochar composite to uranium-contaminated soil (1000 mg kg^{-1}) decreased cumulative U loss by 53% and leaching concentration by 54% compared to the control.

22.5 Physical modification of biochar and their effect on contaminant bioremediation from soil and water

Physical modifications of biochar [ball milling (Lyu et al., 2018), microwave pyrolysis (Mohamed et al., 2016), steam activation (Chia et al., 2015), gas purging (Rajapaksha et al., 2016), etc]. are comparatively simple and less expensive than chemical treatments. It changes surface area, porosity, surface functional groups, polarity, and hydrophobicity of biochar which helps in inorganic contaminant removal from soil and water. However, physical modification techniques need higher energy input and longer activation times to enhance their performance. So, finding the right balance between improving biochar through physical modifications and minimizing drawbacks is essential in scientific and practical applications (Rajapaksha et al., 2016).

22.5.1 Steam/CO_2-activated biochar

Steam activation transforms raw biochar into an activated form by increasing its porosity, and surface area (Medeiros et al., 2022) and also incorporating oxygen-containing surface functional groups, like carbonyl and hydroxyl (Ahmed, Zhou, Ngo, Guo, & Chen, 2016). In steam activation, water vapor interacts with labile organic matter in biomass-producing gas, and the carbon is consumed as precursor, resulting in increased porosity and surface area of engineered biochar (Anderson, Gu, & Bergman, 2021). Moreover, CO_2 can be used to produce engineered biochar, creating a microporous structure (Medeiros et al., 2022; Kumar et al., 2020a). Prior studies reported that purging CO_2 during the pyrolysis process has proven effective in producing highly porous biochar from barley straw (Pallarés, González-Cencerrado, & Arauzo, 2018), orange peel (Yek et al. 2020), rice straw, sludge (Islam et al. 2021), and waste timber (Sørmo et al. 2021). These modifications have been reported to effectively enhance the performance of engineered biochar in removing contaminants. Steam-activated biochar effectively removed Cu(II) and tetracycline, indicating its potential for simultaneous removal of antibiotics and heavy metals in both simulated and real water samples (Wang et al., 2020). Phosphate adsorption was also reported to increase after steam activation of biochar (Han et al. 2020; Wen et al. 2021).

22.5.2 Microwave-activated biochar

Chen et al. (2022a) reported that microwave activation triggers the dipole rotation of atoms, resulting in the production of heat within the materials which concomitantly enhances porosity, surface area and functional groups of biochar (Chen et al., 2022a). So, this microwave-activated biochar could be used for contaminant removal from soil and water.

Microwave-derived wheat straw biochar, produced at 600 W (WS600) and 500 W (WS500), exhibited excellent specific surface areas (156.09 m^2 g^{-1}) and total pore volumes (0.079 cm^3 g^{-1}). WS500 demonstrated significant adsorption capacities for Pb(II), Cd(II), and Cu(II) as 139.44, 52.92, and 31.25 mg g^{-1}, respectively (Qi et al., 2023). Chen et al. (2023a) explored that biochar derived from microwave pyrolysis of rape stalk revealed a notable Cd(II) adsorption capacity of 53.17 mg g^{-1}. Nevertheless, Kong et al. (2022) evaluated the efficiency of microwave-modified engineered biochar in the adsorption of heavy metal from landfill leachate. They reported that microwave-modified palm kernel shells biochar recorded high surface area of 539.75 mg g^{-1} making it a durable adsorbent. The engineered biochar demonstrated remarkable efficiency in removing Cd(II) and Pb(II) from landfill leachate, with maximum removal capacities of 116.0 mg g^{-1} for Cd(II) and 59.9 mg g^{-1} for Pb(II) under optimized conditions, highlighting its cost-effective potential as an effective adsorbent. So, microwave-activated biochar holds significant promise, however, further research is needed to optimize the energy efficiency during the microwave heating process to further enhance its sustainability.

22.5.3 Ball milling

Ball-milled biochar exhibits enhanced surface properties and increased reactivity, making it a valuable material for various applications such as adsorption and catalysis (Rajapaksha et al., 2016). Ball milling enhanced specific surface area, pore volume, acidic oxygen-containing functional groups, water dispersion and diffusion rate, and negative zeta potential which concomitantly increase the removal of contaminants by electrostatic attraction, surface complexation, pore filling, and $\pi-\pi$ interactions (Zhao et al., 2022).

Wang et al. (2016) combined electrochemical exfoliation with ball mill to create a novel form of ball-milled GO which significantly enhanced U(VI) sorption capabilities compared to unmilled GO. In separate studies, Gao, Wang, Rondinone, He, and Liang (2015) utilized the ball milling process in the production of carbon-based nanocomposites, incorporating activated carbon as a feedstock material, resulting in improved properties for applications in the removal of contaminants. From a metaanalysis, Harindintwali et al. (2023) reported that ball-milled biochar increased inorganic contaminants sorption to the tune of 69.9% as compared to pristine biochar. The same metaanalysis recommended ball milling with a speed of 300–400 rpm for 12–24 hours with a biochar: ball mass ratio of 1:100 could be optimal for contaminant removal. However, it is imperative to explore cost-effective and energy-efficient approaches to produce ball-milled biochar to foster broader adoption and economically sustainable applications (Amusat, Kebede, Dube, & Nindi, 2021).

22.6 Biological modification of biochar and their effect on contaminant bioremediation from soil and water

Biological methods for tailoring biochar have received less attention in comparison to physical and chemical approaches due to their operational complexity (Chen et al., 2022a). Anaerobic bacteria have been employed to convert biomass into biogas, and the resulting digested residue has been integrated into the biochar production process. Tao et al. (2019) reported that biologically modified biochar, produced from the digestion residue of corn straw silage displayed increased surface area, oxygen-containing functional groups, and mineral components compared to pristine corn straw-derived biochar. Notably, the engineered biochar exhibited a threefold higher sorption capacity for Cd(II) removal from water compared to pristine biochar through mineral precipitation, ion exchange, and complexation with oxygen-containing functional groups. Prior studies also reported that biologically modified biochar improved the removal of inorganic contaminants from soil and water (Ho, Yang, Nagarajan, Chang, & Ren, 2017; Inyang et al., 2012; Ni et al., 2019; Yao et al., 2011)

Biological modification of biochar also involves the introduction of microorganisms onto biochar surfaces or their co-pyrolysis during biochar production. This method encompasses various approaches, including impregnation, inoculation, and immobilization of specific microbial species. This biologically modified biochar serves as a conducive habitat and carbon source for the introduced microorganisms, augmenting their potential to remediate contaminants in both soil and water matrices (Chen et al., 2022a). Moreover, according to biofilm theory, living microorganisms secrete different polymers and polysaccharides, and attach themselves to the surface of biochar. This microbial biofilm facilitates the adsorption/degradation of pollutants (An et al. 2022). Furthermore, the high surface area, substantial pore volume, and enduring stability of biochar provide an advantageous niche for microbial colonization and sustained activity (Zhou et al. 2021).

TZ5 strain of Cd immobilizing plant growth promotion bacteria (PGPB) was loaded onto biochar which significantly reduced the percentage of acetic acid-extractable Cd in soil by 11.3%, while also boosting ryegrass dry weight by 77.8% and reduced Cd concentration in ryegrass by 48.49% compared to the control in pot experiments (Ma et al., 2020). Nevertheless, Qi et al. (2021) conducted a study affirming the efficacy of biochar enriched with three bacterial strains—*Bacillus subtilis, B. cereus,* and *Citrobacter* sp.—in immobilizing U(VI) and Cd(II) within contaminated soil. Remarkably, when compared to the control group, this approach resulted in a substantial reduction of 69% in the concentration of DTPA-extractable U and a 56% decrease in Cd within the soil. Furthermore, the biochar loaded with these bacteria inhibited metal uptake and fostered the growth of celery plants. Nevertheless, biochar beads embedded with freshwater algae (*Chlorella pyrenoidosa*) were investigated for their efficiency in pollutant removal, including nutrients and heavy metals from aquatic environments (Guo et al., 2021). This innovative combination of biochar and algae offered a promising approach for enhanced water remediation as the beads exhibited mechanical strength, algae growth promotion, and pollutant removal efficiency. The removal efficiency of pollutants, including ammonia nitrogen (69.2%), total nitrogen (43.0%), total phosphorus (73.8%), zinc (74.4%), and copper (81.0%), underscores their effectiveness (Guo et al., 2021).

Furthermore, An et al. (2022) explored the effectiveness of *Pseudomonas hibiscicola* strain L1, PBC, and a co-system of strain L1 immobilized on PBC for removing Ni(II), Cr(VI), Cu(II), and nitrate from mixed wastewater. Strain L1 removed 15.51%–32.55% of Ni(II), while the co-system showed higher removal rates, particularly for Ni(II) (81.17%), followed by Cu(II) (45.84%) and Cr(VI) (38.21%). Characterization indicated effective immobilization of strain L1 on PBC, leading to the formation of metal hydroxide crystals on the surface, with various functional groups involved in the removal process (An et al. 2022). Moreover, *P. stutzeri* strain XL-2 and modified walnut shell biochar exhibited individual capabilities for ammonium removal. When combined, strain XL-2 and biochar complex demonstrated remarkable synergistic effects, with an average ammonium removal rate of $4.40 \text{ mg} \cdot \text{L}^{-1} \cdot \text{h}^{-1}$, significantly outperforming pure bacteria and biochar alone (Yu et al., 2019).

Despite its potential, the biological modification of biochar faces several challenges, including the selection of compatible microbial species, optimization of growth conditions within the biochar matrix, and the long-term stability and performance of the biochar-microbe system. Ongoing research aims to refine these approaches and broaden the utility of biologically modified biochar for efficient and sustainable contaminant bioremediation in terrestrial and aquatic ecosystems (An et al. 2022).

22.7 Regeneration of used modified biochar

The modified biochar has a finite adsorption capacity, so prolonged exposure to heavy metal ions leads to a thermodynamic equilibrium between the biochar and the sorbates.

Therefore, it becomes necessary to regenerate the biochar sorbents to enable their reuse or proper disposal. Wang et al. (2015b) reported that an affordable and efficient desorption method can greatly reduce the cost of biochar and improve its reuse as the sorption-desorption cycle can be repeated multiple times. Following saturation, we can achieve desorption by utilizing solutions with different concentrations of KNO_3, HNO_3, or $NaNO_3$, which supply cations to displace adsorbed heavy metal ions. Acidic solutions are effective at desorbing heavy metal ions due to their low pH (Sounthararajah, Loganathan, Kandasamy, & Vigneswaran, 2015). Commonly used solvents for regenerating modified biochars encompass acetic acid, NaOH, HCl, NaCl, and EDTA (Godwin, Pan, Xiao, & Afzal, 2019). Several methods can also be employed to regenerate fixed bed columns, including pressure reduction, heat treatment, inert gas or fluid purging, and pH adjustment (Godwin et al., 2019).

22.8 Future perspective

The future of engineered biochar research for inorganic contaminant removal from soil and water holds immense promise. As scientists continue to explore innovative chemical treatments and modifications of biochar, its adsorption and catalytic properties can be tailored to target-specific inorganic pollutants effectively. Through advanced surface functionalization techniques, engineered biochar could exhibit enhanced affinity and selectivity for various contaminants, making it a versatile and eco-friendly solution for soil and water remediation. Moreover, as the quest for sustainable and cost-effective approaches intensifies, engineered biochar could serve as a key component in hybrid systems, combining its pollutant removal capabilities with other emerging technologies like nanomaterials or advanced filtration methods. Depending on the target-specific contaminant, chemical treatment, feedstock type, and pyrolysis temperature can be tailored, requiring further research. Additionally, for mixed pollutant system remediation, engineered biochar can be prepared with simultaneous reactions with CEC and anion exchange capacity enhancing chemicals to remove both cations and anions from soil and water systems. However, information on this approach is limited, necessitating further research. By harnessing the potential of engineered biochar, we may witness a revolutionary breakthrough in combating inorganic contaminants, ultimately leading to a cleaner, healthier future for our planet.

22.9 Conclusion

In conclusion, this chapter underscores the immense potential of engineered biochar, as a potent solution for efficiently mitigating pollutants from soil and water environments. Through these modifications, the physicochemical properties of biochar are strategically altered, resulting in enhanced functional groups, surface charge, porosity, and surface area. These transformative changes equip biochar with the capacity to adsorb contaminants through various mechanisms, encompassing electrostatic attraction, surface precipitation, coprecipitation, ion exchange, complex formation, and hydrogen bonding (as visualized in Fig. 22.2). Among various treatments, chemical modification emerged as one of the promising techniques of tailoring biochar. The diversity in chemical treatments extends a versatile toolkit, where some modifications amplify the sorption of cationic pollutants, while others tailor biochar to remove anionic contaminants. This nuanced approach facilitates the customization of biochar's adsorption capabilities to align with the specific characteristics of the pollutants targeted for removal. A synergistic "win−win" strategy emerges by strategically combining treatments designed for cationic and anionic pollutant sorption. This approach optimizes biochar's adsorption potential, enabling the simultaneous and effective removal of contaminants of diverse charge types However, laboratory-based investigations have provided foundational insights into the potential of chemically treated biochar, the successful translation of these findings into practical applications hinges on rigorous field studies. Such field studies are imperative to validate the efficacy and practicality of these treatments under real-world conditions. The chapter underscores that the synergy between engineered biochar and chemical treatments holds promise for tackling environmental pollution. Through the careful integration of theoretical insights and practical field studies, the transformative potential of chemically treated biochar can be harnessed effectively to address the critical challenge of pollutant removal, offering a sustainable path toward cleaner soils and waters.

References

Ahmed, M. B., Zhou, J. L., Ngo, H. H., Guo, W., & Chen, M. (2016). Progress in the preparation and application of modified biochar for improved contaminant removal from water and wastewater. *Bioresource Technology, 214*, 836−851. Available from https://doi.org/10.1016/j.biortech.2016.05.057.

Amonette, J. E., & Joseph, S. (2009). *Physical properties of biochar. Biochar for Environmental Management* (pp. 13−29). London Sterling, VA: Earthscan.

Amusat, S. O., Kebede, T. G., Dube, S., & Nindi, M. M. (2021). Ball-milling synthesis of biochar and biochar−based nanocomposites and prospects for removal of emerging contaminants: A review. *Journal of Water Process Engineering, 41*, 101993.

An, Q., Jin, N., Deng, S., Zhao, B., Liu, M., Ran, B., & Zhang, L. (2022). Ni(II), Cr(VI), Cu(II) and nitrate removal by the co-system of Pseudomonas hibiscicola strain L1 immobilized on peanut shell biochar. *Science of The Total Environment, 814*, 152635.

An, Q., Zhang, C., Zhao, B., Li, Z., Deng, S., Wang, T., & Jin, L. (2021). Insight into synergies between Acinetobacter sp. AL-6 and pomelo peel biochar in a hybrid process for highly efficient manganese removal. *Science of The Total Environment, 793*, 148609.

Anderson, N., Gu, H., & Bergman, R. (2021). Comparison of novel biochars and steam activated carbon from mixed conifer mill residues. *Energies, 14*(24), 8472. Available from https://doi.org/10.3390/en14248472.

Bai, M., Chai, Y., Chen, A., Yuan, J., Shang, C., Peng, L., & Peng, C. (2023). Enhancing cadmium removal efficiency through spinel ferrites modified biochar derived from agricultural waste straw. *Journal of Environmental Chemical Engineering, 11*(1), 109027.

Chai, Q., Lu, L., Lin, Y., Ji, X., Yang, C., He, S., & Zhang, D. (2018). Effects and mechanisms of anionic and nonionic surfactants on biochar removal of chromium. *Environmental Science and Pollution Research, 25*(19), 18443−18450.

Chandra, S., Medha, I., & Bhattacharya, J. (2020). Potassium-iron rice straw biochar composite for sorption of nitrate, phosphate, and ammonium ions in soil for timely and controlled release. *Science of the Total Environment, 712*, 136337. Available from https://doi.org/10.1016/j.scitotenv.2019.136337.

Chen, H., Gao, Y., Li, J., Fang, Z., Bolan, N., Bhatnagar, A., ... Wang, H. (2022a). Engineered biochar for environmental decontamination in aquatic and soil systems: A review. *Carbon Research, 1*(1), 1−25.

Chen, H., Guo, H., Jiang, D., Cheng, S., Xing, B., Meng, W., ... Xia, H. (2023a). Microwave-assisted pyrolysis of rape stalk to prepare biochar for heavy metal wastewater removal. *Diamond and Related Materials, 134*, 109794.

Chen, H., Yang, X., Liu, Y., Lin, X., Wang, J., Zhang, Z., ... Zhang, Y. (2021). KOH modification effectively enhances the Cd and Pb adsorption performance of N-enriched biochar derived from waste chicken feathers. *Waste Management*, *130*, 82–92. Available from https://doi.org/10.1016/j.wasman.2021.05.015.

Chen, Q., Cheng, Z., Li, X., Wang, C., Yan, L., Shen, G., & Shen, Z. (2022b). Degradation mechanism and QSAR models of antibiotic contaminants in soil by MgFe-LDH engineered biochar activating urea-hydrogen peroxide. *Applied Catalysis B: Environmental*, *302*, 120866.

Chen, S., Zhong, M., Wang, H., Zhou, S., Li, W., Wang, T., & Li, J. (2023b). Study on adsorption of Cu^{2+}, Pb^{2+}, Cd^{2+}, and Zn^{2+} by the $KMnO_4$ modified biochar derived from walnut shell. *International Journal of Environmental Science and Technology*, *20*(2), 1551–1568.

Cheng, L., Ji, Y., Liu, X., Mu, L., & Zhu, J. (2021). Sorption mechanism of organic dyes on a novel self-nitrogen-doped porous graphite biochar: Coupling DFT calculations with experiments. *Chemical Engineering Science*, *242*, 116739.

Cheng, S., Meng, W., Xing, B., Shi, C., Wang, Q., Xia, D., ... Xia, H. (2023). Efficient removal of heavy metals from aqueous solutions by Mg/Fe bimetallic oxide-modified biochar: Experiments and DFT investigations. *Journal of Cleaner Production*, *403*, 136821.

Chia, C. H., Downie, A., & Munroe, P. (2015). Characteristics of biochar: physical and structural properties. In J Lehmann, & H Joseph (Eds.), Biochar for Environmental Management (pp. 89–109). Taylor & Francis e book series.

Cibati, A., Foereid, B., Bissessur, A., & Hapca, S. (2017). Assessment of Miscanthus × giganteus derived biochar as copper and zinc adsorbent: Study of the effect of pyrolysis temperature, pH and hydrogen peroxide modification. *Journal of Cleaner Production*, *162*, 1285–1296.

Dey, S., Purakayastha, T. J., Sarkar, B., Rinklebe, J., Kumar, S., Chakraborty, R., ... Shivay, Y. S. (2023). Enhancing cation and anion exchange capacity of rice straw biochar by chemical modification for increased plant nutrient retention. *Science of The Total Environment*, *886*, 163681.

Du, H., Xi, C., Tang, B., Chen, W., Deng, W., Cao, S., & Jiang, G. (2022). Performance and mechanisms of NaOH and ball-milling co-modified biochar for enhanced the removal of Cd^{2+} in synthetic water: A combined experimental and DFT study. *Arabian Journal of Chemistry*, *15*(6), 103817.

Duan, W., Oleszczuk, P., Pan, B., & Xing, B. (2019). Environmental behavior of engineered biochars and their aging processes in soil. *Biochar*, *1*, 339–351.

El-Nemr, M. A., Abdelmonem, N. M., Ismail, I., Ragab, S., & El Nemr, A. (2020). Ozone and ammonium hydroxide modification of biochar prepared from *Pisum sativum* peels improves the adsorption of copper(II) from an aqueous medium. *Environmental Processes*, *7*(3), 973–1007.

El-Sheekh, M., Abdel-Daim, M. M., Okba, M., Gharib, S., Soliman, A., & El-Kassas, H. (2021). Green technology for bioremediation of the eutrophication phenomenon in aquatic ecosystems: A review. *African Journal of Aquatic Science*, *46*(3), 274–292.

Fan, Y., Wang, B., Yuan, S., Wu, X., Chen, J., & Wang, L. (2010). Adsorptive removal of chloramphenicol from wastewater by NaOH modified bamboo charcoal. *Bioresource Technology*, *101*(19), 7661–7664.

Feng, Y., Lu, H., Liu, Y., Xue, L., Dionysiou, D. D., Yang, L., & Xing, B. (2017). Nano-cerium oxide functionalized biochar for phosphate retention: preparation, optimization and rice paddy application. *Chemosphere*, *185*, 816–825. Available from https://doi.org/10.1016/j.chemosphere.2017.07.017.

Fu, C., Zhang, H., Xia, M., Lei, W., & Wang, F. (2020). The single/co-adsorption characteristics and microscopic adsorption mechanism of biochar-montmorillonite composite adsorbent for pharmaceutical emerging organic contaminant atenolol and lead ions. *Ecotoxicology and Environmental Safety*, *187*, 109763.

Gao, J., Wang, W., Rondinone, A. J., He, F., & Liang, L. (2015). Degradation of trichloroethene with a novel ball milled Fe–C nanocomposite. *Journal of Hazardous Materials*, *300*, 443–450.

Godwin, P. M., Pan, Y., Xiao, H., & Afzal, M. T. (2019). Progress in preparation and application of modified biochar for improving heavy metal ion removal from wastewater. *Journal of Bioresources and Bioproducts*, *4*(1), 31–42.

Gong, Y. P., Ni, Z. Y., Xiong, Z. Z., Cheng, L. H., & Xu, X. H. (2017). Phosphate and ammonium adsorption of the modified biochar based on *Phragmites australis* after phytoremediation. *Environmental Science and Pollution Research*, *24*, 8326–8335.

Guo, Q., Bandala, E. R., Goonetilleke, A., Hong, N., Li, Y., & Liu, A. (2021). Application of *Chlorella pyrenoidosa* embedded biochar beads for water treatment. *Journal of Water Process Engineering*, *40*, 101892.

Hadjittofi, L., Prodromou, M., & Pashalidis, I. (2014). Activated biochar derived from cactus fibres—preparation, characterization and application on Cu(II) removal from aqueous solutions. *Bioresource Technology*, *159*, 460–464.

Hameed, R., Li, G., Son, Y., Fang, H., Kim, T., Zhu, C., ... Du, D. (2023). Structural characteristics of dissolved black carbon and its interactions with organic and inorganic contaminants: A critical review. *Science of The Total Environment*, 872, 162210.

Han, T., Lu, X., Sun, Y., Jiang, J., Yang, W., & Jönsson, P. G. (2020). Magnetic bio-activated carbon production from lignin via a streamlined process and its use in phosphate removal from aqueous solutions. *Science of the Total Environment*, 708, 135069.

Harindintwali, J. D., He, C., Xiang, L., Dou, Q., Liu, Y., Wang, M., ... Wang, F. (2023). Effects of ball milling on biochar adsorption of contaminants in water: A meta-analysis. *Science of The Total Environment*, 882, 163643.

Hazrati, S., Farahbakhsh, M., Cerdà, A., & Heydarpoor, G. (2021). Functionalization of ultrasound enhanced sewage sludge-derived biochar: Physicochemical improvement and its effects on soil enzyme activities and heavy metals availability. *Chemosphere*, 269, 128767.

Ho, S. H., Yang, Z. K., Nagarajan, D., Chang, J. S., & Ren, N. Q. (2017). High-efficiency removal of lead from wastewater by biochar derived from anaerobic digestion sludge. *Bioresource Technology*, 246, 142–149.

Huang, Z. H., Zhang, F., Wang, M. X., Lv, R., & Kang, F. (2012). Growth of carbon nanotubes on low-cost bamboo charcoal for Pb(II) removal from aqueous solution. *Chemical Engineering Journal*, 184, 193–197.

Inyang, M., Gao, B., Yao, Y., Xue, Y., Zimmerman, A. R., Pullammanappallil, P., & Cao, X. (2012). Removal of heavy metals from aqueous solution by biochars derived from anaerobically digested biomass. *Bioresource Technology*, 110, 50–56.

Islam, M. S., Kwak, J. H., Nzediegwu, C., Wang, S., Palansuriya, K., Kwon, E. E., ... Chang, S. X. (2021). Biochar heavy metal removal in aqueous solution depends on feedstock type and pyrolysis purging gas. *Environmental Pollution*, 281, 117094. Available from https://doi.org/10.1016/j.envpol.2021.117094.

Jin, H., Capareda, S., Chang, Z., Gao, J., Xu, Y., & Zhang, J. (2014). Biochar pyrolytically produced from municipal solid wastes for aqueous As (V) removal: Adsorption property and its improvement with KOH activation. *Bioresource Technology*, 169, 622–629.

Kharel, G., Sacko, O., Feng, X., Morris, J. R., Phillips, C. L., Trippe, K., ... Lee, J. W. (2019). Biochar surface oxygenation by ozonization for super high cation exchange capacity. *ACS Sustainable Chemistry & Engineering*, 7(19), 16410–16418.

Kong, S. H., Chin, C. Y. J., Yek, P. N. Y., Wong, C. C., Wong, C. S., Cheong, K. Y., ... Lam, S. S. (2022). Removal of heavy metals using activated carbon from microwave steam activation of palm kernel shell. *Environmental Advances*, 9, 100272.

Kumar, R., Sharma, P., Gupta, R. K., Kumar, S., Sharma, M. M. M., Singh, S., & Pradhan, G. (2020a). Earthworms for eco-friendly resource efficient agriculture. In Kumar, et al. (Eds.), *Resources Use Efficiency in Agriculture* (pp. 47–84). Singapore: Springer Nature.

Kumar, S., Meena, R. S., Datta, R., Verma, S. K., Yadav, G. S., Pradhan, G., Molaei, A., Rahman, G. K. M. M., & Mashuk, H. A. (2020). Legumes for Carbon and Nitrogen Cycling: An organic approach. In R. Datta, et al. (Eds.), *Carbon and nitrogen cycling in soil* (pp. 337–375). Singapore: Springer Nature, Carbon and Nitrogen Cycling in Soil. Available from https://doi.org/10.1007/978-981-13-7264-3_10.

Kumari, A., Rajput, S., Singh, S. V., et al. (2024). New dimensions into the removal of pesticides using an innovative ecofriendly technique: nanoremediation. In *Nano-Bioremediation for Water and Soil Treatment*, Apple Academic Press.

Kumawat, A., Bamboriya, S. D., Meena, R. S., Yadav, D., Kumar, A., Kumar, S., ... Pradhan, G. (2022). Legume-based inter-cropping to achieve crop, soil, and environmental health security. In R. S. Meena, & S. Kumar (Eds.), *Advances in legumes for sustainable intensification* (pp. 307–328). Elsevier. Available from https://doi.org/10.1016/B978-0-323-85797-0.00005-7.

Kwak, J. H., Islam, M. S., Wang, S., Messele, S. A., Naeth, M. A., El-Din, M. G., & Chang, S. X. (2019). Biochar properties and lead(II) adsorption capacity depend on feedstock type, pyrolysis temperature, and steam activation. *Chemosphere*, 231, 393–404.

Li, B., Yang, L., Wang, C. Q., Zhang, Q. P., Liu, Q. C., Li, Y. D., & Xiao, R. (2017). Adsorption of Cd(II) from aqueous solutions by rape straw biochar derived from different modification processes. *Chemosphere*, 175, 332–340.

Li, C., Zhang, L., Gao, Y., & Li, A. (2018). Facile synthesis of nano ZnO/ZnS modified biochar by directly pyrolyzing of zinc contaminated corn stover for Pb(II), Cu(II) and Cr(VI) removals. *Waste Management*, 79, 625–637.

Li, J. H., Lv, G. H., Bai, W. B., Liu, Q., Zhang, Y. C., & Song, J. Q. (2016). Modification and use of biochar from wheat straw (*Triticum aestivum* L.) for nitrate and phosphate removal from water. *Desalination and Water Treatment*, 57(10), 4681–4693.

Li, Y., Yin, H., Cai, Y., Luo, H., Yan, C., & Dang, Z. (2023). Regulating the exposed crystal facets of α-Fe$_2$O$_3$ to promote Fe$_2$O$_3$-modified biochar performance in heavy metals adsorption. *Chemosphere, 311*, 136976.

Liang, H., Wang, W., Liu, H., Deng, X., Zhang, D., Zou, Y., & Ruan, X. (2023). Porous MgO-modified biochar adsorbents fabricated by the activation of Mg (NO$_3$)$_2$ for phosphate removal: Synergistic enhancement of porosity and active sites. *Chemosphere, 324*, 138320.

Liatsou, I., Pashalidis, I., Oezaslan, M., & Dosche, C. (2017). Surface characterization of oxidized biochar fibers derived from *Luffa cylindrica* and lanthanide binding. *Journal of environmental chemical engineering, 5*(4), 4069–4074.

Lin, L., Song, Z., Huang, Y., Khan, Z. H., & Qiu, W. (2019). Removal and oxidation of arsenic from aqueous solution by biochar impregnated with Fe-Mn oxides. *Water, Air, & Soil Pollution, 230*(5), 1–13.

Liu, L., Yang, X., Ahmad, S., Li, X., Ri, C., Tang, J., ... Song, Z. (2023). Silicon (Si) modification of biochars from different Si-bearing precursors improves cadmium remediation. *Chemical Engineering Journal, 457*, 141194.

Liu, R., Wang, H., Han, L., Hu, B., & Qiu, M. (2021a). Reductive and adsorptive elimination of U (VI) ions in aqueous solution by SFeS@ Biochar composites. *Environmental Science and Pollution Research, 28*(39), 55176–55185.

Liu, S., Xie, Z., Zhu, Y., Zhu, Y., Jiang, Y., Wang, Y., & Gao, H. (2021b). Adsorption characteristics of modified rice straw biochar for Zn and in-situ remediation of Zn contaminated soil. *Environmental Technology & Innovation, 22*, 101388.

Liu, Z., Xu, Z., Xu, L., Buyong, F., Chay, T. C., Li, Z., ... Wang, X. (2022). Modified biochar: Synthesis and mechanism for removal of environmental heavy metals. *Carbon Research, 1*(1), 1–21.

Liu, Z., Xue, Y., Gao, F., Cheng, X., & Yang, K. (2016). Removal of ammonium from aqueous solutions using alkali-modified biochars. *Chemical Speciation & Bioavailability, 28*(1–4), 26–32.

Lyu, H., Gao, B., He, F., Zimmerman, A. R., Ding, C., Huang, H., & Tang, J. (2018). Effects of ball milling on the physicochemical and sorptive properties of biochar: Experimental observations and governing mechanisms. *Environmental Pollution, 233*, 54–63.

Lyu, H., Xia, S., Tang, J., Zhang, Y., Gao, B., & Shen, B. (2020). Thiol-modified biochar synthesized by a facile ball-milling method for enhanced sorption of inorganic Hg2 + and organic CH3Hg + . *Journal of Hazardous Materials, 384*, 121357.

Lyu, P., Wang, G., Cao, Y., Wang, B., & Deng, N. (2021). Phosphorus-modified biochar cross-linked Mg–Al layered double-hydroxide composite for immobilizing uranium in mining contaminated soil. *Chemosphere, 276*, 130116.

Ma, H., Wei, M., Wang, Z., Hou, S., Li, X., & Xu, H. (2020). Bioremediation of cadmium polluted soil using a novel cadmium immobilizing plant growth promotion strain Bacillus sp. TZ5 loaded on biochar. *Journal of Hazardous Materials, 388*, 122065.

Meena, R. S., Pradhan, G., Kumar, S., & Lal, R. (2023). Using industrial wastes for rice-wheat cropping and food-energy-carbon-water-economic nexus to the sustainable food system. *Renewable and Sustainable Energy Reviews, 187*, 113756. Available from https://doi.org/10.1016/j.rser.2023.113756.

Meena, R. S., & Pradhan, G. (2023). Industrial garbage-derived biocompost enhances soil organic carbon fractions, CO$_2$ biosequestration, potential carbon credits, and sustainability index in a rice-wheat ecosystem. *Environmental Research*, 116525. Available from https://doi.org/10.1016/j.envres.2023.116525.

Medeiros, D. C. C. S., Nzediegwu, C., Benally, C., Messele, S. A., Kwak, J. H., Naeth, M. A., ... El-Din, M. G. (2022). Pristine and engineered biochar for the removal of contaminants co-existing in several types of industrial wastewaters: A critical review. *Science of The Total Environment, 809*, 151120.

Meena, R. S., Kumawat, A., Kumar, S., Prasad, S. K., Pradhan, G., Jhariya, M. K., ... Raj, A. (2022). Effect of legumes on nitrogen economy and budgeting in South Asia. In R. S. Meena, & S. Kumar (Eds.), *Advances in legumes for sustainable intensification* (pp. 619–638). Academic Press. Available from https://doi.org/10.1016/B978-0-323-85797-0.00001-X.

Mohamed, B. A., Kim, C. S., Ellis, N., & Bi, X. (2016). Microwave-assisted catalytic pyrolysis of switchgrass for improving bio-oil and biochar properties. *Bioresource Technology, 201*, 121–132.

Murad, H. A., Ahmad, M., Bundschuh, J., Hashimoto, Y., Zhang, M., Sarkar, B., & Ok, Y. S. (2022). A remediation approach to chromium-contaminated water and soil using engineered biochar derived from peanut shell. *Environmental Research, 204*, 112125.

Nazari, S., Rahimi, G., & Nezhad, A. K. J. (2019). Effectiveness of native and citric acid-enriched biochar of Chickpea straw in Cd and Pb sorption in an acidic soil. *Journal of Environmental Chemical Engineering, 7*(3), 103064.

Ni, B. J., Huang, Q. S., Wang, C., Ni, T. Y., Sun, J., & Wei, W. (2019). Competitive adsorption of heavy metals in aqueous solution onto biochar derived from anaerobically digested sludge. *Chemosphere, 219*, 351–357.

Pallarés, J., González-Cencerrado, A., & Arauzo, I. (2018). Production and characterization of activated carbon from barley straw by physical activation with carbon dioxide and steam. *Biomass and Bioenergy, 115*, 64−73. Available from https://doi.org/10.1016/j.biombioe.2018.04.015.

Paria, S. (2008). Surfactant-enhanced remediation of organic contaminated soil and water. *Advances in Colloid and Interface Science, 138*(1), 24−58.

Peng, Y., Sun, Y., Fan, B., Zhang, S., Bolan, N. S., Chen, Q., & Tsang, D. C. (2021). Fe/Al (hydro) oxides engineered biochar for reducing phosphorus leaching from a fertile calcareous soil. *Journal of Cleaner Production, 279*, 123877.

Peng, Y., Sun, Y., Sun, R., Zhou, Y., Tsang, D. C., & Chen, Q. (2019). Optimizing the synthesis of Fe/Al (Hydr) oxides-biochars to maximize phosphate removal via response surface model. *Journal of Cleaner Production, 237*, 117770.

Peter, A., Chabot, B., & Loranger, E. (2021). Enhanced activation of ultrasonic pre-treated softwood biochar for efficient heavy metal removal from water. *Journal of Environmental Management, 290*, 112569.

Pradhan, G., & Meena, R. S. (2023). Interaction impacts of biocompost on nutrient dynamics and relations with soil biota, carbon fractions index, societal value of CO_2 equivalent, and ecosystem services in the wheat-rice farming. *Chemosphere*, 139695. Available from https://doi.org/10.1016/j.chemosphere.2023.139695.

Pradhan, G., & Meena, R. S. (2022). Diversity in the rice−wheat system with genetically modified zinc and iron-enriched varieties to achieve nutritional security. *Sustainability, 14*, 9334. Available from https://doi.org/10.3390/su14159334.

Pradhan, G., & Meena, R. S. (2023a). Utilizing waste compost to improve the atmospheric CO_2 capturing in the rice-wheat cropping system and energy-cum-carbon credit audit for a circular economy. *Science of the Total Environment*, 164572. Available from https://doi.org/10.1016/j.scitotenv.2023.164572.

Pradhan, G., Meena, R. S., Kumar, S., & Lal, R. (2023). Utilizing industrial wastes as compost in wheat-rice production to improve the above and below-ground ecosystem services. *Agriculture, Ecosystems and Environment, 358*, 108704. Available from https://doi.org/10.1016/j.agee.2023.108704.

Pradhan, G., Meena, R. S., Kumar, S., Jhariya, M. K., Khan, N., Shukla, U. N., . . . Kumar, S. (2022). Legumes for eco-friendly weed management in an agroecosystem. In R. S. Meena, & S. Kumar (Eds.), *Advances in legumes for sustainable intensification* (pp. 133−154). Elsevier. Available from https://doi.org/10.1016/B978-0-323-85797-0.00033-1.

Premarathna, K. S. D., Rajapaksha, A. U., Sarkar, B., Kwon, E. E., Bhatnagar, A., Ok, Y. S., & Vithanage, M. (2019). Biochar-based engineered composites for sorptive decontamination of water: A review. *Chemical Engineering Journal, 372*, 536−550.

Qi, G., Pan, Z., Zhang, X., Chang, S., Wang, H., Wang, M., & Gao, B. (2023). Microwave biochar produced with activated carbon catalyst: Characterization and adsorption of heavy metals. *Environmental Research, 216*, 114732.

Qi, X., Gou, J., Chen, X., Xiao, S., Ali, I., Shang, R., . . . Luo, X. (2021). Application of mixed bacteria-loaded biochar to enhance uranium and cadmium immobilization in a co-contaminated soil. *Journal of Hazardous Materials, 401*, 123823.

Qian, L., & Chen, B. (2014). Interactions of aluminum with biochars and oxidized biochars: Implications for the biochar aging process. *Journal of Agricultural and Food Chemistry, 62*(2), 373−380.

Qin, Y., Zhu, X., Su, Q., Anumah, A., Gao, B., Lyu, W., . . . Wang, B. (2020). Enhanced removal of ammonium from water by ball-milled biochar. *Environmental Geochemistry and Health, 42*, 1579−1587.

Qu, J., Zhang, B., Tong, H., Liu, Y., Wang, S., Wei, S., . . . Zhang, Y. (2023). High-efficiency decontamination of Pb (II) and tetracycline in contaminated water using ball-milled magnetic bone derived biochar. *Journal of Cleaner Production, 385*, 135683.

Raj, A., Jhariya, M. K., Banerjee, A., Meena, R. S., Nema, S., Khan, N., . . . Pradhan, G. (2021). Agroforestry a model for ecological sustainability. In M. K. Jhariya, R. S. Meena, A. Raj, & S. N. Meena (Eds.), *Natural resources conservation and advances for sustainability* (pp. 289−308). Elsevier. Available from https://doi.org/10.1016/B978-0-12-822976-7.00002-8.

Rajapaksha, A. U., Chen, S. S., Tsang, D. C., Zhang, M., Vithanage, M., Mandal, S., . . . Ok, Y. S. (2016). Engineered/designer biochar for contaminant removal/immobilization from soil and water: Potential and implication of biochar modification. *Chemosphere, 148*, 276−291.

Regmi, P., Moscoso, J. L. G., Kumar, S., Cao, X., Mao, J., & Schafran, G. (2012). Removal of copper and cadmium from aqueous solution using switchgrass biochar produced via hydrothermal carbonization process. *Journal of Environmental Management, 109*, 61−69.

Rizwan, M., Lin, Q., Chen, X., Li, Y., Li, G., Zhao, X., & Tian, Y. (2020). Synthesis, characterization and application of magnetic and acid modified biochars following alkaline pretreatment of rice and cotton straws. *Science of The Total Environment, 714*, 136532.

Roy, A., Chaturvedi, S., Singh, S., Govindaraju, K., Dhyani, V. C., & Pyne, S. (2022). Preparation and evaluation of two enriched biochar-based fertilizers for nutrient release kinetics and agronomic effectiveness in direct-seeded rice. *Biomass Conversion and Biorefinery*. Available from https://doi.org/10.1007/s13399-022-02488-z.

Roy, O., Meena, R. S., Kumar, S., Jhariya, M. K., & Pradhan, G. (2021). Assessment of land use systems for CO_2 sequestration, carbon credit potential, and income security in Vindhyan Region, India. *Land Degradation and Development, 33*(4), 670–682. Available from https://doi.org/10.1002/ldr.4181.

Sheoran, S., Ramtekey, V., Kumar, D., Kumar, S., Meena, R. S., Kumawat, A., ... Shukla, U. N. (2022). Grain legumes: Recent advances and technological interventions. In R. S. Meena, & S. Kumar (Eds.), *Advances in legumes for sustainable intensification* (pp. 507–532). Elsevier. Available from https://doi.org/10.1016/B978-0-323-85797-0.00025-2.

Shin, J., Choi, M., Go, C. Y., Bae, S., Kim, K. C., & Chon, K. (2022). NaOH-assisted H_2O_2 post-modification as a novel approach to enhance adsorption capacity of residual coffee waste biochars toward radioactive strontium: Experimental and theoretical studies. *Journal of Hazardous Materials, 435*, 129081.

Singh, S., Chaturvedi, S., & Datta, D. (2019). Biochar: An eco-friendly residue management approach. *Indian Farming, 69*(08), 27–29.

Singh, S., Chaturvedi, S., Dhyani, V. C., & Govindaraju, K. (2020). Pyrolysis temperature influences the characteristics of rice straw and husk biochar and sorption/desorption behavior of their biourea composite. *Bioresource Technology, 314*, 123674. Available from https://doi.org/10.1016/j.biortech.2020.123674.

Singh, S. V., Luthra, N., Bhattacharya, S., et al. (2024). Occurrence of emerging contaminants in soils and impacts on rhizosphere. In *In Emerging Contaminants Sustainable Agriculture and the Environment*, Elsevier Sciences.

Singh, S. V., Sharma, R., Balan, P, et al. (2024). Nanomaterials for inorganic pollutants removal from contaminated water. In *Nano-Bioremediation for Water and Soil Treatment*, Apple Academic Press.

Sizmur, T., Fresno, T., Akgül, G., Frost, H., & Moreno-Jiménez, E. (2017). Biochar modification to enhance sorption of inorganics from water. *Bioresource Technology, 246*, 34–47.

Sørmo, E., Silvani, L., Bjerkli, N., Hagemann, N., Zimmerman, A. R., Hale, S. E., ... Cornelissen, G. (2021). Stabilization of PFAS-contaminated soil with activated biochar. *Science of the Total Environment, 763*, 144034. Available from https://doi.org/10.1016/j.scitotenv.2020.144034.

Sountharajah, D. P., Loganathan, P., Kandasamy, J., & Vigneswaran, S. (2015). Adsorptive removal of heavy metals from water using sodium titanate nanofibres loaded onto GAC in fixed-bed columns. *Journal of Hazardous Materials, 287*, 306–316.

Sui, L., Tang, C., Du, Q., Zhao, Y., Cheng, K., & Yang, F. (2021). Preparation and characterization of boron-doped corn straw biochar: Fe (II) removal equilibrium and kinetics. *Journal of Environmental Sciences, 106*, 116–123.

Sun, T., Sun, Y., Xu, Y., Wang, L., & Liang, X. (2023). Effective removal of $Hg2+$ and $Cd2+$ in aqueous systems by Fe–Mn oxide modified biochar: A combined experimental and DFT calculation. *Desalination, 549*, 116306.

Tao, Q. I., Chen, Y., Zhao, J., Li, B., Li, Y., Tao, S., ... Wang, C. (2019). Enhanced Cd removal from aqueous solution by biologically modified biochar derived from digestion residue of corn straw silage. *Science of the Total Environment, 674*, 213–222.

Tu, C., Wei, J., Guan, F., Liu, Y., Sun, Y., & Luo, Y. (2020). Biochar and bacteria inoculated biochar enhanced Cd and Cu immobilization and enzymatic activity in a polluted soil. *Environment International, 137*, 105576.

Uchimiya, M., Bannon, D. I., & Wartelle, L. H. (2012). Retention of heavy metals by carboxyl functional groups of biochars in small arms range soil. *Journal of Agricultural and Food Chemistry, 60*(7), 1798–1809.

Vithanage, M., Rajapaksha, A. U., Zhang, M., Thiele-Bruhn, S., Lee, S. S., & Ok, Y. S. (2015). Acid-activated biochar increased sulfamethazine retention in soils. *Environmental Science and Pollution Research, 22*(3), 2175–2186.

Vu, M. T., Chao, H. P., Van Trinh, T., Le, T. T., Lin, C. C., & Tran, H. N. (2018). Removal of ammonium from groundwater using NaOH-treated activated carbon derived from corncob wastes: Batch and column experiments. *Journal of Cleaner Production, 180*, 560–570.

Wang, B., Gao, B., & Wan, Y. (2018). Entrapment of ball-milled biochar in Ca-alginate beads for the removal of aqueous Cd(II). *Journal of industrial and engineering chemistry, 61*, 161–168.

Wang, J., & Wang, S. (2019). Preparation, modification and environmental application of biochar: A review. *Journal of Cleaner Production, 227*, 1002–1022.

Wang, L., Wang, J., Wang, Z., He, C., Lyu, W., Yan, W., & Yang, L. (2018). Enhanced antimonate (Sb (V)) removal from aqueous solution by La-doped magnetic biochars. *Chemical Engineering Journal, 354,* 623–632.

Wang, R. Z., Huang, D. L., Liu, Y. G., Zhang, C., Lai, C., Wang, X., ... Xu, P. (2020). Synergistic removal of copper and tetracycline from aqueous solution by steam-activated bamboo-derived biochar. *Journal of Hazardous Materials, 384,* 121470.

Wang, S., Gao, B., Zimmerman, A. R., Li, Y., Ma, L., Harris, W. G., & Migliaccio, K. W. (2015b). Removal of arsenic by magnetic biochar prepared from pinewood and natural hematite. *Bioresource Technology, 175,* 391–395.

Wang, W., Ma, X., Sun, J., Chen, J., Zhang, J., Wang, Y., ... Zhang, H. (2019). Adsorption of enrofloxacin on acid/alkali-modified corn stalk biochar. *Spectroscopy Letters, 52*(7), 367–375.

Wang, Z., Shen, D., Shen, F., & Li, T. (2015a). Phosphate adsorption on lanthanum loaded biochar. *Chemosphere, 150,* 1–7.

Wang, Z., Wang, Y., Liao, J., Yang, Y., Liu, N., & Tang, J. (2016). Improving the adsorption ability of graphene sheets to uranium through chemical oxidation, electrolysis and ball-milling. *Journal of Radioanalytical and Nuclear Chemistry, 308,* 1095–1102.

Wen, Y., Zheng, Z., Wang, S., Han, T., Yang, W., & Jönsson, P. G. (2021). Magnetic bio-activated carbons production using different process parameters for phosphorus removal from artificially prepared phosphorus-rich and domestic wastewater. *Chemosphere, 271,* 129561.

Wongrod, S., Simon, S., van Hullebusch, E. D., Lens, P. N., & Guibaud, G. (2018). Changes of sewage sludge digestate-derived biochar properties after chemical treatments and influence on As(III and V) and Cd(II) sorption. *International Biodeterioration & Biodegradation, 135,* 96–102.

Wu, L., Zhang, S., Wang, J., & Ding, X. (2020). Phosphorus retention using iron (II/III) modified biochar in saline-alkaline soils: Adsorption, column and field tests. *Environmental Pollution, 261,* 114223.

Xia, Y., Yang, T., Zhu, N., Li, D., Chen, Z., Lang, Q., ... Jiao, W. (2019). Enhanced adsorption of Pb (II) onto modified hydrochar: Modeling and mechanism analysis. *Bioresource Technology, 288,* 121593.

Xiao, J., Hu, R., & Chen, G. (2020). Micro-nano-engineered nitrogenous bone biochar developed with a ball-milling technique for high-efficiency removal of aquatic Cd(II), Cu(II) and Pb(II). *Journal of Hazardous Materials, 387,* 121980.

Xue, Y., Gao, B., Yao, Y., Inyang, M., Zhang, M., Zimmerman, A. R., & Ro, K. S. (2012). Hydrogen peroxide modification enhances the ability of biochar (hydrochar) produced from hydrothermal carbonization of peanut hull to remove aqueous heavy metals: Batch and column tests. *Chemical Engineering Journal, 200,* 673–680.

Yakout, S. M. (2015). Monitoring the changes of chemical properties of rice straw–derived biochars modified by different oxidizing agents and their adsorptive performance for organics. *Bioremediation Journal, 19*(2), 171–182.

Yang, F., Zhang, S., Cho, D. W., Du, Q., Song, J., & Tsang, D. C. (2019). Porous biochar composite assembled with ternary needle-like iron-manganese-sulphur hybrids for high-efficiency lead removal. *Bioresource Technology, 272,* 415–420.

Yao, Y., Gao, B., Inyang, M., Zimmerman, A. R., Cao, X., Pullammanappallil, P., & Yang, L. (2011). Removal of phosphate from aqueous solution by biochar derived from anaerobically digested sugar beet tailings. *Journal of Hazardous Materials, 190*(1–3), 501–507.

Yek, P. N. Y., Peng, W., Wong, C. C., Liew, R. K., Ho, Y. L., Mahari, W. A. W., ... Lam, S. S. (2020). Engineered biochar via microwave CO_2 and steam pyrolysis to treat carcinogenic Congo red dye. *Journal of Hazardous Materials, 395,* 122636. Available from https://doi.org/10.1016/j.jhazmat.2020.122636.

Yin, G., Song, X., Tao, L., Sarkar, B., Sarmah, A. K., Zhang, W., ... Wang, H. (2020). Novel Fe-Mn binary oxide-biochar as an adsorbent for removing Cd (II) from aqueous solutions. *Chemical Engineering Journal, 389,* 124465.

Yin, Y., Guo, X., & Peng, D. (2018). Iron and manganese oxides modified maize straw to remove tylosin from aqueous solutions. *Chemosphere, 205,* 156–165.

Yu, J., Jiang, C., Guan, Q., Ning, P., Gu, J., Chen, Q., ... Miao, R. (2018). Enhanced removal of Cr (VI) from aqueous solution by supported ZnO nanoparticles on biochar derived from waste water hyacinth. *Chemosphere, 195,* 632–640.

Yu, W., Lian, F., Cui, G., & Liu, Z. (2018). N-doping effectively enhances the adsorption capacity of biochar for heavy metal ions from aqueous solution. *Chemosphere, 193,* 8–16.

Yu, Y., An, Q., Zhou, Y., Deng, S., Miao, Y., Zhao, B., & Yang, L. (2019). Highly synergistic effects on ammonium removal by the co-system of Pseudomonas stutzeri XL-2 and modified walnut shell biochar. *Bioresource Technology, 280,* 239–246.

Zhang, C., Zeng, G., Huang, D., Lai, C., Chen, M., Cheng, M., ... Wang, R. (2019). Biochar for environmental management: Mitigating greenhouse gas emissions, contaminant treatment, and potential negative impacts. *Chemical Engineering Journal, 373*, 902–922.

Zhang, X., Xiang, W., Miao, X., Li, F., Qi, G., Cao, C., & Gao, B. (2022). Microwave biochars produced with activated carbon catalyst: Characterization and sorption of volatile organic compounds (VOCs). *Science of the Total Environment, 827*, 153996.

Zhao, L., Zhang, Y., Wang, L., Lyu, H., Xia, S., & Tang, J. (2022). Effective removal of Hg (II) and MeHg from aqueous environment by ball milling aided thiol-modification of biochars: Effect of different pyrolysis temperatures. *Chemosphere, 294*, 133820.

Zheng, Y., Wan, Y., Chen, J., Chen, H., & Gao, B. (2020). MgO modified biochar produced through ball milling: A dual-functional adsorbent for removal of different contaminants. *Chemosphere, 243*, 125344.

Zhong, M., Li, M., Tan, B., Gao, B., Qiu, Y., Wei, X., ... Zhang, Q. (2021). Investigations of Cr (VI) removal by millet bran biochar modified with inorganic compounds: Momentous role of additional lactate. *Science of The Total Environment, 793*, 148098.

Zhou, L., Huang, Y., Qiu, W., Sun, Z., Liu, Z., & Song, Z. (2017). Adsorption properties of nano-MnO2–biochar composites for copper in aqueous solution. *Molecules (Basel, Switzerland), 22*(1), 173.

Zhou, Y., Qin, S., Verma, S., Sar, T., Sarsaiya, S., Ravindran, B., ... Awasthi, M. K. (2021). Production and beneficial impact of biochar for environmental application: A comprehensive review. *Bioresource Technology, 337*, 125451.

Zhu, S., Wang, S., Yang, X., Tufail, S., Chen, C., Wang, X., & Shang, J. (2020). Green sustainable and highly efficient hematite nanoparticles modified biochar-clay granular composite for Cr (VI) removal and related mechanism. *Journal of Cleaner Production, 276*, 123009.

Zhu, Y., Li, H., Zhang, G., Meng, F., Li, L., & Wu, S. (2018). Removal of hexavalent chromium from aqueous solution by different surface-modified biochars: Acid washing, nanoscale zero-valent iron and ferric iron loading. *Bioresource Technology, 261*, 142–150.

CHAPTER 23

Bioremediation of organic pollutants soil and water through biochar for a healthy ecosystem

Diksha Pandey[1], Nikhil Savio[1], Nishtha Naudiyal[1], R.K. Srivastava[1], Prayasi Nayak[1], Beatriz Cabañas[2], Andrés Moreno[2] and Shiv Vendra Singh[3]

[1]College of Basic Sciences and Humanities, Govind Ballabh Pant University of Agriculture & Technology, Pantnagar, Uttarakhand, India [2]Universidad de Castilla-La Mancha, Instituto de Combustión y Contaminación Atmosférica (ICCA). Ciudad Real, Spain [3]Department of Agronomy, College of Agriculture, Rani Lakshi Bai Central Agricultural University, Jhansi, Uttar Pradesh, India

23.1 Introduction

Water and soil serve as important reservoirs for biota like bacteria, fungi, algae, and protozoa. However, due to inadequacies and unavoidable anthropogenic activities, such as industrialization, urbanization and poor agricultural practices, they dynamically standardize ecosystem functioning and ultimately lead to various environmental health risks (Kumari, Rajput, & Singh, 2024; Singh, Sharma, & Balan, 2024). Although a single factor cannot be criticized for these issues, numerous other activities contribute significantly to environmental pollution both directly and indirectly. The ecological environment and human health are severely threatened by organic pollutants, which constitute an essential part of environmental pollutants (Jiang, Pi, & Cai, 2018). The scope and complexity of organic pollution exceed that of heavy metal pollution. Organic environmental pollution is the product of a variety of sources functioning together. Pesticide containers, mulch used after cultivation and wastewater discharge from industrial dyes are significant sources of pollution (Vithanage, Mayakaduwa, Herath, Ok, & Mohan, 2016). Due to their recalcitrant nature, several highly hazardous wastes containing organic compounds in soil and wastewater play a crucial role in bioremediation. Anthropogenic activities can cause soil contamination that varies from 1 to 100 μg kg^{-1} through mining, oil spills, refuse disposal, and other domestic activities (Pradhan, Meena, Kumar, & Lal, 2023; Zhou et al., 2021). Like soil, many organic pollutants have possible reservoirs in water (B.-L. Liu et al., 2021; X. Liu, Wang, Tang, Liu, & Wang, 2021; Meena, Pradhan, Kumar, & Lal, 2023). Because of the severe and long-term effects on human health, the exponential rise in the concentration of organic contaminants in terrestrial and aquatic habitats has alarmed scientific and regulatory groups worldwide (Singh, Luthra, Bhattacharya, & Bera, 2024; Joseph et al., 2009). Compared to traditional physical and chemical remediation techniques, bioremediation is one of the most effective remediation strategies due to its simplicity of use, environmental friendliness, and low expense (Malla et al., 2018; Pradhan & Meena, 2023a). As a result, a sustainable, cost-effective, carbon-rich by-product is developed with unique and multifunctional sorbent called "biochar." In both the soil and water environments, biochar is well recognized for its ability to adsorb various organic and inorganic contaminants (Bolan, Naidu, Syers, & Tillman, 1999; Meena & Pradhan, 2023). The pollutant removal efficacy is highly dependent upon the qualities of biochar (Fig. 23.1). Over the past few decades, study into biochar has contributed to the development and use of bioremediation methodologies. However, this field of study is just now beginning, and it will take more time to scale up such up-and-coming techniques for use in environmental bioremediation in the real world.

Contaminant discharge from commercial, residential, and industrial sources affects the soil and water ecosystems. The primary cause of soil and water contamination is anthropogenic activity, viz., pervasive industrial activities that seep organic contaminants into the land and water and pose serious health risks to individuals, increased use of insecticides, herbicides, pesticides, agricultural fertilizers, antibiotics, and fossil fuels. Researchers and decision-makers have tried to develop new ways to tackle land and water contamination brought on by organic compounds (Pradhan & Meena, 2023b; Zama et al., 2018). In view of improving soil quality, encouraging plant development, easing drought and salinity stresses, and interacting with organic pollutants and heavy metals to stop plants from absorbing those pollutants from the soil, biochar has shown some properties as a soil remediation substance (Guo et al., 2020;

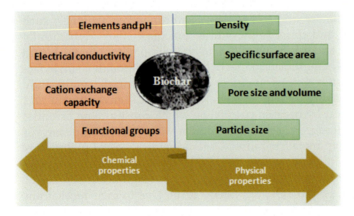

FIGURE 23.1 Chemical and physical properties of biochar for environmental management.

Pradhan & Meena, 2022). The crop yields, the safety of agricultural products and human health have all been severely threatened by soil organic pollution, which has developed from simple pollution to complex pollution through soil-plant system migration and accumulation (Bloom, Micu, & Neamtiu, 2016; Pradhan et al., 2022; Roy, Meena, Kumar, Jhariya, & Pradhan, 2021). With the added advantages of soil fertilization and climate change mitigation, biochar is emerging as a remedy to lower environmental pollutants' bioavailability (Singh et al., 2023; Sohi, 2012). Researchers are interested in using biochar to remove organic contaminants from water and soil because of its appealing properties such as high surface area, more functional groups, high porosity, low bulk density and higher cation exchange capacity (Kumawat et al., 2022; Pandey, Chhimwal, & Srivastava, 2022; Singh, Chaturvedi, Dhyani, & Govindaraju, 2020; Zhang et al., 2013). Despite scientists employing various techniques including biodegradation, adsorption, stripping, hydrolysis, photolysis, and so on, to remove organic pollutants from the environment (Ali, Asim, & Khan, 2012), no comprehensive answer has yet been discovered. In environmental science, research on the adsorption of environmental pollutants by biochar has recently gained attention. To remove various contaminants from soil and water systems, biochar is used as a sustainable substitute for activated carbon. Various studies have demonstrated that biochar can effectively bind to organic contaminants (Ahmed et al., 2017; Cao et al., 2016; Li, Dong, et al., 2017; Li, Zhang, et al., 2017; Zhelezova, Cederlund, & Stenström, 2017; Zhong et al., 2018). The greatest concern of organic contaminants has been focused on pesticides, herbicides, polycyclic aromatic hydrocarbons (PAHs), dyes, and antibiotics (Qiu, Zheng, Zhou, & Sheng, 2009; Teixidó, Pignatello, Beltrán, Granados, & Peccia, 2011; Xu et al., 2012; Zheng, Guo, Chow, Bennett, & Rajagopalan, 2010). Compared to water remediation, there are fewer studies on using biochar to clean up soils polluted with organic pollutants. Soil remediation is the act of eliminating or reducing the availability of contaminants. Recently, environmental remediation has gained attention as a potential field in which biochar can be used (Ahmad et al., 2014; Cao, Ma, Liang, Gao, & Harris, 2011; Sheoran et al., 2022). Accordingly, cheaper and less time-consuming methods of remediation are highly desirable. Blenis, Hue, Maaz, and Kantar (2023) examined the addition of biochar as an effective soil remediation method. The characteristics of biochar, combined with biological activity can help bind and promote the degradation process of these contaminants without using hazardous chemicals or removing a large amount of soil.

Thus this chapter hopes to increase interest in biochar research to fill the missing information gaps that could make biochar production cheaper and more consistent as it offers a more environmentally friendly way to clean up environmental pollutants. However, fewer of these studies apply to soil, and a larger percentage of them concentrated on the remediation of aqueous organic contaminants.

23.2 Bioremediation of organic contaminants in water using biochar

23.2.1 Removal of antibiotics

The discovery of antibiotics has profoundly impacted veterinary and human health. However, due to their widespread use in humans and animals, some pharmaceutical wastewater containing antibiotics progressively changes the environment and persists through a tricky cycle of bioaccumulation and transformation (Ben et al., 2019; Meena et al., 2022). Consequently, antibiotics have been identified as an emerging environmental pollutant (Carvalho & Santos, 2016). The removal of antibiotics has become the study focus of adsorption by biochar. In a study by Liu and coworkers conducted to adsorb doxycycline hydrochloride in a water environment, copper nitrate was added to modify biochar made from peanut shells. Researchers found that the removal rate of doxycycline hydrochloride by copper nitrate-modified biochar is twice as high as that of raw biochar (Liu, Wang, Liu, & Zhu, 2017). Another comprehensive research was conducted using iron and zinc-mixed sawdust biochar to remove tetracycline. According to research, this particular form of biochar has great potential for tetracycline removal in aqueous solution. Tetracycline was separated from water, and after three rounds, the removal rate was still higher than 89% (Zhou et al., 2017). In another research, for the adsorption of tetracycline (TC), rice straw biochar (RCA) and pig manure biochar were used (SCA) after modification by acid treatment. The outcomes demonstrated that modified biochar could improve TC adsorption by RCA and SCA, and even improve SCA's adsorption impact by about 25%. Both can be used as effective materials to remove TC from water as each has greater adsorption capacities (Chen, Luo, et al., 2018; Chen, Yang, et al., 2018; Chen, Zhou, et al., 2018). Magnetic pine sawdust biochar (MPSB), a magnetic adsorbent is produced when $FeCl_2$ is oxidatively hydrolyzed in an alkaline solution. The ultimate goal was to determine how well sulfamethoxazole (SMX) could be removed from an aqueous solution using magnetic pine sawdust biochar as an adsorbent. It has been discovered that MPSB has superior SMX sorption results and a faster sorption rate under low pH (Foo & Hameed, 2012). Research on biochar made from eucalyptus wood revealed that it was a powerful adsorbent for removing nitroimidazole antibiotics. The removal efficiency of biochar for metronidazole and dimetridazole files with concentrations of 20 and 1000 mg/L within 2 hours of the optimum preparation conditions was 97.10% and 96.40%, respectively (Wan, Hua, Sun, Bai, & Liang, 2016).

Researchers need to conduct an extensive amount of study in this field as biochar adsorption is a very efficient method of removing antibiotics from the environment. However, most studies on the biochar adsorption of antibiotics have only been conducted in aqueous solutions; there have been very few investigations in soil environments (Raj et al., 2021; Zhong et al., 2018). Therefore researchers must look into how biochar interacts with the soil environment to bind antibiotics. Additionally, the majority of studies have merely looked at a single antibiotic in a static test, which lacks environmental authenticity because antibiotics are typically found as complex pollutants in real-world settings. In the future, scientists can cope with the environment of antibiotics with multiple pollution by simulating the actual environment using dynamic adsorption experiments (Xu, Shi, & Wang, 2018).

23.2.2 Removal of dyes

Dye wastewater constitutes a sizable part of industrial wastewater and has become a significant source of pollution due to the textile industry's quick growth. Statistics show that more than 100,000 different types of dyes are used in commercial dyeing, and most of the wastewater generated in this industry is discharged. In water, dye is noticeable and harmful even at low concentrations (Hao, Kim, & Chiang, 2000). Dye removal from wastewater can be done in a variety of methods. Compared to other technologies, adsorption technology has a wide range of potential applications in these techniques (Kumar, Meena, et al., 2020; Tang, Li, Wang, & Daroch, 2017). Most organic dye components, including azo dyes and anthraquinone dye, are poisonous or cancer-causing to humans (Donadelli, Carlos, Arques, & Einschlag, 2018; Huang et al., 2018). Adsorption has been tried and applied in wastewater treatment and is thought to be the most acceptable and effective dye removal method to effectively remove these refractory contaminants (Adeyemo, Adeoye, & Bello, 2015; Ersan, Apul, Perreault, & Karanfil, 2017). Adsorbents with many active sites are essential for highly effective and economical adsorption. In-depth research on biochar-based materials has been sparked by biochars (BCs), which have been regarded as a cost-effective adsorbent matrix to provide large amounts of active sites for binding organic dyes via hydrogen bond interaction, n-couple action, electrostatic adsorption, ion exchange process, or other interactions. Researchers investigated the synthesis, characterization, and adsorption characteristics of ferric oxide biochar nano-composites made from pulp and paper sludge in the instance of methyl orange (MO) adsorption. The outcomes demonstrated that Fe_2O_3-BC can adsorb effluent containing MO, indicating that they can be used as dye adsorbents (Chaukura, Murimba, & Gwenzi, 2017; Kumar, Sharma, et al., 2020). Further the sorption of methylene blue (MB) by three types of biochar made from pinewood (BC-PW), pig manure (BC-PM), and cardboard (BCPD) under various pyrolysis conditions were also carried out. They found that all three types of biochar had strong adsorption capacities, with BC-PM having the best adsorption effect because it has the highest ash content (Lonappan et al., 2016). As a low-cost sorbent to remove Reactive Red 141 from aqueous solutions, some researchers assessed biochar made from pecan nutshells. They found that, compared to other adsorbents, the new biochar has the advantages of replacement, low cost, and environmental protection (Zazycki et al., 2018). A thorough investigation into the crystal violet (CV) adsorption capacity was conducted, and the woody tree *Gliricidia sepium* was chosen for producing biochar with the designations GBC300, GBC500, and GBC700 at respective temperatures of 300°C, 500°C, and 700°C. When they removed the CV at a high temperature, they discovered it was very efficient (Wathukarage, Herath, Iqbal, & Vithanage, 2019). Fan, Chen, Li, and Xiong (2017) examined a novel BC made from sewage sludge as an adsorbent to remove MB and found that the removal efficiency of MB increased to almost 100% in just 10 hours. The large surface area, coarse, and porous texture structure were placed for the high removal effectiveness (Fan et al., 2016; Sewu, Boakye, & Woo, 2017).

However, to a certain extent, the adsorption capacity of pristine BCs is insufficient to satisfy the practical requirements, possibly due to the constrained functionalities. Biochar-based hybrid materials are gaining a lot of focus in this regard over time. By fusing the benefits of foreign materials with BCs, these hybrid materials offer chances to meet the demand for low-cost adsorbents with improved removal efficiencies (Chen, Luo, et al., 2018; Chen, Yang, et al., 2018; Chen, Zhou, et al., 2018; Güzel, Sayğılı, Sayğılı, Koyuncu, & Yılmaz, 2017; Wang, Lian, Sun, Ma, & Ning, 2018; Wang, Yan, et al., 2018). Similarly, BCs were created by Kalderis, Kayan, Akay, Kulaksız, and Gözmen (2017) to get rid of the organic dye malachite green. Compared to pristine biochar, the obtained Fe-modified BCs had a larger surface area and pore volume, leading to a greater removal percentage for malachite green. For the photocatalytic degradation of dye contaminants, several excellent catalysts have been discovered, including TiO_2, ZnO, CdS, and Ag nanoparticles (Khan, Lee, & Cho, 2014; Kong, Zhao, Zheng, & Qu, 2017; Wang, Wang, et al., 2017; Wang, Zhang, et al., 2017). For the degradation of Reactive Red 84 (RR84), Khataee, Gholami, Kalderis, Pachatouridou, and Konsolakis (2018) developed a new BC with ceria nanoparticle incorporation called CeO_2-H@BC nanocomposite. When compared to isolated CeO_2-H and BCs samples in an ultrasonic bath, the obtained CeO_2-H@BC nanocomposite demonstrated significantly greater catalytic degradation efficiency (Wang, Wang, et al., 2017; Wang, Zhang, et al., 2017; Zhang, Guo, et al., 2018; Zhang, Xue, Chen, & Li, 2018; Cai et al., 2017; Park et al., 2018). Through a variety of characterization techniques and batch tests, the improvements in dye removal efficiencies and associated removal mechanisms have been verified. However, current research has stayed in single biochar adsorption dyes and there are few studies on various types of biochar composites.

23.2.3 Removal of pesticides

The primary tool for managing agricultural and pest diseases is using pesticides in agricultural output. The widespread use of pesticides can boost crop yields and farm production's financial gains (Chen, Luo, et al., 2018; Chen, Yang, et al., 2018; Chen, Zhou, et al., 2018). Pesticides, however, are migratory and extremely poisonous. Overuse of pesticides may result in some degree of pollution of the air, soil, and water, toxic effects on organisms that are not intended targets, destruction of the ecological equilibrium and risk to human health (Zhong et al., 2018). Due to its high porosity, high surface area, high pH, numerous functional groups, and high aromatic structure, biochar is a good adsorbent (Khorram et al., 2017). Researchers have investigated the ability of biochar to adsorb pesticides from water and soil. At 600°C, imidacloprid was tested for adsorption using biochar produced from pig manure. According to the experimental findings, pore-filling is probably one of the primary adsorption processes for controlling polar chemicals (Jin et al., 2016). To investigate the adsorption impact on thiacloprid (THI), researchers used maize straw as the raw material to produce biochar at 300°C, 500°C, and 700°C. They revealed during the trial that hydrophobic interaction, pore-filling, and p—p interaction are probably responsible for the biochar's ability to bind THI (Zhang, Guo, et al., 2018; Zhang, Xue, et al., 2018). In another investigation, researchers used sugarcane bagasse to adsorb herbicides with an organophosphorus, and they looked at how different variables affected the adsorption. According to studies, adding more bagasse carbon increases the quantity of dimethoate that can be adsorbed and removed (Sun et al., 2018). According to Uchimiya, Wartelle, Lima, and Klasson (2010), heating biochar from broiler litter to 700°C increased the biochar's aromaticity and deisopropyl atrazine sorption capability. Similar findings for trichloroethylene sorption on biochar made from soybean stover and peanut shells at 700°C compared to 300°C were obtained (Ahmad et al., 2012). Contrarily, Sun, Keiluweit, Kleber, Pan, and Xing (2011) discovered that high polarity biochars produced at 400°C were more successful at saturating norflurazon and fluoridone. Variations in the structure of the organic compounds explain these various results. While nonpolar compounds, such as trichloroethylene, access hydrophobic sites on biochar surfaces in the absence of

H-bonding between water and O-containing functional groups, polar compounds, such as norflurazon and fluridone, are adsorbed by H-bonding between the compounds and the O-containing moieties of the biochars (Ahmad et al., 2012; Sun et al., 2011). The adsorption and degradation of two representative neonicotinoid insecticides in typical Chinese paddy soil and red soil by six different kinds of biochars were studied by Zhang, Hou, et al. (2020). The findings showed that adding charcoal increased each soil type's surface area, pH value, total organic carbon, and dissolved organic carbon. Because of the greater surface area and lower H/C, the adsorption of the two pesticides on biochar-soil mixed systems improved by over 4.3 times when the pyrolyzing temperature of biochar was raised from 300°C to 700°C.

Extensive investigation considering the type of biochar and soil properties is needed to optimize biochar technology. Biochar has a complex effect on the behavior of pesticides in soil. Pesticides are widely and ineffectively used to treat crop diseases and pests, which pollute agricultural soils and associated ecosystems (Khalid et al., 2020). Biochar can significantly decrease the health risks associated with pesticide contamination because it is a cheap and environmentally friendly adsorbent. The behavior of herbicides in the soil can be significantly changed by adding biochar to the soil. The bioavailability mechanisms of soil contaminants, such as adsorption, desorption, degradation, and leaching, are specifically impacted by biochar. The literature that can be used to reference the impact of pesticides in aqueous solutions is not very extensive and more research is required (Xiao et al., 2018).

23.2.4 Removal of pharmaceutically active compounds by biochars

Removing harmful pharmaceutically active compounds has demonstrated great promise for biochar-based materials (Zhu et al., 2018). Kemmou and coworkers pyrolyzed malt rootlets to produce BC, which was then used to initiate sodium persulfate (SPS) to oxidize SMX. After 30 minutes of reaction, the degradation efficacy of the compound SMX (250 g L^{-1}) in the presence of 250 mg L^{-1} SPS and 90 mg L^{-1} BC was 94% (Kemmou, Frontistis, Vakros, Manariotis, & Mantzavinos, 2018). Zhu et al. (2018) conducted research very similar to this one, using the wetland plant reed BC as a catalyst for activating peroxydisulfate (PDS) and the degradation of SMX in water. In similar research, the doping of N gave many adsorption sites for the electrostatic binding of organic contaminants and the binding of PDS molecules to form metastable surface-confined reactive species that could effectively oxidize the co-adsorbed organic contaminants. N-BC900 removed 100% of SMX at 20 mg L^{-1} in 20 minutes, which is similar to the removal rates of many well-known carbon materials (Lee et al., 2015). Various studies revealed that using BCs combined with metal oxides or metal nanoparticles could increase the removal efficacy of pharmaceutically active compounds (Shan et al., 2016; Taheran et al., 2017). Tetracycline (TET) has been eliminated from an aqueous solution using a novel iron and zinc codoped sawdust BC described by Zhou et al. (2017). The Fe/Zn BC performed better than purified biochar in the TET removal. This was ascribed to having more active sites to bind TET molecules via hydrogen bonding, electron donor–acceptor (EDA), and Fe (III)-TET due to the higher surface area and larger pore size. Fe/Zn BC was calculated to have a maximal adsorption capacity of 102.0 mg g^{-1}. To eliminate TET, Wang, Yan, et al. (2018), Wang, Lian, et al. (2018) immobilized a new nanoscale zerovalent iron (NZVI) on BC with a polydopamine surface modification (PDA/NZVI@BC) composite. After being loaded with NZVI, TET had 55.9% higher removal effectiveness. Due to its high reducibility, the NZVI could successfully induce the TET by causing electrons to react with H_2O to produce reactive species like H^*. Similarly, a BC-Co_3O_4 compound with cobalt oxide bound was created for the catalytic degradation of the fluoroquinolone antibiotic ofloxacin (OFX) (Chen, Luo, et al., 2018; Chen, Yang, et al., 2018; Chen, Zhou, et al., 2018). According to the degradation tests, the BC-Co_3O_4/Oxone system was much more effective at removing over 90% of OFX within 10 minutes than using only Oxone, only BCs, or only the Co_3O_4/Oxone system. Furthermore, there is efficient inhibition of its aggregation by Co_3O_4 and BCs. Furthermore, the BC-Co_3O_4 composite maintained a high degradation efficiency ($>90\%$) even after three cycles under the same circumstances, evidencing its outstanding catalytic capacity and reusability in the treatment of wastewater containing pharmaceutically active compounds.

23.2.5 Removal of persistent organic pollutants

Persistent organic pollutants (POPs) are extremely difficult to break down in the natural environment, and they last a very long time in the human body and in the water environment, soil, and food chain (Rafiquel et al., 2018). POPs have steadily grown to rank among the most concerning chemical pollutants worldwide due to their long-term bioavailability (Liu et al., 2017). POPs have a high level of toxicity and bioaccumulation, and they can remain in the environment. It poses a severe threat to human health. Studying the removal of POPs from the atmosphere is crucial. Therefore a growing amount of focus is being paid to polybrominated diphenyl ethers, halohydrocarbons, PAHs, and organochlorine pesticides (Ren et al., 2018). POP usage can be decreased using sorbents, and biochar has caught the interest of researchers as a particularly effective adsorbent. When interactions between the POP and the biochar happened that were hydrophobic, electrochemical, and EDA, the toxicity of POPs like PCBs, hexachlorobenzene and polychlorinated dibenzo-p-dioxins was reduced (Hu et al., 2020; Tian, Wang, Liu, & Ma, 2020). Additionally, it has been observed that biochar composites supported by $TiO_2/C_3N_4/CoFe_2O_4/Ag_3PO_4$ can photocatalytically degrade POPs, reducing their toxicity. Scientists employed wood and straw to prepare various types of biochar and evaluate their ability to remove imidazolium-type ionic liquids (ITILS) from wastewater. The outcomes show that biochar made from wood and grass was very effective at eliminating ionic liquids that are water-soluble. In one investigation, researchers used sawdust, bamboo, and straw as raw materials to create biochar at temperatures between 300°C and 700°C. These biochars can be used to absorb aqueous solutions of N-nitrosodimethyl amine (NDMA), and they also suggested the best possibility for adsorption and the mechanism for three different biochars. Following the experimental findings, NDMA was removed from an aqueous solution using bamboo charcoal biochar that was created at 500°C. High nitrosamine removal capacity was achieved with biochar. To understand how biochars affect the buildup of PAHs in rice under anaerobic conditions, systematic research was conducted. Corn stems and bamboo were chosen by scientists as biochar materials and pyrolyzed at temperatures of 300°C and 700°C, respectively. According to research, various kinds of biochar impact how PAHs are absorbed (Ni et al., 2018). Scholars examined the effect of solid organic carbon (SOC) on the biochar's ability to bind polychlorinated biphenyls in the soil during the 120-day experimental time. The empirical findings demonstrated that the presence of organic carbon may cause several interactions, such as competitive sorption, cosorption, and cumulative sorption, to influence adsorption (Huang et al., 2018).

Many studies have been done recently by researchers showing that the use of biochar for organic contaminant removal is an efficient and eco-friendly technique. Researchers have made progress toward its elimination, but the majority of their work has concentrated on the adsorption kinetics and isotherms of various biochars on POPs. There is little systematic theoretical research on the specific adsorption process and little research on the effectiveness of composite biochar in adsorbing POPs (Wu & Liu, 2018; Yang, Chen, Wu, & Xu, 2018). Future research has a lot of scope regarding the POP remediation potential of the above biochar-supported catalysts in the presence of a POP-degrading microbial consortium. Researchers need to continue delving into these issues.

23.2.6 Removal of other contaminants

Constructed wetlands (CWs) are systems that simulate natural wetlands and can be used to treat wastewater from various pollution sources through physical, chemical, and biological purification processes Table 23.1. A critical review of the updated literature on CWs that integrate biochar into the substrate has recently been conducted (El Barkaoui, Mandi, Aziz, Del Bubba, & Ouazzani, 2023). The study focuses on the characteristics of biochar that are generally integrated into this treatment ecotechnology and the processes used to prepare the materials, including thermal conversion conditions and the type of feedstock used (e.g., agricultural, food and wood wastes, sewage sludge, and algal marine feedstock). Biochar quality is affected by biochar preparation conditions (e.g., pyrolysis temperature, heating, and carbonization time). Different CW configurations integrating biochar into the wetland as a filling medium have been described and compared, with the optimal position for the biochar substrate being intermediate between two layers of inert materials, to improve filtration or flotation of the biochar. Also, the mechanisms and applications of (modified) biochar in CWs and its improving effects on typical pollutants (such as nitrogen, phosphorus, complex organics, and heavy metals) removal were summarized (Feng et al., 2023).

Research in the use of magnetic biochar in wastewater treatment has recently focused on removing heavy metals and organic complex organic pollutants (Li et al., 2023). For the future development of magnetic biochar, an environmentally friendly approach is necessary, economic feasibility and joint technology are the directions that need further exploration. Finally, this paper provides an overview of the adsorption mechanisms of magnetic biochar and several common modification methods, aiming to help researchers in their efforts.

On the other hand, steroidal estrogens (SEs) remain one of the notable endocrine-disrupting chemicals that pose a significant threat to the aquatic environment in this era owing to their interference with the normal metabolic functions of the human body systems. They are currently identified as emerging contaminants of water sources. The sources of SEs are either natural or synthetic active ingredients in oral contraceptive and

TABLE 23.1 Removal of organic contaminants in soil and water ecosystem through biochar.

Biochar		Organic pollutants removed	Removal efficiency	References
Pyrolyzed feedstocks	Pyrolysis temperature (°C)			
Pinewood	525	Methylene blue	98.6	Lonappan et al. (2016)
Tomato processing waste	600	Metanil yellow	91	Sayğılı and Güzel (2016)
Sulfidated nano zerovalent iron	–	Nitrobenzene	90	Gao et al. (2022)
Sugarcane bagasse	400	Imidacloprid	92	Chen et al. (2023)
Wood	450	Thiamethoxam	22.8	You et al. (2020)
Sewage sludge	700	PAHs	74.0	Godlewska and Oleszczuk (2022)
Rice husk	700	Pcbs	91.0	Silvani et al. (2019)
Loofah sponges	900	PAHs	31.9	Hao, Wang, Yan, and Jiang (2021)
Rice straw	600	PAHs	58.8	Zhang, He et al. (2020)
Pig manure	700	Clothianidin	90.5	Zhang, Min, Tang, Rafq, and Sun (2020)
Rice straw	700	Tetracycline	75–95	Chen, Yang et al. (2018), Chen, Zhou et al. (2018), Chen, Luo et al. (2018)
Fe/Zn	600	Tetracycline	99	Zhou et al. (2017)
Crab shell		Chlortetracycline hydrochloride	22	Xu et al. (2020)
Eucalyptus sawdust	500	Dimetridazole	97	Wan et al. (2016)
Peanut shells	450	Doxycycline	65	Li, Zhang et al. (2017), Li, Dong et al. (2017)
Sewage sludge	–	Sulfamerazine	95.12	Xing et al. (2022)
Iron-doped biochar	–	Tetracycline	92	Li et al. (2023)
Eucalyptus sawdust	500	Metronidazole	95.1	Wan et al. (2016)
Spent mushroom substrate	–	Crystal violet	99.6	Sewu et al. (2017)
FeCo-biochar	–	Peroxymonosulfate	57.1	Wang and Wang (2023)
Rice straw		Malachite green	55	Hameed and El-Khaiary (2008)
Pecan nutshell	800	Reactive red	85	Zazycki et al. (2018)
Straw	500	Sunset Yellow	–	Ji, Lü, Chen, and Yang (2016)

(Continued)

TABLE 23.1 (Continued)

Biochar		Organic pollutants removed	Removal efficiency	References
Pyrolyzed feedstocks	Pyrolysis temperature (°C)			
Poplar catkins	800	Methylene blue	83	Li, Zhang et al. (2017), Li, Dong et al. (2017)
Maize straw	300	Thiacloprid	—	Zhang, Guo et al. (2018), Zhang, Xue et al. (2018)
Swine manure	600	Imidacloprid	45	Jin et al. (2016)
Bamboo sawdust	600	Sulfamethazine	30	Huang, Wang et al. (2017), Huang, Liu et al. (2017)
Bagasse	600	Sulfapyridine	99	Yao, Zhang, Gao, Chen, and Wu (2017)
Corn straw	450	Atrazine	64	Zhao et al. (2013)
Switchgrasss	425	Metribuzin herbicide	91	Essandoh, Wolgemuth, Pittman, Mohan, and Mlsna (2017)
Loofah sponge	800	Acid Orange 7	96	Zhao et al. (2022)

hormonal replacement therapy drugs and enter the environment primarily from excretes in the form of active free conjugate radicals, resulting in numerous effects on organisms in aquatic habitats and humans. The removal of SEs from water sources is important because of their potential adverse effects on marine ecosystems and human health. Thus adsorption techniques have gained considerable attention as effective methods for removing these contaminants. The adsorption mechanism and factors affecting the removal efficiency, such as pH, temperature, initial concentration, contact time, and adsorbent properties of different adsorbents such as biochar, were investigated in this review (Bayode, Olisah, Emmanuel, Adesina, & Koko, 2023). The analysis provides a comprehensive overview of the adsorption technique for removing SEs from other sources, serving as a valuable resource for researchers involved in developing efficient and sustainable strategies to mitigate the effects of these emerging contaminants.

In addition, the concern for protecting water quantity and quality is one of the most severe challenges of the 21st century, since the demand for water resources is growing as the population and its needs increase. In addition, most human activities produce wastewater containing undesirable pollutants. On the other hand, the generation of agricultural waste and its improper disposal cause other problems. Current wastewater treatment methods involve a combination of physical and chemical processes, technologies, and operations to remove pollutants from effluents; adsorption is an excellent example of an effective way for wastewater treatment, and thus biochar is currently one of the most valuable adsorbents. This review focuses on new research on the application of biochar produced from agricultural residues as a low-cost and environmentally friendly method to remove organic and inorganic compounds from aqueous solutions (Pantoja, Sukmana, Beszdes, & Laszlo, 2023).

23.3 Mechanisms for organic pollutant removal using biochar

Due to high porosity, high surface area and large functional groups availability, biochar is a common substance for adsorbing organic pollutants (Liu et al., 2018; Singh et al., 2023). The chemical characteristics of the adsorbent surface and the type of contaminants determine the adsorption process for removing organic pollutants (Rosales, Meijide, Pazos, & Sanromán, 2017). The various adsorption processes are primarily caused by variations in the organic structure, surface electrical characteristics, and surface functional groups of biochar (Inyang & Dickenson, 2015).

The specific mechanism governing pollutants immobilization is presented in Table 23.2 and Fig. 23.2. Electrostatic attraction, pore-filling, $\pi - \pi$ EDA interaction, H bonding, complex adsorption, hydrophobic interactions, partition on uncarbonized portion, and spectrometer exchange can be used to classify the mechanisms of adsorbing organic contaminants (Singh et al., 2023). Organic contaminants adsorb more biochar as more functional groups contain oxygen, partly because of the $\pi - \pi$ EDA interaction. As electron acceptors, the carboxylic acid, nitro, and ketonic groups on the biochar surface interact with aromatic molecules to create EDA interactions, improving aromatic molecules' adsorption. As p-electron donor sites, biochar can also use various hydroxyl and amine groups (Ahmed et al., 2018). Biochar typically has a negatively charged surface, which attracts positively charged organic molecules via electrostatic forces. The amount of each atomic charge and the separation between two atoms determine the strength of electrostatic attraction (Rosales et al., 2017). To remove the contaminants, biochar interacts hydrophobically with hydrophilic organic substances (Inyang & Dickenson, 2015). Another potential adsorption process involves electrostatic attraction or repulsion between organic contaminants and biochar. Negatively charged biochar surfaces could make positively charged cationic organic molecules more electrostatically attracted to them. Xu, Xiao, Yuan, and Zhao (2011) and Qiu et al. (2009) reported on this electrostatic attraction in the research on the adsorption of cationic dyes, such as methyl violet and rhodanine from water. Highly polar biochars generated at 400°C have abundant electron-withdrawing functional groups in their aromatic $\pi -$ systems (Keiluweit, Nico, Johnson, & Kleber, 2010). They frequently lack electrons and might behave as p-acceptors to electron sources. Because high-temperature-derived biochars contain both electron-rich and electron-poor functional groups, they can potentially interact with electron donors and electron acceptors (Sun et al., 2011). The relationship between biochar's graphene surface, which is π-electron-rich and positively charged organics that lack π electrons is improved (Qiu et al., 2009; Teixidó et al., 2011). However, H-bonding and adsorption could be induced by an electrostatic attraction between negatively charged anionic organic molecules and biochars. In studying the adsorption of biochar on thiaccloprid (THI) using swine manure and maize straw, it was revealed in the findings that a negative correlation between H/C and (O&N)/C and a positive correlation between THI and biochar's sorption affinity, surface area, and aromatic carbon concentration. It shows that hydrophobic interaction, pore-filling, partition on uncarbonized fraction, and $\pi - \pi$ EDA interaction are the primary mechanisms by which biochar is adsorbed (Zhang, Guo, et al., 2018; Zhang, Xue, et al., 2018). In another investigation, scientists used Cu (II) to impregnate biochar to adsorb

TABLE 23.2 Mechanisms involved in removing organic contaminants in soil and water by biochar.

Feedstocks	Pyrolysis temp (°C)	Modifications	Types of organic pollutants	Medium	Mechanism involved	References
Rice straw	700 ± 50	H_3PO_4	Tetracycline	Water	H-bonding, π–π electron donor–acceptor (EDA) interaction	Chen, Yang et al. (2018), Chen, Zhou et al. (2018), Chen, Luo et al. (2018)
Switchgrass	425	Fe^{2+}/Fe^{3+}	Metribuzin herbicide	Water	Electrostatic attraction and hydrogen bonds	Essandoh et al. (2017)
Peanut shells	500 ± 50	H_2O_2	Doxycycline hydrochloride	Water	Strong complexation, electrostatic interactions	Li, Zhang et al. (2017), Li, Dong et al. (2017)
Pine needles	300–700	Pristine	Trichloroethylene	Soil	Chemisorption and pore diffusion	Ahmad et al. (2012)
Corncob	500	Pristine	Bisphenol A	Soil	H-bonding, π–π EDA	Li, Zhang et al. (2017), Li, Dong et al. (2017)
Woody tree	700 ± 50	Pristine	Crystal violet	Water	Chemisorption	Wathukarage et al. (2019)
Bamboo		Pristine	Fuel oil	Water	π-π EDA and acid-base interaction	Yang et al. (2018)
Sawdust	600	Fe-Zn	Tetracycline	Water	Site recognition, bridge enhancement, and site competition	Zhou et al. (2017)
Reed	600	Magnetic biochar	Florfenicol	Water	Pore-filling effect and π – π EDA interaction	Zhao and Lang (2018)
Maple wood	300–700	Oil-imbibed biochar	Crude oil	Water	Pore-filling	Nguyen and Pignatello (2013)
Herb	300 ± 50	Pristine	Metolachlor	Soil	Pore-filling and polar functional groups	Wei, Wang, Fultz, White, and Jeong (2021)
Paper and pulp sludge	700 ± 50	Fe_2O_3	Methyl orange	Water	Complexation	Chaukura et al. (2017)
Pig manure		Fe	Tetracycline	Soil	H-bonding and π−π EDA	Chen, Yang et al. (2018), Chen, Zhou et al. (2018), Chen, Luo et al. (2018)

(Continued)

TABLE 23.2 (Continued)

Feedstocks	Pyrolysis temp (°C)	Modifications	Types of organic pollutants	Medium	Mechanism involved	References
Sewage sludge	500 ± 50	Pristine	2,4-Dichlorophenol	Water	π-π EDA	Kalderis et al. (2017)
Sewage sludge	550	Pristine	Fluoroquinolone	Water	Positive correlation with volatile solid content	Yao et al. (2017)
Paper mill sludge	300°C ± 50°C	Ni–ZVI	Pentachlorophenol	Water	Partition on uncarbonized organic matter	Devi and Saroha (2015)
Hardwood	600°C	Pristine	Sulfamethazine	Water	Adsorption; negative charge assisted H-bonding	Teixidó et al. (2011)
Crop residue	350°C	Pristine	Methyl violet	Water	Water Electrostatic attraction; interaction between dye and carboxylate and phenolic hydroxyl groups; surface precipitation	Xu et al. (2011)
Hardwood, softwood, and grass	250°C, 400°C, and 650°C	Pristine	Catechol and humic acid	Water	Adsorption due to presence of nano-pores	Kasozi, Zimmerman, Nkedi-Kizza, and Gao (2010)
Rice husk	450°C–500°C	H_2SO_4 and KOH	Tetracycline	Water	Formation of $\pi - \pi$ interactions between ring structure of tetracycline molecule and graphite-like sheets of biochars	Liu et al. (2012)
Dairy manure	200°C	Pristine	Atrazine	Water	Partitioning into organic C/sorption	Cao and Harris (2010)
Pine sawdust		Pristine	Sulfamethoxazole	Water	Hydrophobic interaction	Foo and Hameed (2012)

(Continued)

TABLE 23.2 (Continued)

Feedstocks	Pyrolysis temp (°C)	Modifications	Types of organic pollutants	Medium	Mechanism involved	References
Eucalyptus sawdust	500 ± 50	H_3PO_4	Dimetridazole	Water	Physisorption and chemisorptions	Wan et al. (2016)
Crab shell	800°C	Pristine	Chlortetracycline hydrochloride	Water	Cation bridging, electrostatic interaction, hydrogen bonding and $\pi - \pi$ interaction	Xu et al. (2020)
Soybean stover	300°C and 700°C	Pristine	Trichloroethylene	Water	Sorption	Ahmad et al. (2012)
Bamboo sawdust	600°C	Graphene oxide	Sulfamethazine	Soil	$\pi - \pi$ EDA interaction, pore-filling, cation exchange, hydrogen bonding interaction	Huang, Wang, et al. (2017), Huang, Liu, et al. (2017)
Bamboo	600°C	Pristine	Pentachlorophenol	Soil	Reduced leaching due to diffusion and partition	Xu et al. (2012)

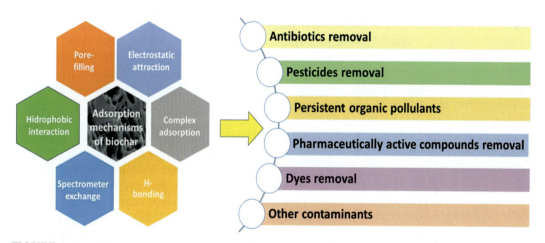

FIGURE 23.2 Contaminants adsorption mechanisms of biochar.

doxycycline in an experiment. This was done because the biochar's surface has a significant number of functional groups that can combine with Cu (II) surface complexes and allow for pH-Zeta potential measurements. Therefore researchers believe that strong complexation and electrostatic interactions might be the adsorption process. It is simple to attract the positively charged cationic organic substance electrostatically when the biochar's surface is negatively charged (Li, Dong, et al., 2017; Li, Zhang, et al., 2017). The researchers discovered that oxygen-containing functional groups, such as -OH and -COOH, can interact with N- and F-containing groups on the surface of florfenicol (FF) molecules to form H bonds when they used magnetic reed biochar (MRBC) to investigate the adsorption mechanism of FF. In addition, the FF surface's high porosity and partly aromatic characteristics show the importance of pore-filling and p–p EDA interaction in the FF's adsorption on MRBC (Zhao & Lang, 2018). Moreover, biochars have different surface-active groups and functional groups, so different biochars have diverse mechanisms for absorbing organic contaminants.

23.4 Factors affecting biochar and its impact on soil and water management

Soil improvement, waste management, climate change mitigation, and energy generation are the four main applications of biochar in environmental management (Lehmann & Joseph, 2015). Biochar can act as a soil conditioner to enhance the physicochemical and biological characteristics of soils because of its high organic C content. With an increase in organic C, soil water retention capacity gets better. The water-holding capacity of soil containing biochar increased by about 18% (Glaser, Lehmann, & Zech, 2002). The hydrophobicity, surface area of biochar, and the improved soil structure after biochar application are all linked to the soil's ability to retain water besides affecting the pollutants immobilization. A reduction in nutrient leakage because of applying biochar has also been noted (Sohi, Krull, Lopez-Capel, & Bol, 2009; Sohi, Lopez-Capel, Krull, & Bol, 2009).

The type of feedstock and the thermochemical production method are two variables that affect the pH of biochar (Singh, Luthra, Bhattacharya, & Bera, 2024). The liming impact, that biochar's alkaline pH causes on acidic soils may increase plant productivity. The ability of biochar to neutralize acids, which varies with the feedstock and pyrolysis temperature, determines how much liming impact it has. Paper mill waste-derived biochar that was pyrolyzed at 550°C had a liming value of about 30% that of $CaCO_3$. In soils amended with biochar, notable improvements in seed germination, plant development, and crop yields have been noted (Glaser et al., 2002). Crop harvests can even be increased by mixing biochar with organic or inorganic fertilizers (Lehmann & Joseph, 2015). Additionally, increased microbial activity and density have been seen in soils that have been amended with biochar (Lehmann et al., 2011; Verheijen, Jeffery, Bastos, Van der Velde, & Diafas, 2010). Biogeochemical processes in soils are significantly impacted by shifts in soil microbial communities and enzyme activities (Awad, Blagodatskaya, Ok, & Kuzyakov, 2012). Only a few studies on earthworm activity in soil have looked at the impacts of biochar on soil fauna. According to Weyers and Spokas (2011), earthworm populations in soils amended with biochar experienced short-term detrimental effects that eventually changed to long-term null effects. The population of earthworms was not affected at all or in an effective means by biochars made from wood. According to Li, Hockaday, Masiello, and Alvarez (2011), a wet biochar application to soil could lessen earthworm avoidance by avoiding desiccation. The priming impact of applying biochar on soil native C was also notable (Wardle, Nilsson, & Zackrisson, 2008). By raising soil pH and enhancing microbial populations, biochar expedites chemical hydrolysis and the breakdown of native soil C (Kuzyakov, Subbotina, Chen, Bogomolova, & Xu, 2009; Yu, Tang, Zhang, Wu, & Gong, 2013). On the other hand, research has shown that biochar speeds up the adsorption of dissolved organic C, slowing down its breakdown (Kwon & Pignatello, 2005). The toxic properties of biochar have been ascribed to this undesirable priming effect, which has decreased microbial activity (Zimmerman, 2010). As a result, it is important to fully understand the characteristics of biochar before using it in soil. Managing the waste stream coming from plants or animals has a lot of promise for reducing the pollution load on the ecosystem. Producing biochar using waste biomass is advantageous and affordable (Barrow, 2012). Variable-charged (or pH-dependent) surfaces are present in biochar. On these surfaces, a rise in pH causes a rise in the negative charge (Xu et al., 2011). The pH-dependent is the proportional impact of ionic strength on the adsorption onto these surfaces. Based on the pH or the biochar's point of zero charge, the impact of ionic strength on adsorption onto biochar can generally be positive or negative (Bolan et al., 1999).

23.5 Challenges, opportunities, and future research directions

This chapter summarizes recent developments in biochar and its role in removing organic contaminants from soil and water environments. A variety of BCs and BC-based hybrid materials have been produced in recent years, driven by their high efficiency and low cost. The broad range of biomasses, including residues from agriculture and forestry operations, industrial by-products and municipal wastes give BCs various chemical compositions and structures. To adsorb organic contaminants via hydrogen bond interaction, n-couple action, electrostatic adsorption, ion exchange process, or other interactions, BCs have a large specific surface area, high porosity, and a lot of oxygen functional groups. Many BCs serve as well or better than commercial activated carbon. However, several obstacles must be surmounted to employing biochar-mediated interactions to achieve industrially relevant contaminant bioremediation. Since contamination frequently results from a combination of contaminants, bioremediation may not always result in a sustainable restoration of contaminated soil. Even though a great deal of study has been done on the development and application of biochar in wastewater treatment, there are still gaps in the knowledge base that need to be filled. Additional studies are still needed to: (1) develop the new low-cost and high-efficiency modification technology of biochar, (2) increase the practical application of biochar in wastewater treatment. Therefore long-term studies are needed to investigate the effect of biochar on contaminant-bound mixed contaminants. Additionally, since the most current study is conducted in the short term, the biochar aging issue caused by long-term experiments may impact the adsorption effect. The use of tailored biochar for contaminant remediation is still in its early stages. To structurally modify biochar, a mix of different material and environmental engineering approaches must be used, which could be further supported by cutting-edge methods like synchrotron-based characterization methods. Additionally, BCs' porous structure and many functional groups serve as a useful base for combining with other functional materials. More effective adsorbents and catalysts with various functions can be produced by coating nanomaterials with nanometal/nanometallic oxides/hydroxide, adding chemically active substances to nanomaterials or combining these materials. In addition, numerous studies have shown that BC-based materials work excellently in removing organic contaminants from aqueous solutions. However, comprehensive information about the actual wastewater treatment process is still lacking. Future studies should focus on the concurrent competing adsorption of numerous organic contaminants by BC-based substances. The combined removal of organic and inorganic pollutants from wastewater and the treatment of BCs and BC-based materials after the preconcentration of various pollutants must all be considered for their use in wastewater treatment. Finally, it is important to consider how BCs affect aquatic organisms and soil microflora, including their potential biotoxicity. Laboratory studies have shown that biochar can remove pollutants from stormwater, municipal sewage, farm sewage, and industrial wastewater. Further research is necessary regarding its utility for onsite applications.

23.6 Conclusion

Water and soil contamination is an increasingly prevalent global issue. Using sustainable and renewable methods to eliminate pollutants in water and soil has become the pursuit of researchers. Thus there is an unending demand for novel, efficient techniques and materials to eradicate organic pollutants from nature, including dyes, antibiotics, and pesticides. Applying biochar to elevate the quality of soil and water by removing contaminants has been regarded as a feasible, cost-effective green strategy. The performance characteristics of biochar are affected by the type of feedstock materials, residence time, and pyrolysis temperature. The mechanisms involved in the removal of organic contaminants include H-bonding, pore-filling, electrostatic interactions, hydrophobic interactions, and cation/p/$\pi-\pi$ interactions. In contrast with aquatic systems, the intricate nature of soil systems has limited biochar applications. This chapter presents a comprehensive understanding of biochar and its potential applications in remediating organic contaminants from water and soil where the paramount focus is on future prospects and opportunities.

References

Adeyemo, A. A., Adeoye, I. O., & Bello, O. S. (2015). Adsorption of dyes using different types of clay: A review. *Applied Water Science*, 7(2), 543–568. Available from https://doi.org/10.1007/s13201-015-0322-y.

Ahmad, M., Lee, S. S., Dou, X., Mohan, D., Sung, J.-K., Yang, J. E., & Ok, Y. S. (2012). Effects of pyrolysis temperature on soybean stover- and peanut shell-derived biochar properties and TCE adsorption in water. *Bioresource Technology*, 118, 536–544. Available from https://doi.org/10.1016/j.biortech.2012.05.042.

Ahmad, M., Lee, S. S., Lim, J. E., Lee, S.-E., Cho, J. S., Moon, D. H., ... Ok, Y. S. (2014). Speciation and phytoavailability of lead and antimony in a small arms range soil amended with mussel shell, cow bone and biochar:

References

Exafs Spectroscopy and chemical extractions. *Chemosphere, 95*, 433–441. Available from https://doi.org/10.1016/j.chemosphere.2013.09.077.

Ahmed, M. B., Zhou, J. L., Ngo, H. H., Guo, W., Johir, M. A., & Sornalingam, K. (2017). Single and competitive sorption properties and mechanism of functionalized biochar for removing sulfonamide antibiotics from water. *Chemical Engineering Journal, 311*, 348–358. Available from https://doi.org/10.1016/j.cej.2016.11.106.

Ahmed, M. B., Zhou, J. L., Ngo, H. H., Johir, M. A. H., Sun, L., Asadullah, M., & Belhaj, D. (2018). Sorption of hydrophobic organic contaminants on functionalized biochar: Protagonist role of π-π electron-donor-acceptor interactions and Hydrogen Bonds. *Journal of Hazardous Materials, 360*, 270–278. Available from https://doi.org/10.1016/j.jhazmat.2018.08.005.

Ali, I., Asim, M., & Khan, T. A. (2012). Low cost adsorbents for the removal of organic pollutants from wastewater. *Journal of Environmental Management, 113*, 170–183. Available from https://doi.org/10.1016/j.jenvman.2012.08.028.

Awad, Y. M., Blagodatskaya, E., Ok, Y. S., & Kuzyakov, Y. (2012). Effects of polyacrylamide, biopolymer, and biochar on decomposition of soil organic matter and plant residues as determined by ^{14}C and enzyme activities. *European Journal of Soil Biology, 48*, 1–10. Available from https://doi.org/10.1016/j.ejsobi.2011.09.005.

Barrow, C. J. (2012). Biochar: Potential for countering land degradation and for improving agriculture. *Applied Geography, 34*, 21–28. Available from https://doi.org/10.1016/j.apgeog.2011.09.008.

Bayode, A. A., Olisah, C., Emmanuel, S. S., Adesina, M. O., & Koko, D. T. (2023). Sequestration of steroidal estrogen in aqueous samples using an adsorption mechanism: A systemic scientometric review. *RSC Advances, 13*, 22675–22697.

Ben, Y., Fu, C., Hu, M., Liu, L., Wong, M. H., & Zheng, C. (2019). Human health risk assessment of antibiotic resistance associated with antibiotic residues in the environment: A Review. *Environmental Research, 169*, 483–493. Available from https://doi.org/10.1016/j.envres.2018.11.040.

Blenis, N., Hue, N., Maaz, M. T., & Kantar, M. (2023). Biochar production, modification, and its uses in soil remediation: A review. *Sustainability, 15*(4), 3442.

Bloom, M. S., Micu, R., & Neamtiu, I. (2016). Female infertility and "emerging" organic pollutants of concern. *Current Epidemiology Reports, 3*(1), 39–50.

Bolan, N. S., Naidu, R., Syers, J. K., & Tillman, R. W. (1999). Surface charge and solute interactions in soils. *Advances in Agronomy*, 87–140. Available from https://doi.org/10.1016/s0065-2113(08)60514-3.

Cai, X., Li, J., Liu, Y., Yan, Z., Tan, X., Liu, S., … Jiang, L. (2017). Titanium dioxide-coated biochar composites as adsorptive and photocatalytic degradation materials for the removal of aqueous organic pollutants. *Journal of Chemical Technology & Biotechnology, 93*(3), 783–791. Available from https://doi.org/10.1002/jctb.5428.

Cao, X., & Harris, W. (2010). Properties of dairy-manure-derived biochar pertinent to its potential use in remediation. *Bioresource Technology, 101*(14), 5222–5228.

Cao, X., Ma, L., Liang, Y., Gao, B., & Harris, W. (2011). Simultaneous immobilization of lead and atrazine in contaminated soils using dairy-manure biochar. *Environmental Science & Technology, 45*(11), 4884–4889. Available from https://doi.org/10.1021/es103752u.

Cao, Y., Yang, B., Song, Z., Wang, H., He, F., & Han, X. (2016). Wheat straw biochar amendments on the removal of polycyclic aromatic hydrocarbons (PAHs) in contaminated soil. *Ecotoxicology and Environmental Safety, 130*, 248–255.

Carvalho, I. T., & Santos, L. (2016). Antibiotics in the aquatic environments: A review of the European scenario. *Environment International, 94*, 736–757.

Chaukura, N., Murimba, E. C., & Gwenzi, W. (2017). Synthesis, characterisation and methyl orange adsorption capacity of ferric oxide–biochar nano-composites derived from pulp and paper sludge. *Applied Water Science, 7*, 2175–2186. Available from https://doi.org/10.1007/s13201-016-0392-5.

Chen, L., Yang, S., Zuo, X., Huang, Y., Cai, T., & Ding, D. (2018). Biochar modification significantly promotes the activity of Co_3O_4 towards heterogeneous activation of peroxymonosulfate. *Chemical Engineering Journal, 354*, 856–865.

Chen, S., Zhou, M., Wang, H. F., Wang, T., Wang, X. S., Hou, H. B., & Song, B. Y. (2018). Adsorption of reactive brilliant red X-3B in aqueous solutions on clay–biochar composites from bagasse and natural attapulgite. *Water, 10*(6), 703. Available from https://doi.org/10.3390/w10060703.

Chen, T., Luo, L., Deng, S., Shi, G., Zhang, S., Zhang, Y., … Wei, L. (2018). Sorption of tetracycline on H_3PO_4 modified biochar derived from rice straw and swine manure. *Bioresource Technology, 267*, 431–437.

Chen, Y., Hassan, M., Nuruzzaman, M., Zhang, H., Naidu, R., Liu, Y., & Wang, L. (2023). Iron-modified biochar derived from sugarcane bagasse for adequate removal of aqueous imidacloprid: Sorption mechanism study. *Environmental Science and Pollution Research*, 30(2), 4754–4768. Available from https://doi.org/10.1007/s11356-022-22357-6.

Devi, P., & Saroha, A. K. (2015). Effect of pyrolysis temperature on polycyclic aromatic hydrocarbons toxicity and sorption behaviour of biochars prepared by pyrolysis of paper mill effluent treatment plant sludge. *Bioresource Technology*, 192, 312–320.

Donadelli, J. A., Carlos, L., Arques, A., & Einschlag, F. S. G. (2018). Kinetic and mechanistic analysis of azo dyes decolorization by ZVI-assisted Fenton systems: pH-dependent shift in the contributions of reductive and oxidative transformation pathways. *Applied Catalysis B: Environmental*, 231, 51–61. Available from https://doi.org/10.1016/j.apcatb.2018.02.057.

El Barkaoui, S., Mandi, L., Aziz, F., Del Bubba, M., & Ouazzani, N. (2023). A critical review on using biochar as constructed wetland substrate: Characteristics, feedstock, design and pollutants removal mechanisms. *Ecological Engineering*, 190, 106927.

Ersan, G., Apul, O. G., Perreault, F., & Karanfil, T. (2017). Adsorption of organic contaminants by graphenenanosheets: A review. *Water Research*, 126, 385–398.

Essandoh, M., Wolgemuth, D., Pittman, C. U., Mohan, D., & Mlsna, T. (2017). Adsorption of metribuzin from aqueous solution using magnetic and nonmagnetic sustainable low-cost biochar adsorbents. *Environmental Science and Pollution Research*, 24, 4577–4590. Available from https://doi.org/10.1007/s11356-016-8188-6.

Fan, C., Chen, H., Li, B., & Xiong, Z. (2017). Biochar reduces yield-scaled emissions of reactive nitrogen gases from vegetable soils across China. *Biogeosciences*, 14(11), 2851–2863. Available from https://doi.org/10.5194/bg-14-2851-2017.

Fan, S., Tang, J., Wang, Y., Li, H., Zhang, H., Tang, J., ... Li, X. (2016). Biochar prepared from co-pyrolysis of municipal sewage sludge and tea waste for the adsorption of methylene blue from aqueous solutions: Kinetics, isotherm, thermodynamic and mechanism. *Journal of Molecular Liquids*, 220, 432–441. Available from https://doi.org/10.1016/j.molliq.2016.04.107.

Feng, L. K., Gao, Z. L., Hu, T. Y., He, S. F., Liu, Y., Jiang, J. Q., ... Wei, L. L. (2023). Performance and mechanisms of biochar-based materials additive in constructed wetlands for enhancing wastewater treatment efficiency: A review. *Chemical Engineering Journal*, 471, 144772. Available from https://doi.org/10.1016/j.cej.2023.144772.

Foo, K. Y., & Hameed, B. H. (2012). Microwave-assisted regeneration of activated carbon. *Bioresource Technology*, 119, 234–240. Available from https://doi.org/10.1016/j.biortech.2012.05.061.

Gao, F., Ahmad, S., Tang, J., Zhang, C., Li, S., Yu, C., ... Sun, H. (2022). Enhanced nitrobenzene removal in soil by biochar supported sulfidated nano zerovalent iron: Solubilization effect and mechanism. *Science of The Total Environment*, 826, 153960. Available from https://doi.org/10.1016/j.scitotenv.2022.153960.

Glaser, B., Lehmann, J., & Zech, W. (2002). Ameliorating physical and chemical properties of highly weathered soils in the tropics with charcoal - A review. *Biology and Fertility of Soils*, 35(4), 219–230. Available from https://doi.org/10.1007/s00374-002-0466-4.

Godlewska, P., & Oleszczuk, P. (2022). Efect of biomass addition before sewage sludge pyrolysis on the persistence and bioavailability of polycyclic aromatic hydrocarbons in biochar-amended soil. *Chemical Engineering, J*, 429, 132143.

Guo, F., Bao, L., Wang, H., Larson, S. L., Ballard, J. H., Knotek-Smith, H. M., ... Han, F. (2020). A simple method for the synthesis of biochar nanodots using hydrothermal reactor. *MethodsX*, 7, 101022. Available from https://doi.org/10.1016/j.mex.2020.101022.

Güzel, F., Sayğılı, H., Sayğılı, G. A., Koyuncu, F., & Yılmaz, C. (2017). Optimal oxidation with nitric acid of biochar derived from pyrolysis of weeds and its application in removal of hazardous dye methylene blue from aqueous solution. *Journal of Cleaner Production*, 144, 260–265. Available from https://doi.org/10.1016/j.jclepro.2017.01.029.

Hameed, B. H., & El-Khaiary, M. I. (2008). Kinetics and equilibrium studies of Malachite Green adsorption on rice straw-derived char. *Journal of Hazardous Materials*, 153(1–2), 701–708. Available from https://doi.org/10.1016/j.jhazmat.2007.09.019.

Hao, O. J., Kim, H., & Chiang, P. C. (2000). Decolorization of wastewater. *Critical reviews in environmental science and technology*, 30(4), 449–505.

Hao, Z., Wang, Q., Yan, Z., & Jiang, H. (2021). Novel magnetic loofah sponge biochar enhancing microbial responses for the remediation of polycyclic aromatic hydrocarbons-contaminated sediment. *Journal of Hazardous Material*, 401, 123859.

Hu, B., Ai, Y., Jin, J., Hayat, T., Alsaedi, A., Zhuang, L., & Wang, X. (2020). Efficient elimination of organic and inorganic pollutants by biochar and biochar-based materials. *Biochar*, 2(1), 47–64. Available from https://doi.org/10.1007/s42773-020-00044-4.

Huang, D., Wang, X., Zhang, C., Zeng, G., Peng, Z., Zhou, J., ... Qin, X. (2017). Sorptive removal of ionizable antibiotic sulfamethazine from aqueous solution by graphene oxide-coated biochar nanocomposites: Influencing factors and mechanism. *Chemosphere*, 186, 414–421.

Huang, Q., Liu, M., Wan, Q., Jiang, R., Mao, L., Zeng, G., ... Wei, Y. (2017). Preparation of polymeric silica composites through polydopamine-mediated surface initiated ATRP for highly efficient removal of environmental pollutants. *Materials Chemistry and Physics*, 193, 501–511. Available from https://doi.org/10.1016/j.matchemphys.2017.03.016.

Huang, R., Tian, D., Liu, J., Lv, S., He, X., & Gao, M. (2018). Responses of soil carbon pool and soil aggregates associated organic carbon to straw and straw-derived biochar addition in a dryland cropping mesocosm system. *Agriculture, Ecosystems & Environment*, 265, 576–586. Available from https://doi.org/10.1016/j.agee.2018.07.013.

Inyang, M., & Dickenson, E. (2015). The potential role of biochar in the removal of organic and microbial contaminants from potable and reuse water: A review. *Chemosphere*, 134, 232–240. Available from https://doi.org/10.1016/j.chemosphere.2015.03.072.

Ji, X. Q., Lü, L., Chen, F., & Yang, C. P. (2016). Sorption properties and mechanisms of organic dyes by straw biochar. *Acta Sci. Circumstantiae*, 36(5), 1648–1654.

Jiang, J., Pi, J., & Cai, J. (2018). The advancing of zinc oxide nanoparticles for biomedical applications. *Bioinorganic Chemistry and Applications*, 3, 1–18.

Jin, J., Li, Y., Zhang, J., Wu, S., Cao, Y., Liang, P., ... Christie, P. (2016). Influence of pyrolysis temperature on properties and environmental safety of heavy metals in biochars derived from municipal sewage sludge. *Journal of Hazardous Materials*, 320, 417–426. Available from https://doi.org/10.1016/j.jhazmat.2016.08.050.

Joseph, S., Peacocke, C., Lehmann, J., & Munroe, P. (2009). Developing a biochar classification and test methods. *Biochar for environmental management: Science and technology*, (vol. 1, pp. 107–126). eBook ISBN9781849770552.

Kalderis, D., Kayan, B., Akay, S., Kulaksız, E., & Gözmen, B. (2017). Adsorption of 2,4-dichlorophenol on paper sludge/wheat husk biochar: Process optimization and comparison with biochars prepared from wood chips, sewage sludge and hog fuel/demolition waste. *Journal of environmental chemical engineering*, 5(3), 2222–2231.

Kasozi, G. N., Zimmerman, A. R., Nkedi-Kizza, P., & Gao, B. (2010). Catechol and humic acid sorption onto a range of laboratory-produced black carbons (biochars). *Environmental Science & Technology*, 44(16), 6189–6195. Available from https://doi.org/10.1021/es1014423.

Keiluweit, M., Nico, P. S., Johnson, M. G., & Kleber, M. (2010). Dynamic molecular structure of plant biomass-derived black carbon (biochar). *Environmental Science & Technology*, 44(4), 1247–1253. Available from https://doi.org/10.1021/es9031419.

Kemmou, L., Frontistis, Z., Vakros, J., Manariotis, I. D., & Mantzavinos, D. (2018). Degradation of antibiotic sulfamethoxazole by biochar-activated persulfate: Factors affecting the activation and degradation processes. *Catalysis Today*, 313, 128–133.

Khalid, S., Shahid, M., Murtaza, B., Bibi, I., Naeem, M. A., & Niazi, N. K. (2020). A critical review of different factors governing the fate of pesticides in soil under biochar application. *Science of the Total Environment*, 711, 134645.

Khan, M. M., Lee, J., & Cho, M. H. (2014). Au@ TiO_2 nanocomposites for the catalytic degradation of methyl orange and methylene blue: An electron relay effect. *Journal of Industrial and Engineering Chemistry*, 20(4), 1584–1590.

Khataee, A., Gholami, P., Kalderis, D., Pachatouridou, E., & Konsolakis, M. (2018). Preparation of novel CeO_2-biochar nanocomposite for sonocatalytic degradation of a textile dye. *Ultrasonics Sonochemistry*, 41, 503–513.

Khorram, M. S., Lin, D., Zhang, Q., Zheng, Y., Fang, H., & Yu, Y. (2017). Effects of aging process on adsorption–desorption and bioavailability of fomesafen in an agricultural soil amended with rice hull biochar. *Journal of Environmental Sciences*, 56, 180–191.

Kong, R. M., Zhao, Y., Zheng, Y., & Qu, F. (2017). Facile synthesis of ZnO/CdS@ ZIF-8 core–shell nanocomposites and their applications in photocatalytic degradation of organic dyes. *RSC advances*, 7(50), 31365–31371. Available from https://doi.org/10.1039/C7RA03918B.

Kumar, R., Sharma, P., Gupta, R. K., Kumar, S., Sharma, M. M. M., Singh, S., & Pradhan, G. (2020). Earthworms for Eco-friendly resource efficient agriculture. In Kumar, et al. (Eds.), *Resources use efficiency in agriculture* (pp. 47–84). Singapore: Springer Nature. Available from https://doi.org/10.1007/978-981-15-6953-1_2.

Kumar, S., Meena, R. S., Datta, R., Verma, S. K., Yadav, G. S., Pradhan, G., Molaei, A., Rahman, G. K. M. M., & Mashuk, H. A. (2020). Legumes for carbon and nitrogen cycling: An organic approach. In R. Datta, et al. (Eds.), *Carbon and nitrogen cycling in soil* (pp. 337–375). Singapore: Springer Nature. Available from https://doi.org/10.1007/978-981-13-7264-3_10.

Kumari, A., Rajput, S., & Singh, S. V. (2024). New Dimensions into the Removal of Pesticides Using an Innovative Ecofriendly Technique: Nanoremediation. In V. D. Rajput (Ed.), *Nano-Bioremediation for Water and Soil Treatment*. Apple Academic Press.

Kumawat, A., Bamboriya, S. D., Meena, R. S., Yadav, D., Kumar, A., Kumar, S., ... Pradhan, G. (2022). Legume-based inter-cropping to achieve crop, soil, and environmental health security. In R. S. Meena, & S. Kumar (Eds.), *Advances in legumes for sustainable intensification* (pp. 307–328). Elsevier. Available from https://doi.org/10.1016/B978-0-323-85797-0.00005-7.

Kuzyakov, Y., Subbotina, I., Chen, H., Bogomolova, I., & Xu, X. (2009). Black carbon decomposition and incorporation into soil microbial biomass estimated by 14C labeling. *Soil Biology and Biochemistry*, 41(2), 210–219. Available from https://doi.org/10.1016/j.soilbio.2008.10.016.

Kwon, S., & Pignatello, J. J. (2005). Effect of natural organic substances on the surface and adsorptive properties of environmental black carbon (char): Pseudo pore blockage by model lipid components and its implications for N2-probed surface properties of natural sorbents. *Environmental Science & Technology*, 39(20), 7932–7939.

Lee, H., Lee, H. J., Jeong, J., Lee, J., Park, N. B., & Lee, C. (2015). Activation of persulfates by carbon nanotubes: Oxidation of organic compounds by nonradical mechanism. *Chemical Engineering Journal*, 266, 28–33. Available from https://doi.org/10.1016/j.cej.2014.12.065.

Lehmann, J., & Joseph, S. (Eds.), (2015). *Biochar for environmental management: Science, technology and implementation*. Routledge. Available from https://doi.org/10.4324/9780203762264.

Lehmann, J., Rillig, M. C., Thies, J., Masiello, C. A., Hockaday, W. C., & Crowley, D. (2011). Biochar effects on soil biota—A review. *Soil biology and biochemistry*, 43(9), 1812–1836. Available from https://doi.org/10.1016/j.soilbio.2011.04.022.

Li, C., Zhang, L., Xia, H., Peng, J., Cheng, S., Shu, J., ... Jiang, X. (2017). Analysis of devitalization mechanism and chemical constituents for fast and efficient regeneration of spent carbon by means of ultrasound and Microwaves. *Journal of Analytical and Applied Pyrolysis*, 124, 42–50. Available from https://doi.org/10.1016/j.jaap.2017.02.025.

Li, C. Y., Zhang, C. B., Zhong, S., Duan, J., Li, M., & Shi, Y. (2023). The removal of pollutants from wastewater using magnetic biochar: A scientometric and visualization analysis. *Molecules (Basel, Switzerland)*, 28(15), 5840. Available from https://doi.org/10.3390/molecules28155840.

Li, D., Hockaday, W. C., Masiello, C. A., & Alvarez, P. J. (2011). Earthworm avoidance of biochar can be mitigated by wetting. *Soil Biology and Biochemistry*, 43(8), 1732–1737.

Li, H., Dong, X., da Silva, E. B., de Oliveira, L. M., Chen, Y., & Ma, L. Q. (2017). Mechanisms of metal sorption by biochars: Biochar characteristics and modifications. *Chemosphere*, 178, 466–478. Available from https://doi.org/10.1016/j.chemosphere.2017.03.072.

Liu, B.-L., Fu, M.-M., Xiang, L., Feng, N.-X., Zhao, H.-M., Li, Y.-W., ... Wong, M.-H. (2021). Adsorption of microcystin contaminants by biochars derived from contrasting pyrolytic conditions: Characteristics, affecting factors, and mechanisms. *Science of The Total Environment*, 763, 143028.

Liu, P., Liu, W. J., Jiang, H., Chen, J. J., Li, W. W., & Yu, H. Q. (2012). Modification of bio-char derived from fast pyrolysis of biomass and its application in removal of tetracycline from aqueous solution. *Bioresource Technology*, 121, 235–240.

Liu, S., Wang, X., Liu, M., & Zhu, J. (2017). Towards better analysis of machine learning models: A visual analytics perspective. *Visual Informatics*, 1(1), 48–56.

Liu, S.-J., Liu, Y.-G., Tan, X.-F., Zeng, G.-M., Zhou, Y.-H., Liu, S.-B., ... Wen, J. (2018). The effect of several activated biochars on CD immobilization and microbial community composition during in-situ remediation of heavy metal contaminated sediment. *Chemosphere, 208*, 655–664.

Liu, X., Wang, D., Tang, J., Liu, F., & Wang, L. (2021). Effect of dissolved biochar on the transfer of antibiotic resistance genes between bacteria. *Environmental Pollution, 288*, 117718.

Lonappan, L., Rouissi, T., Das, R. K., Brar, S. K., Ramirez, A. A., Verma, M., ... Valero, J. R. (2016). Adsorption of methylene blue on biochar microparticles derived from different waste materials. *Waste Management, 49*, 537–544.

Malla, M. A., Dubey, A., Yadav, S., Kumar, A., Hashem, A., & Abd_Allah, E. F. (2018). Understanding and designing the strategies for the microbe-mediated remediation of environmental contaminants using omics approaches. *Frontiers in Microbiology, 9*, 1132.

Meena, R. S., Pradhan, G., Kumar, S., & Lal, R. (2023). Using industrial wastes for rice-wheat cropping and food-energy-carbon-water-economic nexus to the sustainable food system. *Renewable and Sustainable Energy Reviews, 187*, 113756. Available from https://doi.org/10.1016/j.rser.2023.113756.

Meena, R. S., & Pradhan, G. (2023). Industrial garbage-derived biocompost enhances soil organic carbon fractions, CO_2 biosequestration, potential carbon credits, and sustainability index in a rice-wheat ecosystem. *Environmental Research*, 116525. Available from https://doi.org/10.1016/j.envres.2023.116525.

Meena, R. S., Kumawat, A., Kumar, S., Prasad, S. K., Pradhan, G., Jhariya, M. K., ... Raj, A. (2022). Effect of legumes on nitrogen economy and budgeting in South Asia. In Meena, R.S., & Kumar, S. (eds.), *Advances in legumes for sustainable intensification* (pp. 619–638). https://doi.org/10.1016/B978-0-323-85797-0.00001-X.

Nguyen, H. N., & Pignatello, J. J. (2013). Laboratory tests of biochars as absorbents for use in recovery or containment of marine crude oil spills. *Environmental Engineering Science, 30*(7), 374–380. Available from https://doi.org/10.1089/ees.2012.0411.

Ni, N., Wang, F., Song, Y., Bian, Y., Shi, R., Yang, X., ... Jiang, X. (2018). Mechanisms of biochar reducing the bioaccumulation of PAHs in rice from soil: Degradation stimulation vs immobilization. *Chemosphere, 196*, 288–296.

Pandey, D., Chhimwal, M., & Srivastava, R. K. (2022). Engineered biochar as construction material. *Engineered biochar: Fundamentals, preparation, characterization and applications* (pp. 303–318). Singapore: Springer Nature Singapore.

Pantoja, F., Sukmana, H., Beszdes, S., & Laszlo, Z. (2023). Removal of ammonium and phosphates from aqueous solutions by biochar produced from agricultural waste. *Journal of Material Cycles and Waste Management, 25*, 1921–1934.

Park, J. H., Wang, J. J., Xiao, R., Tafti, N., DeLaune, R. D., & Seo, D. C. (2018). Degradation of Orange G by Fenton-like reaction with Fe-impregnated biochar catalyst. *Bioresource Technology, 249*, 368–376. Available from https://doi.org/10.1016/j.biortech.2017.10.030.

Pradhan, G., & Meena, R. S. (2022). Diversity in the rice–wheat system with genetically modified zinc and iron-enriched varieties to achieve nutritional security. *Sustainability, 14*, 9334. Available from https://doi.org/10.3390/su14159334.

Pradhan, G., & Meena, R. S. (2023a). Interaction impacts of biocompost on nutrient dynamics and relations with soil biota, carbon fractions index, societal value of CO_2 equivalent, and ecosystem services in the wheat-rice farming. *Chemosphere*, 139695. Available from https://doi.org/10.1016/j.chemosphere.2023.139695.

Pradhan, G., & Meena, R. S. (2023b). Utilizing waste compost to improve the atmospheric CO_2 capturing in the rice-wheat cropping system and energy-cum-carbon credit audit for a circular economy. *Science of the Total Environment*, 164572. Available from https://doi.org/10.1016/j.scitotenv.2023.164572.

Pradhan, G., Meena, R. S., Kumar, S., & Lal, R. (2023). Utilizing industrial wastes as compost in wheat-rice production to improve the above and below-ground ecosystem services. *Agriculture, Ecosystems and Environment, 358*, 108704. Available from https://doi.org/10.1016/j.agee.2023.108704.

Pradhan, G., Meena, R.S., Kumar, S., Jhariya, M.K., Khan, N., Shukla, U.N., ... Kumar, S. (2022). Legumes for eco-friendly weed management in an agroecosystem. In Meena, R.S., & Kumar, S. (eds.), *Advances in legumes for sustainable intensification* (pp. 133–154). Available from https://doi.org/10.1016/B978-0-323-85797-0.00033-1.

Qiu, Y., Zheng, Z., Zhou, Z., & Sheng, G. D. (2009). Effectiveness and mechanisms of dye adsorption on a straw-based biochar. *Bioresource Technology, 100*(21), 5348–5351.

Rafiquel, I., Sazal, K., Joyanto, K., Kamruzzaman, M., Aminur, R., Nirupam, B., ... Mohammad, M. R. (2018). Bioaccumulation and adverse effects of persistent organic pollutants (POPs) on ecosystems and human exposure: A review study on Bangladesh perspectives. *Environmental Technology & Innovation, 12*, 115–131.

Raj, A., Jhariya, M. K., Banerjee, A., Meena, R. S., Nema, S., Khan, N., ... Pradhan, G. (2021). Agroforestry a model for ecological sustainability. In M. K. Jhariya, R. S. Meena, A. Raj, & S. N. Meena (Eds.), *Natural resources conservation and advances for sustainability* (pp. 289–308). Elsevier. Available from https://doi.org/10.1016/B978-0-12-822976-7.00002-8.

Ren, X., Zeng, G., Tang, L., Wang, J., Wan, J., Liu, Y., ... Deng, R. (2018). Sorption, transport and biodegradation – An insight into bioavailability of persistent organic pollutants in soil. *Science of The Total Environment*, *610–611*, 1154–1163. Available from https://doi.org/10.1016/j.scitotenv.2017.08.089.

Rosales, E., Meijide, J., Pazos, M., & Sanromán, M. A. (2017). Challenges and recent advances in biochar as low-cost biosorbent: From batch assays to continuous-flow systems. *Bioresource Technology*, *246*, 176–192. Available from https://doi.org/10.1016/j.biortech.2017.06.084.

Roy, O., Meena, R. S., Kumar, S., Jhariya, M. K., & Pradhan, G. (2021). Assessment of land use systems for CO_2 sequestration, carbon credit potential, and income security in Vindhyan Region, India. *Land Degradation and Development*, *33*(4), 670–682. Available from https://doi.org/10.1002/ldr.4181.

Sayğılı, H., & Güzel, F. (2016). High surface area mesoporous activated carbon from tomato processing solid waste by zinc chloride activation: Process optimization, characterization and dyes adsorption. *Journal of Cleaner Production*, *113*, 995–1004.

Sewu, D. D., Boakye, P., & Woo, S. H. (2017). Highly efficient adsorption of cationic dye by biochar produced with Korean cabbage waste. *Bioresource Technology*, *224*, 206–213.

Shan, D., Deng, S., Zhao, T., Wang, B., Wang, Y., Huang, J., ... Wiesner, M. R. (2016). Preparation of ultrafine magnetic biochar and activated carbon for pharmaceutical adsorption and subsequent degradation by Ball Milling. *Journal of Hazardous Materials*, *305*, 156–163. Available from https://doi.org/10.1016/j.jhazmat.2015.11.047.

Sheoran, S., Ramtekey, V., Kumar, D., Kumar, S., Meena, R. S., Kumawat, A., ... Shukla, U. N. (2022). Grain legumes: Recent advances and technological interventions. In R. S. Meena, & S. Kumar (Eds.), *Advances in legumes for sustainable intensification* (pp. 507–532). Elsevier. Available from https://doi.org/10.1016/B978-0-323-85797-0.00025-2.

Silvani, L., Hjartardottir, S., Bielska, L., Skulcova, L., Cornelissen, G., Nizzetto, L., & Hale, S. E. (2019). Can polyethylene passive samplers predict polychlorinated biphenyls (PCBs) uptake by earthworms and turnips in a biochar amended soil? *Science of Total Environment*, *662*, 873–880. Available from https://doi.org/10.1016/j.scitotenv.2019.01.202.

Singh, S., Chaturvedi, S., & Datta, D. (2019). Biochar: An eco-friendly residue management approach. *Indian Farming*, *69*(08), 27–29.

Singh, S., Chaturvedi, S., Dhyani, V. C., & Govindaraju, K. (2020). Pyrolysis temperature influences the characteristics of rice straw and husk biochar and sorption/desorption behavior of their biourea composite. *Bioresource Technology*, *314*, 123674. Available from https://doi.org/10.1016/j.biortech.2020.123674.

Singh, S., Luthra, N., Mandal, S., Kushwaha, D. P., Pathak, S. O., Datta, D., ... Pramanik, B. (2023). Distinct behavior of biochar modulating biogeochemistry of salt-affected and acidic soil: A review. *Journal of Soil Science and plant Nutrition*. Available from https://doi.org/10.1007/s42729-023-01370-9.

Singh, S. V., Luthra, N., Bhattacharya, S., Bera, A., et al. (2024). Occurrence of emerging contaminants in soils and impacts on rhizosphere. In *Emerging Contaminants Sustainable Agriculture and the Environment*, Elsevier Sciences.

Singh, S. V., Sharma, R., Balan, P., et al. (2024). Nanomaterials for Inorganic Pollutants Removal from Contaminated Water. In V. D. Rajput, et al. (Eds.), *Nano-Bioremediation for Water and Soil Treatment*. Apple Academic Press.

Sohi, S. P. (2012). Carbon storage with benefits. *Science (New York, N.Y.)*, *338*(6110), 1034–1035.

Sohi, S. P., Krull, E., Lopez-Capel, E., & Bol, R. (2009). A review of biochar and its use and function in soil. *Advances in Agronomy*, *105*, 47–82. Available from https://doi.org/10.1016/S0065-2113(10)05002-9.

Sohi, S., Lopez-Capel, E., Krull, E., & Bol, R. (2009). Biochar, climate change and soil: A review to guide future research. *CSIRO Land and Water Science Report*, *5*(09), 17–31.

Sun, K., Keiluweit, M., Kleber, M., Pan, Z., & Xing, B. (2011). Sorption of fluorinated herbicides to plant biomass-derived biochars as a function of molecular structure. *Bioresource Technology*, *102*(21), 9897–9903. Available from https://doi.org/10.1016/j.biortech.2011.08.036.

Sun, Y., Qi, S., Zheng, F., Huang, L., Pan, J., Jiang, Y., ... Xiao, L. (2018). Organics removal, nitrogen removal and N_2O emission in subsurface wastewater infiltration systems amended with/without biochar and sludge. *Bioresource Technology*, *249*, 57–61. Available from https://doi.org/10.1016/j.biortech.2017.10.004.

Taheran, M., Naghdi, M., Brar, S. K., Knystautas, E. J., Verma, M., & Surampalli, R. Y. (2017). Degradation of chlortetracycline using immobilized laccase on polyacrylonitrile-biochar composite nanofibrous membrane. *Science of The Total Environment, 605*, 315–321.

Tang, J., Li, Y., Wang, X., & Daroch, M. (2017). Effective adsorption of aqueous Pb^{2+} by dried biomass of Landoltiapunctata and Spirodelapolyrhiza. *Journal of Cleaner Production, 145*, 25–34. Available from https://doi.org/10.1016/j.jclepro.2017.01.038.

Teixidó, M., Pignatello, J. J., Beltrán, J. L., Granados, M., & Peccia, J. (2011). Speciation of the ionizable antibiotic sulfamethazine on black carbon (biochar). *Environmental Science & Technology, 45*(23), 10020–10027. Available from https://doi.org/10.1021/es202487h.

Tian, S. Q., Wang, L., Liu, Y. L., & Ma, J. (2020). Degradation of organic pollutants by ferrate/biochar: Enhanced formation of strong intermediate oxidative iron species. *Water Research, 183*, 116054. Available from https://doi.org/10.1016/j.watres.2020.116054.

Uchimiya, M., Wartelle, L. H., Lima, I. M., & Klasson, K. T. (2010). Sorption of deisopropylatrazine on broiler litter biochars. *Journal of Agricultural and Food Chemistry, 58*(23), 12350–12356. Available from https://doi.org/10.1021/jf102152q.

Verheijen, F., Jeffery, S., Bastos, A. C., Van der Velde, M., & Diafas, I. (2010). Biochar application to soils. A critical scientific review of effects on soil properties, processes, and functions. *EUR, 24099*(162), 2183–2207. Available from https://doi.org/10.2788/472.

Vithanage, M., Mayakaduwa, S. S., Herath, I., Ok, Y. S., & Mohan, D. (2016). Kinetics, thermodynamics and mechanistic studies of carbofuran removal using biochars from tea waste and rice husks. *Chemosphere, 150*, 781–789.

Wan, S., Hua, Z., Sun, L., Bai, X., & Liang, L. (2016). Biosorption of nitroimidazole antibiotics onto chemically modified porous biochar prepared by experimental design: Kinetics, thermodynamics, and equilibrium analysis. *Process Safety and Environmental Protection, 104*, 422–435. Available from https://doi.org/10.1016/j.psep.2016.10.001.

Wang, B., Wang, Z., Jiang, Y., Tan, G., Xu, N., & Xu, Y. (2017). Enhanced power generation and wastewater treatment in sustainable biochar electrodes based bioelectrochemical system. *Bioresource Technology, 241*, 841–848.

Wang, L., Yan, W., He, C., Wen, H., Cai, Z., Wang, Z., ... Liu, W. (2018). Microwave-assisted preparation of nitrogen-doped biochars by ammonium acetate activation for adsorption of acid red 18. *Applied Surface Science, 433*, 222–231. Available from https://doi.org/10.1016/j.apsusc.2017.10.031.

Wang, Q., Zhang, S., He, J., Wang, L., Shi, Y., Zhou, W., ... Wang, F. (2017). Surface-enhanced palygorskite coated CDS: Synthesis, characterization and highly improved photocatalytic degradation efficiency of organic dyes. *Journal of Materials Science: Materials in Electronics, 28*(14), 10464–10471. Available from https://doi.org/10.1007/s10854-017-6819-4.

Wang, S., & Wang, J. (2023). Bimetallic and nitrogen co-doped biochar for peroxymonosulfate (PMS) activation to degrade emerging contaminants. *Separation and Purification Technology, 307*, 122807. Available from https://doi.org/10.2139/ssrn.4273735.

Wang, X., Lian, W., Sun, X., Ma, J., & Ning, P. (2018). Immobilization of NZVI in polydopamine surface-modified biochar for adsorption and degradation of tetracycline in aqueous solution. *Frontiers of Environmental Science & Engineering, 12*, 1–11.

Wardle, D. A., Nilsson, M. C., & Zackrisson, O. (2008). Fire-derived charcoal causes loss of forest humus. *Science (New York, N.Y.), 320*(5876), 629. Available from https://doi.org/10.1126/science.1154960.

Wathukarage, A., Herath, I., Iqbal, M. C. M., & Vithanage, M. (2019). Mechanistic understanding of crystal violet dye sorption by woody biochar: Implications for wastewater treatment. *Environmental Geochemistry and Health, 41*, 1647–1661.

Wei, Z., Wang, J. J., Fultz, L. M., White, P., & Jeong, C. (2021). Application of biochar in estrogen hormone-contaminated and manure-affected soils: Impact on soil respiration, microbial community and enzyme activity. *Chemosphere, 270*, 128625.

Weyers, S. L., & Spokas, K. A. (2011). Impact of biochar on earthworm populations: A review. *Applied and Environmental Soil Science, 2011*.

Wu, L. R., & Liu, Y. D. (2018). Application status of biochar as adsorbent in water treatment. *Modern Food, 10*, 46–47.

Xiao, X., Chen, B., Chen, Z., Zhu, L., & Schnoor, J. L. (2018). Insight into multiple and multilevel structures of biochars and their potential environmental applications: A critical review. *Environmental Science & Technology, 52*(9), 5027–5047.

Xing, L., Wei, J., Zhang, Y., Xu, M., Pan, G., Li, J., ... Li, Y. (2022). Boosting active sites of protogenetic sludge-based biochar by boron doping for electro-Fenton degradation towards emerging organic contaminants. *Separation and Purification Technology*, *294*, 121160. Available from https://doi.org/10.1016/j.seppur.2022.121160.

Xu, R. K., Xiao, S. C., Yuan, J. H., & Zhao, A. Z. (2011). Adsorption of methyl violet from aqueous solutions by the biochars derived from crop residues. *Bioresource Technology*, *102*(22), 10293−10298. Available from https://doi.org/10.1016/j.biortech.2011.08.089.

Xu, T., Lou, L., Luo, L., Cao, R., Duan, D., & Chen, Y. (2012). Effect of bamboo biochar on pentachlorophenol leachability and bioavailability in agricultural soil. *Science of the Total Environment*, *414*, 727−731. Available from https://doi.org/10.1016/j.scitotenv.2011.11.005.

Xu, Y., Liu, J., Cai, W., Feng, J., Lu, Z., Wang, H., ... Xu, J. (2020). Dynamic processes in conjunction with microbial response to disclose the biochar effect on pentachlorophenol degradation under both aerobic and anaerobic conditions. *Journal of Hazardous Materials*, *384*, 121503. Available from https://doi.org/10.1016/j.jhazmat.2019.121503.

Xu, Y., Shi, G. Q., & Wang, S. G. (2018). Application of biochar in contaminated-soil remediation. *Journal of Anhui Agricultural University*, *46*, 120−122.

Yang, X., Chen, Z., Wu, Q., & Xu, M. (2018). Enhanced phenanthrene degradation in river sediments using a combination of biochar and nitrate. *Science of the total environment*, *619*, 600−605. Available from https://doi.org/10.1016/j.scitotenv.2017.11.130.

Yao, Y., Zhang, Y., Gao, B., Chen, R., & Wu, F. (2017). Removal of sulfamethoxazole (SMX) and sulfapyridine (SPY) from aqueous solutions by biochars derived from anaerobically digested bagasse. *Environmental Science and Pollution Research*, *25*, 25659−25667.

You, X., Jiang, H., Zhao, M., Suo, F., Zhang, C., Zheng, H., ... Li, Y. (2020). Biochar reduced Chinese chive (Allium tuberosum) uptake and dissipation of thiamethoxam in an agricultural soil. *Journal of Hazardous Material*, *390*, 121749. Available from https://doi.org/10.1016/j.jhazmat.2019.121749.

Yu, L., Tang, J., Zhang, R., Wu, Q., & Gong, M. (2013). Effects of biochar application on soil methane emission at different soil moisture levels. *Biology and Fertility of Soils*, *49*, 119−128. Available from https://doi.org/10.1007/s00374-012-0703-4.

Zama, E. F., Reid, B. J., Arp, H. P. H., Sun, G. X., Yuan, H. Y., & Zhu, Y. G. (2018). Advances in research on the use of biochar in soil for remediation: A review. *Journal of Soils and Sediments*, *18*, 2433−2450. Available from https://doi.org/10.1007/s11368-018-2000-9.

Zazycki, M. A., Godinho, M., Perondi, D., Foletto, E. L., Collazzo, G. C., & Dotto, G. L. (2018). New biochar from pecan nutshells as an alternative adsorbent for removing reactive red 141 from aqueous solutions. *Journal of Cleaner Production*, *171*, 57−65.

Zhang, G., Guo, X., Zhu, Y., Liu, X., Han, Z., Sun, K., ... Han, L. (2018). The effects of different biochars on microbial quantity, microbial community shift, enzyme activity, and biodegradation of polycyclic aromatic hydrocarbons in soil. *Geoderma*, *328*, 100−108. Available from https://doi.org/10.1016/j.geoderma.2018.05.009.

Zhang, G., He, L., Guo, X., Han, Z., Ji, L., He, Q., ... Sun, K. (2020). Mechanism of biochar as a biostimulation strategy to remove polycyclic aromatic hydrocarbons from heavily contaminated soil in a coking plant. *Geoderma*, *375*, 114497.

Zhang, H., Xue, G., Chen, H., & Li, X. (2018). Magnetic biochar catalyst derived from biological sludge and ferric sludge using hydrothermal carbonization: Preparation, characterization and its circulation in Fenton process for dyeing wastewater treatment. *Chemosphere*, *191*, 64−71. Available from https://doi.org/10.1016/j.chemosphere.2017.10.026.

Zhang, J., Hou, D., Shen, Z., Jin, F., O'Connor, D., Pan, S., ... Alessi, D. S. (2020). Effects of excessive impregnation, magnesium content, and pyrolysis temperature on MgO-coated watermelon rind biochar and its lead removal capacity. *Environmental Research*, *183*, 109152.

Zhang, P., Min, L., Tang, J., Rafq, M. K., & Sun, H. (2020). Sorption and degradation of imidacloprid and clothianidin in Chinese paddy soil and red soil amended with biochars. *Biochar*, *2*(3), 329−334. Available from https://doi.org/10.1007/s42773-020-00060-4.

Zhang, X., Wang, H., He, L., Lu, K., Sarmah, A., Li, J., ... Huang, H. (2013). Using biochar for remediation of soils contaminated with heavy metals and organic pollutants. *Environmental Science and Pollution Research*, *20*(12), 8472−8483. Available from https://doi.org/10.1007/s11356-013-1659-0.

Zhao, H., & Lang, Y. (2018). Adsorption behaviors and mechanisms of florfenicol by magnetic functionalized biochar and reed biochar. *Journal of the Taiwan Institute of Chemical Engineers, 88*, 152–160. Available from https://doi.org/10.1016/j.jtice.2018.03.049.

Zhao, X., Ouyang, W., Hao, F., Lin, C., Wang, F., Han, S., & Geng, X. (2013). Properties comparison of biochars from corn straw with different pretreatment and sorption behavior of atrazine. *Bioresource Technology, 147*, 338–344.

Zhao, Y., Dai, H., Ji, J., Yuan, X., Li, X., Jiang, L., & Wang, H. (2022). Resource utilization of Luffa Sponge to produce biochar for effective degradation of organic contaminants through persulfate activation. *Separation and Purification Technology, 288*, 120650. Available from https://doi.org/10.1016/j.seppur.2022.120650.

Zhelezova, A., Cederlund, H., & Stenström, J. (2017). Effect of biochar amendment and ageing on adsorption and degradation of two herbicides. *Water, Air, & Soil Pollution, 228*, 1–13.

Zheng, W., Guo, M., Chow, T., Bennett, D. N., & Rajagopalan, N. (2010). Sorption properties of greenwastebiochar for two triazine pesticides. *Journal of Hazardous Materials, 181*(1–3), 121–126. Available from https://doi.org/10.1016/j.jhazmat.2010.04.103.

Zhong, J., Li, L., Zhong, Z., Yang, Q., Zhang, J., & Wang, L. (2018). Advances on the research of the effect of biochar on the environmental behavior of antibiotics. *Journal of Safety and Environment, 18*(2), 657–663.

Zhou, H., Jiang, L., Li, K., Chen, C., Lin, X., Zhang, C., & Xie, Q. (2021). Enhanced bioremediation of diesel oil-contaminated seawater by a biochar-immobilized biosurfactant-producing bacteria Vibrio sp. LQ2 isolated from cold seep sediment. *Science of The Total Environment, 793*, 148529. Available from https://doi.org/10.1016/j.scitotenv.2021.148529.

Zhou, Y., Liu, X., Xiang, Y., Wang, P., Zhang, J., Zhang, F., … Tang, L. (2017). Modification of biochar derived from sawdust and its application in removal of tetracycline and copper from aqueous solution: Adsorption mechanism and modelling. *Bioresource Technology, 245*, 266–273. Available from https://doi.org/10.1016/j.biortech.2017.08.178.

Zhu, S., Huang, X., Ma, F., Wang, L., Duan, X., & Wang, S. (2018). Catalytic removal of aqueous contaminants on N-doped graphitic biochars: Inherent roles of adsorption and nonradical mechanisms. *Environmental Science & Technology, 52*(15), 8649–8658.

Zimmerman, A. R. (2010). Abiotic and microbial oxidation of laboratory-produced black carbon (biochar). *Environmental Science & Technology, 44*(4), 1295–1301.

CHAPTER 24

Briquetting and pelleting of pine biochar: an income source for small-scale farmers of the hilly area

Hemant Kumar Sharma, T.P. Singh and T.K. Bhattacharya

Department of Farm Machinery and Power Engineering, G. B. Pant University of Agriculture and Technology, Pantnagar, Uttarakhand, India

24.1 Introduction

After China and the United States, India was the third-largest energy consumer in the world in terms of crude oil and natural gas in 2022 (EIA, 2021). Currently, 57.9% of India's primary energy consumption comes from conventional commercial energy sources such as coal, oil, natural gas, nuclear power, and hydropower. Approximately 42.1% of India's overall energy demand is met by renewable energy sources. Biomass is regarded as a natural fuel source that is carbon-neutral, sustainable, and renewable (Goyal, Seal, & Saxena, 2008; Kaygusuz, 2009). Numerous studies are currently being conducted to determine the feasibility of various biomass fuel sources on various routes (Efika, Wu, & Williams, 2012; IEA, 2006). The dominant source of energy in the future is biomass. In India, agricultural waste and residues such as rice husk, soybean stalks, cotton stalks, pigeon pea stalks, safflower residue, groundnut shells, etc. are readily available in large amounts. The world's main energy source by 2050 will be biomass energy, which will account for 15%–50% of global energy consumption. Currently, biomass energy accounts for 11% of global energy consumption. About 500 million metric tons (MMT) of biomass are thought to be available annually in India. There were 120–150 million metric tons of surplus biomass available, which is equivalent to 18,000 MW of potential electric energy (Vyas, Sayyad, Khardiwar, & Kumar, 2015). Crop residue waste had a 4.15 EJ (4.15×10^{18} J) estimated annual bioenergy potential, which corresponds to 17% of India's total primary energy consumption (Hiloidhari, Das, & Baruah, 2014). Globally, 146 billion metric tons (MT) of biomass are produced annually by wild plant evolution. A total of 2740 quads (2.740×10^{15} BTU)

(2.289×10^{15} MJ) of energy are produced from biomass globally each year according to estimates from various sources (Balat & Ayar, 2005).

24.2 Pine needle biomass

24.2.1 Introduction of pine needle

Pine trees (genus *Pinus*), members of the conifer family, are evergreen and can be found all over the world. However, they are indigenous to northern temperate zones and produce cones that contain reproductive seeds. It is difficult to find pines south of the equator that are native, despite the fact that there are many different pines that can endure harsh climates (deserts, rainforests). These settings are suitable for them to exist; however, they prefer a hilly area with regular rainfall and good soils. Older needles on pine trees fall off in the fall, although they still have some foliage throughout the year. To maintain a stable population growth, several pine species need fire. Worldwide, there are about 126 different varieties of pine trees. Pines, in general, are related to conifers such as spruces, firs, and cedars. The most prevalent coniferous tree species in the world is the pine tree. Because their needles endure for two years, most varieties of pine trees are evergreen. Both male and female pines have cones, and male cones and female cones both produce pollen and seeds, respectively. Under ideal circumstances, the majority of pine tree species can reach a century old, and nearly all of them are found in the Northern Hemisphere (Sandborn, 2017). Table 24.1 lists 40 of the most well-known varieties of pine trees that grow in various climates and growing zones (Myers, 2022).

24.2.2 Problems associated with pine needles

Pine forests, which are found practically everywhere on Earth, produce a huge amount of biomass called pine needles. In Himalayan states like Uttarakhand, Himachal Pradesh, and Jammu & Kashmir, Chir Pine (*Pinus roxburghii*) forests are mostly located in the mountainous region and cover a significant amount of forest land. In the Himalayan region, which includes India, Nepal, and Bhutan, there is 7.62 Mha of land covered with pine forests. In Uttarakhand, there is a reserve pine forest that covers about 0.343 Mha (Singh, Gumber, Tewari, & Singh, 2016). About 2.058 MMT of pine needle wastes are accessible each year in Uttarakhand alone (UREDA, 2010). The jungles still held 1.8 MT of untapped pine needles, which can reduce carbon dioxide emissions significantly. For instance, 1.3 tons of biomass briquettes can replace 1 ton of coal used in brick kilns, saving 1.81 tons of carbon dioxide emissions. Likewise, 2.5 tons of biomass briquettes can replace 1 ton of liquefied petroleum gas, saving 2.98 tons of carbon dioxide emissions (Manoj Chandran, 2011). Rural communities can receive 5 hours day^{-1} of electricity from a 789 MW power station using pine needle fuel. It is capable of supporting 165 MW of plants fueled by pine needles for year-round generation (Kala & Subbarao, 2017). Thus, there is a huge potential for using pine needle biomass as an alternative energy source.

Pine needles make a thick layer of dry leaves on the forest land every year between April and May, resulting in fires and slowing the establishment of the grass. Since pine

TABLE 24.1 Species of pine needle worldwide.

S. no.	Type of pine tree	Native region	USDA growing zone
1.	Aleppo Pine	Mediterranean region	USDA Growing Zone: Zone 8–10
2.	Austrian Pine	Southern Europe, Northern Africa, Cyprus, Turkey	USDA Growing Zone: Zone 5–8
3.	Bristlecone Pine	Southern mountainous regions of the US.	USDA Growing Zone: Zone 4–8
4.	Canary Island Pine	Canary Islands	USDA Growing Zone: Zone 9–11
5.	Chir Pine	Himalayan regions of Asia—Afghanistan, Bhutan, China, India, Myanmar, Nepal	USDA Growing Zone: Zone 9–11
6.	Coulter Pine	California, Mexico	USDA Growing Zone: Zone 7–9
7.	Eastern White Pine	US and Canada	USDA Growing Zone: Zone 4–9
8.	Foxtail Pine	California	USDA Growing Zone: Zone 5–8
9.	Gray Pine	California	USDA Growing Zone: Zone 8–9
10.	Italian Stone Pine	Southern regions of Europe, Lebanon, Turkey	USDA Growing Zone: Zone 9–10
11.	Jack Pine	Northern US, Canada	USDA Growing Zone: Zone 3–8
12.	Japanese Black Pine	Japan, South Korea	USDA Growing Zone: Zone 5–8
13.	Japanese White Pine	Japan, South Korea	USDA Growing Zone: Zone 6–9
14.	Jeffrey Pine	California, Nevada, Oregon, Mexico	USDA Growing Zone: Zone 6–8
15.	Lacebark Pine	China	USDA Growing Zone: Zone 5–9
16.	Limber Pine	US, Canada	USDA Growing Zone: Zone 4–7
17.	Loblolly Pine	US	USDA Growing Zone: Zone 6–9
18.	Lodgepole Pine	North America	USDA Growing Zone: Zone 6–8
19.	Longleaf Pine	Southern US	USDA Growing Zone: Zone 7–9

(Continued)

TABLE 24.1 (Continued)

S. no.	Type of pine tree	Native region	USDA growing zone
20.	Luchu Pine	Okinawa, Japan	USDA Growing Zone: Zone 9–11
21.	Maritime Pine	Southern Europe, Morocco	USDA Growing Zone: Zone 7–9
22.	Mexican Weeping Pine	Mexico	USDA Growing Zone: Zone 8–9
23.	Monterey Pine	California, Mexico	USDA Growing Zone: Zone 7–9
24.	Mugo Pine	Europe	USDA Growing Zone: Zone 3–7
25.	Pitch Pine	Eastern US, Canada	USDA Growing Zone: Zone 5–7
26.	Pond Pine	Eastern US	USDA Growing Zone: Zone 7–9
27.	Ponderosa Pine	US, British Columbia, Canada	USDA Growing Zone: Zone 5–8
28.	Red Pine	Northern US, Canada	USDA Growing Zone: Zone 2–7
29.	Sand Pine	Southern US—Alabama and Florida	USDA Growing Zone: Zone 7–10
30.	Scots Pine	Europe, Asia	USDA Growing Zone: Zone 3–7
31.	Single-Leaf Pinyon Pine	Western US, Mexico	USDA Growing Zone: Zone 5–9
32.	Sugar Pine	California, Nevada, Oregon, Mexico	USDA Growing Zone: Zone 6–7
33.	Japanese Red Pine	Asia	USDA Growing Zone: Zone 3–7
34.	Tenasserim Pine	Southeast Asia	USDA Growing Zone: Zone 9–10
35.	Torrey Pine	Southern California	USDA Growing Zone: Zone 8–10
36.	Turkish Pine	Western Asia—Bulgaria, Greece, Italy, Turkey, Ukraine	USDA Growing Zone: Zone 8–11
37.	Two-Needle Pinyon Pine	Western and Central US, Mexico	USDA Growing Zone: Zone 5–8
38.	Virginia Pine	Eastern US	USDA Growing Zone: Zone 4–8

(Continued)

TABLE 24.1 (Continued)

S. no.	Type of pine tree	Native region	USDA growing zone
39.	Western White Pine	Western US, Canada	USDA Growing Zone: Zone 5–7
40.	Whitebark Pine	Western US, Canada	USDA Growing Zone: Zone 4–8

needles can't be utilized as animal feed and don't even disintegrate like other biomass, it can be challenging to incorporate them with other types of forest debris. Nevertheless, they could be a reliable source of biomass for promising, green, and renewable energy sources. Diverse thermal conversion techniques can be employed to transform biomass into different energy products, including combustion, gasification, liquefaction, hydrogenation, and pyrolysis (Font, Conesa, Moltó, & Muñoz, 2009; Haykiri-Acma & Yaman, 2007). Briquetting, gasification, and pyrolysis are three processes that have an edge over others, with respective energy conversion efficiencies of 88%, 52%, and 74% (Mahmood, Parshetti, & Balasubramanian, 2016; Mandal, Bhattacharya, & Tanna, 2017).

24.2.3 Characteristics of pine needle biomass

Table 24.2 depicts the properties of pine needles and pine wood. It is an ideal material for thermo-chemical conversion because of its extremely low ash level and high volatile content. Pine needles are a lignocellulosic biomass with a lignin content of 29.15% and cellulose contents of 27.17%. It is suitable for high-pressure briquetting and thermo-chemical conversion because the high lignin concentration improves compatibility and adhesion. Pine needles have a calorific value that is equivalent to other agricultural waste biomass, making them a potential source of energy. Pine needles' carbon and oxygen contents, which were determined by elemental analysis, were 44.99% and 48.55%, respectively (Hassan, Steele, & Ingram, 2009; Mandal, Prasanna Kumar, Bhattacharya, Tanna, & Jena, 2019).

24.3 Pine needle biochar

A carbon-rich byproduct of biomass (such as wood, manure, or leaves) called biochar is created when the biomass is heated in a closed container with little to no airflow. The resulting biochar is a highly porous, carbon-rich material that can be used as a soil amendment, as a means of carbon sequestration, or as a source of renewable energy generation (Roy et al., 2022).

24.3.1 Preparation of pine needle biochar

Pine needles are high in lignin and carbon content, which are the primary components of wood biomass and other plant materials that are commonly used for biochar

TABLE 24.2 Properties of pine needle biomass.

S. no.	Property	Pine needle	Pine wood
1.	Moisture content, % (wb)	7.78	6.34
2.	Volatile matter, % (db)	71.58	78.54
3.	Ash content, % (db)	2.08	0.46
4.	Fixed carbon, % (db)	26.34	14.66
5.	Extractives, % (db)	17.45	-
6.	Cellulose, % (db)	27.17	-
7.	Hemicellulose, % (db)	24.15	68.1
8.	Lignin, % (db)	29.15	27.7
Elemental analysis (weight, %)			
9.	C	44.99	51.30
10.	H	5.46	5.83
11.	N	0.99	0.07
12.	S	Not traceable	0.01
13.	O	48.55	35.99
14.	H/C	1.46	1.36
15.	O/C	0.81	0.52
16.	Empirical formula	$CH_{1.46}N_{0.02}O_{0.81}$	$CH_{1.36}O_{0.52}$
17.	Calorific value (MJ kg^{-1})	17.67	18.6

production. There exist traditional and modern types of methodologies for the production of pine needle biochar generally employed by small-scale farmers of hilly areas.

24.3.1.1 Traditional approach to pine biochar preparation

The conventional method of producing pine biochar involves heating pine needle biomass in a pit. The small-scale farmers dig a pit in the ground and use it as a fire pit to produce pine biochar. Traditionally it is typically done in a kiln or retort, which can be operated between 400°C and 700°C, which causes the organic material to break down into biochar and other byproducts such as gases and oils. This conventional method of biochar production has less efficiency and can be scaled up to produce large quantities of biochar (Chaturvedi, Singh, Dhyani, Govindaraju, & Mandal, 2021; Singh et al., 2023). Traditionally there are different types of biochar preparation techniques and some of them are discussed in brief.

- Top-lit updraft (TLUD) kiln: The TLUD kiln is a simple and widely used traditional method for biochar production. It consists of a container filled with biomass material,

which is ignited from the top. As the biomass burns, air enters from the bottom, creating an updraft that leads to controlled combustion. This controlled combustion produces biochar as a byproduct.
- Drum or barrel kilns: Drum or barrel kilns involve the use of sealed metal containers filled with biomass. The biomass is burned with limited oxygen supply, resulting in pyrolysis, which converts the biomass into biochar. The sealed design prevents excessive air from entering the chamber, enhancing the biochar yield.
- Flame curtain kilns: Flame curtain kilns are vertical kilns with a central burner that creates a moving flame curtain. The biomass feedstock is loaded from the top, and the controlled flame gradually moves downward through the feedstock, pyrolyzing it into biochar.
- In-drum pyrolysis units: In-drum pyrolysis units are simple, small-scale systems in which biomass is placed in a sealed container and heated to produce biochar. The lack of oxygen inside the drum promotes pyrolysis, and the resulting biochar is collected.
- Batch kilns: Batch kilns are traditional designs where a batch of biomass is loaded into the kiln, pyrolyzed, and then unloaded. They are often used in small-scale operations or for research purposes to produce biochar.
- Hearth-type kilns: Hearth-type kilns involve a combustion chamber where biomass is burned with restricted airflow. This controlled combustion leads to the production of biochar, and the design allows for adjustments to the combustion process to achieve desired biochar characteristics.
- Beehive kilns: Beehive kilns, also known as mound kilns, are dome-shaped traditional kilns used for small-scale biochar production. The dome structure allows for efficient heat retention and uniform pyrolysis.
- Terra Preta kilns: Terra Preta kilns aim to mimic ancient Amazonian practices for creating biochar-rich soils. The kiln design is optimized to produce biochar with properties conducive to enhancing soil fertility and carbon sequestration.

24.3.1.2 Modern approach of pine biochar preparation

To overcome the challenges associated with the traditional biochar production techniques, there are some modern techniques of biochar production and some of them are discussed in brief.

- Retort kiln: Retort kilns are closed containers designed to ensure controlled and efficient pyrolysis and are suitable for pine needles. The limited air ingress during pyrolysis helps improve biochar yields and preserve more of the feedstock's energy content.
- Continuous pyrolysis systems: Continuous pyrolysis systems are larger-scale units with conveyor belts or augers that move biomass through a high-temperature chamber continuously. This design allows for more automated and efficient biochar production. However, pine needles need to be ground for this reactor.
- Gasifier reactors: Gasifier reactors not only produce biochar but also generate syngas (a mixture of carbon monoxide and hydrogen) through partial biomass combustion. This dual production of biochar and syngas makes it an attractive option for integrated bioenergy systems. Recovery in this kind is low.

- Microwave pyrolysis units: Microwave pyrolysis units use microwave energy to rapidly and efficiently heat biomass, leading to faster pyrolysis and biochar production. This technology offers quick processing times and precise control over temperature.
- Fluidized bed reactors: In fluidized bed reactors, biomass is suspended on a stream of hot gas, ensuring uniform heating and fast pyrolysis. These reactors are suitable for large-scale biochar production and only can handle very fine particles of various feedstocks.
- Electric pyrolysis reactors: Electric pyrolysis reactors use electrical heat to generate the heat required for pyrolysis, allowing for precise temperature control and adjustable processing conditions.
- Solar pyrolysis units: Solar pyrolysis units utilize concentrated solar energy to heat biomass and initiate the pyrolysis process, making them environmentally friendly and suitable for regions with ample sunlight.
- Industrial rotary kilns: Industrial rotary kilns are large-scale units designed for the continuous processing of biomass. These rotating cylindrical kilns allow for high-volume biochar production in an efficient manner.
- Mobile pyrolysis units: Mobile pyrolysis units are transportable and can be used on-site, enabling decentralized biochar production and reducing the need to transport biomass over long distances.
- Biomass gasification with biochar capture: This technology combines gasification with biochar capture, allowing for the simultaneous production of syngas for energy use and the recovery of biochar as a valuable co-product.
- Kiln arrays: Kiln arrays involve multiple smaller kilns operated in parallel, offering greater production flexibility and redundancy in case of maintenance or breakdowns. This approach enhances efficiency and production capacity.

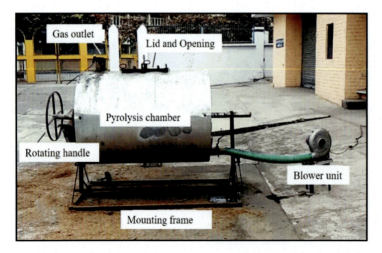

FIGURE 24.1 Biochar production unit.

24.3.1.2.1 Portable rotating biochar drum

A biochar drum was developed by G. B. Pant University of Agriculture and Technology, Pantnagar, India to produce biochar from pine needle biomass. It is a horizontally constructed drum-type biochar production device that turns pine needles into biochar and gas products as illustrated in Fig. 24.1. For the production of pine needle biochar, the unit is filled with pine needles with appropriate moisture content. A small ignition is started and biomass is loaded up to the capacity of the drum. A centrifugal type air blower attached with a 700 W motor and 120 $m^3\ h^{-1}$ bowing capacity for the supply of air and initial burning. The biochar production unit is positioned in the open area and the unit is rotated for the proper burning of biomass (Mandal, Bhattacharya, Verma, & Haydary, 2018).

The biochar production drum has a capacity of 50–60 kg of pine needle biomass. It takes a burning time of 120 minutes for the producing biochar, with an airflow rate of around 40–50 $m^3\ h^{-1}$. The initial moisture content of pine needles should be 12%–15%. The drum has a conversion efficiency of 90% and pine biochar recovery of around 33%. The price of this device is INR 65,000 (US$800). Another small portable biochar kiln was developed by Birsa Agriculture University, Ranchi, India to produce biochar from any biomass. It is a vertical cylindrical drum-type structure having a feeding capacity of 10 kg and a conversion efficiency of 43%. The biochar recovery from the device is 25% and the price of the device is INR 10,200 (US$125).

TABLE 24.3 Properties of pine needle biochar.

Parameter	Pine needle biochar	References
Proximate analysis		
Moisture content (wb%)	4.65	Sharma, Kumain, and Bhattacharya (2020)
Volatile matter (db.%)	25.44	Kumain, Bhattacharya, and Sharma (2020)
Ash content (db.%)	6.64	Sharma et al. (2020)
Fixed Carbon (db.%)	63.30	Sharma et al. (2020)
Elemental analysis (wt.%)		
C	65.1	Li, chen, Zhang, Du, and Wang (2020)
H	4.3	Li et al. (2020)
N	1.8	Li et al. (2020)
Oxygen (by difference)	34.46	Li et al. (2020)
BET surface area ($m^2\ g^{-1}$)	389.4	Li et al. (2020)
Calorific Value, ($MJ\ kg^{-1}$)	29.31	Sharma, Bhattacharya, Singh, and Verma (2022)

24.4 Characteristics of pine needle biochar

Table 24.3 shows the properties of pine needle biochar significantly indicating that the pine needle biochar has great potential to be utilized as a renewable energy source. It has low moisture content, and high calorific value to be used as a fuel source with high energy efficiency. Its high fixed carbon content makes it resistant to combustion, which is important for safety and performance. Low ash content means that it produces little ash when burned, which reduces the need for cleaning and maintenance. The high BET surface area gives it a large surface area to interact with other substances, making it useful for applications such as water filtration and adsorption. Pine needle biochar is also a good source of carbon and other nutrients, which makes it a valuable soil amendment. It can be used to improve soil fertility, water retention, and drainage. It can also help to reduce greenhouse gas emissions by sequestering carbon in the soil.

24.5 Briquetting and pelleting of pine biochar

24.5.1 Briquetting and pelleting

Slack biomass is densified by briquetting or pelleting to improve the slack biomass's fuel-plus-handling properties. It raises raw biomass's calorific value while lowering the transit cost and requiring less storage space. The bulk density of slack biomass can be increased from 40–200 to 600–1200 kg m^{-3} using the briquetting or pelleting process. These briquettes and pellets have been employed as renewable fuels in industries and traditional cooking applications because of their consistent size, high density, and enhanced fuel characteristic qualities. As an alternative to burning forestry and agricultural waste, briquetting or pelleting is also pursued to reduce environmental pollution and carbon emissions into the atmosphere (Mandal et al., 2019). There are two different kinds of briquettes, including biochar and biomass briquettes. In contrast to biochar briquettes, biomass briquettes are created by compacting loose biomass in a high-pressure briquette-making machine.

Briquettes can be produced from any lignocellulosic biomass. Densification of lignocellulosic biomass is a complex process and is influenced by factors such as densification pressure, particle size, moisture content, temperature and type of biomass. Process parameters for the production of briquettes from different biomass materials have varied widely. Further, lignin melts at high temperatures and pressure, and they both affect the strength characteristics of briquettes made from lignocellulosic materials. Densification is influenced by compaction pressure, compaction temperature, compaction velocity, moisture content and particle size distribution.

24.5.2 Methods of briquetting and pelleting

There are several methods for the production of briquettes and pellets from biomass as listed below.

1. Piston press briquetting: This method involves compressing the material using a reciprocating piston and a die. The material is fed into the piston press, which then compresses it into the die. The resulting briquette is then ejected from the die.
2. Screw extrusion briquetting: This method involves feeding the material into a screw extruder, which compresses the material and extrudes it through a die to form briquettes.
3. Hydraulic press briquetting: This method uses a hydraulic press to compress the material into a die to form briquettes. The hydraulic press generates high pressure, which compresses the material into the die.
4. Roller press briquetting: This method involves compressing the material between two rollers that rotate in opposite directions. The rollers apply pressure to the material, forcing it into a die to form briquettes.
5. Adhesive briquetting: Adhesive or binder briquetting is a method of producing briquettes that involve the use of binders or additives to hold the raw materials together. The binder can be a natural substance, such as starch, molasses, or lignin, or a synthetic material, such as polyvinyl alcohol or polyethylene glycol.
6. Binderless briquetting: This method involves compressing the material without the use of any binders or additives. The material is simply compressed under high pressure to form briquettes. This method is often used for biomass and agricultural waste materials.

24.5.3 Pine needle biochar briquettes (beehive shape)

A manually operated beehive-briquette (block) making tool (Fig. 24.2A) was designed and developed for small-scale farmers of the hilly regions of India (Sharma et al., 2022). The ergonomically designed machine can produce four briquettes simultaneously with a production capacity of 48 briquettes per hour. The tool is 90 cm high and has a 150 cm long handle to compact the briquettes for greater mechanical advantage. Molds have a diameter of 12.7 cm and a height of 17 cm. The tool produces biochar briquettes with a 12.7 cm diameter and an 8 cm height (Fig. 24.2B). The briquettes have seven holes, each measuring 1 cm in diameter. This machine is priced at INR 10000 (US$123).

24.5.3.1 Pine needle pellet production

A commercially available screw extrusion machine (Fig. 24.2C) was used to produce pine biochar pellets of large size with a capacity of 100 kg day^{-1}. The machine has a hopper where the raw material is loaded. A feeding mechanism, often a screw conveyor or auger, continuously moves the raw material from the hopper into the compression chamber. Inside the chamber, a large screw, also known as an extruder, rotates with the help of an electric motor. The screw's continuous rotation pushes the raw material forward, compressing it along the way to produce pellets of an average height of 100 mm and diameter of 20 mm as shown in Fig. 24.2D.

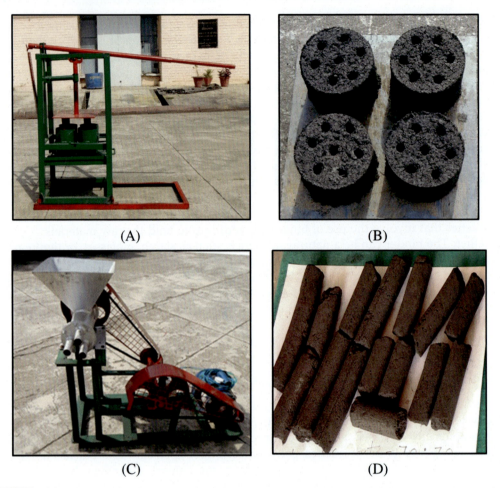

FIGURE 24.2 (A) Manually operated pine biochar beehive briquetting machine, (B) pine needle biochar beehive briquettes, (C) screw type large pelleting machine, and (D) pine needle biochar pellets.

24.5.3.2 Binders for pine needle biochar brqieuttes and pellets

Briquette production, storage, and transportation are all required steps in the pine biochar briquetting and pelleting process. Additional binding material is also needed. The quality of briquettes depends on the binders. Numerous forms of binding substances, including organic binders, inorganic binders, and compound binders, have been used by various researchers.

Four different types of binding materials which included soil, cattle dung, cement, and lime were applied for briquetting and pelleting of pine biochar. Inorganic binders like soil have advantages over organic binders. Since clay soil was the first documented binder used in briquette manufacture, it is also known as civilian briquette binder. The benefit of good hydrophilicity and extensive availability comes with soil binder. Several briquette makers employ cattle dung as an organic binder. The briquettes produced utilizing the cattle dung binder exhibit excellent compressive strength as well as good binding performance. At high temperatures, the organic binders rapidly disintegrate.

An inorganic binder with benefits including strong adhesion, no pollution, low cost, and good hydrophilicity is cement. The industrial briquettes contain cement, which is commonly accessible. Another inorganic binder used in commercial briquettes is lime. It is a readily available, inexpensive, environmentally friendly, and thermally stable binder. It has been claimed that using inorganic binders in the manufacture of briquettes not only lowers the emissions of dangerous gases into the environment but also increases the efficiency with which energy is used (Kaur, Singh, Sharma, Singh, & Singh, 2021; Sharma et al., 2022). Multiple composition combinations were examined as mentioned in Table 24.4 for the manufacturing of pine biochar briquettes and pellets in order to assess their performance in terms of their physical qualities, chemical properties, and thermal properties.

24.5.4 Property assessment of biochar briquettes and pellets

The characterization of pine needle biochar briquettes and pellets can be done on the basis of an assessment of their physical, chemical, and thermal properties. The physical properties of biochar briquettes comprise their size, weight, density, degree of densification, shattering resistance, water absorption resistance, and compressive strength. The chemical property includes proximate analysis and the calorific value of biochar briquettes.

24.5.4.1 Physical properties

The average length and diameter of the briquettes and pellets can be measured with a Vernier caliper to establish their size. Using an electronic weighing balance, the average weight of the prepared briquette samples can be measured to determine the weight of the briquettes. The bulk density of the briquettes and pellets samples can be calculated using the standard test (ASTM E873–82, 2013). Once the mass of an empty, cylindrical container with known capacity had been established, the container was filled with the sample and weighed once more. Bulk density is estimated as

$$\text{Bulk density (kg/m3)} = \frac{\text{Mass of the sample (kg)}}{\text{Volume of the container (m}^3\text{)}} \quad (24.1)$$

The relative size of the briquettes' stability and friability are both determined by the shattering test. It shows how tough the briquettes are and how well they can withstand cracking while being handled. In accordance with the standard test procedure for drop shatter analysis (ASTM D440–86, 2002), sample briquettes with known weights and lengths are twice dropped onto a concrete floor from a height of 1.83 m. The weights of the broken briquettes are recorded, and the amount of material lost as a percentage is computed. Shattering resistance of briquettes is determined by

TABLE 24.4 Properties of pine needle biochar beehive briquettes.

S. no.	Composition	Description	Avg. length (cm)	Avg. Weight (g)	Density (kg m^{-3})	Degree of densification (%)	Shattering Resistance (%)	Water Resistance (%)	MC (%)	VM (%)	AC (%)	FC (%)	Calorific value (MJ/kg)	Compressive strength (MPa)
1.	B70S30	Biochar 70% + soil 30%	9	481	441	42	41	2	4	31	23	43	29	0.2
2.	B60S40	Biochar 60% + soil 40%	9	615	568	70	53	29	5	20	19	57	27	0.3
3.	B50S50	Biochar 50% + soil 50%	9	630	582	74	75	75	5	24	33	34	21	0.4
4.	B70D30	Biochar 70% + dung 30%	8	568	561	40	44	16	9	9	27	56	25	0.1
5.	B60D40	Biochar 60% + dung 40%	8	588	581	40	68	49	10	3	3	43	24	0.1
6.	B50D50	Biochar 50% + dung 50%	8	603	595	39	77	67	8	17	3	43	22	0.3
7.	B70C30	Biochar 70% + cement 30%	9	562	522	44	32	11	4	56	33	7	23	0.3
8.	B60C40	Biochar 60% + cement 40%	9	607	536	50	50	52	6	57	25	13	19	0.3
9.	B50C50	Biochar 50% + cement 50%	9	651	605	56	83	75	6	50	42	2.	19	0.5
10.	B70L30	Biochar 70% + lime 30%	8	500	493	55	44	5	4	62	22	12	25	0.2
11.	B60L40	Biochar 60% + lime 40%	8	570	562	73	57	38	3	57	27	13	23	0.3
12.	B50L50	Biochar 50% + lime 50%	8	687	680	93	82	76	4	54	34	9	19	0.5

B: Biochar; *S*: soil; *D*: dung; *C*: cement; *L*: lime.

$$\% \text{ weight loss} = \frac{W_1 - W_2}{W_1} \times 100 \qquad (24.2)$$

$$\% \text{ shattering resistance} = 100 - \% \text{ weight loss} \qquad (24.3)$$

where W_1 is the mass in (g) of briquettes sample before shattering and W_2 is the mass in (g) of briquette sample after shattering.

A test called the "water absorption test" determines how much water the briquettes actually absorb when submerged entirely. The briquettes and pellets are submerged in room-temperature water for 60 seconds (NAIP, 2014). The formula used to determine the resistance to water absorption and compute the amount of water absorbed by each briquette is as

$$\% \text{ water gained by briquette} = \frac{W_1 - W_2}{W_1} \times 100 \qquad (24.4)$$

$$\% \text{ Water absorption resistance} = 100 - \% \text{ water gained,} \qquad (24.5)$$

where W_1 is the mass in (g) of briquette sample before the test and W_2 is the mass in (g) of briquette sample after the test.

The percentage increase in the density of raw materials produced by the briquetting process is referred to as the degree of densification. This designates the strength of the biomass to get confined and is evaluated using the formula given below (Birwatkar, Khandetod, Mohod, & Dhande, 2014)

$$\text{Degree of densification (\%)} = \frac{\text{Density of the briquette} - \text{Density of the raw material}}{\text{Density of the raw material}} \qquad (24.6)$$

Compressive strength refers to the briquettes' capacity to withstand breaking when subjected to compression force. Utilizing a universal testing machine, the compressive strength is determined. Briquette samples are placed on the platform of the machine and a plunger presses it. In order to cause failure on the briquettes, the machine compresses their surface. The formula for calculating compressive strength is

$$F = \frac{P}{A} \qquad (24.7)$$

where F is the sample's compressive strength (MPa), P is the greatest load that can be applied to it (N), and A is its cross-sectional area (mm^2).

24.5.4.2 Chemical properties

As per the established test procedure for moisture analysis, (ASTM D3173–03, 2008), a hot-air oven at 105°C is used to keep 1 g of the sample for 1 hour. The dried sample is then weighed and the moisture content is calculated. In the analysis of the sample free of moisture content, volatile matter establishes the percentage of gaseous products that are emitted under the particular test conditions. According to the accepted test procedure (ASTM D3175–01, 2007), the oven-dried sample is put inside a crucible with a lid and kept in a muffle furnace at 950°C ± 20°C for 7 minutes. In order to determine weight loss, the crucible is first allowed to cool in the ambient air before being placed in a desiccator.

The standardized test procedure for ash content analysis (ASTM D3174−02, 2002) claims, the leftover sample residue from the volatile matter test is heated in the muffle furnace for 4 hours at 700°C ± 50°C. Based on standard test procedure (ASTM D3172−07a, 2007), the fixed carbon value is determined; it is the sum of the percentages of moisture, ash, and volatile matter after being deducted from 100. The same moisture reference base will be used for all percentages. The conventional formula used to calculate the fixed carbon is

$$FC(\%) = 100 - \{MC(\%) + VM(\%) + AC(\%)\}, \qquad (24.8)$$

where FC is the fixed carbon, MC is the moisture content, VM is the volatile matter and AC is the ash content.

24.5.4.3 Thermal properties

According to the standard test procedure for calorific value, the bomb calorimeter is used to assess the calorific value of briquettes (ASTM E711−87, 2004). The calorific value is calculated using the formula

$$\text{Calorific value (kcal/kg)} = \frac{W \times \Delta T}{M} \qquad (24.9)$$

where M is the mass (g) of fuel placed in the crucible, W is the water equivalent of the bomb calorimeter or heat capacity (cal/°C), $\Delta T = t_2 - t_1$, where t_1 is the initial temperature of water in the calorimeter (°C) and t_2 is the final temperature of water in the calorimeter (°C).

24.5.5 Properties of pine needle briquettes

The results of the above assessment are presented in Table 24.4. The biochar briquettes had a moisture content in the range of 8%−10%. (Mandal, Kumar, Singh, & Ngachan, 2012). For beehive briquettes, it ranges from 5.56 to 10.29% (Mandal, Kumar, Singh, Ngachan, & Kundu, 2014). Briquettes with a higher moisture content burn more slowly and emit more smoke. According to reports, charcoal has an average volatile matter content of 3%−30% (Rominiyi, Olaniyi, Azeez, Eiche, & Akinola, 2017). The pine biochar briquettes' volatile matter content was found to be between 3.02% and 62.08%.

The findings also show a correlation between the biochar content of the briquettes and their shattering resistance. This is because briquettes with a higher binder concentration may be more cohesive in nature. The compressive strength, shattering strength, and water absorption resistance of the biochar briquettes all increase with an increase in the binder level, while the calorific value of the briquettes falls as the binder level rises. A drop in the calorific value of fuels occurs as the amount of binder in the composition rises. The added value of the binder resulted in a higher ash content and a lower calorific value. As a result, the composition with 30% binder had a larger calorific value than the other formulations. Each binder group underwent the optimization of the binder level. Briquettes with

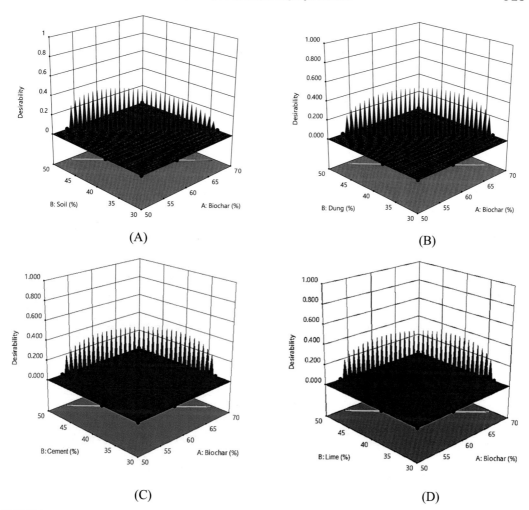

FIGURE 24.3 (A–D) Response surface curve of optimum pine needle biochar and binder ratio.

sufficient strength and heating capabilities can be produced using biochar at a ratio of 60% and binder at 40% as shown in Fig. 24.3A–D.

The response surface curves of optimum pine needle biochar and binder ratio represent the composition of pine needle biochar and binder on the horizontal plane while the significance of desirability on the vertical plane is based on the physical, chemical, and thermal properties of briquettes. The optimum desirability of pine biochar and binder was found to lie at the coordinate of 60% biochar and 40% binder for different types of binders.

24.6 Briquetting and pelleting as an income source

24.6.1 Production cost of pine biochar briquettes and pellets

According to the standardized procedure outlined in IS 9164:1979, the cost of producing pine biochar briquettes and pellets can be assessed in terms of fixed cost and variable cost (IS 9164, 1979). The fixed cost comprises depreciation cost, annual interest cost, insurance/taxes, and storage/housing cost. The depreciation cost is assessed on the basis of decline in the worth of the machine with wear and time. The straight-line technique was used to estimate the depreciation. The annual interest costs are calculated using the actual interest rate that is due. A 2% annual percentage of the machine's average purchase price is used to compute the amount paid in insurance and taxes. The anticipated cost of the housing is calculated at 1.5% of the typical cost of the briquette-making equipment.

On the other hand, the variable cost includes the material cost, repair and maintenance and labor cost. The cost of raw materials used to produce the briquettes is included in the material cost. The pricing of the local market can be surveyed to determine this. The per kilogram price of biochar, soil, dung, cement, and lime is Rs. 4.6, 2, 2, 10, and 10, respectively (Sharma et al., 2020). The expense of repair and maintenance helps to keep a machine in good working order and prevent any accidents or wear-and-tear breakdowns. It represents 6% of the machine's initial cost. The wages or labor costs of an operator are included in the cost of operating a machine on an hourly basis. The labor costs are Rs. 275.00 for each 8-hour shift. The total cost is calculated by adding the hourly rates for fixed and variable costs.

24.6.2 Economic analysis

Based on the above facts, the cost analysis of the briquette manufacturing operation was estimated and presented in Table 24.5.

The cost of production of pine biochar briquette was ₹7 kg^{-1} for soil and dung and ₹11 kg^{-1} for cement and lime. The biochar briquette-making machine has a capacity of producing 48 briquettes per hour and the pellet making machine has a capacity of 100 kg day^{-1}.

24.6.3 Advantages and disadvantages of pine biochar briquettes and pellets

Pine needle biochar briquettes have several advantages, including:

Sustainability: Pine needle biochar briquettes are made from waste materials that would otherwise be discarded, making them a sustainable and environmentally friendly source of fuel.

High calorific value: Pine needle biochar briquettes have a high calorific value, meaning they release a large amount of heat energy when burnt. This makes them an efficient fuel source that can provide a lot of heat for a small amount of material.

TABLE 24.5 Cost analysis of pine biochar beehive briquetting.

Parameter		Cost (₹)	Cost (US$)
Fixed cost	Initial cost of briquetting machine	8000	97.15
	Depreciation (h)	0.32	0.004
	Interest (h)	0.21	0.003
	Insurance and taxes (h)	0.01	0.0001
	Housing charges (h)	0.01	0.0001
Variable cost	Material cost for briquetting, for (50:50) briquettes, material capacity is 24 kg h^{-1}, i.e., 12 kg biochar + 12 kg binder		
	Biochar cost (h)	56	0.7
	Soil cost (h)	24	0.3
	Dung cost (h)	24	0.3
	Cement cost (h)	120	1.5
	Lime cost (h)	120	1.5
	Repair and maintenance cost (h)	0.02	0.0003
	Two Labor charges (h)	70	1
Sum of fixed cost (h)		0.54	0.01

Parameter	Cost (₹)	Cost (US$)
Sum of variable cost (h)		
Biochar:soil	150	1.8
Biochar:dung	150	1.8
Biochar:cement	246	3
Biochar:lime	246	3
Total cost (h)		
Biochar:soil	151	1.83
Biochar:dung	151	1.83
Biochar:cement	247	3
Biochar:lime	247	3
Cost of briquette (kg)		
Biochar:soil	7	0.10
Biochar:dung	7	0.10
Biochar:cement	11	0.13
Biochar:lime	11	0.13

Note: US$1 = ₹82.35 (March 25, 2023 at 12:44 am UTC).

Long burning time: Pine needle biochar briquettes have a longer burn time than traditional wood or charcoal, meaning they can be used for longer periods without needing to be replaced.

Low smoke emissions: Pine needle biochar briquettes produce less smoke than traditional wood or charcoal, which can be beneficial for indoor and outdoor use.

Improved soil health: Biochar, the main component of these briquettes, can be used to improve soil health by increasing nutrient retention and water-holding capacity in the soil. This makes them a valuable resource for agriculture and gardening.

Cost-effective: Pine needle biochar briquettes can be a cost-effective alternative to traditional fuels, as they are made from low-cost and readily available waste materials.

Overall, pine needle biochar briquettes offer a sustainable, efficient, and environmentally friendly alternative to traditional fuels, with the added benefit of improving soil health.

While pine needle biochar briquettes have several advantages, they also have a few disadvantages, including:

Limited availability: Pine needle biochar briquettes may not be widely available in all areas, as they are typically produced in hilly areas and may be difficult to find in some regions.

Higher production costs: The production of pine needle biochar briquettes can be more expensive than traditional wood or charcoal, due to the additional processing steps required to create the biochar.

24.7 Rural development and employment for hilly region farmers

The briquettes and pellets can be used as a fuel source, replacing traditional fuels like coal, wood, or charcoal, and will help in rural development with several benefits, including:

- Economic development: Briquetting can provide an additional source of income for rural communities by creating a market for locally sourced biomass materials. This can help to reduce poverty and increase economic development in rural areas.
- Environmental conservation: Briquetting can help to reduce the use of traditional fuels like coal, wood, and charcoal, which can lead to deforestation and greenhouse gas emissions. Using locally sourced biomass materials for briquetting can also help to reduce transportation emissions.
- Energy security: Briquetting can provide a local source of renewable energy, reducing dependence on imported fuels and increasing energy security for rural communities.
- Waste management: Briquetting can help to manage agricultural and other organic wastes, reducing pollution and improving sanitation in rural areas.
  To promote rural development through briquetting, governments, NGOs, and other organizations can provide technical assistance and training to rural communities on the briquetting process. They can also help to create markets for the briquettes, ensuring that the communities have a sustainable source of income. Briquetting and pelleting can be an effective way to generate employment in rural areas. Here are some ways in which briquetting can create jobs:
- Collection of biomass materials: To produce briquettes, biomass materials such as sawdust, wood chips, and agricultural wastes need to be collected from the surrounding area. This can provide employment opportunities for local people who can collect, sort, and transport these materials.

- Briquette production: The production of briquettes requires labor to operate machines, mix materials, and package the finished product. This can provide employment opportunities for people in the local area.
- Marketing and sales: Briquettes need to be marketed and sold, which can create jobs for people involved in sales and marketing.
- Supporting services: Briquetting can also create jobs in supporting services, such as maintenance and repair of machines, transportation of materials and products, and administration.

Overall, the employment generated through briquetting can help to alleviate poverty, promote economic development, and improve the standard of living for rural communities, small-scale farmers of hilly regions.

24.8 Conclusion

Utilizing pine needle biomass helps the hilly areas by lowering the risk of forest fires and other dangers. Pine needle biomass and biochar have a low moisture content of 8% and 4.3% while having high calorific values of 18 and 29 MJ kg^{-1}, respectively. Pine biochar can be briquetted and pelleted to assist create self-employment and prove successful in halting the emigration of rural farmers from hilly regions. The manually operated pine biochar briquetting machine can enhance the entrepreneurship of small-scale farmers in hilly areas. The beehive briquetting and pelleting machine has a capacity of 48 briquettes per hour and 100 kg day^{-1}, respectively. The per kg cost of production of pine biochar briquette was in the range of INR 7–11 (US$0.10–0.13) The effective and high-quality briquettes and pellets can be produced with this low-cost machine to generate an income source.

Pine needle biochar briquettes can be produced in large quantities and sold to customers as a sustainable alternative to traditional charcoal or firewood. This can provide a steady source of income for those involved in the production process. By producing biochar briquettes from pine needles, rural communities can create a new source of income. This could include selling the briquettes themselves, as well as any byproducts that result from the production process, such as liquid biochar. The production of pine needle biochar briquettes can create employment opportunities for people in rural communities, especially women and youth. This can help to stimulate local economies and reduce poverty. Biochar production can generate carbon credits, which can be sold on the global carbon market. This can provide a significant source of income for communities that produce biochar briquettes, as carbon credits are often sold at a premium price.

References

ASTM D3172-07a (2007). *Standard practice for proximate analysis of coal and coke.* ASTM International, West Conshohocken, PA, USA. https://www.astm.org/d3172-07a.html

ASTM D3173 (2008). *Standard* test method for moisture in the analysis sample of coal and coke. ASTM International, West Conshohocken, PA, USA. https://pdfcoffee.com/astm-d3173-standard-test-method-for-moisture-in-the-analysis-sample-of-coal-and-coke-pdf-free.html

ASTM D3174-02 (2002). *Standard* test method for ash in the analysis sample of coal and coke from coal. ASTM International, West Conshohocken, PA, USA. https://standards.globalspec.com/std/3810237/ASTM%20D3174-02

ASTM D3175-01 (2007). *Standard* test method for volatile matter in the analysis sample of coal and coke. ASTM International, West Conshohocken, PA, USA. https://standards.globalspec.com/std/3823724/ASTM%20D3175-01

ASTM D440-86 (2002). *Standard test method of drop shatter test for coal*. ASTM International. West Conshohocken, PA, USA. http://www.astm.org

ASTM E711-87 (2004). *Standard test method for gross calorific value of refuse-derived fuel by the bomb calorimeter*. ASTM International, West Conshohocken, PA, USA. http://www.astm.org

ASTM E873-82 (2013). *Standard* test method for bulk density of densified particulate biomass fuels. ASTM International, West Conshohocken, PA, USA. https://webstore.ansi.org/standards/astm/astme873822013

Balat, M., & Ayar, G. (2005). biomass energy in the world, use of biomass and potential trends. *Energy Sources, 27*(10), 931–940. Available from https://doi.org/10.1080/00908310490449045.

Birwatkar, V. R., Khandetod, Y. P., Mohod, A. G., & Dhande, K. G. (2014). Physical and thermal properties of biomass briquetted fuel. *Interntional Journal of Scientific Research and Techniques, 2*(4). Available from http://www.indjsrt.com.

Chaturvedi, S., Singh, S., Dhyani, V. C., Govindaraju, K., & Mandal, S. (2021). Characterization, bioenergy value and thermal stability of biochar derived from diverse agriculture and forestry lignocellulosic waste. *Biomass Conversion and Biorefinery*. Available from https://doi.org/10.1007/s13399-020-01239-2.

Efika, C. E., Wu, C., & Williams, P. T. (2012). Syngas production from pyrolysis–catalytic steam reforming of waste biomass in a continuous screw kiln reactor. *Journal of Analytical and Applied Pyrolysis, 95*, 87–94. Available from https://doi.org/10.1016/j.jaap.2012.01.010.

EIA (Energy Information Administration) (2021). *Country analysis executive summary: India*. Washington DC, USA. <https://www.eia.gov>.

Font, R., Conesa, J. A., Moltó, J., & Muñoz, M. (2009). Kinetics of pyrolysis and combustion of pine needles and cones. *Journal of Analytical and Applied Pyrolysis, 85*(1–2), 276–286. Available from https://doi.org/10.1016/J.JAAP.2008.11.015.

Goyal, H. B., Seal, D., & Saxena, R. C. (2008). Bio-fuels from thermochemical conversion of renewable resources: A review. *Renewable and Sustainable Energy Reviews, 12*(2), 504–517. Available from https://doi.org/10.1016/j.rser.2006.07.014.

Hassan, E. B. M., Steele, P. H., & Ingram, L. (2009). Characterization of fast pyrolysis bio-oils produced from pretreated pine wood. *Applied Biochemistry and Biotechnology, 154*(1–3), 182–192. Available from https://doi.org/10.1007/S12010-008-8445-3/TABLES/6.

Haykiri-Acma, H., & Yaman, S. (2007). Interpretation of biomass gasification yields regarding temperature intervals under nitrogen-steam atmosphere. *Fuel Processing Technology, 88*(4), 417–425. Available from https://doi.org/10.1016/J.FUPROC.2006.11.002.

Hiloidhari, M., Das, D., & Baruah, D. C. (2014). Bioenergy potential from crop residue biomass in India. *Renewable and Sustainable Energy Reviews, 32*, 504–512. Available from https://doi.org/10.1016/j.rser.2014.01.025.

IEA (2006). *Wind energy annual report*. International Energy Association. https://www.ewea.org/news/detail/2006/11/27/iea-wind-energy-annual-report-published/

IS 9164 (1979). Guide for estimating cost of farm machinery operation: Bureau of Indian Standards. Internet Archive. https://archive.org/details/gov.in.is.9164.1979

Kala, L. D., & Subbarao, P. M. V. (2017). Pine needles as potential energy feedstock: Availability in the Central Himalayan State of Uttarakhand, India. *E3S Web of Conferences, 23*, 04001. Available from https://doi.org/10.1051/e3sconf/20172304001.

Kaur, L., Singh, H., Sharma, H. K., Singh, T. P., & Singh, J. (2021). Effect of storage on properties of pine needle cattle dung briquettes. *Indian Journal of Engineering and Materials Sciences, 28*(6). Available from https://nopr.niscpr.res.in/bitstream/123456789/59268/3/IJEMS-591-601.pdf.

Kaygusuz, K. (2009). Biomass as a renewable energy source for sustainable fuels. *Energy Sources, Part A: Recovery, Utilization, and Environmental Effects, 31*(6), 535–545. Available from https://doi.org/10.1080/15567030701715989.

Kumain, A., Bhattacharya, T. K., & Sharma, H. K. (2020). Physicochemical and thermal characteristics of pine needle biochar briquetted fuel using soil, lime and cement as a binder. *International Journal of Current Microbiology and Applied Sciences*, 9(10), 3675–3690. Available from https://doi.org/10.20546/ijcmas.2020.910.425.

Li, W., chen, T., Zhang, Z., Du, C., & Wang, G. (2020). Study on adsorption performance of pine needle biochar on rhodamine B. *IOP Conference Series: Materials Science and Engineering*, 729(1), 012076. Available from https://doi.org/10.1088/1757-899X/729/1/012076.

Mahmood, R., Parshetti, G. K., & Balasubramanian, R. (2016). Energy, exergy and techno-economic analyses of hydrothermal oxidation of food waste to produce hydro-char and bio-oil. *Energy*, 102, 187–198. Available from https://doi.org/10.1016/j.energy.2016.02.042.

Mandal, S., Bhattacharya, T. K., & Tanna, H. R. (2017). Energy harnessing routes of rice straw. *Current Science*, 113(1), 21–23. Available from https://www.currentscience.ac.in/Volumes/113/01/0021.pdf.

Mandal, S., Bhattacharya, T. K., Verma, A. K., & Haydary, J. (2018). Optimization of process parameters for bio-oil synthesis from pine needles (*Pinus roxburghii*) using response surface methodology. *Chemical Papers*, 72(3), 603–616. Available from https://doi.org/10.1007/s11696-017-0306-5.

Mandal, S., Kumar, A., Singh, R. K., & Ngachan, S. V. (2012). Evaluation of composition, burn rate and economy beehive charcoal briquettes. *International Journal of Agricultural Engineering*, 5(2), 158–162. Available from http://researchjournal.co.in/upload/assignments/5_158-162.pdf.

Mandal, S., Kumar, A., Singh, R. K., Ngachan, S. V., & Kundu, K. (2014). Drying, burning and emission characteristics of beehive charcoal briquettes: An alternative household fuel of Eastern Himalayan Region. *Journal of Environmental Biology*, 35(3), 543. Available from http://www.jeb.co.in/journal_issues/201405_may14/paper_14.pdf.

Mandal, S., Prasanna Kumar, G. V., Bhattacharya, T. K., Tanna, H. R., & Jena, P. C. (2019). Briquetting of pine needles (*Pinus roxburgii*) and their physical, handling and combustion properties. *Waste and Biomass Valorization*, 10(8), 2415–2424. Available from https://doi.org/10.1007/s12649-018-0239-4.

Manoj Chandran, Sinha, A.R., & Rawat, R.B. S.. (2011). Replacing controlled burning practice by Alternate methods of reducing fuel load in the Himalayan long leaf pine (*Pinus roxburghii* Sarg.) forests. In *5th international wildland fire conference* (p. 5). South Africa. https://www.researchgate.net/publication/268202279_Replacing_controlled_burning_practice_by_Alternate_methods_of_reducing_fuel_load_in_the_Himalayan_Long_leaf_PinePinus_roxburghii_Sarg_forests.

Myers, V. R. (2022). 40 species of pine trees you can grow. *The Spruce*. Available from https://www.thespruce.com/pine-trees-from-around-the-world-3269718.

NAIP (2014). *Value chain on biomass based decentralized power generation for agro enterprises*. Final Report, National Agricultural Innovation Project, ICAR-Central Institute of Agricultural Engineering, Bhopal. <https://naip.icar.gov.in/download/c2-209001.pdf>.

Rominiyi, O., Olaniyi, T., Azeez, T., Eiche, J., & Akinola, S. (2017). Synergetic effect of proximate and ultimate analysis on the heating value of municipal solid waste of Ado – Ekiti, Metropolis, Southwest Nigeria. *Current Journal of Applied Science and Technology*, 22(1), 1–12. Available from https://doi.org/10.9734/CJAST/2017/32953.

Roy, A., Chaturvedi, S., Singh, S., Govindaraju, K., Dhyani, V. C., & Pyne, S. (2022). Preparation and evaluation of two enriched biochar-based fertilizers for nutrient release kinetics and agronomic effectiveness in direct-seeded rice. *Biomass Conversion and Biorefinery*. Available from https://doi.org/10.1007/s13399-022-02488-z.

Sandborn, D. (2017). *Fun facts about pine cones*. Michigan State University Extension. <https://www.canr.msu.edu/news/fun_facts_about_pine_cones>.

Sharma, H. K., Bhattacharya, T. K., Singh, R. P., & Verma, A. K. (2022). Traditional vs. modified approach of pine needle char beehive block production. *Biomass Conversion and Biorefinery*, 12(12), 5799–5812. Available from https://doi.org/10.1007/s13399-020-01008-1.

Sharma, H. K., Kumain, A., & Bhattacharya, T. K. (2020). Characteristic properties of pine needle biochar blocks with distinctive binders. *In Current science*, 118(12). Available from https://doi.org/10.18520/cs/v118/i12/1959-1967.

Singh, S., Luthra, N., Mandal, S., Kushwaha, D. P., Pathak, S. O., Datta, D., ... Pramanik, B. (2023). Distinct behavior of biochar modulating biogeochemistry of salt-affected and acidic soil: A review. *Journal of Soil Science and plant Nutrition*. Available from https://doi.org/10.1007/s42729-023-01370-9.

Singh, R. D., Gumber, S., Tewari, P., & Singh, S. P. (2016). Nature of forest fires in Uttarakhand: Frequency, size and seasonal patterns in relation to pre-monsoonal environment. *Current Science, 111*(2), 398–403. Available from https://doi.org/10.18520/CS/V111/I2/398-403.

UREDA (2010). *Uttarakhand Renewable Energy Development Agency*, Department of Renewable Energy, 2010, Government of Uttarakhand. <https://ureda.uk.gov.in/>.

Vyas, D., Sayyad, F., Khardiwar, M., & Kumar, S. (2015). Physicochemical properties of briquettes from different feed stock. *Current World Environment, 10*(1), 263–269. Available from https://doi.org/10.12944/cwe.10.1.32.

CHAPTER 25

Engineered biochar: potential application toward agricultural and environmental sustainability

Asik Dutta[1], Abhik Patra[2,3], Pooja Nain[4], Surendra Singh Jatav[2], Ram Swaroop Meena[5], Sayon Mukharjee[2], Ankita Trivedi[6], Kiran Kumar Mohapatra[7] and Chandini Pradhan[2]

[1]Crop Production Division, ICAR—Indian Institute of Pulse Research, Kanpur, Uttar Pradesh, India [2]Department of Soil Science and Agricultural Chemistry, Institute of Agricultural Sciences, Banaras Hindu University, Varanasi, Uttar Pradesh, India [3]Krishi Vigyan Kendra, Narkatiaganj, West Champaran, Dr. Rajendra Prasad Central Agricultural University, Samastipur, Pusa, Bihar, India [4]Department of Soil Science, Govind Ballabh Pant University of Agriculture & Technology, Pantnagar, Uttarakhand, India [5]Department of Agronomy, Institute of Agricultural Sciences, Banaras Hindu University, Varanasi, Uttar Pradesh, India [6]Division of Soil Science and Agricultural Chemistry, ICAR—Indian Agricultural Research Institute, Pusa, New Delhi, India [7]Department of Soil Science and Agricultural Chemistry, College of Agriculture, Odisha University of Agriculture and Technology, Bhubaneswar, Odisha, India

25.1 Introduction

Large-scale industrialization and intensive agriculture led to massive emission of greenhouse gases (GHGs) and continuous dwindling of soil health, including a decline in soil fertility and increased erosion (Ding et al., 2016). Therefore, restoring the degraded land using straightforward ecological techniques is imperative. Organic manure and organic waste (sewage sludge and municipal solid waste) and their co-composts contain a numerous harmful pathogens, toxic heavy metals and pharmaceuticals, which may cause the long-term contamination of cultivation land. Moreover, manures and co-composts have the potential to emit ammonia and methane, causing a rise in GHGs and nutrient loss. Engineered biochar is a charcoal-rich solid material produced through heating biomass (thermal decomposition) in an oxygen-depleted environment. This thermal decomposition is accomplished through pyrolysis, in which the biomass is heated to high temperatures, causing changes in its molecular structure. The end products of pyrolysis include syngas, bio-oil, and biochar, a solid, black, extremely porous, and lightweight compound made of 70% carbon (C) in a stable state (Cha et al., 2016; Meena, Pradhan, Kumar, & Lal, 2023). Due to its economic, feasibility and environmental benefits, engineered biochar is a promising approach for sustaining soil fertility and investigating environmental concerns. It contains ample nitrogen (N) both ammonical (NH_4^+) and nitrate (NO_3^-) and phosphate (PO_4^{3-}) could be used as a slow-release organic fertilizer for revamping soil fertility (Ding et al., 2016; Jatav, Singh, Singh, & Kumar, 2018; Pradhan & Meena, 2023a; Pradhan & Meena, 2023b).

The effectiveness of engineered biochar is determined by its chemical and physical properties; however, the interactions between these parameters are highly complex and difficult to comprehend (Sohi, Krull, Lopez-Capel, & Bol, 2010). The seven (7) most crucial characteristics for evaluating engineered biochar are pH, volatile chemical content, water retention, ash content, bulk density, specific surface area (SSA), and volume of pores (Sohi, Krull, Lopez-Capel, & Bol, 2010). The nature of the feedstock, heating rate and time are decisive factors that govern the properties of the end product. Engineered biochar is thus a storehouse of C, although, the entire manufacturing process is a C-negative activity that lowers carbon-di-oxide (CO_2) levels in the atmosphere (Smith, Hatcher, Kumar, & Lee, 2016). Even so, the thermal treatment procedures are frequently maneuvered to customize the properties like SSA, and pore function for the end application. Mechanical, chemical and other miscellaneous activations are the most common biochar engineering approaches (Fig. 25.1). Engineered biochar also can improve soil physical properties such as soil aggregation and macro pore formation (Jatav et al., 2018; Pradhan & Meena, 2023a; Pradhan & Meena, 2023b). These benefits of engineered biochar drew the attention of agricultural scientists to develop the same from various biomass sources and use it as a soil amendment for improving nutrient use efficiency (NUE) and GHG mitigation (Jatav et al., 2018). However, the beneficial effect of engineered biochar on soil quality, crop development, and crop productivity is variable because of the contrasting biochemical nature of s of feedstock, pyrolysis temperature, and other miscellaneous reasons (Pradhan & Meena, 2022; Sohi, Krull, Lopez-Capel, & Bol, 2010). Different literature discussed the utilization of engineered products as a soil amendment to increase soil fertility and prevent loss (El-Naggar et al., 2019; Mohan et al., 2018). The impact of biochar application on soil microorganisms has been observed to vary but is generally positive (Bera, Collins, Alva, Purakayastha, & Patra, 2016; Demisie, Liu, & Zhang, 2014). Jatav et al., 2020 showed the beneficial impact of engineered biochar on microbial activity, potassium (K^+) availability, and reclaiming acid soil. The varying impact of engineered biochar on nutrient availability may correspond primarily to biochar's diverse nutritional composition, underlying soil fertility condition, soil pH, amount of biochar uses, and interaction with soil microbiota (Jatav et al., 2020; Pradhan et al., 2022). As a result, in this particular book, a comprehensive discussion has been made about engineered biochar starting from production technology, physiochemical characteristics, and utilization as an amendment to improve soil fertility. Also, in brief, the role of engineered biochar in soil C-sequestration and mitigating GHGs has been discussed.

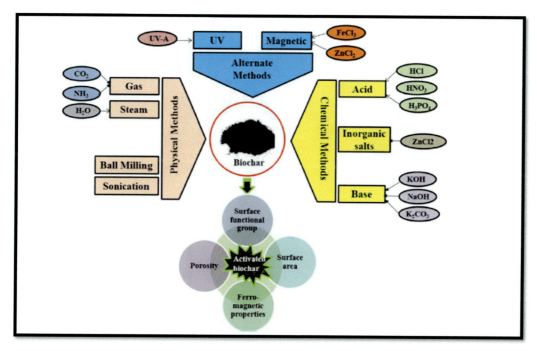

FIGURE 25.1 Methods of biochar engineering using different physical, chemical and other (UV and magnetic) agents.

25.2 Production technology of engineered biochar

Biochar is a carbon-rich, highly porous solid material with higher aromaticity and antidecomposition properties produced by decomposing biomass under limited oxygen (O_2) (Lehmann, Gaunt, & Rondon, 2006). Various raw materials are used to produce biochar, and it has higher porosity and anionic surface functional groups compared to the pristine one. Types of feedstocks used and pyrolysis conditions (temperature, residence time, heating rate and reactor types) determine the physicochemical properties of biochar. Generally, biochar produced at high temperatures (600°C–700°C) had a positive correlation with aromaticity that possesses well-organized C-layers. Still, due to dehydration and deoxygenation at high temperatures, it lacks H and O functional groups (Ahmad, Rajapaksha, & Lim, 2014) which subsequently lowers the ion exchange capacity of biochar. Whereas lower temperature (300°C–400°C) during pyrolysis results in the production of biochar having the advantage of diversified organic characters, including aliphatic and cellulose type structures with more C=O and C–H functional groups (Glaser, Lehmann, & Zech, 2002; Novak et al., 2009; Rajapaksha et al., 2016).

Biochar engineering is the process of the creation of modified or activated biochar. Pristine biochar obtained after pyrolysis of biomass is subjected to physicochemical and biological modification which results in improvement of overall properties [cation exchange capacity (CEC), pH, SSA, surface functional group, porosity, adsorption capacity,

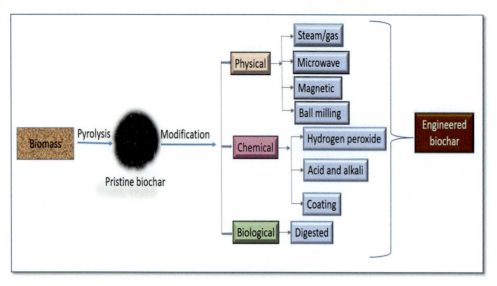

FIGURE 25.2   Process of engineered biochar production.

etc.] regarded as engineered biochar (Mohamed, Ellis, Kim, Bi, & Emam, 2016; Rajapaksha et al., 2016; Yao, Gao, Chen, & Yang, 2013). The process of engineered biochar production is illustrated in Fig. 25.2.

25.2.1 Physical modification

Steam/gas activation, microwave modification, magnetic modification and ball milling come under the physical modification method for engineered biochar production.

Steam gas activation involves pyrolysis of biochar with water vapor, CO_2 air, etc. at a particular temperature. This results in the formation of engineered biochar having increasing porosity and surface area. However, the biochar obtained from this method is heterogeneous because of nonuniform activation, difficulty in controlling the temperature during the reaction and overheating at localized positions (Foo and Hameed, 2011; Raj et al., 2021).

Microwave modification uses high-frequency (350 MHz to 300 GHz) microwave radiation for pyrolysis. In this method, the radiation penetrates biomass and energy is rapidly sent to the reactants' functional groups. Microwave modification can produce biochar with more functional groups and higher surface area. In microwave modification, the pyrolysis temperature can be as low as 200°C with less residence time to get the same result as conventional biochar, thereby saving energy and time. But the downside is, that the raw material has to be mixed with C and certain metal oxides that are effective receptors of microwave radiation (Kumar et al., 2020; Menéndez, Inguanzo, & Pis, 2002; Menéndez, Domınguez, Inguanzo, & Pis, 2004).

Magnetic modification of biochar is attained by combining biochar via co-precipitation with a magnetic material such as ferric chloride and ferrous chloride. Such biochar has higher sorption affinity and is widely used for environmental decontamination, water purification, and many allied fields. However, as biochar is small and low density, it is very difficult to remove from water, limiting its applicability. To remove biochar from liquid, it can be combined with a magnetic medium (Wang et al., 2015). Iron oxide in biochar increases CEC and causes stronger metal binding.

Ball milling: During this process, a mill is employed which has very high energy and along with the milling medium, a specific powder charge is placed (Lin et al., 2017). In this process, moving balls generate kinetic energy applied to the powder charges, breaking down chemical bonds of involved molecules, reducing particles' size, and ultimately breaking the lattice structure of the material. To tailor specific properties, different chemicals (melamine, carboxylic acid, CO_2, ammonium bicarbonate, etc.) can be used to mill together with carbon powder. Gao, Wang, Rondinone, He, and Liang (2015) reported ball milling treated carbon-based materials to have an advantage over unmilled counterparts in terms of adsorption capability, O_2-containing functional group, and efficiency in environmental application. However, certain limitations are involved in this process. The purity and homogeneity of ball-milled products vary depending on the characteristics of feedstock (hardness and particle size) and milling balls in the container. Therefore, ball milling can be regarded greener and cost-effective technology to fabricate novel biochar-based material with improved qualities, which has multiple applications (Kumar et al., 2020; Kumar et al., 2020).

25.2.2 Chemical modification

25.2.2.1 Treatment with H_2O_2

Hydrogen peroxide (H_2O_2) is a strong oxidant with the potential to modify biochar and it has gained importance for being cost-effective and environment friendly. It can increase functional groups containing oxygen in biochar which helps in efficient removal of contaminats from water (Huff and Lee, 2016). The main advantage of the process is the low cost of H_2O_2 compared to other oxidizing agents in modifying biochar and avoiding using other interfering material, which is safer when we use such modified biochar for purification of drinking water and soil amendment.

25.2.2.2 Acid and alkali treatment

Acid modification of biochar causes an extensive increase of the surface area which might be due to an increase in micropores ultimately increasing the adsorption capacity of biochar. The use of organic acid for biochar modification improved and increased the surface functional group of biochar which exhibited enhanced adsorption potential toward diclofenac. In general, alkali-treated biochars have been found to have a larger surface area than those of raw and acid-treated biochar (Ding et al., 2016). Most researchers have used NaOH and KOH for modification of pristine biochar and found increased oxygen-containing functional group, porosity, surface area and adsorption capacity in the modified biochar.

25.2.2.3 Coating

In this method, the surface area of the biochar surface is expanded, functional groups in the surface are increased, CEC and porosity of biochar is increased by coating or impregnating different metal oxides or functional nanoparticles to biochar in different ways at different pyrolysis periods. The nano-metal oxide/hydroxide-biochar composites are prepared through (1) supporting functional nanoparticles, (2) pretreating biomass using metal salt, and (3) mixing metal oxide nanoparticles after pyrolysis.

With large SSA and multiple functional groups, nanoparticles can provide high adsorption capacity to adsorb various pollutants. But the problem with nanomaterials is poor solubility, and ease of gathering. However, nanoparticle-coated biochar can remove contaminants because of improved surface functional group, surface area, porosity and thermal stability of modified biochar after coating of pristine biochar. Another method can be pretreatment of biomass with either chemical reagents, bioaccumulation in biomass or clay before pyrolysis. Metal salts like magnesium chloride ($MgCl_2$), manganese chloride ($MnCl_2.4H_2O$), aluminum chloride ($AlCl_3$), and zinc chloride ($ZnCl_2$) have been used for biochar pretreatment. For example, $MgCl_2$ pretreatment of biomass resulted in the formation of a magnesium oxide (MgO) biochar nanocomposite which showed excellent removal efficiency for nitrate and phosphate.

Another method involves the synthesis of biochar-based composites with metal oxide nanoparticles attached to the biochar matrix. After pyrolysis, iron oxide (Fe_3O_4), ferric nitrate [$Fe(NO_3)_3, 9H_2O$], and potassium permanganate ($KMnO_4$) are used to make metal-loaded biochar composite through either evaporative method, heat treatment, conventional wet impregnation or direct pyrolysis method. Fe-based engineered biochar composites showed greater adsorption capacity, better contaminant removal ability and magnetism in lieu with the pristine one. However, biochar-coated nanomaterials can be good adsorbent, but their ability and existing potential environmental risk cannot be ignored.

25.2.3 Biological modification

In this method of engineered biochar production, feedstock is pretreated biologically through bacterial conversion or anaerobic digestion. Biological modification of biochar is to produce engineered biochar from biologically pretreated feedstocks through anaerobic digestion or bacterial conversion. Biological modification has been identified as a recent approach for the production of economic and efficient biochar-based sorbents. Besides pretreatment of feedstock, anaerobic digestion can also be applied post-modification of biochar for making highly efficient carbon-based sorbents that can sorb heavy metals and cationic methylene-based dye from the environment (Inyang et al., 2012). Biological modification through anaerobic digestion followed by pyrolysis improved physiochemical properties of biochar by increasing cation and anion exchange capacity (AEC), surface area, hydrophobicity and pH as compared to pristine biochar because of alteration of redox potential and pH of feedstock postanaerobic digestion (Inyang, Gao, Pullammanappallil, Ding, & Zimmerman, 2010). Improvement in both AEC and CEC of biologically modified engineered biochar can absorb both positively and negatively charged ions. So, it can be used efficiently for environmental remediation.

25.3 Properties of engineered biochar

25.3.1 Physical properties

The characteristics of biochar are entirely different from its parent material and can be used for a sustainable future. Aspects like biomass retention and nature of feedstock, heating temperature as well as pressure, type of reactor, gas flow rate and postmodification processes like handling etc. (Kazemi Shariat Panahi et al., 2020) influence biochar properties. Pyrolysis-mediated volume reduction in materials causes physical attributes like crystallinity, porosity and SSA modification. Biochemical nature and ash content also determine the physical characteristics of biochar. More precisely attrition, formation of cracks and micro-structural rearrangements also influence the makeup of outcome (Downie, Crosky, & Munroe, 2009). Physical properties like density, pore and particle size distribution and mechanical strength are described here. Among the most properties pore size distribution in the end product is very important to determine its industrial usage. There are three kinds of pores depending upon internal radius namely, macro (>50 nm), meso (2–50 nm) and micro (<2 nm) pores (Rouquerol, Rouquerol, Llewellyn, Maurin, & Sing, 2013). Macropores act as a connecting channel for the other two kinds of pores and as of high value of the materials having agricultural importance (Downie et al., 2009; Palansooriya et al., 2019). Mesopores are significantly involved in the liquid–solid adsorbing process, but micropores are responsible for surface area determination and efficiently absorb nanoscopic materials (Downie et al., 2009). Pre and post processing factors like heating temperature, heating rate, handling and stiffness of the materials during pyrolysis determine the biochar particle size and in the general end product has a comparatively smaller size than initial biomass (Cetin, Moghtaderi, Gupta, & Wall, 2004). Higher particle size can be achievable via a slower heating rate (5–30°C min^{-1}) with higher temperature (450°C–700°C) and the reason is a higher chance of attrition and lesser tensile strength (Kazemi Shariat Panahi et al., 2020). Nonetheless, pressurized system and conglomeration of particles form biochar with larger particle sizes (Cetin et al., 2004). Similarly, density bulk density (0.3–0.4 g cc^{-1}) is lesser than solid/particle density (1.4–1.7 g cc^{-1}) (Brown, Kercher, Nguyen, Nagle, & Ball, 2006). Density also affects mechanical resistance: the ability to withstand environmental stress and acceptability on an industrial scale (Kazemi Shariat Panahi et al., 2020). Biomass quality also influences the strength like biochar from virgin wood is lower than monolithic wood (Byrne and Nagle, 1997). The heating rate is inversely related to mechanical strength forming cracks on the surface (Kazemi Shariat Panahi et al., 2020; Weber and Quicker, 2018).

25.3.2 Chemical properties

Chemical properties of the biochar are highly correlated with precursor material quality and reaction (pyrolysis) temperature and properties of the end product can be predicted via data synthesis technique (Li, Song, Singh, & Wang, 2019). Ion exchange capacities (CEC and AEC), functional group and pH in the biochar alter greatly during pyrolysis. Generally, at higher temperatures, pH and CEC respond oppositely. However, biochar derived from biosolids has higher CEC than others due to the presence of metal cation and oxygen (O⁻) containing functional groups in the surface. Microchemical condition of the biochar depends on the biochemical nature of the solid phase, electrochemical properties, presence of entrapped oils and different functional groups all of these are modified a lot under limited oxygen conditions (Amonette and Joseph, 2009). Temperature between 300°C and 600°C, degradation and volatilization of different anhydrous sugars determine the chemical condition of biochar material, whereas, at temperature >600°C formation of free radicals augments the generation of pyrophoric biochar (Feng, Zheng, & Maciel, 2004). The thermal decomposition of the biochar is largely a four-step temperature-dependent process. In the first step at temperature <250 °C there is continuous mass reduction with concomitant C-enrichment (Paris, Zollfrank, & Zickler, 2005). In the next step, at 250°C–350°C aromatic rings form with complete depolymerization of cellulose and loss of almost 80% biomass (Baldock and Smernik, 2002). Typically, in a biochar C-content could reach up to 90% due to carbonization with concomitant loss of hydrogen (H), and oxygen (O). As discussed, high temperatures cause enrichment in metal ions with concomitant reduction in C, O, and H. The mineral matter in biochar depends on reaction temperature and pressure, steam, air and biomass nature (Bridgwater and Boocock, 2006). Depending on the temperature loss of the minerals, like, highly mobile ions K, chlorine (Cl) gets vaporized easily with lower temperatures. The minerals highly resistant to high temperatures are N, secondary and microelements, and silicon (Si) because they form complex organo-mineral structures, development of phenoxides, etc. (Amonette and Joseph, 2009; Bourke, 2007). Sometimes mineral elements affect the properties of biochar but, pretreatment of biomass materials improves aromatization and strengthens the C-skeleton. Lastly, retaining some mineral acids (phosphoric acid) in the biochar structure is important depending upon its industrial and environmental application. The amount of alkali and alkali earth metal could improve the efficiency of bio-crude oil separation. The presence of bio-crude oils or their derivatives alters the biochar use efficiency and applicability (Amonette and Joseph, 2009).

Chemical nature of the biochar may be acidic, basic, hydrophobic, and hydrophilic, depending on biomass nature and pyrolysis. The main reason behind surface chemical heterogeneity is the charge difference between various inorganic atoms and C-atoms (Brennan, Bandosz, Thomson, & Gubbins, 2001). The functional groups are majorly of two types: (1) Electron acceptors [NO_2, (C = O) H and (C = O) OH], and (2) electron donors [−OR, NH_2, −OH] having α and π electrons (Amonette and Joseph, 2009). Diverse N- and S-containing groups present on the biochar derived from manures, bio-solids, sewage sludge, and high-protein materials (Leng et al., 2020). Lower temperature causes the accumulation of sulfates and sulfonates while theophoric groups form majorly at higher temperatures (Koutcheiko, Monreal, Kodama, McCracken, & Kotlyar, 2007). Pyrone groups form on the surface due to a reaction with molecular oxygen (O_2^-) or its disassociated atoms (O^-) (Swiatkowski, Pakula, Biniak, & Walczyk, 2004). So, the surface chemistry with reaction conditions preliminarily determines the engineered biochar's surface chemical configuration. At temperatures over 400°C organo-chemical properties like H:C-ratio and O:C-ratio change during pyrolysis (Graetz and Skjemstad, 2003). Besides, heating temperature and biomass composition the quality of biochar can be changed with aging in anaerobic conditions and treatment with acids like nitric acid treatment increase oxygen content in the final product while the high temperature promotes C-enrichment (Cheng, Lehmann, & Engelhard, 2008; Trompowsky et al., 2005).

25.4 Critical factors to maximize the advantages of engineered biochar

25.4.1 Quality of feedstock biomass

Biochar originated from various biomasses, including crops, forest leftovers, municipal green trash, paper mills, sawmill waste, pig and poultry waste, and occasionally human waste with peculiar advantages based on nature of parent materials. The advantages of applying biochar to soils can vary such as dessertified terrain or the unproductive agricultural lands of India, South Africa, and Australia. However, not all feedstock materials are suitable for every kind of soil. The quantities and types of nutrients depend on the biomass being used. For example, when wood-based substrates are pyrolyzed, coarse, refractory biochars containing high C levels are produced because the parent material's hard lignolytic character is preserved in the biochar remnant (Winsley, 2007). According to a 2007 Australian study by Chan, Van Zwieten, Meszaros, Downie, and Joseph (2007), biochar derived from poultry manure exhibited greater EC, N, P, and pH values than biochar from organic garden trimmings. These studies show that the advantages of biochar increase with the amount of nutrient-rich organic waste.

25.4.2 Optimum temperature for biochar production

Superior grade biochar has higher C-content and CEC which are closely associated with pyrolysis temperature. Biochar's C-content is decreased by pyrolysis at elevated temperature, which lowers its ability to sequester C and diminishes all of its additional advantages. Similarly, handling small size biochars (<1 mm size), which are generated at extreme temperature pyrolysis, proves to be challenging because doing so necessitates palletization or slurring, both of which come at considerable expense. As a result, 500°C or there abouts is the ideal temperature for pyrolysis (Lehmann, 2007).

25.4.3 Soil carbon level

Soil C content is one of the major determinant for achieving optimum effectivity of applied biochar. In a ten-year study conducted under three different forest types in northern Sweden, charcoal was manufactured, blended with the soil, and remained undisturbed. This resulted in a significant increase in soil fungi and bacterial activity. As a result, native soil organic matter began to mineralize (decompose), which exacerbated CO_2 emissions (Wardle, Nilsson, & Zackrisson, 2008). This demonstrated that using biochar on fields with higher C content might partially cancel out the advantages of reducing GHG. Consequently, biochar ought to be used on soils that are low in C to maximize its merits. Since 25% of South Africa's and 45% of India's cultivable areas are degraded, employing biochar in these agricultural soils may be more effective (FAO, 2008; Hatrack, 2008).

25.4.4 Soil types and soil moisture

The fact that biochar improves water quality and plant-accessible water capacity (PAWC) is one of its main benefits. This would represent a huge advantage in dry nations like Australia and India where both water supply and quality are highly changeable. This kind of soil exerts a big impact on PAWC. There isn't any assurance that adding biochar will enhance the amount of accessible water in loam and clayey soils, even though it does improve the water retention capacity and plant-accessible moisture in sandy soils. By its large surface area plus expanded micropores, biochar helps water retention capacity of sandy soils. However, no modifications were seen in loamy soils, while in clay-rich soils, the quantity of available soil moisture diminished with increased charcoal inputs, perhaps due to the charcoal's hydrophobic nature (Glaser et al., 2002). In farming areas where, the opportunity cost of water is particularly high, for instance, the sandy soils found in the Western Australian wheat belt alongside moisture-scarce soils of India, the favorable effects of biochar on soil moisture retention are therefore improved in sandy soils.

25.4.5 Soil pH and soil contamination

The majority of macronutrients are found in topsoil under neutral pH; however, it is very difficult to draw clear -cut correlation between soil pH and nutrient availability. However, improper nitrogen-containing fertilizer use, crop waste removal, nitrogen leaching, and particularly the existence of calcium sulfate ($CaSO_4$) as a parent material have led to soil acidity becoming a significant global concern. Acid conditions boost soil contamination by elevating Al and Fe cation levels, reducing the number of symbiotic microbes available for efficient plant growth (Shuji, Tanaka, & Ohata, 2007). Farmers use thousands of tonnes of liming materials to neutralize their agricultural soils neutralize acidity. Furthermore, to the above direct expense, lime is produced, packaged, transported, and applied, all of which result in large GHG emissions (Maraseni, 2007). Applying biochar to soils with acidity can significantly reduce intangible costs (such as reducing GHG emissions).

25.5 Application of engineered biochar

25.5.1 Effect of biochar as a soil conditioner

Biochar is a stable carbon matrix material. It will effectively retain water, air, water-soluble nutrients, metals and organic chemicals. At the same time, it can be utilized for the enchantment of quality of irrigation water, soil fertility and productivity, minimizing GHG emissions and increasing agricultural productivity. Nowadays, Australian Farmers use biochar as a stock for feeders mixed with molasses. Due to its good water retention capacity fertilizer industry used biochar as a carrier material. In Brazil, farmers are switching fire-fallow cultivation to slash and char.

As engineered biochar is produced through pyrolysis of plant biomass, they are expected to be rich in C and macro (N, P, K, Ca, Mg, and S) and micronutrients (Fe, Mn, Zn, and Cu) (Table 25.1). The addition of engineered biochar in the soil can enhance the availability of soil moisture, water holding capacity (WHC), oxygen diffusion rate (ODR), organic matter (OM) content, microbial biomass C (MBC), microbial biomass nitrogen (MBN) and microbial metabolic activity, enzymatic activity, and nutrient absorption and mineralization, which resulted in reduced chemical fertilizer demands and also decreases nutrients losses (El-Naggar et al., 2019). In general, engineered biochar can be an alternative option over chemical fertilizer as an organic source of fertilizer for soil amendment. There is still a rising demand to improve the effectiveness of biochar to enhance soil fertility and carbon sequestration by incorporating modern tools and technologies in engineered biochar production (Mandal et al., 2016; Rajapaksha et al., 2016). The potential for engineered biochar to increase soil fertility has been recorded (Khan, Reid, Li, & Zhu, 2014). Soil fertility is the function of numerous factors such as soil physical and biogeochemical cycles, soil structure, BD, soil reaction, CEC, buffering capacity available major and macronutrient, microbial activity and soil organic carbon (SOC). This segment addresses the various manners in which engineered biochar may help to enhance soil quality.

25.5.1.1 Soil physical properties

The addition of engineered biochar directly or indirectly influenced the soil physical properties. The addition of engineered biochar increases WHC by an average of 10%, depending on the type of soil (Dai, Zheng, Jiang, & Xing, 2020). According to a meta-analysis of 37 studies, applying engineered biochar increased the permanent wilting point (PWP), field capacity (FC), and soil moisture content by an average of 16.7%, 20.4%, and 28.5%, respectively (Edeh, Mašek, & Buss, 2020). Therefore, the basic surface area, particle size distribution, and porosity of the engineered biochar were the primary determinants of these biochar-induced alterations (independent of soil type). Therefore, biochar may be added to sand, sandy loam, or silty soil to increase water retention capacity. For this purpose, 30–70 t ha^{-1} of biochar application in coarse-textured soils is best suited (Edeh et al., 2020).

25.5.1.2 Physicochemical properties

Application of engineered biochar improved soil physicochemical properties by optimizing soil reaction (pH), which is ambient for higher crop production. More importantly, in acidic soils engineered biochar will dramatically increase soil reaction and enhance the nutrient's bioavailability (Cornelissen et al., 2018; Raboin et al., 2016). The meta-analysis study found that amendment with enriched biochar increases soil pH by 9% compared to control (Dai et al., 2020). The high surface area of biochar helped increase the CEC in addition to the pH, especially in sandy soils. Similarly, result of meta-analysis study reported, the application of engineered biochar capable of enhancing CEC of soil by 194% (Dai et al., 2020). As a result, modified biochar reduced the leaching losses of available basic cations and enhanced the nutrient availability for plant growth (Fig. 25.3) (Laird et al., 2010).

25.5.1.3 Available nutrients

Engineered biochar is a nutrient-enriched organic amendment for soil health. The application of enhanced biochar enhances N bioavailability and decreases different N losses such as leaching and volatilization losses of N fertilizer (Mia, Singh, & Dijkstra, 2019; Zhao, Cao, Mašek, & Zimmerman, 2013). In soil, enriched biochar acts like a buffering agent by absorbing inorganic nitrogenous compounds (NH_4^+ and NO_3^-), releasing them

TABLE 25.1 Impact of engineered biochar addition as soil fertilizer on crop yield.

Engineered biochar	Materials used for producing biochar	Location	Crop	Yield	Remarks/comments	References
Cow dung-derived EBC	Cow dung, $MgCl_2$	Shanghai, China	Lettuce	16.77 gm pot^{-1}	25.3% yield increased and over ordinary BC	Chen et al. (2018)
Biochar-based fertilizer	200 g wheat straw, 15 g urea, 15 g bentonite clay, 15 g rock phosphate, 5 g Fe_2O_3 and 5 g $FeSO_4.7H_2O$	Nanjing	Rice	3.02 g per plant	83.2% dry weight increased over RDF	Chew et al. (2020)
Iron-modified biochar	Platanus orientalis Linn, $FeCl_3.6H_2O$	Zhejiang A&F University in Hangzhou City	Rice	Straw yield 55 g pot^{-1}, grain yield 14 g pot^{-1}	FeBC-treated soils increased uptake and translocation of the metals and increased the grain yield but not significantly with RBC	Wei et al. (2021)
Sulfur biochar	[A 5:100 (w/w) mixture of elemental sulfur (S) and citrus wood biochar (BCH)]	Fayoum, Egypt	(Capsicum annuum L., cv. "Omiga")	42–67 g 10 m^{-2} plot	RSBC might be a promising and an eco-friendly approach to maximizing crop production in saline environments.	Abd El-Mageed et al. (2020)
Paper mill EBC	Paper mill sludge		Wheat	Increase in wheat height by 30%–40% in acid soil but not in alkaline soil	Mainly liming value	Van Zwieten (2018)

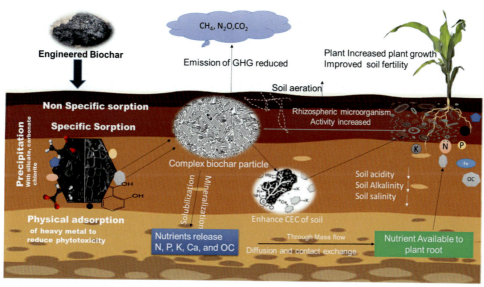

FIGURE 25.3 Mechanisms of nutrient release through engineered biochar.

slowly to the plant root by reducing different losses (Haider, Steffens, Moser, Müller, & Kammann, 2017; Solaiman & Anawar, 2015). Engineered biochar has a 20% greater NH_4^+-N adsorption potential than ordinary biochar (Chen et al., 2017). The mechanism of the N cycle in soil is also disturbed by biochar-induced N-fixation, nitrification, and/or denitrification processes. Engineered biochar increased the total available soil P and K by up to 33% and 50% respectively (Namasivayam and Sangeetha, 2004; Zheng, Wang, Deng, Herbert, & Xing, 2013) and improved the P-adsorb. It has been reported that $ZnCl_2$-engineered biochar improved the P adsorption potential by 5 and 30 mg g^{-1}. Mg-enriched engineered biochar synthesized from tomato leaf had the maximum P-adoption potential (100 mg g^{-1}) compared to ordinary biochar (Yao et al., 2013). Overall, engineered biochar enhances the slow release of available essential nutrients (N, P, K, Zn, Fe, Cu, and Mn), fixes environmental problems like runoff and eutrophication, and enhances seed germination and crop growth.

25.5.1.4 Soil biological properties

Several studies have demonstrated that engineered biochar affects soil microbial structure and diversity (Quilliam, Glanville, Wade, & Jones, 2013). This neutral to negative impact may be clarified by the fact that the effects of biochar on the soil microbial population are a function of the soil type, soil moisture content, type of biochar used, and rate of biochar amendment. Among the different types of biochar, EB, having a highly porous nature, created a microenvironment that was conducive to microbial colonization and their growth and development in soil (Awad, Lee, Kim, Ok, & Kuzyakov, 2018). In contrast, biochar stimulates rhizobial nodulation by adsorbing soil N, releasing Ca, P, K, Mo, and B, and more effectively retaining signaling molecules like nod factors (Biederman & Harpole, 2013). For example, the amendment of engineered biochar at 10–20 t ha^{-1} in P-deficient forest soil has been reported to enhance the bacterial/fungal population ratio while reducing the ratio of Gram-negative/Gram-positive bacterial community (Mitchell, Simpson, Soong, & Simpson, 2015). Amendments of engineered biochar can restore deteriorated soils by encouraging soil biomass carbon and soil enzyme activity such as N-acetyl-β-glucosaminidase, β-D-cellobiosidase and β-glucosidase (Wang et al., 2015), which participate in the nutrient's transformation (Kazemi Shariat Panahi et al., 2020).

25.5.1.5 Crop response

An overview of the impact of the engineered biochar amendment on crop yield is given in Table 25.1. For example, by adding 2% of engineered biochar to the soil, crop productivity increased significantly as compared to the untreated one (Kammann et al., 2015). Total C and CEC content in soil was increased by 1.5%, and crop productivity was 70.8% to 309% higher than control with the amendment of engineered biochar (Busch and Glaser, 2015; Liu et al., 2016; Schulz, Dunst, & Glaser, 2013).

25.5.1.6 Reduce soil erosion

One of the major issues associated with climate change is soil erosion due to deforestation and degradation of soil. It has been reported that when biochar was added at different rates of 33.75 and 65.5 t ha^{-1} to extremely weathered soils, there was a 50% and 64% reduction in soil erosion compared to control (Palansooriya et al., 2019). Also, the use of biochar reduced runoff ratios, total runoff, and nutrient losses on sloping uplands. The desirable impact of biochar on reducing soil erosion is due to various factors, including the development of micro-aggregates and the coherence of biochar and soil particles resulting in an increase in soil water stable aggregates, water retention and decrement in soil crusting.

25.5.2 Engineered biochar for climate resilience agriculture

25.5.2.1 Carbon sequestration

Biochar has been found to contain a significant amount of recalcitrant C due to which it is ideal and capable of C sequestration, thereby lowering soil-based emissions of GHGs (Borchard et al., 2019; Han et al., 2018; Sui et al., 2016). The mean residence time of biochar in the soil varies from a few decades to many thousands of years. The C in biochar exists in two forms: i. condensed or recalcitrant pool ii. labile pool. The efficiency of biochar in carbon sequestration depends on the stability of biochar and structural or quantitative changes in native SOC. Short-term biochar application may lead to a rise in CO_2 emission because of rapid release of the labile-C from biochar and further deterioration of the native SOC. Whereas, in the long run, C- sequestration potential has been increased due to biochar application, and the reasons are (1) development of stable biochar-organic matter macro-aggregates derived from the interaction between native SOC and biochar. (2) modification in the size and composition of the microbial community (Borchard et al., 2019). Engineered biochar amendment will increase the concentration of C (48%) in low mineralizable N soils simultaneously it helps sequester atmospheric C and also mitigates greenhouse emissions to the atmosphere. In addition to other organic fertilizers such as farmyard manure, vermicompost, and compost, biochar has a stable long C chain with a 102–107-year half-life period depending on the type of biomass and pyrolysis temperature (Sánchez-García, Sánchez-Monedero, & Cayuela, 2020). Finally, these characteristics of EB help enhance soil fertility and crop productivity (El-Naggar et al., 2018).

25.5.2.2 *Mitigation of methane emission*

It has been widely accepted that biochar has the ability to reduce the emission of methane. The capacity of biochar to minimize methane emission depends on attributes of biochar and environmental conditions. For instance, structural properties of biochar like porosity and surface area are one of the most important properties which reduce methane emission. Biochar can be engineered to develop materils with higher pores which increase the adsorption and oxidation of methane. Adding biochar improves the aeration of soil and supplements suitable habitat, inhibiting methanogenic archaea and enhancing methanotrophic bacteria, eventually decreasing methane production (Qin et al., 2016). Moreover, biochars have high SSA and porosity, so they can absorb a greater amount of soil-derived organic or inorganic materials, such as dissolved organic C, which decreases the availability of substrate resources for methane production. Apart from physical properties, chemical properties such as the pH of biochar which is present in a wide range (pH 4–12), can be specifically modified or engineered for a specific pH to create a favorable condition for reducing methane emission. According to studies, adding alkaline biochar to acidic soil raises the soil's pH level by significantly reducing methane emissions (Feng, Xu, Yu, Xie, & Lin, 2012; Jeffery, Verheijen, Kammann, & Abalos, 2016). Feng et al. (2012) observed that adding biochar to acidic paddy soil increased soil pH, which enhanced the numbers of methanotrophic bacteria; this led to a 59%–63% reduction in methane emission.

25.5.2.3 *Reduction of NO_2 emission*

It is widely observed that the addition of biochar lessens the emission of oxides of N (NOx). There are two main reasons for this: firstly, adding biochar prohibits soil denitrification and secondly, due to greater surface area and porosity, biochar has an impressive capacity for adsorption. This leads to the adsorption of NH_4^+ and NO_3^- which immobilizes N compounds due to which there is a decrease in NH_3 volatilization and reduction in inorganic N pool for nitrogen di oxide (NO_2)-generating microorganisms through denitrification process (Clough, Condron, Kammann, & Müller, 2013). Biochar can be engineered to have more surface area and improved porosity. This will ensure greater absorbance of dissolved organic C and mineral N, which will result in decreased availability of substrate resources for NO_2 production. Under a similar context Ameloot, Maenhout, De Neve, and Sleutel (2016) found that applying biochar can decrease NO_2 emission by 50%–90% through a reduction in the bioavailability of C substrate to denitrifies.

25.5.3 Carbon capture and utilization

Global SOC stocks in the top 1 m of soil account for 1505 Pg C out of the 3170 Pg C in terrestrial ecosystems. Strategies for increasing SOC include conservation tillage, organic amendments, crop rotations, and cover cropping (Lal, 2018; Van Zwieten, 2018; White, Brennan, & Cavigelli, 2020). Increasing SOC for climate change mitigation is complex due to the depletion of soil carbon in intensive farming, and it hinders agricultural potential (Lal, 2018). Organic amendments, depending on quality, increase microbial growth, releasing nutrients for microbes and plants. Fresh organic matter increases microbial growth, activating microorganisms and enhancing SOM degradation through co-metabolism and enzymatic production (Blagodatskaya & Kuzyakov, 2008). Chen et al. (2018) from a meta-analysis study concluded that SOC content increased by an average of 29%–50% compared to controls, more significant with prolonged organic matter application and higher amendment doses.

The key is to increase soil carbon storage, minimizing positive priming from organic matter application, and reduce the organic amendment's mineralization rate. Biochar adds stable carbon directly to various SOC pools (Kookana, Sarmah, & Van Zwieten, 2017) and decreases the mineralization rate of native SOC and rhizodeposits through negative priming (Weng, Van Zwieten, & Singh, 2018a; Weng, Van Zwieten, & Singh, 2018b). Although it may exhibit positive priming in the initial years. Higher C:N ratio of biochar, pyrolytic time, and soil clay content increase negative priming (benefiting soil C stabilization). Biochar also increases aggregation and mineral protection of rhizodeposits and SOM within macroaggregates (250–2000 μm) (Wang, Fonte, Parikh, Six, & Scow, 2017). Key mechanisms for biochar stabilization in soil includes high residence time, inherent resistance to decomposition (due to aromaticity, amorphous structures, turbostatic crystallites, and rounded structures) (Bourke et al., 2007; Paris et al., 2005), reduced accessibility, biochar's particulate nature, and interactions with mineral surfaces (Sollins, Homann, & Caldwell, 1996). Additionally, the persistence of organic matter within soil aggregates and organo-mineral associations contributes to the durability of biochar in the soil.

25.5.4 Bioremediation of contaminated aqueous and soil ecosystem

Numerous studies suggest that biochar effectively immobilizes organic contaminants and potentially toxic elements (PTEs) in soil and water due to its unique properties influenced by feedstock, pyrolytic conditions, and modification methods (Ahmad et al., 2014; Dai, Zhang, & Xing, 2019). Key mechanisms for biochar's immobilization of organic contaminants include partitioning by noncarbonized fractions, pore filling, electrostatic attraction, π–π electron donor–acceptor interactions, and the hydrophobic effect (Ahmad et al., 2014; Dai et al., 2019). Biochar's effectiveness depends on factors like properties (e.g., aromatic nature, surface functional groups, micro-porosity, high SSA, polarity, high buffering capacity, high ash content, alkalinity and aromatic properties, etc), contaminant characteristics (e.g., hydrophobicity, size, chemical nature etc.), soil properties (e.g., pH, texture, organic matter), and other environmental factors (Ahmad et al., 2014; Ahmad, Ok, & Rajapaksha, 2016; Khalid, Shahid, & Murtaza, 2020).

Modification with 50% phosphoric acid (H_3PO_4) increased surface area, and potential atrazine absorption ability via Van der Waals forces, hydrogen bonding, electrostatic interaction, and pore filling (Zeng et al., 2022). Likewise, biochar-based zero-valent iron aided atrazine degradation using SO_4^{-2} and $HO \cdot$ as reactive species (Jiang, Li, & Jiang, 2020). Combining biochar with Fe-Mn binary oxides boosted polycyclic aromatic hydrocarbon (e.g., naphthalene) adsorption and activated H_2O_2 (Li, Lai, & Huang, 2019).

Biochar interacts with PTEs through electrostatic interaction, ion exchange, and surface complexation with functional groups. Divalent metal cations are immobilized through cation exchange, precipitation, surface complexation, and electrostatic interaction (Bandara, Franks, & Xu, 2020a; Bandara, Xu, & Potter, 2020b; Pellera & Gidarakos, 2015; Rechberger, Kloss, & Wang, 2019). Alkali metals like Ca and Mg on biochar's surface can exchange with positively charged Pb(II) ions (Bandara et al., 2020a). For Cr(VI), immobilization occurs through reduction, surface complexation, or precipitation (Bandara et al., 2020a). Mercury (Hg) immobilization by engineered biochar primarily involves forming Hg-sulfide and Hg-dissolved organic carbon complexes, as well as Hg precipitation (Bandara et al., 2020a; Xing, Wang, & Shaheen, 2019). Binding Hg with thiols (e.g., cysteine) in biochar reduces Hg solubility in treated soils (Xing et al., 2019). Another mechanism is electrostatic outer-sphere complexation due to metal exchange with Ca^{++}, Mg^{++}, K^+, and Na^+ present in the biochar. Chemisorption, surface complex formation, and electrostatic adsorption are key controls on PTEs adsorption onto biochar (Dong, Ma, & Gress, 2014; Hu, Ding, & Zimmerman, 2015; Tong, Li, & Yuan, 2011).

25.5.5 Engineered biochar as catalyst

Biochar is a cost-effective and eco-friendly alternative to carbon-based supports and be an effective catalyst support and catalyst due to its carbon-rich nature and porosity-controlling abilities. The inclusion of intrinsic heteroatoms (N, O, S, and P) and metallic compounds enhances its catalytic potential. Biochar's surface properties can be engineered through physical or chemical activation processes (Jung, Oh, & Park, 2019; Lee, Kim, & Kwon, 2017).

Internal combustion engines and power systems emit harmful NO_x. Activated carbon and biochar, structurally similar to activated carbon, reduce NOx at lower temperatures. Selective catalytic reduction (SCR) employs NH_3 as the reducing agent and utilizes over-activated biochars (Wu et al., 2016), heteroatom-doped biochars (Klinik et al., 2011), and metal/metal oxide-impregnated biochars (Shen et al., 2015). For example, MnOx supported on rice straw biochar achieved 85% and 84% NOx removal efficiency at 250°C and 50°C, respectively (Cha et al., 2022). Mn–CeOx supported on cotton stalk biochar exhibited over 80% NO removal efficiency at 280°C (Shen et al., 2015). However, the NOx removal efficiency of biochar-based catalysts remains below that of conventional SCR catalysts, which attain up to 99% efficiency at temperatures below or equal to 200°C (Wang et al., 2019).

Biodiesel production involves esterifying and trans-esterifying vegetable or animal oils using various catalysts, including both homogeneous (e.g., KOH, NaOH, HCl, and H_2SO_4) and heterogeneous materials (e.g., CaO, zeolite, amberlyst resins, SiO_2, TiO_2, and Al_2O_3) (Ahmad et al., 2023). The high cost of metal precursors limits metal/metal oxide catalyst availability.

Homogeneous catalysts raise concerns about product separation and environmental impact. Biochar, known for its outstanding acid density, surface area, catalytic activity, and reusability, achieves an esterification efficiency of 90%–100% ((Qian, Kumar, Zhang, Bellmer, & Huhnke, 2015). Its thermal stability up to 600°C enhances liquid oil quality during plastic waste pyrolysis by removing impurities (Ding et al., 2016). Like sulphonated rice husk biochar, engineered biochar yields 88% biodiesel through waste cooking oil transesterification at 100°C (Li, Zheng, Chen, & Zhu, 2014). Chicken manure-derived biochars yield over 95% biodiesel from waste cooking oil transesterification due to the presence of $CaCO_3$, expediting transesterification kinetics (Jung, Oh, & Baek, 2018; Jung, Park, & Kwon, 2019).

25.5.6 Application in Livestock science

Biochar in animal feed boosts health by absorbing toxins, reducing pathogens, enhancing feed efficiency, nutrient absorption, and productivity, and reducing nutrient losses and GHGs. When integrated with good farming practices, it enhances animal husbandry sustainability by increasing feed intake, weight gain, egg quality in poultry, boosting the immune system, improving meat quality, reducing odors, diseases, and veterinary costs. Historically known as charcoal, biochar has been used for centuries to combat the harmful effect of bacteria such as *Clostridium tetani* and *C. botulinum* (Gerlach, Gerlach, & Schrödl, 2014), and reduce oocyst excretion in animals (Volkmann, 1935).

When combined with silage, biochar curbs mycotoxin formation, inhibits butyric acid production, and increases lactic bacteria populations (Calvelo Pereira et al., 2014). It also absorbs pesticides such as organochlorides (Fries, Marrow, & Gordon, 1970). Moreover, biochar reduces milk's somatic cell count (SCC), indicating fewer harmful bacteria, while increasing milk protein and fat content (Gerlach & Schmidt, 2012). Such absorption by activated carbon and charcoal toward proteins, amino acids, enzymes, and toxins, is contributed to their high surface area, surface functional group richness and porosity (O'Toole et al., 2016; Schirrmann, 1984; Steinegger & Menzi, 1955; Toth & Dou, 2016). Biochar also reduces greenhouse gases like methane from ruminant guts by acting as an electron acceptor (Cabeza, Waterhouse, Sohi, & Rooke, 2018; Saleem et al., 2018). Germany and Switzerland pioneered biochar use in animal feed, catalyzing European biochar market growth. Research by Kammann et al. (2017), O'Toole et al. (2016), underscores that a significant portion of European biochar production now targets animal feed, bedding, manure treatment, and soil enhancement. In 2016, the European Biochar Foundation (EBC) introduced a specialized certification standard for animal feed biochar (European Biochar Foundation EBC, 2018) to ensure quality and regulatory alignment.

25.5.7 Construction material

The comprehensive exploration of biochar's potential to improve mortar and concrete properties is a relatively recent development. Especially emphasizing how biochar particles acting as fillers significantly enhance mortar's mechanical properties. As reinforcement fibers weaken concrete, biochar coating strengthens the interface of the cementitious matrix and polypropylene fibers, reducing the latter's water sorptivity and water penetration (Akhtar & Sarmah, 2018; Gupta, Kua, & Cynthia, 2017). Biochar highers the compression strength and lowers the water absorption by means of "reservoir effect" of presoaked biochar—a phenomenon in which absorbed water is released from the pores of biochar back into the surrounding cementitious matrix — promotes cement hydration under air-curing conditions (Gupta & Kua, 2018; Gupta & Kua, 2019). They also explored and confirmed the concept of immobilizing carbonate-precipitating bacterial spores with biochar across three healing cycles. Once internal cracks were sealed, mortar containing bacteria immobilized in biochar exhibited 7.7%–8.8% higher compressive strength compared to mortar where bacteria were directly mixed in. Biochar mixing increases the toughness and flexural strength of plain cement paste (Ahmad, Khushnood, & Jagdale, 2015). Recent research focuses is biochar in sediment-based construction and ultra-high-performance concrete (UHPC). Wang, Chen, and Tsang (2019) demonstrated that adding biochar to sediment enhances cement hydration and effectively traps potential toxic elements and organic pollutants within its pores. Incorporating 5 wt.% of biochar into UHPC mixtures resulted in a 17% boost in cement hydration while maintaining similar 28-day compressive strength (Dixit, Gupta, & Dai Pang, 2019)

25.6 Conclusion

Biochar is a C-rich material produced from numerous materials like crop biomass, sewage sludge, and virgin wood by pyrolysis under a limited supply of oxygen. The physical and chemical properties of the precursor material and the end product must be known to get maximum benefits. Depending on the need, the biochar can be engineered for multipurpose usage by improvising the original biomass composition and the reacting environment. Manipulating techniques like steam gas activation, microwave modification, ball milling, chemical treatments with hydrogen peroxide, acids and alkali and biological digestion greatly modifies the original characteristics like pore and particle size distribution, density, functional group composition and other organo-chemical characteristics thereby making it suitable for diversified usage like improving NUE and mitigating greenhouse gas emission, Bioremediation of contaminated aqueous and soil ecosystem, increasing the efficiency of fuel and construction industry. Engineered biochar can be an important option for removing all loopholes in the pristine biochar; but its techno-economic viability must be worked out. Future research must be based on biomass type, pyrolysis conditions, and technological advancement. Efficient and cost-effective biochar synthesis and characterization from various biowaste sources is a critical research focus. Inventing modification techniques to modify biochar production for specific applications while monitoring the long-term effects of biochar among diverse agroecosystems are must. Furthermore, the reusability of spent biochar for remediating contaminated water sources is needed to investigate understanding the long-term stability in immobilizing contaminants. Additionally, ongoing work is to develop highly efficient biochar-based catalysts for industrial catalytic reactions. These endeavors may collectively contribute to a broader understanding and utilization of biochar in a variety of environmental and industrial contexts.

References

Abd El-Mageed, T. A., Rady, M. M., Taha, R. S., Abd El Azeam, S., Simpson, C. R., & Semida, W. M. (2020). Effects of integrated use of residual sulfur-enhanced biochar with effective microorganisms on soil properties, plant growth and short-term productivity of Capsicum annuum under salt stress. *Scientia Horticulturae, 261*, 108930.

Ahmad, J., Awais, M., Rashid, U., Ngamcharussrivichai, C., Naqvi, S. R., & Ali, I. (2023). A systematic and critical review on effective utilization of artificial intelligence for bio-diesel production techniques. *Fuel, 338*, 127379.

Ahmad, M., Lee, S. S., Lim, J. E., Lee, S. E., Cho, J. S., Moon, D. H., ... Ok, Y. S. (2014). Speciation and phytoavailability of lead and antimony in a small arms range soil amended with mussel shell, cow bone and biochar: EXAFS spectroscopy and chemical extractions. *Chemosphere, 95*, 433–441.

Ahmad, M., Ok, Y. S., Rajapaksha, A. U., et al. (2016). Lead and copper immobilization in a shooting range soil using soybean stover- and pine needle-derived biochars: Chemical, microbial and spectroscopic assessments. *Journal of Hazardous Materials, 301*, 179–186.

Ahmad, M., Rajapaksha, A. U., Lim, J. E., et al. (2014). Biochar as a sorbent for contaminant management in soil and water: A review. *Chemosphere, 99*, 19–33.

Ahmad, S., Khushnood, R. A., Jagdale, P., et al. (2015). High performance self-consolidating cementitious composites by using micro carbonized bamboo particles. *Materials & Design, 76*, 223–229.

Akhtar, A., & Sarmah, A. K. (2018). Novel biochar-concrete composites: Manufacturing, characterization and evaluation of the mechanical properties. *The Science of the Total Environment, 616*, 408–416.

Ameloot, N., Maenhout, P., De Neve, S., & Sleutel, S. (2016). Biochar-induced N_2O emission reductions after field incorporation in a loam soil. *Geoderma, 267*, 10–16.

Amonette, J. E., & Joseph, S. (2009). Characteristics of biochar: Microchemical properties. *Biochar for Environmental Management: Science and Technology, 33*..

Awad, Y. M., Lee, S. S., Kim, K. H., Ok, Y. S., & Kuzyakov, Y. (2018). Carbon and nitrogen mineralization and enzyme activities in soil aggregate-size classes: Effects of biochar, oyster shells, and polymers. *Chemosphere, 198*, 40–48.

Baldock, J. A., & Smernik, R. J. (2002). Chemical composition and bioavailability of thermally altered *Pinus resinosa* (Red pine) wood. *Organic Geochemistry, 33*, 1093–1109.

Bandara, T., Franks, A., Xu, J., et al. (2020a). Chemical and biological immobilization mechanisms of potentially toxic elements in biochar-amended soils. *Critical Review Environment Science & Technology., 50*, 903–978, 2020.

Bandara, T., Xu, J., Potter, I. D., et al. (2020b). Mechanisms for the removal of Cd(II) and Cu(II) from aqueous solution and mine water by biochars derived from agricultural wastes. *Chemosphere, 254*, 126745.

Bera, T., Collins, H. P., Alva, A. K., Purakayastha, T. J., & Patra, A. K. (2016). Biochar and manure effluent effects on soil biochemical properties under corn production. *Applied Soil Ecology, 107*, 360–367.

Biederman, L. A., & Harpole, W. S. (2013). Biochar and its effects on plant productivity and nutrient cycling: A meta-analysis. *Global Change Biology: Bioenergy, 5*, 202–214.

Blagodatskaya, E., & Kuzyakov, Y. (2008). Mechanisms of real and apparent priming effects and their dependence on soil microbial biomass and community structure: critical review. *Biology and Fertility of Soils, 45*, 115–131.

Borchard, N., Schirrmann, M., Cayuela, M. L., Kammann, C., Wrage-Mönnig, N., Estavillo, J. M., & Novak, J. (2019). Biochar, soil and land-use interactions that reduce nitrate leaching and N2O emissions: A meta-analysis. *Science of the Total Environment, 651*, 2354–2364.

Bourke, J. (2007). *Preparation and properties of natural, demineralized, pure, and doped carbons from biomass; Model of the chemical structure of carbonized charcoal* (Doctoral dissertation). The University of Waikato.

Bourke, J., Manley-Harris, M., Fushimi, C., Dowaki, K., Nonoura, T., & Antal, M. J. (2007). Do all carbonized charcoals have the same chemical structure? 2. A model of the chemical structure of carbonized charcoal. *Industrial and Engineering Chemistry Research, 46*, 5954–5967.

Brennan, J. K., Bandosz, T. J., Thomson, K. T., & Gubbins, K. E. (2001). Water in porous carbons. *Colloids and surfaces A: Physicochemical and engineering aspects, 187*, 539–568.

Bridgwater, A. V., & Boocock, D. G. B. (Eds.), (2006). *Science in thermal and chemical biomass conversion*. Thatcham, UK: CPL press.

Brown, R. A., Kercher, A. K., Nguyen, T. H., Nagle, D. C., & Ball, W. P. (2006). Production and characterization of synthetic wood chars for use as surrogates for natural sorbents. *Organic Geochemistry, 37*, 321–333.

Busch, D., & Glaser, B. (2015). Stability of co-composted hydrochar and biochar under field conditions in a temperate soil. *Soil Use and Management, 31*, 251–258.

Byrne, C. E., & Nagle, D. C. (1997). Carbonization of wood for advanced materials applications. *Carbon, 35*, 259–266.

Cabeza, I., Waterhouse, T., Sohi, S., & Rooke, J. A. (2018). Effect of biochar produced from different biomass sources and at different process temperatures on methane production and ammonia concentrations in vitro. *Animal Feed Science and Technology, 237*, 1–7.

Calvelo Pereira, R., Muetzel, S., Camps Arbestain, M., Bishop, P., Hina, K., & Hedley, M. (2014). Assessment of the influence of biochar on rumen and silage fermentation: A laboratory-scale experiment. *Animal Feed Science and Technology, 196*, 22–31.

Cetin, E., Moghtaderi, B., Gupta, R., & Wall, T. F. (2004). Influence of pyrolysis conditions on the structure and gasification reactivity of biomass chars. *Fuel, 83*, 2139–2150.

Cha, J. S., Kim, Y. M., Choi, Y. J., Rhee, G. H., Song, H., Jeon, B. H., ... Park, Y. K. (2022). Mitigation of hazardous toluene via ozone-catalyzed oxidation using MnOx/Sawdust biochar catalyst. *Environmental Pollution, 312*, 119920.

Cha, J. S., Park, S. H., Jung, S. C., Ryu, C., Jeon, J. K., Shin, M. C., & Park, Y. K. (2016). Production and utilization of biochar: A review. *Journal of Industrial and Engineering Chemistry, 40*, 1–15.

Chan, K.Y., Van Zwieten, L., Meszaros, I., Downie, A., & Joseph, S. (2007). Assessing the agronomic values of contrasting char materials on an Australian hard setting soil. In *Paper presented in International Agrichar Initiative (IAI) 2007 Conference*, Terrigal. New South Wales, Australia, 27 April–2 May 2007.

Chen, Y., Camps-Arbestain, M., Shen, Q., Singh, B., & Cayuela, M. L. (2018). The long-term role of organic amendments in building soil nutrient fertility: a meta-analysis and review. *Nutrient Cycling in Agroecosystems, 111*, 103–125.

Chen, Y., Liu, Y., Li, Y., Wu, Y., Chen, Y., Zeng, G., & Li, H. (2017). Influence of biochar on heavy metals and microbial community during composting of river sediment with agricultural wastes. *Bioresource Technology, 243*, 347–355.

Cheng, C. H., Lehmann, J., & Engelhard, M. H. (2008). Natural oxidation of black carbon in soils: Changes in molecular form and surface charge along a climosequence. *Geochimica et Cosmochimica Acta, 72*(6), 1598–1610.

Chew, J., Zhu, L., Nielsen, S., Graber, E., Mitchell, D. R., Horvat, J., & Fan, X. (2020). Biochar-based fertilizer: supercharging root membrane potential and biomass yield of rice. *Science of the Total Environment, 713*, 136431.

Clough, T. J., Condron, L. M., Kammann, C., & Müller, C. (2013). A review of biochar and soil nitrogen dynamics. *Agronomy, 3*, 275–293.

Cornelissen, G., Nurida, N. L., Hale, S. E., Martinsen, V., Silvani, L., & Mulder, J. (2018). Fading positive effect of biochar on crop yield and soil acidity during five growth seasons in an Indonesian Ultisol. *Science of the Total Environment, 634*, 561–568.

Dai, Y., Zhang, N., Xing, C., et al. (2019). The adsorption, regeneration and engineering applications of biochar for removal organic pollutants: A review. *Chemosphere, 223*, 12–27.

Dai, Y., Zheng, H., Jiang, Z., & Xing, B. (2020). Combined effects of biochar properties and soil conditions on plant growth: A meta-analysis. *Science of the Total Environment, 713*, 136635.

Demisie, W., Liu, Z., & Zhang, M. (2014). Effect of biochar on carbon fractions and enzyme activity of red soil. *Catena, 121*, 214–221.

Ding, Y., Liu, Y., Liu, S., Li, Z., Tan, X., Huang, X., & Zheng, B. (2016). Biochar to improve soil fertility. A review. *Agronomy for sustainable development, 36*(2), 1–18.

Ding, Z., Wan, Y., Hu, X., Wang, S., Zimmerman, A. R., & Gao, B. (2016). Sorption of lead and methylene blue onto hickory biochars from different pyrolysis temperatures: Importance of physicochemical properties. *Industrial & Engineering Chemistry Research, 37*, 261e267.

Dixit, A., Gupta, S., Dai Pang, S., et al. (2019). Waste valorization using biochar for cement replacement and internal curing in ultra-high performance concrete. *Journ Clean Prod., 238*, 117876.

Dong, X., Ma, L. Q., Gress, J., et al. (2014). Enhanced Cr(VI) reduction and As(III) oxidation in ice phase: Important role of dissolved organic matter from biochar. *Journal of Hazardous Materials, 267*, 62–70.

Downie, A, Crosky, A, & Munroe, P (2009). Chapter 2. Physical properties of biochar. In J Lehmann, & S Joseph (Eds.), *Biochar for Environmental Management: Science and Technology*. London: Earthscan.

Edeh, I. G., Mašek, O., & Buss, W. (2020). A meta-analysis on biochar's effects on soil water properties—New insights and future research challenges. *Science of the Total Environment, 714*, 136857.

El-Naggar, A., El-Naggar, A. H., Shaheen, S. M., Sarkar, B., Chang, S. X., Tsang, D. C., & Ok, Y. S. (2019). Biochar composition-dependent impacts on soil nutrient release, carbon mineralization, and potential environmental risk: A review. *Journal of Environmental Management, 241*, 458–467.

El-Naggar, A., Lee, S. S., Awad, Y. M., Yang, X., Ryu, C., Rizwan, M., ... Ok, Y. S. (2018). Influence of soil properties and feedstocks on biochar potential for carbon mineralization and improvement of infertile soils. *Geoderma, 332*, 100–108.

El-Naggar, A., Lee, S. S., Rinklebe, J., Farooq, M., Song, H., Sarmah, A. K., ... Ok, Y. S. (2019). Biochar application to low fertility soils: A review of current status, and future prospects. *Geoderma., 337*, 536–554.

European Biochar Foundation (EBC). (2018). Guidelines for EBC-feed certification. https://www.european-biochar.org/en (Accessed 14.08.23).

FAO. (2008). Preliminary map of land degradation, Food and Agriculture Organization of the United Nations, 2008.

Feng, J. W., Zheng, S., & Maciel, G. E. (2004). EPR investigations of the effects of inorganic additives on the charring and char/air interactions of cellulose. *Energy & fuels, 18*(4), 1049–1065.

Feng, Y., Xu, Y., Yu, Y., Xie, Z., & Lin, X. (2012). Mechanisms of biochar decreasing methane emission from Chinese paddy soils. *Soil Biology & Biochemistry, 46*, 80–88.

Foo, K. Y., & Hameed, B. H. (2011). Preparation and characterization of activated carbon from pistachio nut shells via microwave-induced chemical activation. *Biomass Bioenergy, 35*, 3257–3261.

Fries, G., Marrow, G., Jr, Gordon, C., et al. (1970). Effect of actilvated carbon on elimination of organochlorine pestilcides from rats and cows. *Journal of Dairy Science, 53*, 1632–1637.

Gao, J., Wang, W., Rondinone, A. J., He, F., & Liang, L. (2015). Degradation of trichloroethene with a novel ball milled Fe–C nanocomposite. *The Journal of Hazardous Materials, 300*, 443–450.

Gerlach, A., & Schmidt, H. P. (2012). Pflanzenkohle in der Rinderhaltung. *Ithaka Journal, 1*, 80–84.

Gerlach, H., Gerlach, A., Schrödl, W., et al. (2014). Oral appliIcation of charcoal and humic acids to dairy cows influences *Clostridium botulinum* blood serum antibody level and glyphosate excretion in urine. *Journal of Clinical Toxicology, 4*(2), 1000186.

Glaser, B., Lehmann, J., & Zech, W. (2002). Ameliorating physical and chemical properties of highly weathered soils in the tropics with charcoal—A review. *Biology and Fertility of Soils, 35*(4), 219–230.

Graetz, R. D., & Skjemstad, J. O. (2003). The charcoal sink of biomass burning on the Australian continent. *CSIRO Atmospheric Research, 64*, 1–61.

Gupta, S., & Kua, H. W. (2018). Effect of water entrainment by pre-soaked biochar particles on strength and permeability of cement mortar. *Construction Building Materials., 159*, 107–125.

Gupta, S., & Kua, H. W. (2019). Carbonaceous micro-filler for cement: Effect of particle size and dosage of biochar on fresh and hardened properties of cement mortar. *The Science of the Total Environment, 662*, 952–962.

Gupta, S., Kua, H. W., & Cynthia, S. Y. T. (2017). Use of biocharcoated polypropylene fibers for carbon sequestration and physical improvement of mortar. *Cement Concrete Composites., 83*, 171–187.

Haider, G., Steffens, D., Moser, G., Müller, C., & Kammann, C. I. (2017). Biochar reduced nitrate leaching and improved soil moisture content without yield improvements in a four-year field study. *Agriculture, Ecosystems & Environment, 237*, 80–94.

Han, L., Ro, K. S., Wang, Y., Sun, K., Sun, H., Libra, J. A., & Xing, B. (2018). Oxidation resistance of biochars as a function of feedstock and pyrolysis condition. *The Science of the Total Environment, 616–617*, 335–344.

Hatrack, L. (2008). Forty-five percent of Indian agricultural land degraded; particulate air pollution spiking in all major cities. Available online at: http://www.zeenews.com/news554697.

Hu, X., Ding, Z., Zimmerman, A. R., et al. (2015). Batch and column sorption of arsenic onto iron-impregnated biochar synthesized through hydrolysis. *Water Research, 68*, 206–226.

Huff, M. D., & Lee, J. W. (2016). Biochar-surface oxygenation with hydrogen peroxide. *Journal of Environmental Management, 165*, 17–21.

Inyang, M., Gao, B., Pullammanappallil, P., Ding, W., & Zimmerman, A. R. (2010). Biochar from anaerobically digested sugarcane bagasse. *Bioresour Technology, 101*, 8868–8872.

Inyang, M., Gao, B., Yao, Y., Xue, Y., Zimmerman, A. R., Pullammanappallil, P., & Cao, X. (2012). Removal of heavy metals from aqueous solution by biochars derived from anaerobically digested biomass. *Bioresour Technology, 110*, 50–56.

Jatav, H. S., Singh, S. K., Jatav, S. S., Rajput, V. D., Parihar, M., Mahawer, S. K., & Singhal, R. K. (2020). Importance of biochar in agriculture and its consequence. *Applications of Biochar for Environmental Safety*, 109.

Jatav, H. S., Singh, S. K., Singh, Y., & Kumar, O. (2018).). Biochar and sewage sludge application increases yield and micronutrient uptake in rice (*Oryza sativa* L.). *Communications in Soil Science and Plant Analysis, 49*(13), 1617–1628.

Jeffery, S., Verheijen, F. G. A., Kammann, C., & Abalos, D. (2016). Biochar effects on methane emissions from soils: A meta-analysis. *Soil Biology & Biochemistry, 101*, 251–258.

Jiang, Z., Li, J., Jiang, D., et al. (2020). Removal of atrazine by biochar-supported zero-valent iron catalyzed persulfate oxidation: Reactivity, radical production and transformation pathway. *Environmental Research, 184*, 109260.

Jung, J.-M., Oh, J.-I., Baek, K., et al. (2018). Biodiesel production from waste cooking oil using biochar derived from chicken manure as a porous media and catalyst. *Energy Conversion Management., 165*, 628–633.

Jung, J.-M., Oh, J.-I., Park, Y.-K., et al. (2019). CO_2-mediated chicken manure biochar manipulation for biodiesel production. *Environmental Research, 171*, 348–355.

Jung, S., Park, Y.-K., & Kwon, E. E. (2019). Strategic use of biochar for CO_2 capture and sequestration. *Journal of CO_2 Utilization., 32*, 128–139.

Kammann, C., Ippolito, J., Hagemann, N., Borchard, N., Cayuela, M. L., Estavillo, J. M., ... Novak. (2017). Biochar as a tool to reduce the agricultural greenhouse-gas burden—knowns, unknowns and future research needs. *Journal of Environmental Engineering and Landscape Management, 25*(2), 114–139.

Kammann, C. I., Schmidt, H. P., Messerschmidt, N., Linsel, S., Steffens, D., Müller, C., & Joseph, S. (2015). Plant growth improvement mediated by nitrate capture in co-composted biochar. *Scientific Reports, 5*(1), 1–13.

Kazemi Shariat Panahi, H., Dehhaghi, M., Ok, Y. S., Nizami, A. S., Khoshnevisan, B., Mussatto, S. I., ... Lam, S. S. (2020). A comprehensive review of engineered 604 biochar: Production, characteristics, and environmental applications. *Journal of Cleaner Production, 605*, 122462.

Khalid, S., Shahid, M., Murtaza, B., et al. (2020). A critical review of different factors governing the fate of pesticides in soil under biochar application. *The Science of the Total Environment, 711*, 134645.

Khan, S., Reid, B. J., Li, G., & Zhu, Y. G. (2014). Application of biochar to soil reduces cancer risk via rice consumption: A case study in Miaoqian village, Longyan, China. *Environment International, 68*, 154–161.

Klinik, J., Samojeden, B., Grzybek, T., Suprun, W., Papp, H., & Gläser, R. (2011). Nitrogen promoted activated carbons as DeNOx catalysts. 2. The influence of water on the catalytic performance. *Catalysis Today, 176*(1), 303–308.

Kookana, R. S., Sarmah, A. K., Van Zwieten, L., et al. (2017). Biochar application to soil: Agronomic and environmental benefits and unintended consequences. *Advanced Agronomy, 2011*, 103–143.

Koutcheiko, S., Monreal, C. M., Kodama, H., McCracken, T., & Kotlyar, L. (2007). Preparation and characterization of activated carbon derived from the thermo-chemical conversion of chicken manure. *Bioresource Technology, 98*(13), 2459–2464.

Kumar, M., Xiong, X., Wan, Z., Sun, Y., Tsang, D. C., Gupta, J., ... Ok, Y. S. (2020). Ball milling as a mechanochemical technology for fabrication of novel biochar nanomaterials. *Bioresour Technology, 1*, 123613.

Kumar, R., Sharma, P., Gupta, R. K., Kumar, S., Sharma, M. M. M., Singh, S., & Pradhan, G. (2020). Earthworms for eco-friendly resource efficient agriculture. In Kumar, et al. (Eds.), *Resources use efficiency in agriculture* (pp. 47–84). Singapore: Springer Nature. Available from 10.1007/978-981-15-6953-1_2.

Kumar, S., Meena, R. S., Datta, R., Verma, S. K., Yadav, G. S., Pradhan, G., ... Mashuk, H. A. (2020). Legumes for carbon and nitrogen cycling: An organic approach. carbon and nitrogen cycling in soil. In R. Datta, et al. (Eds.), *Carbon and nitrogen cycling in soil* (pp. 337–375). Singapore: Springer Nature. Available from https://doi.org/10.1007/978-981-13-7264-3_10.337-375.

Laird, D. A., Fleming, P., Davis, D. D., Horton, R., Wang, B., & Karlen, D. L. (2010). Impact of biochar amendments on the quality of a typical Midwestern agricultural soil. *Geoderma, 158*(3–4), 443–449.

Lal, R. (2018). Digging deeper: A holistic perspective of factors affecting soil organic carbon sequestration in agroecosystems. *Global Change Biology, 24*, 3285–3301.

Lee, J., Kim, K.-H., & Kwon, E. E. (2017). Biochar as a catalyst. *Renewable Sustainable Energy Review, 77*, 70–79.

Lehmann, J. (2007). Bio-energy in the black. *Frontiers in Ecology and the Environment, 5*(7), 381–387.

Lehmann, J., Gaunt, J., & Rondon, M. (2006). Bio-char sequestration in terrestrial ecosystems—A review. *Mitigation and Adaptation Strategies for Global Change, 11*, 403–427.

Leng, L., Xu, S., Liu, R., Yu, T., Zhuo, X., Leng, S., ... Huang, H. (2020). Nitrogen containing functional groups of biochar: An overview. *Bioresource Technology, 298*, 122286.

Li, L., Lai, C., Huang, F., et al. (2019). Degradation of naphthalene with magnetic bio-char activate hydrogen peroxide: Synergism of bio-char and Fe-Mn binary oxides. *Water Research, 160*, 238–248.

Li, M., Zheng, Y., Chen, Y., & Zhu, X. (2014). Biodiesel production from waste cooking oil using a heterogeneous catalyst from pyrolyzed rice husk. *Bioresource Technology, 154*, 345–348.

Li, Z., Song, Z., Singh, B. P., & Wang, H. (2019). The impact of crop residue biochars on silicon and nutrient cycles in croplands. *Science of the Total Environment, 659*, 673–680.

Lin, X., Liang, Y., Lu, Z., Lou, H., Zhang, X., Liu, S., ... Wu, D. (2017). Mechanochemistry: A green, activation-free and top-down strategy to high-surface-area carbon materials. *ACS Sustainable Chemistry & Engineering, 5*, 8535–8540.

Liu, C., Wang, H., Tang, X., Guan, Z., Reid, B. J., Rajapaksha, A. U., ... Sun, H. (2016). Biochar increased water holding capacity but accelerated organic carbon leaching from a sloping farmland soil in China. *Environmental Science and Pollution Research, 23*, 995–1006.

Mandal, S., Thangarajan, R., Bolan, N. S., Sarkar, B., Khan, N., Ok, Y. S., & Naidu, R. (2016). Biochar-induced concomitant decrease in ammonia volatilization and increase in nitrogen use efficiency by wheat. *Chemosphere, 142*, 120–127.

Maraseni, T.N. (2007). *Re-evaluating land use choices to incorporate carbon values: A case study in the South Burnett region of Queensland, Australia* (Doctoral dissertation). University of Southern Queensland.

Meena, R. S., Pradhan, G., Kumar, S., & Lal, R. (2023). Using industrial wastes for rice-wheat cropping and food-energy-carbon-water-economic nexus to the sustainable food system. *Renewable and Sustainable Energy Reviews, 187*, 113756. Available from https://doi.org/10.1016/j.rser.2023.113756.

Menéndez, J. A., Domínguez, A., Inguanzo, M., & Pis, J. J. (2004). Microwave pyrolysis of sewage sludge: Analysis of the gas fraction. *The Journal of Analytical and Applied Pyrolysis, 71*, 657–667.

Menéndez, J. A., Inguanzo, M., & Pis, J. J. (2002). Microwave-induced pyrolysis of sewage sludge. *Water Research, 36*, 3261–3264.

Mia, S., Singh, B., & Dijkstra, F. A. (2019). Chemically oxidized biochar increases ammonium-15 N recovery and phosphorus uptake in a grassland. *Biology and Fertility of Soils, 55*(6), 577–588.

Mitchell, P. J., Simpson, A. J., Soong, R., & Simpson, M. J. (2015). Shifts in microbial community and water-extractable organic matter composition with biochar amendment in a temperate forest soil. *Soil Biology & Biochemistry, 81*, 244–254.

Mohamed, B. A., Ellis, N., Kim, C. S., Bi, X., & Emam, A. E. (2016). Engineered biochar from microwave-assisted catalytic pyrolysis of switch grass for increasing water-holding capacity and fertility of sandy soil. *The Science of the Total Environment, 566*, 387–397.

Mohan, D., Abhishek, K., Sarswat, A., Patel, M., Singh, P., & Pittman, C. U. (2018). Biochar production and applications in soil fertility and carbon sequestration a sustainable solution to crop-residue burning in India. *RSC Advances, 8*(1), 508–520.

Namasivayam, C., & Sangeetha, D. (2004). Equilibrium and kinetic studies of adsorption of phosphate onto $ZnCl_2$ activated coir pith carbon. *Journal of Colloid and Interface Science, 280*(2), 359–365.

Novak, J. M., Lima, I., Xing, B., Gaskin, J. W., Steiner, C., Das, K. C., ... Schomberg, H. (2009). Characterization of designer biochar produced at different temperatures and their effects on a loamy sand. *Annals of Environmental Science and Toxicology*.

O'Toole, A., Andersson, D., Gerlach, A., Glaser, B., Kammann, C. I., Kern, J., ... Srocke Franziska Stenrød, M. (2016). Current and future applications for biochar. In S. Shackley, G. Ruysschaert, K. Zwart, & B. Glaser (Eds.), *Biochar in European soils and agriculture: Science and practice* (pp. 253–280). Abington: Taylor & Francis Group.

Palansooriya, K. N., Ok, Y. S., Awad, Y. M., Lee, S. S., Sung, J. K., Koutsospyros, A., & Moon, D. H. (2019). Impacts of biochar application on upland agriculture: A review. *Journal of Environmental Management, 234*, 52–64.

Palansooriya, K. N., Ok, Y. S., Awad, Y. M., Lee, S. S., Sung, J. K., Koutsospyros, A., & Moon, D. H. (2019). Impacts of biochar application on upland agriculture: A review. *Journal of Environmental Management, 234*, 52–64.

Paris, O., Zollfrank, C., & Zickler, G. A. (2005). Decomposition and carbonisation of wood biopolymers – A microstructural study of softwood pyrolysis. *Carbon, 43*, 53–66.

Pellera, F.-M., & Gidarakos, E. (2015). Effect of dried olive pomace-derived biochar on the mobility of cadmium and nickel in soil. *Journal of Environmentl Chemistry & Engineering., 3*, 1163–1176.

Pradhan, G., & Meena, R. S. (2022). Diversity in the rice–wheat system with genetically modified zinc and iron-enriched varieties to achieve nutritional security. *Sustainability, 14*, 9334. Available from https://doi.org/10.3390/su14159334.

Pradhan, G., & Meena, R. S. (2023a). Interaction impacts of biocompost on nutrient dynamics and relations with soil biota, carbon fractions index, societal value of CO_2 equivalent, and ecosystem services in the wheat-rice farming. *Chemosphere*, 139695. Available from https://doi.org/10.1016/j.chemosphere.2023.139695.

Pradhan, G., & Meena, R. S. (2023b). Utilizing waste compost to improve the atmospheric CO_2 capturing in the rice-wheat cropping system and energy-cum-carbon credit audit for a circular economy. *Science of the Total Environment*, 164572. Available from https://doi.org/10.1016/j.scitotenv.2023.164572.

Pradhan, G., Meena, R. S., Kumar, S., Jhariya, M. K., Khan, N., Shukla, U. N., ... Kumar, S. (2022). Legumes for eco-friendly weed management in an agroecosystem. In R. S. Meena, & S. Kumar (Eds.), *Advances in legumes for sustainable intensification* (pp. 133–154). . Available from https://doi.org/10.1016/B978-0-323-85797-0.00033-1.

Qian, K., Kumar, A., Zhang, H., Bellmer, D., & Huhnke, R. (2015). Recent advances in utilization of biochar. *Renewable and Sustainable Energy Reviews, 42*, 1055–1064.

Qin, X., Li, Y., Wang, H., Liu, C., Li, J., Wan, Y., ... Liao, Y. (2016). Long-term effect of biochar application on yield-scaled greenhouse gas emissions in a rice paddy cropping system: A four-year case study in south China. *The Science of the Total Environment, 569–570*, 1390–1401.

Quilliam, R. S., Glanville, H. C., Wade, S. C., & Jones, D. L. (2013). Life in the 'charosphere'—Does biochar in agricultural soil provide a significant habitat for microorganisms? *Soil Biology & Biochemistry, 65*, 287–293.

Raboin, L. M., Razafimahafaly, A. H. D., Rabenjarisoa, M. B., Rabary, B., Dusserre, J., & Becquer, T. (2016). Improving the fertility of tropical acid soils: Liming versus biochar application? A long term comparison in the highlands of Madagascar. *Field Crops Research, 199*, 99–108.

Raj, A., Jhariya, M. K., Banerjee, A., Meena, R. S., Nema, S., Khan, N., ... Pradhan, G. (2021). Agroforestry a model for ecological sustainability. In M. K. Jhariya, R. S. Meena, A. Raj, & S. N. Meena (Eds.), *Natural resources conservation and advances for sustainability* (pp. 289–308). Elsevier. Available from https://doi.org/10.1016/B978-0-12-822976-7.00002-8.

Rajapaksha, A. U., Chen, S. S., Tsang, D. C., Zhang, M., Vithanage, M., Mandal, S., & Ok, Y. S. (2016). Engineered/designer biochar for contaminant removal/immobilization from soil and water: Potential and implication of biochar modification. *Chemosphere, 148*, 276–291.

Rechberger, M. V., Kloss, S., Wang, S.-L., et al. (2019). Enhanced Cu and Cd sorption after soil aging of woodchip-derived biochar: What were the driving factors? *Chemosphere, 216*, 463–471.

Rouquerol, J., Rouquerol, F., Llewellyn, P., Maurin, G., & Sing, K. S. (2013). *Adsorption by powders and porous solids: Principles, methodology and applications*. Academic press.

Saleem, A. M., Ribeiro, G. O., Yang, W. Z., Ran, T., Beauchemin, K. A., McGeough, E. J., ... McAllister, T. A. (2018). Effect of engineered biocarbon on rumen fermentation, microbial protein synthesis, and methane production in an artificial rumen (RUSITEC) fed a high forage diet1. *Journal of Animal Science*, 3121–3130.

Schirrmann, U. (1984). *Aktivkohle und ihre Wirkung auf bakterien und deren toxine im gastrointestinaltrakt*. Technische-Universität München.

Schulz, H., Dunst, G., & Glaser, B. (2013). Positive effects of composted biochar on plant growth and soil fertility. *Agronomy for Sustainable Development, 33*(4), 817–827.

Shen, B., Li, G., Wang, F., Wang, Y., He, C., Zhang, M., & Singh, S. (2015). Elemental mercury removal by the modified bio-char from medicinal residues. *Chemical Engineering Journal, 272*, 28–37.

Shuji, Y., Tanaka, S., & Ohata, M. (2007). Preservation of woods in forest acid soil by addition of biomass charcoal. In: *Proceedings of the International Agrichar Initiative (IAI) 2007 Conference* (Vol. 27).

Smith, C. R., Hatcher, P. G., Kumar, S., & Lee, J. W. (2016). Investigation into the sources of biochar water soluble organic compounds and their potential toxicity on aquatic microorganisms. *ACS Sustainable Chemistry & Engineering, 4*(5), 2550–2558.

Sohi, S. P., Krull, E., Lopez-Capel, E., & Bol, R. (2010). A review of biochar and its use and function. *Advances in Agronomy, 105*, 47–82.

Solaiman, Z. M., & Anawar, H. M. (2015). Application of biochars for soil constraints: challenges and solutions. *Pedosphere, 25*(5), 631–638.

Sollins, P., Homann, P., & Caldwell, B. A. (1996). Stabilization and destabilization of soil organic matter: Mechanisms and controls. *Geoderma, 74*, 65–105.

Steinegger, P., & Menzi, M. (1955). Versuche über die Wirkung von Vitamin-Zusätzen nach Verfütterung von Adsorbentien an mastpoulets. *Verlag Nicht Ermittelbar, 18*, 165–176.

Sui, Y., Gao, J., Liu, C., Zhang, W., Lan, Y., Li, S., ... Tang, L. (2016). Interactive effects of straw-derived biochar and N fertilization on soil C storage and rice productivity in rice paddies of Northeast China. *The Science of the Total Environment, 544*, 203–210.

Swiatkowski, A., Pakula, M., Biniak, S., & Walczyk, M. (2004). Influence of the surface chemistry of modified activated carbon on its electrochemical behaviour in the presence of lead (II) ions. *Carbon, 42*(15), 3057–3069.

Sánchez-García, M., Sánchez-Monedero, M. A., & Cayuela, M. L. (2020). N_2O emissions during *Brassica oleracea* cultivation: Interaction of biochar with mineral and organic fertilization. *European Journal of Agronomy, 115*, 126021.

Tong, X.-j, Li, J.-y, Yuan, J.-h, et al. (2011). Adsorption of Cu(II) by biochars generated from three crop straws. *Chemical Engineering Journl, 172*, 828–834.

Toth, J. D., & Dou, Z. (2016). Use and impact of biochar and charcoal in animal production systems. In M. Guo, Z. He, & S. M. Uchimiya (Eds.), *Agricultural and environmental applications of biochar: Advances and barriers* (pp. 199–224). Madison: Soil Science Society of America, Inc.

Trompowsky, P. M., de Melo Benites, V., Madari, B. E., Pimenta, A. S., Hockaday, W. C., & Hatcher, P. G. (2005). Characterization of humic like substances obtained by chemical oxidation of eucalyptus charcoal. *Organic Geochemistry, 36*(11), 1480–1489.

Van Zwieten, L. (2018). The long-term role of organic amendments in addressing soil constraints to production. *Nutritional Cycle Agroecosystem, 111*, 99–102.

Volkmann, A. (1935). *Behandlungsversuche der Kaninchenlbzw. Katzencoccidiose mit Viscojod and Carbo medlicinalis*. Edited by Edelmann. Leipzig; 1.

Wang, B., Lee, X., Theng, B. K., Zhang, L., Cheng, H., Cheng, J., & Lyu, W. (2019). Biochar addition can reduce NOx gas emissions from a calcareous soil. *Environmental Pollutants and Bioavailability, 31*(1), 38–48.

Wang, D. Y., Fonte, S. J., Parikh, S. J., Six, J., & Scow, K. M. (2017). Biochar additions can enhance soil structure and the physical stabilization of C in aggregates. *Geoderma, 303*, 110–117.

Wang, H., Gao, B., Wang, S., Fang, J., Xue, Y., & Yang, K. (2015). Removal of Pb (II), Cu (II), and Cd (II) from aqueous solutions by biochar derived from $KMnO_4$ treated hickory wood. *Bioresource Technology, 197*, 356–362.

Wang, L., Chen, L., Tsang, D. C., et al. (2019). The roles of biochar as green admixture for sediment-based construction products. *Cement Concrete Composites., 104*, 103348.

Wang, S., Gao, B., Zimmerman, A. R., Li, Y., Ma, L., Harris, W. G., & Migliaccio, K. W. (2015). Removal of arsenic by magnetic biochar prepared from pinewood and natural hematite. *Bioresource Technology, 175*, 391–395.

Wardle, D. A., Nilsson, M. C., & Zackrisson, O. (2008). Fire-derived charcoal causes loss of forest humus. *Science (New York, N.Y.), 320*(5876), 629, -629.

Weber, K., & Quicker, P. (2018). Properties of biochar. *Fuel, 217*, 240–261.

Wei, C., Lu, H. H., Wang, Y. Y., Liu, Y. X., He, L. L., Yang, X. Y., & Yang, S. M. (2021). Adsorption effectiveness of ammonium nitrogen by iron-modified rice husk biochars. *Journal of Plant Nutrition and Fertilizers, 27*(4), 595–609.

Weng, Z. H., Van Zwieten, L., Singh, B. P., et al. (2018a). The accumulation of rhizodeposits in organo-mineral fractions promoted biochar-induced negative priming of native soil organic carbon in ferralsol. *Soil Biology & Biochemistry, 118*, 91–96.

Weng, Z.H., Van Zwieten, L., Singh, B.P., et al. (2018b) Biochar built soil carbon over a decade by stabilizing rhizodeposits. *Nature Climate Change*.7:371–376.

White, K. E., Brennan, E. B., Cavigelli, M. A., et al. (2020). Winter cover crops increase readily decomposable soil carbon, but compost drives total soil carbon during eight years of intensive, organic vegetable production in california. *PLoS One, 15*, e0228677.

Winsley, P. (2007). Biochar and bioenergy production for climate change mitigation. *New Zealand Science Review, 64*(1), 5–10.

Wu, B., Cheng, G., Jiao, K., Shi, W., Wang, C., & Xu, H. (2016). Mycoextraction by Clitocybe maxima combined with metal immobilization by biochar and activated carbon in an aged soil. *Science of the Total Environment, 562*, 732–739.

Xing, Y., Wang, J., Shaheen, S. M., et al. (2019). Mitigation of mercury accumulation in rice using rice hull-derived biochar as soil amendment: A field investigation. *Journal of Hazardous Materials, 388*, 121747.

Yao, Y., Gao, B., Chen, J., & Yang, L. (2013). Engineered biochar reclaiming phosphate from aqueous solutions: Mechanisms and potential application as a slow-release fertilizer. *Environmental Science & Technology, 47*(15), 8700–8708.

Zeng, X. Y., Wang, Y., Li, R. X., Cao, H. L., Li, Y. F., & Lü, J. (2022). Impacts of temperatures and phosphoric-acid modification to the physicochemical properties of biochar for excellent sulfadiazine adsorption. *Biochar, 4*(1), 14.

Zhao, L., Cao, X., Mašek, O., & Zimmerman, A. (2013). Heterogeneity of biochar properties as a function of feedstock sources and production temperatures. *Journal of Hazardous Materials, 256*, 1–9.

Zheng, H., Wang, Z., Deng, X., Herbert, S., & Xing, B. (2013). Impacts of adding biochar on nitrogen retention and bioavailability in agricultural soil. *Geoderma, 206*, 32–39.

Index

Note: Page numbers followed by "*f*" and "*t*" refer to figures and tables, respectively.

A

ABA. *See* Abscisic acid (ABA)
Abelmoschus esculentus. *See* Okra (*Abelmoschus esculentus*)
Abiotic stress(es), 227–228, 255
  biochar-mediated suppression of, 255–259
    improving drought tolerance, 255–258
    improving heavy metals tolerance, 259
    improving salt tolerance, 258–259
  resilience
    impact of biochar for sustainable agriculture, 229*f*
    drought, 229–232
    flood, 233–234
    future perspective, 242
    heat stress, 238–240
    heavy metals, 240–242
    salinization, 235–238
  use of biochar amendment for abiotic stress mitigation, 233*f*
Ablative reactors, 31–32
  vortex ablative bed reactor, 31*f*
Abscisic acid (ABA), 258
AC. *See* Ash content (AC)
Acacia, 325
Acid treatment, 535
Acid-treated biochar, 459–461
Acidic environment, 98–99
Acidic soil(s), 122–123, 188–190, 192–193, 303
  biochar
    interactions in, 188–195, 189*t*
    properties, 187
    and soil carbon sequestration, 195–196, 196*t*
  conditions, 438
  extent of, 188
  future challenges and perspective, 197–198
Acidobacteria, 290–291
Actinobacteria, 287–288, 290–291
Actinomycetes, 292–293
Actual density, 85–86
AD. *See* Anaerobic digestion (AD)
Adhesive briquetting, 517
Adsorption, 214, 339
AEC. *See* Anion exchange capacity (AEC)

Aerobic conversion, 193–194
Aging process, 152
Agricultural arable land, uses and limitations of biochar in, 175–177
Agricultural biochar, 140
Agricultural CO_2 emissions, 394–395
Agricultural management, 117–118
Agricultural manure, 241–242
Agricultural plants, 168–169
Agricultural plastic films, 57–58
Agricultural practices, 195–196
  for C footprint management, 394–395
Agricultural production, 400
Agricultural residues, 7, 191–192
Agricultural sectors, 394–395
Agricultural soils, 422–423, 453
Agricultural systems, 137–138
Agricultural wastes, 57–58, 71, 228, 361–362
Agricultural-based industries, 62
Agriculture, 100–102, 168, 207–208, 389–390, 395–397, 415–416
  GHG
    emissions and role of agriculture's production system, 395–400
    removal potential of agriculture practices, 402*f*
  impact of heavy metal on, 240–241
  productivity, 162–163
    biochar's on soil properties and diverse applications in, 165*t*
    influence of biochar on, 164–168
  systems, 300–301
  water resources, 235–236
Agro-ecological hazards from problematic wastes, 61–63
  from agro-industrial products, 62
  from e-waste, 63
  from municipal solid waste, 62
  from plastic waste, 62–63
Agro-industrial waste, 59
Agroforestry, 423
Agronomic measures, different, 400
  agronomic practices and input details, 400*t*
Air pollution, 3

Airborne pollutants, 64
Aliphatic groups, 193
Alkali treatment, 535
Alkali-treated biochar, 461–462
Alkaline, 188–190
   biochar, 210
   soils, 172–173
Aluminum chloride ($AlCl_3$), 536
Aluminum toxicity alleviation, 193
   mechanism of biochar for ameliorating acidic soil, 194f
Amending biochar, 251
Amending growth substrate, 255
Amending maize straw biochar, 257–258
AMF. See Arbuscular mycorrhizal fungi (AMF)
Ammonia (NH_3), 269–270, 316
Ammonia-oxidizing bacteria (AOB), 193–194
Ammonia-oxidizing genes (amoA), 194–195
Ammonium, 193–194, 273, 314, 317
amoA. See Ammonia-oxidizing genes (amoA)
Amorphous silicon, 104
Anaerobic digestion (AD), 63–64, 66–67
Anethum graveolens. See Dill (*Anethum graveolens*)
Animal(s)
   body waste, 241–242
   dung, 140
   genetic enhancement of, 399
Anion exchange capacity (AEC), 87, 536
Anthropogenic carbon dioxide emissions, 128–129
Anthropogenic origin, 1–2
Anthropogenic sources, 191–192
Antibiotics, 15–17, 148–150, 436, 443–444, 480–481
   removal of, 481–482
AOB. See Ammonia-oxidizing bacteria (AOB)
Aquaculture, 398
Aqueous system, biochar application for removal of ECs from, 438–442
Arachis hypogaea, 438–440
Arbuscular mycorrhizal fungi (AMF), 169, 236–238, 257–258
Archaea, 287
Argon (Ar), 67
Aromatic carbon framework, 169–170
Aromatic groups, 193
Aromatic polymer structure, 46
Aromatic ring, 104
Arrhenius equation, 47
Ascorbate peroxidase, 236–238
Ash content (AC), 190–191, 522
ATR-FTIR spectroscopy. See Attenuated total reflection FTIR spectroscopy (ATR-FTIR spectroscopy)
Attenuated total reflection FTIR spectroscopy (ATR-FTIR spectroscopy), 461–462

Auger reactors, 32–33
Auxins, 143–144

B

Bacillus
   B. cereus, 469
   B. subtilis, 469
Bacteria, 9, 285–286, 290–293
Bacterial populations, 290–291
Ball milling, 455–459, 468, 535
Bamboo-based biochar, 290–291
Bambusa vulgaris, 443–444
Barrel kilns, 513
Batch kilns, 513
Batch reactor, 67–68
BBFs. See Biochar-based fertilizers (BBFs)
BBNFs. See Biochar-based N fertilizers (BBNFs)
BC. See Biochar (BC)
BC-PM. See Biochar-pig manure (BC-PM)
BC-PW. See Biochar-pinewood (BC-PW)
Bean (*Phaseolus vulgaris*), 258–259
Beehive kilns, 513
Binder briquetting, 517
Binder-free biochar-based SRFs, 276
Binderless briquetting, 517
Binders, 276
Bio-energy production, 120
Bio-hydrogen, 71
Bio-oil, 33, 276
   encapsulated biochar-based SRFs, 276
Biochar (BC), 1, 6, 28, 50–52, 81–82, 98, 118–119, 152, 161, 188–190, 195–196, 208–209, 212, 214–215, 218, 228, 230–232, 234–236, 241–242, 249–250, 253, 258–259, 270–271, 275, 285–292, 295–301, 311–312, 336–339, 360–363, 394–395, 403–405, 414, 422, 435, 440–442, 453, 479–483, 511, 533, 539, 548
   activation process, 144–145
   amendment, 241–242
      use for abiotic stress mitigation, 233f
   application for removal of ECs, 8–18
      antibiotics, 443–444
      effect on crop yield and quality, 14–15
      effect on soil health parameters, 125–128
      from aqueous system, 438–442, 439t
      from soil system, 442–445
      heavy metals, 438–440, 443
      impact on carbon footprint under different landforms, 421–424
      nanoparticles to enhance surface chemistry of, 105–107
      organic pollutants, 440–442
      pesticides, 444–445

polycyclic aromatic hydrocarbons, 444
synthetic dyes, 443
BC-amended soils, 288–291
    BC-induced changes in microbial activity and diversity, 288–293, 289t
    biochar impact on soil microbial communities, 287–288
    mechanisms behind biochar impacts on soil microorganisms, 293–295
    microbial community in soils, 287, 288f
    microbial dynamics in, 287–295
BC-based carbon crediting, 424–427
    BC in enhancing C sequestration in different land use systems, 425t
BC-based carbon sequestration, 366
BC-based N-P-K SRFs, 273
BC-based nano-composite, 273–274
BC-based S fertilizers, 320–321
BC-mediated carbon cycle, 365f
as biofilter, 17
bioremediation of organic contaminants in water using, 481–489
as biostimulant, 17
C-content, 539
and carbon sequestration, 13
carbon storage potential of, 364–368
    long-term stability of, 365–366
as catalyst, 17
challenges and future prospects, 382
characterization, 81–82
and climate change, 415–421
    application for carbon stabilization and sequestration in soil, 415–418
    indirect impacts of BC on CO_2-equivalent emissions, 419–421
contribution of BC in economics of carbon farming, 129–130
control of greenhouse gas emissions, 15
crop residue availability in India, 362f
to cut down potential on greenhouse effects, 400–403
determinants of BC stability, 122–123
    BC's inherent chemical characteristics, 122
    interaction with minerals and SOM, 122–123
different salinity levels and effects on crops, 206t
drum, 515
and effect on contaminant bioremediation from soil and water
    acid-treated BC, 459–461
    alkali-treated BC, 461–462
    ball milling, 468
    biological modification of, 468–469
    carbon nanotube-coated BC, 466
    chemical modification of, 459–466
    clay-modified BC, 465–466
    layered double hydroxide–modified BC, 466
    metal oxide-loaded BC, 463–464
    microwave-activated BC, 467
    nonmetallic heteroatom–doped BC, 464–465
    oxidizing agent–treated BC, 462–463
    physical modification of, 466–468
    steam/CO_2-activated BC, 467
    surface active agent–treated BC, 465
effect
    as oil conditioner, 540–544
    available nutrients, 541–543
    co-application on soil health, 145
    crop response, 544
    on crop yield, 124t
    on microorganism in salt-effected soil, 216–217
    on mitigating drought stress, 230–232
    on physicochemical properties of saline soil, 213–215
    on plant facing salinity and heavy metal stress, 237t
    on plants facing drought and flood stress, 231t
    physicochemical properties, 541, 542t
    production process parameters on BC yield and quality, 6–7
    reducing soil erosion, 544
    soil biological properties, 543–544
    soil physical properties, 541
emerging applications, 17–18
engineering, 533–534
enhanced seed germination using, 261–262
for enriching soil biological health, 126–127
factors affecting
    ability of BC to fix carbon in soil, 415
    BC and its impact on soil and water management, 494–495
factors influencing effectiveness of, 151–152
    biochar dosage and application methods, 151
    climate and environmental conditions, 151
    interactions with microbial communities, 152
    long-term effects and aging, 152
    soil type, texture, and mineralogy, 151
    synergies with soil amendments, 152
as filler material for cement, 18
fortified BC, 363–364
and heavy metal contamination, 10–13
for horticultural rooting media, 250–254
    as alternative to peat media, 254
    as container substrate, 253
    as nursery substrate, 254
    as soil-less substrate, 251–252
impact on heat stress, 239–240

Biochar (BC) (*Continued*)
   impact on SMBC and carbon fractions, 367
      decomposition rate of, 367–368
   inaccessibility of BC to decomposers, 366
   interactions in acidic soils, 188–195, 189t
      and aluminum toxicity alleviation, 193
      and nutrient availability, 191–193
      and soil nitrification, 193–195
      and soil pH, 190–191
   as matrix to produce sustainable fertilizers, 312–314
      effects on crop productivity, 313–314
      enrichment techniques, 312–313
   mechanisms for organic pollutant removal using, 490–494
   mediated carbon and nutrient dynamics
      characteristics and production of, 163–164, 164f
      effects on different soil nutrient dynamics, 170–175
      effects on soil carbon dynamics and soil fertility, 168–169
      influence of BC on agriculture productivity, 164–168
      mechanisms of BC-mediated carbon sequestration, 169–170
      uses and limitations of BC in agricultural arable land, 175–177
   meta-analysis of BC's effects on biogenic GHG fluxes, 402–403
   mitigating greenhouse gases emission, 368–382, 369t
      and CO_2 sequestration in carbon pools, 368–380
      role in mitigation of CH_4 emission, 380
      role in mitigation of N_2O emission, 380–382
   for mitigation of climate change in different soils, 128–129
   modification techniques
      for improving surface chemistry of, 99–100
      scanning electron micrograph image of surface of, 102f
      in surface functional groups of, 98–99
      transmitting electron micrograph of surface of, 103f
   option for carbon farming, 119–122
   for overall soil health improvement, 127–128
   phenotype, 104
   physicochemical properties of BC relevant to methane mitigation, 348–350
      carbon content, 350
      electro-chemical properties, 350
      high porosity, 349
      large specific surface area, 349
      strong ion exchange capacity, 349–350
   pores, 318–319
   porous pores, 232
   potential advantages of BC and application in enhancing soil health and promoting plant growth, 142–143
   potential drawbacks and limitations, 152–153
   potential to mitigate methane emissions, 338–340
   poultry-manure compost, 236–238
   preparation methods, and properties, 138–139
   production, 71–72, 130, 276
      and application goals to upscale BC technology, 119f
      feedstock for, 7–8
      optimum temperature for, 539
      pyrolysis, 3–6
      technologies, 3–6
   properties and characterization of, 2, 82–92, 82f, 187
      high heating value, 88–89
      morphology and surface functionality, 89–92
      physiochemical characterization, 85–88
      proximate analysis, 82–84
      ultimate analysis and elemental composition, 84–85
   properties of BC affecting removal of emerging contaminants, 437–438
   property assessment of BC briquettes, 519–522
      chemical properties, 521–522
      physical properties, 519–521
      thermal properties, 522
   for protecting soil physical health, 126
   pyrolysis systems, 4
   qualities associated with microbial diversity, 139–140
   quality control and standardization, 153
   reclaiming health of disturbed soil, 127
   regulatory and policy issues regarding BC application, 153
   related favorable impacts of BC on climate change mitigation, 403–406
      alteration in plant growth rates after BC application, 405–406
      increasing soil C stock reduce soil surface albedo, 404–405
      negative priming effect of BC on mineralization of soil organic carbon, 403–404
   role
      in flood management, 234
      in management of soil salinity, 235–238
      in plant growth in salt-effected soil, 217–219
      of feedstock in enhancing surface chemistry of, 100–103
      on nutrient status of salt-affected soils, 209
      on soil properties, 209–219
   in salt-affected soils, 210–213
   samples, 437–438

serve as viable amendment for soil with high salinity, 207–209
and significant properties, 362–364
on soil
  biological properties, 10
  physical and chemical properties, 8–9, 11t
stimulates methanogenic and methanotrophic microbial shift, 345–348
surplus crop residue vis-a-vis biochar production, 361–362
synthesis, 228, 361–362
as tool for sequestering C in soil, 401–402
wastewater treatment, 15–17
yield, 50
for yield sustenance, 123–125
Biochar-based fertilizers (BBFs), 274–275, 312, 321
  slow-release mechanisms of, 314–323
    nitrogen, phosphorus, and sulfur dynamics, 314–321
    potassium release dynamics, 321–323
Biochar-based N fertilizers (BBNFs), 316–317
Biochar-based slow-release fertilizers, 270–271, 272t
  applications of, 277–278
    carbon sequestration, 278
    metal adsorption, 277–278
    nitrogen use efficiency, 277
    water retention, 277
  biochar-based SRFs, 270–273
biochar-C, 297
biochar-led methanogenesis and methanotrophy, 343–344
  biochar's impact on methanobacteria in crop systems, 345t
biochar-mediated enhanced root growth of horticultural crops, 262
biochar-mediated suppression
  of abiotic stresses, 255–259
  biotic stresses, 259–261
  effect on foliar and soil borne disease suppression, 260t
biochar-mineral fertilizer composites, 318–319
biochar-mineral urea composites, 275
biochar-soil interactions, 176–177
biochar–microbe interactions, 140–141, 294–295
future perspectives, 278–279
preparation methods of, 271–276, 273f
  encapsulation, 276
  granulation, 275–276
  impregnation, 273–274
  pyrolysis, 274–275
removal of pharmaceutically active compounds by, 485–486
Biochar-pig manure (BC-PM), 482–483

Biochar-pinewood (BC-PW), 482–483
Biochemical technology of waste management, 66–67
  anaerobic digestion, 66–67
  fermentation, 66
Biodiesel, 71
  production, 547–548
Bioethanol, 71
Biofuels, 71
  production, 71
Biogas, 71
Biogenic greenhouse gas fluxes, meta-analysis of biochar's effects on, 402–403
Biogeochemical processes, 286
Biomass, 50–52, 129, 186, 362–364, 424–427, 507–508
  biochar, 85–86
  biomass-based biochar, 285–286
  energy-based industrial research, 43
  feedstock, 7–8
  fragments, 32
  gasification with biochar capture, 514
  pyrolysis, 37, 129
  slurry, 49
Bioremediation, 241, 438
  biochar
    bioremediation of organic contaminants in water using, 481–489
    chemical and physical properties of, 480f
    mechanisms for organic pollutant removal using, 490–494
  challenges, opportunities, and future research directions, 495–496
  of contaminated aqueous and soil ecosystem, 546–547
  factors affecting biochar and impact on soil and water management, 494–495
Biotic stress, 251
  biochar-mediated suppression of, 259–261
Birchwood-based biochar, 232
Black carbon, 1, 161
Black liquor-based hydrogel, 277
Bomb calorimeter, 88–89
Boron (B), 210
Botrytis cinerea, 260–261
Brachiaria humidicola, 343–344
Bradyrhizobiaceae, 291
Bradyrhizobium, 232
Brassica
  B. napus, 438–440
  B. nigra, 240–241
Brassica oleracea var. *capitata*. See Cabbage (*Brassica oleracea* var. *capitata*)
Brinjal (*Solanum melongena* L), 258–259

Briquetting, 508–511, 516, 519–521
  as income source, 524–526
    advantages and disadvantages of pine biochar briquettes and pellets, 524–526
    economic analysis, 524
    production cost of pine biochar briquettes and pellets, 524
  of pine biochar, 516
Broad mite pest (*Polyphagotarsonemus latus*), 260–261
Brunauer–Emmett–Teller, 90
Bubbling fluidized bed, 29, 29f
Bulk density, 8–9, 519
  of slack biomass, 516

C

Cabbage (*Brassica oleracea* var. *capitata*), 262
Cadmium (Cd), 241, 454
  adsorption capacity, 100
Cadmium ion (Cd (II)), 270–271
Calcareous loam soil, 214
Calcite ($CaCO_3$), 207
Calcium (Ca), 139, 173–175, 206–207, 217
Calcium sulfate ($CaSO_4$), 540
Calculation process, 121
Calorific value, 522
Capsicum annuum L. *See* Pepper (*Capsicum annuum* L.)
Carbon (C), 1, 84–85, 186, 195–196, 295–296, 366, 413–414, 423–427, 435, 532
  biochar impact on carbon fractions, 367
  capture and utilization, 546
  carbon-based biowastes, 285–286
  carbon-based compounds, 216–217
  carbon-rich biochar, 215–216
  carbon-rich byproduct of biomass, 511
  carbon-rich material, 15, 138
  carbon-rich organic matter, 161
  content, 2, 118, 187, 350, 511–512
  cycling, 153, 286
  dynamics, 169
  farming, 117–119, 129
    biochar option for, 119–122
    carbon credit gain from biochar, 121t
    contribution of biochar in economics of, 129–130
    positive outcomes of carbon farming for sustainable C cycle, 120f
  functional groups, 90–91
  improvement of plant yield and improved C–sequestration in plant biomass, 420–421
  mineralization, 300, 365–366
  negative, 416–417
  pools, 1
    biochar and CO_2 sequestration in, 368–380
  sequestration, 13, 28, 43, 97–98, 121–122, 278, 299–300, 394–395, 399, 415–417, 544–545
  stabilization, 299, 301–302
    biochar application for carbon stabilization and sequestration in soil, 415–418
  storage potential of biochar, 364–368
Carbon dioxide (CO_2), 1, 15, 117–118, 162, 278, 285–286, 335–336, 338–339, 359–360, 368, 399, 413–414, 423, 455–459, 532
  CO_2-activated biochar, 467
  emissions, 125–126, 300–301, 364–365
    global CO_2 emissions, 392f
    status of, 393f
    trend and analysis of, 392–393
  sequestration in carbon pools, 368–380
Carbon dioxide equivalent (CO_2-e), 389–392, 424–427
  indirect impacts of biochar on, 419–421
    improvement of plant yield and improved C-sequestration in plant biomass, 420–421
    methane, 419
    nitrous oxide, 420
Carbon footprint (CF), 299–300, 390–395
  agricultural practices for CF management, 394–395
  biochar application impact on CF under different landforms, 421–424
    agricultural soil, 422–423
    agroforestry, 423
    forestry, 423–424
    mining spoil, 423
  diversifying crop rotations to reduce, 399
  and significance in changing global environment, 390–391
  trend and analysis of current CO_2 emission, 392–393
Carbon nanotubes (CNTs), 466
  CNT-coated biochar, 466
Carbon stability, 286
  and association with soil health restoration, 215–216
  biochar and, 295–301
    biochar role in soil carbon sequestration, 295–296
    biochar-induced changes in soil carbon stability, 299–301
    factors affecting soil carbon stability, 297–299, 298t
    interactions between microbial dynamics and, 301–303, 302f
Carbonaceous material, 118
Carbonic oxide (CO), 278
Carbonization process, 35, 90, 100–102, 163, 186
Carbonizing biomass, 28
Carboxyl, 322–323
Carboxylic groups, 87
Catalyst, 5–6, 45–46, 49–50
  addition, 49–50
  engineered biochar as, 547–548

Catalytic microwave pyrolysis, 69
Catalytic pyrolysis, 5–6
Cation exchange capacity (CEC), 87, 97–98, 139, 169, 187, 209, 228, 253, 312, 349–350, 390, 454, 533–534
Cationic methylene blue dye, 440–442
Cations, 169
CEC. *See* Cation exchange capacity (CEC)
Cellulose, 102–103, 123, 186
Centaurea cyanus L. *See* Cornflower (*Centaurea cyanus* L.)
Centrifugal force, 31–32
CeO_2-H@BC nanocomposite, 483–484
CF. *See* Carbon footprint (CF)
CFB reactor. *See* Circulating fluidized bed reactor (CFB reactor)
Charcoal, 1, 14–15, 150, 163, 548
Chemical degradation, 241
Chemical fertilizers, 274, 312
Chemical pesticides, 436
Chemical processes, 288–290
Chickpea (*Cicer arietinum*), 230
Chir Pine (*Pinus roxburghii*), 508
Chlorella pyrenoidosa. *See* Freshwater algae (*Chlorella pyrenoidosa*)
Chlorine (Cl), 258, 537–538
Cicer arietinum. *See* Chickpea (*Cicer arietinum*)
Circulating fluidized bed reactor (CFB reactor), 30, 30*f*
Citrobacter sp., 469
Civilian briquette binder, 518–519
Clay, 297
  clay-modified biochar, 465–466
  clay-rich soils, 170
  minerals, 128–129
  soils, 297
Climate change, 27, 185–186, 188–190, 227, 233, 238–239, 359–360, 364–365, 389–390, 413–421, 423–424
  biochar for mitigation climate change in different soils of, 128–129
  mitigation, 480–481, 494
    related favorable impacts of biochar on, 403–406
Climate conditions, 151
Climate innovative farming practices for reducing GHG emission, 398–399
  mitigation of climate change, 398*f*
Climate resilience agriculture, engineered biochar for, 544–545
  carbon sequestration, 544–545
  mitigation of methane emission, 545
  reduction of NO_2 emission, 545
Closed soilless systems, 258
Closed-cycle soilless methods, 258

Closed-cycle systems, 258
Clostridium
  C. botulinum, 548
  C. tetani, 548
CNTs. *See* Carbon nanotubes (CNTs)
Co-pyrolysis, 274–276
Coal-based power plants, 43
Coating, 536
Cocomposted biochar (COMBI), 312
  biochar as matrix to produce sustainable fertilizers, 312–314
  as enriched nutrient source, 323–326
    and sustainable ecological effects on soil, 324*f*
  slow-release mechanisms of biochar-based fertilizers, 314–323
Coconut husk biochar, 176–177
Colletotrichum acutatum, 260–261
COMBI. *See* Cocomposted biochar (COMBI)
Combustion process, 28, 83–84
Compost fertilizers, 152
Condensation reactions, 48
Conferences of the Parties (COP27), 392–393
Conservation agriculture, 399
Constructed wetlands (CWs), 487
Construction material, 548–549
Container capacity, 253
Container plant production, 254
Container substrate, biochar as, 253
Contaminant, 480–481
  adsorption mechanisms of biochar, 493*f*
  bioremediation from soil and water
    biological modification of biochar and effect on, 468–469
    chemical modification of biochar and effect on, 459–466
    physical modification of biochar and effect on, 466–468
  removal of, 487–489, 488*t*
Contaminated aqueous, bioremediation of, 546–547
Continuous pyrolysis systems, 513
Conventional heating, 44
Conventional methods, 82
Conventional thermochemical conversion techniques, 27–28
COP27. *See* Conferences of the Parties (COP27)
Copper (Cu), 236–238, 241, 277–278
Cornflower (*Centaurea cyanus* L.), 253
Cosmetics, 436
COVID-19 pandemic, 60
Cow dung, 274
Cow urine-enriched biochar, 234
Cowpea (*Vigna unguiculata* (L.)), 258–259, 262

Crop(s), 249, 345–346
　effect of biochar application on crop yield and quality, 14–15
　different salinity levels and effects on, 206t
　diversifying crop rotations to reduce carbon footprint, 399
　diversity, 399
　effect on crops growth and productivity, 150
　genetic enhancement of, 399
　management techniques, 13
　production, 271
　productivity, 123–125, 229–230, 312
　　effects on, 313–314
　residue, 360–362, 367–368
　　availability in India, 362f
　　burning, 398
　　crop residue-derived biochar, 122
　　waste, 507–508
　response, 544
　waste biochar, 241–242
　waste conversion
　　applications of BC, 8–18
　　effect of BC production process parameters on BC yield and quality, 6–7
　　BC production technologies, 3–6
　　current BC markets, 18
　　feedstock for BC production, 7–8
　　policy and legislative framework, 18
　　properties of BC, 2
　yield, 125, 167–168
Cropping systems, 300–301
　methanogenesis and methane emissions from, 340–342
　　percent methane contribution of countries from paddy cultivation, 341f
　　substrate consumption-based methanogenesis classification, 340t
　methanotrophs and methane oxidation from, 342–343
Cunninghamia lanceolate, 291–292
Cupric ion (Cu (II)), 270–271
Cutting-edge method, 193
CWs. *See* Constructed wetlands (CWs)
Cytokinins, 143–144

D

Dairy, 398
Dark fermentation, 66
Decarboxylation reactions, 48
Decomposition, 141–142
Dehydration reactions, 46, 48
Deionized water, 85–86
Densification, 516

Depreciation cost, 524
Dill (*Anethum graveolens*), 258–259
Disc granulation, 275
Disposable plastics, 60
Dissolution process, 207
Dissolved organic carbon (DOC), 210–211, 345–346, 403–404
Dissolved organic matter (DOM), 299–300
Diverse carbon-rich materials, 161
Diverse thermal conversion techniques, 508–511
Diversification of problematic wastes through thermal degradation, 71–72
DOC. *See* Dissolved organic carbon (DOC)
DOM. *See* Dissolved organic matter (DOM)
Domestic food waste, 62
Draining peat bogs, 254
Drought, 229–232, 255
　impact of, 229–230
　biochar effect on mitigating drought stress, 230–232
　drought-stressed cabbage plants, 257–258
　drought-stressed soils, 232
　improving drought tolerance, 255–258
　　effect of biochar on abiotic stresses in horticultural crops, 255t
　stress, 229–230, 257–258
Drum kilns, 513
Dry spells, 227
Dye(s), 480–481
　removal of, 482–484
　wastewater, 482–483
Dynamic bed behaviour, 68–69

E

EBC. *See* European Biochar Certificate (EBC)
EC. *See* Electrical conductivity (EC)
Ecosystem, occurrence and fates of ECs in, 436–437
ECs. *See* Emerging contaminants (ECs)
EDA. *See* Electron donor–acceptor (EDA)
EDAX. *See* Energy dispersive X-ray spectroscopy (EDAX)
EDCs. *See* Endocrine-disrupting substances (EDCs)
EIA. *See* Environmental Impact Assessment (EIA)
Electric pyrolysis reactors, 514
Electrical conductivity (EC), 86–87, 145–146, 193, 205–206, 253
　impact on soil pH and, 145–146
Electricity, 43
Electro-chemical properties of biochar, 350
Electromagnetic fields, 52
Electromagnetic radiation, 44–45
Electron donor–acceptor (EDA), 485–486
Electronic waste (E-waste), 61
　hazards from, 63

Elemental composition, 84–85
Elemental metals, 105
Emerging contaminants (ECs), 433–434. *See also* Organic contaminants
   biochar application for removal of ECs
      from aqueous system, 438–442
      from soil system, 442–445
   challenges and future perspective, 445
   in environment and sources, 436f
   occurrence and fates of ECs in ecosystem, 436–437
   properties of biochar affecting removal of, 437–438
Emerging volatile pollutants, 437
Encapsulation methods, 276
Endocrine-disrupting substances (EDCs), 436
Energy
   conservation, 121–122
   crops, 7
   efficiency, 52–53
   energy-rich solid carbonaceous compound, 186
   generation, 27, 494
   recovery, 53
   storage, 100–102
Energy dispersive X-ray spectroscopy (EDAX), 463
Engineered biochar, 531–532, 540–543. *See also* Tailored biochar
   application of, 540–549
      application in Livestock science, 548
      effect of biochar as oil conditioner, 540–544
      bioremediation of contaminated aqueous and soil ecosystem, 546–547
      carbon capture and utilization, 546
      as catalyst, 547–548
      for climate resilience agriculture, 544–545
      construction material, 548–549
   biological modification, 536
   chemical modification, 535–536
      acid and alkali treatment, 535
      coating, 536
      treatment with H_2O_2, 535
   critical factors to maximize advantages of, 539–540
      optimum temperature for biochar production, 539
      quality of feedstock biomass, 539
      soil carbon level, 539
      soil pH and soil contamination, 540
      soil types and soil moisture, 539–540
   different methodologies for preparation of, 459f
   modification methods for preparation of, 455–459
   physical modification, 534–535
   production technology of, 533–536, 534f
   properties of, 537–538
      chemical properties, 537–538
      physical properties, 537
   recent developments in inorganic contaminant removal by, 456t
Environment conservation practices, 100–102
Environmental conditions, 151
Environmental elements, 151
Environmental Impact Assessment (EIA), 18
Environmental management, 117–118
Enzymatic activity, 540–541
Enzymes, 148–150
   activities, 294–295
ESP. *See* Exchangeable sodium percentage (ESP)
Esterification processes, 193
Ethanol drop test, 88
Ethylene glycol, 49
Eucalyptus wastes, 170
Eupatorium, 325
European Biochar Certificate (EBC), 81–82
European Biochar Foundation, 548
E-waste. *See* Electronic waste (E-waste)
Ex situ remediation techniques, 241
Exchangeable sodium percentage (ESP), 206, 213–214
Extruder, 517

F

Faba beans (*Vicia faba* L.), 257–258, 262
Farming practices effect on soil greenhouse gas emission, 396–400
Fast pyrolysis, 5, 186
FC. *See* Field capacity (FC); Fixed carbon (FC)
Feedstock, 6–7, 100–102
   bed height, 37–38
   materials, 103–104
   for microwave-assisted hydrothermal carbonization, 45–46
   quality of feedstock biomass, 539
Fermentation, 66
Ferric nitrate (Fe(NO$_3$)$_3$, 9H$_2$O), 536
Fertilizer(s), 145–146, 269–271, 275, 312, 362–363, 380–382, 400
Field capacity (FC), 541
Firmicutes, 287–288, 290–291
Fish farming, 398
Fixed carbon (FC), 83, 522
Fixed-bed reactor, 67–68
Flame curtain kilns, 513
Flash pyrolysis, 5
Flavobacterium, 210–211
Flood, 233–234
   impact of, 233–234
   biochar role in flood management, 234
   flood-stressed soil, 234
Floodwaters, 234
Fluid catalytic cracking, 72–73

Fluidized bed reactors, 5–6, 29, 68, 514
Food
  production chain, 59
  security, 230, 269
Forestry, 423–424
  residues, 7
Forests, 300–301
Fossil fuels, 58–59
  combustion, 97–98
  excessive uneconomical use of, 397
Fossil-based power plants, 43
Fourier transform infrared spectroscopy (FTIR spectroscopy), 87–88, 461
FR. *See* Fuel ratio (FR)
Fragaria X ananassa. *See* Strawberry (*Fragaria X ananassa*)
Freshwater algae (*Chlorella pyrenoidosa*), 469
FTIR spectroscopy. *See* Fourier transform infrared spectroscopy (FTIR spectroscopy)
Fuel cells, 58–59
Fuel ratio (FR), 83–84
Fungi, 142, 285–287, 291–293

G

Galaxolide, 436
Gas
  generation, 5
  phase, 185–186
Gaseous compounds, 50
Gaseous discharge, 52–53
Gasification, 4, 28, 64–65, 138, 508–511
Gasifier reactors, 513
GCV. *See* Gross calorific value (GCV)
Gemmatimonadetes, 290–291
GeoPyc 1360 Envelope Density Analyzer, 85–86
Geranium (*Pelargonium peltatum*), 254
GHGs. *See* Greenhouse gases (GHGs)
Gibberellins, 143–144
Ginseng (*Panax ginseng*), 261
Gliricidia sepium, 482–483
Global environment, carbon footprint and significance in changing, 390–391
Global GHG emissions, 395
Global warming, 57–58, 238–239, 359–360, 380, 390–391
Glutathione reductase levels, 236–238
Glycine max, 438–440
Glycine max L. *See* Soybeans (*Glycine max* L.)
GOs. *See* Graphene oxides (GOs)
Gram-negative bacteria, 169–170
Gram-positive bacteria, 169–170
Granulation methods, 275–276
Graphene oxides (GOs), 466
Grasslands, 300–301
Green energy, 416–417
Greenhouse effects, potential of biochar to cut down on, 400–403
Greenhouse emissions, 197
Greenhouse gases (GHGs), 1–2, 43, 57–58, 97–98, 117–118, 141–142, 191–192, 250, 311–312, 335–336, 359–360, 389–390, 531–532
  emission, 394–395, 413–414
    agricultural sectors and greenhouse gas inventories, 395
    control of, 15
    effect of land use and farming practices on soil GHG emission, 396–400
    and role of agriculture's production system, 395–400
  productions, 169
  removal potential of agriculture practices, 402f
Gross calorific value (GCV), 47–48
Grow bag production systems, 253

H

Haplic Luvisol soil, 191
Harmful toxic metals, 259
Harvest trash, 361–362
Hazardous waste, 59
Hearth-type kilns, 513
Heat conversion techniques, 4
Heat Flow Meter (HFM), 88
Heat radiation, 44
Heat stress, 238–240
  impact, 238–239
    of biochar on heat stress, 239–240
Heat transfer rate, 30
Heating rate, 7, 37
Heavy metals (HMs), 61, 104, 140, 146, 161, 240–242, 259, 277–278, 293–294, 311–312, 438–440, 443
  adsorption capacity, 15–17
  biochar and HM interaction, 241–242
  contamination, 10–13
  impact of HM on agriculture, 240–241
  improving HMs tolerance, 259
  stresses, 255
Hemicellulose, 35, 102–103, 123, 186
Herbaceous plants, 236
Herbicides, 15–17, 394–395, 480–481
HFCs. *See* Hydrofluorocarbons (HFCs)
HFM. *See* Heat Flow Meter (HFM)
HHV. *See* Higher heating value (HHV)
High fixed carbon, 52
High porosity, 349

High salinity, biochar serve as viable amendment for soil with high, 207–209
High-input fertigation systems, 250–251
High-temperature pyrolysis, 207–208
High-temperature treatment (HTT), 104
Higher heating value (HHV), 88–89
Hilly region farmers, rural development and employment for, 526–527
HMs. *See* Heavy metals (HMs)
Hormones, 436
Horticultural charcoal, 253
Horticultural crops, 250, 262
  effect of biochar on abiotic stresses in, 255t
  biochar-mediated enhanced root growth of, 262
Horticultural rooting media, biochar for, 250–254
Hot gas filtration system, 29
HTC. *See* Hydrothermal carbonization (HTC)
HTT. *See* High-temperature treatment (HTT)
Hydraulic conductivity, 254
Hydraulic press briquetting, 517
Hydro-pyrolysis process, 5–6
Hydrochloric acids (HCl), 455
Hydrofluorocarbons (HFCs), 15
Hydrogen (H), 1, 84–85, 122, 537–538
Hydrogen peroxide (H_2O_2), 455, 462, 535
  treatment with H_2O_2, 535
Hydrogen sulfide (H_2S), 66–67
Hydrolysis, 46
Hydroponic production systems, 251
Hydrothermal carbonization (HTC), 43–44
Hydrothermal process, 45, 47
Hydroxyapatite, 100
Hyphomicrobiaceae, 291

I

IBI. *See* International Biochar Initiative (IBI)
Imidazolium-type ionic liquids (ITILS), 486–487
Impregnation method, 273–274
In situ crop residue decomposition by microbes, 396
In situ pyrolysis, 274
In situ remediation techniques, 241
In situ surface modification, 49–50
In-drum pyrolysis units, 513
Inceptisol, 301
Incineration process, 64
Incombustible ash, 64
Indo-Gangetic rice-wheat system, 299–300
Induced resistance (IR), 259–260
Industrial goods, 228, 436
Industrial rotary kilns, 514
Industrial vacuum pyrolysis reactor, 32
Industrial wastes, 57–58
Industrialization, 43

Innovative method, 31–32
Inorganic binder, 519
Inorganic components, 100
Inorganic compounds, 139
Inorganic contaminants, 453
Inorganic fertilizers, 210, 394–395
Inorganic ions, 241–242
Inorganic minerals, 103–104, 190–191
Inorganic nutrients in biochar, 363–364
Inorganic pollutants, 454–455
  occurrence and fates of inorganic pollutants in soil and water ecosystem, 454–455
Inorganic solutes, 262
Intergovernmental Panel on Climate Change (IPCC), 359–360, 416–417
International Biochar Initiative (IBI), 81
Ion chromatography system, 87
Ion exchange capacities, 537–538
Ionic stress, 217
Ionized gas, 70
IPCC. *See* Intergovernmental Panel on Climate Change (IPCC)
IR. *See* Induced resistance (IR)
Iron (Fe), 207, 236–238, 277–278
Iron oxide (Fe_3O_4), 536
Iron sulfide (FeS), 463
ITILS. *See* Imidazolium-type ionic liquids (ITILS)

K

K-rich biochar fertilizers, 274
Kalsilite ($KAlSiO_4$), 323
Kaolinite, 465–466
Kiln arrays, 514
Kinetic equation, 46

L

Land use effect on soil greenhouse gas emission, 396–400
Lantana camara, 230–232, 423
Lanthanum (La), 464–465
Layered double hydroxides (LDHs), 466
  LDHs—modified biochar, 466
LDHs. *See* Layered double hydroxides (LDHs)
Leaching method, 235–236
Lead (Pb), 259, 454
Life cycle assessment, 279, 391–392
Lignin, 3, 35, 102–103, 186, 511–512
Lignocellulosic biomass, 45, 207–208, 516
Linear regression functions, 87
Liquid discharge, 52–53
Liquid phase, 185–186
Liquid products, 28

Livestock science, application in, 548
Loamy soil, 240
Lolium perenne, 444–445
Long-chain hydrocarbons, 35
Long-term stability of biochar, 365–366
Low-temperature pyrolysis, 207–208
Luffa cylindrica, 460–461
Lupinus angustifolius, 232

M

Macronutrients, 125, 173–175, 217, 273–274, 363–364, 540
   in salt-affected soils, 210–213
Magnesium (Mg), 139, 173–175, 206
   Mg-enriched biochar, 271
Magnesium chloride ($MgCl_2$), 536
Magnesium oxide (MgO), 274–275, 536
Magnetic modification, 534–535
Magnetic reed biochar (MRBC), 490–494
Malondialdehyde, 236–238
Manganese, 236–238
Manganese chloride ($MnCl_2O \cdot 4H_2O$), 536
Mango wood-based biochar, 240
Manure application of soil greenhouse gas emission, 397
Manure fertilizers, 152
MAP. *See* Monoammonium phosphate (MAP)
MB. *See* Methylene blue (MB)
MBN. *See* Microbial biomass nitrogen (MBN)
MC. *See* Moisture content (MC)
Mean residence time (MRT), 300, 417
Mechanical extrusion, 275
Mercury (Hg), 454
   porosimetry method, 85
Merthanosarcina, 346–347
Mesopores, 537
Mesoporosity, 140
Meta-analysis of biochar's effects on biogenic greenhouse gas fluxes, 402–403
Metabolic processes, 216–217, 255
Metal adsorption, 277–278
Metal oxide-loaded biochar, 463–464
Metal oxyhydroxides, 87
Metal salts, 536
Metals, 277–278
Methane (CH_4), 15, 128–129, 164–167, 335–336, 338–339, 359–360, 368, 413–414, 419, 421
   Emission, 346–347
      biochar role in mitigation of, 380
      from cropping systems, 340–342, 340*t*
      from different sectors, 336*f*
      mitigation of, 545
      munchers, 341–342

oxidation from cropping systems, 342–343
physicochemical properties of biochar relevant to CH_4 mitigation, 348–350
production, 292–293
Methane monooxygenase (MMO), 342–343
Methanobacterium, 336–337, 341–342
Methanocella, 336–337, 341–342
Methanogenesis from cropping systems, 340–342, 340*t*
Methanogenic and methanotrophic microbial shift, biochar stimulates, 345–348
   abundance of methanotrophs and oxidation activity, 347–348
   promotes methanogenic microbial shift and inhibits methanogenesis activity, 346–347
Methanogens, 340–341
Methanomassiliicoccus, 336–337
Methanoregula, 341–342, 346–347
Methanosaeta, 336–337, 341–342
Methanosarcina, 336–337
Methanospirillum, 341–342
Methanotrophic bacteria, 342–343
Methanotrophs
   abundance of, 347–348
   from cropping systems, 342–343
Methyl orange (MO), 482–483
Methylene blue (MB), 482–483
Methylobacter, 342–343
Methylocaldum, 342–343
Methylocapsa, 347–348
Methylocella, 347–348
Methylococcus, 342–343
Methylocystis, 342–343, 347–348
Methylomicrobium, 342–343
Methylomonas, 342–343
Methylosarcina, 342–343
Methylosinus, 342–343, 347–348
MHTC. *See* Microwave-assisted hydrothermal carbonization (MHTC)
Micro nutrients, 273–274
Microbes, 152, 285–286, 301–302, 362–363
   in situ crop residue decomposition by, 396
Microbial activity, 137–138
   actinomycetes, 292
   bacteria, 290–291
   biochar-induced changes in microbial activity and diversity, 288–293, 289*t*
   fungi, 291–292
Microbial biomass carbon, 301–302, 540–541
Microbial biomass nitrogen (MBN), 148–150, 540–541
Microbial communities, 197
   interactions with, 152
   in soils, 287, 288*f*
Microbial decomposition, 417

Microbial diversity, 343–344
  biochar qualities associated with, 139–140
    agricultural biochar, 140
    municipal solid waste biochar, 140
    woody biochar, 140
Microbial dynamics, 303
  in biochar-amended soils, 287–295
  interactions between microbial dynamics and carbon stability, 301–303, 302f
Microbial enzymes, 301–302
  activity, 301–302
Microbial metabolic activity, 540–541
Microbial metabolites, 301–302
Microcline ($KAlSi_3O_8$), 323
Micromonosporaceae, 291
Micronutrients, 173–175, 217
  in salt-affected soils, 210–213
Microorganisms, 10, 62–63, 140–141, 290–292, 367
  effect of biochar on microorganism in salt-effected soil, 216–217
Microplastics (MPs), 57–58, 436
Micropore fraction, 187
Microscope, 89–90
Microwave
  activation, 455–459
  energy, 44–45
  heating, 44–45
  microwave-activated biochar, 467
  microwave-assisted heating technique, 44
  microwave-assisted pyrolysis, 64–65, 163
  modification, 534
  power, 50
  pyrolysis, 275
    reactor, 69
    units, 514
  reactors, 69
Microwave-assisted hydrothermal carbonization (MHTC), 44–50
  applications of, 50–52
  chemistry of, 46, 46f
  feedstock for, 45–46
  limitations of, 52–53
  reaction profile, 46–50, 47f
    catalyst, 49–50
    factors, 50, 51t
    process water recycling, 50
    reaction time, 48, 48f
    selection of solvent, 49
    solids loading, 49
    temperature, 47–48
Mineral fertilizers, 152, 319
Mineral nutrients, 208, 297–299
Mineralization process, 120

Minerals and SOM, interaction with, 122–123
Mining spoil, 423
Miscanthus giganteas, 462–463
Mitigating methane emissions
  biochar stimulates methanogenic and methanotrophic microbial shift, 345–348
  biochar-led methanogenesis and methanotrophy, 343–344
  biochar's potential to mitigate methane emissions, 338–340
  global rice cultivated area and production levels, 337f
  long term methane emissions from major rice cultivated countries, 338f
  methane emissions from different sectors, 336f
  methanogenesis and methane emissions from cropping systems, 340–342
  methanotrophs and methane oxidation from cropping systems, 342–343
  physicochemical properties of biochar relevant to methane mitigation, 348–350
MMO. See Methane monooxygenase (MMO)
MO. See Methyl orange (MO)
Mobile pyrolysis units, 514
Modified biochar, regeneration of, 469–470
Moisture content (MC), 522
Molybdenum (Mo), 210
Monoammonium phosphate (MAP), 313
Montmorillonite, 465–466
Montmorillonite-biochar SRFs, 274–275
Morphology, 89–92
MPs. See Microplastics (MPs)
MRBC. See Magnetic reed biochar (MRBC)
MRT. See Mean residence time (MRT)
MSW. See Municipal soil waste (MSW); Municipal solid waste (MSW)
Multistress tolerant crops, 227–228
Municipal plastic debris, 57–58
Municipal soil waste (MSW), 140
Municipal solid waste (MSW), 57–59
  biochar, 140
  hazards from, 62
  sources and types, 60t
Mycorrhiza, 143–144, 293–294
Mycorrhizal fungi, 140–141, 212, 291–292

N

Nano zerovalent iron (nZVI), 463–464, 485–486
Nanomaterials, 436
Nanoparticles (NPs), 105, 436
  application of NPs to enhance surface chemistry of biochar, 105–107

National Biomass Action Plan (NBAP), 18
Natural degradation process, 301
Natural gas, 400
Natural materials, 270–271
Natural Resource Conservation Services, 119–120
Natural resources, 433–434
NBAP. See National Biomass Action Plan (NBAP)
NDMA. See Nitrosodimethyl amine (NDMA)
Net zero emission, 389–390
Neutral reaction, 123–125
Nickel stress, 240–241
Nitrates (NO_3^-), 105–107, 170–172, 193–194, 230, 273, 531–532
Nitric acids (HNO_3), 455
Nitrification process, 193–194
Nitrogen (N), 1, 66–67, 84–85, 167, 207, 212, 217, 269, 277, 314–321, 396, 531–532
   cycling, 128–129
   fertilizer, 125, 277
      injudicious use of, 396–397
      N-concentrated biochar, 98–99
      N-NDMA, 486–487
   use efficiency, 277
Nitrogen oxide emissions (NOx emissions), 84–85
Nitrogen trifluoride (N_2F_3), 359–360
Nitrogen use efficiency (NUE), 316–317
Nitrosodimethyl amine (NDMA), 486–487
Nitrous oxide (N_2O), 15, 142, 164–167, 269–270, 316, 335–336, 359–360, 368, 413–414, 420–421
   emissions, 128–129
      biochar role in mitigation of, 380–382
      reduction of, 545
NMR spectroscopy, 90–92
Non-calcareous loamy sand soil, 214
Nonbiodegradable nature, 60
Nonmetallic heteroatom–doped biochar, 464–465
Novel polymer-coated urea SRFs, 276
NPs. See Nanoparticles (NPs)
NUE. See Nitrogen use efficiency (NUE); Nutrient use efficiency (NUE)
Nursery substrate, biochar as, 254
Nutrient use efficiency (NUE), 312, 317, 532
Nutrient(s), 9, 141, 164–167, 207, 274–275, 292–293
   absorption and mineralization, 540–541
   availability, 191–193, 209
   available, 541–543
   biochar on nutrient dynamics, 192t
   cycling, 148–150, 162–163
      biochar for improving nutrient cycling in soil, 126
   dynamics, 126
   elements, 125
   enriched biochar
      biochar as matrix to produce sustainable fertilizers, 312–314
      cocomposted biochar as enriched nutrient source, 323–326
      slow-release mechanisms of biochar-based fertilizers, 314–323
   impact on availability of nutrients in soil, 146–147
   management, 167
   nutrient-enriched biochar fertilizers, 363–364
   nutrient-rich feedstock materials, 170
   retention ability, 144–145
   role of biochar on nutrient status of salt-affected soils, 209
   sources, 313
nZVI. See Nano zerovalent iron (nZVI)

O

Ocimum basilicum. See Sweet basil (*Ocimum basilicum*)
ODR. See Oxygen diffusion rate (ODR)
Ofloxacin (OFX), 485–486
OFX. See Ofloxacin (OFX)
Oidiopsis sicula, 260–261
Oil, 129
   effect of biochar as oil conditioner, 540–544
Okra (*Abelmoschus esculentus*), 257–259
OM. See Organic matter (OM)
Optimal water retention, 176
Optimum reaction, 48
Optimum temperature for biochar production, 539
Opuntia ficus indica, 460–461
Orchid industry, 253
Organic acids, 187, 535
Organic amendments, 218, 300, 546
Organic biomass, 10
Organic biowastes, 45
Organic carbon, 9, 297, 299–301
Organic compounds, 139, 186, 293–294
Organic contaminants, 15–17, 490–494. See also Emerging contaminants (ECs)
   bioremediation in water using biochar, 481–489
   removal
      of antibiotics, 481–482
      of contaminants, 487–489
      of dyes, 482–484
      of pesticides, 484–485
      of pharmaceutically active compounds by biochars, 485–486
      of POPs, 486–487
   removing organic contaminants in soil and water by biochar, 491t
Organic environmental pollution, 479–480
Organic feedstocks, 81
Organic fertilizers, 319

Organic manure, 531–532
Organic materials, 32, 365–366
Organic matter (OM), 3, 128–129, 141–142, 227–228, 316, 540–541
Organic nitrogen (O-N), 170–172
  in biochar, 317–318
  forms, 170
Organic pollutants, 140, 440–442
  mechanisms for organic pollutant removal using biochar, 490–494
Organic pollution, 479–480
Organic solutes, 262
Organic waste, 531–532
  decomposition, 62
  materials, 7, 210
Organo-mineral combinations, 434–435
Osmotic stress, 217–218
Oxidation activity, abundance of, 347–348
Oxidation reactions, 194–195
Oxidative stress, 207
Oxidizing agent–treated biochar, 462–463
Oxisol, 297, 301
Oxycarbide, 278
Oxygen (O_2), 1, 66–67, 88–89, 122, 533, 537–538
  O_2-free cylindrical tube, 33
Oxygen diffusion rate (ODR), 540–541
Ozone (O_3), 335–336, 455, 462

P

Paddy
  percent methane contribution of countries from paddy cultivation, 341f
  soil, 126–127
PAHs. *See* Polycyclic aromatic hydrocarbons (PAHs)
Panax ginseng. *See* Ginseng (*Panax ginseng*)
Paper mill biochar, 15–17
Particle size of biomass feedstock, 7
Particulate organic carbon (POC), 422–423
PAWC. *See* Plant-accessible water capacity (PAWC)
PBC. *See* Peanut shell biochar (PBC)
PCBs. *See* Polychlorinated biphenyls (PCBs); Printed circuit boards (PCBs)
PCPs. *See* Personal care products (PCPs)
PDS. *See* Peroxydisulfate (PDS)
Peanut shell biochar (PBC), 464
Peat, 250, 253
  biochar as alternative to peat media, 254
  bogs, 250–251
  mining, 250
  moss, 254
  peat-based root substrate, 254
  peat-based substrate, 253
  substitution, 250–251

Pelargonium peltatum. *See* Geranium (*Pelargonium peltatum*)
Pellets
  as income source, 524–526
    advantages and disadvantages of pine biochar briquettes and pellets, 524–526
    economic analysis, 524
    production cost of pine biochar briquettes and pellets, 524
  binders for, 518–519
  of pine biochar, 516
  property assessment of, 519–522
    chemical properties, 521–522
    physical properties, 519–521
    thermal properties, 522
Pepper (*Capsicum annuum* L.), 260–261
Perfluorocarbons (PFCs), 15
Peri-urban fruit, 259
Permanent wilting point (PWP), 541
Peroxydisulfate (PDS), 485–486
Persistent organic pollutants (POPs), 486–487
  removal of, 486–487
Personal care items, 436
Personal care products (PCPs), 436
Pesticides, 15–17, 395, 444–445, 480–481
  containers, 479–480
  removal of, 484–485
Petunia (*Petunia hybrida*), 254
Petunia hybrida. *See* Petunia (*Petunia hybrida*)
PFCs. *See* Perfluorocarbons (PFCs)
PGPB. *See* Plant growth promotion bacteria (PGPB)
PGPRs. *See* Plant growth-promoting rhizobacteria (PGPRs)
Pharmaceuticals
  compounds, 436
  removal of pharmaceutically active compounds by biochars, 485–486
  in sewage water, 435
Phaseolus vulgaris. *See* Bean (*Phaseolus vulgaris*)
Phenol oxidase, 301–302
Phenolic groups, 87
Phosphates (PO_4^{3-}), 105–107, 273, 531–532
  phosphates-solubilizing bacteria, 210–211
Phosphoric acid (H_3PO_4), 274–275, 455
Phosphorus (P), 143–144, 167, 206–207, 210–211, 217, 269, 314–321
  concentrations, 236–238
  content, 172
  P-enriched biochar, 271
Photosynthesis, 168–169
Physiochemical characterization of biochar, 85–88
  cation and anion exchange capacity, 87
  density, 85–86

Physiochemical characterization of biochar (*Continued*)
  electrical conductivity, 86–87
  FTIR spectroscopy, 87–88
  pH, 86
  porosity, 85
  thermal conductivity, 88
  water-holding capacity, 88
Phytohormones, 148–150
Phytoremediation, 241
Pine biochar, 516–523
  briquettes
    advantages and disadvantages of, 524–526
    cost analysis of, 525*t*
    production cost of, 524
  briquetting and pelleting of, 516
    methods of, 516–517
  pine needle biochar briquettes, 517–519
  preparation
    modern approach of, 513–515
    portable rotating biochar drum, 515
    traditional approach to, 512–513
  properties of pine needle briquettes, 522–523
  property assessment of biochar briquettes and pellets, 519–522
Pine forests, 508
Pine needle biochar, 511–516
  briquettes, 517–519
    binders for pine needle biochar brqieuttes and pellets, 518–519
    pine needle pellet production, 517
    properties of, 520*t*
  characteristics of, 516
  pine biochar preparation
    modern approach of, 513–515
    traditional approach to, 512–513
  preparation of, 511–515
  properties of, 512*t*, 515*t*
Pine needle biomass, 508–511
  characteristics of, 511
  problems associated with, 508–511
    species of pine needle worldwide, 509*t*
Pine needles, 508–511
  properties pine needle briquettes of, 522–523
Pine sawdust biochar, 481–482
Pine trees (*Pinus*), 443–444, 508
Pinus. *See* Pine trees (*Pinus*)
Pinus roxburghii. *See* Chir Pine (*Pinus roxburghii*)
Piston press briquetting, 517
Pisum sativum, 463
Plant growth, 139, 141–142, 228, 235–236, 253
  alteration in plant growth rates after BC application, 405–406
  hormones, 144–145
  role of BC in plant growth in salt-effected soil, 217–219
  potential advantages of BC and application in enhancing soil health and promoting, 142–143
    reduces contaminant, 143
    reduces greenhouse gas emissions, 142
    soil fertility and plant growth, 142
  regulators, 143–144
  stimulation
    mechanism by plant growth promoting rhizobacteria, 143–144
    by plant growth promoting rhizobacteria, 144*f*
Plant growth promotion bacteria (PGPB), 469
Plant growth-promoting rhizobacteria (PGPRs), 137–138, 143–144
  effect on crops growth and productivity, 150
  effect on soil microbial communities, 148–150, 149*t*
  effect on water holding capacity of soil, 147–148
  effects of plant growth promoting rhizobacteria and biochar co-application on soil health, 145
  impact on availability of nutrients in soil, 146–147
  impact on soil pH and electrical conductivity, 145–146, 147*f*
  interaction of biochar and, 144–150
Plant-accessible water capacity (PAWC), 539–540
Plant(s), 211–212, 235, 271
  glycine receptor, 143–144
  improvement of plant yield and improved C-sequestration in plant biomass, 420–421
  materials, 123, 297–299
  nutrients, 206, 362–363
  quality, 208
  roots, 287
    growth, 126–127
  stress, 142
Plasma, 70
  processing, 173
  reactor, 70
Plastic waste, 60, 62–63, 68–69
  hazards from, 62–63
Plasticizers, 433–434
POC. *See* Particulate organic carbon (POC)
Podosphaera apahanis, 260–261
Pollutant removal efficacy, 479–480
Pollution mitigation processes, 97–98
Polychlorinated biphenyls (PCBs), 63
Polycyclic aromatic hydrocarbons (PAHs), 153, 417, 442, 444, 480–481
Polyethylene terephthalate (PET), 64–65
Polymerization reactions, 48
Polyphagotarsonemus latus. *See* Broad mite pest (*Polyphagotarsonemus latus*)
Polyvinylpyrrolidone (PVP), 276

POPs. *See* Persistent organic pollutants (POPs)
Porosity, 85
Porous carbon material, 147–148
Portable rotating biochar drum, 515
  biochar production unit, 514f
Postpyrolysis, 313
Potassium (K), 139, 167, 206–207, 211–212, 217–218, 236–238, 269
  content, 173
  release dynamics, 321–323
    slow-release dynamics of K from biochar-based fertilizers, 322f
  solubilizing enzymes, 143–144
Potassium dihydrogen phosphate (KH_2PO_4), 278
Potassium permanganate ($KMnO_4$), 455, 462–463, 536
Potato spindle tuber viroid (PSTVd), 261
Potentially toxic elements (PTEs), 546
Prepyrolysis acidification, 170
Printed circuit boards (PCBs), 61
Problematic wastes
  agro-ecological hazards from, 61–63
  challenges of pyrolysis of, 72–73
  composting of, 72
  current technologies for waste management, 63–67
  diversification of problematic wastes through thermal degradation, 71–72
  types, 58–61
    agro-industrial waste, 59
    E-waste, 61
    municipal solid waste, 59
    of pyrolysis reactors used for treating problematic wastes, 67–70
    plastic waste, 60
Process water recycling, 50
Protozoa, 287
Pseudomonas, 210–211
  P. hibiscicola, 469
PSTVd. *See* Potato spindle tuber viroid (PSTVd)
PTEs. *See* Potentially toxic elements (PTEs)
Pumpkin (*Cucurbita pepo*), 257–258
PWP. *See* Permanent wilting point (PWP)
Pyrolysis, 3–6, 28, 64–66, 83, 138, 197, 249–250, 274–275, 343–344, 348, 424–427, 508–511
  challenges of pyrolysis of problematic wastes, 72–73
  co-pyrolysis, 274–275
  liquids, 72–73
  optimizing product yield, 4f
  process, 69, 90, 117–118, 210–211, 215, 285–286, 293–296
  reactors, 67
  in situ pyrolysis, 274
  temperature, 6

  role in enhancing surface chemistry of biochar, 103–105, 106t
  types, 4–6
    catalytic pyrolysis and hydro-pyrolysis, 5–6
    fast pyrolysis, 5
    flash pyrolysis, 5
    slow pyrolysis, 4
Pyrolysis reactors, 7, 37–38
  factors affecting design of, 35–38
    feedstocks bed height, 37–38
    heating rate, 37
    particle size, 36–37
    residence time, 35–36
    temperature, 35, 36t
  types, 28–35
    ablative reactors, 31–32
    bubbling fluidized bed, 29, 29f
    circulating fluidized bed, 30, 30f
    rotary klin, 33–35
    screw reactors, 32–33
    vacuum pyrolysis reactors, 32, 33f
  used for treating problematic wastes, types of, 67–70
    fixed-bed reactor and batch reactor, 67–68
    fluidized-bed reactor, 68
    microwave pyrolysis reactor, 69
    plasma reactor, 70
    rotary kiln reactor, 69
    screw auger pyrolysis reactor, 70
    solar reactor, 70
    spouted bed reactor, 68–69

R

Ralstonia solanacearum, 261
Raman spectroscopy, 87–88
Rapid pyrolysis. *See* Fast pyrolysis
Raw materials, 274, 521, 533
Razort loam soil, 192–193
RCA. *See* Rice straw biochar (RCA)
Reaction time, 48, 48f
Reactive oxygen species (ROS), 227–228, 257–258
Reactive Red 84 (RR84), 483–484
Reactor designs, 28
Recirculating systems, 258
Recycling waste biomass, 274
Reducing GHG emission, climate innovative farming practices for, 398–399
Regeneration of used modified biochar, 469–470
Remediation, 480–481
Renewable energy, 43, 50–52
Repolymerization process, 35–36
Residence time, 7
Residual materials, 27–28
Retort kilns, 513

Rhizophagus irregularis, 232
Rhizospheric microorganisms, 232
Rice
   cultivation, 336–337, 398
   rice-based cropping systems, 336–337, 361–362
   rice-based systems, 341–342
   straw-based biochar, 230–232
   straw-derived biochar, 230–232
Rice straw biochar (RCA), 126, 481–482
Roller press briquetting, 517
Rooting media, 250
ROS. *See* Reactive oxygen species (ROS)
Rotary drum, 33–34
Rotary kiln reactor, 33–35, 34f, 69
RR84. *See* Reactive Red 84 (RR84)
RT-qPCR method, 261
Rural development and employment for hilly region farmers, 526–527

S

16S rRNA gene, 287
SA. *See* Surface area (SA)
Salicaceous feedstocks, 104
Saline soil, 127, 205–206, 235
   effect of biochar on physicochemical properties of, 213–215
Saline-sodic soils, 205–207
Salinity, 205–206, 235, 255
   stress, 218
   stressed-maize, 236–238
Salinization, 235–238
   impact of, 235
   role of biochar in management of soil salinity, 235–238
   salinization-exposed soils, 236–238
   salinization-tolerance, 235–236
Salt, 235
   adsorption capacity, 235–236
   improving salt tolerance, 258–259
   salt-affected soils, 206, 214
      biochar and micronutrients and macronutrients in, 210–213
      effect of biochar on microorganism in, 216–217
      role of biochar in plant growth in, 217–219
      role of biochar on nutrient status of, 209
Sand, 29
Sandy soils, 151, 176, 240, 297, 302–303
Satureja hortensis L. *See* Summer savory (*Satureja hortensis* L.)
Sawdust biochar, 104
SC. *See* Soil carbon (SC)
Scanning electron microscopy (SEM), 89–90, 104, 461

SCC. *See* Somatic cell count (SCC)
Scholars examined effect of organic carbon (SOC), 486–487
SCR. *See* Selective catalytic reduction (SCR)
Screw auger pyrolysis reactor, 70
Screw extrusion briquetting, 517
Screw pyrolysis reactors, 32–33
Screw reactors, 32–33
   screw or screw pyrolysis reactor, 34f
Sea level, 235
Secondary pyrolysis processes, 37
Seed germination using biochar, enhanced, 261–262
Selective catalytic reduction (SCR), 547
SEM. *See* Scanning electron microscopy (SEM)
Sequestered carbon, 169
Sequestering C in soil, biochar as tool for, 401–402
SEs. *See* Steroidal estrogens (SEs)
Sewage sludge (SS), 170, 321
Sewage sludge biochar (SSB), 320–321
Siderophores, 143–144
Silica–carbon complexes, 169–170
Silicon (Si), 104, 261–262, 537–538
Slack biomass, 516
Slow pyrolysis, 4, 163–164
Slow-release fertilizers (SRFs), 269–270
Slow-release phosphorus fertilizers, 275, 277
Slow–release mechanisms of biochar-based fertilizers, 314–323
SMBC. *See* Soil microbial biomass carbon (SMBC)
SMX. *See* Sulfamethoxazole (SMX)
SOC. *See* Scholars examined effect of organic carbon (SOC); Soil organic carbon (SOC)
Sodic soils, 205–207, 209
Sodium (Na), 207, 214, 217–218, 258
Sodium persulfate (SPS), 485–486
Soil amendment, 100–102, 128–129, 151, 162, 300
   BC, preparation methods, and properties, 138–139
   BC qualities associated with microbial diversity, 139–140
   BC–microbe interactions, 140–141
   considerations and difficulties, 152–153
      biochar quality control and standardization, 153
      biochar's potential drawbacks and limitations, 152–153
      ecological implications and unintended consequences, 153
      regulatory and policy issues regarding biochar application, 153
   factors influencing effectiveness of BC, 151–152
   interaction of BC and plant growth promoting rhizobacterias, 144–150

mechanism of plant growth stimulation by plant growth promoting rhizobacteria, 143–144
potential advantages of BC and application in enhancing soil health and promoting plant growth, 142–143
significance of soil microbial diversity on soil–plant–atmospheric continuum, 141–142
synergies with, 152
Soil carbon (SC), 297, 299–301, 390
  biochar effects on SC dynamics and soil fertility, 168–169
  level, 539
  mineralization, 303
  sequestration, 117–118, 195–196
    biochar role in, 295–296
    contribution of biochar to soil carbon stable pool, 296f
  stability, 303
    biochar-induced changes in, 299–301
    factors affecting, 297–299, 298t
  storage, 164–167
Soil erosion, reducing, 544
Soil health, 125–126, 137–138
  biochar
    application effect on, 125–128
    for enriching soil biological health, 126–127
    for improving nutrient cycling in soil, 126
    for overall soil health improvement, 127–128
    for protecting soil physical health, 126
    reclaiming health of disturbed soil, 127
  carbon stability and association with, 215–216
  effects of plant growth promoting rhizobacteria and biochar co-application on, 145
Soil microbial biomass carbon (SMBC), 360–361
  biochar impact on, 367
Soil organic carbon (SOC), 9, 196, 215–216, 285–286, 295–297, 299–301, 361–362, 364–365, 390, 417, 540–541
  negative priming effect of biochar on mineralization of, 403–404
Soil organic matter (SOM), 122–123, 141–142, 168–170, 190–191, 367
  interaction with minerals and, 122–123
Soil quality index (SQI), 127–128
Soil(s), 87, 152, 185–186, 197, 205–206, 208, 210–212, 230–232, 271, 277, 287, 293–294, 479–480
  absorb energy, 239–240
  acidity, 188
  aggregates, 169
  amelioration, 97–98
  bacteria, 147, 287, 301
  biochar, 347–348, 361–362, 420
    and effect on contaminant bioremediation

application for carbon stabilization and sequestration in, 415–418
application for removal of ECs from soil system, 442–445
as soil-less substrate, 251–252
as tool for sequestering C in, 401–402
biological modification of, 468–469
chemical modification of, 459–466
contaminants removal from soil ecosystem, 442t
effect on growth and development in different horticultural crops, 251t
effect on microorganism in salt-effected soil, 216–217
effect on physicochemical properties of saline soil, 213–215, 213f
effects on different soil nutrient dynamics, 170–175
and micronutrients and macronutrients in salt-affected soils, 210–213, 211f
for mitigation of climate change in different, 128–129
role in plant growth in salt-effected soil, 217–219, 219t
role on nutrient status of salt-affected soils, 209
role on soil properties, 209–219
serve as viable amendment for soil with high salinity, 207–209
effect of biochar soil physical and chemical properties on, 8–9
biological processes, 288–290
biological properties, 543–544
  effect of biochar on, 10
biota, 10
carbon, nitrogen, potassium contents percentage in different biochar product, 174t
carbon stability and association with soil health restoration, 215–216
carbon transformation and migration in, 416f
contamination, 480–481, 540
ecosystem
  bioremediation of, 546–547
  occurrence and fates of inorganic pollutants in, 454–455
effect on water holding capacity of, 147–148
erosion, 167
factors affecting ability of biochar to fix carbon in, 415
facts and prospects from different, 123–125
fertility, 28, 139, 142, 162–163, 421–422, 540–541
fertilization, 480–481
flushing techniques, 241
impact on availability of nutrients in, 146–147
improvement, 494

Soil(s) (*Continued*)
   increasing soil C stock reduce soil surface albedo, 404–405
   insight of soil physical-fertility-biological factors, 125–128
   land use and farming practices effect on soil GHG emission, 396–400
      carbon sequestration, 399
      climate innovative farming practices for reducing GHG emission, 398–399
      conservation agriculture, 399
      crop residue burning, 398
      dairy, aquaculture and fish farming, 398
      different agronomic measures, 400
      diversifying crop rotations to reduce carbon footprint, 399
      excessive uneconomical use of fossil fuels, 397
      genetic enhancement of crops and animals, 399
      increasing input-use efficiency, 400
      injudicious use of N fertilizers, 396–397
      land use changes, 397–398
      manure application, 397
      in situ crop residue decomposition by microbes, 396
   management
      factors affecting biochar and impact on, 494–495
      techniques, 13
   matrix, 295–296
   mechanism and regulations of carbon and nitrogen between soil and plant, 171*f*
   microbes, 126–127, 145
   microbial communities
      biochar impact on, 287–288
      effect on, 148–150
   microbial community in, 287, 288*f*
   microbial diversity, 141–142
   microbial dynamics, 286
   microbial respiration, 216
   microorganisms, 147–148, 216, 286, 301
      mechanisms behind biochar impacts on, 293–295
      potential mechanism underlying biochar on microbial dynamics, 294*f*
   minerals, 169–170
   moisture, 539–541
   nitrification, 193–195
      biochar for reducing Al toxicity in soil, 195*f*
   nitrogen, 300
   organic mixtures, 169–170
   organic pollution, 480–481
   penetration resistance, 126
   pH, 190–191, 287–288, 299, 540
      impact on soil pH and electrical conductivity, 145–146
   physical modification of biochar and effect on contaminant bioremediation from, 466–468
   physical properties, 541
   pollutants, 143
   pores function, 8
   porosity, 8
   processes, 287
   quality, 228
   remediation, 480–481
   salinity, 207, 212
      role of biochar in management of, 235–238
   salinization, 235–236
   salts, 212
   soil-applied biochar of different feedstocks, 418*t*
   soil-based production system, 249
   soil-DOC, 345–346
   soil-dwelling bacteria, 143–144
   soil-dwelling fungi, 291–292
   soil–plant systems, 228–229
   soil–plant–atmospheric continuum functions, 141–142
   texture, 297
      and composition, 176
   types, 539–540
      texture, and mineralogy, 151
   washing techniques, 241
Soilless cultivation
   biochar for horticultural rooting media, 250–254
   biochar-mediated enhanced root growth of horticultural crops, 262
   biochar-mediated suppression
      of abiotic stresses, 255–259
      of biotic stresses, 259–261
   enhanced seed germination using biochar, 261–262
   future perspectives, 262–263
Solanum lycopersicum. *See* Tomato (*Solanum lycopersicum*)
Solanum melongena L. *See* Brinjal (*Solanum melongena* L)
Solar pyrolysis units, 514
Solar reactor, 70
Solid biochar, 43–44
Solid biofuel, 43–44
Solid char, 43–44
Solid material, 311–312
Solid phase, 185–186
Solid products, 28
Solid Waste Management Rules, 18
Solidification, 241
Solids loading, 49
Soluble organic C, 100
Soluble salts, 206, 209

Solvents, 49
  selection of, 49
Solvothermal synthesis, 49
SOM. *See* Soil organic matter (SOM)
Somatic cell count (SCC), 548
Soybeans (*Glycine max* L.), 218, 230
  leaf photosynthetic rate, 230
Specific surface area (SSA), 349, 390, 532
Spinacea oleracia. *See* Spinach (*Spinacea oleracia*)
Spinach (*Spinacea oleracia*), 258–259
Spouted bed reactor, 68–69
SPS. *See* Sodium persulfate (SPS)
SQI. *See* Soil quality index (SQI)
Sr. *See* Strontium (Sr)
SRF encapsulation, 276
SRFs. *See* Slow-release fertilizers (SRFs)
SS. *See* Sewage sludge (SS)
SSA. *See* Specific surface area (SSA)
SSB. *See* Sewage sludge biochar (SSB)
Stabilization, 241
Stable carbon molecules, 295–296
Starch, 71
Steam activation transforms, 467
Steam gas activation, 534
Steroidal estrogens (SEs), 487–489
Strawberry (*Fragaria* X *ananassa*), 262
Streptosporangineae, 291
Stress hormones, 235
Strong ion exchange capacity, 349–350
Strontium (Sr), 462–463
Substrate amendments, 259–260
Sugarcane wastes, 170
Sulfamethoxazole (SMX), 481–482
Sulfonated biochar, 98–99
Sulfur (S), 1, 173–175, 319–320
  dynamics, 314–321
Sulfur hexafluoride (SF_6), 15
Sulfur oxides (SOx), 84–85
Sulfuric acids (H_2SO_4), 49–50, 455
Summer savory (*Satureja hortensis* L.), 262
Summer squash (*Cucurbita. pepo* L.), 259
Supercapacitors, 58–59, 100–102
Surface active agent–treated biochar, 465
Surface adsorption capacity, 98
Surface albedo, 404–405
Surface area (SA), 163, 187
Surface chemistry of biochar, modification techniques for improving, 99–100, 99f, 101t
Surface functionality, 89–92
  Brunauer–Emmett–Teller, 90
  NMR spectroscopy, 90–92
  scanning electron microscopy, 89–90
  thermogravimetric analysis, 90

Surface hydrophobic-hydrophilic interactions, 239
Surfactants, 436
Surplus biomass waste, 360–361
Surplus crop residue vis-a-vis biochar production, 361–362
Sustainability, 197, 279
Sustainable agricultural practices, 128–129
Sustainable agriculture, 269
Sustainable C cycle, 120
Sustainable fertilizers, biochar as matrix to produce, 312–314
Sustainable waste treatment, 70
Sweet basil (*Ocimum basilicum*), 261–262
Synergistic interaction, 105–107
Syngas, 129
Synthetic dyes, 443
Synthetic fertilizers, 72

T

Tailored biochar. *See also* Engineered biochar
  biochar and effect on contaminant bioremediation from soil and water
    biological modification of, 468–469
    chemical modification of, 459–466
    physical modification of, 466–468
  modification methods for preparation of engineered biochar, 455–459
  occurrence and fates of inorganic pollutants in soil and water ecosystem, 454–455
  regeneration of used modified biochar, 469–470
TC. *See* Tetracycline (TC)
Terra Preta de Indio, 1–2, 285–286
Terra Preta kilns, 513
Tetracycline (TC), 481–482, 485–486
TGA. *See* Thermogravimetric analysis (TGA)
Thermal conductivity, 68, 88
Thermal conversion process, 365–366
Thermal cracking, 4
Thermal decomposition, 186
Thermal degradation, diversification of problematic wastes through, 71–72
  biochar production, 71–72
  biofuel production, 71
  composting of problematic wastes, 72
Thermal gasification, 64–66
  strategies for problematic waste management, 65f
Thermochemical conversion process, 1, 67, 362–363
Thermochemical techniques, 1–2, 32, 63–66, 163
  incineration, 64
  pyrolysis and thermal gasification, 64–66
Thermogravimetric analysis (TGA), 90

Thermomonosporaceae, 291
Thermos-chemical conversion technologies, 3
THI. *See* Thiacloprid (THI)
Thiacloprid (THI), 484–485
Thiobacillus, 210–211
TLUD kiln. *See* Top-lit updraft kiln (TLUD kiln)
Tomato (*Solanum lycopersicum*), 176–177, 230–232, 236, 258–262, 405–406
Tonalide, 436
Top-lit updraft kiln (TLUD kiln), 512–513
Total soil C (TSC), 367–368
Toxic compounds, 279
Toxic gases, 60
Traditional approach to pine biochar preparation, 512–513
Traditional energy resources, 27
Trichoderma, 150
True particle density. *See* Actual density
TSC. *See* Total soil C (TSC)

U

UHPC. *See* Ultra-high-performance concrete (UHPC)
Ultisol, 297
Ultra-high-performance concrete (UHPC), 548–549
Ultraviolet light, 435
UNFCC. *See* United Nations Framework Convention on Climate Change (UNFCC)
United Nations Framework Convention on Climate Change (UNFCC), 392–393
Urban fruit, 259
Urea fertilizers, 275

V

Vacuum pyrolysis reactors, 32, 33f
Vaporization, 32
Vegetable farming, 259
Vermiculite, 465–466
Vertisol, 297
Vicia faba L. *See* Faba beans (*Vicia faba* L.)
Vigna unguiculata (L.)). *See* Cowpea (*Vigna unguiculata* (L.))
VM. *See* Volatile matter (VM)
VOCs. *See* Volatile organic compounds (VOCs)
Volatile compounds, 50, 83–84
Volatile matter (VM), 83, 522
Volatile organic compounds (VOCs), 17
Volatile organic vapors, 186
Volatile solids (VSs), 83–84
Volatilization-induced nitrogen, 170
Vortex ablative bed reactor, 31
VSs. *See* Volatile solids (VSs)

W

Waste
  absorption, 238–239
    test, 521
  biomass product, 97–98
  contamination, 480–481
  disposal, 57–58
  generation, 57–58
  management, 494
    biochemical technology, 66–67
    current technologies for, 63–67
    thermochemical technology, 64–66
  materials, 64, 270–271
  occurrence and fates of inorganic pollutants in water ecosystem, 454–455
  waste-derived resources, 228
Waste printed circuit boards (WPCBs), 61
Waste stream of electrical and electronic equipment (WEEE), 61
Wastewater treatment, 15–17, 28
  adsorption of heavy metals and organic contaminants by biochar in, 16t
Water, 164–167, 479–480
  biochar and effect on contaminant bioremediation
    biological modification of, 468–469
    chemical modification of, 459–466
    physical modification of, 466–468
  bioremediation of organic contaminants in water using biochar, 481–489
  factors affecting biochar and impact on water management, 494–495
  quality, 539–540
  recirculation, 50
  retention, 277
    capacity, 129–130
  scarcity, 167–168
  stress, 232
  water-soluble ions, 86
  water-soluble salts, 206
  water-stressed regions, 235
  water-stressed tomato plants, 262
Water holding capacity (WHC), 187, 240, 419, 540–541
  effect on water holding capacity of soil, 147–148
Water use efficiency (WUE), 240
Water-holding capacity (WHC), 88, 139, 232
Waterborne polymers, 276
Waterlogged conditions, 233–234
WEEE. *See* Waste stream of electrical and electronic equipment (WEEE)
Wet spells, 227
WHC. *See* Water holding capacity (WHC); Water-holding capacity (WHC)

Wheat, 218, 236
    plants, 236–238
    wheat-maize cropping sequence, 127–128
Wood, 297–299
    pellets biochar, 257–258
    wood-based biochar, 295–296
    wood-derived biochar, 209
Woody biochar, 140
Woody biomass, 140
Woody plant material, 241–242
WPCBs. *See* Waste printed circuit boards (WPCBs)
WUE. *See* Water use efficiency (WUE)

X

X-ray diffraction (XRD), 461–462
X-ray photoelectron spectroscopy (XPS), 320–321, 461–462
XPS. *See* X-ray photoelectron spectroscopy (XPS)
XRD. *See* X-ray diffraction (XRD)

Z

Zeolite, 465–466
Zinc (Zn), 147, 207, 217, 236–238, 277–278
    nanoparticles, 261–262
Zinc chloride ($ZnCl_2$), 536

Printed in the United States
by Baker & Taylor Publisher Services